STEVENSON

Refrigeration and Air Conditioning Technology

Refrigeration and Air Conditioning Technology

by
REX MILLER
Professor of Industrial Arts and Vocational Education
State University College
Buffalo, New York

Bennett Publishing Company
Peoria, Illinois 61615

Copyright © 1983

by Rex Miller

All rights reserved

83 84 85 86 87   V-H   5 4 3 2 1

ISBN 87002-379-9

Library of Congress Catalog No. 81-71181

Printed in the United States of America

Preface

An introduction to the basic principles and practices of the air conditioning and refrigeration industry is more than just a review of the facts and figures. It requires a complete look at the industry. This text presents the basics of refrigeration. It explains the equipment that makes it possible for us to live comfortably in air conditioned spaces.

Up-to-date methods of equipment maintenance are stressed. The latest tools are shown. The applications of the newer types of units are emphasized. The field of air conditioning technology is still growing and will continue to grow far into the future. New technicians will need to be aware of the fact that change is inevitable. They will have to continue to keep up with the latest developments as long as they stay in the field.

This textbook has been prepared to aid in instructional programs in high schools, technical schools, trade schools, and community colleges. Adult evening classes and apprenticeship programs also may find it useful. This book provides a thorough knowledge of the basics and a sound foundation for anyone entering the air conditioning and refrigeration field.

Acknowledgments

No author works without being influenced and aided by others. Every book reflects this fact. This book is no exception. A number of people cooperated in providing technical data and illustrations. For this I am grateful.

I would like to thank those organizations that so generously contributed information and illustrations. The following have been particularly helpful.

Admiral Group of Rockwell International
Air Conditioning and Refrigeration Institute
Air Temp Division of Chrysler Corporation
Americold Compressor Corporation
Amprobe Instrument Division of SOS Consolidated, Inc.
Arkla Industries, Inc.
Bryant Manufacturing Company
Buffalo Evening News
Calgon Corporation
Carrier Air Conditioning Company
E.I. DuPont de Nemours & Co., Inc.
Dwyer Instruments, Inc.
Ernst Instruments, Inc.
General Controls Division of ITT
General Electric Co. (Appliance Division)
Haws Drinking Faucet Company
Hubbell Corporation
Hussman Refrigeration, Inc.
Johnson Controls, Inc.
Kari-Kold, Inc.
Kodak Corporation
Lennox Industries, Inc.
Lima Register Company
Marley Company
Marsh Instrument Company, Division of General Signal
Mueller Brass Company
Packless Industries, Inc.
Parker-Hannifin Corporation
Penn Controls, Inc.
Schaefer Corporation
Sears, Roebuck and Company
Snap-on Tools, Inc.
Sporlan Valve Company
Superior Electric Company
Tecumseh Products Company
Thermal Engineering Company
Trane Company
Turner Division of Cleanweld Products, Inc.
Tuttle & Bailey Division of Allied Thermal Corporation
Tyler Refrigeration Company
Union Carbide Company, Linde Division
Universal-Nolin Division of UMC Industries, Inc.
Virginia Chemicals, Inc., with particular appreciation to Mr. Paul Nelson, Mr. Gene Abrams, and Mr. Kenneth Scribner.
Wagner Electric Motors
Weksler Instrument Corporation
Westinghouse Electric Corporation
Worthington Compressors

**DEDICATION
To Sylvia Loffler Navara**

Table of Contents

Preface 5

Acknowledgments 6

Chapter 1—Mathematics 11
    Basic Mathematics, 11; Fractions, 13; Decimals, 15; Square Root, 15; Areas and Volumes, 16; Equations, 19; Formulas, 20; The Metric System of Measurement, 20; Reading Graphics, 21; Measuring Temperature, 21; Review Questions, 24.

Chapter 2—Air Conditioning and Refrigeration Tools and Instruments 25
    Tools and Equipment, 25; Gages and Instruments, 32; Superheat Measurement Instruments, 39; Halide Leak Detectors, 43; Electrical Instruments, 45; Other Instruments, 50; Service Tools, 52; Special Tools, 53; Vacuum Pumps, 54; Evacuating a System, 57; Charging Cylinder, 58; Charging Oil, 58; Changing Oil, 59; Mobile Charging Stations, 59; Tubing, 59; Cutting Copper Tubing, 60; Flaring Copper Tubing, 62; Constricting Tubing, 63; Swaging Copper Tubing, 64; Forming Refrigerant Tubing, 64; Fitting Copper Tubing By Compression, 65; Soldering, 65; Silver Soldering or Brazing, 67; Testing for Leaks, 67; Cleaning and Degreasing Solvents, 67; Review Questions, 69.

Chapter 3—Refrigeration 70
    Development of Refrigeration, 70; Structure of Matter, 70; Pressure, 72; Temperature and Heat, 75; Refrigeration Systems, 76; Review Questions, 79.

Chapter 4—Electricity and Refrigeration Controls 80
    Voltage, Current, and Resistance, 80; Ohm's Law, 80; Series Circuits, 81; Parallel Circuits, 82; AC and DC Power, 83; Power Rating of Equipment, 85; Capacitors, 85; Inductance, 92; Transformers, 93; Semiconductors, 95; Diodes, 96; Transistors, 97; Silicon-Controlled Rectifier (SCR), 97; Bridge Circuits, 98; Sensors, 99; Controllers, 101; Actuators, 104; Auxiliary Devices, 106; Review Questions, 106.

Chapter 5—Electric Motors and Controls 107
    DC Motors, 107; Right-Hand Rule, 109; Basic DC Motor Theory, 110; Multiloop Motors, 110; Shunt Motor, 110; Starting an Electric Motor, 110; Armature Speed, 111; Series Rheostat, 111; Armature Reaction, 112; Series Motor, 112; Compound Motor, 112; DC Motor Applications, 113; AC Motors, 113; Troubleshooting Electric Motors with a Volt-Ammeter, 123; Split-Core AC Volt-Ammeter, 124; Using the Megohmmeter for Troubleshooting, 128; Insulation Resistance Testing, 129; AC Motor Control, 131; Temperature Controls, 136; Defrost Controls, 138; Motor Burnout Cleanup, 140; Reading a Schematic, 141; Review Questions, 144.

Chapter 6—Refrigerants 145
    Classification of Refrigerants, 145; Common Refrigerants, 146; Refrigerant Properties, 156; Detecting Leaks, 159; Review Questions, 161.

TABLE OF CONTENTS

Chapter 7—Compressors 162
Refrigeration Compressors, 162; Condensers, 162; Hermetic Compressors, 163; Hermetic Compressor Motor Types, 176; Compressor Motor Relays, 180; Compressor Terminals, 181; Motor Mounts, 182; Crankcase Heaters, 183; Electrical Systems for Compressor Motors, 184; Compressor Connections and Tubes, 195; Rotary Compressors, 198; Review Questions, 201.

Chapter 8—Condensers and Cooling Towers 202
Condensers, 202; Chillers, 207; Cooling Towers, 211; Evaporative Condensers, 213; New Developments, 213; Review Questions, 214.

Chapter 9—Cooling Water Problems 215
Fouling, Scaling, and Corrosion, 215; Corrosion, 218; The Problem of Scale, 220; How To Clean Cooling Towers and Evaporative Condensers, 221; How to Clean Shell (Tube or Coil) Condensers, 223; Safety, 224; Solvents and Detergents, 225; Review Questions, 225.

Chapter 10—Evaporators 226
Coiled Evaporator, 226; Application of Controls for Hot-Gas Defrost of Ammonia Evaporators, 229; Valves and Controls for Hot-Gas Defrost of Ammonia-Type Evaporators, 246; Back-Pressure Regulator Applications of Controls, 249; Valve Troubleshooting, 251; Review Questions, 253.

Chapter 11—Refrigerant Flow Control 254
Metering Devices, 254; Fittings and Hardware, 255; Driers, Line Strainers, and Filters, 259; Liquid Indicators, 261; Thermostatic Expansion Valve (TEV), 263; Crankcase Pressure Regulating Valves, 269; Evaporator Pressure Regulating Valves, 271; Head Pressure Control Valves, 273; Discharge Bypass Valves, 278; Level Control Valves, 283; Level-Master Control, 285; Other Types of Valves, 290; Accumulators, 291; Rating Data, 292; Installation of the Accumulator, 292; Review Questions, 293.

Chapter 12—Refrigerators 294
Features of the Single-Door Refrigerator, 295; Cycle Defrost Refrigerator/Freezer, 297; No-Frost Top-Mount Refrigerator/Freezer, 298; No-Frost Side-By-Side Refrigerator/Freezer, 300; Icemaker, 301; Troubleshooting, 311; Defrosting System Analysis, 313; Rapid Electrical Diagnosis, 319; Energy-Saver Switch, 320; Condenser Fan Motor, 320; Run Capacitor, 320; Types of Compressors, 321; Use and Care of a Refrigerator/Freezer, 322; Touch-Up and Refinishing Procedure, 323; Troubleshooting the Electrical Components of a Refrigerator, 325; Identification of Refrigerator Parts in Troubleshooting, 327; Light Switches, 330; Tube Heaters, 331; Perimeter Tube and Drier Coil, 332; Freezer Divider Crossrail, 333; Defrost Timers, 333; Defrost Heaters, 335; Fan and Motor Assemblies, 337; Air Circulation, 338; Air Shutters, 338; Drain-Trough Heater, 339; Light Bulb Sockets, 339; Review Questions, 340.

Chapter 13—Servicing 341
Safety, 341; Servicing the Refrigerator Section, 343; Compressor Replacement, 344; Troubleshooting Compressors, 345; Troubleshooting Refrigerator Components, 345; Compressor Motor Burnout, 349; Cleaning System after Burnout, 350; Replacing the Filter Drier, 350; Replacing the Condenser, 352; Replacing the Heat Exchanger, 352; Repairing the Perimeter Tube (Fiberglass Insulated), 352; Replacing the Evaporator-Heat Exchanger Assembly, 356;

TABLE OF CONTENTS

Adding Refrigerant, 357; Low-Side Leak or Slight Undercharge, 357; High-Side Leak or Slight Undercharge, 358; Overcharge of Refrigerant, 358; Testing for Refrigerant Leaks, 358; Service Diagnosis, 359; Start and Run Capacitors, 362; Permanent Split-Capacitor (PSC) Compressor Motors, 363; Field Testing Hermetic Compressors, 363; Review Questions, 368.

Chapter 14—Freezers 369
Types of Freezers, 369; Installing a Freezer, 369; Freezer Components, 370; Complete Recharge of Refrigerant, 378; Overcharge of Refrigerant, 378; Testing for Refrigerant Leaks, 378; Portable Freezers, 378; Review Questions, 387.

Chapter 15—Commercial Refrigeration Systems 388
Refrigeration Cycle, 388; Air-Cooled Refrigeration Equipment, 389; Refrigeration Units, 391; Self-Contained Walk-in Coolers, 392; Self-Contained Unit Corner Section, 393; Sectional Walk-in Storage Coolers, 396; Commercial Refrigeration Units, 400; Open Produce Cases, 403; Ice Cream Display, 412; Outdoor Installation, 417; Reclaiming Heat and Air Conditioning, 417; Heat Reclaimer, 419; Review Questions, 423.

Chapter 16—Psychrometrics and Air Movement 424
Temperature, 424; Converting Temperatures, 424; Pressures, 425; Psychrometric Chart, 429; Air Movement, 433; Comfort Conditions, 434; Designing a Perimeter System, 436; Selection of Diffusers and Grilles, 439; Return Grilles, 442; Types of Registers and Grilles, 443; Fire and Smoke Dampers, 443; Ceiling Supply Grilles and Registers, 443; Ceiling Diffusers, 444; Linear Grilles, 447; Review Questions, 447.

Chapter 17—Comfort Air Conditioning 448
Window Units, 448; Evaporators for Add-On Residential Use, 451; Remote Systems, 455; Single-Package Rooftop Units, 455; Refrigerant Pipe Sizes, 458; Mobile Homes, 463; Review Questions, 467.

Chapter 18—Commercial Air-Conditioning Systems 468
Expansion-Valve Air Conditioning System, 468; Troubleshooting, 469; Packaged Cooling Units, 469; Direct Multizone System, 473; Evaporative Cooling System, 475; Absorption-Type Air Conditioning Systems, 476; Chilled Water Air Conditioning, 477; Chillers, 481; Console-Type Air Conditioning Systems, 483; Review Questions, 487.

Chapter 19—Heat Pumps, Gas-Fired Air Conditioners, and Solar Air Conditioners 488
Gas Air Conditioning, 488; Gas-Fired Chillers, 489; Chiller-Heater, 490; Absorption Refrigeration Machine, 494; Heat Pumps, 500; Review Questions, 504.

Chapter 20—Load Estimation and Insulation 505
Refrigeration and Air Conditioning Load, 505; Running Time, 505; Calculating Cooling Load, 506; Calculating Heat Leakage, 507; Calculating Product Cooling Load, 507; Air Doors, 509; Insulation, 509; Review Questions, 512.

Chapter 21—Wire Sizes and Voltage Requirements for Air Conditioning and Refrigeration Equipment 513
Choosing Wire Size, 513; Wire Size and Low Voltage, 514; The Effects of Voltage Variations on

TABLE OF CONTENTS

AC Motors, 515; Selecting Proper Wire Size, 516; Unacceptable Motor Voltages, 517; Calculating Starting Current Values and Inrush Voltage Drops, 519; Code Limitations on Amperes per Conductor, 520; Heat Generated Within Conductors, 521; Circuit Protection, 521; Fuses, 522; Review Questions, 523.

Chapter 22—Careers **524**

Industries that Employ Air-Conditioning and Refrigeration Mechanics, 524; Job Qualifications, 525; The Future, 527; Pay and Benefits, 527; Teaching as a Career, 528; More Information, 529; Review Questions, 529.

Index **530**

1

Mathematics

BASIC MATHEMATICS

Work on any air conditioning system requires a knowledge of liquids and gases. These liquids and gases must be moved from one place to another through pipes and tubing. In many cases, the pipes and tubing must be cut to the proper length in order to be fitted into the system. For instance, a length of pipe may be ten feet long. Unions, tees, elbows, and valves must be subtracted from this length to get the proper fit. Here, and in other cases, a knowledge of basic mathematics is necessary.

Numbers are composed of digits. Consider, for example, the number 527. The 5, the 2, and the 7 are all *digits*. Numbers between 1 and 9 are *one*-digit numbers. Numbers from 10 to 99 are made up of *two* digits. Numbers between 100 and 999 are composed of *three* digits. Any number can be defined in terms of the number of digits. Examine the number 1,245,683,967.

Note that the number is marked off by commas in *groups* of three, starting from the right, or with the number 7. Then, read from right to left. The digits are called *units, tens, hundreds, thousands*, and so on. Read the number now. It becomes one billion, two-hundred forty-five million, six hundred eighty-three thousand, nine hundred sixty-seven. Fig. 1-1.

Addition

Addition is the process of totaling two or more numbers. Like objects can be added. Frequently it may be necessary to add the number of values or the length of pieces of pipe or tubing. However, it is necessary to place the numbers in a column according to their values. Use the plus sign (+) to indicate addition. For instance;

```
   27
  950
 8281
    4
 ----
 9262
```

1. Add the units first, or 7 + 0 + 1 + 4. They total 12.

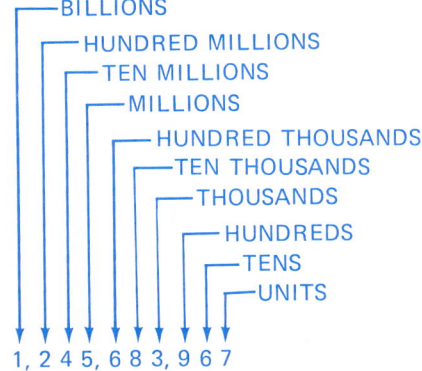

1-1. Reading a number.

2. Place the 2 in the unit column under the horizontal line. *Carry* the 1 to the next column, (the tens column).

3. Total the tens column (2 + 5 + 8). Add the 1 carried from the units column. The total is 16.

4. Place the 6 in the tens column beneath the line. Carry the 1 to the hundreds column.

5. Add the hundreds column (9 + 2). Add the 1 carried over from the tens column. The result is 12.

6. Enter the 2 in the hundreds column beneath the line. Carry the 1 to the thousands column.

7. Add the 8 and the 1 carried over from the hundreds column. The result is 9.

8. Counting from the right, mark off each group of three digits with a comma.

9. The number is read as nine thousand, two hundred and sixty-two. Check your work by adding from the bottom of the column to the top or:

```
    4
 8281
  950
   27
 ----
 9262  or  9,262.
```

If the same total is obtained using bottom to top and top to bottom addition, then the answer is correct.

11

Subtraction

Subtraction is a process whereby the *difference* between two numbers is found. The minus sign (−) is used to indicate the process of subtraction. The larger of the two numbers is written on top. Make sure the digits line up in the proper columns. For instance:

$$\begin{array}{r} 983 \\ -231 \\ \hline 752 \end{array}$$

Start from the right. One from 3 equals 2. Place the 2 under the horizontal line and go to the next column. Subtracting 3 from 8 gives 5. Place the 5 under the horizontal line. Then, go to the next column and subtract 2 from 9 to produce 7. Enter the 7 in that column under the 2. The result is 752, or as it is read, seven hundred and fifty-two. Check your work by adding the answer: 752 and 231. The result should be 983.

Another type of subtraction problem will give further exercise. For example:

$$\begin{array}{r} 358 \\ -279 \\ \hline 79 \end{array}$$

Start with the right-hand column, where 9 is subtracted from 8. Since 8 is smaller than 9, we have to *borrow* one from the next column. This means the 8 is now 18 and the 5 in the other column becomes 4. Now, subtract 9 from 18. The answer is 9. Place the 9 in the proper column under the horizontal line. Move to the next column. Note that the 4 is smaller than the 7. Thus, 1 must be *borrowed* from the next column. The 5 now becomes 14. Subtract 7 from 14. The result is 7. Place the 7 under the horizontal line and in line with the 7 above the line. Now, move to the hundreds column. The 3 in this column is now 2, since 1 has been borrowed from it. This means you subtract 2 from 2, giving an answer of 0. Therefore, no number is entered. The answer for the problem is 79. To check your answer, add 79 and 279 to see if they produce 358.

$$\begin{array}{r} 79 \\ +279 \\ \hline 358 \end{array}$$

If so, 79 is the correct answer to this subtraction problem.

Multiplication

Multiplication is another method of addition. It can, however, be done more quickly. A multiplication sign (×) is used to indicate multiplication. The × is read as "times." Thus, 2 times 2 (2 × 2) is 4. The *equals* (=) sign is used to separate the operation and the answer. The answer, or result of multiplication, is called the *product.* Thus the product of 2 times 2 is 4.

Multiplication tables can help you solve a problem rapidly. However, you should memorize the multiplication tables for numbers up to 12. Then the more complicated multiplication problems can be solved rather easily.

In multiplying two numbers, place the numbers as follows:

$$\begin{array}{r} 35 \\ \times\ 2 \\ \hline 70 \end{array}$$

First, multiply 2 by 5. The answer is 10. Place the 0 under the horizontal line. Add the 1 to the next column for adding to the next step.

Second, multiply 2 by 3 to get 6. Add the 1 from the previous operation to obtain 7. Enter the 7 to the left of the 0. The answer is 70. To check this add 35 and 35:

$$\begin{array}{r} 35 \\ +35 \\ \hline 70 \end{array}$$

Note how the columns line up when numbers of three digits are multiplied.

$$\begin{array}{r} 321 \\ \times 219 \\ \hline 2889 \\ 321 \\ 642 \\ \hline 70299 \end{array}$$

To check the answer, reverse the two and multiply again.

$$\begin{array}{r} 219 \\ \times 321 \\ \hline 219 \\ 438 \\ 657 \\ \hline 70299 \end{array}$$

Division

The process used to determine the number of times one number is contained in another is called *division.* For instance, 9 has three 3's and 10 has two 5's. The terms *divisor, dividend,* and *quotient* are all used in division. A divisor is the number by which the dividend is divided. The dividend is the number that is divided. The quotient is the result of division.

$$5\overline{\smash{)}10}^{\,2}$$

In the above problem, 5 is the divisor, 10 is the dividend, and 2 is the quotient.

The sign used to indicate division is ÷. The ÷ sign is read as "divided by."

Check your division skill by dividing 9,856 by 2.

12

$$\begin{array}{r}4\\2\overline{)9856}\\-8\\\hline 1\end{array}$$

Two is contained in 9 four times. The 4 is placed above the 9. Now, multiply 4 by 2. The result is 8. This 8 is subtracted from 9 to produce 1.

$$\begin{array}{r}4\\2\overline{)9856}\\-8\\\hline 18\end{array}$$

Next, bring down the 8 to make 18. Determine the number of times 2 is contained in 18. The answer is 9. Place the 9 over the 8:

$$\begin{array}{r}49\\2\overline{)9856}\\-8\\\hline 18\end{array}$$

Multiply 9 by the divisor (2). The product is 18. Place this 18 under the one already found previously:

$$\begin{array}{r}49\\2\overline{)9856}\\-8\\\hline 18\\-18\end{array}$$

Subtract this 18 from the existing 18. This produces 0. Move on to the next step. Bring down the 5 from the dividend:

$$\begin{array}{r}49\\2\overline{)9856}\\-8\\\hline 18\\18\\\hline 5\end{array}$$

The number 5 contains two 2s, so the 2 is placed above the line and over the 2:

$$\begin{array}{r}492\\2\overline{)9856}\\8\\\hline 18\\18\\\hline 5\\-4\\\hline 1\end{array}$$

Multiply the 2 in the quotient times the 2 in the divisor to produce 4. Place the 4 under the 5 and subtract. The answer is 1. Next, bring down the 6 from the dividend to produce 16:

$$\begin{array}{r}492\\2\overline{)9856}\\8\\\hline 18\\18\\\hline 5\\4\\\hline 16\end{array}$$

Next, determine that the 16 contains 8 2s. Place the 8 above the 6 in the dividend.

$$\begin{array}{r}4928\\2\overline{)9856}\\8\\\hline 18\\18\\\hline 5\\4\\\hline 16\\-16\\\hline\end{array}$$

Now, multiply the 8 in the quotient by the 2 in the divisor to get 16. Place the 16 under the existing 16 and subtract. Subtraction produces a 0. Since no remainder exists, the problem is solved.

The number of 2s in 9,856 is 4,928.

To check for accuracy, multiply the quotient by the divisor. *The result should be the dividend.*

Thus, 9,856 ÷ 2 is 4,928.

FRACTIONS

Fractions come in all sizes. They are used to indicate some part of a whole. For example, the whole may be a dollar. The fractional parts of a dollar are one-half ($\frac{1}{2}$), one-quarter ($\frac{1}{4}$), one-tenth ($\frac{1}{10}$) usually referred to as a dime, a nickel ($\frac{1}{20}$), and the penny, or cent ($\frac{1}{100}$). These are everyday uses of fractions. There are other fractional units we use in daily activities: $\frac{1}{4}$ mile, $\frac{1}{2}$ bushel, $\frac{3}{4}$ of an inch. Fractions will be used until the metric system of measurement is used for all things. The metric system measures in decimals, instead of fractions.

Numerators and Denominators

The upper number in a fraction is called the *numerator*. The lower part is called the *denominator*. The denominator tells you how many parts the whole is divided into. For instance, $\frac{3}{8}$, or three-eighths, tells us there are eight parts in the whole. The numerator means we are referring to three parts of that whole.

Note here how the denominator indicates size:

$\frac{1}{2}$ is *more* than $\frac{1}{4}$, or $\frac{1}{4}$ is *less* than $\frac{1}{2}$.

If the numerators are the same, the larger the denominator, the smaller the fraction.

If the denominators are the same, the larger the numerator, the larger the fraction. For example, $\frac{3}{8}$ is larger than $\frac{1}{8}$.

Proper Fractions

Fractions that indicate less than *one* are called *proper fractions*. For example, $\frac{1}{2}$, $\frac{1}{3}$, $\frac{1}{4}$, $\frac{1}{8}$,

$1/16$ are proper fractions. In proper fractions, the numerator is *equal to or smaller than* the denominator.

If the numerator and denominator are the same ($4/4$ or $5/5$), the number equals 1. However, if the numerator is less than the denominator ($1/2$, $3/4$, or $5/16$) then the fraction is less than 1.

Improper Fractions

A fraction is called *improper* when it is equal to one ($5/5$) or *more* than one ($5/4$, $3/2$, $9/8$). This means the numerator is equal to or larger than the denominator.

Mixed Numbers

Mixed numbers are made up of a whole number and a fraction. For instance, $1\frac{1}{2}$, $2\frac{3}{4}$, and $5\frac{1}{8}$ are mixed numbers.

Lowest Terms

Fractions are usually reduced to their lowest terms. This means that both the numerator and denominator are numbers that cannot be evenly divided by any other number. For example, $3/6$ can be reduced to $1/2$ since both the 3 and 6 can be divided by 3 to produce $1/2$:

$$\frac{3 \div 3}{6 \div 3} = \frac{1}{2}$$

One-half ($1/2$) is the *lowest term* for $3/6$, since the 1 and the 2 cannot be further reduced by dividing to produce an even number without a remainder.

Reducing fractions to their lowest terms is accomplished by dividing both numerator and denominator by a divisor common to both numbers. Thus, $14/21$ is reduced by 7.

$$\frac{14 \div 7}{21 \div 7} = \frac{2}{3}$$

To have a *correct solution*, the answer must be stated with fractions in their lowest terms.

Converting Fractions

It is often necessary to convert fractions and whole numbers to improper fractions to multiply them. Three steps are usually needed to change a mixed number to an improper fraction:

Step 1. Multiply the denominator of the fraction by the whole number.
Step 2. Add the numerator and the result of Step 1.
Step 3. Place the result over the denominator.

Example:

$3\frac{3}{4}$ is changed to an improper fraction by the following three steps:

Step 1. The *whole number* (3) is multiplied by the denominator (4) to give a result of 12.
Step 2. The *numerator* (3) is added to the result of Step 1 (12) to give a result of 15.
Step 3. The result of Step 2 (15) is placed above the denominator.

Thus, $\frac{15}{4}$ is the improper fraction of $3\frac{3}{4}$.

Multiplication of Fractions

If you have a piece of pipe in your hand how long must it be should you have to produce three pieces, each $3/8''$ long?

$$\frac{3 \times 3}{8 \times 1} = \frac{9}{8}$$

The answer is an improper fraction.

(Note: The whole number is placed over 1)

Reduce the improper fraction by dividing the numerator by the denominator:

$\frac{9}{8}$ equals 1 with a remainder of 1.

The remainder goes over the denominator or $1\frac{1}{8}''$ is the answer. A length of pipe $1\frac{1}{8}''$ long will produce three pieces, each $3/8''$ long.

Division of Fractions

Sometimes it becomes necessary to divide fractions. For instance, if you had a piece of tubing $7/8''$ long but wanted to use it to make four bushings, how long would the four equal bushings be?

Step 1. State problem: $\frac{7}{8} \div 4$

Step 2. Place 1 beneath the whole number: $\frac{7}{8} \div \frac{4}{1}$

Step 3. Invert the last fraction and multiply:

$$\frac{7}{8} \times \frac{1}{4} = \frac{7}{32}$$

The result is four bushings, each $7/32''$ long.

Cut a piece of pipe $15/16''$ long into three equal parts. What would be the length of each part?

$$\frac{15}{16} \div \frac{3}{1}$$

$$\frac{15}{16} \times \frac{1}{3} = \frac{15}{48}$$

Further reduce $15/48$ by dividing both numerator and denominator by 3, or,

$$\frac{15 \div 3}{48 \div 3} = \frac{5}{16}''$$

Thus, three $5/16''$ lengths of pipe are contained in one piece $15/16''$ long.

CHAPTER 1—MATHEMATICS

DECIMALS

Fractions have a numerator and a denominator. Fractions can be expressed as a *decimal*. A decimal is a fraction with ten, or some multiple of ten, acting as a denominator.

Decimals are indicated by placing a *decimal point* (.), before the number. If 10 is written as a decimal, the point is placed after it (10.0). The unit one-tenth ($1/10$) can be written as 0.1. The location of the decimal determines the value of the unit. Fig. 1-2.

```
Decimal     Fraction
0.1      = 1/10         (one-tenth)
0.01     = 1/100        (one-one hundredth)
0.001    = 1/1000       (one-one thousandth)
0.0001   = 1/10,000     (one-ten thousandth)
0.00001  = 1/100,000    (one-hundred thousandth)
0.000001 = 1/1,000,000  (one-millionth)
```

Examples:
0.2 = two-tenths
0.20 = twenty hundredths
0.234 = two hundred thirty four thousandths

Combinations of whole numbers and decimals are the same as fractions. Such numbers are called *mixed* decimals. For example, 5.1 is five and one-tenth.

One important use of the decimal is in our money system. The *dollar* is divided into hundredths (cents). The *mixed* decimal is used to express amounts of money. Take, for instance, $12.57. Move the decimal one point to the right and it becomes $125.70. If the decimal is moved one decimal point to the left, $12.57 becomes $1.257 or one dollar twenty five and seven-tenths cents.

Addition of Decimals

To add decimals, it is very important to line up the decimal points. Note how the whole numbers and decimals are aligned.

```
   1.15          1.15
  13.47       2308.62
 103.21        103.21
+2308.62      + 13.47
────────     ────────
 2426.45      2426.45
```

Subtraction of Decimals

In subtracting decimals, make sure the decimal points are aligned. Some trouble may occur if the top number is a whole number and the bottom number is a mixed decimal. For example,

```
  45.
−21.673
```

Add the correct number of zeros to the whole number. This will balance each decimal part of the lower number.

```
 45.000
−21.673
```

Subtract as usual.

```
 45.000
−21.673
───────
 23.327
```

SQUARE ROOT

Many formulas use square root. For instance, square root can be used to determine the length of one side of a triangle when the lengths of the other sides are known.

The *root* of a number is that number which, when multiplied by itself a given number of times, will equal the given number. For example, the square root of 25 is 5, since 5 × 5, or 5^2, equals 25.

The symbol used to indicate square root is $\sqrt{}$. This symbol is called a *radical sign*.

Finding the Square Root of a Number

In some instances, the square root can be determined mentally. Thus, a knowledge of multiplica-

1-2. A decimal fraction.

MILLIONS	HUNDRED THOUSANDS	TEN THOUSANDS	THOUSANDS	HUNDREDS	TENS	UNITS	DECIMAL	TENTHS	HUNDREDTHS	THOUSANDTHS	TEN THOUSANDTHS	HUNDRED THOUSANDTHS	MILLIONTHS
5	3	2	8	7	6	4	.	2	1	3	4	5	6

5,328,764.213456

tion is important. For example, $\sqrt{4} = 2$, or $2 \times 2 = 4$; $\sqrt{9} = 3$, or $3 \times 3 = 9$; $\sqrt{16} = 4$, or $4 \times 4 = 16$; $\sqrt{25} = 5$, or 5×5 is 25. Similarly, $\sqrt{144} = 12$, since $12 \times 12 = 144$.

In most cases, the square root of a number must be determined by a mathematical process. If the number is a perfect square, such as 4 is 2×2, the square root will be an integral number. If the number is not a perfect square (3398.89), the square root will be a *continued decimal*.

Find the square root of 3398.89.

Step 1. Starting at the decimal point, mark off the digits in pairs in both directions. Leave a space between each pair of digits.

$$\sqrt{33\ 98.89}$$

Step 2. Place the decimal point for the answer directly above the decimal point that appears under the radical sign.

$$\sqrt{33\ 98.89}$$

Step 3. Determine by inspection the largest number that can be squared without exceeding the first pair of digits—33. The answer is 5, since the square of any number larger than 5 will be greater than 33. Place the 5 above the first pair of digits.

$$\begin{array}{r} 5\quad . \\ \sqrt{33\ 98.89} \end{array}$$

Step 4. Square 5 to obtain 25, and place it under 33. Subtract 25 from 33 and obtain 8. Bring down the next pair of digits—98.

$$\begin{array}{r} 5\quad . \\ \sqrt{33\ 98.89} \\ -25 \\ \hline 8\ 98 \end{array}$$

Step 5. Double the answer, 5, to obtain a trial divisor of 10. Divide the trial divisor into all but the last digit of the modified remainder. It will go into 89 eight times. Place the 8 above the second pair of digits. Also place the 8 to the right of the trial divisor. Thus, the true divisor is 108. Multiply 108 by 8 and obtain 864. Subtract 864 from 898 to obtain 34. Bring down the next pair of digits—89.

$$\begin{array}{rl} & 5\ \boxed{8}. \\ & \sqrt{33\ 98.89} \\ 2 \times 5 = 10\boxed{8} & -25 \\ & 8\ 98 \\ 8 \times 108 = 864 & -8\ 64 \\ & 34\ 89 \end{array}$$

With each new successive digit in the answer:
1. Place the digit in the answer above the pair of digits involved.
2. Place the same digit to the right of the trial divisor to obtain the true divisor.
3. Multiply the digit by the true divisor. (Do not use the square boxes in actual problems.)

Step 6. Double the answer, 58, to obtain a trial divisor of 116. Divide the trial divisor into all but the last digit of the remainder. It will go into 348 three times. Place the 3 above the third pair of digits. Also place the 3 to the right of the trial divisor. Thus, the true divisor is 1163. Multiply 1163 by 3 to obtain 3489. Subtract 3489 from 3489. There is no remainder. Therefore, 3398.89 is a perfect square and its square root is 58.3.

$$\begin{array}{rl} & 5\quad 8.\boxed{3} \\ & \sqrt{33\ 98.89} \\ & 25 \\ 2 \times 58 = 116\boxed{3} & 8\ 98 \\ & 8\ 64 \\ & 34\ 89 \\ 3 \times 1163 = 3489 & 34\ 89 \end{array}$$

Step 7. Check the answer by squaring 58.3, or $58.3^2 = 3398.89$.

AREAS AND VOLUMES

When working with air conditioning and refrigeration, it is necessary to know the volume of a room. This indicates the amount of heat that must be removed to cool the room. A refrigerator is usually rated according to its cubic foot capacity. Volume calculations are also needed to find the capacity of a pipe or any system. Volume is concerned with the amount of air, liquid, or gas that can be contained within an object. The volume of air that must be moved must be known to calculate the size of a fan or refrigeration unit.

Area is the measure of a surface. Some customary units of area measurement are square inches, square feet, square yards, and square miles. In the metric system of measurement, square centimeters, square meters, and

CHAPTER 1—MATHEMATICS

square kilometers are used. In most cases you will be working with square inches and square feet. A square foot is 12 inches by 12 inches, or 12″ × 12″ = 144 square inches = 1 square foot. To convert square feet to square inches, multiply by 144. To convert square inches to square feet, divide by 144.

Houses, factories, and commercial buildings are designed in several different shapes. You may have to be able to figure square footage for each shape.

Areas

RECTANGLE

A *rectangle* is an object with four sides. Opposite sides are parallel. All corners are 90°. Most rooms are rectangles.

The *area* of a rectangle is found by multiplying its height (altitude) by its width (base). Expressed as a formula this is A (area) = a (height or altitude) × b (width or base). For example, the altitude, or height, of a rectangle is 3. The base, or width, is 12. If 3 is multiplied by 12, the area of the rectangle is 36. If the measurement is in inches, then the area is 36 square inches. If the measurement is in meters, then the area is 36 square meters (m²). Take a closer look at Fig. 1-3. There are 36 squares, each of the same size. In this case, the squares represent inches.

The area was found by multiplying the altitude by the base. Thus, we can find the altitude if the area and base are known:

$$\text{Altitude} = \frac{\text{Area}}{\text{Base}} \quad \text{or} \quad \frac{36}{12} = 3$$

The base can be found if the altitude and the area are known:

$$\text{Base} = \frac{\text{Area}}{\text{Altitude}} \quad \text{or} \quad \frac{36}{3} = 12$$

PARALLELOGRAM

A *parallelogram* is a four-sided figure that has opposite sides parallel and equal in length.

The *area* of a parallelogram may be found by multiplying its base by its altitude, or A = ab. The altitude is measured on a perpendicular line from the base to the opposite side. Fig. 1-4.

TRIANGLE

A *triangle* is a three-sided figure that has three lines intersecting by twos in three points that form angles.

The *area* of a triangle is found by using one-half of a parallelogram. Note that Fig. 1-5 is one-half of Fig. 1-4. Therefore, the area of a triangle is one-half that of a parallelogram, or A = ½ ab.

CIRCLE

A *circle* is a closed plane curve such that all its points are equally distant from a point within, the center.

The *area* of a circle is found by multiplying the square of the *radius* by pi. A = πr². Pi (π) is a Greek letter used to indicate a constant number, 3.141592654+. The radius is equal to one-half the diameter. Fig. 1-6.

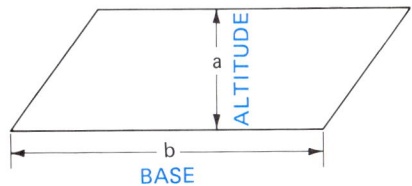

1-4. The area of a parallelogram equals its altitude times its base.

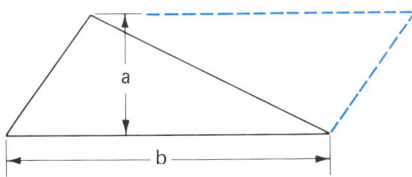

1-5. The area of a triangle equals ½ ab. Or, one-half the area of a parallelogram.

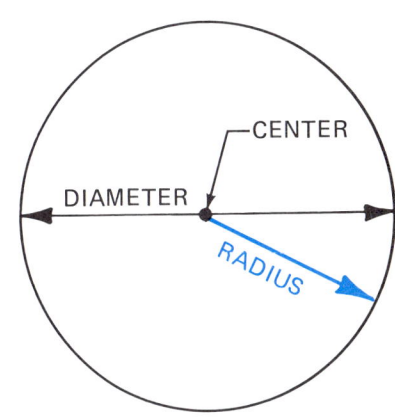

1-6. The area of a circle is equal to the square of the radius times pi. Pi equals 3.141592654+.

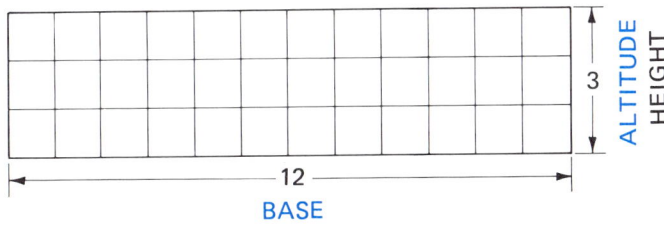

1-3. The area of a rectangle equals altitude times base.

17

REFRIGERATION AND AIR CONDITIONING TECHNOLOGY

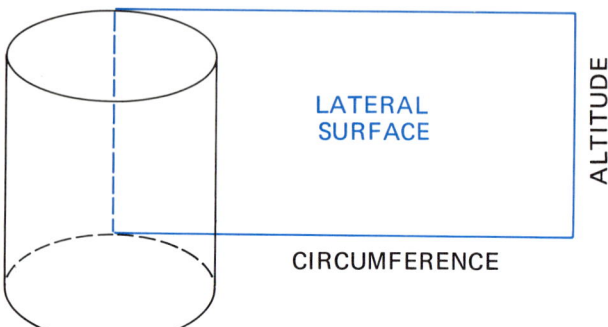

1-7. The area of a cylinder. The area of the lateral surface of a cylinder equals 2πra. Top and bottom can be found in terms of using a circle's formula or 2 × πr².

CYLINDER

A *cylinder* is a modified version of a rectangle. Figure 1-7 shows that the lateral surface of a cylinder, if it could be peeled off (as when a label is removed from a can) would form a rectangle. The altitude of a cylinder is the height. The base is equal to the circumference. Circumference is equal to pi multiplied by the diameter, or C = πd.

The area of the base of the cylinder is that of a circle, or πr². There is also a top to the cylinder. The top is a circle, too. Its area is also equal to πr². To find the area of a cylinder, you must find the area of the lateral surface and then *add* the top and bottom areas. The formula becomes:

A = 2πra + (2 × πr²)

(2πr² represents area of top and bottom circles.)

A = area
π = 3.14159
r = radius
a = altitude

Volume

Volume is the measure of space. The unit of measure is the cube, such as cubic inches, cubic feet, or cubic centimeters (cm³). A cubic foot has 1,728 cubic inches (12″ × 12″ × 12″ = 1,728 *cubic* inches). To convert cubic inches to cubic feet, divide by 1728.

RECTANGULAR SOLID

Figure 1-8 shows a number of cubes or blocks stacked up to form one large block. The number of blocks in the large block can be determined by multiplying the number of blocks in the length by the number in the width by the number in the height. If the length is called the base, then use *b* to represent it. If the width is called the thickness, then use *t* to represent it. If the height is called the altitude, then use *a*. Thus, the formula becomes V = a b t.

Sometimes, it is necessary to cool a solid mass. This formula determines the volume of a mass so you can determine the number of Btu needed for cooling.

CYLINDER

The volume of a cylinder is

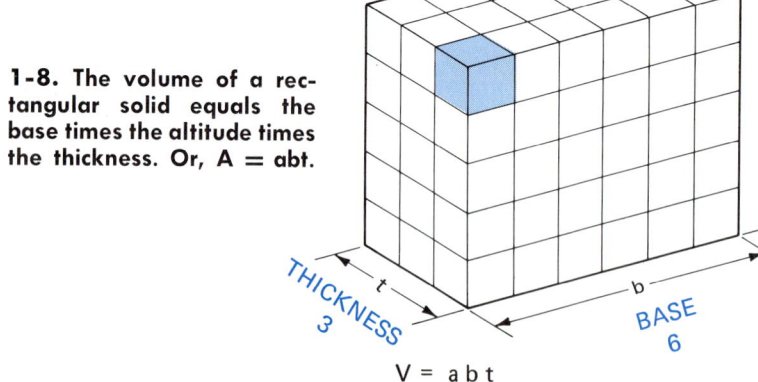

1-8. The volume of a rectangular solid equals the base times the altitude times the thickness. Or, A = abt.

V = a b t
V = 5×6×3 = 90
THERE ARE 90 SMALL BLOCKS IN THE LARGE BLOCK.

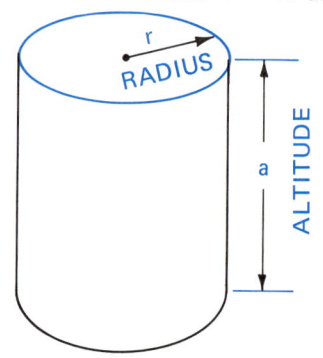

1-9. The volume of a cylinder is equal to the area of the base or top times the altitude. V = πr²a.

18

formed in a manner similar to that of the rectangular solid. Fig. 1-9. The area of the end of a cylinder is found by using the formula for a circle. The volume is then found by multiplying this area by the cylinder's altitude. Or, $V = \pi r^2 a$.

V = volume
π = 3.14159
r = radius
a = altitude

EQUATIONS

An *equation* is a statement of equality. For example, A = B is an equation. The expression to the left of the equals sign is called the *left-hand member* of the equation. The expression to the right of the equals sign is called the *right-hand member*. Finding the values of the unknown quantities of an equation is known as "solving the equation." The answer is called the *solution*. If only one unknown is involved, the solution is called the *root*.

To solve an equation means to determine the value of the unknown number. There are rules to be used to help solve equations.

Rule: *If equal numbers are added to equal numbers, the sums are equal.*

For example, use this equation: X − 6 = 10
Add 6 to each side.

$$X - 6 + 6 = 10 + 6$$

Collect terms.

$$X = 10 + 6 \quad \text{or} \quad X = 16$$

Rule: *If equal numbers are subtracted from equal numbers, the remainders are equal.*
Equation: X + 2 = 7
Subtract 2 from each side.

$$X + 2 (-2) = 7 (-2)$$

Collect terms.

$$X = 5$$

Rule: *If equal numbers are multiplied by equal numbers, the products are equal.*
Equation: ½X = 5
Multiply each side by 2.
  a. Left side of equal sign: 2 × ½ = 1. Therefore, 2 × ½X = 1X, or X.
  b. Other side of equal sign: 2 × 5 = 10
  c. Therefore, X = 10

Rule: *If equal numbers are divided by equal numbers, not by zero, the quotients are equal.*
Equation: 6X = 12
Divide both sides by 6.
  a. $\frac{6X}{6} = X$.
  b. $\frac{12}{6} = 2$.
  c. Therefore, X = 2

Rule: *Like roots of equal numbers are equal.*
Equation: $X^2 = 25$
Take the *square root* of both sides.
  a. $\sqrt{X^2} = X$
  b. $\sqrt{25} = 5$
  c. Therefore, X = 5

Rule: *Like powers of equal numbers are equal.*
Equation: X = 8
Square both sides.
  a. $X \times X = X^2$
  b. 8 × 8 = 64
  c. Therefore, $X^2 = 64$

Transposing

If it is desired that the unknown is always on the left and the known on the right, then transposing can be used.

When a term is added to or subtracted from both sides of an equation, it is *transposed* from one side to the other. Its sign is changed. For example,

$$X - 5 = 2$$

the −5 on the left side becomes +5 on the right side:

$$X = 2 + 5$$
$$X = 7$$

This places the unknown on the left and the known on the right. X is an unknown value and 7 is a known value.

Proportions

When an equation is made up of two fractions, there is a proportion. For instance,

$$\frac{3}{4} = \frac{6}{8} \quad \text{or} \quad \frac{3}{4} = \frac{3}{4}$$

when $\frac{6}{8}$ is reduced to $\frac{3}{4}$

Another form of proportion is 3:4 = 6:8. This is read "three is to 4 as 6 is to 8."

In solving problems, always use the fractional form:

$$\frac{3}{4} = \frac{6}{8}$$

Cross multiply:

3 × 8 = 24   and
    4 × 6 = 24   or   24 = 24.

Translating Word Problems into Equations

Many practical problems are stated in words. These must be translated into symbols before the rules of algebra can be applied. There are no specific rules for the translation of a written problem into an equation of numbers, signs, and symbols. Here are two general suggestions that may prove helpful in developing equations:

REFRIGERATION AND AIR CONDITIONING TECHNOLOGY

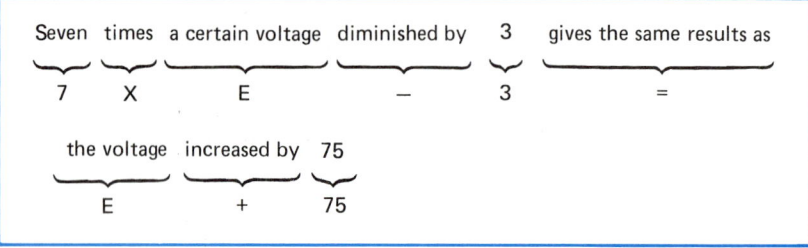

1-10. Translating a mathematical problem from words to symbols.

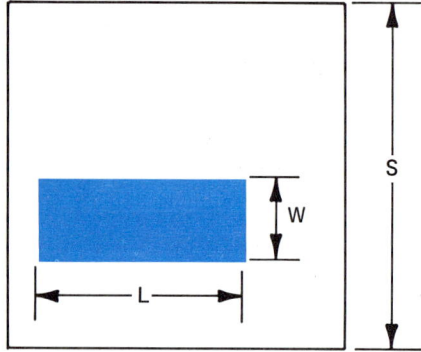

1-11. The area of a square with a rectangle removed is found by using the formula $A = S^2 - LW$.

1. From the worded statement of the problem, select the unknown quantity (or one of the unknown quantities) and represent it by a letter, such as X. Write the expression, stating exactly what X represents and the units in which it is measured.

2. If there is more than one unknown quantity in the problem, try to represent each unknown in terms of the first unknown.

EXAMPLE

In simple problems, an equation may be written by an almost direct translation into algebraic symbols. Fig. 1-10.

Solving the equation:

$$7E - 3 = E + 75$$
$$7E - E = 75 + 3$$
$$6E = 78$$
$$E = 13$$

Solution: Voltage is 13 volts.

FORMULAS

To solve a problem in mathematics it is often necessary to check for rules that state conditions of operation. It is sometimes hard to tell from a rule just how different values enter into the solution. It is best to set down a statement of the values concerned. A simple method for doing this is to represent each value by a letter or other mark. An example of this is the use of the Greek letter pi (π) to represent 3.141592654. In some cases, Greek letters and our own alphabet are mixed in a formula. By writing down the values in symbol form, you develop a formula.

PROBLEM: What is the volumetric efficiency of a compressor that has a volume of 10 cubic inches, but pumps only 7 cubic inches of gas each stroke of the piston?

First, develop a series of letters to represent the various values to be considered, as below:

V_E = volumetric efficiency
V_G = volume of gas pumped per stroke
V_C = volume of compressor cylinder

$V_E = \dfrac{V_G}{V_C}$, or volumetric efficiency is equal to the volume of gas pumped per stroke divided by the volume of the compressor cylinder.

This is a *formula*, or a fixed rule, indicating relationships of the various quantities.

Secondly, place the known values in the formula and solve mathematically:

$$V_E = \frac{7}{10} \quad \text{or} \quad V_E = .7 \text{ or } 70\%$$

efficient.

A formula can also be used for finding the area of a square with a rectangle removed. Fig. 1-11.

Develop a formula for solving the problem. Or, represent the length of the rectangle by L and the width with W. The side of the square can be represented by S. The area of the square is now found. The area of the rectangle is found and its value subtracted from the area of the square. If everything is measured in inches, then the answer would be in square inches. If the area had measurements in feet, then the answer would be in square feet. However, if the metric system of measurement were used, the answer would be in square centimetres.

If the area is represented by the letter A, then the formula becomes:

$$A = S^2 - LW$$

THE METRIC SYSTEM OF MEASUREMENT

Two systems of measurement are in use in the United States today—the English, or customary system, and the metric system. The English system is based on the foot and the pound. The metric system is based on the centimetre (or metre) and the gram (or kilogram). Both systems are used in electronics and in air conditioning work. Table 1-A

CHAPTER 1—MATHEMATICS

lists the metric base units and their symbols.

Metric prefixes are used in electronics in combination with the basic units of measurement. Table 1-B gives the meaning of those prefixes with respect to various units of measurement. Thus, for example, a megavolt is one million volts. A microvolt is one-millionth of a volt.

In the metric system of measurement, in numbers of five digits and more, a space (rather than a comma) is inserted every three digits, counting from the decimal point. Thus, 10,000 is written 10 000. The same rule applies to decimals: 0.000001 is written 0.000 001.

Conversion Factors

Table 1-C lists the common customary and metric units of measurement and the factors by which these units must be multiplied to convert them to other units. Table 1-B is very useful when dealing with a single measuring system or with two.

READING GRAPHICS

Engineering graphs of operational or experimental data are constructed in the same manner as graphs of equations. First, a chart is compiled of the data available. The data is plotted on an axis. The independent variable (the variable to which values are assigned) usually is plotted on the X axis. The dependent variable is plotted on the Y axis. The scales on the axes should be as large as practicable. At the same time, keep the graph within the space available.

Figure 1-12 is an example of a graph constructed from information gathered earlier. In fact, the information as it was gathered is shown in Table 1-D.

After a study of Table 1-D, it will be found that it is more convenient to use a much smaller unit of measurement on the X axis than on the Y axis. Also, the entire graph falls in the first quadrant, since all values are positive. The resulting graph, Fig. 1-12, is the current-resistance characteristic of the circuit. Note that the current decreases as the resistance increases. The current for any value of the variable resistance can be found by reading the graph.

The process of finding properties of a function by inspection of the graph representing it is called *reading the graph*.

MEASURING TEMPERATURE

Temperature Scales

There are two standard temperature scales used by the refrigeration technician. Both of these scales use water as the reference for freezing and boiling.

Water freezes at a constant temperature. It also turns to steam at a constant temperature. The thermometer is marked according to these two points.

The customary, or English, system of measurement measures temperature in degrees Fahrenheit. On this scale, water freezes at 32° Fahrenheit and turns to steam at 212° Fahrenheit. The

Table 1-A.
SI Metric Base Units

Quantity	Unit	Symbol
Length	metre	m
Mass	kilogram	kg
Time	second	s
Electric current	ampere	A
Temperature	kelvin	K
Luminous intensity	candela	cd
Amount of substance	mole	mol

Table 1-B.
Metric Prefixes

Metric prefix		Meaning		Associated with
Mega	M	Million	(1 000 000)	Volts, ohms, hertz, amperes
Kilo	k	Thousand	(1 000)	Volts, watts, hertz, metres, amperes, ohms
Hecto	h	Hundred	(100)	Metres
Deka	da	Ten	(10)	Metres
Deci	d	One-tenth	(0.1)	Metres
Centi	c	One-hundredth	(0.01)	Metres
Milli	m	One-thousandth	(0.001)	Volts, amperes, metres, henrys, watts, ohms
Micro	μ	One-millionth	(0.000 001)	Volts, amperes, farads, henrys, ohms
Pico	p	One-millionth of one millionth (0.000 000 000 001)		Volts, amperes, farads, coulombs

REFRIGERATION AND AIR CONDITIONING TECHNOLOGY

scale is extended below 32° and above 212° to increase its range.

Table 1-C.
Conversions

To convert—	Into—	Multiply by—	Conversely, multiply by—
Inches	Centimetres	2.54	0.3937
Inches	Mils	1000	0.001
Joules	Foot-pounds	0.7376	1.356
Joules	Ergs	10^7	10^{-7}
Kilogram-calories	Kilojoules	4.186	0.2389
Kilograms	Pounds (avoirdupois)	2.205	0.4536
Kg per sq metre	Pounds per sq foot	0.2048	4.882
Kilometres	Feet	3281	3.048×10^{-4}
Kilowatt-hours	Btu	3413	2.93×10^{-4}
Kilowatt-hours	Foot-pounds	2.655×10^6	3.766×10^{-7}
Kilowatt-hours	Joules	3.6×10^6	2.778×10^{-7}
Kilowatt-hours	Kilogram-calories	860	1.163×10^{-3}
Kilowatt-hours	Kilogram-metres	3.671×10^5	2.724×10^{-6}
Litres	Cubic metres	0.001	1000
Litres	Cubic inches	61.02	1.639×10^{-2}
Litres	Gallons (liq US)	0.2642	3.785
Litres	Pints (liq US)	2.113	0.4732
Metres	Yards	1.094	0.9144
Metres per min	Feet per min	3.281	0.3048
Metres per min	Kilometres per hr	0.06	16.67
Miles (nautical)	Kilometres	1.853	0.5396
Miles (statute)	Kilometres	1.609	0.6214
Miles per hr	Kilometres per min	2.682×10^{-2}	37.28
Miles per hr	Feet per minute	88	1.136×10^{-2}
Miles per hr	Kilometres per hr	1.609	0.6214
Poundals	Dynes	1.383×10^4	7.233×10^{-5}
Poundals	Pounds (avoirdupois)	3.108×10^{-2}	32.17
Sq inches	Circular mils	1.273×10^6	7.854×10^{-7}
Sq inches	Sq centimetres	6.452	0.155
Sq feet	Sq metres	9.29×10^{-2}	10.76
Sq miles	Sq yards	3.098×10^6	3.228×10^{-7}
Sq miles	Sq kilometres	2.59	0.3861
Sq millimetres	Circular mils	1973	5.067×10^{-4}
Tons, short (avoir 2000 lb)	Tonnes (1000 kg)	0.9072	1.102
Tons, long (avoir 2240 lb)	Tonnes (1000 kg)	1.016	0.9842
Tons, long (avoir 2240 lb)	Tons, short (avoir 2000 lb)	1.120	0.8929
Watts	Btu per min	5.689×10^{-2}	17.58
Watts	Ergs per sec	10^7	10^{-7}
Watts	Ft-lb per minute	44.26	2.26×10^{-2}
Watts	Horsepower (550 ft-lb per sec)	1.341×10^{-3}	745.7
Watts	Horsepower (metric) (542.5 ft-lb per sec)	1.36×10^{-3}	735.5
Watts	Kg-calories per min	1.433×10^{-2}	69.77

The SI metric system of measurement measures temperature in degrees Celsius. The Celsius temperature scale is divided into 100 degrees between the freezing point and the boiling point of pure water. Thus, water freezes at 0°C and boils at 100°C. Because there are one hundred graduations between the freezing point of water and its boiling point, this temperature scale was once referred to as the centigrade scale. (*Centi* means "one hundredth.")

There are other temperature scales. One is the Rankine scale, based on the point at which molecular movement stops. This point is believed to be 492° below zero on the Fahrenheit scale. To convert degrees Fahrenheit to degrees Rankine, add 492° to the Fahrenheit reading. Measurement in degrees Rankine is used in some industrial applications. However, in most residential and commercial refrigeration, measurements are taken in degrees Fahrenheit, with some conversions to degrees Celsius.

TEMPERATURE CONVERSIONS

Sometimes it is necessary to convert a temperature. To convert degrees Fahrenheit to degrees Celsius, use the formula:

$$°C = \frac{5}{9} \times (°F. - 32)$$

Example: Convert 95° Fahrenheit to degrees Celsius.

$$°C = \frac{5}{9} \times (95 - 32)$$

$$°C = \frac{5}{9} \times 63$$

$$°C = 35°$$

To convert degrees Celsius to degrees Fahrenheit use the formula:

CHAPTER 1—MATHEMATICS

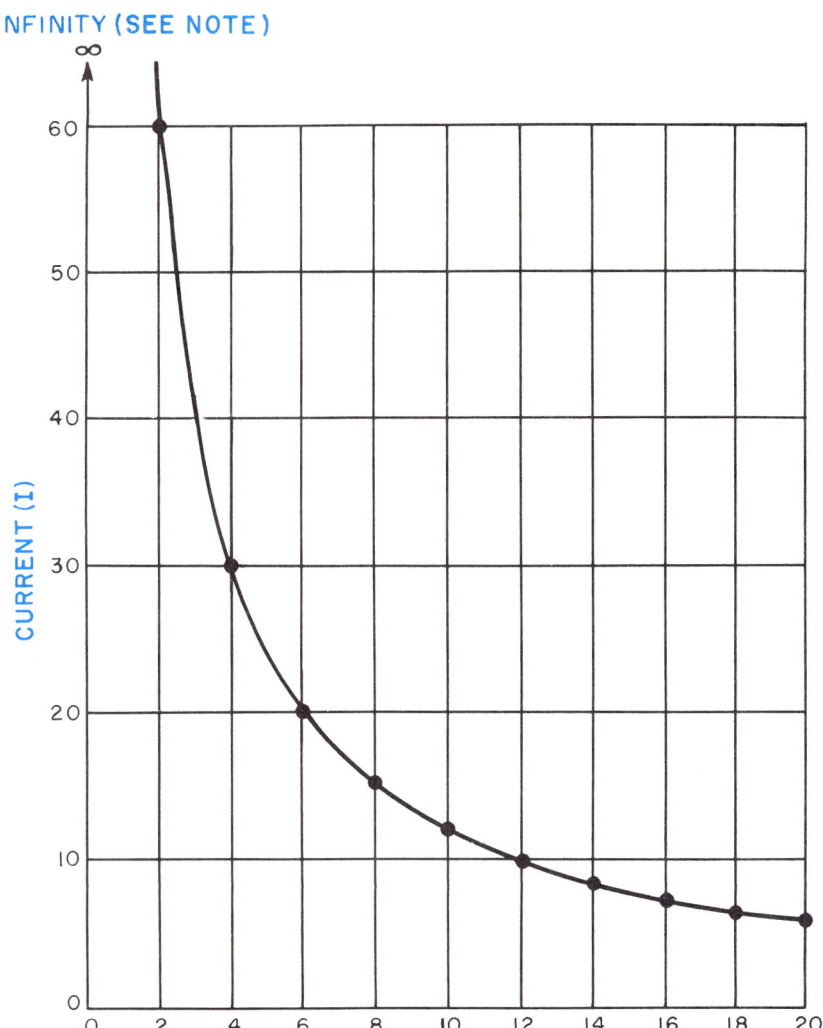

NOTE:
ZERO DIVIDED INTO ANY NUMBER (EXCEPT ZERO) IS REPRESENTED BY THE INFINITY SYMBOL (∞).

1-12. Graph showing the relationship of resistance and current.

$°F. = (°C \times 1.8) + 32$

Example: Convert 35° Celsius to degrees Fahrenheit.

$°F. = (35 \times 1.8) + 32$
$°F. = 63 + 32$
$°F. = 95$

Table 1-D.
The Relationship of Resistance and Current

Resistance	0	2	4	6	8	10	12	14	16	18	20
Current	∞	60	30	20	15	12	10	8.5	7.5	6.6	6

*Some numbers have been rounded to one decimal place. For greater accuracy or for resistance multiplied by current to equal 120 volts, the current would have to be carried past one decimal point.

23

REFRIGERATION AND AIR CONDITIONING TECHNOLOGY

REVIEW QUESTIONS

1. What is a digit?
2. What is a divisor?
3. What is a dividend?
4. What is a quotient?
5. What is a numerator?
6. What is a denominator?
7. What is a proper fraction?
8. What is a parallelogram?
9. What is the value of pi?
10. What is an equation?
11. What is a conversion factor?

2

Air Conditioning and Refrigeration Tools and Instruments

TOOLS AND EQUIPMENT

The air conditioning technician must work with electricity. Equipment that has already been wired may have to be replaced or rewired. In any case, it is necessary to identify and use safely the various tools and pieces of equipment.

Special tools are needed to install and maintain electrical service to air conditioning units. Wires and wiring should be installed according to the National Electrical Code (NEC). However, it is possible that this may not have been done. In such a case, the electrician will have to be called to update the wiring to carry the extra load of the installation of new air conditioning or refrigeration equipment.

This section deals only with interior wiring. Following is a brief discussion of the more important tools used by the electrician.

Pliers and Clippers

Pliers come in a number of sizes and shapes designed for special applications. Pliers are available with either insulated or uninsulated handles. Although pliers with insulated handles are always used when working on or near "hot" wires, they must not be considered sufficient protection alone. Other precautions must be taken. *Long-nose pliers* are used for close work in panels or boxes. *Slip-joint*, or *gas*, *pliers* are used to tighten locknuts or small nuts. Fig. 2-1.

Wire cutters are used to cut wire to size.

Fuse Puller

The *fuse puller* is designed to eliminate the danger of pulling and replacing cartridge fuses by hand. Fig. 2-2. It is also used for bending fuse clips, adjusting loose cutout clips, and handling live electrical parts. It is made of a phenolic material, which is an insulator. Both ends of the puller are used. One end is for large-diameter fuses; the other is for small-diameter fuses.

Screwdrivers

Screwdrivers are made in many sizes and tip shapes. Those used by electricians and refrigeration technicians should have insulated handles. One variation of the screwdriver is the screwdriver bit. It is held in a brace and used for heavy-duty work. For safe and efficient use, screwdriver tips should be kept square and sharp. They should be selected to match the screw slot. Fig. 2-3.

The *Phillips-head screwdriver* has a tip pointed like a star. It is designed to be used with a Phillips screw. These screws are com-

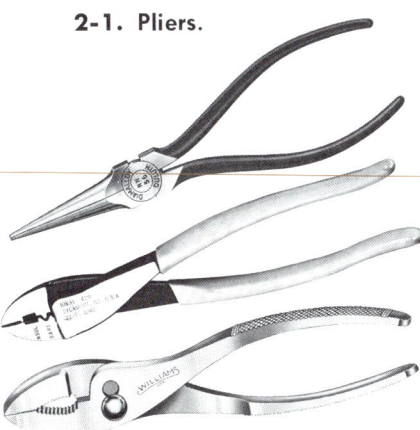

2-1. Pliers.

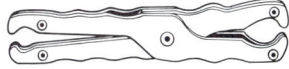

2-2. A fuse puller.

REFRIGERATION AND AIR CONDITIONING TECHNOLOGY

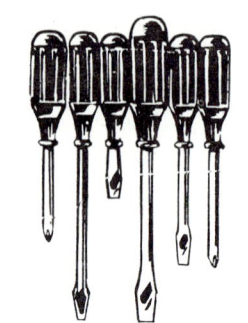

2-3. Screwdrivers.

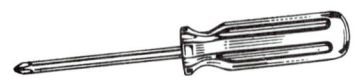

2-4. A Phillips-head screwdriver.

monly found in production equipment. The presence of four slots, rather than two, assures that the screwdriver will not slip in the head of the screw. There are a number of sizes of Phillips-head screwdrivers. They are designated as No. 1, No. 2, and so on. The proper point size must be used to prevent damage to the slot in the head of the screw. Fig. 2-4.

Wrenches

Three types of wrenches are shown in Fig. 2-5. The adjustable open-end wrenches are commonly called *crescent wrenches*. *Monkey wrenches* are used on hexagonal and square fittings such as machine bolts, hexagon nuts, or conduit unions. *Pipe wrenches* are used for pipe and conduit work. They should not be used where crescent or monkey wrenches can be used. Their construction will not permit the application of heavy pressure on square or hexagonal material. Continued misuse of the tool in this manner will deform the teeth on the jaw faces and mar the surfaces of the material being worked.

Soldering Equipment

The standard soldering kit used by electricians consists of the same equipment as the refrigeration mechanic uses. Fig. 2-6. It consists of a nonelectric soldering device—in the form of a torch with propane fuel cylinder or an electric soldering iron, or both. The torch can be used for heating the solid-copper soldering iron or for making solder joints in copper tubing. A spool of solid tin-lead wire solder or flux-core solder is used. *Flux-*

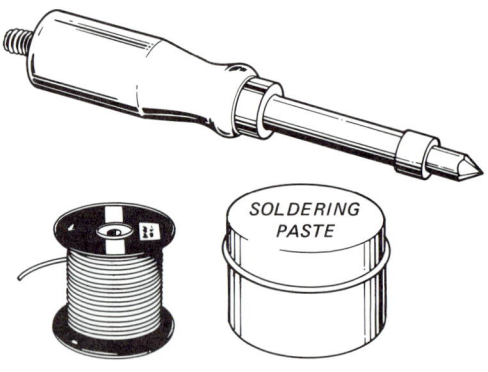

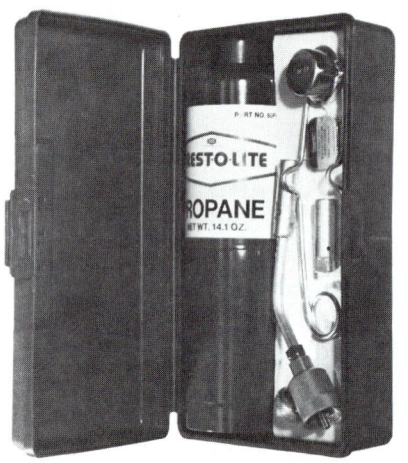

2-6. Soldering equipment.

core solder with a rosin core is used for electrical soldering. *Acid-core solder* is used for soldering metals. It is strongly recommended that acid-core solder not be used with electrical equipment. *Soldering paste* is used with the solid wire solder for solder joints on copper pipe or solid materials.

Drilling Equipment

Drilling equipment consists of a *brace*, a *joint-drilling fixture*, an *extension bit* to allow for drilling into and through thick material, an *adjustable bit*, and a *standard wood bit*. These are required in electrical work to drill holes in building structures for the passage of conduit or wire in new or

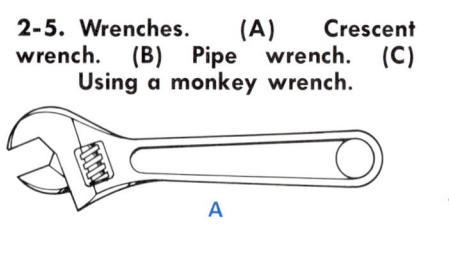

2-5. Wrenches. (A) Crescent wrench. (B) Pipe wrench. (C) Using a monkey wrench.

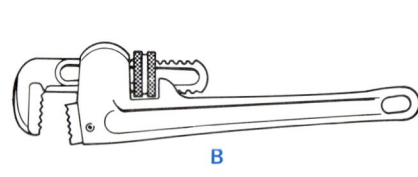

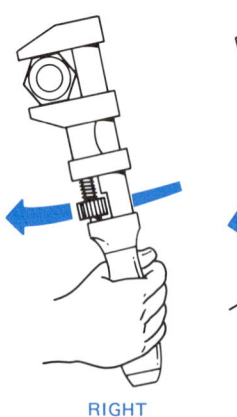

RIGHT

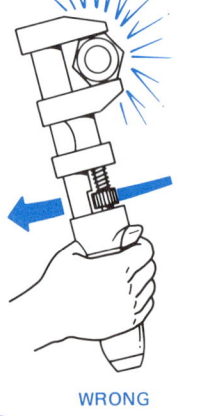

WRONG

C

26

CHAPTER 2—AIR CONDITIONING AND REFRIGERATION TOOLS AND INSTRUMENTS

modified construction. Similar equipment is required for drilling holes in sheet-metal cabinets and boxes. In this case, high-speed or carbide-tipped drills should be used in place of the carbon-steel drills used in wood drilling. Electric power drills are also used. Fig. 2-7.

Woodworking Tools

Crosscut saws, keyhole saws, and *wood chisels* are used by electricians and refrigeration and air conditioning technicians. Fig. 2-8. They are used to remove wooden structural members obstructing a wire or conduit run and to notch studs and joists to take conduit, cable, box-mounting brackets, or tubing. They are also used in the construction of wood-panel mounting brackets. The keyhole saw may again be used when cutting openings in walls of existing buildings where boxes are to be added or tubing is to be inserted for a refrigeration unit.

Metalworking Tools

The *cold chisel* and *center punch* are used when working on steel panels. Fig. 2-9. The *knockout punch* is used either in making or enlarging a hole in a steel cabinet or outlet box. The *hacksaw* is usually used when cutting conduit, cable, or wire that is too large for wire cutters. It is also a handy device for cutting copper tubing or pipe. The *mill file* is used to file the sharp ends of such cutoffs. This is a precaution against short circuits or poor connections in tubing.

Masonry Working Tools

The air conditioning technician should have several sizes of *masonry drills* in the tool kit. These drills normally are carbide-tipped. They are used to drill holes in brick or concrete

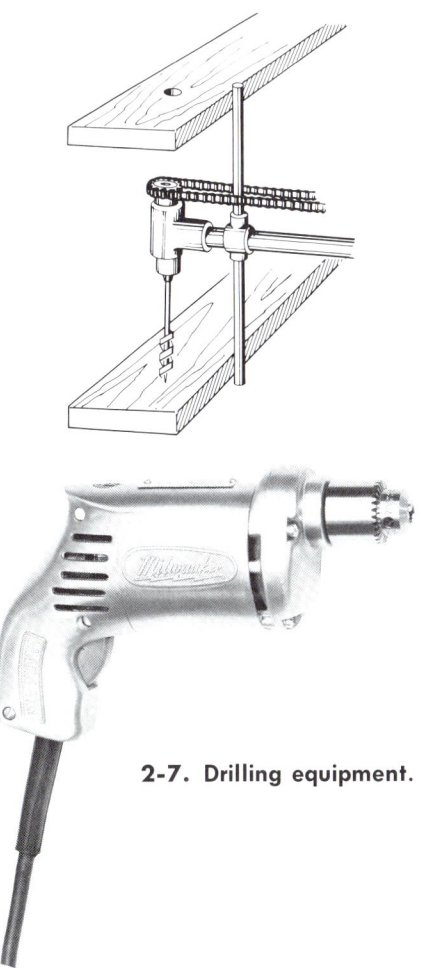

2-7. Drilling equipment.

2-8. Woodworking tools.

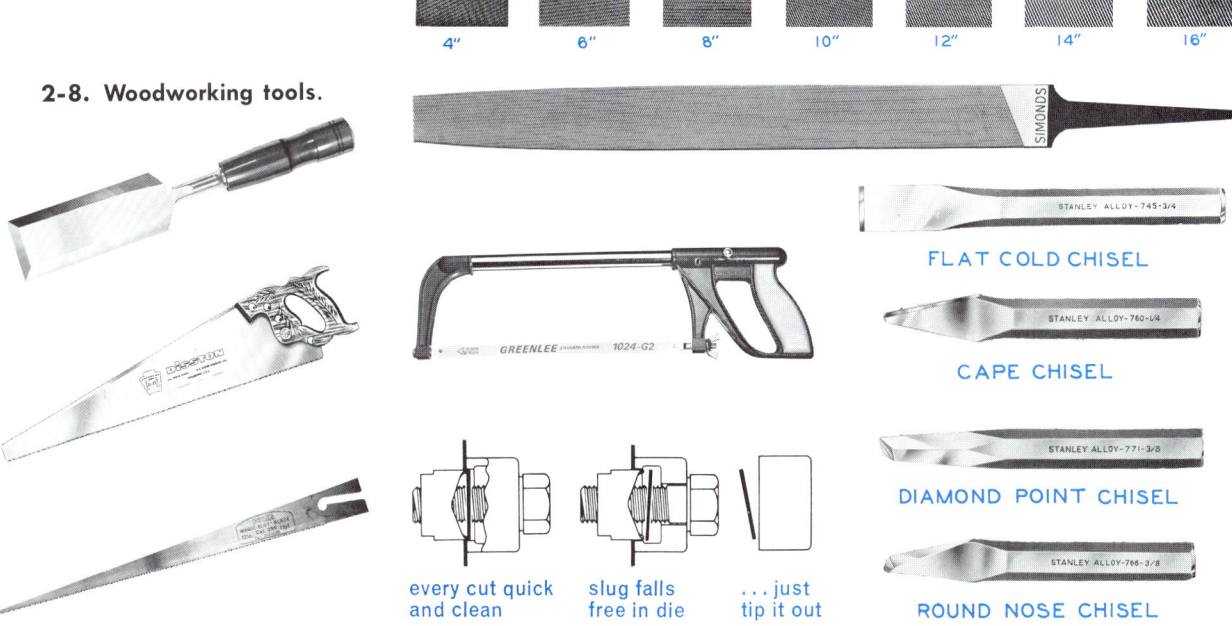

2-9. Metalworking tools.

27

REFRIGERATION AND AIR CONDITIONING TECHNOLOGY

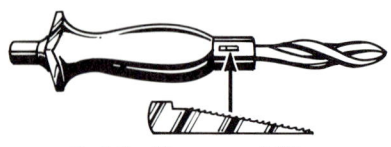

2-10. Masonry drills.

walls. These holes are used for anchoring apparatus with expansion screws or to allow the passage of conduit, cable, or tubing. Figure 2-10 shows the carbide-tipped bit used with a power drill and a hand-operated masonry drill.

Knives and Other Insulation-Stripping Tools

The stripping or removing of wire and cable insulation is accomplished by the use of the tools shown in Fig. 2-11. The *knives* and *patented wire strippers* are used to bare the wire of insulation before making connections. The *scissors* shown are used to cut insulation and tape. The *armored cable cutter* may be used instead of a hacksaw to remove the armor from the electrical conductors at box entry or when cutting the cable to length.

Hammers

Hammers are used either in combination with other tools such as chisels or in nailing equipment to building supports. Figure 2-12 shows a *carpenter's claw hammer* and a *machinist's ball peen hammer*.

Tape

Various tapes are available. They are used for replacing removed insulation and wire coverings. *Friction tape* is a cotton tape impregnated with an insulating adhesive compound. It provides weather resistance and limited mechanical protection to a splice already insulated. *Rubber tape* or *varnished cambric tape* may be used as an insulator when replacing wire covering. *Plastic electrical tape* is made of a

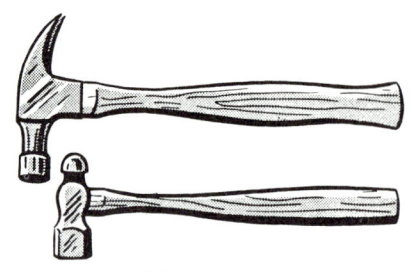

2-12. Hammers.

plastic material with an adhesive on one side of the tape. It has replaced friction and rubber tape in the field for 120- and 208-volt circuits. It serves a dual purpose in taping joints. It is preferred over the former tapes.

Ruler and Measuring Tape

The technician should have a *folding rule* and a *steel tape*. Both of these are aids to cutting to exact size.

Extension Cord and Light

The *extension light* shown in Fig. 2-13 normally is supplied with a long extension cord. It is used by the technician when normal building lighting has not been installed or is not functioning.

Wire Code Markers

Tapes with identifying numbers or nomenclature are available for permanently identifying

2-13. Extension light.

2-11. Tools for cutting and stripping. (A) Electrician's knife. (B) Electrician's scissors. (C) Skinning knife. (D) Stripper. (E) Cable cutter.

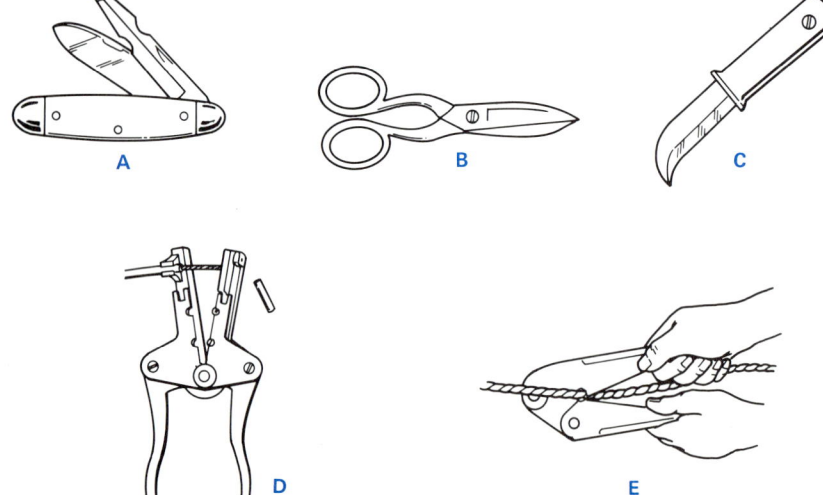

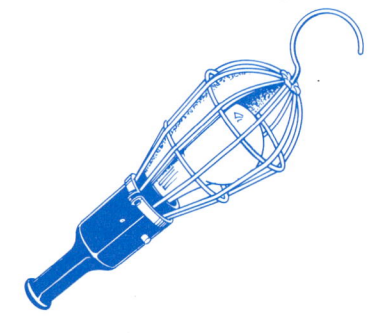

28

CHAPTER 2—AIR CONDITIONING AND REFRIGERATION TOOLS AND INSTRUMENTS

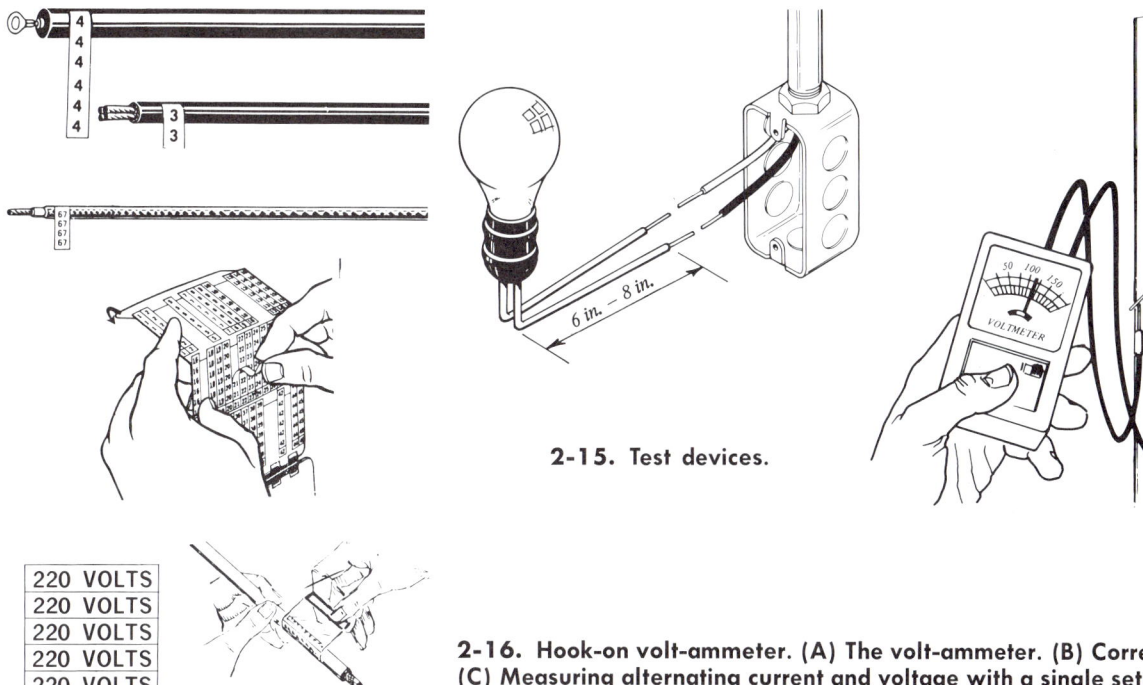

2-15. Test devices.

2-14. Wire code markers.

2-16. Hook-on volt-ammeter. (A) The volt-ammeter. (B) Correct operation. (C) Measuring alternating current and voltage with a single setup. (D) Looping conductor to extend current range of transformer.

wires and equipment. These *wire code markers* are particularly valuable for identifying wires in complicated wiring circuits, in fuse boxes, and circuit breaker panels, or in junction boxes. Fig. 2-14.

Meters and Test Prods

An *indicating voltmeter* or *test lamp* is used when determining the system voltage. It is also used in locating the ground lead and for testing circuit continuity through the power source. They both have a light that glows in the presence of voltage. Fig. 2-15.

A modern method of measuring current flow in a circuit uses the hook-on volt-ammeter. Fig. 2-16. This instrument does not have to be hooked into the circuit. It can be operated with comparative ease. Just remember that it measures only one

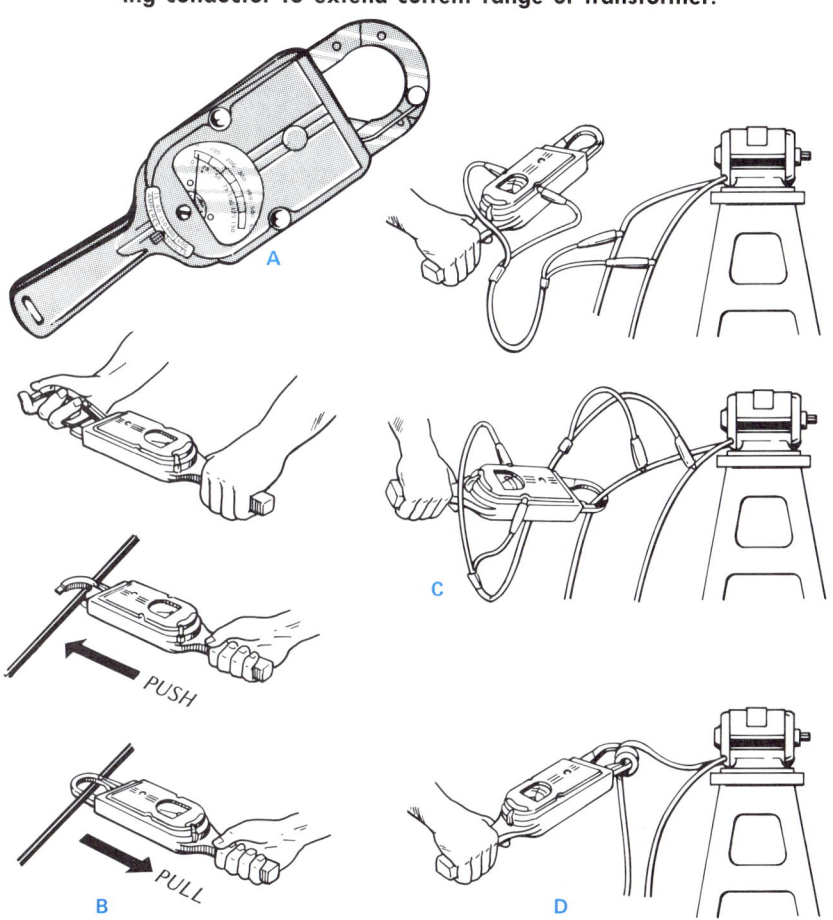

29

REFRIGERATION AND AIR CONDITIONING TECHNOLOGY

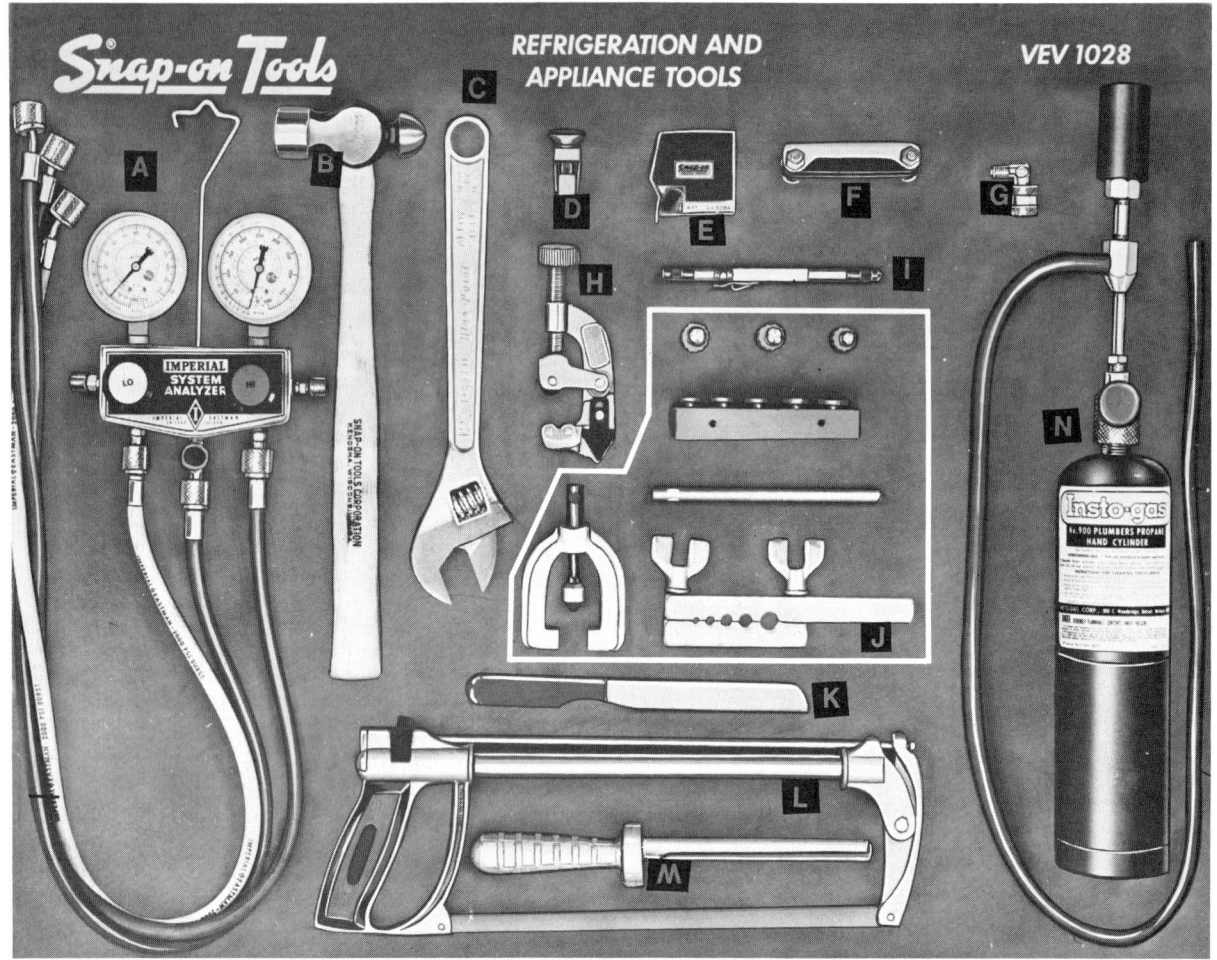

2-17. Refrigeration and appliance tools. (A) Servicing manifold, (B) ball peen hammer, (C) adjustable wrench, (D) tubing tapper, (E) tape measure, (F) Allen wrench set, (G) 90° adapter service part, (H) tubing cutter, (I) thermometer, (J) flaring tool kit, (K) knife, (L) hacksaw, (M) jab saw, (N) halide leak detector.

wire. Do not clamp it over a cord running from the consuming device to the power source. Also, this meter is used only on alternating current circuits. The AC current will cancel the reading if two wires are covered by the clamping circle. Note how the clamp-on part of the meter is used on *one wire* of the motor.

To make a measurement, the hook-on section is opened by hand and the meter is placed against the conductor. A slight push on the handle snaps the section shut. A slight pull on the handle springs open the hook on the C-shaped current transformer and releases the conductor. Applications of this meter are shown in Fig. 2-16. Figure 2-16B shows current being measured by using the hook-on section. Figure 2-16C shows the voltage being measured using the meter leads. An ohmmeter is included in some of the newer models. However, power must be off when the ohmmeter is used. The ohmmeter uses leads to complete the circuit to the device under test.

Use of the volt-ammeter is a quick way of testing if the air conditioning or refrigeration unit has a motor that is drawing too much current. A motor that is drawing too much current will overheat and burn out.

Tool Kits

Some tool manufacturers make

30

Snap-on Tools

2-18. Air conditioning and refrigeration portable tool kit. (A) Air conditioning charging station, (B) excavating/charging valve, (C) 90° adapter service part, (D) O-ring installer, (E) refrigeration ratchet, (F) snap-ring pliers, (G) stem thermometer, (H) seal remover and installer, (I) test light, (J) puller, (K) puller jaws, (L) retainer ring pliers, (M) refrigerant can tapper, (N) dipsticks for checking oil level, (O) halide leak detector, (P) flexible charging hose, (Q) goggles.

2-19. AC and DC voltage probe-voltmeter.

Amprobe

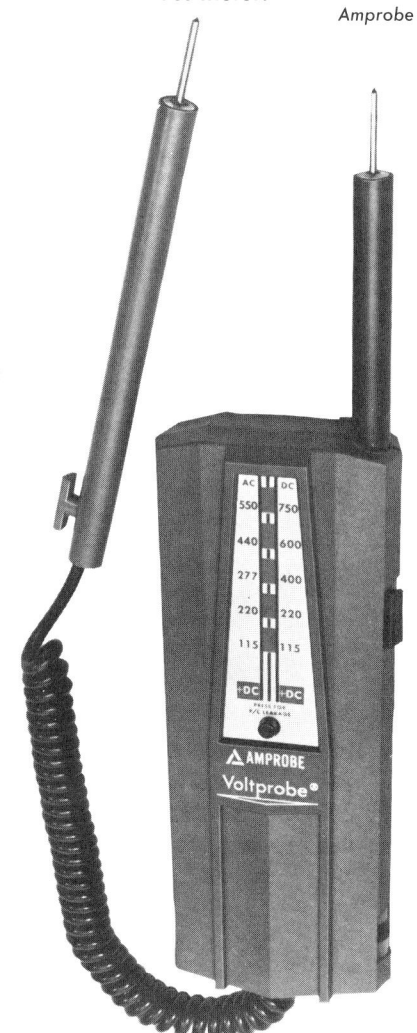

up tool kits for the refrigeration and appliance trade. Fig. 2-17. In the Snap-On® tool kit, the leak detector is part of the kit. The gages are also included. An adjustable wrench, tubing cutter, hacksaw, flaring tool, and ball peen hammer can be hung on the wall and replaced when not in use. One of the biggest problems for any repairperson is keeping track of the tools. Markings on a board will help locate at a glance what is missing.

Figure 2-18 shows a portable tool kit.

Figure 2-18 (J) shows a pulley puller. This tool is used to remove the pulley if necessary to get at the seals. A cart (A) is included so that the refrigerant and vacuum pump can be easily handled in large quantities. The goggles (Q) protect the eyes from escaping refrigerant.

Figure 2-19 shows a voltmeter probe. It detects the presence of 115 to 550 volts AC and 115 to 750 volts DC. The hand-held meter is used to find whether the voltage is AC or DC and what the potential difference is. It is

31

REFRIGERATION AND AIR CONDITIONING TECHNOLOGY

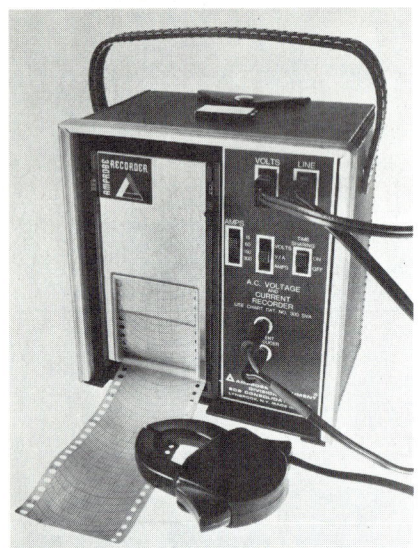

Amprobe
2-20. Voltage and current recorder.

rugged and easy to handle. This meter is useful when working around unknown power sources in refrigeration units.

Figure 2-20 shows a voltage and current recorder. It can be left hooked to the line for an extended period of time. Use of this instrument can help determine the exact cause of a problem, since voltage and current changes can affect the operation of air conditioning and refrigeration units.

GAGES AND INSTRUMENTS

It is impossible to install or service air conditioning and refrigeration units and systems without using gages and instruments.

A number of values must be measured accurately if air conditioning and refrigeration equipment is to operate properly. Refrigeration and air conditioning units must be properly serviced and monitored if they are to give the maximum efficiency for the energy expended. Here, the use of gages and instruments becomes important. It is not possible to analyze a system's operation without the proper equipment and procedures. In some cases, it takes thousands of dollars worth of equipment to troubleshoot or maintain modern refrigeration and air conditioning systems.

Instruments are used to measure and record such values as temperature, humidity, pressure, air flow, electrical quantities, and weight. Instruments and monitoring tools can be used to detect incorrectly operating equipment. They can also be used to check efficiency. Instruments can be used on a job, in the shop, or in the laboratory.

If properly cared for and correctly used, modern instruments are highly accurate.

Pressure Gages

Pressure gages are relatively simple in function. Fig. 2-21. They read positive pressure or negative pressure, or both. Fig. 2-22. Gage components are rela-

2-21. Pressure gage.
Weksler

Marsh
2-22. This gage measures up to 150 lb. per sq. in. pressure and also reads from 0 to 30 for vacuum. The temperature scale runs from —40 to 115° F. [—40 to 46.1°C].

tively few. However, different combinations of gage components can produce literally millions of design variations. Fig. 2-23. One gage buyer may use a gage with 0–250 psi range. Another person with the same basic measurement requirements will order a gage with a range of 0–300 psi. High-pressure gages can be purchased with scales of 0 to 1000, 2000, 3000, 4000, and 5000 psi.

There are, of course, many applications that will continue to require custom instruments, specially designed and manufactured. Most gage manufacturers have both stock items and specially manufactured gages.

Gage Selection

Since 1939, gages used for pressure measurements have been standardized by the American National Standards Institute, formerly called the American Standard Association. Most gage

32

CHAPTER 2—AIR CONDITIONING AND REFRIGERATION TOOLS AND INSTRUMENTS

manufacturers are consistent in case patterns, scale ranges, and grades of accuracy. Industry specifications were reviewed and updated in 1974.

Gage accuracy is stated as the limit that error must not exceed when the gage is used within any combination of rated operating conditions. It is expressed as a percentage of the total pressure (dial) span.

Classification of gages by ANSI standards has a significant bearing on other phases of gage design and specification. As an example, a test gage with ±0.25% accuracy would not be offered in a 2″ dial size. Readability of smaller dials is not sufficient to permit the precision indication necessary for this degree of accuracy. Most gages with accuracy of ±0.5% and better have dials that are at least 4½″. Readability can be improved still further by increasing the dial size.

Accuracy

How much accuracy is enough? That is a question only the application engineer can answer. However, from the gage manufacturer's point of view, increased accuracy represents a proportionate increase in the cost of building a gage. Tolerances of every component must be more exacting as gage accuracy increases. The time needed for factory technicians to calibrate the gage correctly increases accordingly. Nevertheless, a broad selection of precision instruments are available. Grades A (±1%), 2A (±0.5%), and 3A (±0.25%) are examples of the tolerances available.

Medium

In every gage selection, the medium to be measured must be evaluated for potential corrosiveness to the Bourdon tube of the gage.

There is no ideal material for Bourdon tubes. No one material adapts to all applications. Bourdon tube materials are chosen for their elasticity, repeatability, ability to resist "set," and corrosion resistance to the fluid mediums.

Ammonia refrigerants are commonly used in refrigeration. All-steel internal construction is required. Ammonia gages have corresponding temperature scales. A restriction screw protects the gage against sudden impact, shock, or pulsating pressure. A heavy-duty movement of stainless steel and Monel steel prevents corrosion and give extra-long life. The inner arc on the dial shows pressure. The outer arc shows the corresponding temperature. Fig. 2-24.

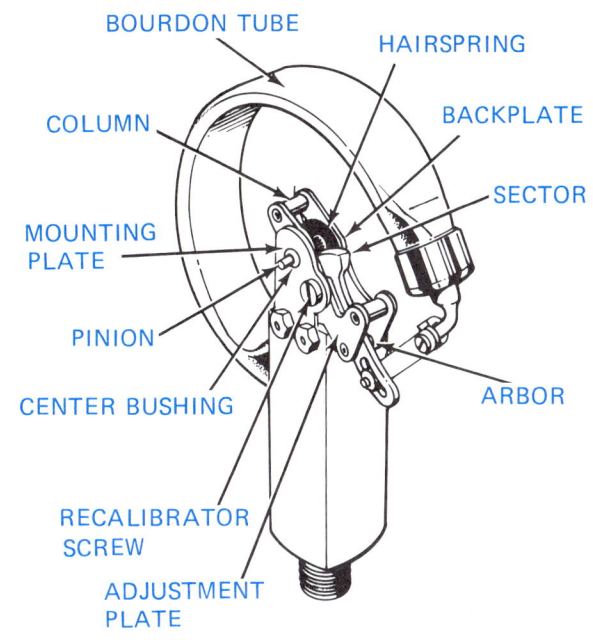

2-23. Bourdon tube arrangement and parts of a gage.
Marsh

2-24. Ammonia gage.
Marsh

Line Pressures

The important consideration regarding line pressures is to determine whether the pressure reading will be fairly constant or whether it will fluctuate. The maximum pressure at which a

33

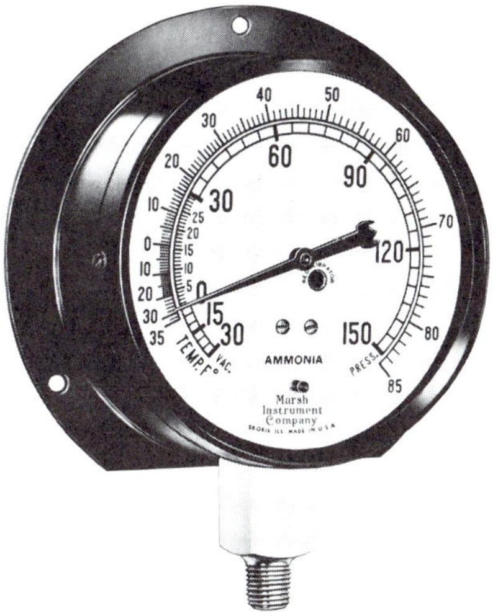

REFRIGERATION AND AIR CONDITIONING TECHNOLOGY

gage is continously operated should not exceed 75% of the full scale range. For the best performance, gages should be graduated to twice the normal system-operating pressure.

This extra margin provides a safety factor in preventing overpressure damage. It also helps avoid a permanent set of the Bourdon tube. For applications with substantial pressure fluctuations, this extra margin is especially important. In general, the lower the Bourdon tube pressure, the greater the overpressure percentage it will absorb without damage. The higher the Bourdon tube pressure, the less overpressure it can safely absorb.

Pulsation causes pointer flutter, which makes reading difficult. Pulsation also can drastically shorten gage life by causing excessive wear of the movement gear teeth. A pulsating pressure is defined as a pressure variation of more than 0.1% full scale per second. Following are conditions often encountered and suggested means of handling them.

The *restrictor* is a low-cost means of combating pulsation problems. This device reduces the pressure opening. The reduction of the opening allows less of the pressure change to reach the Bourdon tube in a given time interval. This dampening device protects the Bourdon tube by retarding overpressure surges. It also improves gage readability by reducing pointer flutter. When specifying gages with restrictors, indicate whether the pressure medium is liquid or gas. The medium determines the size of the orifice. Also, restrictors are not recommended for dirty line fluids. Dirty materials in the line can easily clog the orifice. For

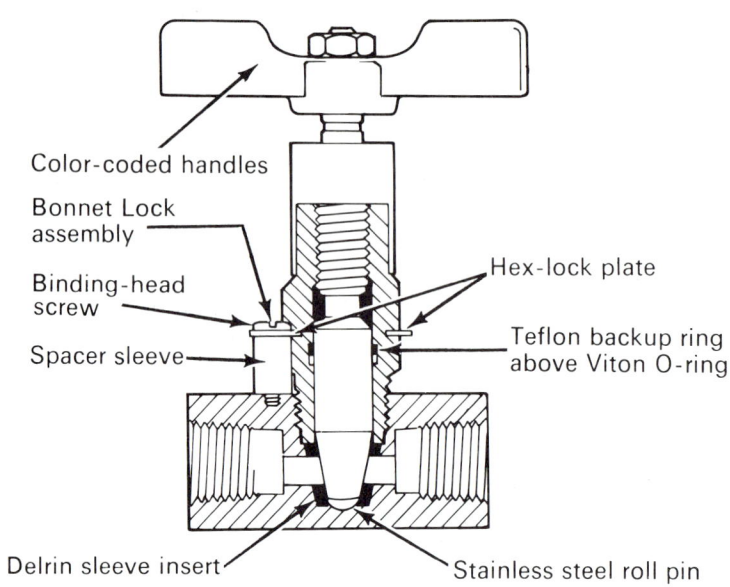

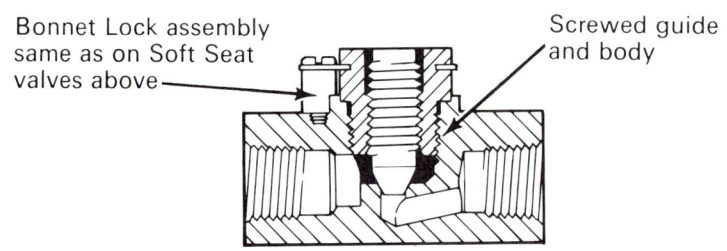

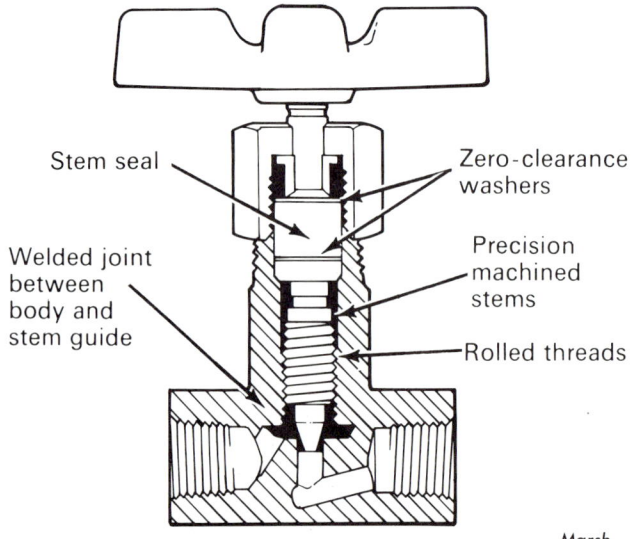

2-25. Different types of needle valves.

34

such conditions, diaphragm seals should be specified.

The *needle valve* is another means of handling pulsation if used between the line and the gage. Fig. 2-25. The valve is throttled down to a point where pulsation ceases to register on the gage.

In addition to the advantage of precise throttling, needle valves also offer complete shutoff, an important safety factor in many applications. Use of a needle valve can greatly extend the life of the gage by allowing it to be used only when a reading is needed.

Liquid-filled gages are another very effective way to handle line pulsation problems. Because the movement is constantly submerged in lubricating fluid, reaction to pulsating pressure is dampened and the pointer flutter is practically eliminated.

Silicone-oil-treated movements dampen oscillations caused by line pressure pulsations and/or mechanical oscillation. The silicone oil, applied to the movement, bearings, and gears, acts as a shock absorber. This extends the gage life while helping to maintain accuracy and readability.

Effects of Temperature on Gage Performance

Because of the effects of temperature on the elasticity of the tube material, the accuracy may change. Gages calibrated at 75° F. [23.9°C] may change by more than ±2% Full Scale (FS) below −30° F. [−34°C] or above 150° F. [65.6°C].

Care of Gages

The pressure gage is one of the serviceperson's most valuable tools. Thus, the quality of the work depends on the accuracy of the gages used. Most are precision-made instruments that will give many years of dependable service if properly treated.

The test gage set should be used primarily to check pressures at the low and high side of the compressor.

The ammonia gage should be used with a steel Bourdon tube tip and socket to prevent damage.

Once you become familiar with the construction of your gages you will be able to handle them more efficiently. The internal mechanism of a typical gage was shown in Fig. 2-23. The internal parts of a vapor tension thermometer are very similar.

Drawn brass is almost always used for case material. It does not corrode. However, some gages now use high-impact plastics. A copper alloy Bourdon tube with a brass tip and socket is used for most refrigerants. Stainless steel is used for ammonia. Engineers have found that moving parts involved in rolling contact will last longer if made of unlike metals. That is why many top-grade refrigeration gages have bronze-bushed movements with a stainless steel pinion and arbor.

The socket is the only support for the entire gage. It extends beyond the case. The extension is long enough to provide a wrench flat for use in attaching the gage to the pressure source. Never twist the case when threading the gage into the outlet. This could cause misalignment or permanent damage to the mechanism.

Note: Keep gages and thermometers separate from other tools in your service kit. They can be knocked out of alignment by a jolt from a heavy tool.

Most pressure gages for refrigeration testing have a small orifice restriction screw. The screw is placed in the pressure inlet hole of the socket. It reduces the effects of pulsations without throwing off pressure readings. If the orifice becomes clogged, the screw can be easily removed for cleaning.

Gage Recalibration

Most gages retain a good degree of accuracy in spite of daily usage and constant handling. Since they are precision instruments, however, you should set up a regular program for checking them. If you have such a regular program, you can be sure that you are working with accurate instruments.

Gages may develop reading errors if they are dropped or subjected to excessive pulsation, vibration, or a violent surge of overpressure. You can restore a gage to accuracy by adjusting the recalibrator screw. Fig. 2-26. If the gage does not have a recalibrator screw, remove the ring and glass. Connect the gage you are testing and a gage of known

2-26. Recalibrating a gage.
Marsh

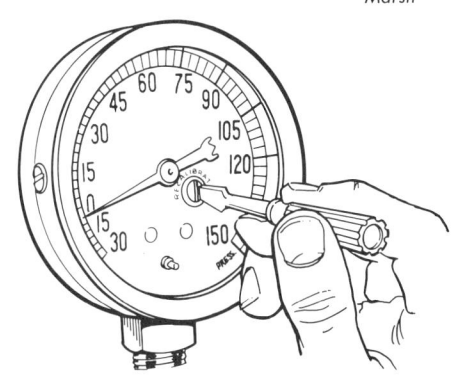

REFRIGERATION AND AIR CONDITIONING TECHNOLOGY

accuracy to the same pressure source. Compare readings at mid-scale. If the gage under test is not reading the same as the test gage, remove the pointer and reset.

This type of adjustment on the pointer acts merely as a pointer setting device. It does not reestablish the original even increment (linearity) of pointer travel. This becomes more apparent as the correction requirement becomes greater.

If your gage has a recalibrator screw on the face of the dial as in Fig. 2-26, remove the ring and glass. Relieve all pressure to the gage. Turn the recalibration screw until the pointer rests at zero.

The gage will be as accurate as when it left the factory if it has a screw recalibration adjustment. Resetting the dial to zero restores accuracy throughout the entire range of dial readings.

If you cannot calibrate the gage by either of these methods, take it to a qualified gage specialist for repair.

Thermometers

Temperature scales were discussed in Chapter 1. A thermometer is used to measure heat. A thermometer should be chosen according to its application. Consider first the kind of installation—direct mounting or remote reading.

If remote readings are necessary, then the vapor tension thermometer is best. It has a closed, filled Bourdon tube. A bulb is at one end for temperature sensing. Changes in temperature at the bulb result in pressure changes in the fill medium. Remote-reading thermometers are equipped with six feet of capillary tubing as

2-27. Thermometers used to measure superheat.

standard. Other lengths are available on special order.

The location, direct or remote, is important when choosing a thermometer. Four common types of thermometers are used to measure temperature: the pocket thermometer, the bimetallic thermometer, the thermocouple thermometer, and the resistance thermometer.

POCKET THERMOMETER

The *pocket thermometer* depends upon the even expansion of a liquid. The liquid may be mercury or colored alcohol. This type of thermometer is versatile. It can be used to measure temperatures of liquids, air, gas, and solids. This can be strapped to the suction line during a superheat measurement. For practical purposes, it can operate wet or dry. This type of thermometer can withstand extremely corrosive solutions and atmospheres.

When the glass thermometer is read in place, temperatures are accurate if proper contact is made between the stem and the medium being measured. Refrigeration service persons are familiar with the need to attach the thermometer firmly to the suction line when taking superheat readings. Fig. 2-27. Clamps are available for this purpose. One thing should be kept in mind. That is the depth at which the thermometer is to be immersed in the medium being measured. Most instruction sheets point out that for liquid measurements the thermometer should be immersed so many inches. When used in a duct, a specified length of the stem should be in the air flow. Dipping only the bulb into a glass of water does not give the same reading as immersing to the prescribed length.

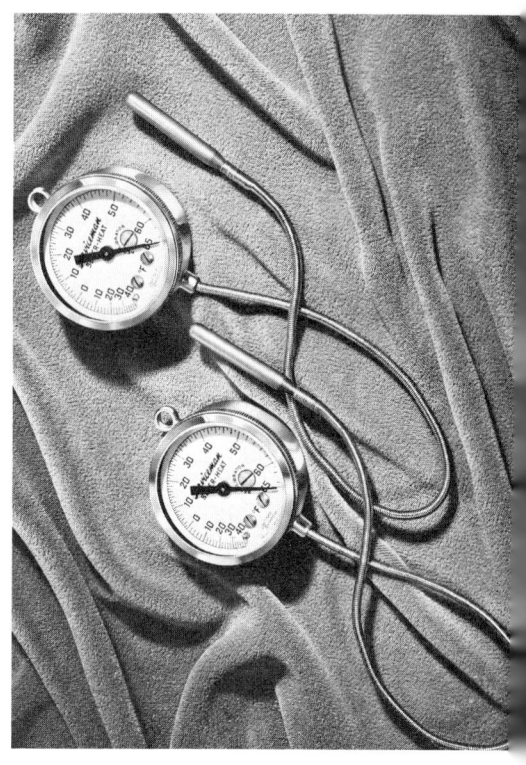

Shielding is frequently overlooked in the application of the simple glass thermometer. The instrument should be shielded from radiated heat. Heating repairpersons often measure air temperature in the furnace bonnet. Do not place the thermometer in a position where it receives direct radiation from the heat exchanger surfaces. This causes erroneous readings.

The greatest error in the use of the glass thermometer is that it is often not read in place. It is removed from the outlet grille of a packaged air conditioner. Then it is carried to eye level in the room

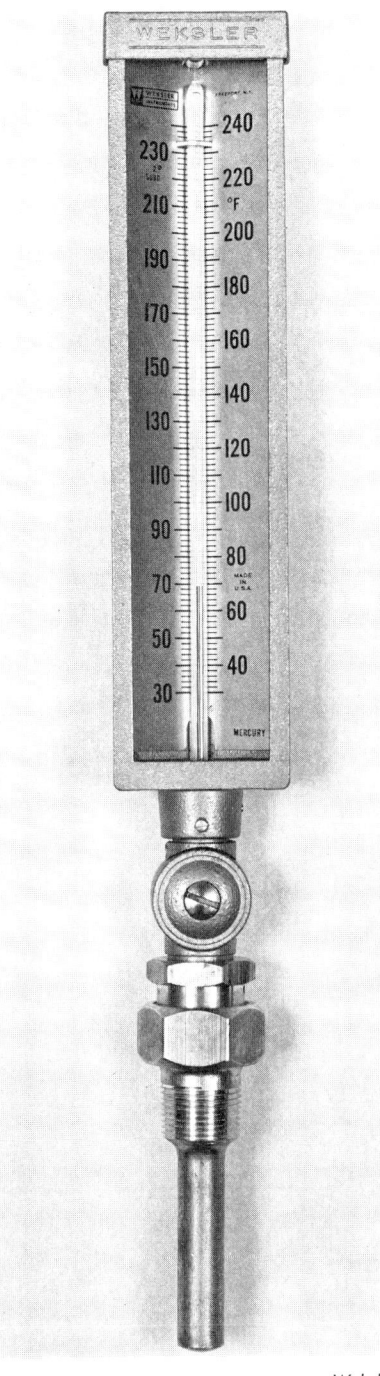

CHAPTER 2—AIR CONDITIONING AND REFRIGERATION TOOLS AND INSTRUMENTS

different from that which it was measuring.

A liquid bath temperature reading is taken with the bulb in the bath. It is left for a few minutes immersed, then raised to be read.

A simple rule helps eliminate incorrect readings:

Read glass thermometers while they are actually in contact with the medium being measured.

If a thermometer must be handled, do so with as little hand contact as possible. Read the thermometer immediately!

A recurring problem with mercury-filled glass thermometers is separation of the mercury column. Fig. 2-28. This results in what is frequently termed a "split thermometer." The cause of the column's splitting is always rough handling. Such handling cannot be avoided at all times in service work. Splitting does not occur in thermometers that *do not* have a gas atmosphere over the mercury. Such thermometers allow the mercury to move back and forth by gravity, as well as temperature change. Such thermometers may not be used in other than vertical positions.

A "split thermometer" can be repaired. Most service thermometers have the mercury reservoir at the bottom of the tube. In this case, cool the thermometer bulb in shaved ice. This draws the mercury to the lower part of the reservoir. Add more ice or salt to lower the temperature, if necessary. With the thermometer in an upright position, tap the bottom of the bulb on a padded piece of paper or cloth. The entrapped gas causing the split column should then rise to the top of the mercury. After the column has been joined, test the service thermometer against a standard thermometer. Do this at several service temperatures.

Weksler
2-28. Mercury thermometer.

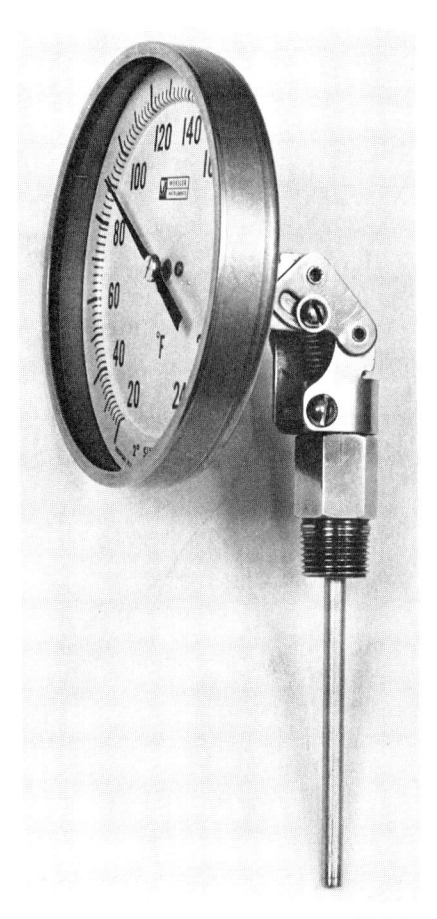

at ambient temperatures. Here it is read a few seconds to a minute later. It is read in a temperature

Weksler
2-29. Dial-type thermometer.

Bimetallic Thermometers

Dial thermometers are actuated by bimetallic coils, by mercury, by vapor pressure, or gas. They are available in varied forms that allow the dial to be used in a number of locations. Fig. 2-29. The sensing portion of the instrument may be located somewhere else. The dial can be read in a convenient location.

Bimetallic thermometers have a linear dial face. There are equal increments throughout any given dial range. Dial ranges are also available to meet higher temperature measuring needs. Ranges are available to 1000° F. [537.8°C]. In four selected ranges, dials giving both Celsius and Fahrenheit readings are available. Bimetallic thermometers are economical. There is no need for a machined movement or gearing. The temperature-sensitive bimetallic element is connected directly to the pointer. This type of thermometry is well adapted to measuring the temperature of a surface. Dome-mounted thermal protectors actually react to the surface temperature of the compressor skin. These thermometers are used where direct readings need to be taken, such as on pipelines, tanks, ovens, ducts, sterilizers, heat exchangers, and laboratory temperature baths.

The simplest type of dial thermometer has a stem. The stem is inserted into the medium to be measured. With the stem immersed 2" in liquids and 4" in gases, this thermometer gives reasonably accurate readings.

Although dial thermometers have many uses, there are some limitations. They are not as universally applicable as the simple glass thermometer. When ordering a dial thermometer, specify the stem length, scale range, and medium in which it will be used.

One of the advantages of bimetallic thermometry is that the thermometer can be applied directly to surfaces. It can be designed to take temperatures of pipes from $\frac{1}{2}$" through 2".

In operation, the bimetallic spiral is closely coupled to the heated surface that is to be measured. The thermometer is held fast by two permanent magnets. One manufacturer claims that their type of thermometer reaches stability within 3 minutes. Its accuracy is said to be plus or minus 2% in working ranges.

A simple and inexpensive type of bimetallic thermometer scribes temperature travel on a load of food in transit. It can be used also to check temperature variations in controlled industrial areas. The replaceable chart gives a permanent record of temperature variations during the test period.

Bimetallic drives are also used in control devices. For example, thermal overload sensors for motors and other electrical devices use bimetallic elements. Other examples will be discussed later.

THERMOCOUPLE THERMOMETERS

Thermocouples are made of two dissimilar metals. Once the metals are heated they give off an EMF (electro-motive force, or voltage). This electrical energy can be measured with a standard type of meter designed to measure small amounts of current. The meter can be calibrated in degrees, instead of amperes, milliamperes, or microamperes.

In use, the thermocouples are placed in the medium that is to be measured. Extension wires run from the thermocouple to the meter. The meter then gives the temperature reading at the remote location.

The extension wires may be run outside closed chests and rooms. There is no difficulty in closing a door, neither will the wires be pinched. On air conditioning work, one thermocouple may be placed in the supply grille and another in the return grille. Readings can then be taken seconds apart without handling a thermometer.

Thermocouples are easily taped onto the surface of pipes to check the inside temperature. It is a good idea to insulate the thermocouple from ambient and radiated heat.

Although this type of thermometer is rugged, it should be handled with care. It should not be handled roughly.

Thermocouples should be protected from corrosive chemicals and fumes. Manufacturer's instructions for protection and use are supplied with the instrument.

RESISTANCE THERMOMETERS

One of the newer ways to check temperature is with a thermometer that uses a resistance sending element. An electrical sensing unit may be made of a *thermistor*. A thermistor is a piece of material that changes resistance rapidly when subjected to temperature changes. When heated, the thermistor lowers its resistance. This decrease in resistance makes a circuit change its current. A meter can be inserted in the circuit. The change in current can be calibrated against a standard thermometer. The scale can be marked to read temperature in degrees Celsius or degrees Fahrenheit.

Another type of resistance thermometer indicates the temperature by an indicating light. The resistance-sensing bulb is placed in the medium to be measured. The bridge circuit is adjusted until the light comes on. The knob that adjusts the bridge circuit is calibrated in degrees Celsius or Fahrenheit.

CHAPTER 2—AIR CONDITIONING AND REFRIGERATION TOOLS AND INSTRUMENTS

The knob then shows the temperature. The sensing element is just one of the resistors in the bridge circuit. The bridge circuit is described in detail in Chapter 4, page 98.

There is the possibility of having practical precision of ±1° F. [0.5°C] in this type of measurement. The range covered is −325 to 250° F. [−198 to 121°C]. A unit may be used for deep freezer testing, for air conditioning units, and for other work.

Response is rapid. Special bulbs are available for use in rooms, outdoors, immersion, surfaces, and ducts.

SUPERHEAT THERMOMETER

The *superheat thermometer* is used to check for correct temperature differential of the refrigerant gas. The inlet and outlet side of the evaporator coil have to be measured to obtain the two temperatures. The difference is obtained by subtracting.

Test thermometers are available in boxes. Fig. 2-30. The box protects the thermometer. It is important to keep the thermometer in operating condition. Several guidelines must be followed. Figure 2-31 illustrates how to keep the test thermometer in good working condition. Preventing kinks in the capillary is important. Keep the capillary clean by removing grease and oil. Clean the case and crystal with a mild detergent.

SUPERHEAT MEASUREMENT INSTRUMENTS

Superheat plays an important role in refrigeration and air conditioning service. For example, the thermostatic expansion valve operates on the principle of superheat. In charging capillary tube systems, the superheat measurement must be carefully watched. The suction line superheat is an indication of whether liquid refrigerant is flooding the compressor from the suction side. A measurement of zero superheat is a definite indicator that liquid is reaching the compressor. A measurement of 6 to 10° F. [−14.4 to −12.2°C] for the expansion valve system and 20° F. [6.7°C] for capillary tube system indicates that all refrigerant is vaporized before entering the compressor.

The superheat at any point in a refrigeration system is found by first measuring the actual refrigerant temperature at that point using an electronic thermometer. Then the boiling point temperature of the refrigerant is found by connecting a compound pressure gage to the system and reading the boiling temperature from the center of the pressure gage. The

2-30. Test thermometer.

Marsh

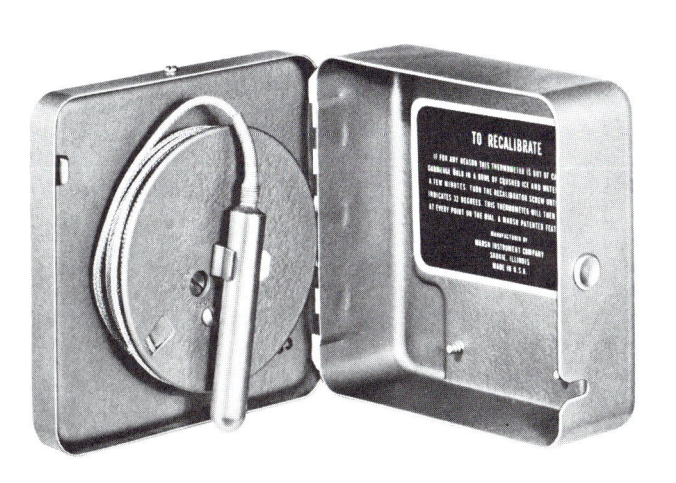

REFRIGERATION AND AIR CONDITIONING TECHNOLOGY

KEEP YOUR THERMOMETERS WORKING BY FOLLOWING THESE STEPS

1. DO NOT CUT, TWIST, OR KINK CAPILLARY.

When capillary becomes kinked, remove the kink by carefully bending the capillary in a direction opposite to the kink.

To straighten twisted capillary, grasp the tubing in both hands and untwist short sections at a time, being careful not to break the fine wire armor.

Cutting the capillary will release the charge and render the instrument useless.

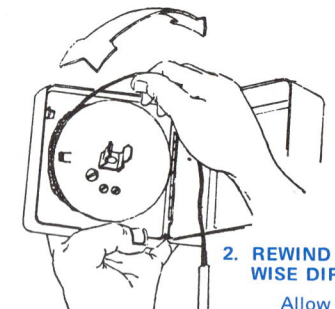

2. REWIND CAPILLARY CAREFULLY IN CLOCKWISE DIRECTION.

Allow bulb to hang free and turn with winding.

Keep bulb in holding clip when thermometer is not in use. Clip will turn in any direction to receive bulb.

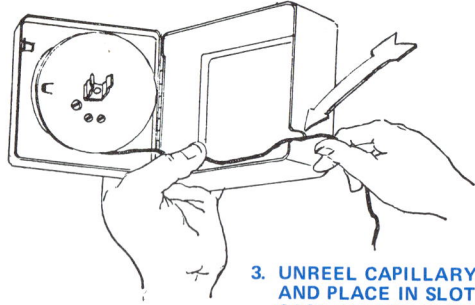

3. UNREEL CAPILLARY CAREFULLY AND PLACE IN SLOT AT SIDE OF CASE BEFORE CLOSING.

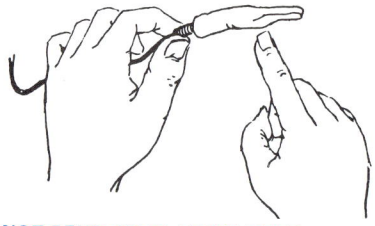

4. DO NOT BEND OR FLATTEN BULB.

Distortion of the bulb will result in false reading.

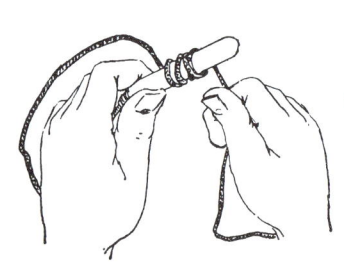

5. DO NOT TWIST CAPILLARY AROUND BULB TO HOLD IN POSITION.

A small piece of tape will usually be adequate to hold bulb in place.

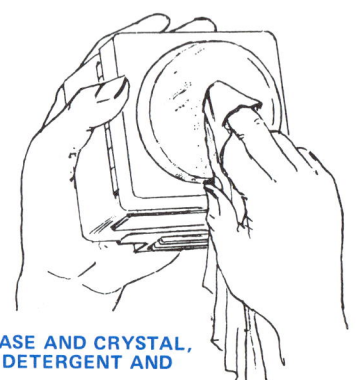

6. TO CLEAN CASE AND CRYSTAL, USE A MILD DETERGENT AND SOFT RAG.

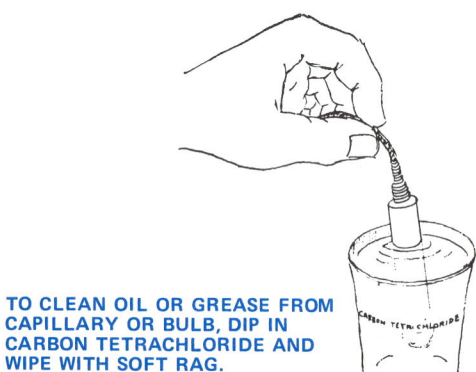

7. TO CLEAN OIL OR GREASE FROM CAPILLARY OR BULB, DIP IN CARBON TETRACHLORIDE AND WIPE WITH SOFT RAG.

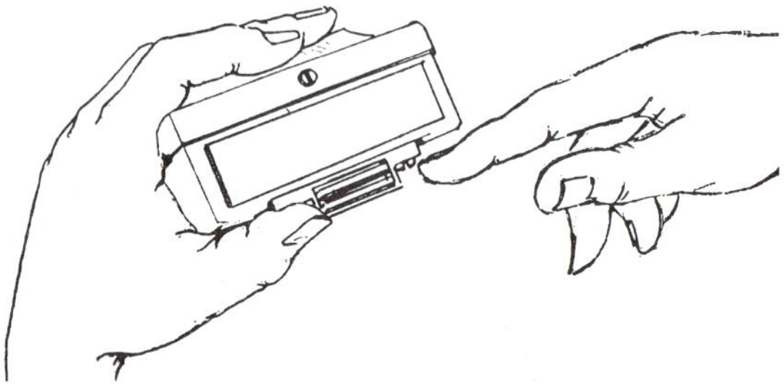

8. MAGNETIC BASE UNIT USED FOR CONVENIENT POSITION MOUNTING OF THERMOMETER.

Marsh

2-31. How to take care of the thermometer.

CHAPTER 2—AIR CONDITIONING AND REFRIGERATION TOOLS AND INSTRUMENTS

Amprobe

2-32. Hand-held electronic thermometer.

Thermal Engineering

2-33. Electronic thermometer for measuring superheat. The probes are made of thermocouple wire. They can be strapped on anywhere with total contact to the surface. This thermometer covers temperatures from −50° F. to +1500° F. on four scales. The temperature differential probe reads the difference in temperature between any two points directly. It reads superheat directly. It is battery operated and has a +2% accuracy on all ranges. Celsius scales are available.

difference between the actual temperature and the boiling point temperature is superheat. If the superheat is zero, the refrigerant must be boiling inside. Then, there is a good chance that some of the refrigerant is still liquid. If the superheat is greater than zero, at least 5° F. [−15°C] or better, then the refrigerant is probably past the boiling point stage and is all vapor.

The method of measuring superheat described here has obvious faults. If there is no attachment for a pressure gage at the point in the system where you are measuring superheat, the hypothetical boiling temperature cannot be found. To determine the superheat at such a point, the following method can be used. This method is particularly useful for measuring the refrigerant superheat in the suction line.

Instead of using a pressure gage, the boiling point of the refrigerant in the evaporator can be determined by measuring the temperature in the line just after the expansion valve where the boiling is vigorous. This can be done with any electronic thermometer, Fig. 2-32. As the refrigerant heats up through the evaporator and the suction line, the actual temperature of the refrigerant can be measured at any point along the suction. Comparison of these two temperatures gives a superheat measurement sufficient for field service unless a distributor metering device is used or the evaporator is very large with a great amount of pressure drop across the evaporator.

Using the meter shown in Fig. 2-33, it is possible to read superheat directly, using the temperature differential feature. Strap one end of the differential probe to the outlet of the metering device. Strap the other end to the point on the suction line where the superheat measure is to be taken. Turn the meter to temper-

41

REFRIGERATION AND AIR CONDITIONING TECHNOLOGY

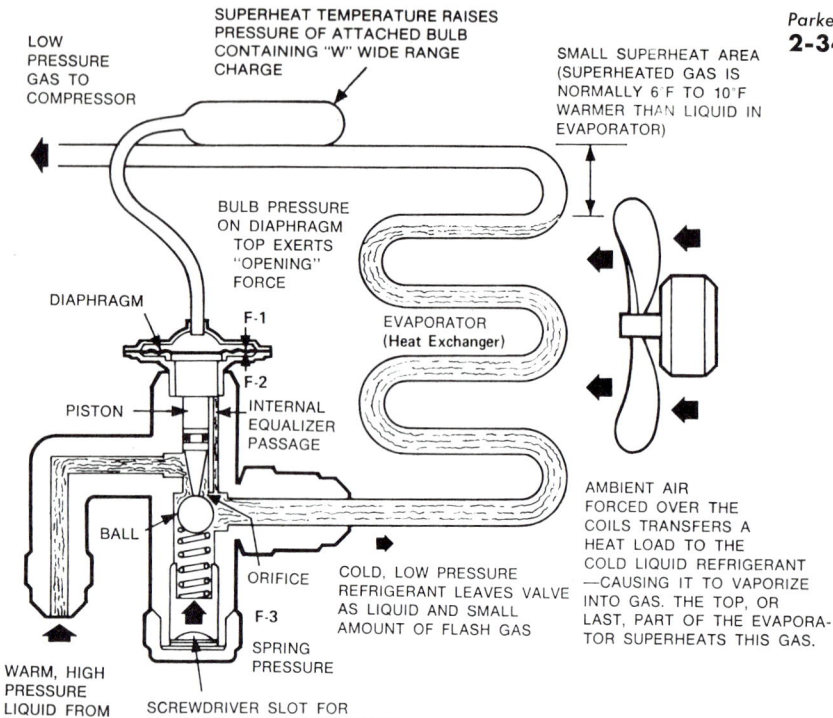

Parker-Hannifin
2-34. How superheat works.

temperature at the bulb location. In the example, the temperature is 37° F.

2. Measure the suction line pressure at the bulb location. In the example, the suction line pressure is 27 psi.

3. Convert the suction line pressure to the equivalent saturated (or liquid) evaporator temperature by using a standard temperature-pressure chart (27 psi = 28° F.).

4. Subtract the two temperatures. The difference is *superheat*. In this case, superheat is found by the following method: 37° F. − 28° F. = 9° F.

ature differential and the superheat will be directly read on the meter.

Figure 2-34 illustrates the way superheat works. The bulb "opening" force (F-1) is caused by bulb temperature. This force is balanced against the system back pressure (F-2), and the valve spring force (F-3) to hold the evaporator pressure within a range that will vaporize all of the refrigerant just before it reaches the upper part or end of the evaporator.

The method of checking superheat is shown in Fig. 2-35. The procedure is as follows:

1. Measure the suction line

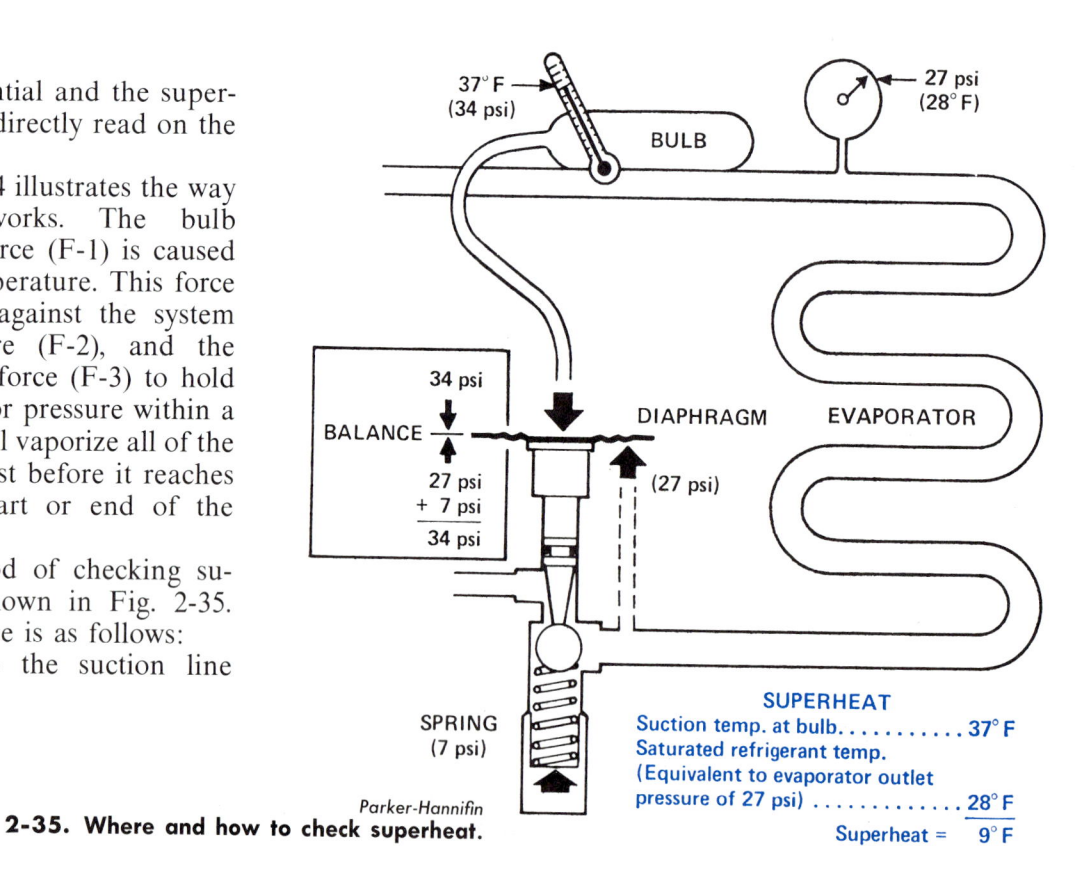

SUPERHEAT
Suction temp. at bulb............37° F
Saturated refrigerant temp.
(Equivalent to evaporator outlet
pressure of 27 psi).............28° F
Superheat = 9° F

Parker-Hannifin
2-35. Where and how to check superheat.

42

CHAPTER 2—AIR CONDITIONING AND REFRIGERATION TOOLS AND INSTRUMENTS

Suction pressure at the bulb may be obtained by either of the following methods:

1. If the valve has an external equalizer line, the gage in this line may be read directly.

2. If the valve is internally equalized, take a pressure gage reading at the compressor base valve. Add to this the estimated pressure drop between the gage and the bulb location. The sum will approximate the pressure at the bulb.

The system should be operating normally when the superheat is between 6 and 10° F. [−14.4 and −12.2°C].

HALIDE LEAK DETECTORS

Not too long ago leaks were detected by using soap bubbles and water. If possible, the area of the suspected leak was submerged in soapy water. Bubbles pinpointed the leak area. If the unit, or suspected area was not easily submerged in water—then it was coated with a soap solution. As the leak was covered with soap, bubbles would be produced. These indicated the location of the leak. These methods are still used today in some cases. However, it is now possible to obtain better indications of leaks with electronic equipment and with halide leak detectors.

Halide leak detectors are used in the refrigeration and air conditioning industry. They are designed for locating leaks of noncombustible halide refrigerant gases. Figs. 2-36 and 2-37.

The supersensitive detector will detect the presence of as little as 20 parts per million of refrigerant gases. Fig. 2-38. Another model will detect 100 parts of halide gas per million parts of air.

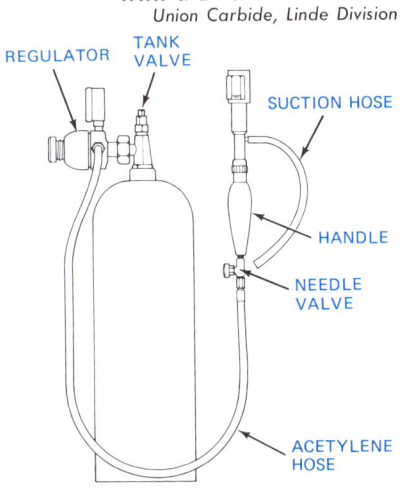

Setting Up

The leak detector is normally used with a standard torch handle. The torch has a shut-off valve. Acetylene can be supplied by a "B" tank (40 cubic feet) or an MC tank (10 cubic feet). In either case, the tank must be equipped with a pressure-reducing regulator. The torch handle is connected to the regulator by a suitable length of fitted acetylene hose. Fig. 2-36.

An alternate setup uses an adapter to connect the leak-detector stem to an MC tank. No regulator is required. The tank must be fitted with a handle. Fig. 2-37.

In making either setup, be sure all seating surfaces are clean before assembling. Tighten all connections securely. Use a wrench to tighten hose and regulator connections. If you use the B tank setup, be sure to follow the instructions supplied with the torch handle and regulator.

Lighting

● Setup with tank, regulator, and torch handle. Refer to Fig. 2-36.

1. Open the tank valve one-quarter turn, using a P-O-L tank key.

2. Be sure the shut-off valve on the torch handle is closed. Then adjust the regulator to deliver 10 psi. Do this by turning in the pressure-adjusting screw until the "C" marking on the flat surfaces of the screw is opposite the face of the front cap. Test for leaks.

3. Open the torch handle shut-off valve and light the gas above the reaction plate. Use a match or taper.

4. Adjust the torch until a steady flame is obtained.

● Setup with MC tank and adaptor. Refer to Fig. 2-37.

1. With the needle valve on the adaptor closed tightly, *just barely open* the tank valve, using a P-O-L tank key. Test for leaks.

2. Open the adaptor needle valve about one-quarter turn.

2-36. Halide leak detector for use with a B tank.
Union Carbide, Linde Division

2-37. Halide leak detector for use with an MC tank.
Union Carbide, Linde Division

43

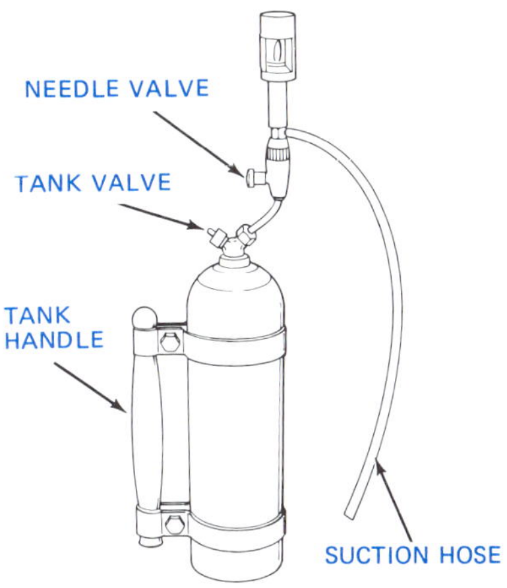

REFRIGERATION AND AIR CONDITIONING TECHNOLOGY

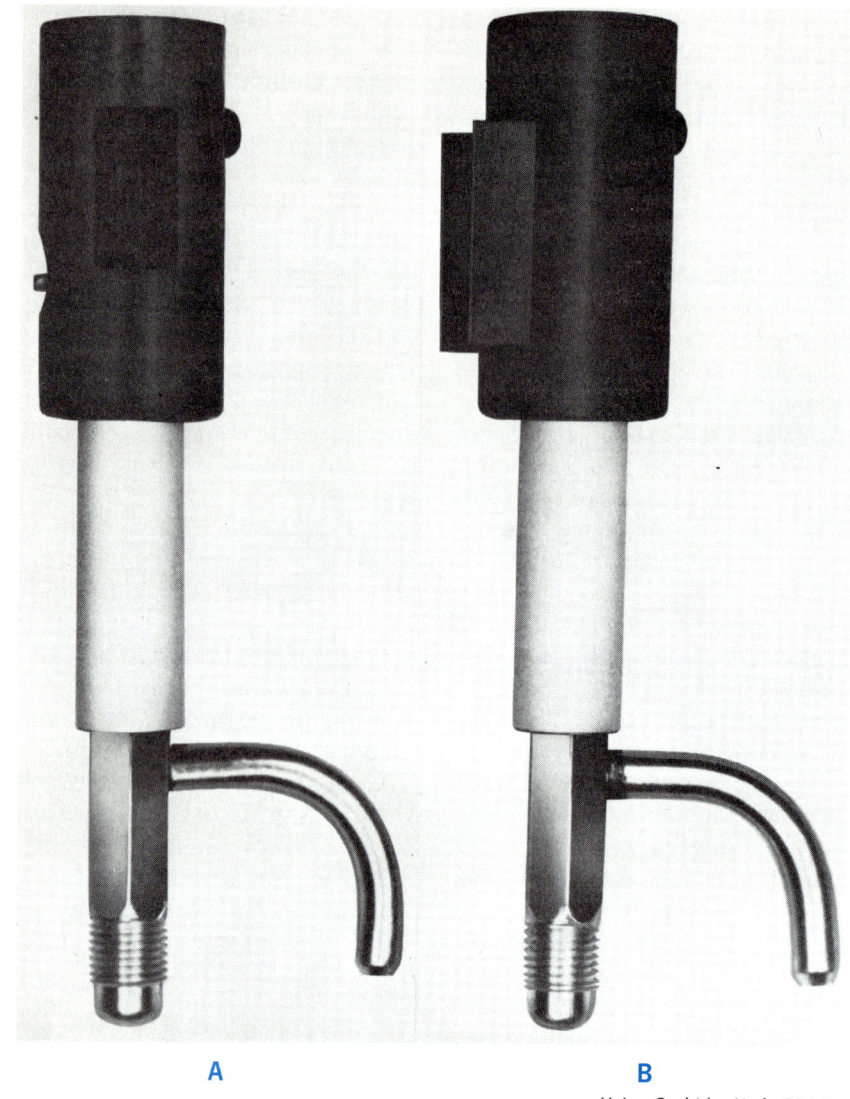

2-38. Detectors. (A) Supersensitive detector of refrigerant gases. This detects 20 parts per million. (B) Standard model detector torch. This detects 100 parts per million.

Union Carbide, Linde Division

Light the gas above the reaction plate. Use a match or taper.

Leak Testing the Setup

Using a small brush, apply a thick solution of soap and water to test for leaks. Check for leaks at the regulator and any connection point. Check the hose to handle connection, hose to regulator connection, and regulator or adaptor connection. If you find a leak, correct it before you light the gas. A leak at the valve stem of a small acetylene tank can often be corrected by tightening the packing nut with a wrench. If this will not stop the leak, remove the tank. Tag it to indicate valve stem leakage. Place it outdoors in a safe spot until you can return it to the supplier.

Adjusting the Flame

Place the inlet end of the suction hose so that it is unlikely to draw in air contaminated by refrigerant vapor. Adjust the needle valve on the adaptor or torch handle until the pale blue outer envelope of the flame extends about 1" above the reaction plate. The inner cone of the flame, which should also be visible above the reaction plate, should be clear and sharply defined.

If the outer envelope of the flame, when of proper length, is yellow, not pale blue, the hose is picking up refrigerant vapors. There may also be some obstruction in the suction hose. Make sure the suction tube is not clogged or bent sharply. If the suction tube is clear, shut off the flame. Close the tank valve. Disconnect the leak detector from the handle or adaptor. Check for dirt in the filter screw or mixer disc. Fig. 2-39. Use a $\frac{1}{8}$" socket key (Allen wrench) to remove or replace the filter screw. This screw retains the mixer disc.

Detecting Leaks

To explore for leaks, move the end of the suction hose around all points where there might be leaks. Be careful not to kink the suction hose.

Watch for color changes in the flame as you move the end of the suction hose. These are the changes which you should look for:

With the model that has a large opening in the flame shield (wings on each side), a small leak will change the color of the outer

CHAPTER 2—AIR CONDITIONING AND REFRIGERATION TOOLS AND INSTRUMENTS

flame to a yellow or an orange yellow hue. As the concentration of halide gas increases, the yellow will disappear. The lower part of the flame will become a bright, light blue. The top of the flame will become a vivid purplish blue.

With the model that has no wings alongside the flame shield opening, small concentrations of halide gas will change the color. A bright blue green outer flame indicates a leak. As the concentration of halide gas increases, the lower part of the flame will lose its greenish tint. The upper portion will become a vivid purplish blue.

Watch for color intensity changes. The location of small leaks can be pinpointed readily. Color in the flame will disappear almost instantly after the intake end of the hose has passed the point of leakage. With larger leaks, you will have to judge the point of leakage. Note the color change from yellow to purple blue or blue green to blue purple, depending upon the model used.

Maintenance

With intensive usage, an oxide scale may form on the surface of the reaction plate. Thus, sensitivity is reduced. Usually, this scale can be easily broken away from the plate surface. If you suspect a loss in sensitivity, remove the reaction plate. Scrape its surface with a knife or screwdriver blade, or install a new plate.

ELECTRICAL INSTRUMENTS

Several electrical instruments are used by the air conditioning serviceperson to see if the equipment is working properly. Studies show that the most trouble calls on heating and cooling equipment are electrical in nature.

The most frequently measured quantities are volts, amperes, and ohms. In some cases, wattage is measured to check for shorts and other malfuctions. A wattmeter is available. However, it must be used to measure volt-amperes instead of watts. To measure watts, it is necessary to use *DC only* or convert the volt-amperes (VA) to watts by using the power factor. The power factor times the volt-amperes produces the actual power consumed in watts. Since most cooling equipment uses AC, it is necessary to convert to watts by this method.

A number of factors can be checked with electrical instruments. For example, electrical instruments can be used to check the flow rate from a centrifugal water pump, the condition of a capacitor, or the qualification of a start or run-winding of an electric motor.

Ammeter

The ammeter is used to measure current. It can measure the amount of current flowing in a circuit. It may use one of a number of different basic meter movements to accomplish this. The most frequently used of the basic meter movements is the D'Arsonval type. Fig. 2-40. It uses a permanent magnet and an electromagnet to determine circuit current. The permanent magnet is used as a standard or basic source of magnetism. As the current flows through the coil of wire it creates a magnetic field around it. This magnetic field is strong or weak, depending upon

2-39. Position of filter screw and mixer disk on Prest-O-Lite halide leak detector standard Model II and supersensitive Model I.

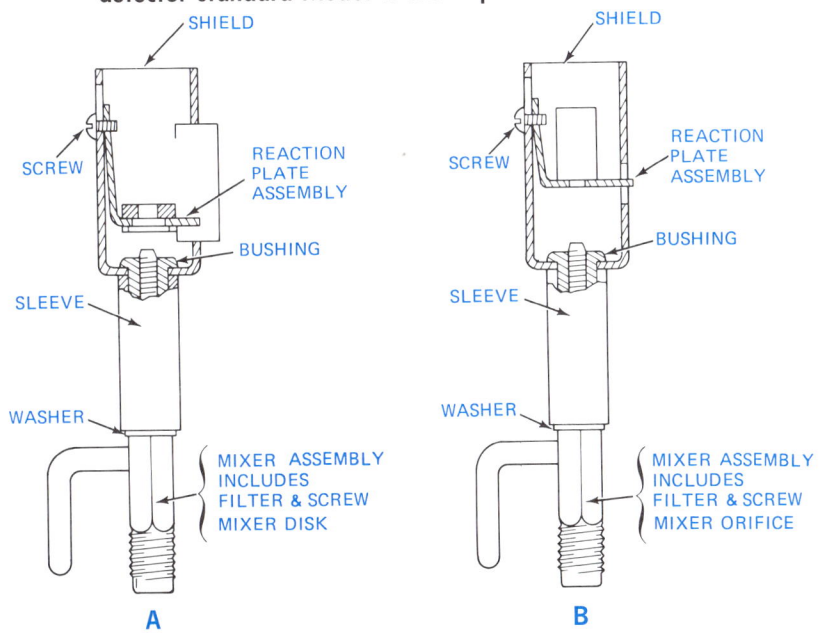

A

B

45

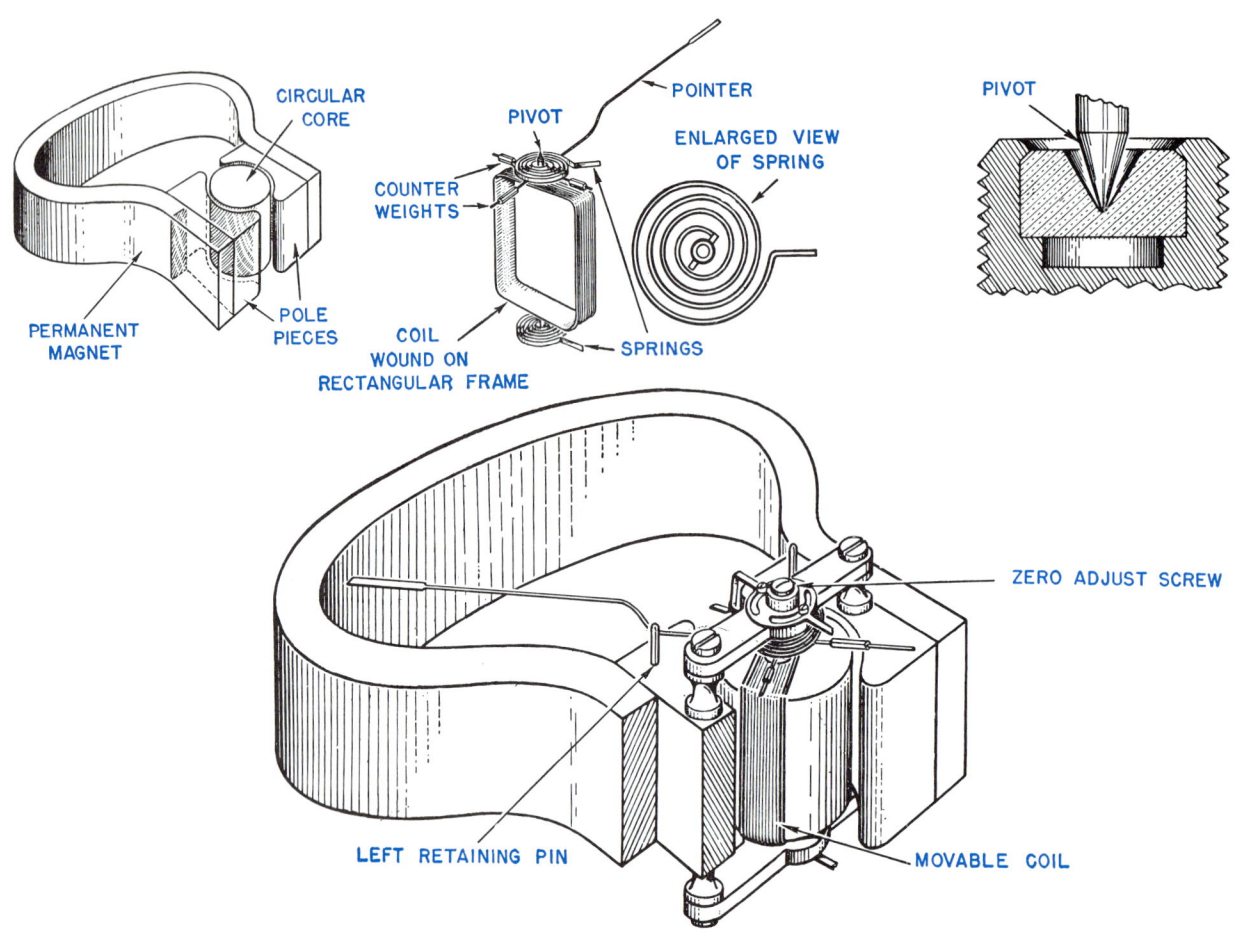

REFRIGERATION AND AIR CONDITIONING TECHNOLOGY

2-40. Moving-coil (D'Arsonval) meter movement.

the amount of current flowing through it. The stronger the magnetic field created by the moving coil, the more it is repelled by the permanent magnet. This repelling motion is calibrated to read amperes, milliamperes (0.001 A), or microamperes (0.000 001 A).

The D'Arsonval meter movement may also be used on alternating current (AC) when a diode is placed in series with the moving coil winding. The diode changes the AC to DC and the meter works as on DC. Fig. 2-41. The dial or face of the instrument is calibrated to indicate the AC readings.

There are other types of AC ammeters. They are not always as accurate as the D'Arsonval, but they are effective. In some moving magnet meters, the coil is stationary and the magnet moves. Although rugged, this type is not as accurate as the D'Arsonval meter.

The moving vane meter is useful in measuring current when AC is used. Fig. 2-42.

The clamp-on ammeter has already been discussed. It has some limitations. However, it does have one advantage in that it can be used without having to break the line to insert it. Most ammeters must be connected in series with the consuming device. That means one line has to be

2-41. Diode inserted in the circuit with a D'Arsonval meter movement to produce an AC ammeter.

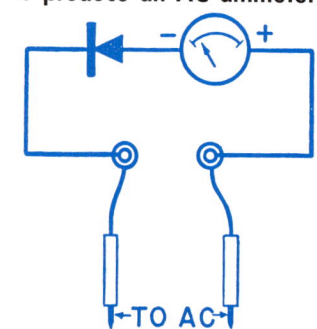

broken or disconnected to insert the meter into the circuit.

The ampere reading can be used to determine if the unit is drawing too much current or insufficient current. The correct current amount is usually stamped on the nameplate of the motor or the compressor.

Starting and running amperes may be checked to see if the motor is running under too much load or is shorted. The flow rate of some pumps can be determined by reading the current the motor pulls. The load on the entire line can be checked by inserting the ammeter in the line. This is done by taking out the fuse and completing the circuit with the meter. Be careful.

If the ammeter has more than one range, it is best to start on the highest range and work down. The reading should be in or near the center of the meter scale.

Make sure you have some idea of what the current in the circuit should be before inserting the meter. Thus, the correct range—or, in some instances, the correct meter—can be selected.

Voltmeter

The voltmeter is used to measure voltage. Voltage is the electrical pressure needed to cause current to flow. The voltmeter is used across the line or across a motor or whatever is being used as a consuming device.

Voltmeters are nothing more than ammeters that are calibrated to read volts. There is, however, an important difference. The voltmeter has a very high internal resistance. Thus, very small amounts of current flow through its coil. Fig. 2-43. This high resistance is produced by multipliers. Each range on the voltmeter has a different resistor to increase the resistance so the line current will not be diverted

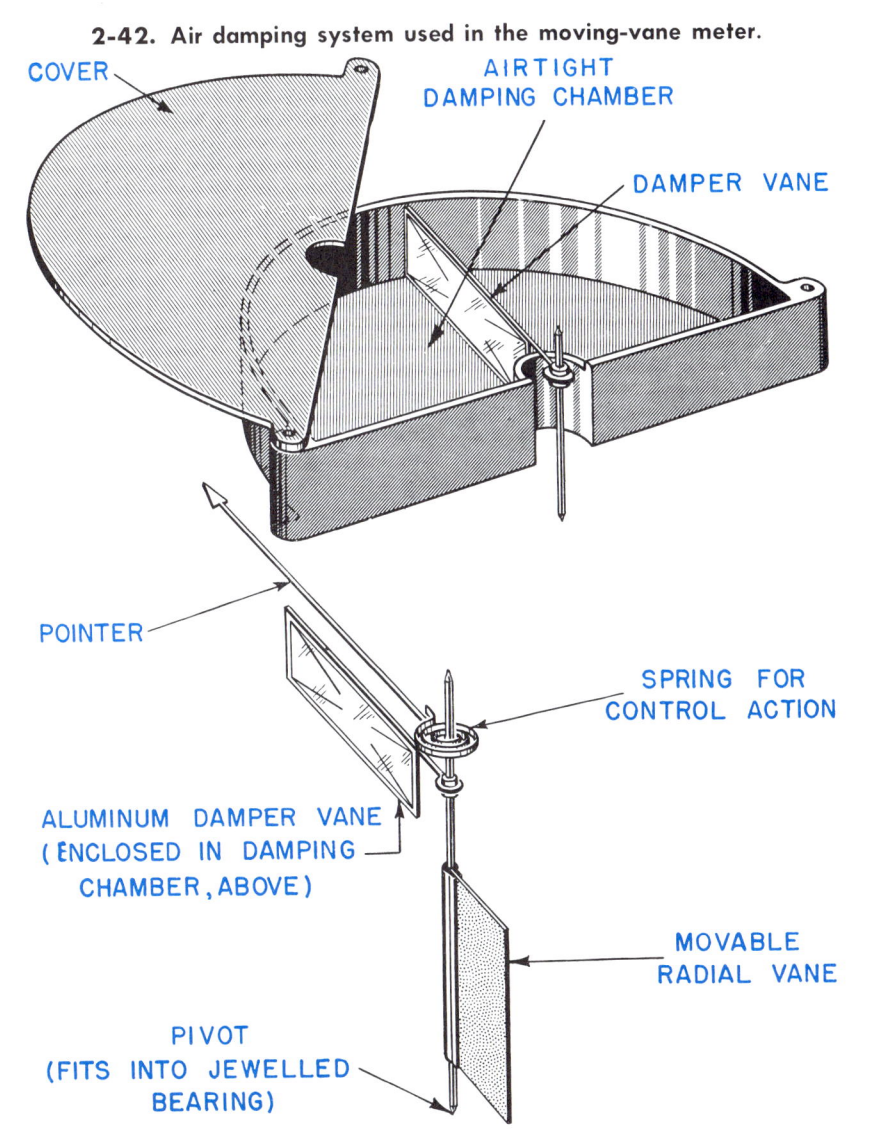

2-42. Air damping system used in the moving-vane meter.

2-43. A voltmeter circuit with high resistance in series with the meter movement allows it to measure voltage.

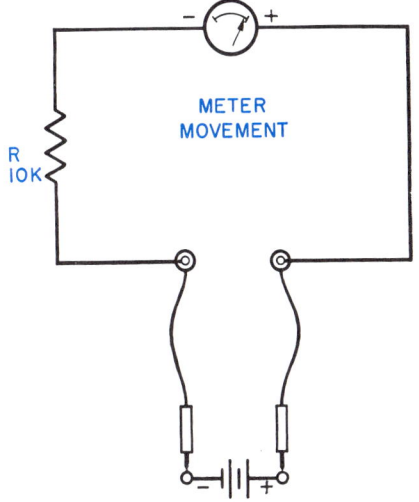

47

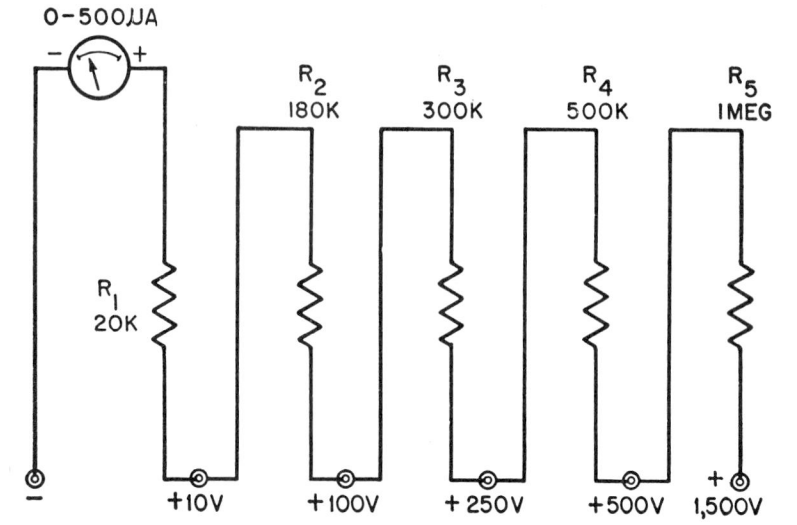

REFRIGERATION AND AIR CONDITIONING TECHNOLOGY

2-44. Different types of multirange voltmeters. This view shows the interior of the meter box or unit.

through it. Fig. 2-44. The voltmeter is placed across the line, whereas the ammeter is placed in series. You do not have to break the line to use the voltmeter. The voltmeter has two leads. If you are measuring DC, you have to observe polarity. The red lead is positive (+) and the black lead is negative (−). However, when AC or alternating current is used, it does not matter which lead is placed on which terminal. Using a D'Arsonval meter movement, voltmeters can be made with the proper diode to change AC to DC. Voltmeters can be made with a stationary coil and a moving magnet. Other types of voltmeters are available. They use various means of registering voltage.

If the voltage is not known, use the highest scale. Turn the range switch to a point where the reading is in the midrange of the meter movement.

Normal line voltage in most locations is 120 V. When line voltage is lower than normal, it is possible for the equipment to draw excessive current. This will cause overheating and eventual failure due to burnout. The correct voltage is needed for the equipment to operate according to its designed specifications. The voltage range is usually stamped on the nameplate of the device. Some state 208 V. This is obtained from a three-phase connection. Most home or residential power is supplied with 120 or 240 V. The range is 220 to 240 V for normal residential service. The size of the wire used to connect the equipment to the line is important. If the wire is too small, it will drop voltage. There will be low voltage at the consuming device. For this reason, a certified electrician with knowledge of the National Electrical Code should wire a new installation.

Ohmmeter

The *ohmmeter* measures resistance. The basic unit of resistance is the ohm (Ω). Every device has resistance. That is why it is necessary to know the proper resistance before trying to troubleshoot a device by using an ohmmeter. The ohmmeter has its own power source. Fig. 2-45. Do not use an ohmmeter on a line that is energized or connected to a power source.

An ohmmeter can read the resistance of the windings of a motor. If the correct reading has been given by the manufacturer, it is then possible to see if the reading has changed. If the reading is much lower, it may indicate a shorted winding. If the reading is infinite (∞), it may mean there is a loose connection or an open circuit.

Ohmmeters have ranges. Fig. 2-46. The R × 1 range means the scale is read as is. If the R × 10 range is used, it means that the scale reading must be multiplied by 10. If the R × 1000 range is selected, then the scale reading must be multiplied by 1000. If the meter has a R × 1 meg range, the scale reading must be multiplied by 1 000 000. A meg is one million.

Multimeter

The *multimeter* is a combination of meters. Fig. 2-47. It may

2-45. Internal circuit of an ohmmeter.

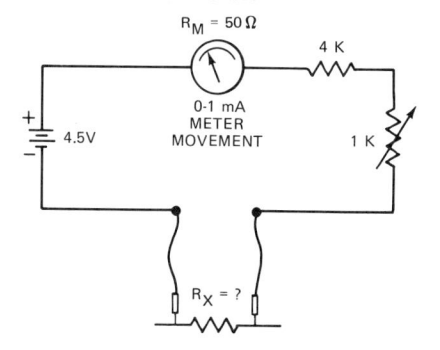

CHAPTER 2—AIR CONDITIONING AND REFRIGERATION TOOLS AND INSTRUMENTS

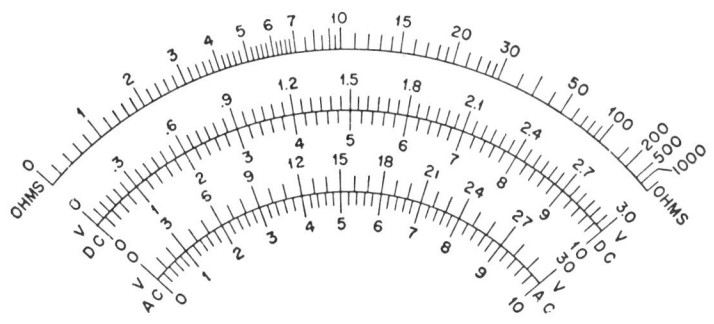

2-46. A multimeter scale. Note the ohms and volt scales.

have a voltmeter, ammeter, and ohmmeter in one case. This is the usual arrangement for fieldwork. This way it is possible to have all three meters in one portable combination. It should be checked for each of its functions.

The snap-around meter uses its scale for a number of applications. It can read current by snapping around the current-carrying wire. If the leads are used, it can be used as a voltmeter or an ohmmeter. Remember that the power must be off to use the ohmmeter. This meter comes in its own case. It should be protected from shock and vibration just as any other sensitive instrument.

Wattmeter

The *wattmeter* is used to measure watts. However, when used on an alternating current line (AC) it measures volt-amperes. If watts are to be measured, the reading must be converted to watts mathematically. Multiply the reading on the wattmeter by the power factor (usually available on the nameplate) to obtain the reading in watts.

Wattmeters use the current and the voltage connections as with individual meters. Fig. 2-48. One coil is heavy wire and is connected in series. It measures the current. The other connection is made in the same way as with a voltmeter—across the line. This coil is made of many turns of fine wire. It measures the voltage. By the action of the two magnetic fields, the current is multiplied by the voltage. Wattage is read on the meter scale.

The volt-ampere is the unit used to measure volts times amperes in an AC circuit. If a device has inductance (as in a motor) or capacitance (some motors have run-capacitors) the true wattage is not given on a wattmeter. The reading is in volt-amperes, instead of watts. It is converted to watts by multiplying the reading by the power factor. A wattmeter reads watts only when it is connected to a DC circuit or to an AC circuit with resistance only.

The power factor is the ratio of true power to apparent power. Apparent power is what is read on a wattmeter on an AC line. True power is the wattage reading on DC. The two can be used to find the power factor. The power factor is the cosine of the phase angle. The power factor can be found by using a mathematical computation or a very delicate meter designed for the purpose. However, the power factor of equipment using alter-

2-47. Two types of multimeters.

49

REFRIGERATION AND AIR CONDITIONING TECHNOLOGY

nating current is usually stamped on the nameplate of the compressor, the motor, or the unit itself.

Wattmeters are also used to test capacitors. Some companies provide charts to convert wattage ratings to microfarad ratings. The wattmeter can test the actual condition of the capacitor. The ohmmeter tells if the capacitor is good or bad. However, it is hard to indicate how a capacitor will function in a circuit with the voltage applied. This is why testing with the wattmeter is preferred.

OTHER INSTRUMENTS

Many types of meters and gages are available to test almost any quantity or condition. For example, air efficiency gages, air measurement gages, humidity measurement devices, moisture analyzers, and Btu meters, are used. Vibration and sound meters and recorders are also available.

Air Filter Efficiency Gages

Air measurements are taken in an air-distribution system. They often reveal the existence and location of unintentionally closed or open dampers. Obstructions, leaks in the ductwork, and sharp bends are located this way.

Air measurements frequently show the existence of a blocked filter. Dirty and blocked filters can upset the balance of either a heating or cooling system. This is important whether it is in the home or in a large building.

Certain indicators and gages can be mounted in air plenums. They can be used to show when

Amprobe

2-48. (A) Wattmeter connection for measuring input power. (B) Alternate wattmeter connection. (C) With load disconnected, uncompensated wattmeter measures own power loss.

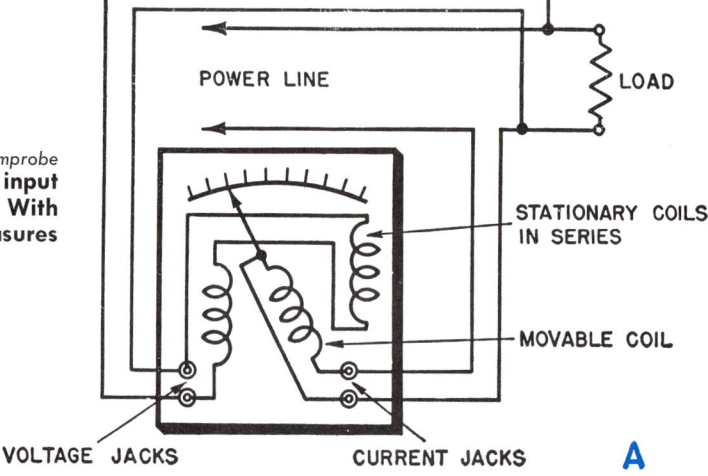

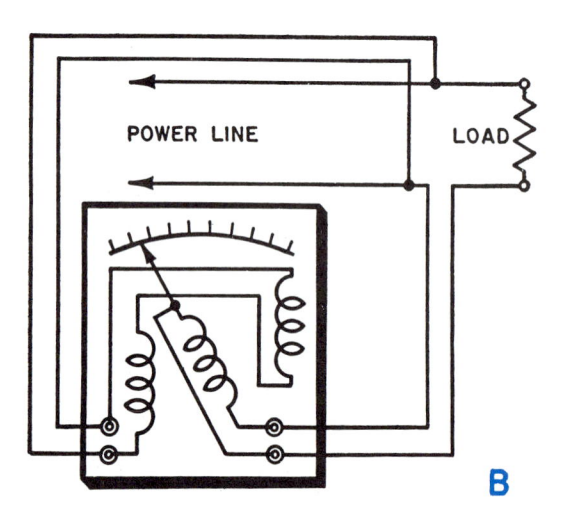

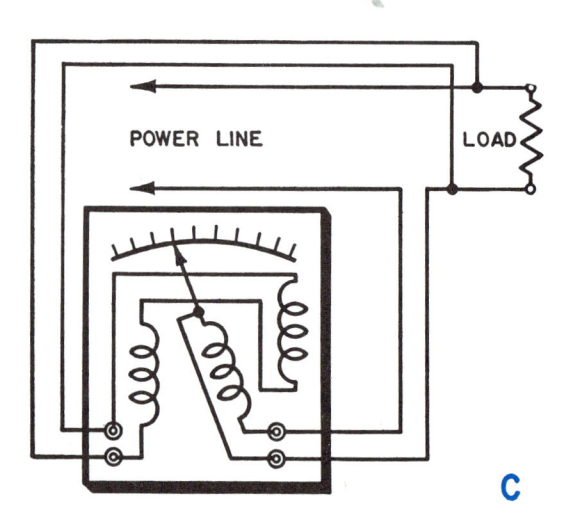

50

CHAPTER 2—AIR CONDITIONING AND REFRIGERATION TOOLS AND INSTRUMENTS

the filter has reached the point that it is restricting the air flow. An air plenum is a large space above the furnace heating or cooling unit.

Air Measurement Instruments

The volume and velocity of air are important measurements in the temperature control industries. Proper amounts of air are indispensable to the best functioning of refrigeration cycles, regardless of the size of the system. Air conditioning units and systems also rely upon volume and velocity for proper distribution of conditioned air.

Only a small number of contractors are equipped to measure volume and velocity correctly. The companies that are doing the job properly are in great demand. Professional handling of air volume and velocity ensure the efficient use of equipment. Large buildings are very much in need of the skills of air balancing teams.

Some people attempt to obtain proper air flow by measuring air temperature. They adjust dampers and blower speeds. However, they usually fail in their attempts to balance the air flow properly.

There are instruments available to measure air velocity and volume. Such instruments can accurately measure the low pressures and differentials involved in air distribution.

Draft gages do measure pressure. However, their specific application to air control makes it more appropriate to discuss them here, rather than under pressure gages. They measure pressure in inches of water. They come in several styles. The most familiar is the slant type. It may be used either in the field or in the shop.

Meter-type draft gages are better for field work. They can be carried easily. They can sample air at various locations, with the meter box in one location.

Besides air pressure, it is frequently necessary to measure air volume (cfm). Air volume is measured in cubic feet per minute (cfm). Air velocity is measured in feet per minute (fpm).

The measure of air flow is still somewhat difficult. However, newer instruments are making more accurate measurements possible.

Humidity Measurement Instruments

Humidity is read in rh, or *relative humidity*. To obtain the rh it is necessary to use two thermometers. One thermometer is a dry bulb; the other is a wet bulb. The device used to measure relative humidity is the *sling psychrometer*. It has two glass-stem thermometers. The wet bulb thermometer is moistened by a wick attached to the bulb. As the dual thermometers are whirled, air passes over them. The dry and wet bulb temperatures are recorded. Relative humidity is determined by graphs, slide rules, or similar devices.

STATIONARY PSYCHROMETERS

Stationary psychrometers take the same measurements as sling-psychrometers. They do not move, however. They use a blower or fan to move the air over the thermometer bulbs.

For approximate rh readings, there are metered devices. They are used on desks and walls. They are not accurate enough to be used in engineering work.

Humidistats, which are humidity controls, are used to control humidifiers. They operate the same way as thermostats in closing contacts to complete a circuit. They do not use the same sensing element, however.

Moisture Analyzers

It is sometimes necessary to know the percentage of water in a refrigerant. The water vapor or moisture is measured in parts per million. The necessary measuring instrument is still used primarily in the laboratory. Instruments for measuring humidity are not used here. Soon, perhaps, moisture analyzers that can be used in the field will be available.

Btu Meters

The *British thermal unit* (*Btu*) is used to indicate the amount of heat present. Meters are specially designed to indicate the Btu in a chilled water line, a hot water line, or a natural gas line. Specially designed, they are used by skilled laboratory personnel at present.

Vibration and Sound Meters

More cities are now prohibiting air conditioning units that make too much noise. In most cases, vibration is the main problem. However, it is not an easy task to locate the source of vibration. However, special meters have been designed to aid in the search for vibration noise.

Portable noise meters are available. The dB, or decibel, is the unit for the measurement of sound. There are a couple of bands on the noise meters. The dB-A scale corresponds roughly to the human hearing range.

REFRIGERATION AND AIR CONDITIONING TECHNOLOGY

Other scales are available for special applications.

More emphasis is now being placed on noise levels in factories, offices, and schools. The Occupational Safety and Hazards Act (OSHA) lays down strict guidelines regarding noise levels. There are penalties for noncompliance. Thus, it will be necessary for all new and previously installed units to be checked for noise.

High-velocity air systems—used in large buildings—are engineered to reduce noise to levels set by OSHA. For example, there are chambers to lower the noise in the ducts. Air engineers are constantly working on high-velocity systems to try to solve some of the problems associated with them.

Recorders

If a permanent or a continuous reading of temperature, humidity, pressure, or voltage is needed, a *recorder* will be needed. Recorders are used to record on paper (or in some instances, tape) the operation of a circuit or piece of equipment. Pressure in lines or temperatures of lines or air flow can also be recorded.

Several service calls may fail to locate the problem in a circuit or piece of equipment. A recorder might then be used to detect variations in performance. Once the variations are known, a solution can be effected.

Recorders are also used for testing equipment in the shop.

The recorder is nothing more than a recording electrical instrument. It does not record pressure, temperature, or humidity. All it records are variations in the current. The main component of the recorder, the transducer, converts changes in temperature, humidity, and pressure into electrical impulses. The marking device on the recorder can be a ballpoint pen, a specially adapted ink pen, or a sensitive paper tape with a pressure-sensitive coating. The pressure-sensitive coating on the recorder paper is scribed by a small point, which makes a pressure line. Small electrical motors drive the paper or tape recording surface. The recorders can be either bat-

Thermal Engineering

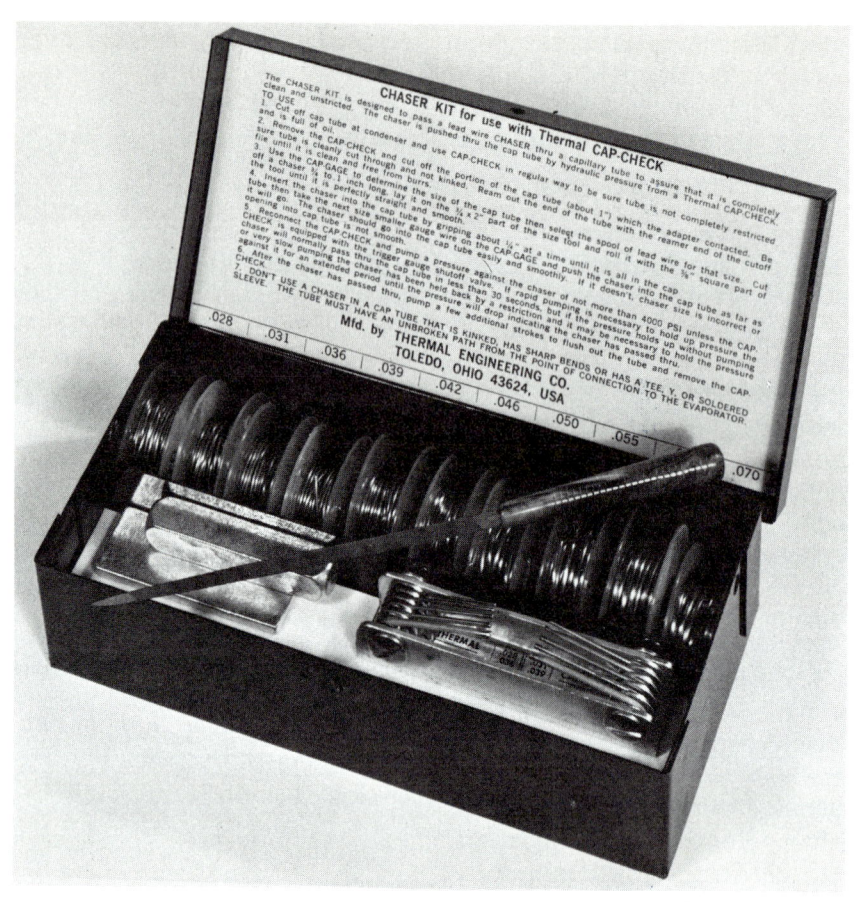

2-49. Cap-Check® chaser kit. This is a means to clean partially plugged cap tubes. It has ten spools of lead alloy wire. These wires can be used as a chaser for the ten most popular sizes of cap tubes. A cap tube gage, set of sizing tools, and a combination file / reamer are included in the kit.

tery operated or plugged into the available power source.

SERVICE TOOLS

Servicepersons use some special devices to help them with repair jobs in the field. One of them is the *chaser kit*. Fig. 2-49. It is used for cleaning partially plugged capillary tubes. The unit includes ten spools of lead alloy wire. These wires can be used as chasers for the ten most popular sizes of capillary tubes. In addition to the wire, a cap tube gage,

52

CHAPTER 2—AIR CONDITIONING AND REFRIGERATION TOOLS AND INSTRUMENTS

Thermal Engineering

2-50. Cap-Check® is a portable, self-contained hydraulic power unit with auxiliary equipment especially adapted to cleaning refrigeration capillary tubes. It is hand operated.

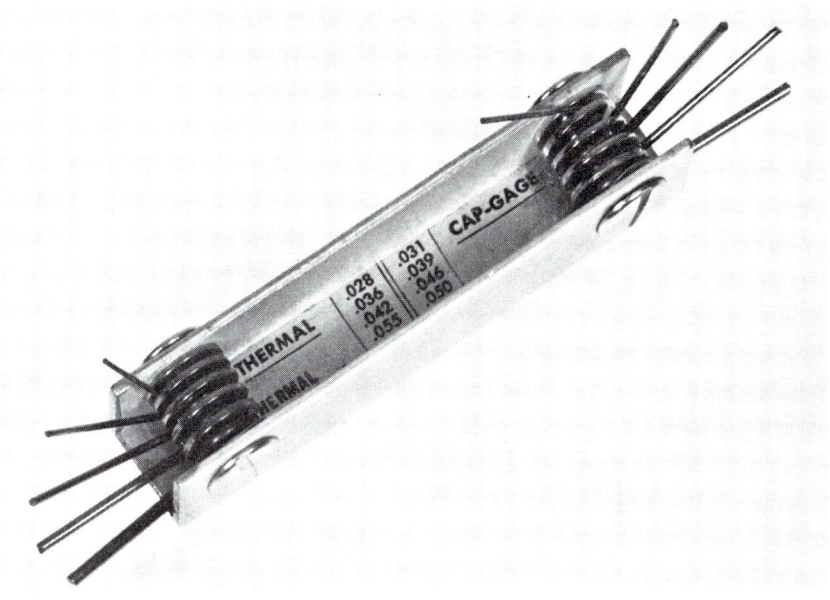

Thermal Engineering

2-51. The Cap-Gage is a pocketknife-type cap tube gage with 10 stainless steel gages to measure the most popular sizes of cap tubes.

a set of sizing tools, and a combination file/reamer are included in the metal case. This kit is used in conjunction with the Cap-Check*.

The Cap-Check is a portable, self-contained hydraulic power unit with auxiliary equipment especially adapted to cleansing refrigeration capillary tubes. Fig. 2-50. A small plug of wire from the chaser kit is inserted into the capillary tube. The wire is a few thousands of an inch smaller than the internal diameter of the capillary tube. This wire is pushed like a piston through the capillary tube with hydraulic pressure from the Cap-Check. A 0-5000 psi gage shows pressure build-up if the capillary tube is restricted. It also shows when the chaser has passed through the tube. A trigger-operated gage shutoff is provided so the gage will not be damaged if pressures greater than 5000 psi are desired. When the piston stops against a partial restriction, high-velocity oil is directed around the piston and against the wall, washing the restriction away and allowing the wire to move through. The lead wire eventually ends up in the bottom of the evaporator, where it remains. The capillary tube is then as clean as when it was originally installed.

A 30" high pressure hydraulic hose with a $\frac{1}{4}$" SAE male flare outlet connects the cap tube to the Cap-Check for simple handling. An adapter comes with the Cap-Check to connect the cap tube directly to the hose outlet without a flared fitting.

The Cap-Gage® is a capillary tube gage. It has ten stainless steel gages to measure the most popular sizes of capillary tubes. Fig. 2-51.

SPECIAL TOOLS

Sooner or later, almost every refrigerant charging job turns into a vapor charging job. Unless the compressor is turned on, liquid can be charged into the high side only so long before the system and cylinder pressures become unfavorable. When this happens, all refrigerant must be taken in the low side in the form of vapor.

Vapor charging is much slower than liquid charging. To create a vapor inside the refrigerant cylinder, the liquid refrigerant must be boiling. Boiling refrigerant absorbs heat. This is the principle on which refrigeration operates.

The boiling refrigerant absorbs heat from the refrigerant

---
*"Cap-Check" is a registered trademark of Thermal Engineering Co.

REFRIGERATION AND AIR CONDITIONING TECHNOLOGY

surrounding it in the cylinder. The net effect is that the cylinder temperature begins to drop soon after you begin charging with vapor. As the temperature drops, the remaining refrigerant will not vaporize as readily. Charging will slow.

To speed charging, servicepersons add heat to the cylinder by immersing part of it in hot water. The cylinder temperature rises. The boiling refrigerant becomes vigorous, and charging returns to a rapid rate. It is not long, though, before all the heat has been taken from the water and more hot water must be added.

The Vizi-Vapr® can remove liquid from a cylinder and apply it to the system in the form of a vapor. Fig. 2-52. No heat is required. This eliminates the hazards of using a torch and hot water. The change from a liquid to a gas or vapor takes place in the Vizi-Vapr®. It restricts the charging line between the cylinder and compressor. This restriction is much like an expansion valve in that it maintains high cylinder pressure behind it to hold the refrigerant as a liquid. However, it has a large pressure drop across it to start evaporation. The heat required to vaporize the refrigerant is taken from the air surrounding the unit, not from the remaining refrigerant. This produces a dense, saturated vapor.

The amount of restriction in the unit is very critical. Too much restriction will slow charging considerably. It also will allow liquid to go through and cause liquid slugging in the compressor. The restriction setting is different for each size system, for different types of refrigerants,

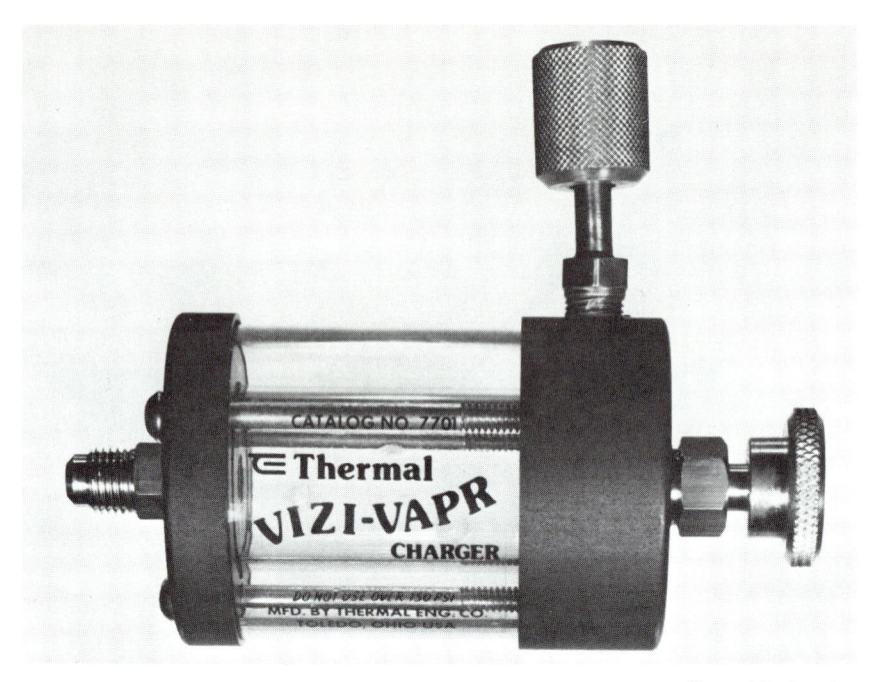

Thermal Engineering
2-52. The Vizi-Vapr® is a device that allows rapid charging of a compressor without heating the cylinder of refrigerant.

and even for different ambient temperatures.

The unit can be set easily for a specific job. The charging refrigerant is visible through the clear plastic case of the unit. If the valve is closed too far, you will see only a fine spray of refrigerant. Opening the metering valve slightly will bring forth a dense, liquid-vapor spray. This will charge rapidly and will not affect the compressor. Opening the valve too far will cause liquid to form in the visible portion of the unit. Eventually, this would flood into the compressor. Open the valve until just a little liquid forms and stays at a constant level in the visible chamber. This will result in the best operating point and the maximum safe charging rate.

The unit can be connected to the low side of your charging manifold. To save refrigerant, it can also be attached directly to the refrigerant cylinder. When using a charging cylinder for exact charge, the unit connects to the liquid discharge line. It settles the refrigerant level in the cylinder so that it is easily read, even when charging with vapor.

VACUUM PUMPS

The use of the vacuum pump may be the single most important development in refrigeration and air conditioning service in the last ten years.

The purpose of a vacuum pump is to remove the undesirable materials that create pressure in a refrigeration system. These include moisture, air (oxygen), hydrochloric acid, and other materials that will vaporize in the low micron range. These, along with a wide variety of solid mate-

CHAPTER 2—AIR CONDITIONING AND REFRIGERATION TOOLS AND INSTRUMENTS

rials, are pulled into the vacuum pump in the same way a vacuum cleaner sucks up dirt.

Evacuation is being routinely performed on almost every service call on which recharging is required. Vacuum levels formerly unheard of for field evacuation are being accomplished daily by servicepersons who are knowledgeable regarding vacuum equipment. These servicepersons have found, through experience, that the two-stage pump is much better than the single-stage pump for deep evacuation. Figs. 2-53 and 2-54. It was devised as a laboratory instrument. With minor alterations, it has been adapted to the refrigeration field. It is the proper tool for vacuum evacuations in the field.

To understand the advantages of a two-stage pump over a single-stage pump, refer to Fig. 2-55. This shows the interior of a two-stage vacuum pump. This is a simplified version of a vacuum stage. It is built on the principle of a Wankel engine. There is a stationary chamber with an eccentric rotor revolving inside. The sliding vanes pull gases through the intake. They compress them and force them into the atmosphere through the exhaust. The vanes create a vacuum section and a pressure section inside the pump. The seal between the vacuum and the pressure sections is made by the vacuum pump oil. These seals are the critical factor in the depth of the vacuum a pump can pull. If the seals leak, the pump will not be able to draw a deep vacuum. Consequently, less gas can be processed. A pump with high leakage across the seal will be able to pull a deep vacuum on a small system. But the leakage will decrease the pumping speed (CFM) in the deep vacuum region. Long pull-down times will result.

There are three oil seals in a

Thermal Engineering
2-53. Single-stage portable vacuum pump.

Thermal Engineering
2-54. Two-stage portable vacuum pump.

55

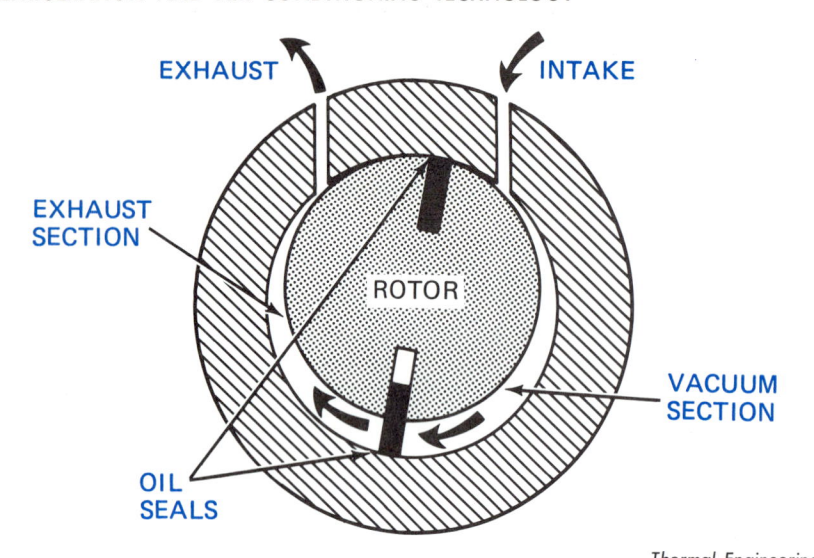

REFRIGERATION AND AIR CONDITIONING TECHNOLOGY

2-55. Two-stage vacuum pump showing seals and intake, exhaust, and vacuum section.

Thermal Engineering

single-stage vacuum pump. Each seal must hold against a high pressure on one side and a deep vacuum on the other side. This places a great deal of strain on the oil seal. A two-stage vacuum pump cuts the pressure strain on an oil seal in half. Such a pump uses two chambers instead of one to evacuate a system. The first chamber is called the deep vacuum chamber. It pulls in the vacuum gases from the deep vacuum and exhausts them into the second chamber at a moderate vacuum. The second chamber, or stage, brings in these gases at a moderate vacuum and exhausts them into the atmosphere. By doing this, the work of a single chamber is split between two chambers. This, in turn, cuts in half the strain on each oil seal, which reduces the leakage up to 90%.

A two-stage vacuum pump is more effective than a single-stage vacuum pump. For example, a single-stage vacuum pump rated for 1.5 CFM capacity will take one and one-half hours to evacuate one drop of water. A two-stage vacuum pump with the same rating will evacuate the drop in twelve minutes.

For evacuation of a five-ton system saturated with moisture, a minimum of fifteen hours evacuation time is required in using a single-stage vacuum pump. A two-stage pump with the same CFM rating could do the job in as little as two hours.

Another advantage of the two-stage pump is reliability. As you can see, if the oil seal is to be effective, the tolerances in these vacuum pumps must be very close between rotor and stator. If the tolerences are not correct, the oil seal will not be effective. Slippage of tolerance due to wear is the major cause of vacuum pump failure. With a single-stage pump, when the tolerance in the stage slips, the pump loses effectiveness. With a two-stage pump, if one stage loses tolerance, the other one will still pull the vacuum of a single-stage pump.

Today, the larger CFM, two-stage vacuum pumps are preferred to the one-stage vacuum pumps. The cost difference between the two is not great. Also, the time saved by using the two-stage pump is evident on the first evacuation.

Vacuum Pump Maintenance

The purpose of vacuum pump oil is to lubricate the pump and act as a seal. To perform this function, the oil must have:

• A low vapor pressure that does not materially increase up to 125° F. [51.7°C]. A viscosity sufficiently low for use at 60° F. [15.6°C] yet fairly constant up to 125° F. [51.7°C].

These requirements are easily met by using a low vapor pressure, paraffinic-base oil having a viscosity of approximately 300 SSU (shearing stress units) at 100° F. [37.8°C] and a viscosity index in the range of 95–100. This type of uninhabited oil is readily obtainable. It is the material provided by virtually all sellers of vacuum pump oil to the refrigeration trade.

Vacuum Pump Oil Problems

The oils used in vacuum pumps are designed to lubricate and seal. Many of the oils available for other jobs are not designed to clean as they lubricate. Neither are they designed to keep in suspension the solids freed by the cleaning action of oil. In addition, the oil is not usually heavily inhibited against the action of oxygen. Therefore, the vacuum pump must be run with flushing oil periodically to clean it. Otherwise, its efficiency will be reduced. The use of flushing oils is recommended by pump manufacturers.

CHAPTER 2—AIR CONDITIONING AND REFRIGERATION TOOLS AND INSTRUMENTS

Thermal Engineering

2-56. This vacuum check gage is designed to be as handy as a charging manifold. This one has a three hose-charging manifold with compound and pressure gage and a third gage. It has a minielectronic vacuum gage. When the manifold is connected to a vacuum pump it becomes a light-duty, leaktight vacuum manifold with the center gage coming active to measure vacuum levels below 29 inches of mercury. The vacuum sensor and all electronics are housed inside a flexible plastic case.

As a charging manifold, it is a conventional front-wheel, three-hose manifold with internal design that doubles the flow capacity for fast charging.

With the center line attached to the vacuum pump, outside lines allow simultaneously evacuation of both high and low sides of the system. High-side and low-side valves act as isolation valves to the system. A vacuum sensor is mounted to the low side of the manifold to automatically record the vacuum in microns. The unit is powered by 2 C-cell batteries.

If hydrochloric acid has been pulled into the pump, water, solids, and oil will bond together to form a sludge or slime that may be acidic. The oil also may deteriorate due to oxidation (action on the oil by oxygen in air pulled through the pump). This results in a pump that will not pull a proper vacuum, may wear excessively, seriously corrode, or rust internally.

Operating Instructions

Use vacuum pump oil in the pump when new. After five to ten hours of running time change the oil. Make sure all of the original oil is removed from the pump. Thereafter, change the oil after every thirty hours of operation or when the oil becomes dark due to suspended solids drawn into the pump. Such maintenance will ensure peak efficiency in the operation of the pump.

If the pump has been operated for a considerable time on regular pump oil, drain the oil and replace with dual-purpose vacuum pump oil. Operate for ten hours and drain. The oil will probably be quite dark due to sludge removed from the pump. Operate the second charge of oil for ten hours and drain again. The second charge of oil may still be dark. However, it will probably be lighter in color than the oil drained after the first ten hours.

Change the oil at thirty-hour intervals after that. Change the oil before such intervals if it becomes dark due to suspended solids pulled into the pump. Be sure to change the oil every thirty hours thereafter to keep the vacuum pump in peak condition.

EVACUATING A SYSTEM

How Long Should It Take?

Present-day techniques of evacuation will clean refrigeration and air-conditioning systems to a degree never before reached. Properly used, a good vacuum pump will eliminate 99.99% of the air and virtually all of the moisture in a system. There is no firm answer regarding the time it will take a pump to accomplish this level of cleanliness. The time required for evacuation depends on many things. Some factors that must be considered are:

- The size of the vacuum pump.
- The type of vacuum pump—single-stage or two-stage.
- The size of the hose connections.
- The size of the system.
- The contamination in the system.
- The application for the system.

Evacuations sometimes take fifteen minutes. They may sometimes take weeks. The only way to know when evacuation is complete is to take micron vacuum readings, using a good electronic vacuum gage. A number of electronic meters are available. Figs. 2-56 and 2-57.

Evacuating down to 29″ eliminates 97% of all air. Moisture removal, however, does not begin until a vacuum below 29″ is reached. This is the micron level of vacuum. It can be measured only with an electronic vacuum gage. Dehydration of a system does not certainly begin until the vacuum gage reads below 5000 microns. If the system will not pump down to this level, something is wrong. There may be a leak in the vacuum connections. The vacuum pump oil may be contaminated. There may be a leak in the system. Vacuum gage readings between 5000 microns and 1000 microns assure

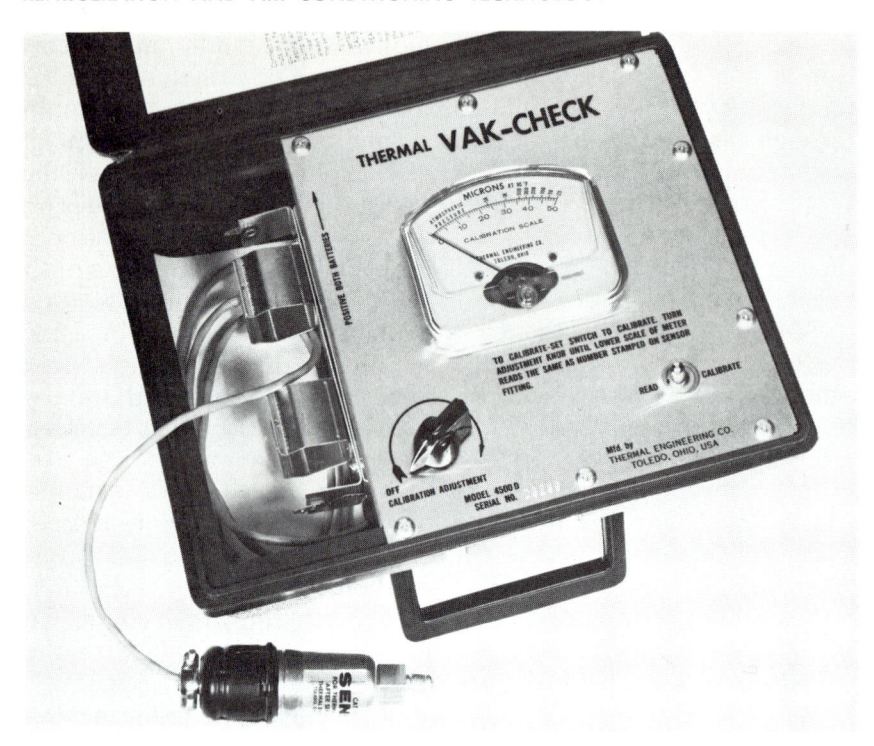

REFRIGERATION AND AIR CONDITIONING TECHNOLOGY

2-57. An electronic high-vacuum gage that reads directly in microns. It is designed specifically for refrigeration and air conditioning evacuation. It reads down to 25 microns. It is powered by 2 D-cell batteries.
Thermal Engineering

that dehydration is proceeding. When all moisture is removed, the micron gage will pull down below 1000 microns.

Pulling a system down below 1000 microns is not a perfect test for cleanliness. If the vacuum pump is too large for the system, it may pull down this level before all of the moisture is removed. Another test is preferred. After the system is pulled down below 1000 microns and will not go any further, the system should be valved off from the vacuum pump and the pump turned off. If the vacuum in the system does not rise over 2000 microns in the next five minutes, evacuation has been completed. If it goes over this level, either the moisture is not completely removed or the system has a slight leak. To find which, reevacuate the system down to its lowest level. Valve it off again and shut off the vacuum pump. If the vacuum leaks back to the same level as before, there is a leak in the system. If the rise is much slower than before, small amounts of moisture are probably left in the system. Reevacuate until the vacuum will hold.

CHARGING CYLINDER

The *charging cylinder* lets you charge with heat to speed up the charging process. This unit, with its heater assembly, allows up to 50 watts of heat to be used in charging. Refrigerant is removed rapidly from the cylinder as a liquid, but injected into the system as a gas with the Vizi-Vapr®. It requires no heat during the charging process. The Extracta-Charge® device allows the serviceperson to carry small amounts of refrigerant to the job. The refrigerant can be bought in large drums and stored at the shop. The Extracta-Charge® comes in a rugged, steel carrying case to protect it from tough use. It provides a method for draining refrigerant even from capillary tube, sealed systems.

It may soon be mandatory to capture the escaping refrigerant. The Extracta-Charge® is the instrument to use. When systems are overcharged, the excess can be transported back to the drum. The amount removed can be measured also. A leak found after the charging operation usually means the loss of the full charge. Using this device, the serviceperson can extract the charge and save it for use after the leak has been found and repaired.

CHARGING OIL

In charging a compressor with oil, there is danger of drawing air and moisture into the refriger-

2-58. Oil charging pump.
Thermal Engineering

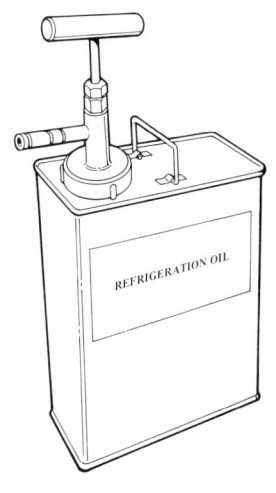

58

CHAPTER 2—AIR CONDITIONING AND REFRIGERATION TOOLS AND INSTRUMENTS

ation system. Use of the pump shown in Fig. 2-58 eliminates this danger. This pump reduces charging time by over 70% without pumping down the compressor. The pump fits the can with a cap seal so the pump need not be removed until the can is empty. It is a piston-type high-pressure pump designed to operate at pressures to 250 psi. It pumps one quart in twenty full strokes of the piston. The pump can be connected to the compressor by a refrigerant charging line or copper tubing from a $\frac{1}{2}''$ male flare fitting.

CHANGING OIL

Whenever it is impossible to drain oil in the conventional manner, it becomes necessary to hook up a pump. Removing oil from refrigeration compressors before dehydrating with a vacuum is a necessity. The pump shown in Fig. 2-59 has the ability to remove one quart of oil with about ten strokes. It is designed for use in pumping oil from refrigeration compressors, marine engines, and other equipment.

2-60. Mobile charging station.
Thermal Engineering

2-59. Oil changing pump.
Thermal Engineering

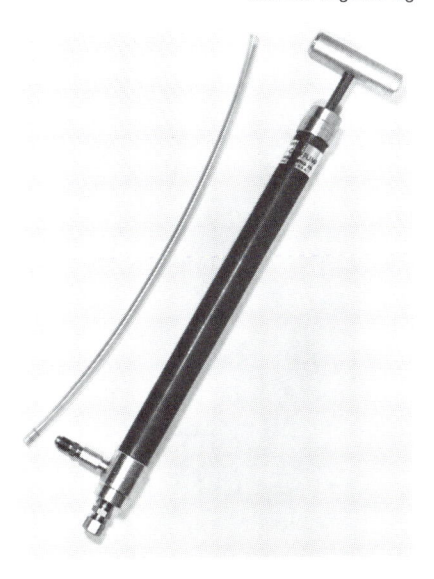

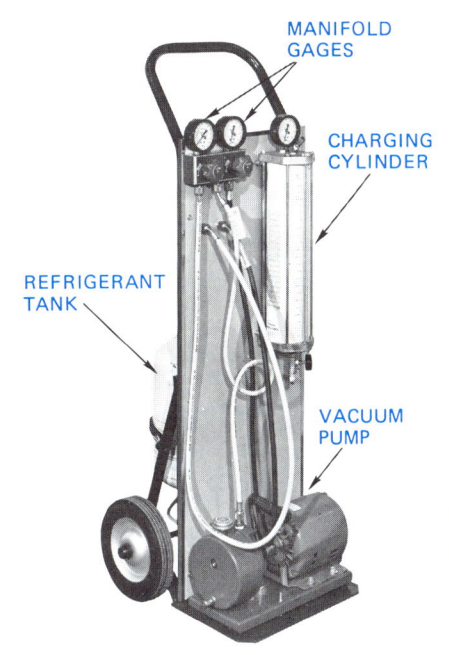

MOBILE CHARGING STATIONS

Mobile charging stations can be easily loaded into a pickup truck, van, or station wagon. They take little space. Fig. 2-60. Stations come complete with manifold gage set, charging cylinder, instrument and tool sack, and vacuum pump. The refrigerant tank can also be mounted on the mobile charging station.

TUBING

Several types of tubing are used in plumbing, refrigeration, and air conditioning work. Air-conditioning and refrigeration, however, use special types of tubing. Copper, aluminum, and stainless steel are used for tubing materials. They ensure that refrigerants do not react with the tubing. Each type of tubing has a special application.

Most of the tubing used in refrigeration and air conditioning is made of copper. This tubing is especially processed to make sure it is clean and dry inside. It is sealed at the ends to make sure the cleanliness is maintained.

Stainless steel tubing is used with R-717 or ammonia refrigerant. Brass or copper tubing should *not* be used in ammonia refrigerant systems.

Aluminum tubing is used in condensers in air-conditioning systems for the home and automobile. This requires a special type of treatment for soldering or welding.

Copper tubing is the type most often used in refrigeration systems. There are two types of copper tubing—hard-drawn and soft copper tubing. Each has a particular use in refrigeration.

Soft Copper Tubing

Some commercial refrigeration systems use soft copper tubing. However, such tubing is most commonly found in domestic systems. Soft copper is annealed. Annealing is the process whereby the copper is heated to a blue surface color and allowed to cool gradually to room temperature. If copper is hammered or bent repeatedly it will become hard. Hard copper tubing is subject to cracks and breaking.

Soft copper comes in rolls and is usually under $\frac{1}{2}''$ in outside diameter (OD). Small-diameter copper tubing is made for capillary use. It is soft drawn and flexible. It comes in random lengths of 90' to 140'. Table 2-A gives the available inside and outside diameters. This type of tubing usually fits in a $\frac{1}{4}''$ OD solder fitting

REFRIGERATION AND AIR CONDITIONING TECHNOLOGY

Table 2-A.
Inside and Outside Diameter of Small Capillary Tubing*

Inside Diameter (ID)	Outside Diameter (OD)
.026″	.072″
.031″	.083″
.036″	.087″
.044″	.109″
.050″	.114″
.055″	.125″
.064″	.125″
.070″	.125″
.075″	.125″
.080″	.145″
.085″	.145″

*Reducing bushing fits in 1/4″ OD solder fitting and takes 1/8″ OD tubing.

Table 2-B.
Dehydrated and Sealed Copper Tubing Outside Diameters, Wall Thicknesses, and Weights*

50-Foot Coils		
Outside Diameter	Wall Thickness	Approximate Weight
1/8″	.030″	1.74 lbs.
3/16″	.030″	2.88 lbs.
1/4″	.030″	4.02 lbs.
5/16″	.032″	5.45 lbs.
3/8″	.032″	6.70 lbs.
1/2″	.032″	9.10 lbs.
5/8″	.035″	12.55 lbs.
3/4″	.035″	15.20 lbs.
7/8″	.045″	22.75 lbs.
1 1/8″	.050″	32.75 lbs.
1 3/8″	.055″	44.20 lbs.

*The standard soft dehydrated copper tubing is made in the wall thickness recommended by the Copper and Brass Research Association to the National Bureau of Standards. Each size has ample strength for its capacity.

that takes a 0.125″ or 1/8″ OD tubing.

There are three types of copper tubing—types K, L, and M. Type-K tubing is heavy duty. It is used for refrigeration, general plumbing, and heating. It can also be used for underground applications.

Type-L tubing is used for interior plumbing and heating. Type-M tubing is used for light-duty waste vents, water, and drainage purposes.

Type-K soft copper tubing that comes in 60-foot rolls is available in outside diameters of 5/8″, 3/4″, 7/8″, and 1 1/8″. It is used for underground water lines. Wall thickness and weight per foot are the same as for hard copper tubing.

Copper tubing used for air-conditioning and refrigeration purposes is marked *ACR*. It is deoxidized and dehydrated to ensure that there is no moisture in it. In most cases, the copper tubing is capped after it is cleaned and filled with nitrogen. Nitrogen keeps it dry and helps prevent oxides from forming inside when it is heated during soldering.

Refrigeration dehydrated and sealed soft copper tubing must meet standard sizes for wall thickness and outside diameter. These sizes are shown in Table 2-B.

Hard and soft copper tubing are available in two wall thicknesses—K and L. The L thickness is used most frequently in air-conditioning and refrigeration systems.

Hard-Drawn Copper Tubing

Hard-drawn copper tubing is most frequently used in refrigeration and air-conditioning systems. Since it is hard and stiff, it does not need the supports required by soft copper tubing. This type of tubing is not easily bent. In fact, it should not be bent for refrigeration work. That is why there are several tubing fittings available for this type of tubing. Hard-drawn tubing comes in 10′ or 20′ lengths. Table 2-C. Remember that there is a difference between hard copper sizes and nominal pipe sizes. Table 2-D shows the differences. Nominal sizes are used in water lines, home plumbing, and drains. They are *never* used in refrigeration systems. Keep in mind that Type K is heavy-wall tubing, Type L is medium-wall tubing, and Type M is thin-wall tubing. The thickness determines the pressure the tubing will safely handle.

CUTTING COPPER TUBING

Copper tubing can be cut with a copper tube cutter or a hacksaw. ACR tubing is cleaned, degreased, and dried before the end is sealed at the factory. The sealing plugs are reusable.

60

Table 2-C.
Outside Diameter, Wall Thickness, and Weight per Foot of Hard Copper Refrigeration Tubing

Type-K Tubing		
Outside Diameter	Wall Thickness	Weight per Foot
3/8″	0.035	0.145
1/2″	0.049	0.269
5/8″	0.049	0.344
3/4″	0.049	0.418
7/8″	0.065	0.641
1 1/8″	0.065	0.839
1 3/8″	0.065	1.040
1 5/8″	0.072	1.360
2 1/8″	0.083	2.060
2 5/8″	0.095	2.930
3 1/8″	0.109	4.000
4 1/8″	0.134	6.510
Type-L Tubing		
3/8″	0.030	0.126
1/2″	0.035	0.198
5/8″	0.040	0.285
3/4″	0.042	0.362
7/8″	0.045	0.445
1 1/8″	0.050	0.655
1 3/8″	0.055	0.884
1 5/8″	0.060	1.114
2 1/8″	0.070	1.750
2 5/8″	0.080	2.480
Type-M Tubing		
1/2″	0.025	0.145
5/8″	0.028	0.204
7/8″	0.032	0.328
1 1/8″	0.035	0.465
1 3/8″	0.042	0.682
1 5/8″	0.049	0.940

Table 2-D.
Comparison of Outside Diameter and Nominal Pipe Size

Outside Diameter	Nominal Pipe Size
3/8″	1/4″
1/2″	3/8″
5/8″	1/2″
3/4″	—
7/8″	3/4″
1 1/8″	1″

To provide further dryness and cleanliness, nitrogen, an inert gas, is used to fill the tube. It materially reduces the oxide formation during brazing. The remaining nitrogen limits excess oxides during succeeding brazing operations. Where tubing will be exposed inside food compartments, tinned copper is recommended.

To uncoil the tube without kinks, hold one free end against the floor or on a bench. Uncoil along the floor or bench to the desired length. The tube may be cut to length with a hacksaw or a tube cutter. In either case, deburr the end before flaring. Bending is accomplished by use of an internal or external bending spring. Lever-type bending tools may also be used. These tools will be shown and explained later.

The hacksaw should have a 32-tooth blade. The blade should have a wave set. No filings or chips can be allowed to enter the tubing. Hold the tubing so that when it is cut the scraps will fall out of the usable end.

Figure 2-61 shows some of the tubing cutters available. The tubing cutter is moved over the spot to be cut. The cutting wheel is adjusted so it touches the copper. A slight pressure is applied to the tightening knob on the cutter to penetrate the copper slightly. Then the knob is rotated around the tubing. Once around, it is tightened again to make a deeper cut. Rotate again to make a deeper cut. Do this by degrees so that the tubing is not crushed during the cutting operation.

After the tubing is cut through it will have a crushed end. The

REFRIGERATION AND AIR CONDITIONING TECHNOLOGY

2-61. Three types of tubing cutters.

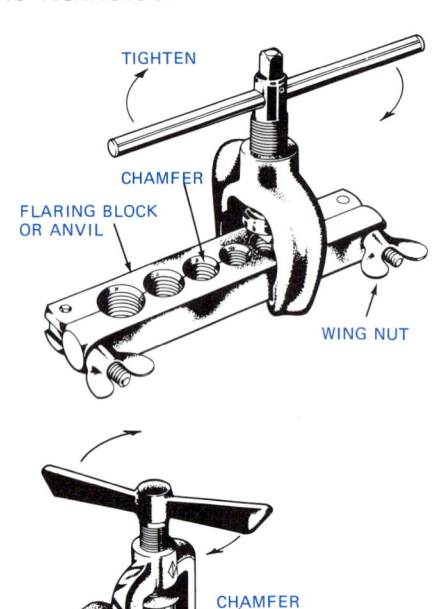

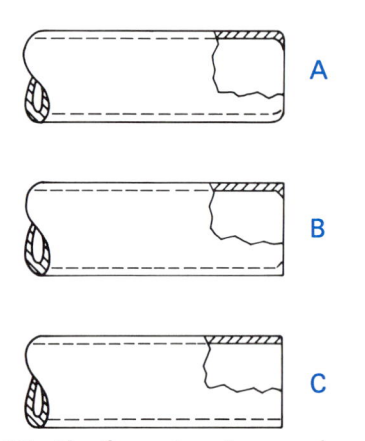

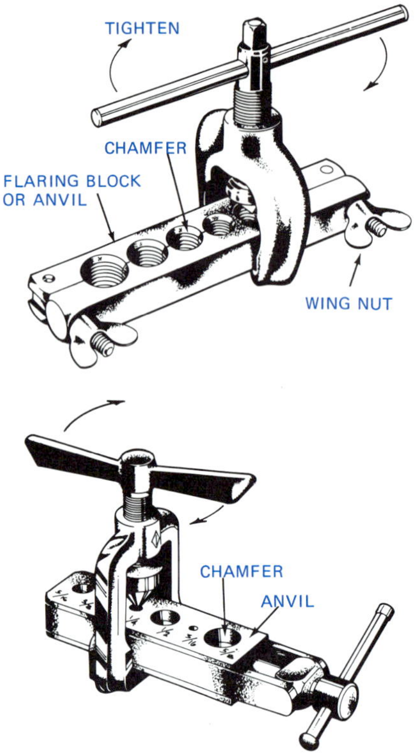

2-62. The three steps in removing a burr after the tubing has been cut with a tubing cutter. (A) The end of the cut tubing. (B) Squaring with a file produces a flat end. (C) The tube has been filed and reamed. It can now be flared.

2-63. Two types of flaring tools for soft copper tubing.

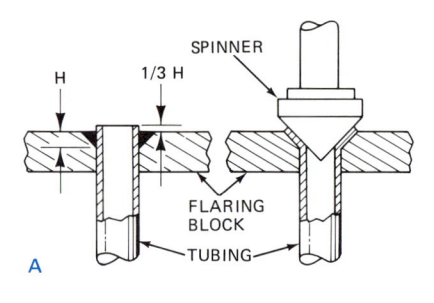

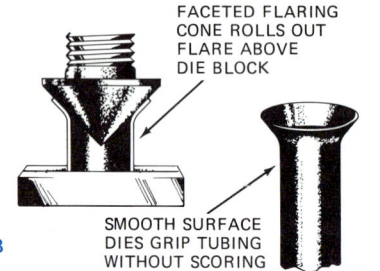

2-64. Flaring tools. (A) This type of tool calls for the tubing to be inserted into the proper size hole with a small amount of the tubing sticking above the flaring block. (B) This type of tool calls for the tubing to stick well above the flaring block. This type is able to maintain the original wall thickness at the base of the flare. The faceted flaring cone smooths out any surface imperfections.

crushed end is prepared for flaring by filing and reaming. Fig. 2-62. A file and the deburring attachment on the cutting tool can also be used.

After the tubing is cut to length it probably will require flaring or soldering.

FLARING COPPER TUBING

A flaring tool is used to spread the end of the cut copper tubing outward. Two types of tools are designed for this operation. Fig. 2-63. The flaring process is shown in Fig. 2-64. Note that the flaring is done by holding the end of the tubing rigid at a point slightly below the protruding part of the tube. This protruding part allows for the stretching of the copper.

A flare is important for a strong, solid, leakproof joint. The flares shown in Fig. 2-64 are single-flares. These are used in most refrigeration systems. The other type of flare is the double flare. Here the metal is doubled over to make a stronger joint. They are used in commercial refrigeration and automobile air-conditioners. Figure 2-65 shows how the double flare is made. The tool used is called a block-and-punch. Adapters can be used with a single-flare tool to produce a double flare. Fig. 2-66.

Figure 2-67 shows joints that use the flare. The flared tubing fits over the beveled ends. The flare tee uses the flare connection on all three ends. The half-union elbow uses the flare at one end and a male pipe thread (MPT) on the other end. A female pipe

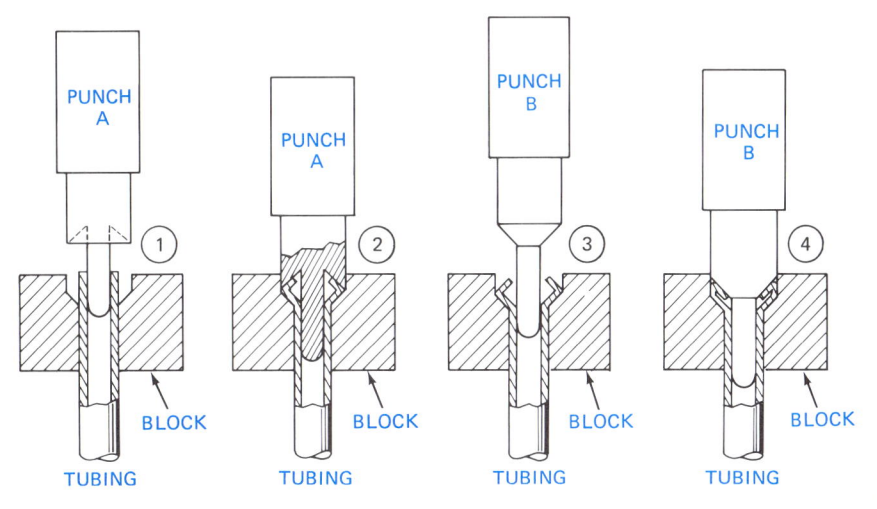

CHAPTER 2—AIR CONDITIONING AND REFRIGERATION TOOLS AND INSTRUMENTS

thread is designated by the abbreviation FPT.

Double flaring is recommended for copper tubing $5/16''$ and over. Double flares are not easily formed on smaller sizes of tubing.

CONSTRICTING TUBING

A tubing cutter adapted with a roller wheel is used to constrict a tubing joint. Two tubes are placed so that one is inserted inside the other. They should be within 0.003″ when inserted. This space is then constricted by a special wheel on the tube cutter. Fig. 2-68. The one shown is a combination tube cutter and constrictor. The wheel tightens

2-65. Double flares formed by the punch-and-block method. (1) Tubing is clamped into the block opening of the proper size. The female punch, Punch A, is inserted into the tubing. (2) Punch A is tapped to bend the tubing inward. (3) The male punch, Punch B, is inserted into the first fold. (4) The male punch is tapped to create the final double flare.

2-66. Making a double flare with an adapter for the single-flare tool. (1) Insert the tubing into the proper size hole in the flaring bar. (2) Place the adapter over the tubing. (3) Place the adapter inside the tubing. Apply pressure with the flaring cone to push the tubing into a doubled-over configuration. (4) Remove the adapter and use the flaring cone to form a double-thickness flare.

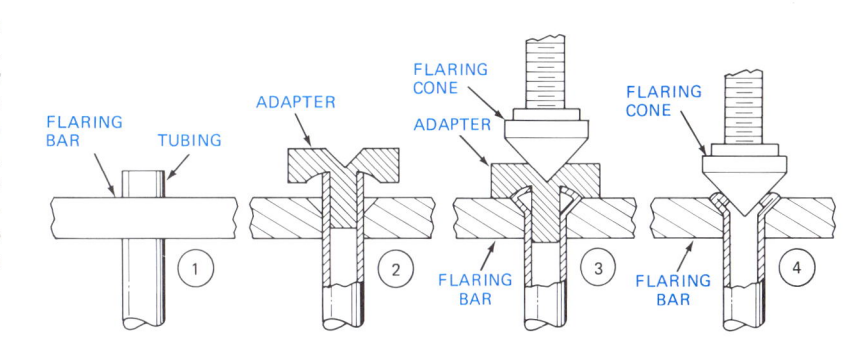

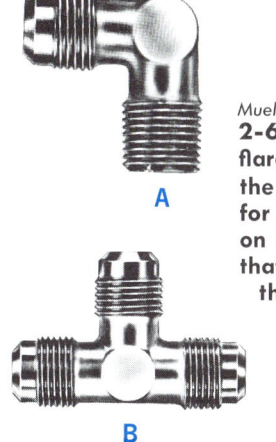

Mueller Brass
2-67. A half-union elbow (A) and a flare tee (B). Note the 45° angle on the end of the half-union elbow fitted for a flare. Also note the 45° angles on both ends of the flare tee. Note that the flared end does not have threads to the end of the fitting.

2-68. Tubing cutter adapted with a roller wheel to work as a tubing constrictor.

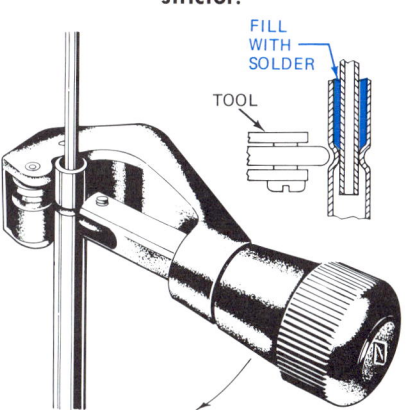

63

REFRIGERATION AND AIR CONDITIONING TECHNOLOGY

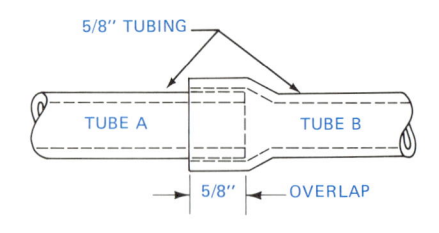

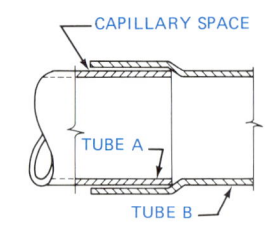

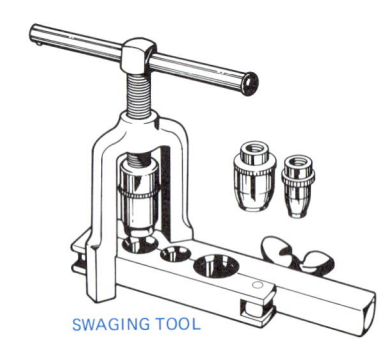

2-69. Swaging tool and swaging techniques. The swaging punches screw into the yoke and are changed for each size of tubing. Swages are available in $\frac{1}{2}''$, $\frac{5}{8}''$, and $\frac{7}{8}''$ OD, or $\frac{3}{8}''$, $\frac{1}{2}''$, and $\frac{3}{4}''$ nominal soft copper and aluminum tubing sizes.

the outside tube around the inside tube. The space between the two is then filled with solder. Of course, proper cleanliness for the solder joint must be observed before attempting to fill the space with solder. Both pieces of tubing must be hot enough to melt the solder. Flux must be used to prevent oxidation during the heating cycle. Place flux only on the tube to be inserted. No flux should be allowed to penetrate the inside of the tubing. It can clog filters and restrict refrigerant flow.

SWAGING COPPER TUBING

Swaging joins two pieces of copper without a coupling. This makes only one joint, instead of the two that would be formed if a coupling were used. With fewer joints, there are fewer chances of leaks. Punch-type swaging tools and screw-type swaging tools are used in refrigeration work. The screw-type swaging tool works the same as the flaring tool.

Tubing is swaged so that one piece of tubing is enlarged to the outside diameter of the other tube. The two pieces of soft copper are arranged so that the inserted end of the tubing is inside the enlarged end by the same amount as the diameter of the tubing used. Fig. 2-69. Once the areas have been properly prepared for soldering, the connection is soldered.

Today, most mechanics use fittings, rather than take the time to prepare the swaged end.

FORMING REFRIGERANT TUBING

There are two types of bending tools made of springs. One fits inside the tubing. The other fits outside and over the tubing being bent. Fig. 2-70. Tubing must be bent so that it does not collapse and flatten. To prevent this, it is necessary to place some device over the tubing to make sure that the bending pressure is not unevenly applied. A tube-bending spring may be fitted either inside or outside the copper tube while it is being bent. Fig. 2-71.

Keep in mind that the minimum safe distance for bending

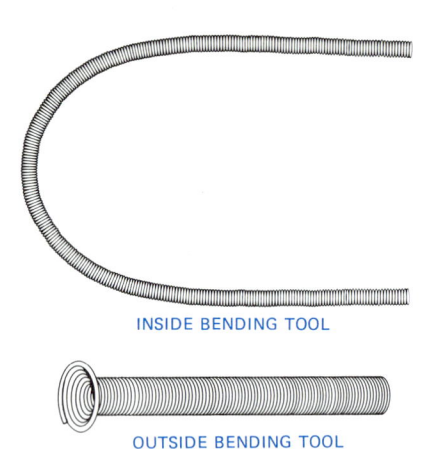

2-70. Bending tools for soft copper tubing.

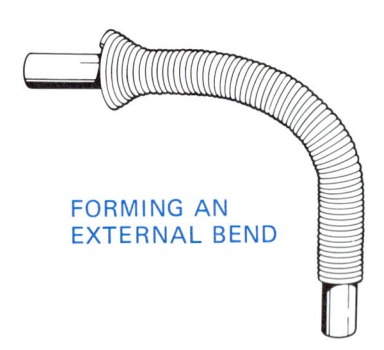

2-71. Using a spring-type tool to bend tubing.

small tubing is five times its diameter. On larger tubing, the minimum safe distance is ten times the diameter. This prevents the tubing from flattening or buckling.

Make sure the bending is done slowly and carefully. Make a large radius bend first, then go on to the smaller bends. Do not try to make the whole bend at one time. A number of small bends will equalize the applied pressure and prevent tubing collapse. When using the internal bending spring, make sure part of it is outside the tubing. This gives you a handle on it when it is time

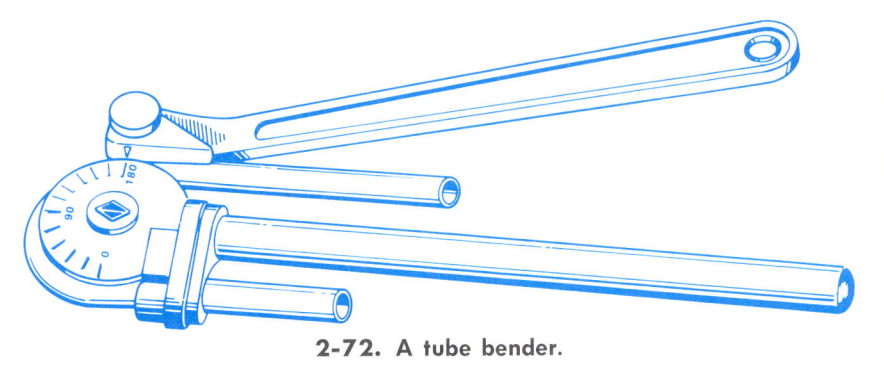

CHAPTER 2—AIR CONDITIONING AND REFRIGERATION TOOLS AND INSTRUMENTS

2-72. A tube bender.

to remove it after the bending. You may have to twist the spring to release it after the bend. By bending it so the spring compresses, it will become smaller in diameter and pull out easily. The external spring is usually used in bending tubing along the midpoint. It is best to use the internal spring when a bend comes near the end of the tubing or close to a flared end.

The lever-type tube bender is also used for bending copper tubing, Fig. 2-72. This one-piece open-side bender makes a neat, accurate bend since it is calibrated in degrees. It can be used to make bends up to 180°. A 180° bend is U-shaped. This tool is to be used when working with hard-drawn copper or steel tubing. It can also be used to bend soft copper tubing. The springs are used only for soft copper, since the hard-drawn copper would be difficult to bend by hand. Hard-drawn copper tubing can be bent, if necessary, using tools that electricians use to bend conduit.

FITTING COPPER TUBING BY COMPRESSION

Making leak-proof and vibration-proof connections can be difficult. A capillary tube connection can be used. Fig. 2-73. This compression fitting is used with a capillary tube. The tube extends through the nut and into the connector fitting. The nose section is forced tightly against the connector fitting as the nut is tightened. The tip of the nose is squeezed against the tubing.

If you service this type of fitting, you must cut back the tubing at the end and replace the soft nose nut. If the nut is reused, it will probably cause a leaky connection.

SOLDERING

Much refrigeration work requires soldering. Brass parts, copper tubing, and fittings are soldered. The cooling unit is also soldered. Thus, the air-conditioning and refrigeration mechanic should be able to solder properly.

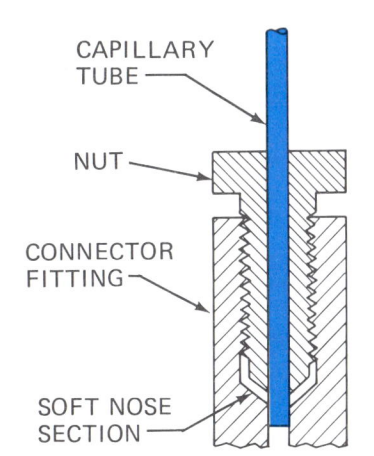

Two types of solder are used in refrigeration and air-conditioning work. Soft solder and silver solder are most commonly used for making good joints. Brazing is really silver soldering.

Brazing requires careful preparation of the products prior to heating for brazing or soldering. This preparation must include steps to prevent contaminants such as dirt, chips, flux residue, and oxides from entering and remaining in an installation. A general-purpose solder for water lines and temperatures below 250° F. [121.1°C] is 50-50. It is made of 50% tin and 50% lead. The 50-50 solder flows at 414° F. [212.2°C].

Another low-temperature solder is 95-5. It flows at 465° F. [240.5°C]. It has a higher resistance to corrosion. It will result in a joint shear strength approximately $2\frac{1}{2}$ times that of a 50-50 joint at 250° F. [121.1°C].

A higher temperature solder is No. 122. It is 45% silver brazing alloy. This solder flows at 1145° F. [618.2°C]. It provides a joining material that is suitable for a joint strength greater than the other two solders. It is recommended for use on ACR copper tubing.

Number 50 solder is 50-50 lead and tin. Number 95 solder is 95% tin and 5% antimony. Silver solder is really brazing rod, instead of solder. The higher temperature requires a torch to melt it.

2-73. A capillary tube connection.

65

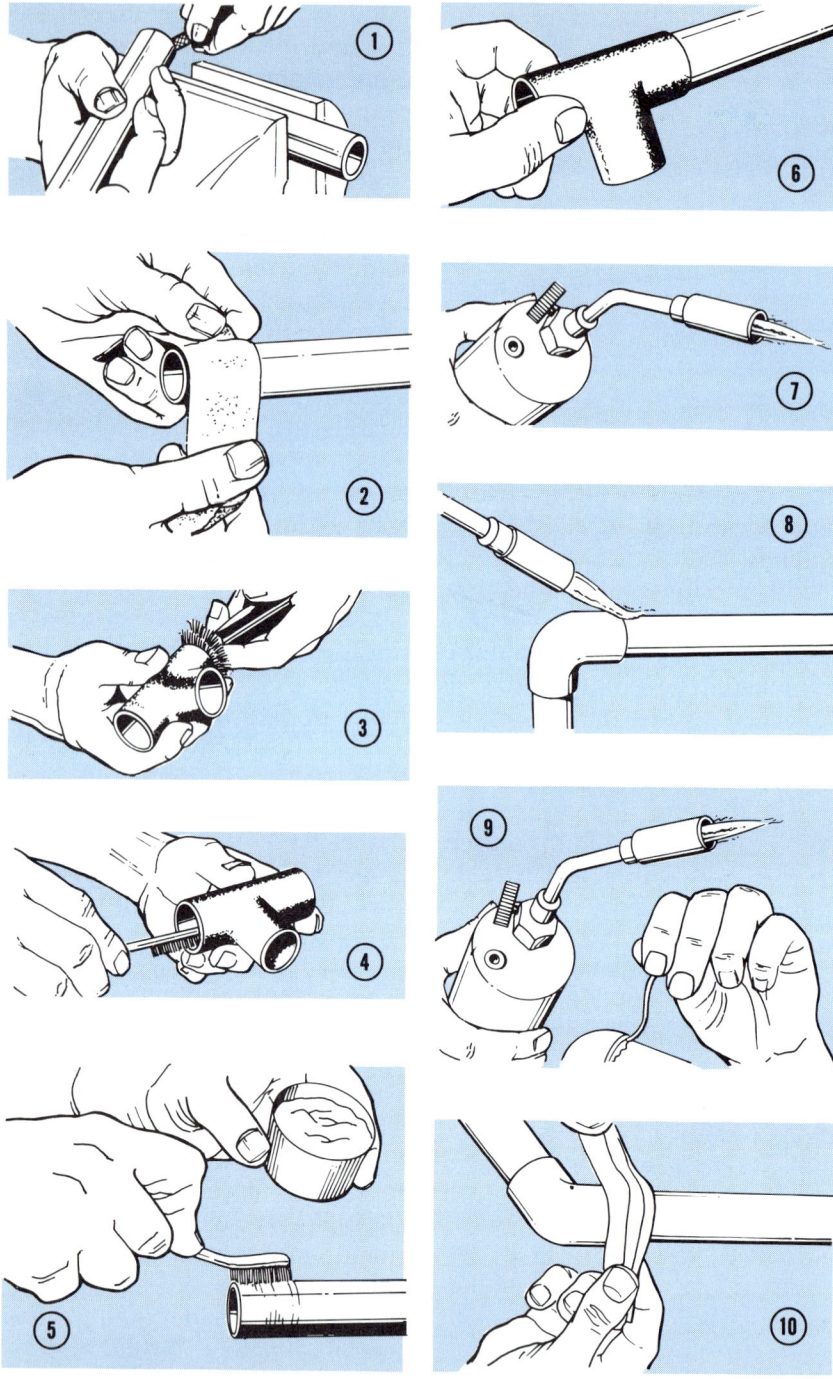

REFRIGERATION AND AIR CONDITIONING TECHNOLOGY

2-74. Soldering procedures. (1) Cut the tubing to length and remove the burrs. (2) Clean the joint area with sandpaper or sandcloth. (3) Clean the inside of the fitting. Use sandpaper, sandcloth, or wire brush. (4) Apply flux to inside of fitting. (5) Apply flux to outside of tubing. (6) Assemble the fitting onto the tubing. (7) Obtain proper tip for the torch and light it. Adjust the flame for the soldering being done. (8) Apply heat to the joint. (9) When solder can be melted by the heat of the copper (not the torch), simply apply solder so it flows around the joint. (10) Clean the joint of excess solder and cool it quickly with a damp rag.

The torch calls for an acetelyene tank and an oxygen tank.

Soft Soldering

Soldering calls for a very clean surface. Sandcloth is used to clean the copper surfaces. Flux must be added to prevent oxidation of the copper during the heating process. A no-corrode solder is absolutely necessary. Fig. 2-6. Acid-core solder must not be used. The acid in the solder will corrode the copper and cause leaks.

Soldering is nothing more than applying a molten metal to join two pieces of tubing or a tubing end and a fitting. It is important that both pieces of metal being joined are at the flow point of the solder being used. Never use the torch to melt the solder. The torch is used to heat the tubing or fitting until it is hot enough to melt the solder.

The steps in making a good solder joint are shown in Fig. 2-74. Cleanliness is essential. Flux can damage any system. It is very important to keep flux out of the lines being soldered.

The use of excessive amounts of solder paste affects the operation of a refrigeration system. This is especially true of R-22 systems. Solder paste will dissolve in the refrigerant at the high liquid line temperature. It is then carried through a drier or strainer and separated out at the colder expansion valve temperature. Generally, R-22 systems will be more seriously affected than those carrying R-12. This is because the solid materials separate out at a higher temperature. Sound practice would indicate the use of only enough solder paste to secure a good joint. The paste should be applied accord-

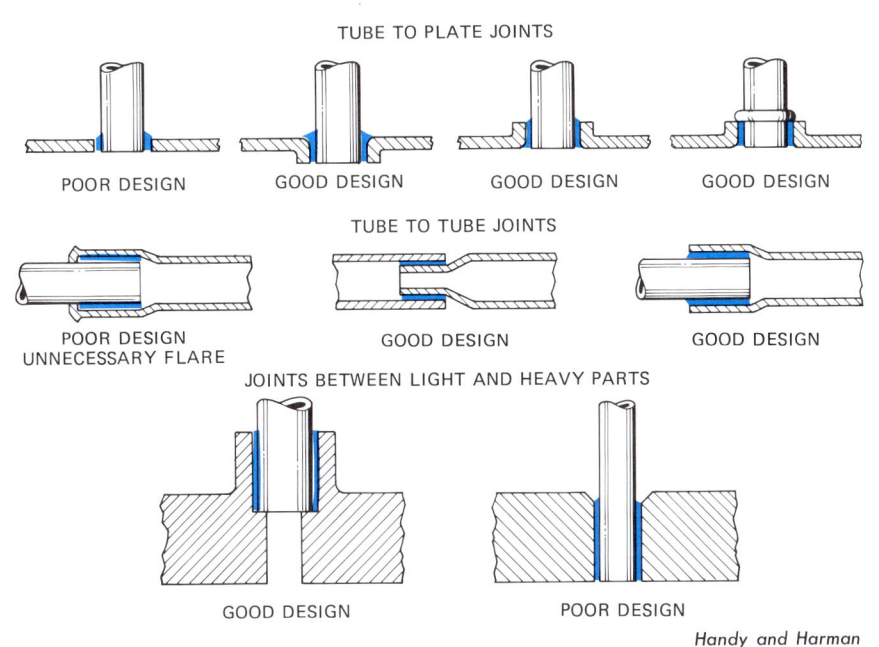

2-75. Designs that are useful in silver soldering copper tubing. Here, the clearances between the copper tubing are exaggerated for the sake of illustration. They should be much less than shown here.

ing to directions specified by its manufacturer.

SILVER SOLDERING OR BRAZING

Silver solder melts at about 1120° F. [604.4°C] and flows at 1145° F. [618°C]. An acetylene torch is needed for the high heat. It is used primarily on hard-drawn copper tubing.

Caution: *Before using silver solder, make sure it does not contain* **cadmium.** *Cadmium fumes are very poisonous. Make sure you work in a very well-ventilated room. The fumes should not contact your skin or eyes. Do not breathe the fumes from the cadmium type of silver solder. Most manufacturers will list the contents on the container.*

Silver soldering also calls for a clean joint area. Use the same procedures as shown previously for soldering (Fig. 2-74). Figure 2-75 shows good and poor design characteristics. No flux should enter the system being soldered. Make your plans carefully to prevent any flux entering the tubing being soldered.

Nitrogen or carbon dioxide can be used to fill the refrigeration system during brazing. This will prevent any explosion or the creation of phosgene when the joint has been cleaned with carbon tetrachloride.

In silver soldering, you need a tip that is several sizes larger than that used for soft soldering. The pieces should be heated sufficiently to have the silver solder adhere to them. Never hold the torch in one place. Keep it moving. Use a slight feather on the inner cone of the flame to make sure you have the proper heat. A large soft flame may be used to make sure the tip does not burn through the fitting or the tubing being soldered.

It is necessary to disassemble sweat-type valves when soldering to the connecting lines. In soldering sweat-type valves where they connect to a line, make sure the torch flame is directed away from the valve. Avoid excessive heat on the valve diaphragm. As an extra precaution, a damp cloth may be wrapped around the diaphragm during the soldering operation. The same is true for soldering thermostatic expansion valves to the distributor.

Either soft or hard solder or silver brazing is acceptable in soldering thermostatic expansion valves. Keep the flame at the fittings and away from the valve body and distributor tube joints. *Do not overheat.* Always solder the outside diameter (OD) of the distributor, never the inside diameter (ID).

TESTING FOR LEAKS

Never use oxygen to test a joint for leaks. Any oil in contact with oxygen under pressure will form an explosive mixture.

Do not use emery cloth to clean a copper joint. Emery cloth contains oil. This may hinder the making of a good soldering joint. Emery cloth is made of silicon carbide, which is a very hard substance. Any grains of this abrasive in the refrigeration mechanism or lines can damage a compressor. Use a brush to help clean the area after sanding.

CLEANING AND DEGREASING SOLVENTS

Solvents, including carbon tetrachloride (CCl_4), are frequently used in the refrigeration industry for cleaning and degreasing equipment. While there is no absolutely safe solvent, there are several that may be used with relative safety. Carbon tetrachloride is *not* one of them. Use of one of the safer solvents will reduce the likelihood of serious illness developing in the course of daily use. Some of these solvents are stabilized methyl chloroform, methylene chloride, trichlorethylene, and perchloroethylene. Some petroleum solvents are available. These are flammable in varying degrees.

Most solvents may be used safely if certain rules are followed.

1. Use no more solvent than the job requires. This helps keep solvent vapor concentrations low in the work area.

2. Use the solvent in a well-ventilated area and avoid breathing the vapors as much as possible. If the solvents are used in shop degreasing, it is wise to have a ventilated degreasing unit to keep the level of solvent vapors as low as possible.

3. Keep the solvents off the skin as much as possible. All solvents are capable of removing the oils and waxes that keep the skin soft and moist. When these oils and waxes are removed, the skin becomes irritated, dry, and cracked. A skin rash may develop more easily.

A word of caution is in order concerning the commonly used solvent, carbon tetrachloride. While this material has many virtues as a solvent, it has caused much illness among those who use it. Each year several deaths result from its use. Usually, these occur in the small shop or the home. Most large industries have discontinued its use or use it only with extreme caution. A measure of its harmful nature is indicated by the fact that it bears a *poison* label. It should never be placed in a container that is not labeled "Poison." It is for industrial use only.

While occasional deaths result from swallowing carbon tetrachloride, the vast majority of deaths are caused by breathing its vapors. When exposure is very great, the symptoms will be headache, dizziness, nausea, vomiting, and abdominal cramping. The person may lose consciousness. While the person seems to recover from breathing too much of the vapor, a day or two later he or she again becomes ill. Now there is evidence of severe injury to the liver and kidneys. In many cases, this delayed injury may develop after repeated small exposures or after a single exposure not sufficient to cause illness at the time of exposure. The delayed illness is much more common and more severe among those who drink alcoholic beverages. In some episodes where several persons were equally exposed to carbon tetrachloride, the only one who became ill or the one who became most seriously ill was the person who stopped for a drink or two on the way home. When overexposure to carbon tetrachloride results in liver and kidney damage, the patient begins a fight for life without the benefit of an antidote. The only sure protection against such serious illness is not to breathe the vapors or allow contact with the skin.

Human responses to carbon tetrachloride are not predictable. A person may occasionally use carbon tetrachloride in the same job in the same way without apparent harm. Then, one day severe illness may result. This unpredictability of response is one factor that makes the use of "carbon tet" so dangerous.

Other solvents will do a good job of cleaning and degreasing. It is much safer to select one of those for regular use rather than to expose yourself to the potential dangers of carbon tetrachloride.

CHAPTER 2—AIR CONDITIONING AND REFRIGERATION TOOLS AND INSTRUMENTS

REVIEW QUESTIONS

1. What does NEC stand for?
2. What type of solder core is preferred for electrical work?
3. What type of tips must masonry drill bits have?
4. What is a thermocouple?
5. What is a thermistor?
6. What is superheat?
7. What symbol identifies infinite resistance on an ohmmeter?
8. What is a draft gage?
9. What is the difference between a sling psychrometer and a stationary psychrometer?
10. Where are humidistats used?
11. What is a British thermal unit (Btu)?
12. What is a capillary tube?
13. Why is vapor charging slower than liquid charging?
14. What is the purpose of a vacuum pump?
15. What is a micron?
16. What type of tubing is needed with R-717 or ammonia refrigerant?
17. Name the three types of copper tubing and describe each.
18. What does ACR on a piece of copper tubing signify?
19. How do you shape or form copper tubing without collapsing it?
20. What is swaging?
21. At what temperature does silver solder melt?

3
Refrigeration

DEVELOPMENT OF REFRIGERATION

Refrigeration is the process of removing heat from where it is not wanted. Heat is removed from food to preserve its quality and flavor. It is removed from room air to establish human comfort. There are innumerable applications in industry in which heat is removed from a certain place or material to accomplish a desired effect.

During refrigeration, unwanted heat is transferred mechanically to an area where it is not objectionable. A practical example of this is the window air conditioner that cools air in a room and exhausts hot air to the outdoors.

The liquid called the *refrigerant* is fundamental to the heat transfer accomplished by a refrigeration machine. Practically speaking, a commercial refrigerant is any liquid that will evaporate and boil at relatively low temperatures. During evaporation or boiling, the refrigerant absorbs the heat. The cooling effect felt when alcohol is poured over the back of your hand illustrates this principle.

In operation, a refrigeration unit allows the refrigerant to boil in tubes that are in contact, directly or indirectly, with the medium to be cooled. The controls and engineering design determine the temperatures reached by a specific machine.

Historical Development

Natural ice was shipped from the New England states throughout the Western world from 1806 until the early 1900s. Although ice machines were patented in the early 1800s, they could not compete with the natural ice industry. Artificial ice was first commercially manufactured in the southern United States in the 1880s.

Domestic refrigerators were not commercially available until about 1920. Fig. 3-1. During the 1920s, the air conditioning industry also got its start with a few commercial and home installations.

The refrigeration industry has now expanded to touch most of our lives. There is refrigeration in our homes, and air conditioning in our place of work, and even in our automobiles. Refrigeration is used in many industries, from the manufacture of instant coffee to the latest hospital surgical techniques.

STRUCTURE OF MATTER

To be fully acquainted with the principles of refrigeration, it is necessary to know something about the structure of matter.

General Electric
3-1. One of the first commercial home refrigerators.

Matter is anything that takes up space and has weight. Thus, matter includes everything but a perfect vacuum.

There are three familiar physi-

70

CHAPTER 3—REFRIGERATION

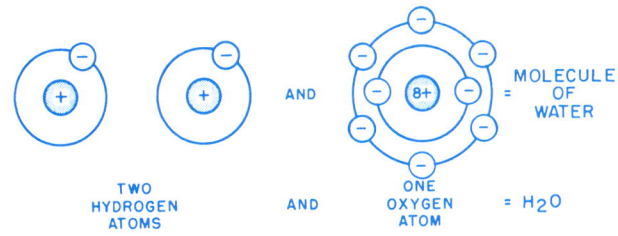

3-2. Two or more atoms linked are called a molecule. Here two hydrogen atoms and one oxygen atom form a molecule of the compound water (H_2O).

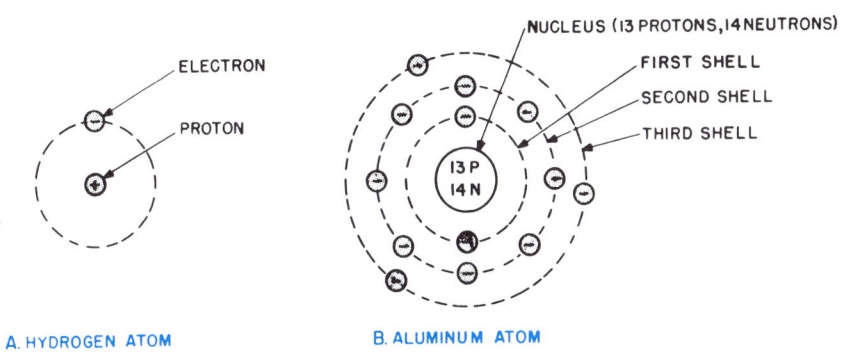

3-3. Atoms contain protons, neutrons, and electrons.

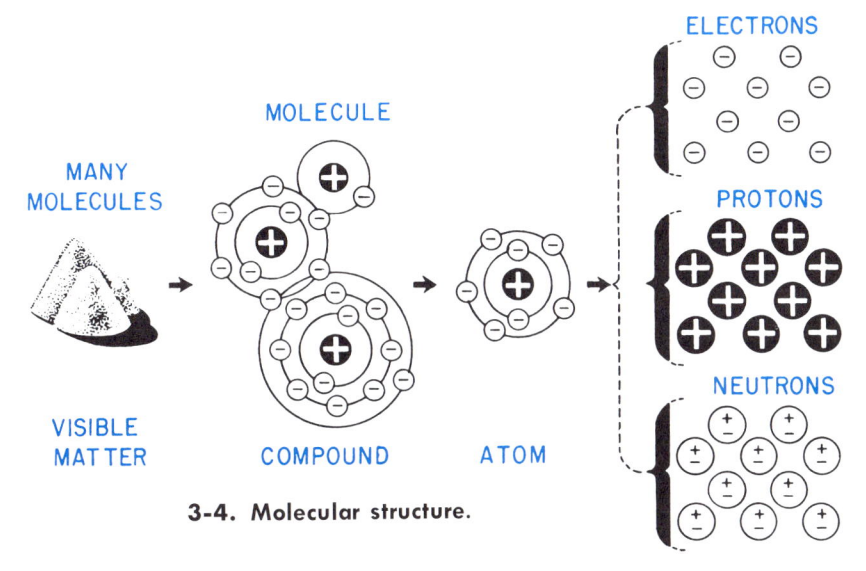

3-4. Molecular structure.

cal states of matter: solid, liquid, and gas or vapor.

A *solid* occupies a definite amount of space. It has a definite shape. The solid does not change in size or shape under normal conditions.

A *liquid* takes up a definite amount of space, but does not have any definite shape. The shape of a liquid is the same as the shape of its container.

A *gas* does not occupy a definite amount of space and has no definite shape. A gas that fills a small container will expand to fill a large container.

Matter can be described in terms of our five senses. We use our senses of touch, taste, smell, sound, and sight to tell us what a substance is. More accurate methods of detecting matter have been developed by scientists. Some of these are discussed below.

Elements

Scientists have discovered 105 building blocks for all matter. These building blocks are called *elements*. Elements are the most basic materials in the universe. Ninety-four elements, such as iron, copper, and nitrogen, have been found in nature. Scientists have made eleven others in laboratories. Every known substance—solid, liquid, or gas—is composed of elements.

It is very rare for an element to exist in a pure state. Nearly always the elements are found in combinations called *compounds*. Compounds contain more than one element. Even such a common substance as water is a compound, rather than an element. Fig. 3-2.

Atom

An *atom* is the smallest particle of an element that retains all the properties of that element. Each element has its own kind of atom. That is, all hydrogen atoms are alike. They are different from the atoms of all other elements. However, all atoms have certain things in common. They all have an inner part, the *nucleus*. This is composed of tiny particles called *protons* and *neutrons*. An atom also has an outer part. It consists of other tiny particles, called

71

REFRIGERATION AND AIR CONDITIONING TECHNOLOGY

electrons, which orbit around the nucleus. Figs. 3-3 and 3-4.

Neutrons have no electrical charge, but protons have a positive charge. Electrons are particles of energy and have a negative charge. Because of these charges, protons and electrons are particles of energy. That is, these charges form an electric field of force within the atom. Stated very simply, these charges are always pulling and pushing each other. This makes energy in the form of movement.

The atoms of each element have a definite number of electrons, and they have the same number of protons. A hydrogen atom has one electron and one proton. An aluminum atom has thirteen of each. The opposite charges—negative electrons and positive protons—attract each other and tend to hold electrons in orbit. As long as this arrangement is not changed, an atom is electrically balanced.

When chemical engineers know the properties of atoms and elements they can then engineer a substance with the properties needed for a specific job. Refrigerants are manufactured in this way.

Properties of Matter

It is important for a refrigeration technician to understand the structure of matter. With this knowledge, the person can understand those factors that affect this structure. These factors can be called the *properties* of matter. These properties are chemical, electrical, mechanical, or thermal (related to heat). Some of these properties are force, weight, mass, density, specific gravity, and pressure.

Force is described as a push or a pull on anything. Force is applied to a given area.

Weight is the force of gravity pulling all matter toward the center of Earth. The unit of weight in the English system is the *pound.* The unit of mass in the metric system is the *gram.*

Mass is the amount of matter present in a quantity of any substance. Mass is not dependent on location. A body has the same mass whether here on Earth, on the moon, or anywhere else. The weight *does* change at other locations.

In the metric system the kilogram (symbol *kg*) is the unit of mass. In the English system the *slug* is the unit of mass.

Density is the mass per unit of volume. Densities are comparative figures. That is, the density of water is used as a base and is set at 1.00. All other substances are either more or less dense than water.

The densities of gases are determined by a comparison of volumes. The volume of one pound of air is compared to the volume of one pound of another gas.

3-5. Pressure sensing element, diaphragm type.

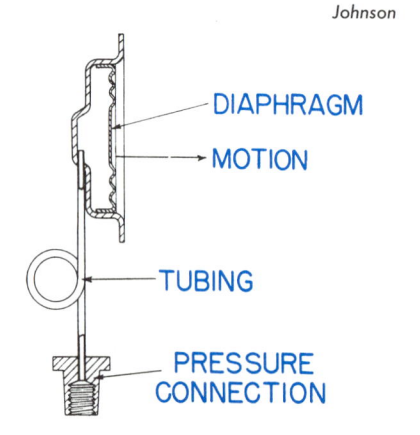

Johnson

Both gases are under standard conditions of temperature and pressure.

The *specific gravity* of a substance is its density compared to the density of water.

Specific gravity has many uses. It can be used as an indicator of the amount of water in a refrigeration system. Testing methods are discussed in later chapters.

PRESSURE

Pressure is a force that acts on an area. Stated in a formula, it becomes:

$$P = \frac{F}{A}$$

where:

F = Force
A = Area
P = Pressure

The unit of measurement of pressure in the English system is the pound per square foot or pounds per square inch (psi). The metric unit of pressure is the kilopascal (kPa). Pressure measuring elements translate changes or differences in pressure into motion. The three types most commonly used are the diaphragm, the bellows, and the Bourdon spring tube.

Pressure Indicating Devices

Pressure indicating devices are most important in the refrigeration field. It is necessary to know the pressures in certain parts of a system to locate troublespots.

The *diaphragm* is a flexible sheet of material held firmly around its perimeter so there can be no leakage from one side to the other. Fig. 3-5. Force applied to one side of the diaphragm will cause it to move or flex. Some

72

CHAPTER 3—REFRIGERATION

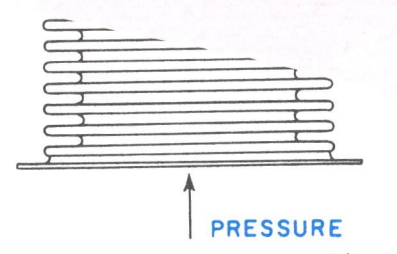

3-6. Pressure sensing element, bellows type.

diaphragms are simply a flat sheet of material with a limited range of motion. Other diaphragms are made with at least one corrugation or fold. This allows more movement at the point where work is produced.

Some types of pressure controllers require more motion per unit change of force applied. To accomplish the desired end result, the diaphragm is joined to the housing by a section with several convolutions or folds called a *bellows*. Thus, the diaphragm moves in response to pressure changes. Each fold flexes only a small amount. Fig. 3-6. The bellows element may be assembled to expand or to compress as pressure is applied. The bellows itself acts as a spring to return the diaphragm section to the original position when the pressure differential is reduced to zero. If a higher spring return rate is required, to match or define the measured pressure range, then an appropriate spring is added.

One of the most widely used types of pressure measuring elements is the *Bourdon spring tube*, applied within the the tube tends to straighten or unwind. This produces motion, which may be used to position an indicator or actuate a controller.

Pressure of Liquids and Gases

Pascal's Law states that when a fluid is confined in a container that is completely filled, the pressure exerted on the fluid is transmitted at equal pressure on all surfaces of the container.

The pressure of a gas is the same on all areas of its container.

ATMOSPHERIC PRESSURE

The layer of air that surrounds the Earth is several miles deep. The weight of the air above exerts pressure in all directions. This pressure is called *atmospheric pressure*. Atmospheric pressure at sea level is 14.7 psi. Converted to metric, the pressure is 1.013×10^5 newtons per square meter.

When fluid eter.

The aneroid barometer has a sealed chamber containing a partial vacuum. As the atmospheric pressure increases, the chamber is compressed—causing the needle to move. As the atmospheric pressure decreases, the chamber expands—causing the needle to move in the other direction. A dial on the meter is calibrated to indicate the correct pressure.

The mercury barometer has a glass tube about 34 inches long. The tube holds a column of mercury. The height of this column reflects the atmospheric pressure. Standard atmospheric pressure at sea level is indicated by 29.92 inches of mercury. That converts to 759.96 mm.

GAGE PRESSURE

Gage pressure is the pressure *above* or *below* atmospheric pressure. This is the pressure measured with most gages. A gage that measures both pressure and vacuum is called a *compound gage*. *Vacuum* is pressure that is

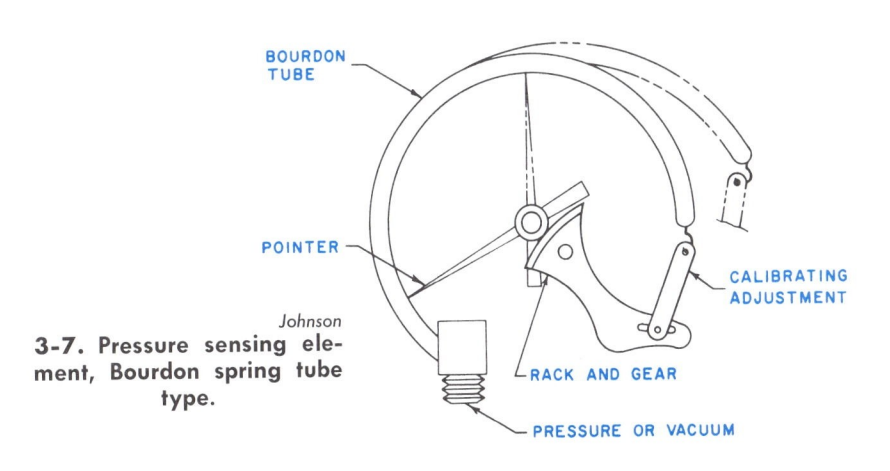

3-7. Pressure sensing element, Bourdon spring tube type.

REFRIGERATION AND AIR CONDITIONING TECHNOLOGY

below atmospheric pressure. A gage indicates zero pressure before you start to measure. It does not take the pressure of the atmosphere into account. In the customary system, gage pressure is measured in pounds per square inch (psi).

ABSOLUTE PRESSURE

Absolute pressure is the *sum* of the gage pressure and atmospheric pressure. This is abbreviated as psia. A good example of this is the pressure in a car tire. This is usually 28 psi. That would be 42.7 psia. For example:

$$\text{psi (gage)} = 28 \text{ psi}$$
$$\text{Atmospheric Pressure} = 14.7 \text{ psi}$$
$$\text{Absolute Pressure} = 42.7 \text{ psi}$$

The abbreviation for *pounds per square inch gage* is psig. The abbreviation for *pounds per square inch absolute* is psia. Absolute is found by adding 14.7 to the psig. However, the atmospheric pressure does vary with altitude. In some cases it is necessary to convert to the atmospheric pressure at the altitude where the pressure is being measured. This small difference can make a tremendous difference in correct readings of psia.

To convert psi to kPa (kilopascals), the metric unit of pressure, multiply psi by 6.9.

COMPRESSION RATIO

Compression ratio is defined as the *absolute* head pressure divided by the *absolute* suction pressure.

$$\text{Compression Ratio} = \frac{\text{Absolute Head Pressure}}{\text{Absolute Suction Pressure}}$$

Example 1.
There is a 0 or above gage reading.
Absolute Head Pressure =
 Gage Reading + 15 pounds (14.7 actually)
Absolute Suction Pressure =
 Gage Reading + 15 pounds (14.7 actually)

Example 2.
The low side is reading in vacuum range.
Absolute Head Pressure =
 Gage Reading + 15 pounds (14.7 actually)
Absolute Suction Pressure =
$$\frac{30 - \text{Gage Reading in Inches}}{2}$$

The calculation of compression ratio can be illustrated by the following:

Example 1.
Head Pressure—160 pounds
Suction Pressure—10 pounds
Compression Ratio =
$$\frac{\text{Absolute Head Pressure}}{\text{Absolute Suction Pressure}} =$$
$$\frac{160 + 15}{10 + 15} = \frac{175}{25} = 7:1$$

Example 2.
Head Pressure—160 pounds
Suction Pressure—10 inches of vacuum
Absolute Head Pressure =
 160 + 15 = 175 pounds
Absolute Suction Pressure =
$$\frac{30 - 10}{2} = \frac{20}{2} = 10$$
Compression Ratio =
$$\frac{\text{Absolute Head Pressure}}{\text{Absolute Suction Pressure}} =$$
$$\frac{175}{10} = 17.5:1$$

The preceding examples show the influence of back pressure on the compression ratio. A change in the head pressure does not produce such a dramatic effect. If the head pressure in both cases were 185 pounds, the compression ratio in Example 1 would be 8:1 and in Example 2 it would be 20:1.

A high compression ratio will make a refrigeration system run hot. A system with a very high compression ratio may show a discharge temperature as much as 150° F. [65.6°C] above normal. The rate of a chemical reaction approximately doubles with each 18° F. [−7.8°C] rise in temperature. Thus, a system running an abnormally high head temperature will develop more problems than a properly adjusted system. The relationship between head pressure and back (suction) pressure, wherever possible, should be well within the accepted industry bounds of a 10:1 compression ratio.

It is interesting to compare, assuming a 175-pound heat pressure in both cases, Refrigerant 12 versus Refrigerant 22 operating at −35° F. [−37°C] coil. At a −35° F. [−37°C] coil, as described above, the R-22 system would show a 10.9:1 compression ratio while the R-12 system would be at 17.4:1. The R-22 system is a borderline case. The R-12 system is not in the safe range and would run very hot with all of the accompanying problems.

A number of other factors will produce serious high temperature conditions. However, high compression ratio alone is enough to cause serious trouble. The thermometer shown in Fig. 3-8 reads temperature as a function of pressure. This device reads the pressure of R-22 and R-12 refrigerant. It also indicates

3-8. Thermometer and pressure gage. *Marsh*

the temperature in degrees Farenheit on the outside scale.

TEMPERATURE AND HEAT

The production of excess heat in a system will cause problems. Normally, matter expands when heated. This is the principle of thermal expansion. The linear dimensions increase, as does the volume. Removing heat from a substance causes it to contract in linear dimensions and in volume. This is the principle of the liquid in a glass thermometer.

Temperature is the measure of hotness or coldness on a definite scale. Every substance has temperature.

Molecules are always in motion. They move faster with a temperature increase and more slowly with a temperature decrease. In theory, molecules stop moving at the lowest temperature possible. This temperature is called *absolute zero*. It is approximately −460° F. [−273°C].

The amount of heat in a substance is directly related to the amount of molecular motion. The absence of heat would occur only at absolute zero. Above that temperature, there is molecular motion. The amount of molecular motion corresponds to the amount of heat.

The addition of heat causes a temperature increase. The removal of heat causes a temperature decrease. This is true except when matter is going through a change of state.

Heat is often confused with temperature. Temperature is the measurement of heat *intensity*. It is not a direct measure of heat *content*. Heat content is not dependent on temperature. Heat content depends on the type of material, the volume of the material, and the amount of heat that has been put into or taken from the material. For example, one cup of coffee at 200° F. [93.3°C] contains less heat than one gallon of coffee at 200° F. [93.3°C]. The cup at 200° F. [93.3°C] can also contain less heat than the gallon at a lower temperature of 180° F. [82.2°C].

Specific Heat

Every substance has a characteristic called *specific heat*. This is the measure of the temperature change in a substance when a given amount of heat is applied to it.

One Btu (British thermal unit) is *the amount of heat required to raise one pound of water 1° F. at 39° F.* With a few exceptions, such as ammonia gas and helium, all substances require less heat per pound than water to raise the temperature one degree Fahrenheit.

Thus, the specific heat scale is based on water, which has a specific heat of 1.0. The specific heat of aluminum is 0.2. This means that 0.2 Btu will raise the temperature of one pound of aluminum one degree Fahrenheit. One Btu will raise the temperature of five pounds of aluminum one degree Fahrenheit, or of one pound, five degrees Fahrenheit.

Heat Content

Every substance theoretically contains an amount of heat equal to the heat energy required to raise its temperature from absolute zero to its temperature at a given time. This is referred to as *heat content*, which consists of *sensible heat* and *latent heat*. *Sensible heat* can be felt because it changes the temperature of the substance. *Latent heat*, which is not felt, is seen as it changes the state of substance from solid to liquid or liquid to gas.

SENSIBLE HEAT

Heat that changes the temperature of a substance without changing its state when added or removed is called sensible heat. Its effect can be measured with a thermometer in degrees as the difference in temperatures of a substance (Delta T, or ΔT).

If the weight and specific heat of a medium are known, the amount of heat added or removed in Btu can be computed by multiplying the sensible change (ΔT) by the weight of the medium and by its specific heat. Thus, the amount of heat required to raise the temperature of one gallon of water (8.34 pounds) from 140° F. to 160° F. is:

Sensible Heat = $\Delta T \times$ Weight $\times$ Specific Heat
= $(160 - 140) \times 8.34 \times 1$
= 20×8.34
= 166.8 Btu

REFRIGERATION AND AIR CONDITIONING TECHNOLOGY

LATENT HEAT

The heat required to change the state of a substance without changing its temperature is called latent heat, or hidden heat. Theoretically, any substance can be a gas, liquid, or solid, depending on its temperature and pressure. It takes heat to change a substance from a solid to a liquid, or from a liquid to a gas.

For example, it takes 144 Btu of latent heat to change one pound of ice at 32° F. to one pound of water at 32° F. It takes 180 Btu of sensible heat to raise the temperature of one pound of water 180° F. from 32° F. to 212° F. It takes 970 Btu of latent heat to change one pound of water to steam at 212° F. When the opposite change is effected, equal amounts of heat are taken out or given up by the substance.

This exchange of heat, or the capability of a medium such as water to take and give up heat, is the basis for most of the heating and air conditioning industry. Most of the functions of the industry are concerned with adding or removing heat at a central point and distributing the heated or cooled medium throughout a structure to warm or cool the space.

Other Sources of Heat

Other heat in buildings comes principally from four sources: electrical energy, the sun, outdoor air temperatures, and the building's occupants.

Every kilowatt of electrical energy in use produces 3413 Btu per hour, whether it is used in lights, the heating elements of kitchen ranges, toasters, or irons.

The sun is a source of heat. At noon, a square foot of surface directly facing the sun may receive 300 Btu per hour on a clear day.

When outdoor air temperatures exceed the indoor space temperature, the outdoors becomes a source of heat. The amount of heat communicated depends on the size and number of windows, among other factors.

The occupants of a building are a source of heat, since body temperatures are higher than normal room temperatures. An individual seated and at rest will give off about 400 Btu per hour in a 74° F. [23.3°C] room. If the person becomes active, this amount of heat may be increased two or three times, depending upon the activity involved. Some of this heat is sensible heat, which the body gives off by convection and radiation. The remainder is latent heat, resulting from the evaporation of visible or invisible perspiration. The sensible heat increases the temperature of the room. The latent heat increases the humidity. Both add to the total heat in the room.

REFRIGERATION SYSTEMS

The refrigerator was not manufactured until the 1920s. Before that time, ice was the primary source of refrigeration. A block of ice was kept in the icebox. The icebox was similar to the modern refrigerator in construction. It was well insulated and had shelves to store perishables. The main difference was the method of cooling. The iceman came about once a week to put a new 50- or 100-pound block of ice in the icebox. How much cooling effect does a 50-pound block of ice produce? The latent heat of melting for one pound of ice is 144 Btu. The latent heat of melting for a 50-pound block is 50×144, or 7,200 Btu. The latent heat of melting for the 100-pound block was 14,400 Btu. The refrigeration was accomplished by convection in the icebox.

One of the first refrigerators is shown in Fig. 3-9. The unit on the top identified it as a refrigerator, instead of an icebox. Some of these units, made in the 1920s, are still operating today.

Refrigeration from Vaporization (Open System)

The perspiration on your body evaporates and cools your body. Water kept in a porous container is cooled on a hot day. The water seeps from the inside to the outside. There is a small amount of water on the outside surface. The surface water is vaporized—it evaporates. Much of the heat required for vaporization comes from the liquid in the container. When heat is removed this way,

3-9. Early modification of the icebox to make it a refrigeration unit.

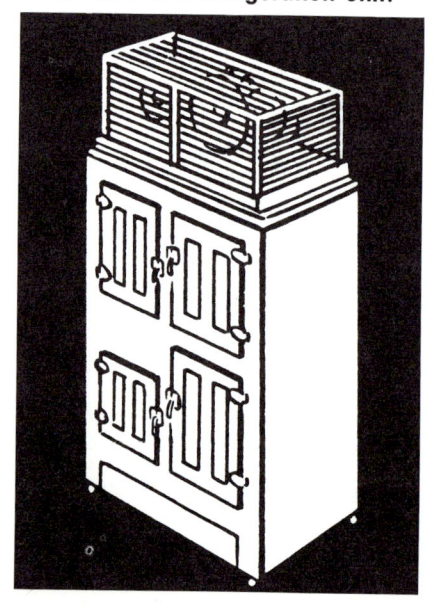

CHAPTER 3—REFRIGERATION

the liquid is cooled. The heat is carried away with the vapor.

Basic Refrigeration Cycle

A substance changes state when the inherent amount of heat is varied. Ice is water in a solid state and steam is a vapor state of water. A solid is changed to a liquid and a liquid to a vapor by applying heat. Heat must be added to vaporize or boil a substance. It must be taken away to liquefy or solidify a substance. The amount of heat necessary will depend on the substance and the pressure changes in the substance.

Consider, for example, an open pan of boiling water heated by a gas flame. The boiling temperature of water at sea level is 212° F. [100°C]. Increase the temperature of the flame and the water will boil away more rapidly, although the temperature of the water will not change. To heat or boil a substance, heat must be removed from another substance. In this case, heat is removed from the gas flame. Increasing the temperature of the flame merely speeds the transfer of heat. It does not increase the temperature of the water.

A change in pressure will affect the boiling point of a substance. As the altitude increases above sea level, the atmospheric pressure and the boiling temperature drop. For example, water will boil at 193° F. [89.4°C] at an altitude of 10,000 feet. At pressures below 100 psi, water has a boiling point of 338° F. [170°C].

The relationship of pressure to refrigeration is shown in the following example. A tank contains a substance that is vaporized at atmospheric pressure but condenses to a liquid when 100 pounds of pressure are applied. The liquid is discharged from the tank through a hose and nozzle into a long coil of tubing to the atmosphere as shown in Fig. 3-10.

As the liquid enters the nozzle, its pressure is reduced to that of the atmosphere. This lowers its vaporization or boiling point. Part of the liquid vaporizes or boils, using heat from itself. The unevaporated liquid is immediately cooled as its heat is taken away. The remaining liquid takes heat from the metal coil or tank and vaporizes—cooling the coil. The coil takes heat from the space around it—cooling the space. This unit would continue to provide cooling or refrigeration for as long as the substance remained under pressure in the tank.

All of the other components of a refrigeration system are merely for reclaiming the refrigeration medium after it has done its job of cooling. The other parts of a refrigeration system in order of assembly are: tank or liquid receiver, expansion valve, evaporator coil, compressor, and condenser.

Figure 3-11 illustrates a typical refrigeration system cycle. The refrigerant is in a tank or liquid receiver under high pressure and in a liquid state. When the refrigerant enters the expansion valve, the pressure is lowered and the liquid begins to vaporize. Complete evaporation takes place when the refrigerant moves into the evaporator coil. With evaporation, heat must be added to the refrigerant. In this case, the heat comes from the evaporator coil. As heat is removed from the coil, the coil is cooled. The refrigerant is now a vapor under low pres-

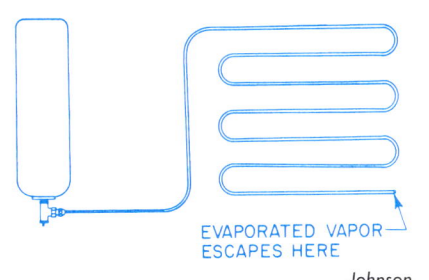

3-10. Basic step of refrigeration.

3-11. High and low sides of a refrigeration system.

sure. The evaporator section of the system is often called the low pressure, back pressure, or suction side. The warmer the coil, the more rapidly evaporation takes place and the higher the suction pressure becomes.

The compressor then takes the low-pressure vapor and builds up the pressure sufficiently to condense the refrigerant. This starts the high side of the system. To return the refrigerant to a liquid state (to condense it), heat picked up in the evaporator coil and the compressor must be removed. This is the function of the condenser used with an air- or water-cooled coil. Being cooler than the refrigerant, the air or water absorbs its heat. As it cools, the refrigerant condenses into a liquid and flows into the liquid receiver or tank. Since the pressure of the refrigerant has been raised, it will condense at a lower temperature.

In some systems, the liquid receiver may be part of another unit such as the evaporator or condenser.

Capacity

Refrigeration machines are rated in tons of refrigeration. This rating indicates the size and ability to produce cooling energy in a given period of time. One ton of refrigeration has cooling energy equal to that produced by one ton of ice melting in twenty-four hours. Since it takes 288,000 Btu of heat to melt one ton of ice, a one-ton machine will absorb 288,000 Btu in a twenty-four hour period.

Refrigerants

Theoretically, any gas that can be alternately liquified and vaporized within mechanical equipment can serve as a refrigerant. Thus, carbon dioxide serves as a refrigerant on many ships. However, the piping and machinery handling it must be very heavy-duty.

Practical considerations have led to the use of several refrigerants that can be safely handled at moderate pressures by equipment having reasonable mechanical strength and with lines of normal size and wall thickness. While no substance possesses all the properties of an ideal refrigerant, the hydrocarbon (Freon®) refrigerants come quite close.

Refrigerant 12 (R-12) is made of carbon (C), chlorine (Cl), and fluorine (F). Its formula is CCl_2F_2. It is made of a combination of elements. Refrigerant 22 (R-22) is made of carbon (C), hydrogen (H), chlorine (Cl), and fluorine (F). Its formula—$CHClF_2$—is slightly different from that of R-12.

Each of these manufactured refrigerants has its own characteristics, such as odor and boiling pressure.

Refrigerants are the vital working fluids in refrigeration systems. They transfer heat from one place to another for the purpose of cooling air or water in air conditioning installations.

Many substances can be used as refrigerants, including water under certain conditions. The following are some common refrigerants:

- *Ammonia.* The oldest commonly used refrigerant, still used in some systems. Very toxic.
- *Sulphur dioxide.* First to replace ammonia and to be used in small domestic machines. Very toxic.
- *Refrigerant 12* (R-12). The first synthetic refrigerant to be used commonly. Used in a large number of reciprocating machines operating in the air-conditioning range. Nontoxic.
- *Refrigerant 22* (R-22). Used in many of the same applications as Refrigerant 12. Its lower boiling point and higher latent heat permit the use of smaller compressors and refrigerant lines. Nontoxic.
- *Refrigerant 40.* Methyl chloride is used in the commercial refrigeration field, particularly in small installations. Today it is no longer used. It will explode when allowed to combine with air. Nontoxic.

REVIEW QUESTIONS

1. Define refrigeration.
2. Define refrigerant.
3. What is a compound?
4. What is an atom?
5. What is a Bourdon spring?
6. What is Pascal's Law?
7. What is the difference between an aneroid barometer and a mercury barometer?
8. Describe gage pressure.
9. What is absolute pressure?
10. What is absolute zero?
11. What is specific heat?
12. What is sensible heat?
13. What is latent heat?
14. How much heat is produced by a kilowatt hour of electrical energy?
15. What amount of heat is needed to melt 1 ton of ice in 24 hours?

4

Electricity and Refrigeration Controls

VOLTAGE, CURRENT, AND RESISTANCE

Every electrical circuit has current, voltage, and resistance. The movement of electrons along a wire or conductor is referred to as current.

Voltage is the electrical pressure that pushes electrons through a resistance. Voltage is measured in volts (V). Electrical pressure, electromotive force (EMF), difference of potential, and voltage are all terms used to designate the difference in electrical pressure or potential. For example, a battery is a common power source. It furnishes the energy needed to cause electrical devices to function. A battery has a difference of potential between its terminals. This difference of potential is called voltage.

Current is the flow of electrons. Current is measured in amperes (A). A *coulomb* is 6.28×10^{18} electrons. When a coulomb is standing still, or static, it is referred to as *static electricity*. Once the coulomb is in motion, it is referred to as *current electricity*. The movement of one coulomb past a given point in one second is one ampere. At times, it is necessary to refer to smaller units of the ampere. The *milli-ampere* is one-thousandth of an ampere (0.001 A-1 mA). A *micro-ampere* is one-millionth of an ampere (0.000 001 A-1 µA). These smaller units are commonly used in working with transistorized circuits.

Resistance is the opposition offered to the passage of electrical current. Resistance is measured in *ohms* (Ω). The ohm is the amount of opposition presented by a substance when a pressure of one volt is applied and one ampere of current flows through it.

OHM'S LAW

Ohm's Law states the relationship among the three factors of an electrical circuit. A *circuit* is a path for the flow of electrons from one side of a power source or potential difference to the other side. Fig. 4-1.

Ohm's Law states that the current (I) in a circuit is equal to the voltage (E) divided by the resistance (R). Ohm's Law is expressed by the following three formulas:

$$I = \frac{E}{R}$$

$$R = \frac{E}{I}$$

$$E = I \times R$$

The best way to become familiar with Ohm's Law is to do a few problems. If two of the factors or quantities are known, it is easy to find the unknown. Since the size of the wire used in a circuit is determined by the amount of current it is to handle, it is necessary to find the current and check

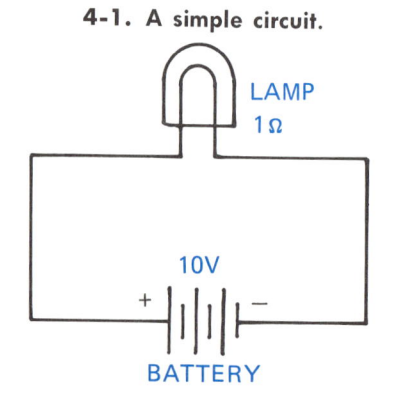

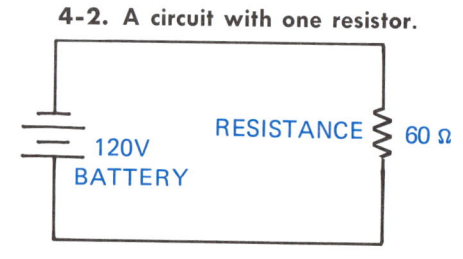

4-1. A simple circuit.

LAMP 1Ω
10V
BATTERY

4-2. A circuit with one resistor.

120V BATTERY
RESISTANCE 60Ω

80

CHAPTER 4—ELECTRICITY AND REFRIGERATION CONTROLS

Table 4-A.
Current-Carrying Ability of Copper Wire with Different Types of Insulation Coating

Wire size	In Conduit or Cable Type RHW* THW*	In Conduit or Cable Type TW, R*	In Free Air Type RHW* THW*	In Free Air Type TW, R*	Weather-proof Wire
14	15A	15	20	20	30
12	20A	20	25	25	40
10	30A	30	40	40	55
8	45A	40	65	55	70
6	65A	55	95	80	100
4	85A	70	125	105	130
3	100A	80	145	120	150
2	115A	95	170	140	175
1	130A	110	195	165	205
0	150A	125	230	195	235
00	175A	145	265	225	275
000	200A	165	310	260	320

* Types "RHW," "THW," "TW," or "R" are identified by markings on outer cover

Actual size of copper conductors. Note the larger the gage number, the smaller the diameter of the wire.

Sears

Table 4-B.
Wire Size and Current-Carrying Capacity

Wire Size A.W.G. (B & S)	Current-Carrying Capacity at 700 C.M. per Ampere
8	23.6
10	14.8
12	9.33
14	5.87
16	3.69
18	2.32
20	1.46
22	.918
24	.577
26	.363
28	.228
30	.144
32	.090
34	.057
36	.036
38	.022
40	.014

a chart to see what size the wire should be. Tables 4-A and 4-B.

PROBLEM: The voltage available is 120 volts. The resistance of the circuit is 60 ohms. What is the current? What size of wire will handle this amount of current? See Fig. 4-2 and Tables 4-A and 4-B.

$$I = \frac{E}{R}$$

$$I = \frac{120}{60}$$

$$I = 2 \text{ amperes}$$

Now that you know the amount of current (2 amperes),

refer to Table 4-B to find the size of wire that would be used to handle 2 amperes. The table says that a No. 18 wire handles 2.32 amperes. Thus, there is a safety factor of 0.32 amperes, or 320 milliamperes.

SERIES CIRCUITS

A series circuit consists of two or more consuming devices connected with one terminal after the other. Fig. 4-3. This shows that the current through the cir-

4-3. A series circuit with two bulbs.

cuit is the same in all devices. However, the total resistance is found by adding the resistances. Thus, $R_T = R_1 + R_2 + R_3 + \ldots$

Therefore, if a resistance of 1

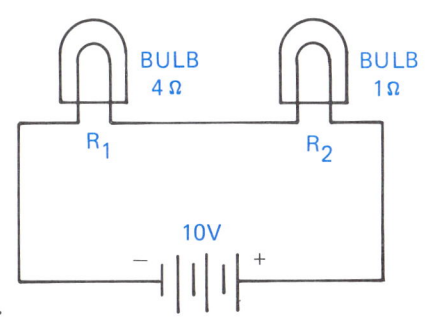

81

ohm and a resistance of 4 ohms are connected in series, the total resistance is 5 ohms. To find the total current, divide the total resistance into the voltage (in this case 10 volts). That gives Ohm's Law another use—finding the total current in the circuit. Since the total current in a series circuit is the current through each resistance, the individual light bulbs will have the same current through them. Or,

$$I = \frac{10 \text{ volts}}{5 \text{ ohms}} = 2 \text{ amperes}$$

There are three basic laws regarding series circuits:

1. Current is the same in all parts of the circuit.
2. Voltage drop across each resistance varies according to the resistance of the individual device.
3. Resistance is added to equal the total. Or, $R_T = R_1 + R_2 + R_3 + \ldots$.

Another example of how series circuit laws and Ohm's Law can be of assistance follows:

In a circuit with 120 volts and a current of 5 amperes, what is the resistance?

$$R = \frac{E}{I}$$

$$R = \frac{120}{5}$$

$$R = 24 \text{ ohms}$$

In a circuit with 20 amperes and 40 ohms, what is the voltage needed for normal operation?

$$E = I \times R$$

$$E = 20 \times 40$$

$$E = 800 \text{ volts}$$

Suppose you have a series circuit for which you know the voltage (120 volts), the current (4 amperes), and the resistance of one of the two resistors (20 ohms). How do you find the value of the other resistor in the circuit?

Use Ohm's Law and the laws of a series circuit:

$$R = \frac{E}{I}$$

$$R = \frac{120}{4}$$

$$R = 30 \text{ ohms}$$

Subtract the known resistance of 20 ohms from the total of 30 ohms. This gives 10 ohms for the missing resistor value.

PARALLEL CIRCUITS

Parallel circuits are the most common type of circuit. They are used for wiring lights in a house or for connecting equipment that must operate on the same voltage as the power source.

A parallel circuit consists of two or more resistors connected as in Fig. 4-4. Both resistors have the same voltage available as furnished by the battery. Thus, if the battery puts out 12 volts, the resistors will have 12 volts across them.

There are three basic laws regarding parallel circuits:

1. The voltage is the same across each resistor.
2. The current divides according to the resistance.
3. There are two ways of finding total resistance.

This formula can be used for *only* two resistors:

$$R_T = \frac{R_1 \times R_2}{R_1 + R_2}$$

This formula can be used for any number of resistors:

$$\frac{1}{R_T} = \frac{1}{R_1} + \frac{1}{R_2} + \frac{1}{R_3} + \ldots$$

Current in a Parallel Circuit

The current divides according to the resistance. For example,

$$R_1 = 10 \text{ ohms}$$
$$R_2 = 20 \text{ ohms}$$
$$R_3 = 30 \text{ ohms}$$

If the voltage is 60 volts, then the following method is used to determine the current through each resistor.

Voltage across each resistor is the same (60 volts). Therefore, the resistance and the voltage are known. Use Ohm's Law and find the current through each resistor:

$$I_{R_1} = \frac{E_{R_1}}{R_1} \text{ or, } I_{R_1} = \frac{60}{10} = 6A$$

$$I_{R_2} = \frac{E_{R_2}}{R_2} \text{ or, } I_{R_2} = \frac{60}{20} = 3A$$

$$I_{R_3} = \frac{E_{R_3}}{R_3} \text{ or, } I_{R_3} = \frac{60}{30} = 2A$$

Since current is divided according to the resistance of the individual resistor, the total current is found by adding the individual currents:

$$I_{R_1} + I_{R_2} + I_{R_3} = I_T$$

Resistance in a Parallel Circuit

As has already been stated, the total resistance of a parallel circuit can be found by two methods. For instance,

$$R_1 = 10 \text{ ohms}$$
$$R_2 = 15 \text{ ohms}$$
$$R_3 = 20 \text{ ohms}$$

$$\frac{1}{R_T} = \frac{1}{R_1} + \frac{1}{R_2} + \frac{1}{R_3}$$

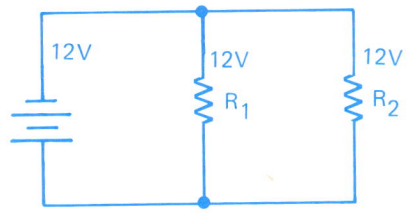

4-4. A parallel circuit.

4-5. A parallel circuit with three resistors.

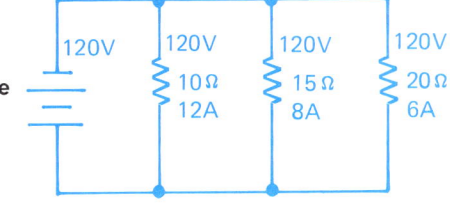

$$\frac{1}{R_T} = \frac{1}{10} + \frac{1}{15} + \frac{1}{20}$$

Find a common denominator for the fractions:

$$\frac{1}{R_T} = \frac{6 + 4 + 3}{60}$$

Add the numerators:

$$\frac{1}{R_T} = \frac{13}{60}$$

Invert:

$$\frac{R_T}{1} = \frac{60}{13}$$

$R_T = 4.6153846$ ohms

Notice that the total resistance is *always* smaller than the resistance of the smallest resistor.

If this circuit with three resistors of 10, 15, and 20 ohms in parallel has 120 volts applied, what is the current through each resistor? Fig. 4-5.

Total resistance is 4.6153846 ohms. Applied voltage is 120 volts. Therefore, using Ohm's Law you can find the total current in the circuit:

$$I_T = \frac{120}{4.6153846} = 26.0 \text{ Amperes}$$

This means that the total current in the circuit divides three ways through each resistor. Now, check to see if each resistor has the proper amount of current so the total of the individual currents is 26.0.

$$I_{R_1} = \frac{120}{10} = 12 \text{ amperes}$$

$$I_{R_2} = \frac{120}{15} = 8 \text{ amperes}$$

$$I_{R_3} = \frac{120}{20} = 6 \text{ amperes}$$

Add these individual currents. The result is 26 amperes.

AC AND DC POWER

Electrical power can be supplied in two different forms—AC and DC. The difference is in the characteristics of the current flow. A power source that causes current to flow in only one direction is referred to as a *direct current (DC) source*. A power source that causes current to flow alternately in one direction and then in the other is referred to as an *alternating current (AC) source*.

Batteries and automobile DC generators are common examples of DC electrical power sources. Normally, a DC current flow is thought of as a continuous *unidirectional* (*uni* means one) flow that is constant in magnitude. However, a pulsating current flow that changes in magnitude, but not direction, is also considered direct current.

The power supplied by power companies in the United States is the most common example of AC power. If the magnitude of the current is recorded as it varies with time, the shape of the resultant curve is called the *waveform*. The waveform produced by the power companies' generators is a sine wave, as shown in Fig. 4-6.

When the waveform of an AC voltage or current passes through a complete set of positive and negative values, it completes a *cycle* (now called a hertz). The frequency of an AC voltage or current is the number of hertz (cycles) that occur in one second. The frequency of the voltage supplied by United States power companies is 60 hertz. In Europe, it is 50 hertz. Hertz is abbreviated as Hz.

All calculations using AC voltage are based on *sine waves*. There are four values of sine waves that are of particular importance.

● *Instantaneous value.* The voltage or current in an AC circuit is continuously changing. The value varies from zero to maximum and back to zero. If you measure the value at any given instant, you will obtain the

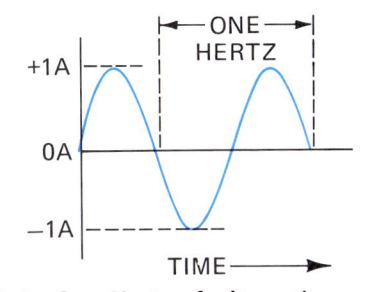

4-6. One Hertz of alternating current.

83

REFRIGERATION AND AIR CONDITIONING TECHNOLOGY

instantaneous voltage or current value.

- *Maximum value.* For two brief instants in each hertz the sine wave reaches a maximum value. One is a positive maximum and the other is a negative maximum. Maximum value is often referred to as *peak value*. The two terms have identical meanings and are interchangeable.

- *Average value.* The positive and negative halves of a sine wave are identical. Thus, the average value can be found by determining the area below the wave and calculating what DC value would enclose the same area over the same amount of time. For either the positive or negative half of the sine wave, the average value is 0.636 times the maximum value.

- *Effective value.* The effective value is often referred to as *rms (root-mean-square)*. The effective value is the same as a DC voltage or current required to provide the same average power or heating effect.

The heating effect is independent of the direction of electron flow. The effective value of an AC sine wave is equal to 0.7071 times the maximum value. Thus, the alternating current of a sine wave having a maximum value of 10 amperes produces in a circuit the amount of heat produced by a DC current of 7.071 amperes.

The heating effect varies as the square of the voltage or current. If we square the instantaneous values of a voltage or current sine wave, we obtain a wave that is proportional to the instantaneous power, or heating effect of the original sine wave. The average of this new waveform represents

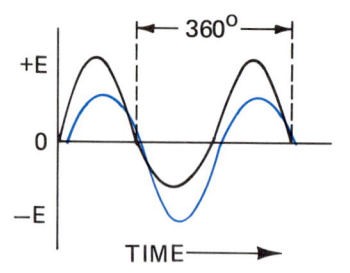

4-7. Two AC waveforms, unequal in amplitude, but in phase.

the average power that will be supplied. The square root of this average value is the voltage or current that represents the heating effect of the original sine wave of voltage or current. This is the effective value of the wave, or the rms value, the square root of the average of the squared waveform.

The effective values of voltage and current are more important than instantaneous, maximum, or average values. Most AC voltmeters and ammeters are calibrated to read in rms values.

Phase

The phase of an AC voltage refers to the relationship of its instantaneous polarity to that of another AC voltage.

Figure 4-7 shows two AC sine waves in phase, but unequal in amplitude. Figure 4-8 shows the two sine waves 45° out of phase.

4-8. Two AC waveforms, out of phase by 45°.

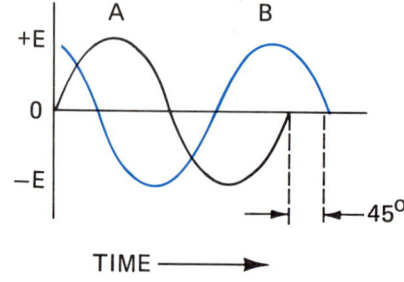

Figure 4-9 shows the two sine waves 180° out of phase.

The length of a sine wave can be measured in angular degrees because each hertz is a repetition of the previous one. One complete cycle (hertz) of a sine wave is 360 angular degrees.

Power in DC Circuits

Whenever a force causes motion, work is performed. Electrical force is expressed as voltage. When voltage causes a movement of electrons (current) from one point to another, energy is expended. The rate of work, or the rate of producing, transforming, or expending energy, is generally expressed in watts or kilowatts. A kilowatt (kW) is 1000 watts. In a DC circuit, one volt forcing a current of one ampere through a one-ohm resistance results in one watt of power being expended. The formula for this is:

P (watts) = E (volts) × I (amperes)

Find the power used by the light bulbs in a series circuit. Use the values given in Fig. 4-3.

$$I = \frac{E}{R_T}$$

$$I = \frac{10 \text{ volts}}{5 \text{ ohms}}$$

4-9. Two AC waveforms, out of phase by 180°.

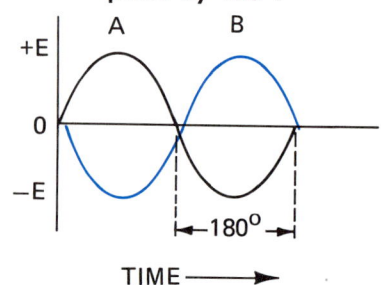

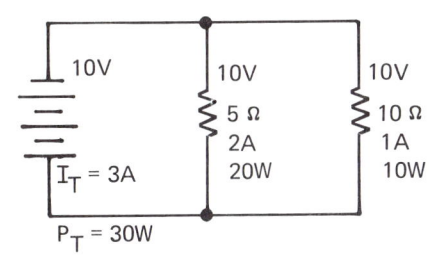

CHAPTER 4—ELECTRICITY AND REFRIGERATION CONTROLS

4-10. Total current is found by adding the individual currents.

$I = 2$ amperes

Then, because $P = E \times I$:

$P = 10$ volts $\times 2$ amperes $= 20$ watts

The same calculations can be made for the bulbs in a parallel connection. See Fig. 4-10. Total current is 3 amperes. Applied voltage is 10 volts. Since it is a parallel circuit, each resistor will have a 10-volt potential across it. Now, use the power formula:

$P = EI$

$P = 10 \text{ V} \times 3 \text{ A}$

$P = 30 \text{ W}$

Power can be computed if any two of the three values of current, voltage, and resistance are known:

When *resistance* is *unknown*:
$P = EI$

When *voltage* is *unknown*:
$P = I^2 R$

When *current* is *unknown*:
$P = \dfrac{E^2}{R}$

POWER RATING OF EQUIPMENT

Most electrical equipment is rated for both voltage and power. Electric lamps rated at 120 volts are also rated in watts. Then, they are commonly identified by their wattage rating, rather than by voltage. The voltage is usually 120 volts in the United States and 240 volts in Europe and certain other countries.

The wattage rating of a light bulb or other electrical device indicates the rate at which electrical energy is changed into other forms of energy, such as light and heat. The greater the amount of electrical power a lamp changes to light, the brighter the lamp will be. Therefore, a 100-watt bulb furnishes more light than a 75-watt bulb.

Similarly, the power ratings of motors, resistors, and other electrical devices indicate the rate at which the devices are designed to change electrical energy into some other form of energy. If the rated wattage is exceeded, the excess energy is usually converted to heat. Then, the equipment will overheat and perhaps be damaged. Some devices will have maximum DC voltage and current ratings instead of wattage. Multiplied, these values give the effective wattage.

Resistors are rated in watts dissipated, in addition to ohms of resistance. Resistors of the same resistance value are available with different wattage ratings. Usually, carbon composition resistors are rated from about $1/10$-watt to 2-watts. Carbon composition resistors have color bands to indicate their resistance. The physical size determines their wattage rating. The fourth band determines the tolerance of the resistor. Fig. 4-11.

Wirewound resistors are used when a higher wattage is needed. Generally the larger the physical size of the resistor, the higher the wattage rating, since a larger amount of surface area exposed to the air is capable of dissipating more heat. Fig. 4-12.

CAPACITORS

A *capacitor* is a device that opposes any change in circuit voltage. It may be used in AC or DC circuits. It does, however, have different uses for different types of current—AC or DC.

Capacitance is that property of a capacitor that opposes any change in circuit voltage. In a

4-11. Wattage rating of carbon composition resistors varies from ¼th to 2 watts. Color bands indicate their ohmic value and tolerance.

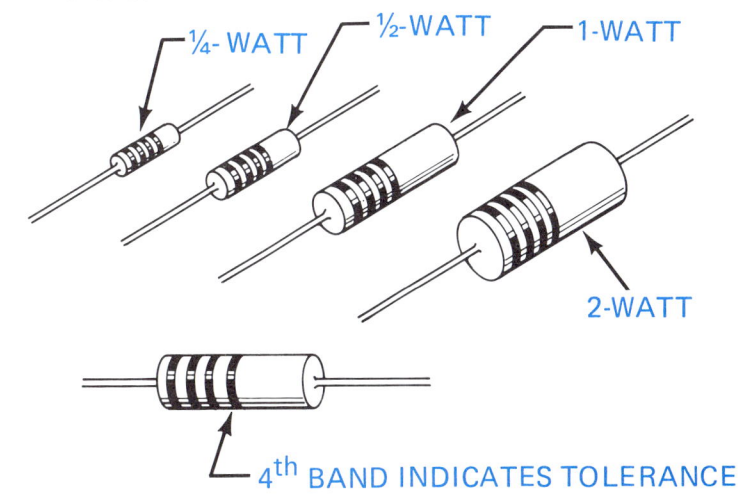

85

REFRIGERATION AND AIR CONDITIONING TECHNOLOGY

How a Capacitor Works

If a capacitor has no electron charge, it is neutral, or uncharged. Fig. 4-13(A). This is the condition when no applied voltage has been connected to the plates. When a source of voltage is connected to the two leads of the capacitor, the difference in potential created by the voltage source causes electrons to be transferred from the positive plate and placed on the negative plate. Fig. 4-13(B). This transfer continues until the accumulated charge equals the potential difference of the applied voltage. Once the voltage source is removed, Fig. 4-13(C), the potential difference remains until a conductor for discharging the excess electrons on the negative plate is connected to the positive or deficient plate. Fig. 4-13(D).

The discharge path for electrons, from one plate to the other, is in the opposite direction of the charge path. This indicates that any change in circuit voltage also results in a minor change in the capacitor charge. Some electrons leave the excess (negative) plate to try to keep the voltage in the circuit constant. This capability of a capacitor to oppose a change in circuit voltage by placing stored electrons back into circulation is called *capacitance*.

Capacitance tries to hold down circuit voltage when it increases and tries to hold it up when circuit voltage decreases.

Since DC voltage varies only when turned on and off, there is little capacitance effect other than at these times. However, AC is continuously changing. Thus, the capacitance effect is continuous in an AC circuit.

The symbols used for devices placed in circuits to produce capacitance are shown in Fig. 4-14.

Capacity of a Capacitor

The plates of a capacitor may be made of any material. A dielectric is made of an insulator such as air, a vacuum, wood, mica, plastic, rubber, Bakelite®,

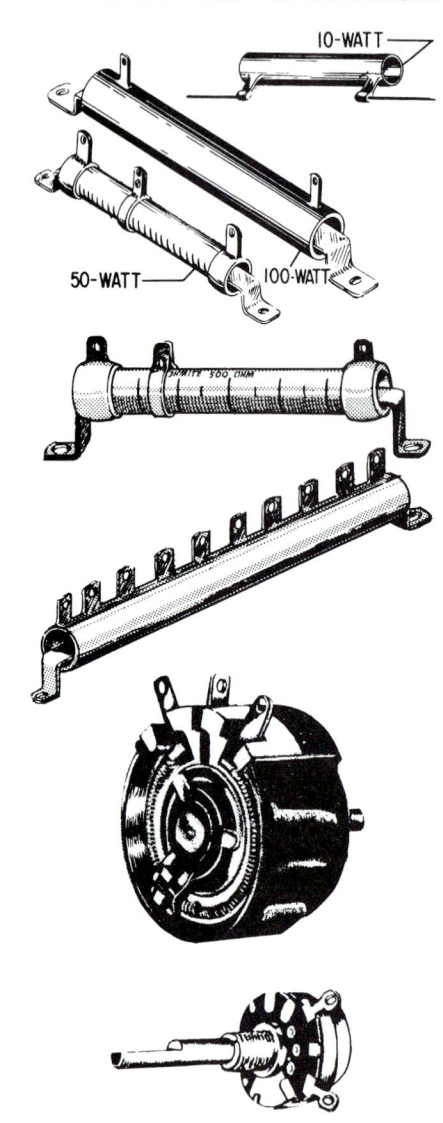

4-12. Wirewound resistors are usually over 2 watts. Shown above are various shapes of wirewound resistors.

capacitor, a device used to obtain capacitance, two plates of a conductor material are isolated from each other by a *dielectric*. A dielectric is a material that does not conduct electrons easily. Electrons are stored on the surface of the two plates. If the surface area is made larger, there is more room to store electrons and more capacitance is produced.

4-13. Capacitor charges. (A) A capacitor with no charge. (B) A capacitor charged by a battery. (C) A capacitor holding its charge after battery is removed. (D) A capacitor discharging, since it has been shorted.

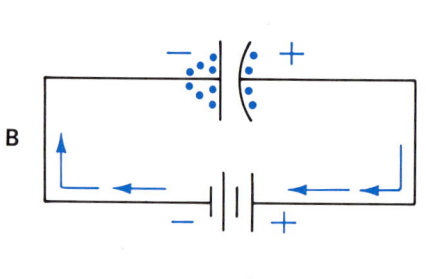

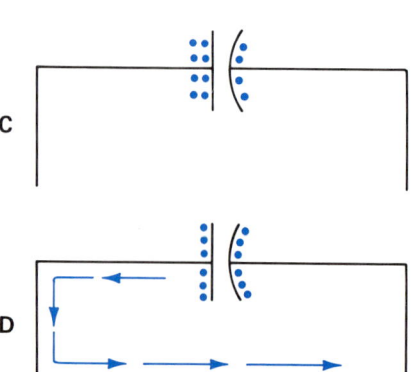

4-14. Capacitor symbols.

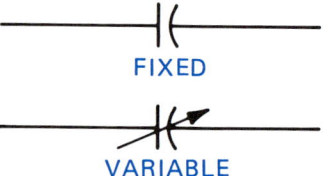

FIXED

VARIABLE

86

paper, or oil. If electrons accumulate on a surface, it has capacitance. The larger the surface area, the larger the capacity of the capacitor.

The following three factors determine the capacity of a capacitor:
1. Area of the plates.
2. Distance between the plates.
3. Material used for the dielectric.

The area of the plates determines the ability of the capacitor to hold electrons. The larger the plate area, the more electrons it can hold.

The distance between the plates determines the effect of the electron charges on one another.

The electrostatic field between opposite plates can store a greater charge when the plates are close together. They also can produce more electron interaction than plates that are farther apart. The capacitance between two plates increases as the plates are brought closer together. Capacitance decreases when plates are separated.

The thinner the dielectric, the closer the plates of a capacitor. This ensures a greater effect of the stored charges on the plates and a greater capacitance.

Some dielectrics have better insulating properties than others. This insulating property is referred to as the *dielectric constant*.

Dielectric Failure

The *breakdown voltage* is the voltage at which the dielectric will break down and allow a path of electrons to flow through it. At this point, the dielectric is no longer an insulator and will short the capacitor. In some cases, the dielectric allows small amounts of electrons to flow at different times. A capacitor in which this happens is referred to as a leaky capacitor.

Basic Units of Capacitance

The *farad* is the basic unit of capacitance. It was named for the English physicist Michael Faraday. The farad is equal to the capacitance of a capacitor that has stored in its dielectric one coulomb of electrons. One coulomb is 6.28×10^{18} electrons, or 6 280 000 000 000 000 000 electrons. Thus, one coulomb of electrons on one plate and no electrons on the other would produce a difference in capacitance of one farad.

As can be seen by the number of electrons, the farad is a large unit. For practical purposes, it is broken down into smaller quantities. The *microfarad* is one-millionth of a farad (.000 001 F) and the *micromicrofarad* is one-millionth of one-millionth of a farad (0.000 000 000 001 F). Micromicrofarad is an old term, but can still be found on older capacitors. The term *micromicro* has been replaced by *pico*. The symbol for *micro* is the Greek letter mu, or M (μ). The symbol for *picofarad* is pF. The symbol MMF was used for micromicrofarad or picofarad. There are several ways to represent capacitor values, but they all use microfarads or picofarads. For instance, MMF, mmf, UUF, uuf, UUFD, and MMFD were all formerly used to designate micromicrofarads. Today, pF is used as the prefix for the symbol for micromicrofarad. The letters MM and UU were used to symbolize *micromicro*. This made it unnecessary to buy a separate font of Greek letters and use just one of them.

The symbols MFD or mfd may also be found on equipment with older component parts. Today, MF is used almost exclusively. Occasionally, the Greek letter mu (μ) will be used with the F to represent microfarads.

Working with Capacitive Values

Sometimes it is necessary to convert the farad to smaller units. It may also be necessary to change the smaller units to larger units. For example, it may become necessary to convert 10 000 picofarads to microfarads or farads. This would mean moving the decimal place. For example, 10 000 picofarads equal 0.01 microfarad or 0.000 000 01 farad.

A schematic may be marked 10K pF, meaning 10 000 picofarads. However, some schematics may call for a 0.01-MF capacitor. This would be equivalent to a 10 000-pF capacitor.

The 10 000 is sometimes abbreviated as 10K on disc ceramic capacitors, and no pF follows. It is assumed that such a large value could only be in the pF range.

Table 4-C lists the methods by which capacitive values can be converted.

Capacitor Types

The following five types of capacitors are available for commercial applications:
1. Air.
2. Mica.
3. Paper.
4. Ceramic.
5. Electrolytic.

The capacitor with polarity markings (+ or −) is called an *electrolytic*. The other four types

REFRIGERATION AND AIR CONDITIONING TECHNOLOGY

Table 4-C.
Capacitive Value Conversion Table

To Convert:	Move Decimal Point:
pF to MF	six places to the left
MF to F	six places to the left
F to MF	six places to the right
MF to pF	six places to the right
pF to F	twelve places to the left
F to pF	twelve places to the right

are not polarized and are not marked + or −. Some of the other capacitors will have a black band around one end. This indicates the terminal or lead that is connected to the outside foil of the capacitor. Fig. 4-15.

• *Air capacitors.* Air capacitors have air as the dielectric separating the plates. These capacitors are usually variable capacitors.

• *Mica capacitors.* Mica is the dielectric separating aluminum foil plates. Mica capacitors are not common today. Many other materials are less expensive. Mica capacitors are usually contained in Bakelite®. They usually have capacitances of 50 to 500 pF.

• *Paper capacitors.* In paper capacitors, paper is the dielectric separating the two plates of aluminum foil. The materials, aluminum foil and paper separators, are rolled in a cylinder. Leads are attached to each foil layer. Fig. 4-15. The cylindrical roll is placed in a container tube made of cardboard and sealed with wax. Paper capacitors usually have a capacitance of 0.001 to 1.0 MF. Some use Teflon® or Mylar®, instead of paper, as a dielectric. These capacitors have the added advantage of high breakdown voltage and low losses. They operate efficiently over a longer period of time than the regular paper capacitors.

• *Oil-filled capacitors.* Oil-filled capacitors are paper capacitors encased in oil. Usually mounted in a metal case, they are referred to as *bathtub capacitors*. Their values are not over one microfarad. Their main advantage is a higher breakdown voltage and ruggedness.

• *Ceramic capacitors.* A ceramic capacitor has a high voltage rating, since ceramic is a good insulator. They are usually small and rugged. They consist of a ceramic disc with a coating of silver on both sides. Leads are soldered to the coatings. The whole assembly is then covered with a ceramic glaze and fired. They are made with values from 1 pF to 0.05 MF. Breakdown voltages can be as high as 10 000 volts. The value and the code for voltage are stamped on the capacitor. Small ceramic capacitors are now made for transistor circuits with very low voltage breakdown capability. Such low voltages are common in transistor circuits.

• *Tubular ceramic capacitors.* Tubular ceramic capacitors are used in electronic circuits where stability of capacitance is required—as in control circuits. Such a capacitor is nothing more than a ceramic tube coated on the inside and outside with a metallic substance—small wires soldered to them and then coated with epoxy or some other coating

4-15. A paper capacitor with the top cover removed.

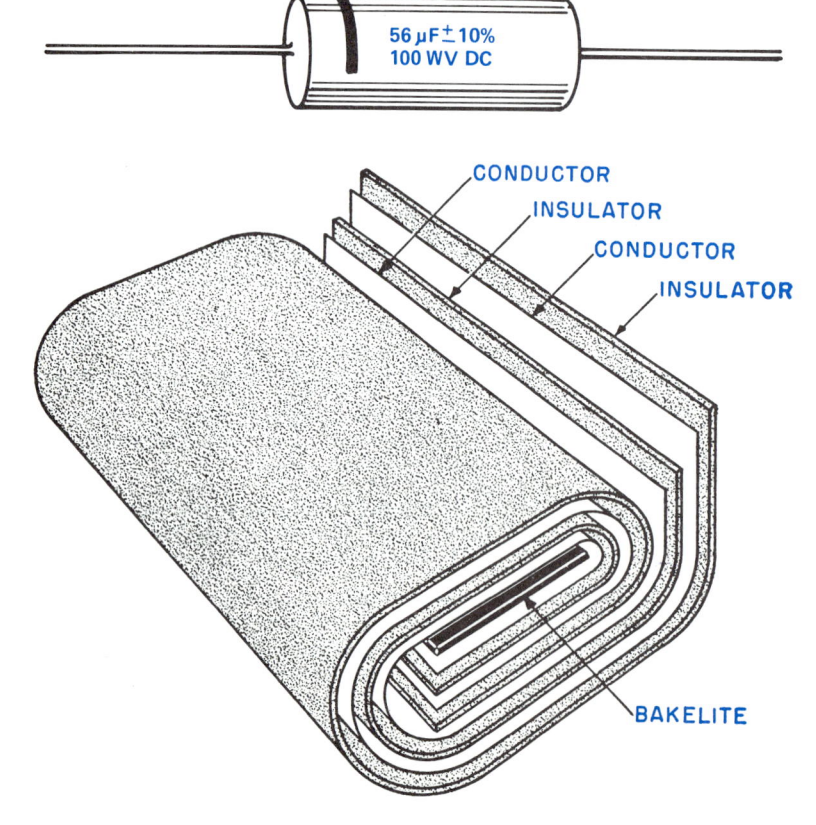

CHAPTER 4—ELECTRICITY AND REFRIGERATION CONTROLS

4-16. How an electrolytic capacitor is made.

to protect the plate surface. Their value is applied by color code. They have been designed to replace mica capacitors. Values of 1 to 500 pF are common. Their physical size is their greatest advantage.

• *Electrolytic capacitors.* An electrolytic capacitor is easily identified, since it will have a — or a + at one end of the tubular case. There are two types of electrolytic capacitors—wet and dry. The dry type is the most common. The wet type is used in heavy-duty electronic equipment, such as transmitters.

ELECTROLYTIC CAPACITORS

Dry electrolytic capacitors are not really dry. They have an electrolyte that is damp. Once the electrolyte dries up, the capacitor is defective. Such drying can occur under a number of conditions. For example, an electrolytic capacitor will dry up if allowed to sit without use for a period of time. In some cases they will become leaky, shorted, or open due to age.

Electrolytic capacitors are often called merely electrolytics. They are available in sizes starting at 1 MF and going up to 1 F. Of course, the working voltage—the point where there is a difference of potential across the plates—is very small at the higher values. The method used to construct electrolytic capacitors is shown in Fig. 4-16.

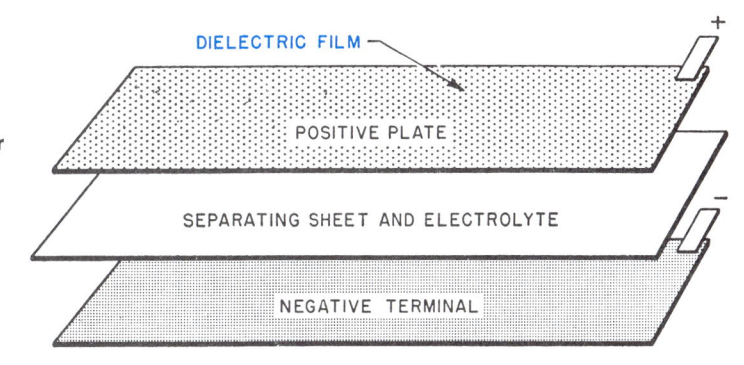

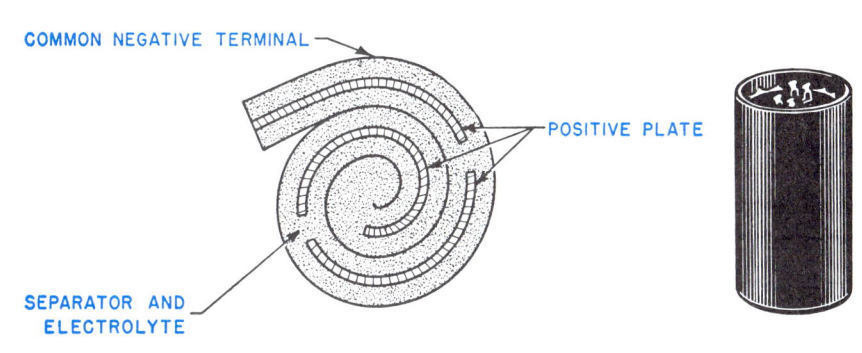

Making an electrolytic. In manufacturing, a DC voltage is applied to the electrolytic. An electrolytic action creates a molecule-sized film of aluminum oxide with a thin layer of gas at the junction between the positive plate and the electrolyte.

The oxide film is a dielectric. There is capacitance between the positive plate and the electrolyte through the film. The negative plate provides a connection to the electrolyte. This thin film allows many layers of foil to be placed into a can or cardboard cover. Larger capacitance values can thus be produced—by having plates closer together. Some electrolytic capacitors have more than one capacitor in a case. Such capacitors are either labeled on the cover or at the bottom of the can. Fig. 4-17.

Connecting electrolytics in circuits. The *polarity* of electrolytics must be observed when they are

4-17. The can capacitor has more than one electrolytic. The tubular capacitor has more than one electrolytic.

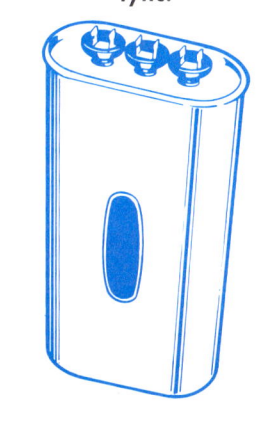

89

connected in a circuit. If they are not connected properly, that is, − to − and + to +, the oxide film that was formed during manufacture will break down and form large amounts of gas under pressure. This can rupture the can or container and cause an explosion. Thus, it is best to make sure polarity is properly connected.

Electrolytics should be used in circuits where at least 75% of their working voltage (WVDC) is available. This will keep the capacitor formed to its rating.

AC and electrolytics. AC electrolytics are found in air conditioning units in connection with the motors that power the compressors. These electrolytics are not polarized. Nonpolarized electrolytics are made by connecting them in series, but back to back. Thus, two capacitors of 50 MF can be used to produce an AC nonpolarized capacitor by placing them in series and connecting the two negative (−) terminals and using the two positive (+) terminals for connections in the circuit. It can also be done by connecting the two positive terminals and using the negative terminals for connection purposes. Using this arrangement, it is possible to arrange two standard electrolytics to substitute for a capacitor. Remember—the placing of the two capacitors in series lowers the capacitance of the combination.

SERIES CAPACITORS

Capacitors placed in series effectively separate the plates. This reduces the total capacitance of the capacitors placed in series. The working voltage, however, is increased by placing the plates farther apart.

$$C_T = \frac{1}{\frac{1}{C_1} + \frac{1}{C_2} + \frac{1}{C_3} + \cdots}$$

or

$$\frac{1}{C_T} = \frac{1}{C_1} + \frac{1}{C_2} + \cdots$$

or, for only two capacitors:

$$C_T = \frac{C_1 \times C_2}{C_1 + C_2}$$

Therefore, two 50-MF capacitors would make a series combination of:

$$\frac{50 \times 50}{50 + 50} = \frac{2500}{100} = 25 \text{ MF}$$

When connecting capacitors in series, consider the working voltage DC rating (WVDC). If two capacitors are connected in series, the outside plates are farther apart. This increase in distance between the plates *increases* the WVDC rating of the capacitor. For example, if one capacitor has a 100 WVDC rating and the other a 50 WVDC rating, then the total WVDC rating would be 150. Just add the two WVDC ratings.

PARALLEL CAPACITORS

Connecting capacitors in parallel increases the capacitance. This is primarily since the plate area is increased by parallel connections. The area for electron storage is increased. Total capacitance is found by adding the individual capacitances. With a parallel connection and with the working volts DC, total working voltage equals the working voltage of the smallest capacitor. For instance,

$C_1 = 50$ MF at 400 WVDC

$C_2 = 25$ MF at 200 WVDC

$C_T = 75$ MF at 200 WVDC

The weakest point in the connection is the 200-WVDC capacitor. That would be the one used to protect the combination from voltage breakdown.

Capacitor Tolerances

Capacitors have a tolerance of ±20%, unless otherwise noted. The manufacturer's specifications must be checked to make sure. In some cases, a ±10% capacitor is available. However, this is *not* the case with electrolytic capacitors. The electrolytic may have a tolerance of −20% and +100%. For instance, the capacitor marked 50 MF may be somewhere between 40 and 100 MF. In the case of AC electrolytics made for use on AC circuits (as opposed to one made from DC electrolytics), the capacitance *range* will be given on the capacitor. For example, it may read 40 to 100 MF at 200 volts AC, 60 Hertz.

If you are working with close-tolerance control equipment, you may encounter the mica or tubular ceramic capacitor. These capacitors have extremely small tolerances. Their tolerance may be ±2% to ±20%. The closer the tolerance, the more expensive the capacitor. If very close tolerances are required, silver-plated mica may be specified with a ±1% tolerance.

The AC Circuit and the Capacitor

Direct current and alternating current affect a capacitor differently. When DC is applied to a capacitor, the capacitor charges to the voltage of the source. Once the voltage source is removed from the capacitor, the capacitor

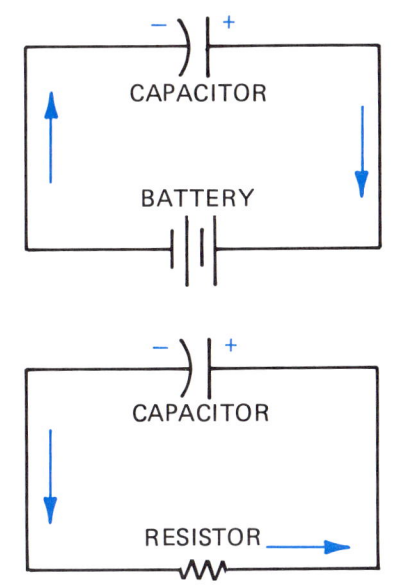

CHAPTER 4—ELECTRICITY AND REFRIGERATION CONTROLS

4-18. Note direction of charge and discharge of a capacitor.

will discharge through the resistor in the opposite direction than that from which it was charged. Fig. 4-18. No current flow takes place once the capacitor is charged to the source voltage level.

In an AC circuit with a capacitor, *capacitive reactance* (X_C) must be considered. Capacitive reactance is the opposition to current flow presented by a given capacitance. Capacitive reactance is determined by the frequency of the alternating current and the capacity of the capacitor. Capacitive reactance is found by the following formula:

$$X_C = \frac{1}{2\pi FC}$$

Where:

X_C = capacitive reactance, measured in ohms

π = 3.14

F = frequency (usually 60 Hz)

C = capacity in Farads

Alternating current *appears* to pass through a capacitor. However, it is blocked. The capacitor is charged in first one direction and then the other as the current alternates. Fig. 4-19. Note that the circuit allows current to flow when the capacitor is charging and discharging. The AC source voltage increases to a maximum, decreases to zero, then increases to a minimum in the opposite direction. Then it drops to zero again. Since the current is alternating, the charging and discharging current moves through the lamp as quickly as the source can change its direction. At 60 Hertz, the bulb increases and decreases its intensity so rapidly (120 times per second) that the human eye is unable to detect the change. However, the bulb *appears* to glow continuously.

A small capacitor will cause the lamp to glow dimly. A larger capacitor will cause the lamp to glow brightly. This change indicates that the same amount of current is not available to make the lamp glow brightly in the dimmer circuit. That means something must have caused the difference in the bulb brightness. Since nothing was changed except the size of the capacitor, it must be surmised that the size of the capacitor affects the brightness of the bulb's glow.

The following problem illustrates the exactness with which this phenomenon can be checked mathematically.

PROBLEM: A circuit has 120 volts, 60 Hertz, AC applied to a 40-watt light bulb in series with a 10-MF capacitor. What will be the current flow through the bulb?

SOLUTION: The capacitive reactance (X_C) is the opposition.

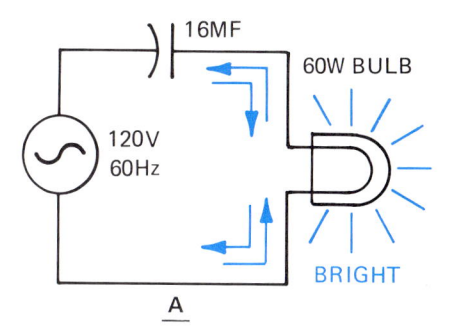

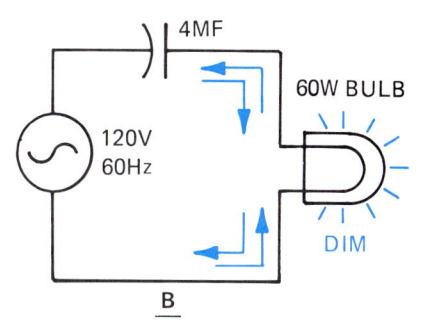

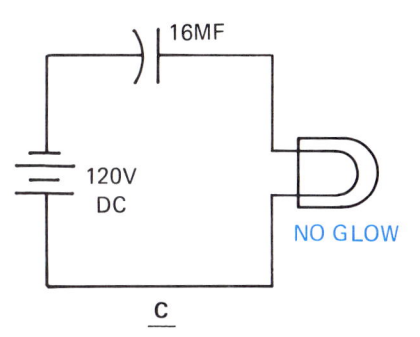

4-19. Alternating current in a capacitor. (A) Large capacitor (16 MF) allows the bulb to glow brightly. (B) Small capacitor (4 MF) allows the bulb to glow dimly. (C) Capacitor in DC circuit will not allow the bulb to glow.

Use it where the resistance is called for in the Ohm's Law formula.

$$X_C = \frac{1}{2\pi FC}$$

or

$$\frac{1}{6.28 \times 60 \times 0.000\,01}$$

REFRIGERATION AND AIR CONDITIONING TECHNOLOGY

(Note: The capacitance must be measured in farads.)

$$X_C = \frac{1}{0.003\,768} = 265.39 \text{ ohms}$$

Since the voltage for the whole circuit is 120 volts, the wattage rating of the bulb tells what the current should be, or:

$$I = \frac{E}{R} = \frac{120}{360} = 0.3333 \text{ amperes}$$

Resistance of the bulb is found by:

$$P = \frac{E^2}{R} \quad \text{or} \quad R = \frac{E^2}{P}$$

In this case,

$$R = \frac{14\,400}{40} = 360 \text{ ohms}$$

E = 120 V; E² = 14 400
R = Resistance of filament
P = Watts (40 W in this case)

I, then, is equal to 0.3333 amperes.

To find the impedance, or total opposition (Z), made up of the capacitive reactance and the resistance of the lamp bulb filament, use the following formula:

$$Z = \sqrt{R^2 + X_C^2}$$
$$Z = \sqrt{360^2 + 265.39^2} \quad \text{or}$$
$$\sqrt{129\,600 + 70\,431.85}$$
$$= 447.2 \text{ ohms}$$

Now that the impedance (Z), has been found, the problem of finding the total current in the circuit with the capacitor and bulb can be found using the following formula:

$$I_T = \frac{E_{Applied}}{Z} \quad \text{or} \quad \frac{120}{447.2}$$
$$= 0.2683 \text{ amperes}$$

The answer—0.2683 amperes—is less than the 0.3333 amperes needed to give the bulb full brightness. Thus, the bulb glows dimmer than it would without the capacitor in the circuit.

The same procedure can be followed with the larger capacitor. If the capacity of the capacitor is increased, it means the capacitive reactance is lower. If the X_C is lower, a larger current value is obtained when the voltage is divided by the capacitive reactance. Thus,

$$\uparrow I_T = \frac{120}{X_C} \downarrow$$

The smaller the capacitive reactance (X_C), the larger the total current (I_T). Thus, the brighter the bulb glows.

Uses of Capacitors

Capacitors are used in electronic circuits for one of three basic purposes:

1. To couple an AC signal from one section of a circuit to another.
2. To block out and/or stabilize any DC potential from a component.
3. To bypass or filter out the AC component of a complex wave.

Capacitors are also used as part of a circuit in an electric motor. They improve the operating characteristics of some motors. It is possible to start a motor under load if it is a *capacitor-start* type. This is very important when an air conditioning unit must start under load.

A capacitor-start, capacitor-run type of motor is also used in air-conditioning units. This type of motor will be discussed later.

INDUCTANCE

Inductors have inductance. *Inductance* is that property of a coil that opposes any change in circuit current. Inductance is measured in henries (H). The symbol for inductance is L. Inductance is sometimes measured in *millihenries*. (*Milli* means 1/1000 or 0.001 H.) The symbol for millihenry is mH. There are occasions when even smaller units of the henry are used, such as *micro-henry*. (*Micro* means one-millionth or 0.000 001.)

Inductors are used in circuits containing audio frequencies (those that *can* be heard) and in circuits containing radio frequencies (those that *cannot* be heard). The symbol for a coil is ⎧⎭⎫⎭ . If the coil has application in a circuit with audio frequencies it will have an iron core. The symbol will be ⎧⎭⎫⎭ .

The symbol for an inductor used in a circuit with radio frequencies is ⎧⎭⎫⎭ . Notice there is

4-20. Counter-electromotive force. (A) Magnetic field builds up and expands when switch is closed. (B) Magnetic field collapses when the switch is opened.

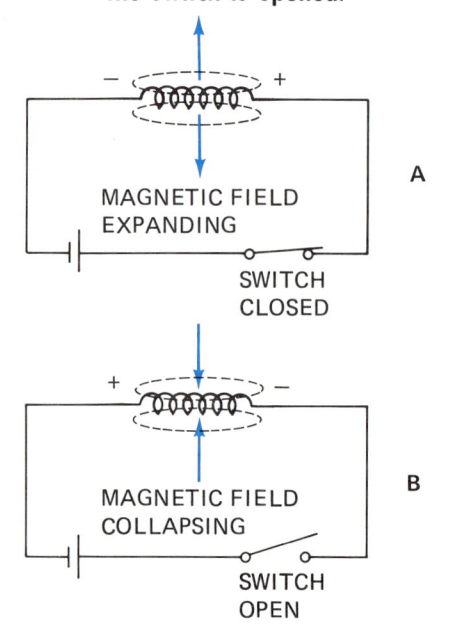

no core in a radio frequency choke. In some cases, a ferrite core is used and then the symbol will be ⌇⌇⌇.

Four Methods of Changing Inductance

The following four factors affect the inductance of a coil:
1. The *number of turns*.
2. The *diameter* of the coil.
3. The *permeability* of the *core material*.
4. The *length* of the coil.

Changing any of these factors will change the inductance of the coil. In a coil with an air core having the turns close-wound, inductance is increased four times by doubling the number of turns. Doubling the diameter of the coil also quadruples the inductance. The length of the coil directly increases the inductance.

Self-Inductance

The capability of a conductor to induce voltage in itself when the current changes is called *self-inductance*, or *inductance*. When a current that is changing in value (such as alternating current) passes through a coil, the moving magnetic field around the windings of the coil produces electromagnetic induction. The magnetic field around each turn of the coil cuts across the remaining turns and a voltage is generated across the coil. Since this induced voltage is generated by the moving magnetic field produced by an increasing or decreasing current, it is generated in the opposite direction to the voltage that caused it. This is referred to as a counter-electromotive force (CEMF). Fig. 4-20.

One henry (H) is the amount of inductance present when a current variation of one ampere per second results in an induced EMF of one volt.

In Fig. 4-20A the current is shown rising from zero to a maximum rather quickly. This causes the magnetic field around the coil to expand. A CEMF is produced by the expanding magnetic field, cutting the windings of the coil ahead of the current. The windings are usually alongside or on top of the energized part of the coil. In Fig. 4-20B the circuit is shown opened by a switch. The magnetic field collapses and the current in the circuit changes from its maximum value to zero. As the field collapses, it induces a voltage across the coil. This opposes the decrease in current and prevents the current from dropping to zero as quickly as it would in a straight wire. Note that the time lag shown in Fig. 4-21 is produced by a coil.

Mutual Inductance

Mutual inductance is concerned with two or more coils. *Mutual inductance* refers to the condition in which two circuits share the energy of one circuit. The energy in one circuit is transferred to the other circuit. The coupling that takes place between the circuits is done by means of magnetic flux. Fig. 4-22. When two coils have a mutual inductance of one henry it means a change in current of one ampere takes place in one second. One coil induces one volt in the other coil.

Inductive Reactance

When alternating current flows through an inductor it creates a certain amount of opposition to its flow. This opposition is called *inductive reactance* (X_L). Inductive reactance is measured in ohms. This type of reactance is not present in a coil when energized by DC. The only opposition encountered by a DC current passing through a coil is the resistance of the copper wire used to wind the coil.

A number of factors determine inductive reactance. Frequency and inductance are the major factors. The formula for inductive reactance is:

$$X_L = 2\pi FL$$

Where:

$2\pi = 6.28$
F = frequency of alternating current
L = inductance (in henrys)

USES OF INDUCTIVE REACTANCES

Inductive reactances are very important in filter circuits. It is sometimes necessary to smooth out the variations in a power source current. The inductor can help make the fluctuations less severe.

Inductive reactance (X_L) becomes very useful when dealing with electronic circuits. When combined with capacitive reactance (X_C), it is possible to obtain a resonant frequency. Inductive reactance and capacitive reactance can have the same value. Under such conditions, they can cause a circuit to resonate at a given frequency and no other. Thus, it is possible to pick out one frequency from a number present. This is helpful in tuning in a radio or television station.

TRANSFORMERS

A *transformer* is a device that transfers energy from one circuit to another without being physi-

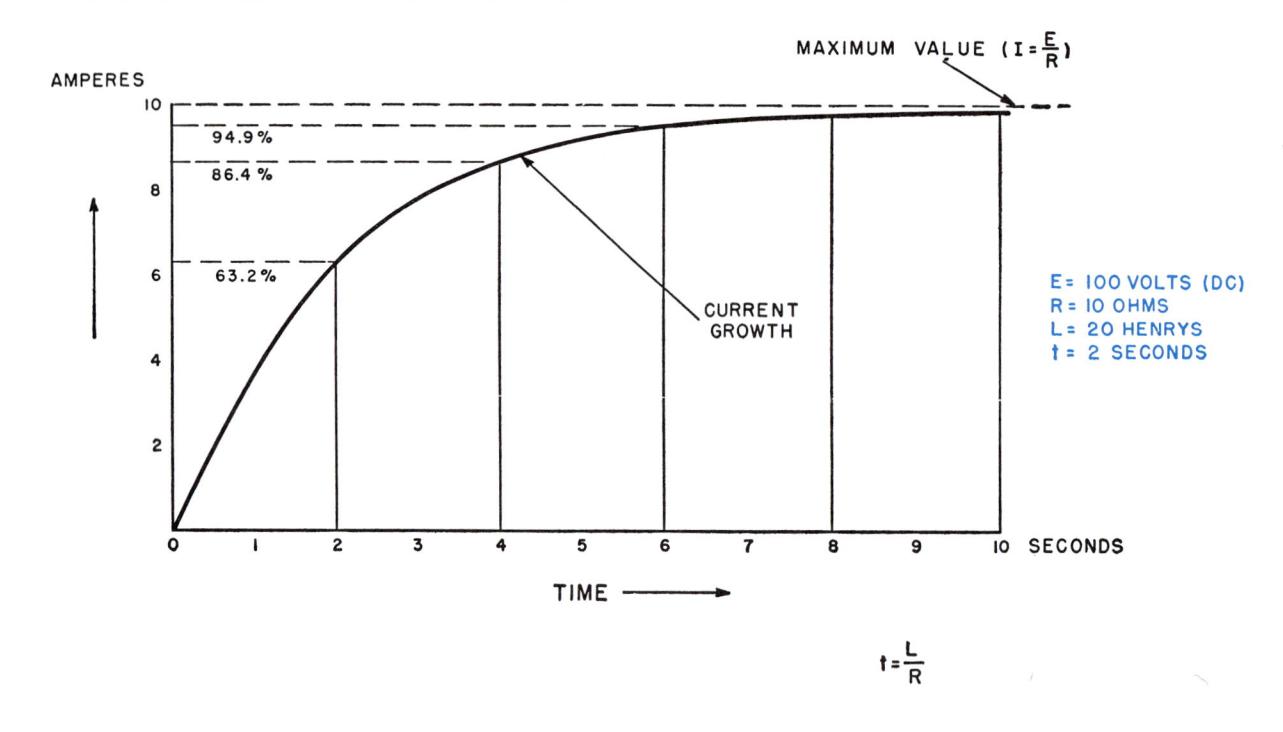

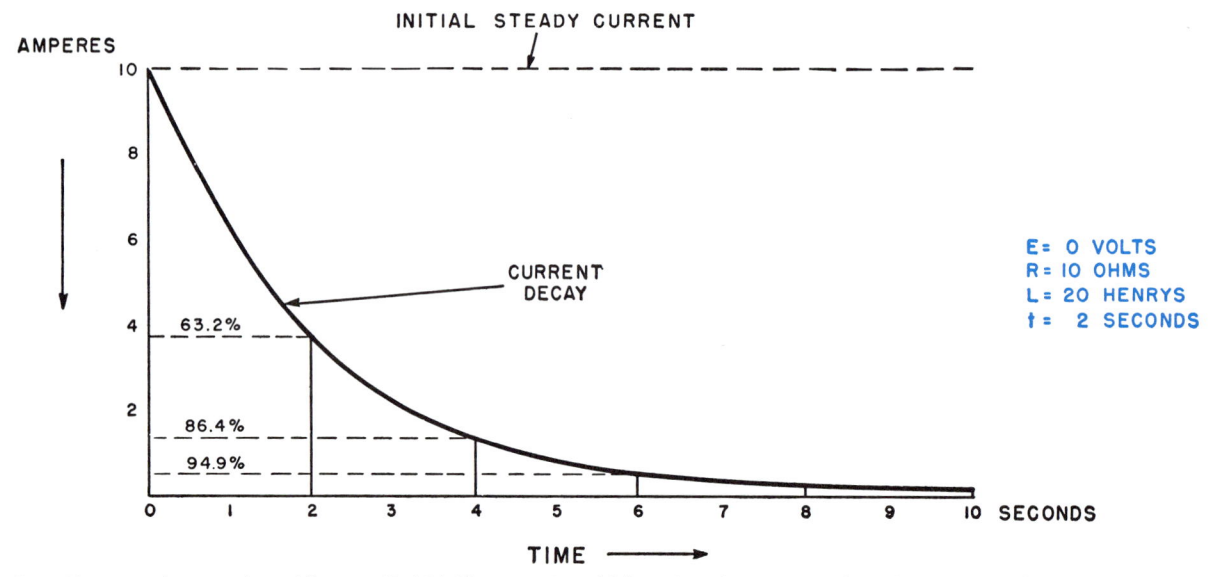

REFRIGERATION AND AIR CONDITIONING TECHNOLOGY

4-21. The time lag produced by a coil. (A) The way in which a time lag is introduced in a circuit by an inductor. (B) It takes time for the magnetic field to collapse.

$$t = \frac{L}{R}$$

Graph A: Current Growth
- E = 100 VOLTS (DC)
- R = 10 OHMS
- L = 20 HENRYS
- t = 2 SECONDS

Graph B: Current Decay
- E = 0 VOLTS
- R = 10 OHMS
- L = 20 HENRYS
- t = 2 SECONDS

Maximum Value $(I = \frac{E}{R})$

4-22. Magnetic flux is the coupling between the primary and secondary transformer circuits.

cally connected to both circuits. A transformer operates on the principle of mutual inductance. Magnetic lines of force (or a *flux field*) are created by the primary side of the transformer. These lines of force (or the *force field*) change as the alternating current changes in polarity. The changing magnetic field creates an induced emf in the secondary side of the transformer. The amount of current available is determined by the size of the wire and the amount of iron in the core of the transformer. The symbols for transformers are shown in Fig. 4-23.

Transformer Construction

Transformers are constructed with a coil in the primary winding and a coil in the secondary winding. The coil in the primary winding is connected to the power source. The secondary coil is connected to the circuit needing its particular voltage and current. The primary coil is the *input*. The secondary coil is the *output*. There is some power loss in the transfer of energy from the input to the output coils. Nevertheless, transformers are very close to being 100% efficient. This is due partly to the fact that there are no moving parts—only the current varies. The core of the transformer may be air (no core) or iron. Air cores are used in radio frequency applications. Iron cores are used in power line frequencies and audio frequency applications. The magnetic path is usually through the iron core. The core makes a difference in the capability of the transformer to transfer large amounts of energy from one coil to the other. The core also represents a power loss potential.

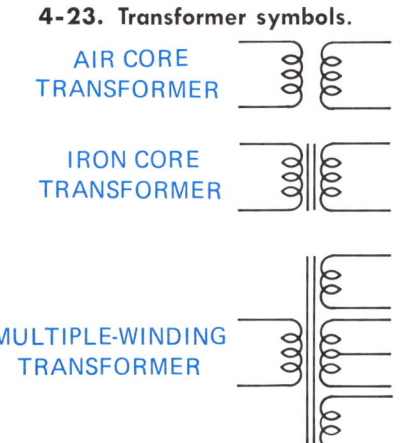

The following three types of losses are encountered in transformers:

1. *Hysteresis losses*—caused by the reluctance of the iron core to change polarity with changes in current direction and resultant changes in magnetic field polarity.

2. *Eddy current losses*—created by small currents induced into the core material by changing magnetic fields.

3. *Copper losses*—due to the copper content of the wire. This copper has resistance as an inherent factor.

Losses can be reduced by the following methods:
- Hysteresis losses are reduced by using silicon steel.
- Eddy current losses are reduced by using laminations.
- Copper losses are reduced by using the correct size of wire.

Turns Ratio

A transformers output voltage is determined by its number of turns as compared to those of the input primary. For example, if the primary has 100 turns and the secondary has 10 turns, then the turn ratio is 10:1. Thus, if 100 volts are applied to the primary, the secondary will put out 10 volts. However, if the input current is 1 ampere, then the available output current would be 10 amperes. The *power in* must equal the *power out* less any inefficiency. Power in (or, $P = E \times I$) is equal to power out (or, $P = E \times I$). Therefore, a step-up transformer refers to the voltage because the current will be the opposite of voltage. The example just mentioned is a *step-down* transformer. In such a transformer, the output voltage is less than the input.

Transformer Applications

Most heating and cooling devices use transformers to step down the voltage for control circuits. Transformers mean you can have the proper voltage for use by any type of equipment from one source voltage. Transformers are used on *AC only*, since DC does not have a moving magnetic field.

Transformers are used in electronic air cleaners to step up the voltage sufficiently to operate the equipment and trap dust particles.

SEMICONDUCTORS

Semiconductors are used in making diodes and transistors. These devices are made primarily of germanium and silicon crystals. Controlled amounts of impurities are placed into a 99.999 999% pure silicon wafer or germanium wafer. When arsenic or antimony are added, the N-type semiconductor material is formed. This means the material has an excess of electrons. Electrons have a negative charge.

When gallium or indium is used as the impurity, a P-type

4-23. Transformer symbols.

AIR CORE TRANSFORMER

IRON CORE TRANSFORMER

MULTIPLE-WINDING TRANSFORMER

REFRIGERATION AND AIR CONDITIONING TECHNOLOGY

semiconductor material is produced. This means it has a positive charge, or is missing an electron.

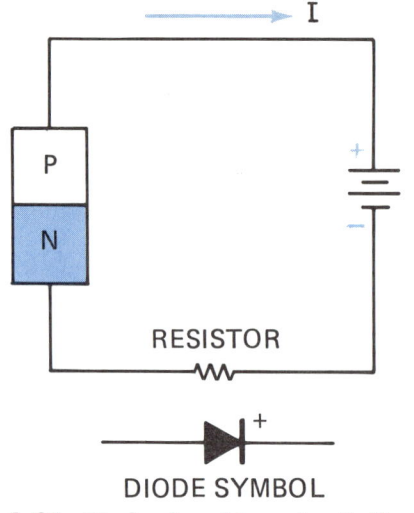

DIODES

When N- and P-type materials are joined, they form a *diode*, also called a *rectifier*. This device is used to change alternating current to direct current. The PN junction (diode) acts as a one-way valve to control the current flow. The forward, or low resistance direction through the junction, allows current to flow through it. The high resistance direction does not allow current to flow. This means that only one-half of an alternating current hertz is allowed to flow in a circuit with a diode. Figure 4-24 shows how a diode is used in the forward biased direction that allows current to flow. Figure 4-25 indicates the arrangement in the reverse bias configuration. No current is allowed to flow under these conditions. Note the polarity of the battery.

Diodes are also used in isolating one circuit from another. A simple rectifier circuit is shown in Fig. 4-26. The output from the transformer is an AC voltage, as shown in Fig. 4-27. However, the rectifier action of the diode blocks current flow in one-half of the sine wave and produces a pulsating DC across the resistor. Fig. 4-27(B).

Zener Diode

When one polarity of voltage is applied to a rectifier (diode), it blocks the current flow. However, if the voltage is raised high enough the diode breaks down. This allows current to flow. Normal diodes would be destroyed by this breakdown. However, a *zener diode* is designed to operate in the breakdown region.

Figure 4-28 shows how a zener diode is connected in a circuit. The breakdown voltage on the diode is 8.2 volts. As long as the battery voltage is 8.2 volts *or lower*, the output across the diode will be 8.2 volts or lower. However, if the battery voltage is more than 8.2 volts the voltage drop across the diode is still 8.2 volts. If the battery voltage is 10 volts, the voltage drop is 1.8 across the series resistor and 8.2 across the diode. If the battery voltage reaches 12 volts, the voltage is 3.8 across the resistor and 8.2 across the diode. As can be seen in Fig. 4-28, the zener diode can be used in a circuit to regulate the voltage and keep it constant—or at least no higher than its rating. That is why the circuit

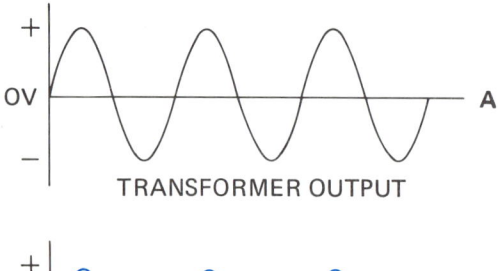

4-24. Diode placed in a circuit. The symbol and the silicon wafer are represented in a circuit.

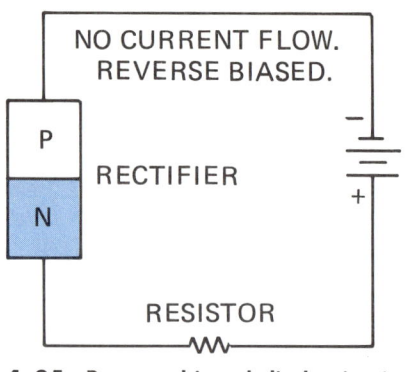

4-25. Reverse-biased diode circuit.

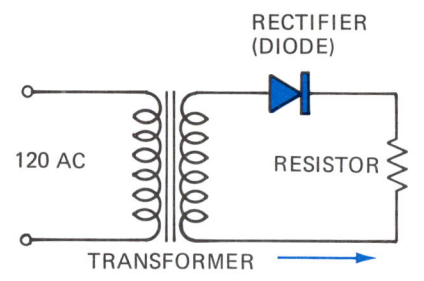

4-26. Rectifier circuit using a diode to produce DC from AC.

4-27. Results of the rectifier circuit. The transformer output is changed to pulsating DC across the resistor.

96

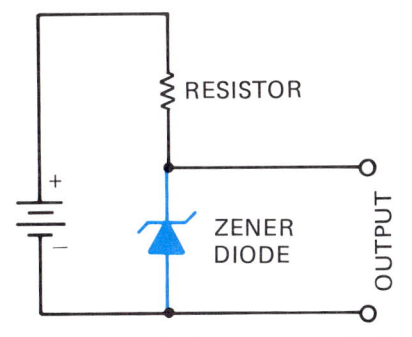

CHAPTER 4—ELECTRICITY AND REFRIGERATION CONTROLS

4-28. Zener diode in a circuit. Resistor is necessary to the proper operation of the circuit.

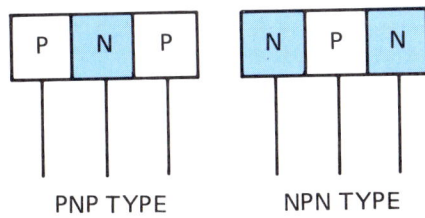

4-29. Arrangement of wafers of silicon or germanium to produce a PNP or NPN transistor.

is called a voltage regulator circuit. Such a circuit is very important when a constant voltage is necessary for sensing equipment to operate accurately.

TRANSISTORS

In 1948, the Bell Telephone research laboratories announced that a crystal could amplify. Such a crystal was called a *transistor*—meaning *trans*fer re*sistor*. The transistor has replaced the vacuum tube in almost all applications. It is made up of three layers of P and N-type semiconductor material arranged in either of two ways. Fig. 4-29.

The transistor is used as a switching device or an amplifier. The advantages of the transistor are well known, since it is used in the transistor radio and the semiconductor television receiver.

Figure 4-30 shows a simple transistor amplifier. Battery 1 (B_1) and adjustable resistor R_1 determine the input current to the transistor. When R_1 is high in resistance, the current flowing from the base to the emitter is very small. When the base-to-emitter current is small, the collector-to-emitter resistance appears as a very high resistance. This limits the current flow from battery 2 (B_2). The result is a limiting of the voltage drop across R_2.

As the resistance of R_1 is lowered, the current flowing through the base-to-emitter junction increases. As the base-to-emitter current is increased, the resistance of the transistor from collector to emitter is decreased. More current is flowing from battery 2 through R_2 and the voltage drop across R_2 is increased. A very small change in the current from battery 1 causes a large change in the current from battery 2. The ratio of the large change to the small change is defined as the *gain of the transistor*.

SILICON-CONTROLLED RECTIFIER (SCR)

The silicon-controlled rectifier (SCR) is a four-layered PNPN device. The SCR can be defined as a high-speed semiconductor switch. It requires only a short pulse to turn it *on*. It remains *on* as long as current is flowing through it.

Look at the circuit shown in Fig. 4-31. Assume that the SCR is off. (It would then have a very high resistance.) No current would be flowing through the resistor. When switch S_1 is closed

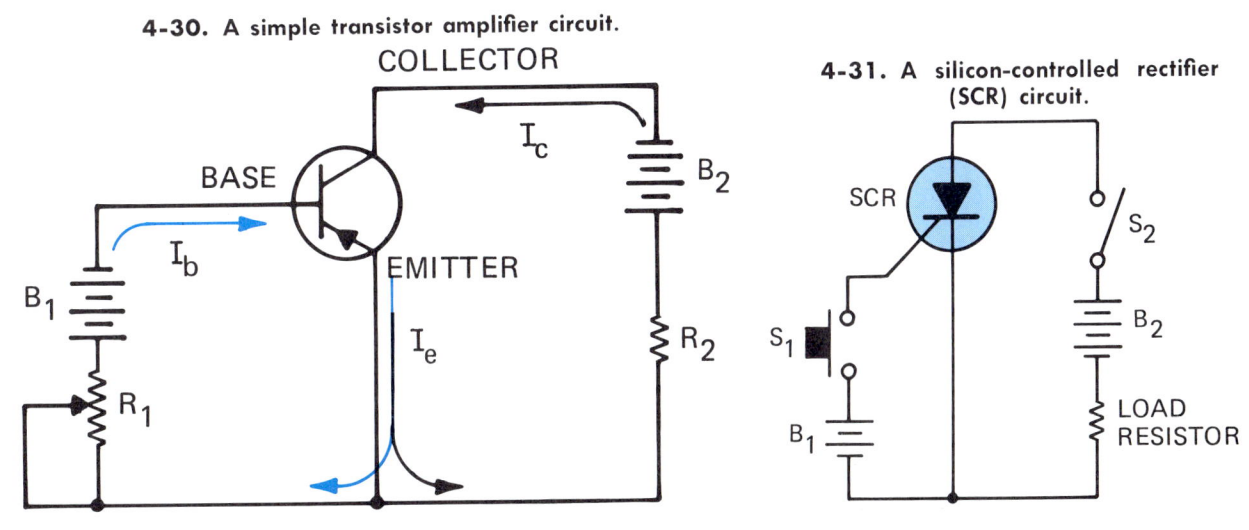

4-30. A simple transistor amplifier circuit.

4-31. A silicon-controlled rectifier (SCR) circuit.

97

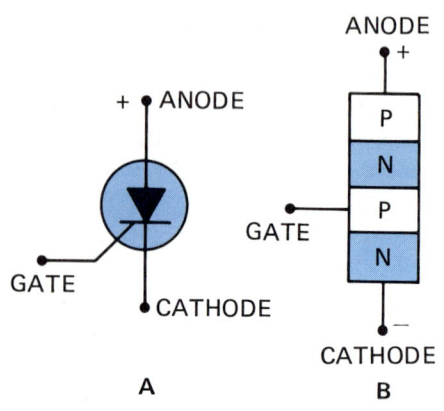

4-32. A silicon-controlled rectifier. (A) Schematic representation of an SCR. (B) Wafer arrangement needed to produce a SCR.

just long enough to turn on the SCR (which then has a very low resistance) a current will flow through the resistor and the SCR. The SCR will remain on until switch S_2 stops the flow of current through the resistor and SCR. Then the SCR will turn off. When S_2 is again closed, the resistance of the SCR remains high. No current will flow through the resistor until S_1 is reclosed.

Figure 4-32(A) shows an SCR represented schematically. Figure 4-32(B) shows the arrangement of the layers of P- and N-type materials that produce the SCR effect. The anode is the positive terminal. The gate is the terminal used to turn on the SCR. The cathode is the negative terminal.

BRIDGE CIRCUITS

Wheatstone Bridges

A *bridge circuit* is a network of resistances and capacitive or inductive impedances. The bridge circuit is usually used to make precise measurements. The most common bridge circuit is the *Wheatstone bridge*. This consists of variable and fixed resistances. Simply, it is a series-parallel circuit. Redrawn, as shown in Fig. 4-33, is a Wheatstone bridge circuit. The branches of the circuit forming the diamond shape are called legs.

If 10 volts DC were applied to the bridge shown in Fig. 4-34, one current would flow through R_1 and R_2 and another through

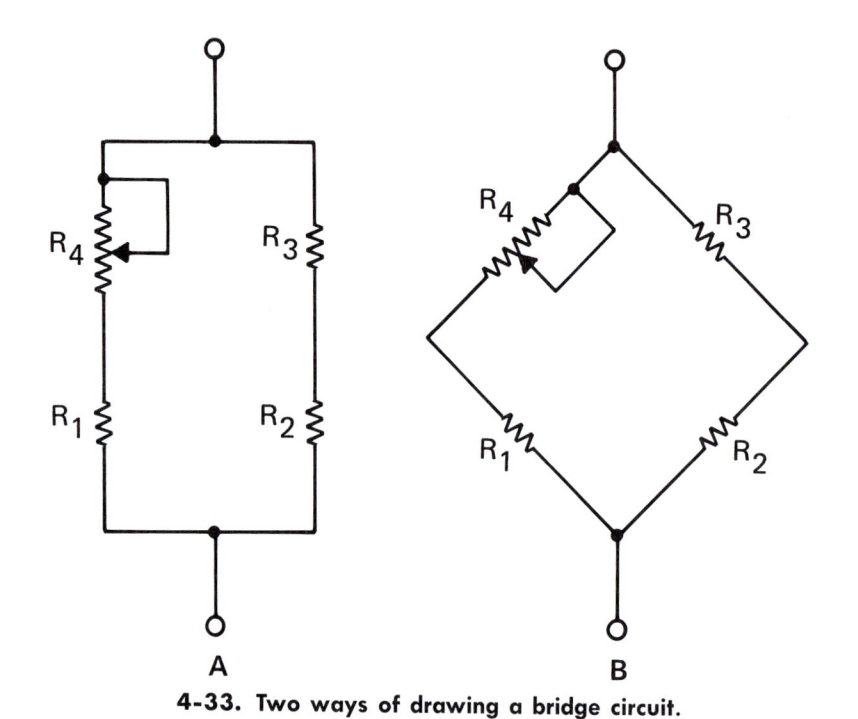

4-33. Two ways of drawing a bridge circuit.

4-34. Operation of a bridge circuit.

98

CHAPTER 4—ELECTRICITY AND REFRIGERATION CONTROLS

4-35. Operation of a bridge circuit.

4-36. Operation of a bridge circuit.

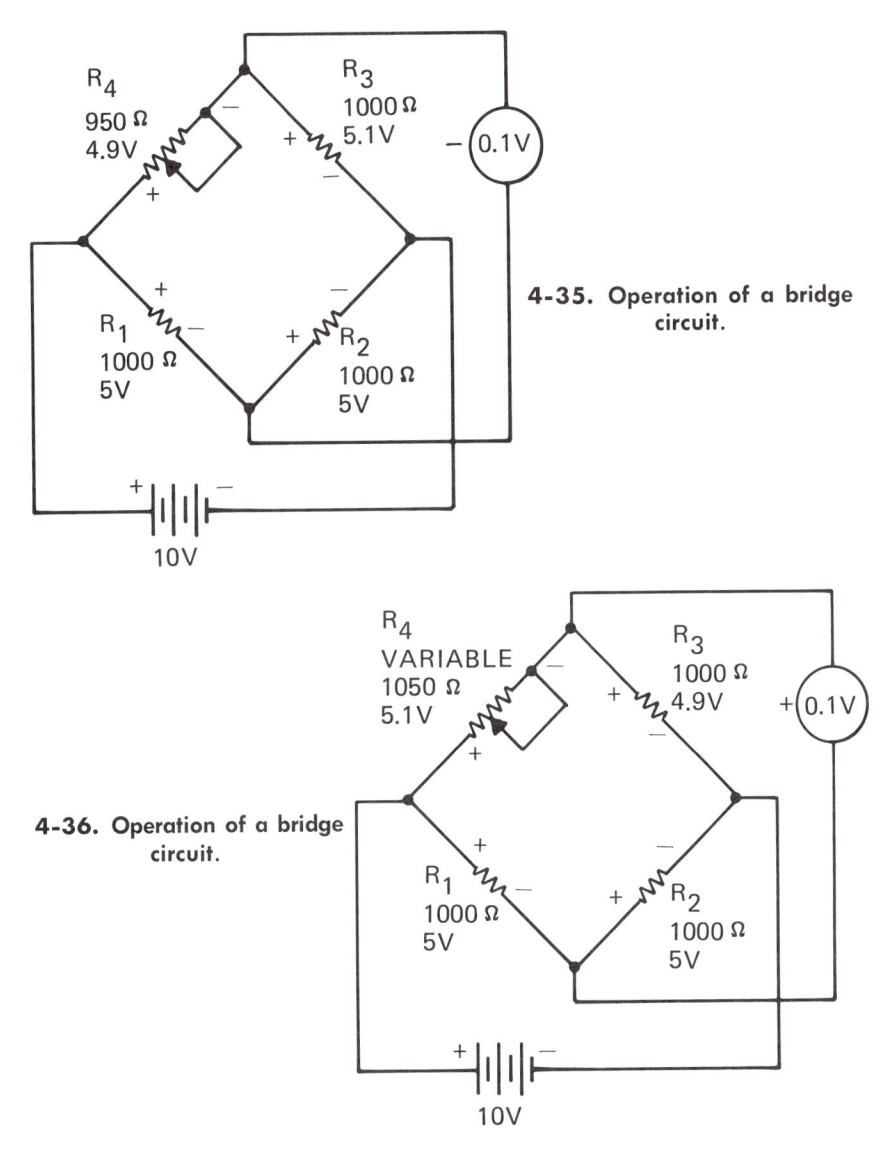

R_3 and R_4. Since R_1 and R_2 are both fixed 1000-ohm resistors, the current through them is constant. Each resistor will drop one-half of the battery voltage, or 5 volts. 5 volts is dropped across each resistor. The voltmeter senses the sum of the voltage drops across R_2 and R_3. Both are 5 volts. However, the R_2 voltage drop is a positive (+) to negative (−) drop. The R_3 drop is a negative to positive drop. They are opposite in polarity and cancel each other. This is called a *balanced bridge*. The relationship is usually expressed as a ratio of:

$$\frac{R_1}{R_2} = \frac{R_3}{R_4}$$

The actual resistance values are not important. What is important is that this ratio is maintained and the bridge is balanced.

Variable Resistor

In Fig. 4-35, the value of the variable resistor R_4 is 950 ohms. The other resistors have the same value. Using Ohm's Law, the voltage drop across R_4 is found to be 4.9 volts. The remaining voltage, 5.1 volts, is dropped across R_3. As shown in Fig. 4-35, the voltmeter measures the sum of the voltage drops across R_2 and R_3 as 5.0 volts (+ to −) and 5.1 volts (− to +). It registers a total of −0.1 volt.

In Fig. 4-36, the converse is true. The value of R_4 is 1050 ohms. The voltage drop across R_3 is 4.9 volts. The voltmeter senses the sum of 5 volts (+ to −) and 4.9 volts (− to +), or +0.1 volt.

When R_4 changes the same amount above or below the balanced bridge resistance, the magnitude of the DC output, measured by the voltmeter, is the same. However, the polarity is reversed.

SENSORS

The *sensor* in a control system is a resistance element that varies in resistance value with changes in the variable it is measuring. These resistance changes are converted into proportional amounts of voltage by a bridge circuit. The voltage is amplified and used to position actuators that regulate the controlled variable.

Temperature Elements

The temperature element used in *cybertronic* devices is a nickel wire winding. (A cybertronic device is an electronic control system.) This wire is very sensitive to temperature changes. It increases its resistance to current flow at the rate of approximately 3 ohms for every degree Fahrenheit increase in temperature. This is called a *positive tempera-*

99

REFRIGERATION AND AIR CONDITIONING TECHNOLOGY

ture coefficient. The length and type of wire give the winding a reference resistance of 1000 ohms at 70° F. A temperature drop decreases the resistance and a temperature rise increases the resistance. The winding is accurate over a range of −40 to 250° F.

Humidity Elements

Improvements in the design of humidity-sensing elements and the materials used in their construction have minimized many of the past limitations of humidity sensors. The humidity sensor used with the electronic controls is a resistance CAB (cellulose acetate butyrate) element. This resistance element is an improvement over other resistance elements. It has greater contamination resistance, stability, and durability. The humidity CAB element is a multilayered humidity-sensitive polymeric film. It consists of an electrically con-

CAB RESISTIVE ELEMENT
Johnson Controls
4-37. CAB resistive element.

4-38. A hydrolyzed humidity element.

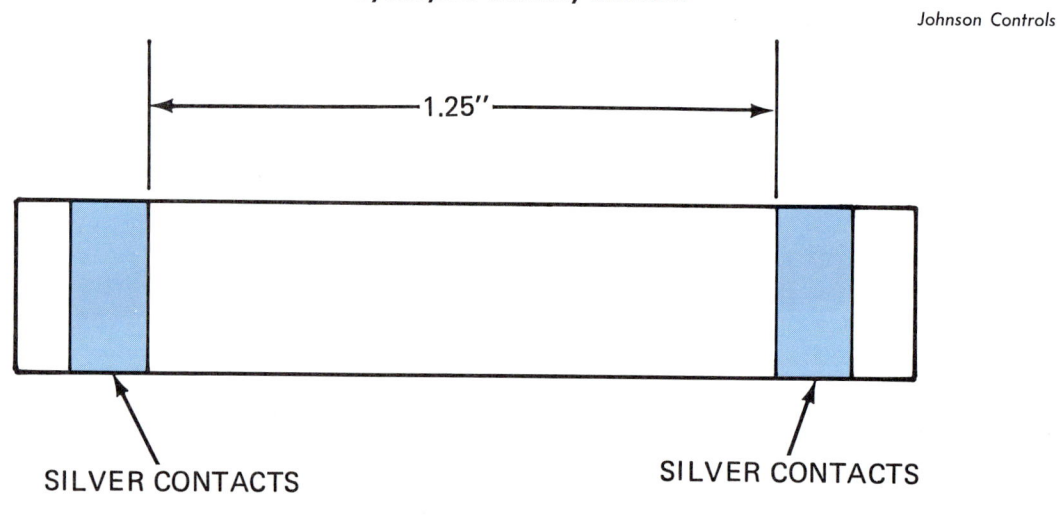

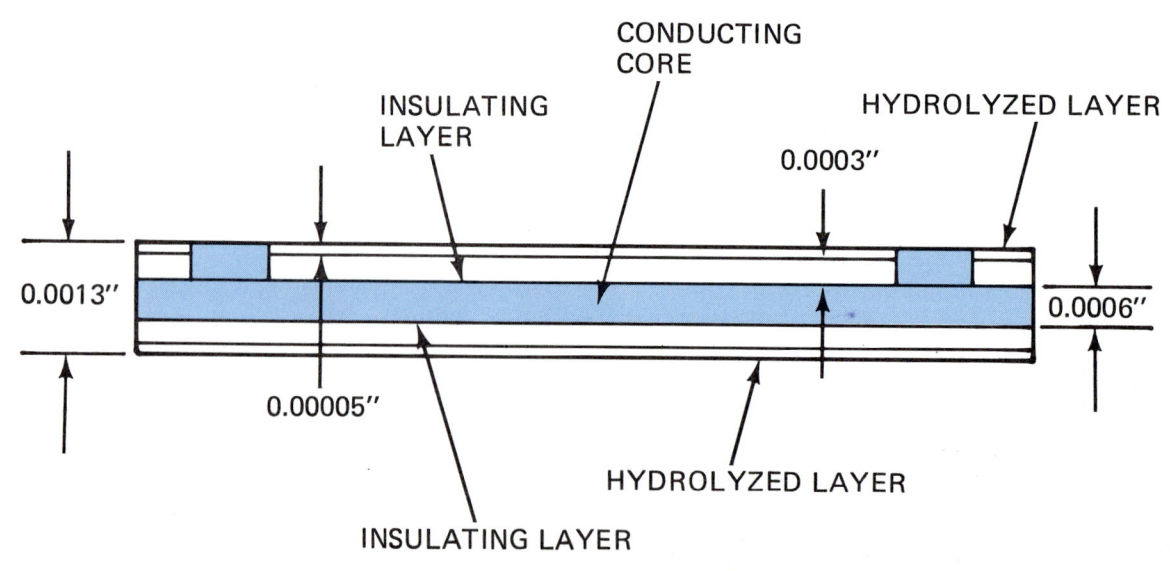

Johnson Controls

100

ductive core and insulating outer layers. These layers are partially hydrolyzed. The element has a nominal resistance of 2500 ohms. It has a sensitivity of 2 ohms per 1% relative humidity (rh) at 50% rh. Humidity sensing range is rated at 0 to 100% rh.

The CAB element consists of a conductive humidity-sensitive film, mounting components, and a protective cover. Fig. 4-37. The principle component of this humidity sensor is the film. The film has five layers of cellulose acetate butyrate in the form of a ribbonlike strip. The CAB material is used because of its good chemical and mechanical stability and high sensitivity to humidity. It also has excellent film-forming characteristics. Fig. 4-38.

The CAB resistance element is a carbon element having the resistance/humidity tolerances shown in Fig. 4-39. With an increase in relative humidity, water is absorbed by the CAB, causing it to swell. This swelling of the polymer matrix causes the suspended carbon particles to move farther apart from each other. This results in an increased element resistance.

When relative humidity decreases, water is given up by the CAB. The contraction of the polymer causes the carbon particles to come closer together. This, in turn, makes the element more conductive, or less resistive.

CONTROLLERS

The sensing bridge is the section of the controller circuit that contains the temperature-sensitive element or elements. The potentiometer for establishing the "set point" is also part of the control system. The bridges are energized with a DC voltage.

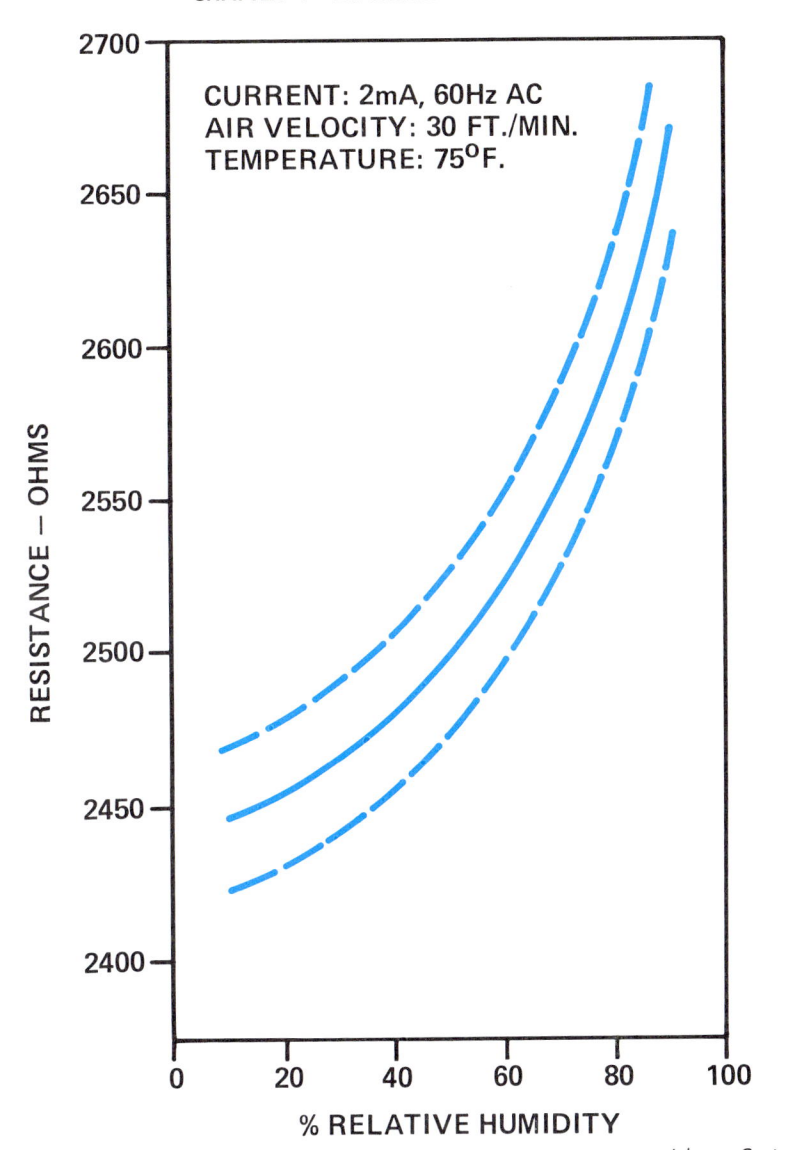

4-39. Operational characteristics of a humidity element.
Johnson Controls

CURRENT: 2mA, 60Hz AC
AIR VELOCITY: 30 FT./MIN.
TEMPERATURE: 75°F.

This permits long wire runs in sensing circuits without the need for compensating wires or for other capacitive compensating schemes.

Both integral (room) and remote-sensing element controllers produce a proportional 0- to 16-volt DC output signal in response to a measured temperature change. Controllers can be wired to provide direct or reverse action. Direct-acting operation provides an increasing output signal in response to an increase in temperature. Reverse-acting operation provides an increasing output signal in response to a decrease in temperature.

Single-Element Controllers

Electronic controllers have

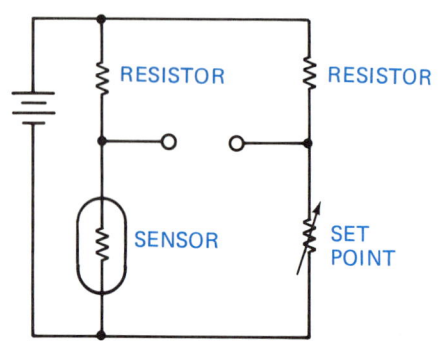

4-40. A bridge arrangement with a sensor and set point.

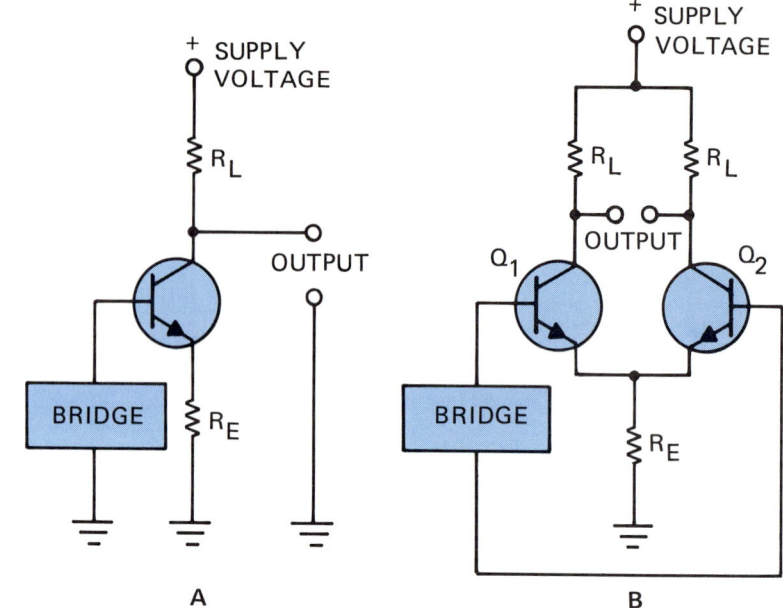

4-42. Amplifier stages. (A) Single transistor amplifier stage. (B) Two-transistor amplifier stage.

three basic parts: the bridge, the amplifier, and the output circuit.

Bridge theory has been covered previously. Two legs of the bridge are variable resistances. Fig. 4-40. The sensor and the set point potentiometer are shown in the bridge circuit. If temperature changes, or if the set point is changed, the bridge is in an *unbalanced* state. This gives a corresponding output result. The output signal, however, lacks power to position actuators. Therefore, this signal is amplified.

4-41. DC differential amplifiers for use in a controller circuit.

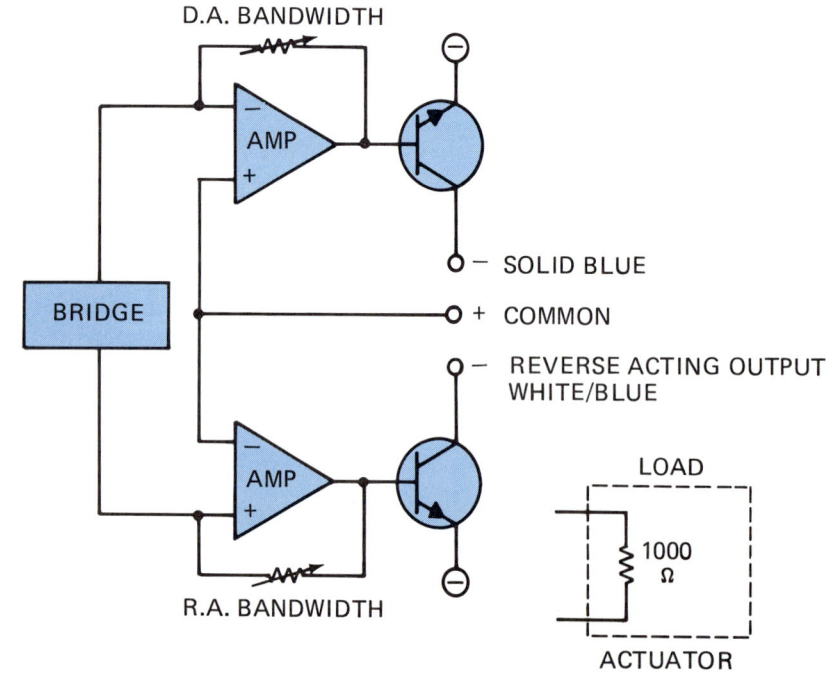

Differential Amplifiers

Controllers utilize direct-coupled DC differential amplifiers to increase the millivolt signal from the bridge to the necessary 0- to 16-volt level for the actuators. There are two amplifiers—one for direct reading and one for reversing signals. Each amplifier has two stages of amplification. This arrangement is shown in block form in Fig. 4-41.

The differential transistor circuits provide gain and good temperature stability. Figure 4-42 compares a single transistor amplifier stage with a differential amplifier. Transistors are temperature sensitive. That is, the current they allow to pass depends upon the voltage at the transistor and its ambient temperature. An increase in ambient temperature in the circuit shown in Fig. 4-42(A) would cause the current through the transistor to increase. The output voltage would, therefore, decrease. The emitter resistor R_E reduces this temperature effect. It also reduces the available voltage gain in the circuit because the signal voltage across the resistor amounts to a negative feedback voltage. That is, it causes a decrease in the voltage difference which was originally produced by the change in temperature at the sensing element.

Since it is desirable for the output voltage of the controller to correspond only to the temperature of the sensing elements and not to the ambient temperature of the amplifier, the circuit shown in Fig. 4-42(B) is used. Here any ambient temperature changes affect both transistors simultaneously. The useful output is taken as the difference in output levels of each transistor and the effects of temperature changes are cancelled. The voltage gain of the circuit shown in Fig. 4-42(B) is much higher than that shown in Fig. 4-42(A). This is because the current variations in the two transistors produced by the bridge signal are equal and opposite. An increase in current through Q_1 is accompanied by a decrease in current through Q_2. The sum of these currents through R_E is constant. No signal voltage appears at the emitters to cause negative feedback as in Fig. 4-42(A).

Output Circuit Connections

The output circuit of the controller has three connections:
- Common positive (+), solid red wire.
- Direct acting negative (−), solid blue wire.
- Reverse acting negative (−), white/blue wire.

A load in the form of an actuator, which is equivalent to 1000 ohms, can be connected to either set of wires or terminals. This depends upon the controller action desired.

The controller's amplifier and output circuits are also designed to provide sequential operation of two actuators. This is accomplished by connecting an actuator to the direct output, as well as to the reverse acting output.

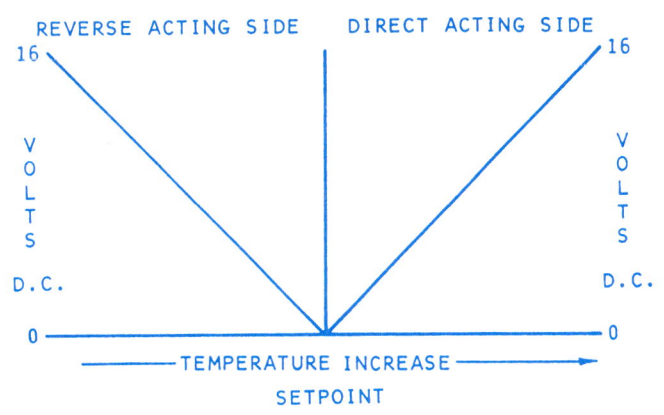

The result is sequentially varying DC signals in response to temperature change at the sensing element. Fig. 4-43. When sequential operation is used, the controller is calibrated so the set point and sensing element provide the bridge with a balanced condition at set point. This means both the direct and reverse acting outputs are zero.

When the temperature is significantly below the set point, a 16-volt DC output is present on the reverse acting side and a 0-volt DC signal is present on the direct acting side. As temperature increases, the reverse acting signal decreases. When the temperature reaches set point, both outputs are 0 volts DC or at "null." On a further increase in temperature, the direct acting signal increases from 0 to 16 volts DC. When the temperature is such that operation is on the reverse acting side of null, only the actuator connected to that side is operating. Similarly, when the temperature is above the set point, operation is on the direct acting side of null and only that actuator is operating. In other words, the actuators operate in sequence, not simultaneously.

Bandwidths

Bandwidths in these controllers are adjusted separately for direct- and reverse-acting signals. This permits optimum settings for both heating and cooling systems. Fig. 4-43.

Bandwidth adjustment of an electronic controller is defined as "the number of desired degree changes at the element needed to cause a full 0- to 16-volt DC change in the output signal." When sequential operation is utilized, the total temperature change at the element, which caused the outputs of both sides of null to vary, must be considered in the evaluation of system control.

Since there are two bandwidth settings, consider individually

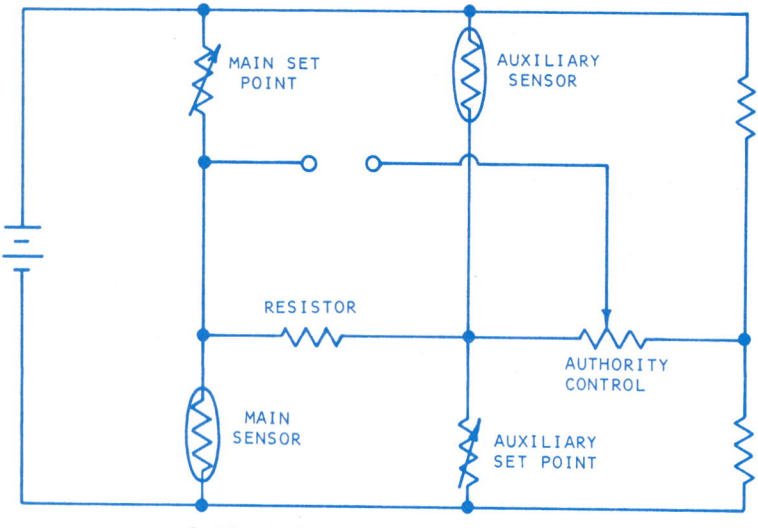

REFRIGERATION AND AIR CONDITIONING TECHNOLOGY

4-44. A dual bridge arrangement.

the temperature change from set point to where the full 16-volt output on each side of null should occur.

Dual-Element Controllers

Dual-element controllers function the same as single-element controllers with one exception. In place of one bridge, two bridges are used. Two bridge controllers are used where temperature effects on one element are to be used to readjust the set point of another element to provide greater accuracy of control and improved comfort for occupants.

Dual-Bridge

A dual-bridge arrangement is shown in Fig. 4-44. Bridge output is proportional to the algebraic sum of the temperature effects on both elements. This algebraic sum is expressed in terms of percentage of *authority*. An authority of 100% simply means that a temperature change (ΔT) on the auxiliary element has the same effect as a temperature change (ΔT) on the main element, except that the temperature change at each element is the opposite in direction. This is referred to as *reverse adjustment*.

Main Element

Determining main and auxiliary assignments is dependent upon the measured temperature span at each element. The *main* element is always the element having the least measured temperature change of the two elements. The auxiliary element is always the element having the greatest measured temperature change. This arrangement is essential, since authority settings are always between 0 and 100%.

A typical system might have a ratio of main to auxiliary sensor effects of 20 to 1. This corresponds to a 5% authority setting. This means that a 20° F. change in temperature at the auxiliary element produces a bridge output equal to that of a 1° F. change at the main element. For a 2 to 1 ratio, authority is 50%. This means a 2° F. change at the auxiliary element has the effect of a 1° F. change at the main element.

Dual-element controllers differ from single-element controllers only in regard to bridge configurations. There is an interacting effect within the bridge circuitry caused by the two elements and the authority setting. The amplifier circuitry and output circuitry cause the signals on both sides of null to be identical to those encountered with single-element controllers.

ACTUATORS

Electro-Hydraulic Actuators

Cybertronic actuators perform the work in an electronic system. They accept a control signal and translate that signal into mechanical movement to position valves or dampers.

The electro-hydraulic actuators are so called because they convert an electric signal into a fluid movement and force. Damper actuators, equipped with linkage for connection to dampers and valve actuators, having a yoke and linkage to facilitate mounting on a valve body, are available.

Operation of Actuators

Two voltages are applied to the actuator. Fig. 4-45. A 0- to 16-volt DC control signal regulates or controls the servo valve. Then a 24- or 120-volt AC signal, depending on the unit, operates the oil pump. The oil pump moves oil from the upper chamber to the lower chamber. The servo valve controls pressure at the diaphragm by varying the return flow from the lower to the upper chamber.

When there is no DC voltage applied to the servo valve, the

CHAPTER 4—ELECTRICITY AND REFRIGERATION CONTROLS

flapper is pushed off the servo port by way of the hydraulic pressure developed by the pump. The open servo port allows the pump to move all the oil through the lower chamber back into the upper chamber.

When the voltage on the servo increases, a magnetic force is developed. This magnetic force holds the flapper down over the servo port. The pump continues to pump oil into the lower chamber. But, the return flow to the upper chamber is stopped by the blocked servo port. Pressure is built up in the lower chamber until the magnetic force on the flapper is overcome and the flapper is pushed away from the servo port. This equalizes the flow through the pump and servo valve, while maintaining a pressure in the lower chamber.

Each increase in DC voltage results in a hydraulic pressure increase in the lower chamber. The increased pressure begins to overcome the opposing pressure from the return spring, and forces out the actuator shaft. Each further increase in DC voltage causes an increased extension of the actuator shaft.

The servo valve represents the load of 1000 ohms required by the controller to cause a variation of 0 to 16 volts DC output signal. Two actuators can be connected in parallel across the output terminals of an electronic controller. However, this will provide only 500 ohms resistance, which the controller also can handle.

Thermal Actuators

Thermal actuators should more properly be called electro-thermal actuators. This is because they take a 0- to 16-volt DC signal and convert the signal into heat. The thermal damper actuator has linkage for connection to a damper. The thermal valve actuator is directly connected to a valve body.

OPERATION OF A THERMAL ACTUATOR

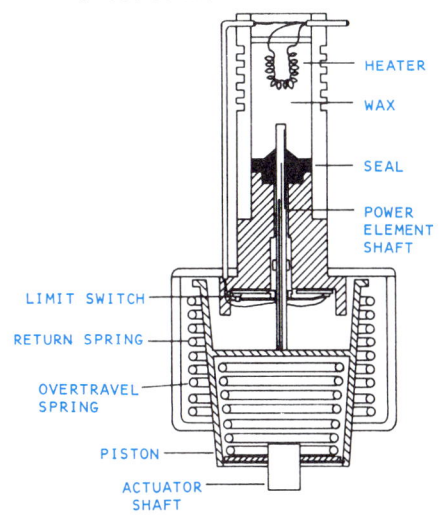

A thermal actuator is shown in Fig. 4-46. A small electrical control circuit is encapsulated in the electrical cable about 12" from the thermal unit. The 0- to 16-volt DC signal from the controller and the 24-volt AC supply voltage are fed into the control circuit. The circuit allows the 0- to 16-volt DC signal to control the amount of current from the 24-volt supply to the actuator.

Inside the actuator, the controlled current from the 24-volt source heats up a small heater that is embedded in wax. When the wax reaches approximately 180° F. it changes from a solid to a liquid. During this change the wax expands. This is the point at which the motion of the device is controlled.

As the wax expands, the power element shaft is forced out to move the piston. This, in turn, compresses the return spring and moves the actuator shaft. After the power element shaft has traveled the full stroke, a limit switch is opened to stop the flow of current to the heater. The wax begins to cool and contract. The

4-45. An electro-hydraulic actuator.

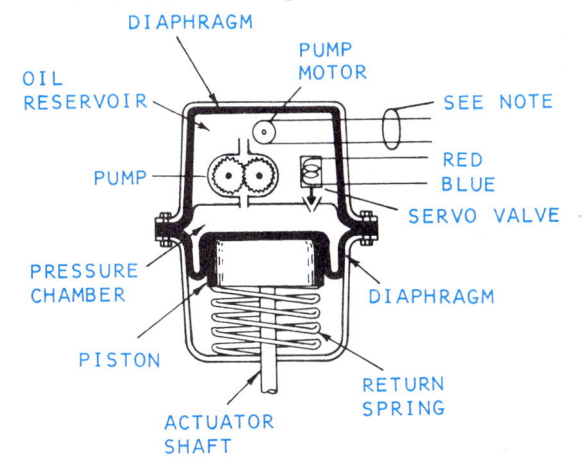

NOTE: POWER WIRES FOR 120V UNIT ARE BLACK AND BLACK/RED; WIRES FOR 24V UNIT ARE YELLOW AND WHITE

4-46. A thermal actuator.

105

REFRIGERATION AND AIR CONDITIONING TECHNOLOGY

power element shaft is forced to retract by the return spring. This closes the limit switch and the sequence is repeated. However, only when the control signal is high enough to hold the actuator at its fully extended position does it take place.

AUXILIARY DEVICES

Low- and high-signal selectors accept several control signals. Such selectors then compare them and pass the lowest or highest. For example, a high-signal selector can be used on a multizone unit to control the cooling coil. The zone requiring the most cooling transmits the highest control signal. This, in turn, will be passed by the high-signal selector to energize the cooling.

Minimum position networks are used to ensure that the outdoor air dampers are positioned to admit a minimum amount of air for ventilation, regardless of the controller demand.

Reversing networks change the action of a controller output signal from direct to reverse or reverse to direct acting.

Sequencing networks amplify a selected portion of an input voltage from a controller. A common application is where two actuators function in sequence.

Two-position power supplies permit two-position override of a proportional control system.

A unison amplifier allows a controller to operate up to eight actuators, where a controller alone will operate only two.

REVIEW QUESTIONS

1. What is Ohm's Law?
2. Describe a parallel circuit.
3. What is the formula for finding resistance in a parallel circuit?
4. What is the basic unit of measurement for electrical power?
5. What is a capacitor?
6. What is a dielectric?
7. What three factors determine the capacitance of a capacitor?
8. What is a microfarad?
9. What makes an electrolytic capacitor different from a standard paper-type capacitor?
10. What is the unit of measurement for inductance?
11. What is the symbol for an audio frequency inductor?
12. What is inductive reactance?
13. What effect does the turns ratio have on the output voltage of a transformer?
14. What is a zener diode?
15. What do the letters SCR stand for?
16. What is a bridge circuit?
17. What are the three parts of electronic controllers?
18. What is a thermal actuator?

5

Electric Motors and Controls

DC MOTORS

Motors are simply a means of transforming energy or power. Motors convert electrical power into mechanical power. The essential parts of a DC motor are the field coils, armature coils, commutator, and brush assembly.

Principle of Operation

An understanding of basic motor action may be obtained by considering the following simple facts concerning magnetic fields.

Figure 5-1 shows the uniform magnetic field that exists between the poles of a magnet when its field coils are connected to a DC source. The lines of flux are directed from the north pole to the south pole.

Figure 5-2 represents a cross-section of a current-carrying conductor when the direction of current flow is away from the observer. By applying the left-hand rule for current carrying conductors, it is found that the direction of the field is counter-clockwise. Arrows indicate the lines of flux. *The left-hand rule states that if a current-carrying conductor were grasped in the left hand with the thumb pointing in the direction of current flow (negative to positive), the fingers*

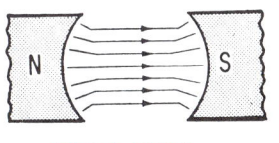

FIELD FLUX

5-1. Magnetic field and flux between poles.

5-2. Magnetic field around a current-carrying wire.

FLUX AROUND CONDUCTORS

5-3. The left-hand rule.

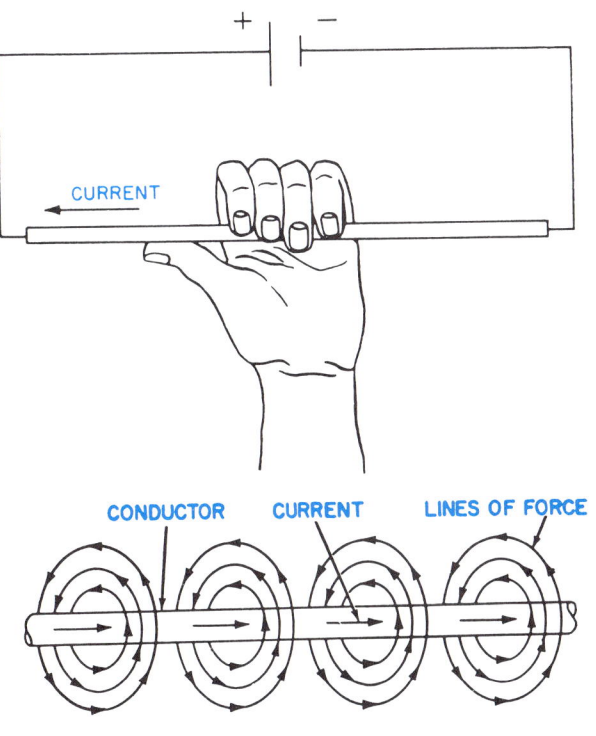

107

REFRIGERATION AND AIR CONDITIONING TECHNOLOGY

would encircle the wire in the direction of the magnetic lines of force. Fig. 5-3.

Figure 5-4 shows the field produced by the action of the current-carrying wire in the presence of the pole pieces and their magnetic field. Above the conductor, the field is weakened because the field produced by the poles and the field produced by the conductor are opposite in direction. They tend to cancel each other. Below the conductor, the two fields run in the same direction and the resultant field is strengthened. Because magnetic lines of force act to push each other apart, those lines below the conductor tend to push the conductor up. Those lines above the conductor tend to push the conductor down. However, because the field below has many more lines and is therefore much stronger, the push upward is greater. The result is that the conductor in Fig. 5-4 is moved upward.

Figure 5-5 indicates the condition when the current flow has been reversed. The direction of the motion of the conductor will also be reversed. This is because, in this case, the field above the conductor is strengthened and the field below is weakened.

Briefly, motors operate on the following principle: *A conductor carrying current in a magnetic field tends to move at right angles to (across) the field.*

Basic DC Motors

A basic DC motor consists of poles that set up a magnetic field, an armature made of a single-turn loop and commutator, and a brush assembly. Each coil side of the loop lies in the magnetic field. Fig. 5-6.

Motor Operation

When a DC voltage is applied to the brushes, a current flows around the loop. The magnetic field produced by the flow of current interacts with the magnetic field produced by the poles of the magnet. The resultant magnetic field is represented by flux lines, as shown in Fig. 5-6.

Small permanent-magnet motors are used to power toys, movie cameras, and other devices. The permanent magnet limits its size. This means its power will be limited to fractional horsepower. It works only on DC. What would happen if AC were applied?

The permanent magnet motor has a permanent magnet for a fixed magnetic field. It has a commutator, brushes, and a wound rotor (armature). The permanent magnet motor shows the basic principles of a DC motor in its simplest form. It also has practical applications.

The loop in the motor rotates clockwise under the influence of the magnetic field. It may be noted that the force acting in one direction (on one side of the

5-4. Note how the wire is pushed upward when the magnetic field is distorted by the presence of a current-carrying wire.

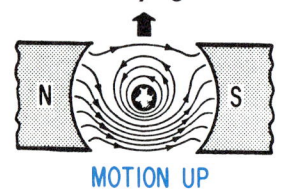

MOTION UP

5-5. Note downward movement of the wire with the current flowing through it.

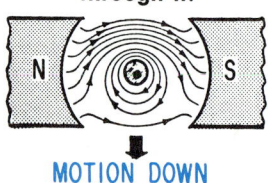

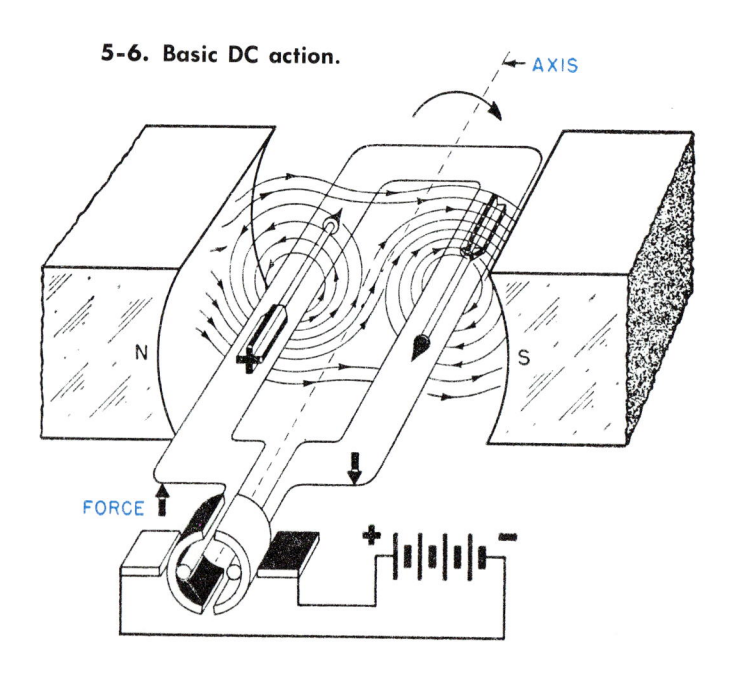

MOTION DOWN

5-6. Basic DC action.

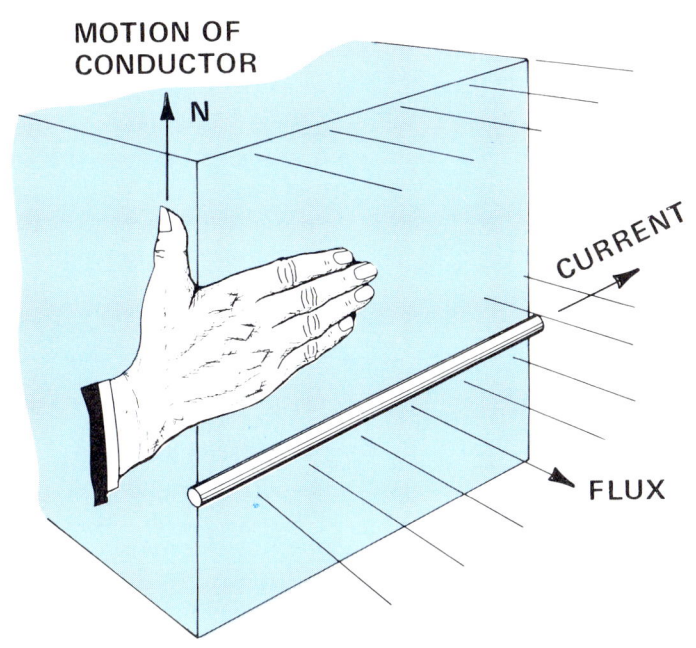

CHAPTER 5—ELECTRIC MOTORS AND CONTROLS

is expressed in the right-hand rule for motors, which states:

Place your right hand in a position so that the lines of force from the north pole of the magnet enter the palm of the hand. Fig. 5-7. Let the extended fingers point in the direction of the current in the conductor. Then the thumb, placed at right angles to the fingers, points in the direction of motion of the conductor.

From this rule, the direction of rotation of a DC motor can be determined, if the direction of the field flux and direction of current flow are known.

5-7. Right-hand rule for motors.

loop) and the force acting in the other direction (on the other side of the loop) combine. This causes the coil to turn on its axis. The loop thus acts as if it were a lever with a turning force, or *torque,* at each end. Because of this lever arrangement, the force at each end is magnified by the distance from the center. Thus, the torque is equal to the combined force on the two sides of the loop, multiplied by the distance of the conductors from the axis about which the loop rotates. The greater the torque, the more pull the motor has to drive a mechanical load.

RIGHT-HAND RULE

There is a definite relationship between the direction of the magnetic field produced by the poles, the direction of the current flow in the conductor, and the direction in which the conductor tends to move. This relationship

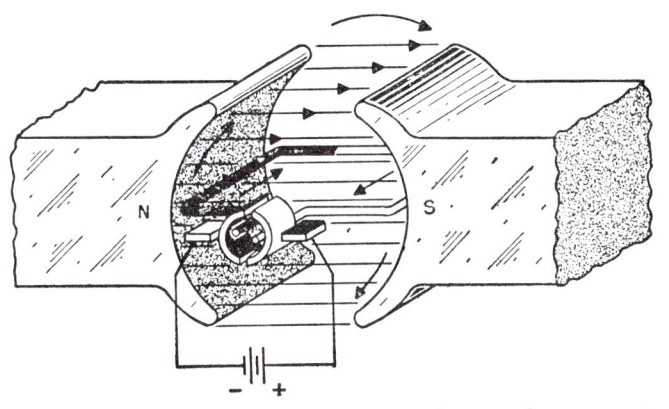

5-8. Note the position of the armature loop and commutator.

5-9. Loop has moved 90°. Note commutator location.

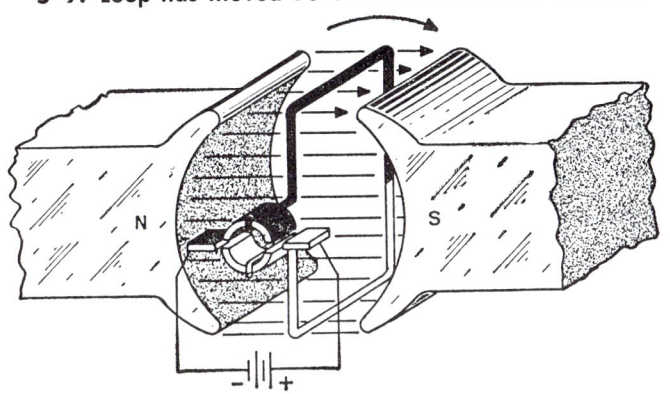

109

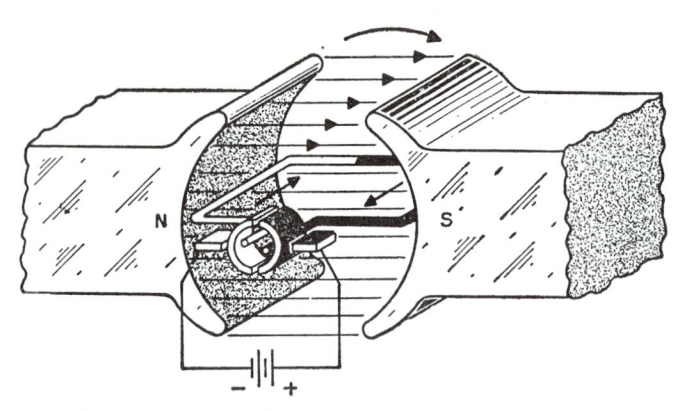

REFRIGERATION AND AIR CONDITIONING TECHNOLOGY

5-10. Loop has made a 180° rotation. Note the commutator segments.

BASIC DC MOTOR THEORY

Figures 5-8 through 5-10 show the continuous rotation of the armature of a DC motor.

MULTILOOP MOTORS

Torque developed by a single-loop motor is too small for practical purposes. It is referred to only in explaining the theory of operation of an electric motor. Each time the loop passes through a neutral plane, the torque is zero, and a jerk is caused in the rotation of the armature. To develop more torque and to eliminate the effect of the zero-torque points on armature rotation, additional loops are added to the armature. Fig. 5-11. Two loops can reduce the zero-torque condition. More loops can reduce the problem even further. They provide a more evenly distributed, smooth rotation of the armature. Both ring- and drum-type armatures are used on practical motors.

SHUNT MOTOR

Structurally a shunt motor is the same as a shunt generator. The field coils of a shunt motor are connected directly across the DC input terminals. When the supply voltage remains constant, the current through the field coils remains constant. As a result, the magnetic field is also constant. The torque of the shunt motor varies only with the current through the armature windings.

Whenever an armature rotates in a magnetic field, an EMF is induced in the armature coils. During rotation, the armature windings cut the lines of force and an EMF is induced in the windings. Therefore, there is motor and generator action at the same time. The direction of the induced EMF is opposite the applied EMF. For this reason, it is known as *counter electromotive force,* or CEMF.

In a shunt motor, the amount of current that flows through the armature depends upon the resistance between the applied voltage and induced voltage (CEMF). The CEMF, in turn, depends solely upon the armature speed because the armature coils are fixed to the armature core and the field does not vary. Therefore, the following formula is used for motor calculations:

$$I_{Armature} = \frac{E_{Applied} - CEMF}{R_{Armature}}$$

Where $R_{Armature}$ is the armature resistance, $I_{Armature}$ is the current through the armature, and $E_{Applied}$ is the voltage applied to the motor.

For example, suppose that a motor develops a CEMF of 100 volts at 1000 rpm, the armature resistance is 1 ohm, and the applied voltage is 105 volts. Find the armature current at the speed of 1000 rpm.

$$I_{Armature} = \frac{105 - 100}{1}$$

$$= 5 \text{ amperes at 1000 rpm}$$

STARTING AN ELECTRIC MOTOR

The CEMF developed is proportional to the speed of the armature. Consequently, no CEMF is developed when the armature is not turning, even though a voltage is applied to the input terminals. Under such a condition, the armature current

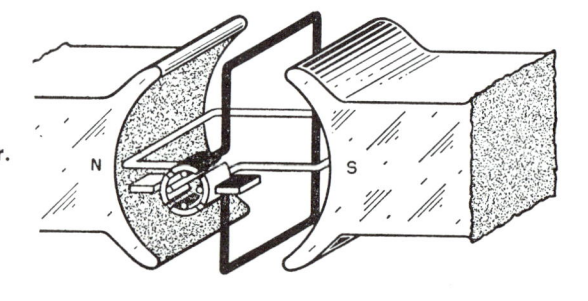

5-11. Two-loop motor.

is equal to the applied voltage divided by the ohmic resistance. For example, if the applied voltage were 100 volts and the ohmic resistance were 1 ohm, the armature current would be 100 divided by 1, or 100 amperes. Such a high current could burn out the armature windings. Also, if the field windings were not excited, no magnetic field would be produced by the poles. This means no flux lines would be cut when the armature rotates. The current again would be limited only by the value of applied voltage and the ohmic resistance of the armature windings. The current would be excessive since no CEMF would be produced.

To prevent such excessive current from flowing in the armature windings, a resistance is placed in series with the windings. The resistance is in the circuit while the armature speed is building to its proper value. This resistance is usually inserted by means of a starting box. In addition to protecting the motor as it starts, the starting box also has provisions for breaking the circuit. It will open the circuit if the field should open or if the power supply should fail.

Motor Starters

The starting box shown in Fig. 5-12 has the starting lever in the *off* position. To start the motor, the operator first moves the lever to the first contact of Coil C, Fig. 5-12. This action closes the circuit to the field coil through the electromagnet, H. It also closes the circuit to the armature windings through high-resistance coils in the starter. As a result, maximum current flows through the field coil, and the magnetic field produced by the poles has maximum strength. Yet, the current through the armature is limited by the series resistance.

As the armature gradually builds up speed, the operator moves the lever successively to other contacts in order to reduce the value of the series resistance.

Finally, the operator moves the lever to the extreme right position. It is held in this position by the electromagnet. At the extreme right position, the series resistance is removed from the circuit. If the power fails, or if the field coils open for any reason, the electromagnet becomes de-energized. That means the lever is returned automatically to the *off* position by the spring, P.

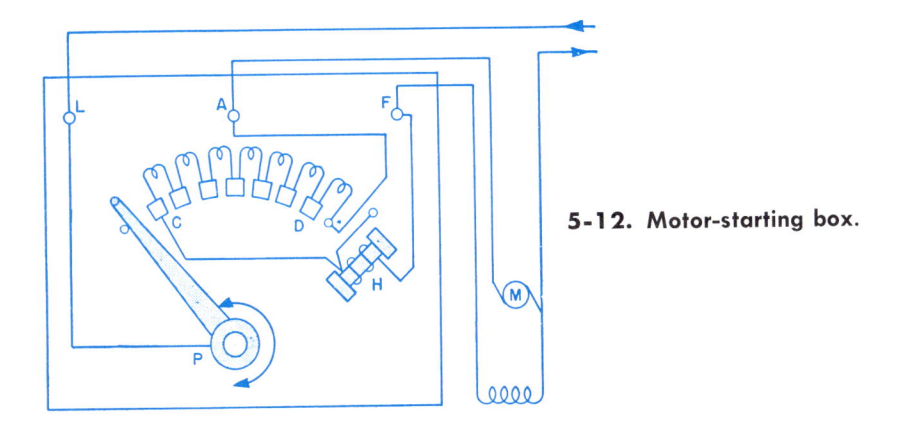

5-12. Motor-starting box.

ARMATURE SPEED

The speed of a shunt motor is fairly constant under conditions of changing load. As a load is applied, the speed of the armature decreases. This decreases the CEMF and increases the current. The increase in current boosts the coupling between field and armature and increases the torque. This causes the motor to resume approximate running speed.

SERIES RHEOSTAT

The speed of a shunt motor may be controlled by a rheostat in series with the armature windings. A rheostat in series with the field winding can also be used. The rheostat in series with the field winding is the most commonly used method.

Adding resistance in series with the armature reduces the armature current and decreases the torque. This causes the motor to slow down until the current increases sufficiently to give the desired torque. When resistance is added in series, it also reduces the field strength, since the current through the windings is reduced. This, in turn, reduces the CEMF. The reduced CEMF makes the armature draw more current. This process increases the speed of the motor, due to the increased torque.

The increasing CEMF produced by the increasing speed causes a current reduction sufficient to produce the torque required for the load.

A rheostat in series with the field winding is preferable to a rheostat in series with the armature. This is because the field current is much lower than the armature current. The loss in the field rheostat is much less than

REFRIGERATION AND AIR CONDITIONING TECHNOLOGY

the power loss in the armature rheostat.

ARMATURE REACTION

The current flow through a motor armature is opposite to the current flow through a generator armature. This armature reaction tends to shift the neutral plane back to the vertical plane, rather than forward as in the case of generators. Therefore, the brushes of a shunt motor are set back of the neutral plane, instead of forward.

A shunt motor will operate only on direct current. An AC voltage applied accidentally to a DC shunt motor causes unpredictable motor operation.

SERIES MOTOR

A series motor is structurally the same as a series generator.

The series motor is adapted for giving a very high starting torque. Actually, the torque of this motor varies approximately as the square of the current. Remember, in a series motor current flows through its series-connected armature and field coils. If the armature current is doubled, the flux is also doubled. Hence the torque, which is proportional to the current times the flux, is increased many times.

When an armature is at rest, the armature current (and therefore the torque) is at maximum. This is because no CEMF is generated in the coils. The current is limited only by the applied voltage and the ohmic resistance of the armature and field coils. As the armature gains speed, the CEMF increases. This decreases the armature current and torque. If additional load is applied to the motor, the armature slows down. This decreases the generated CEMF. Thus, a greater current flows, and a greater torque is produced. The speed of the motor is controlled by the load. If the load is removed, the motor will race. It will gain speed until centrifugal force causes the armature to disintegrate.

Series motors are generally used only where the load is constantly applied, and a good starting torque is required. An example of this is the automobile engine starter where high torque is needed for cranking the engine. Hoists, street cars, and other devices use the series motor.

One of the advantages of the series motor is its capability of operating on either DC or AC. On DC the brushes are set back of the neutral plane to compensate for armature action. For AC operation, both the field and the armature change polarity at the same time. The brushes are set in the vertical or neutral plane. In AC motors, the field core must be laminated to prevent eddy current losses. The theory of operation of AC will be explained later.

Series-type motors are used on electric drills and other small power tools.

Series motors are generally used where the load is constantly applied. A good starting torque is available. A starting box is generally used with large series motors to limit current under starting conditions. This starter has a rheostat that can be connected in series with the motor windings. All the resistance is inserted in the circuit when the motor is being started. The value of resistance is reduced gradually as the speed of the motor increases.

COMPOUND MOTOR

A compound motor differs from the stabilized shunt type. It has a more predominant series field. Like compound generators, compound motors may be divided into two classes—cumulative and differential. The classification depends upon the connection of the series field in relation to the shunt field.

Cumulative Compound Motor

The cumulative compound motor as shown in Fig. 5-13 is connected so that its series and shunt fields *aid* each other. A motor thus connected will have a very strong starting torque, but poor speed regulation. Motors of

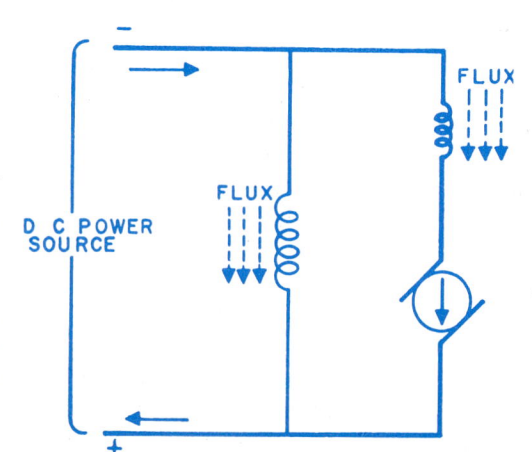

5-13. Connection of a cumulative compound motor.

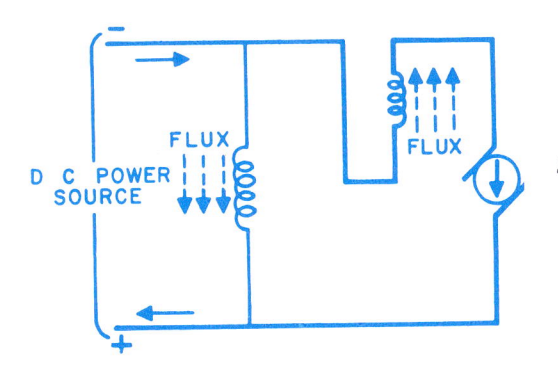

CHAPTER 5—ELECTRIC MOTORS AND CONTROLS

5-14. Connection of a differential compound motor.

Differential Compound Motor

A diagram of a differential compound motor is shown in Fig. 5-14. In this type, the series field *opposes* the connected shunt field. Therefore, this motor operates at a practically constant speed. As the load increases, the armature current increases to provide more torque. The series magnetomotive force increases, thus weakening the shunt field. This reduces the counterelectromotive (CEMF) force in the armature, without causing a reduction in speed.

The operating characteristics of series, shunt, and compound motors are shown in Fig. 5-15.

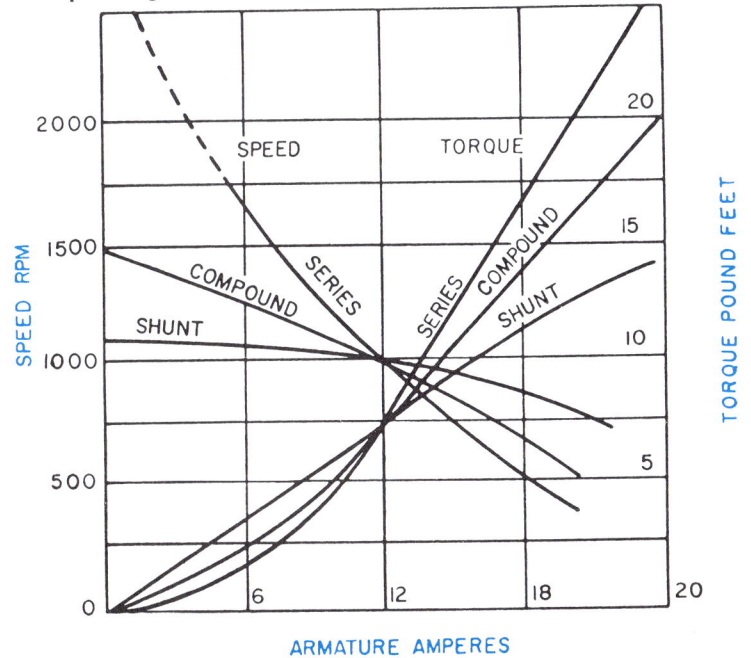

DC MOTOR APPLICATIONS

Direct current motors have an advantage when it comes to reversing. To reverse a DC motor, simply change the polarity of the input power. The DC polarity can be changed with a switch or a relay. Usually, it is done with a relay. In the case of an elevator car that moves up and down, the DC motor is capable of turning first in one direction and then in the other.

Note that the cutaway drawing of the DC motor, Fig. 5-16, shows brushes, commutator segments, and a wound rotor. The unit sticking up on the right end of the motor is a brake.

Heavy-duty DC motors are still used in many areas of the country. An understanding of their workings and the interaction of the magnetic fields serves as a starter for the look at AC motors.

AC MOTORS

The induction motor is the most widely used of several available AC motors. It has a simple design and rugged construction. Fig. 5-17. The induction motor is particularly well adapted for constant speed applications. Because it does not use a commutator, most of the troubles encountered in the operation of DC motors are eliminated. An induction motor can be either a single-phase or a polyphase machine. The operating principle, the same in either case, depends on a revolving, or rotating, magnetic field to produce torque. The key to understanding the induction motor is a thorough comprehension of the rotating magnetic field.

Rotating Magnetic Field

Consider the field structure of A in Fig. 5-18. The poles have

5-15. Operating characteristics of series, shunt, and compound motors.

113

REFRIGERATION AND AIR CONDITIONING TECHNOLOGY

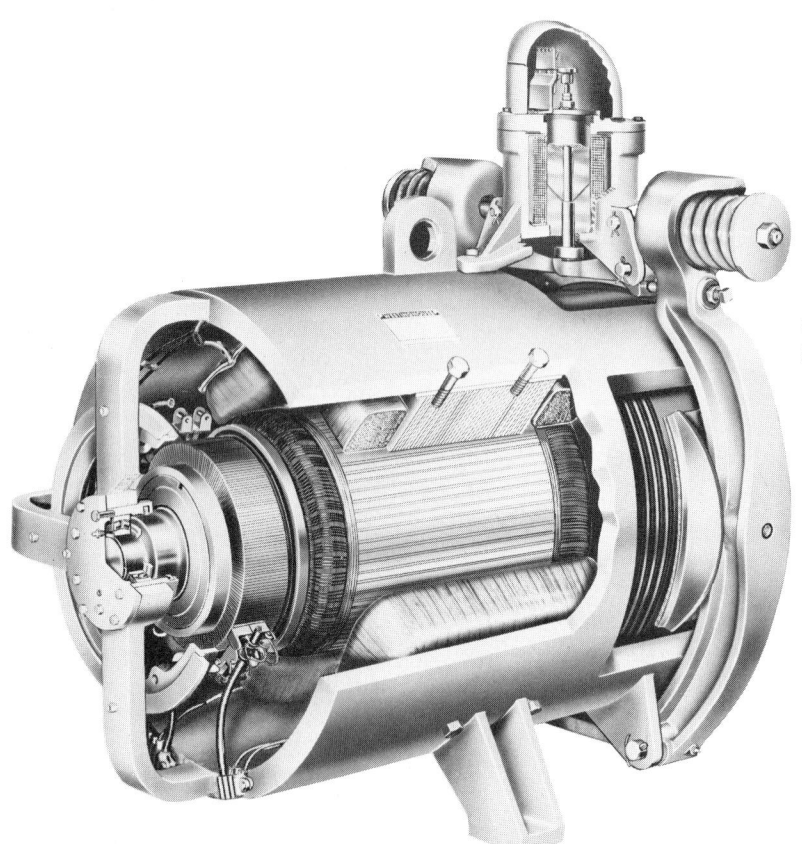

Westinghouse
5-16. Cutaway view of a DC motor used for elevator operation.

Westinghouse
5-17. An assembler attaches the inner bearing cap and ball bearings on a rotor shaft for a 250-horsepower AC motor adapted for high-torque requirements.

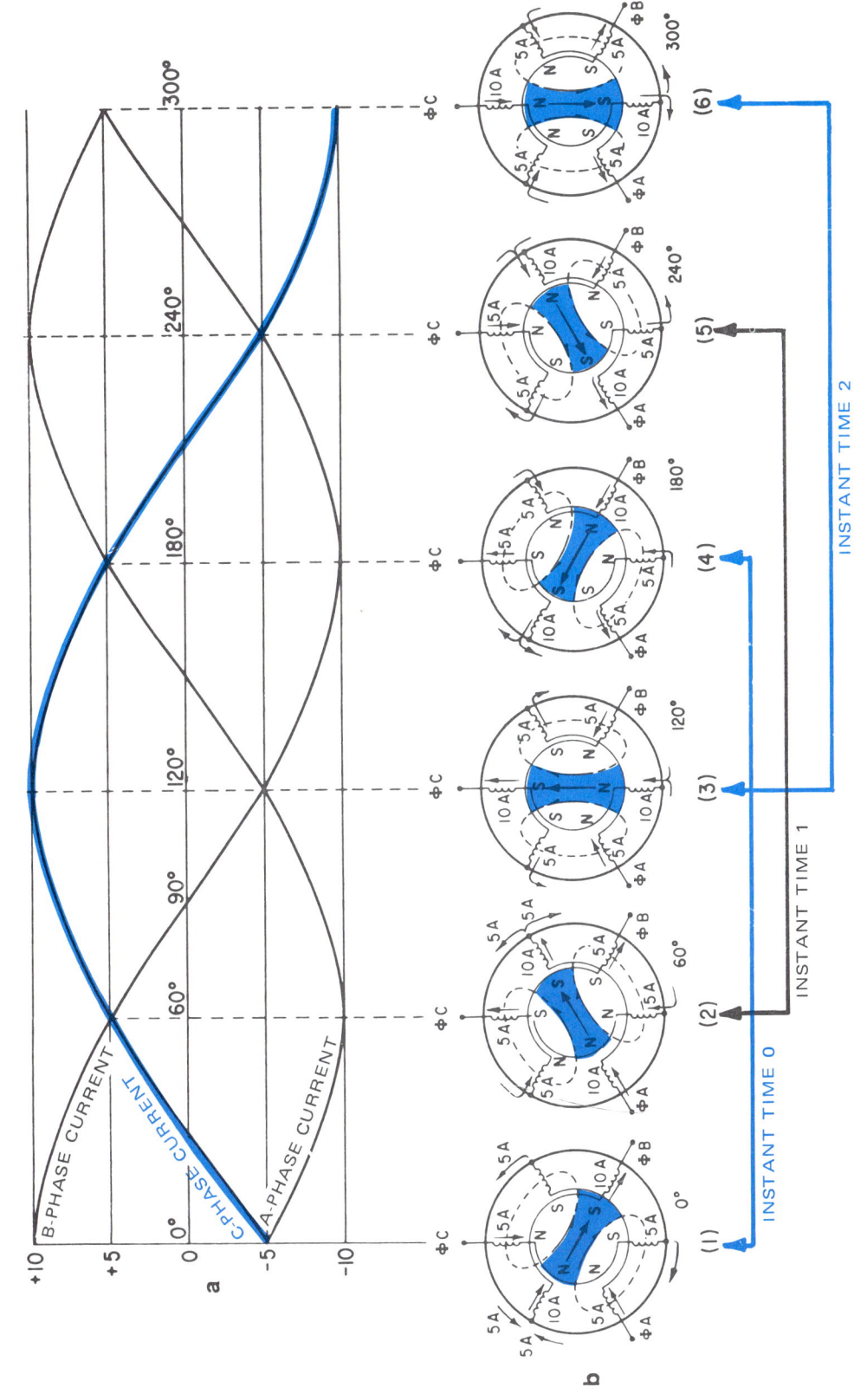

5-18. A rotating field developed by application of 3-phase voltages.

REFRIGERATION AND AIR CONDITIONING TECHNOLOGY

windings that are energized by three AC voltages. Each phase can be represented by a letter: ØA, ØB, and ØC. These voltages have equal magnitude but differ in phase, as shown in B of Fig. 5-18.

At the instant of time shown as Ø, the resultant magnetic field produced by the application of the three voltages has its greatest intensity in a direction extending from pole (1) to pole (4). Under this condition, pole (1) can be considered as a north pole, and pole (4) as a south pole.

At the instant of time shown as 1, the resultant magnetic field will have its greatest intensity in the direction extending from pole (2) to pole (5). In this case, pole (2) can be considered as a north pole and pole (5) as a south pole. See resultant poles shown inside pole (2) and pole (3) on B in Fig. 5-18. The arrowhead shows resultant poles.

Thus, between instant 0 and 1, the magnetic field has rotated clockwise. The magnet shown in A of Fig. 5-18 rotates counterclockwise. However, the magnetic field that caused it rotated clockwise.

At time 2, the resultant magnetic field has its greatest intensity in the direction from pole (3) to pole (6). It is apparent that the resultant magnetic field has continued to rotate clockwise.

At instant 3, poles (4) and (1) can be considered as north and south poles, respectively, and the field has rotated still further.

At later instants of time, the resultant magnetic field rotates to other positions while traveling in a clockwise direction. A single revolution of the field occurs in one cycle. If the exciting voltages have a frequency of 60 hertz, the magnetic field makes 60 revolutions per second, or 3600 revolutions per minute. This speed is known as the synchronous speed of the rotating field.

Construction of an Induction Motor

In an induction motor, the stationary portion of the machine is called a *stator*. The rotating member is called a *rotor*. Instead of salient poles in the stator, as shown in A of Fig. 5-18, distributed windings are used. These are placed in slots around the periphery of the stator. Fig. 5-19.

It is not usually possible to determine the number of poles from visual inspection of an induction motor. A look at the nameplate will usually tell the number of poles. It also gives the rpm, the voltage required, and the current needed. This rated speed is usually less than the synchronous speed because of the *slip*. Slip is due to the inability of the rotor to keep up with the rotating field. To determine the number of poles per phase of the motor, divide 120 times the frequency by the rated speed:

$$P = \frac{120 \times f}{N}$$

P—the number of poles per phase
f—frequency in hertz (Hz)
N—the rated speed in rpm
120 is a constant.

The result is very nearly the number of poles per phase. For example, consider a 60-hertz, 3-phase machine rated with a speed of 1750 rpm. In this case:

$$P = \frac{120 \times 60}{1750} = \frac{7200}{1750} = 4.1$$

Therefore the motor has four

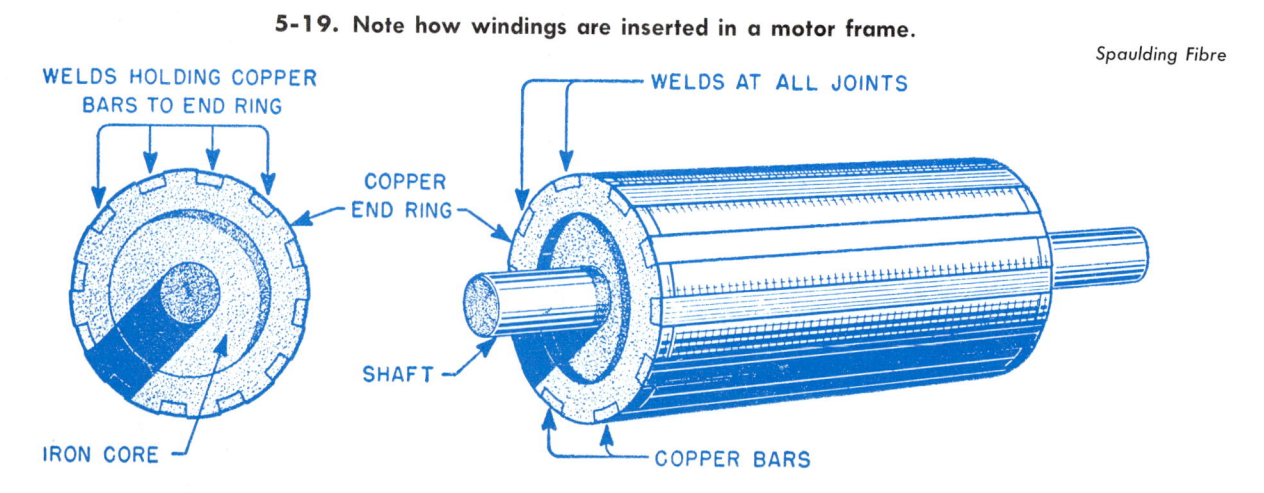

5-19. Note how windings are inserted in a motor frame.

Spaulding Fibre

CHAPTER 5—ELECTRIC MOTORS AND CONTROLS

5-20. Shaded-pole motor.

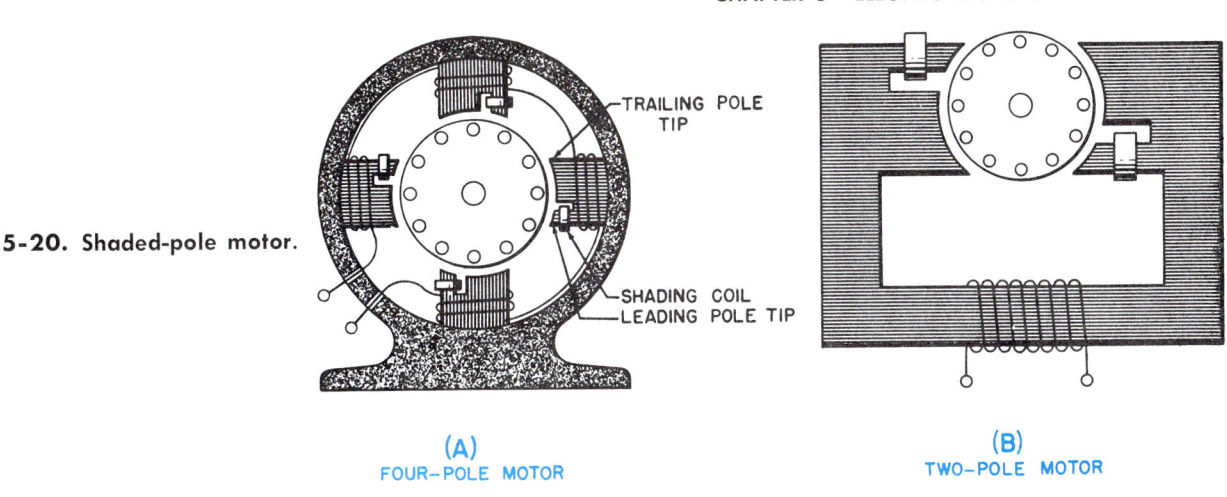

(A) FOUR-POLE MOTOR

(B) TWO-POLE MOTOR

poles per phase. If the number of poles per phase is given on the nameplate, the synchronous speed can be determined. Divide 120 times the frequency by the number of poles per phase. In the example just given, the synchronous speed is equal to 7200 divided by 4, or 1800 rpm.

The rotor of an induction motor consists of an iron core with longitudinal slots around its circumference, in which heavy copper or aluminum bars are embedded in the slots. These bars are welded to a heavy ring of high conductivity on either end. This composite structure is sometimes called a *squirrel cage*. Motors containing such a rotor are called squirrel-cage induction motors. Fig. 5-19.

Single-Phase Motors

The field of a single-phase motor, instead of rotating, merely pulsates. No rotation of the rotor takes place. A single-phase pulsating field may be visualized as two rotating fields revolving at the same speed, but in opposite directions. It follows, therefore, that the rotor will revolve in either direction at nearly synchronous speed—if it is given an initial impetus in either one direction or the other. The exact value of this initial rotational velocity varies widely with different machines. A velocity higher than 15% of the synchronous speed is usually sufficient to cause the rotor to accelerate to the rated or running speed. A single-phase motor can be made self-starting if means can be provided to give the effect of a rotating field.

Shaded-Pole Motor

The shaded-pole motor resulted from one of the first efforts to make a self-starting single-phase motor. Fig. 5-20. This motor has salient poles. A portion of each pole is encircled by a heavy copper ring. The presence of the ring causes the magnetic field through the ringed portion of the pole face to lag behind that through the other portion of the pole face. Fig. 5-20. The effect is the production of a slight component of rotation of the field. That slight component of rotation is sufficient to cause the rotor to revolve. As the rotor accelerates, the torque increases until the rated speed is obtained. Such motors have low starting torque. Their greatest use is in small fans where the initial torque is low. They are also used in clocks, inexpensive record players, and some electric typewriters. Fig. 5-21.

Split-Phase Motor

Many types of split-phase motors have been made. Such motors have a start winding that is displaced 90 electrical degrees from the main or run winding. In some types, the start winding has a fairly high resistance. This causes the current in it to be out of phase with the current in the run winding. This condition produces, in effect, a rotating field and the rotor revolves. A centrif-

5-21. Shaded-pole motors used for fans and clocks.

REFRIGERATION AND AIR CONDITIONING TECHNOLOGY

ugal switch is used to disconnect the start winding automatically after the rotor has attained approximately 75% of its rated speed. Fig. 5-22.

Split-phase motors are used where there is no need to start under load. They are used on grinders, buffers, and other similar devices. They are available in fractional horsepower sizes with various speeds, and are wound to operate on 120 volts AC or 240 volts AC.

Capacitor-Start Motor

With the development of high-quality and high-capacity electrolytic capacitors, a variation of the split-phase motor known as the capacitor-start motor, has been made. Almost all fractional-horsepower motors in use today on refrigerators, oil burners, washing machines, table saws, drill presses, and similar devices are capacitor-start. A capacitor motor has a high starting current and the ability to develop about four times its rated horsepower if it is suddenly overloaded. In this adaptation of the split-phase motor, the start winding and the run winding have the same size and resistance value. The phase shift between currents of the two windings is obtained by means of capacitors connected in series with the start winding. Capacitor-start motors have a starting torque comparable to their torque at rated speed and can be used in places where the initial load is heavy. A centrifugal switch is required for disconnecting the start winding when the rotor speed is up to about 25% of the rated speed. Figure 5-23 shows a disassembled capacitor motor. Note in Fig. 5-24, also a capacitor-start motor, the centrifugal-switch arrangement with the governor mechanism. Figure 5-25 shows the windings, the rotor, and the capacitor housing on top of the motor. Note that the windings overlap.

One of the advantages of the single-value capacitor-start motor is its ability to be reversed easily and frequently. Figs. 5-26 and 5-27. The motor is quiet and smooth running. If a 5- to 20-hp capacitor-start motor is called for, the two-value capacitor motor is used. Fig. 5-28. This motor has two sets of field windings in the stator—an auxiliary winding, called a phase winding, and the main winding. The phase winding is designed for continuous duty; a capacitor remains in series with the winding at all times. A start capacitor is added to the phase current to increase starting torque. However, it is disconnected by a centrifugal switch during acceleration. This

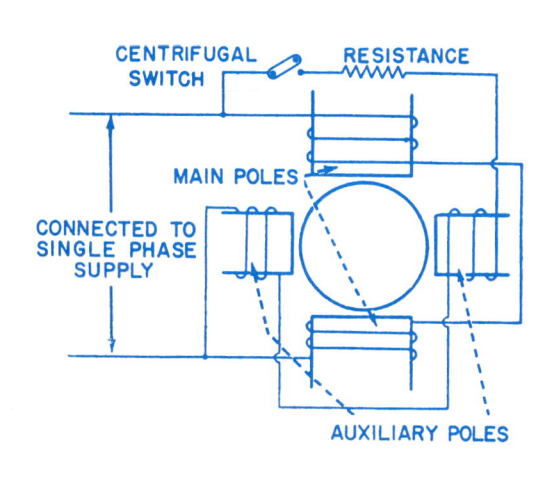

5-22. Split-phase motor.

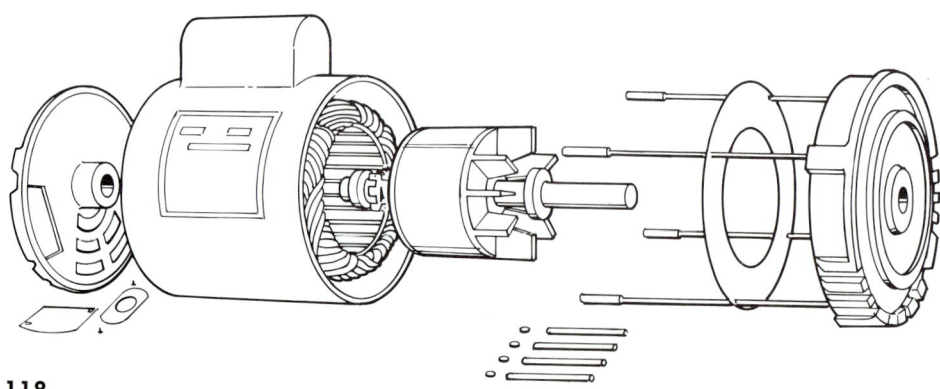

5-23. Disassembled single-phase, capacitor-start motor.

118

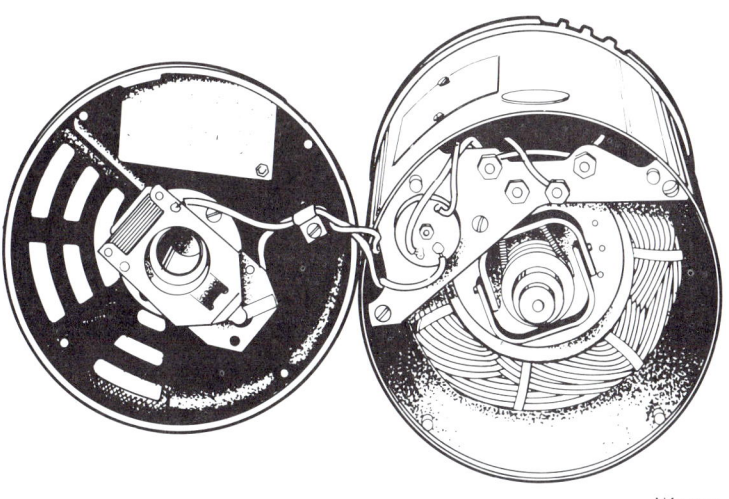

5-24. Single-phase starting switch and governor mechanism.

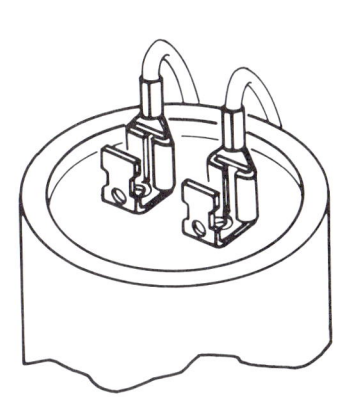

QUICK—CONNECT

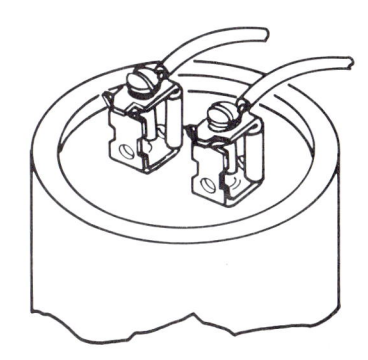

SCREW

SOLDER

5-26. Electrolytic capacitor with three methods of connection.

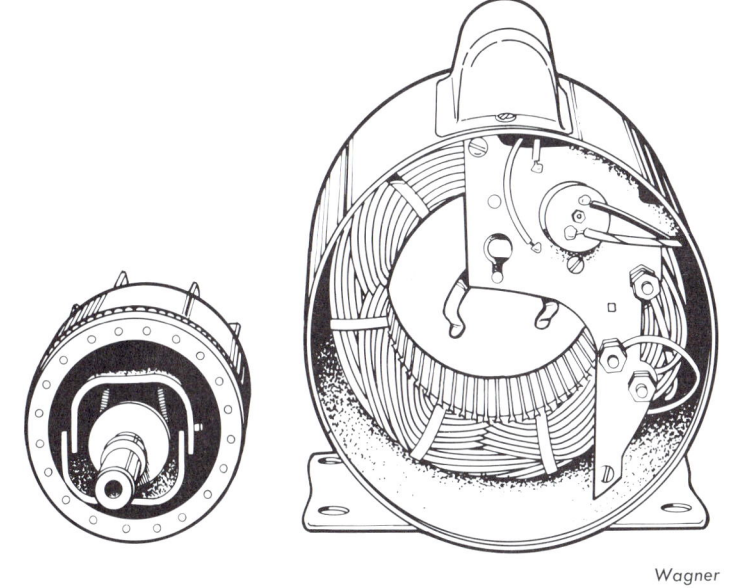

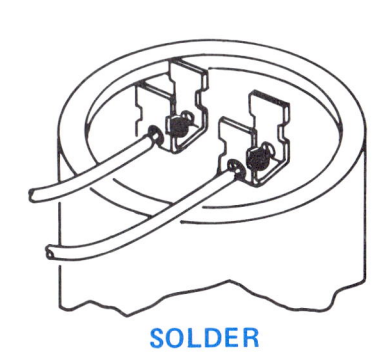

5-25. Single-phase stator and rotor.

5-27. Capacitor-start diagram.

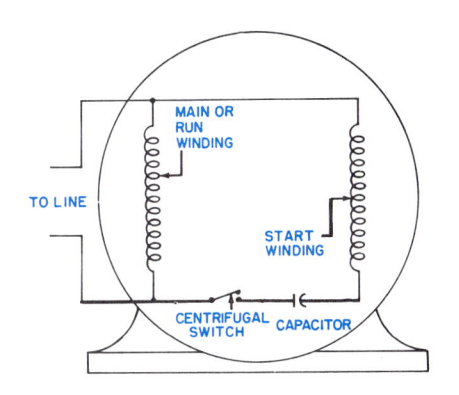

type of motor is used in many air conditioning applications where the unit is 2 to 4 tons.

In general, the single-phase motor is more expensive to purchase and to maintain than the

119

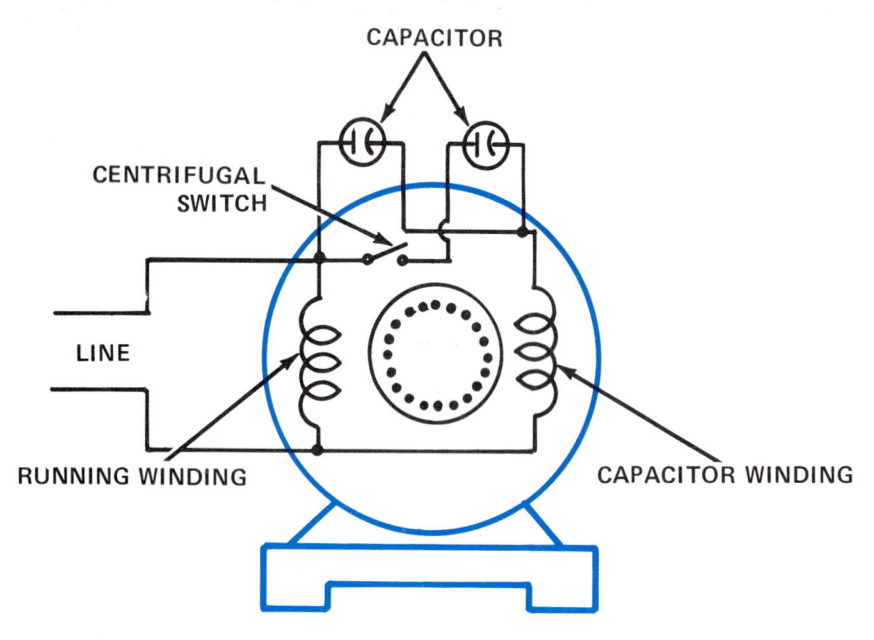

REFRIGERATION AND AIR CONDITIONING TECHNOLOGY

TWO-VALUE CAPACITOR MOTOR
5-28. Two-valve capacitor-start motor.

three-phase motor. It is less efficient, and its starting currents are relatively high. All run at essentially constant speed. Nonetheless, most machines using electric motors around the home, on the farm, or in small commercial plants are equipped with single-phase motors.

Those who select a single-phase motor usually do so because three-phase power is not available to them. Figure 5-29 shows the simple methods used in the construction of a three-phase motor. Note this is a half-etched, squirrel-cage rotor. The bearings are *not* sealed ball bearings. They are a sleeve-type with the oil caps placed so that oil may be added occasionally to keep the bearings lubricated.

Figure 5-30 is a cutaway view of a three-phase motor. Note the simple rotor and fan blades. The windings and the sealed ball bearings make it simple for maintenance. This is an almost maintenance-free motor. Figure 5-31 shows a polyphase motor that has been made explosion-proof.

Sizes of Motors

Some single-phase induction motors are rated as high as 2 hp. The major field of use is 1 hp, or less, at 120 or 240 volts for the smaller sizes. For larger power ratings, polyphase (two-phase, three-phase, etc.) are generally specified, since they have excellent starting torque and are practically maintenance free.

Figure 5-32 is a brush-lifting, repulsion-start, induction-run, single-phase motor. The following should be noted about this type of motor: The rotor is wound (just like that in a DC motor). The brushes can be lifted by centrifugal force once the rotor comes up to speed. This means the rotor can then act as a squirrel-cage type. This type pulls a lot of current in starting, but is capable of starting under full load conditions.

Cooling and Mounting Motors

Figure 5-33 shows an improved motor ventilating system. A large volume of air is directed through the motor to reduce temperatures. The large blower on the right is located behind a baffle that controls air movement to the blower blades. The blower draws outside air through the

5-29. Cutaway view of a three-phase motor with a half-etched, squirrel-cage rotor.
Wagner

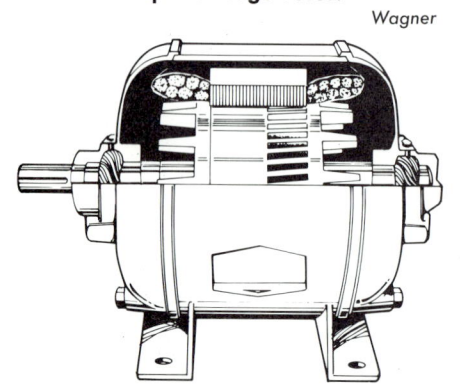

5-30. Cutaway view of a three-phase motor showing the cast rotor.
Wagner

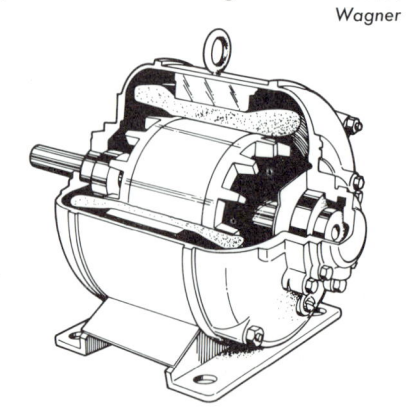

120

CHAPTER 5—ELECTRIC MOTORS AND CONTROLS

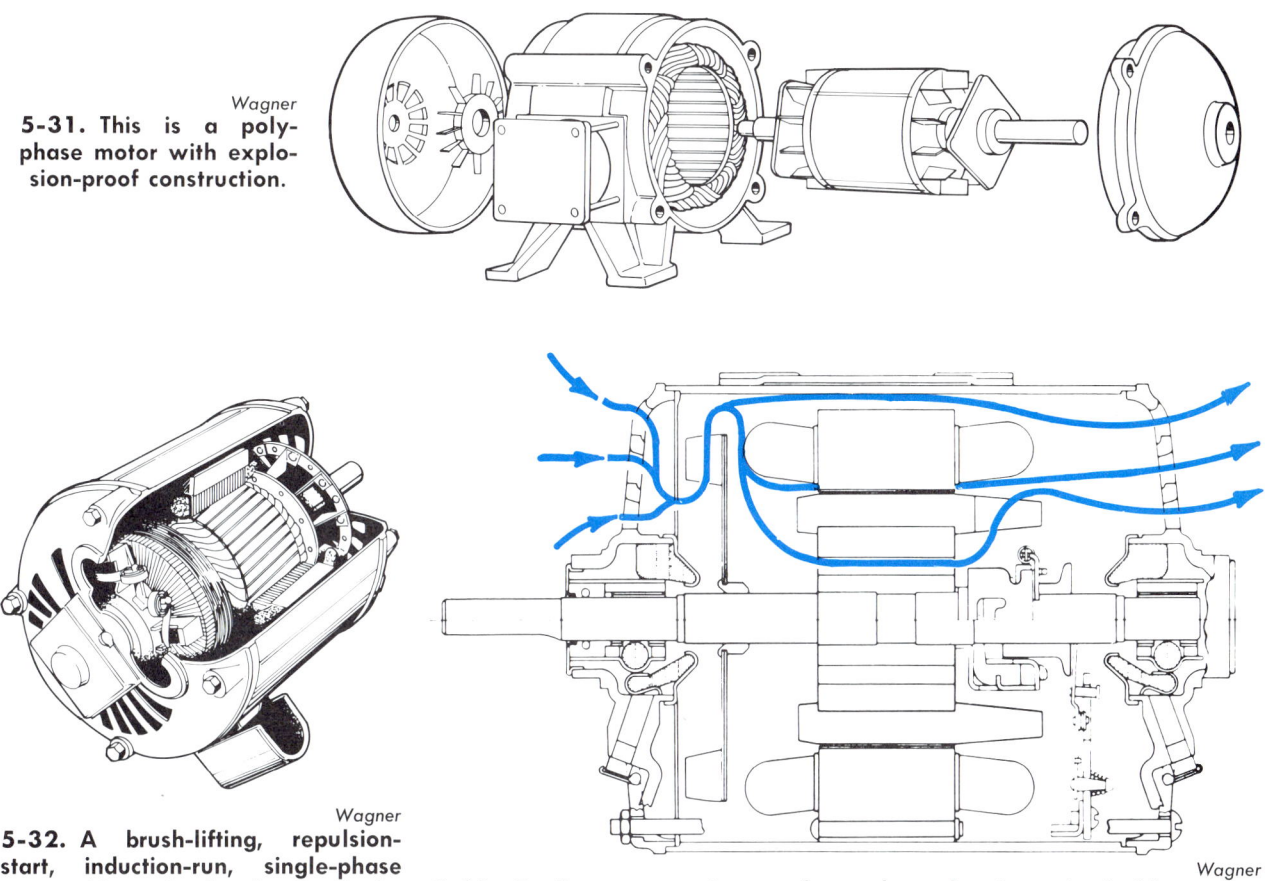

5-31. This is a polyphase motor with explosion-proof construction.

5-32. A brush-lifting, repulsion-start, induction-run, single-phase motor. Note the brushes and the wound rotor.

5-33. Cooling system using two fans to keep the air moving inside an electric motor.

large drip-proof openings in the back end plate. It then forces the cooling air around the back coil extension, through the rotor vent holes, the air gap, and through the passages between the stator core and the frame. A second blower on the front end of the rotor at left, cast as an integral part of the rotor, circulates the air around the inside of the front coil extensions and then speeds the flow of heated air out the motor through the drip-proof openings in the front end-plate.

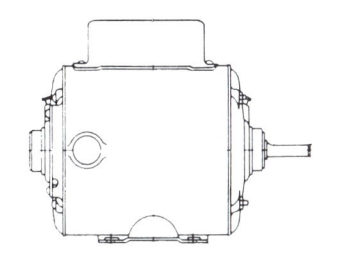

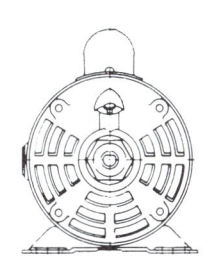

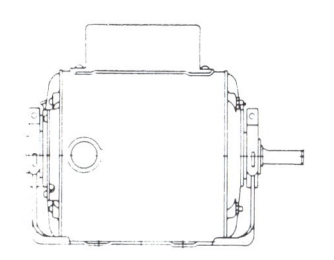

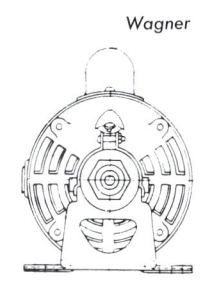

5-34. Rigid-base and resilient-base mountings for electric motors.

RIGID BASE

RESILIENT BASE

121

Figure 5-34 shows the rigid base and the resilient base. Note that the resilient base has a mounting bracket attached to the ends of the rotor with some material used to make it more silent. However, the rigid base has its support mechanism welded to the frame of the motor. If the support mechanism is welded, it can transmit the noise of the running motor to whatever it is attached to in operation.

Direction of Rotation

The direction of rotation of a three-phase induction motor can be changed simply by reversing two of the leads to the motor. The same effect can be obtained in a two-phase motor by reversing connections to one phase. In a single-phase motor, reversing connections to the start winding will reverse the direction of rotation. Most single-phase motors designed for general use have provisions for readily reversing connections to the start winding. Nothing can be done to a shaded-pole motor to reverse the direction of rotation. The direction of rotation is determined by the physical location of the copper shading ring on the shaded-pole.

If, after starting, one connection to a three-phase motor is broken, the motor will continue to run but will deliver only one-third of the rated power. Also, a two-phase motor will run at one-half its rated power if one phase is disconnected. Neither motor will start under these conditions. They can be started by hand in either direction—manually. Once started by hand, they do run. Incidentally, about the only place that a two-phase motor will be found is in Europe, where some two-phase power is distributed for local use. In the United States only single-phase power is available to residential customers. Three-phase is usually available to industry and commercial establishments. Schools usually have three-phase power located within easy connection, if needed. Some parts of the southwestern United States now have three-phase distributed to homes. This is due primarily to the requirements of the air conditioning units they need. The three-phase motor requires fewer service calls and has a long life. Therefore, it is often worth the extra expense to have three-phase power brought in by the power company. It is less expensive if a whole subdivision uses three-phase.

Synchronous Motor

A synchronous motor is one of the principal types of AC motors. Like the induction motor, the synchronous motor is designed to take advantage of a rotating magnetic field. Unlike the induction motor, however, the torque developed does not depend upon the induction of currents in the rotor. Briefly, the principle of operation of the synchronous motor is as follows.

A multiphase source of AC is applied to the stator windings and a rotating magnetic field is produced. A direct current is applied to the rotor windings and another magnetic field is produced. The synchronous motor is so designed and constructed that these two fields react upon each other. They act in such a manner that the rotor is dragged along. It rotates at the *same speed* as the rotating magnetic field produced by the stator windings.

THEORY OF OPERATION

An understanding of the operation of the synchronous motor may be obtained by considering the simple motor shown in Fig. 5-35. Assume that poles A and B are being rotated clockwise by some mechanical means to produce a rotating magnetic field. The rotating poles induce poles of opposite polarity, as shown in the illustration of the soft iron rotor, and forces of attraction exist between corresponding north and south poles. Consequently, as poles A and B rotate, the rotor is dragged along at the same speed. However, if a load is applied to the rotor shaft, the rotor axis will momentarily fall behind that of the rotating field, but will thereafter continue to rotate with the field at the same speed, as long as the load remains constant. If the load is too large, the rotor will pull out of synchronization with the rotating field. As a result, it will no longer rotate with the field at the same speed. The motor is then said to be overloaded.

SYNCHRONOUS MOTOR ADVANTAGES

Some advantages of the synchronous motor are:

1. When used as a synchronous capacitor, the motor is connected on the AC line in parallel with the other motors on the line. It is run either without load or with a very light load. The rotor field is *overexcited* just enough to produce a *leading* current that offsets the lagging current of the line with the motors operating. A unity power factor (1.00) can usually be achieved. This means the load on the generator is the same as though *only* resistance made up the load.

CHAPTER 5—ELECTRIC MOTORS AND CONTROLS

cage type on the rotor. This induction winding brings the rotor almost into synchronous speed. When the DC is connected to the rotor windings, the rotor pulls into step with the field. The latter method is the more commonly used.

Figure 5-36 shows a small synchronous motor that has a number of applications. Because of their stable speed, synchronous motors are used for turntables in stereo equipment. This type is also used in timing devices.

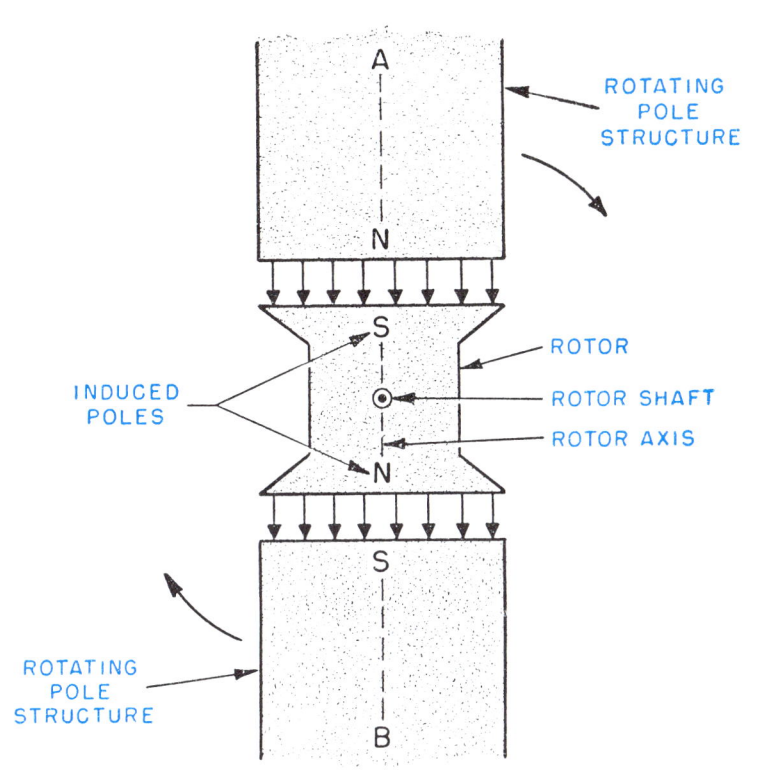

5-35. Simple synchronous motor.

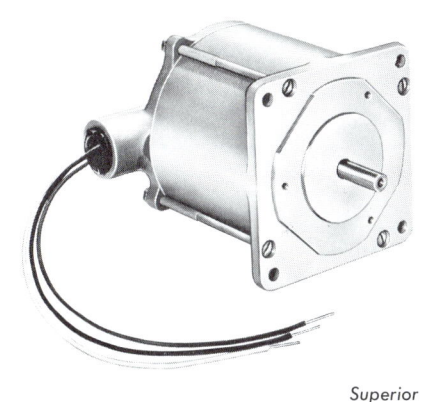

5-36. Synchronous motor. *Superior*

PROPERTIES OF THE SYNCHRONOUS MOTOR

The synchronous motor is not a self-starting motor in most cases. The rotor is heavy. From a dead stop, it is impossible to bring the rotor into magnetic lock with the rotating magnetic field. For this reason, all synchronous motors have a starting device. Such a simple starter is another motor, either AC or DC, which can bring the rotor up to approximately 90% of the synchronous speed. The starting motor is then disconnected and the rotor locks in step with the rotating field.

Another starting method is a second winding of the squirrel-cage type on the rotor. This induction winding brings the rotor almost into synchronous speed. When the DC is connected to the rotor windings, the rotor pulls into step with the field. The latter method is the more commonly used.

2. The synchronous motor can be made to produce as much as 80% leading power factor. However, because a leading power factor on a line is just as detrimental as a lagging power factor, the synchronous motor is regulated to produce just enough leading current to compensate for lagging current in the line.

TROUBLESHOOTING ELECTRIC MOTORS WITH A VOLT-AMMETER

Electrical equipment is designed to operate at a specific voltage and current. Usually the equipment will work satisfactorily if the line voltage differs plus or minus 10% from the actual nameplate rating. In a few cases, however, a 10% voltage drop may result in a breakdown. Such may be the case with an induction motor that is being loaded to its fullest capacity both on start and run. A 10% loss in line voltage will result in a 20% loss in torque.

The full load current rating on the nameplate is an approximate value based on the average unit coming off the manufacturers' production line. The actual current for any one unit may vary as much as plus or minus 10% at rated output. However, a motor whose load current exceeds the rated value by 20% or more will reduce the life of the motor due to higher operating temperatures, and the reason for excessive current should be determined. In many cases it may simply be an overloaded motor. The percentage increase in load will not correspond with percent-

123

age increase in load current. For example, in the case of a single-phase induction motor, a 35% increase in current may correspond to an 80% increase in output torque.

The operating conditions and behavior of electrical equipment can be analyzed only by actual measurement. A comparison of the measured terminal voltage and current will check whether the equipment is operating within electrical specifications.

The measurement of voltage and current requires the use of two basic instruments—a voltmeter and an ammeter. To measure voltage, the test leads of the voltmeter are in contact with the terminals of the line under test. To measure current, the conventional ammeter must be connected in series with the line so that the current will flow through the ammeter.

The insertion of the ammeter means shutting down the equipment, breaking open the line, connecting the ammeter, starting up the equipment, reading the meter and then going through as much work to remove the ammeter from the line. Additional time-consuming work may be involved if the connections at the ammeter have to be shifted to a higher or lower range terminal.

SPLIT-CORE AC VOLT-AMMETER

These disadvantages are practically eliminated by use of the split-core AC volt-ammeter. This instrument combines an AC voltmeter and AC split-core ammeter into a single pocket-size unit with a convenient range switch to select any of the multiple voltage ranges or current ranges. Fig. 5-37. With the split-core ammeter, the line to be tested does not have to be disconnected from its power source.

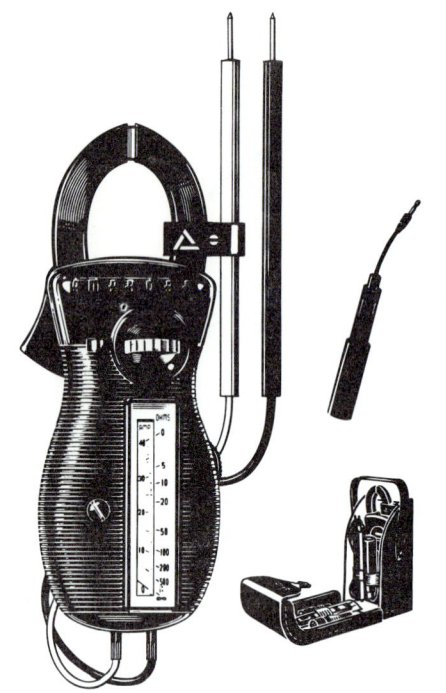

This type of ammeter uses the transformer principle to connect the instrument into the line. Since any conductor carrying alternating current will set up a changing magnetic field around itself, that conductor can be used as the primary winding of the transformer. The split-core ammeter carries the remaining parts of the transformer, which are the laminated steel core and the secondary coil. To get transformer action, the line to be tested is encircled with the split-type core by simply pressing the trigger button. Fig. 5-38. Aside from measuring terminal voltages and load currents, the split-core ammeter-voltmeter can be used to track down electrical difficulties in electric motor repair.

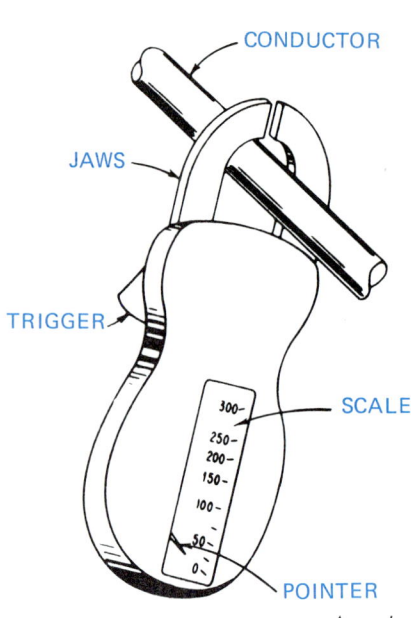

5-37. Clamp-on volt/ampere/ohm-meter with rotary scale.

5-38. The clamp-on volt/ampere/ohmmeter with parts labeled.

Testing for Grounds

To determine whether a winding is grounded or has a very low value of insulation resistance, connect the unit and test leads as shown in Fig. 5-39. Assuming the available line voltage is approximately 120 volts, use the unit's lowest voltage range. If the winding is grounded to the frame, the test will indicate full line voltage.

A high resistance ground is simply a case of low insulation resistance. The indicated reading for a high resistance ground will be a little less than line voltage. A winding that is not grounded will be evidenced by a small or negligible reading. This is due mainly to the capacitive effect between the winding and the steel lamination.

To locate the grounded portion of the winding, disconnect the necessary connection jumpers and test. Grounded sections will be detected by a full line voltage indication.

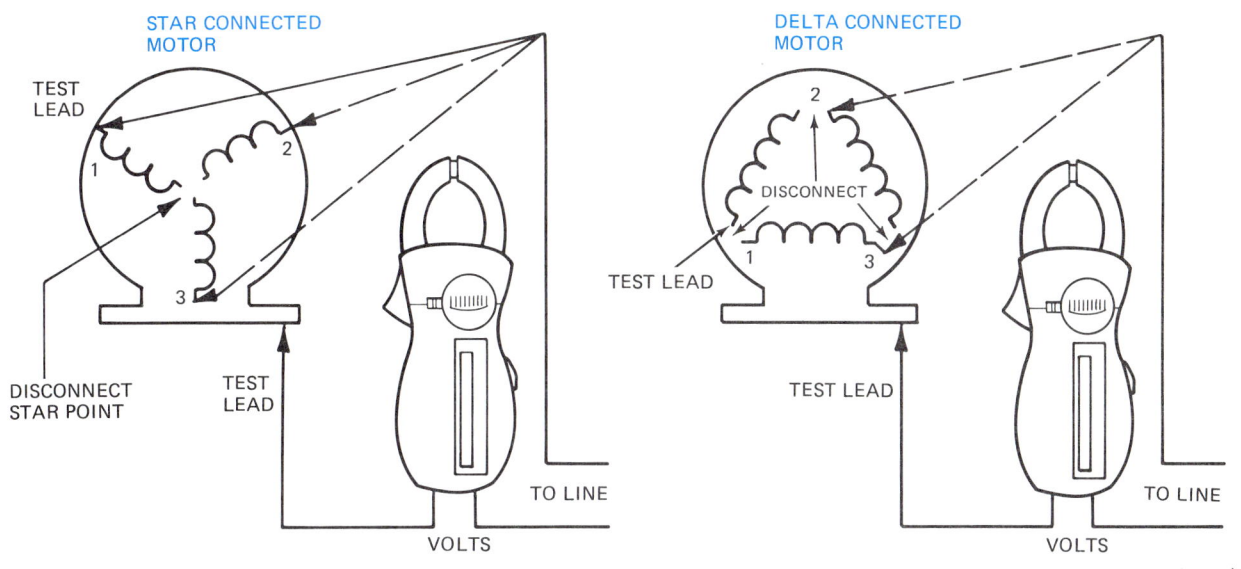

5-39. Find the location of a grounded phase of a motor.

Testing for Opens

To determine whether a winding is open, connect test leads as shown in Figs. 5-40 and 5-41. If the winding is open, there will be no voltage indication. If the circuit is not open, the voltmeter indication will read full line voltage.

Checking for Shorts

Shorted turns in the winding of a motor behave like a shorted secondary of a transformer. A motor with a shorted winding will draw excessive current while running at no load. Measurement of the current can be made without disconnecting lines. This means you engage one of the lines with the split-core transformer of the tester. If the ampere reading is much higher than the full load ampere rating on the nameplate, the motor is probably shorted.

In a two- or three-phase motor, a partially shorted winding produces a higher current reading in the shorted phase. This becomes evident when the current in each phase is measured.

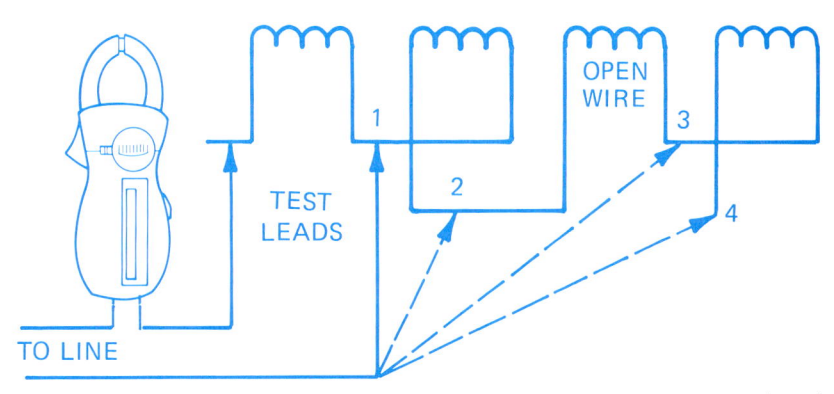

5-40. Isolating an open phase.

Testing Squirrel-Cage Rotors

In some cases, loss in output torque at rated speed in an induction motor may be due to opens in the squirrel-cage rotor. To test the rotor and determine which rotor bars are loose or open, place the rotor in a growler. Engage the split-core ammeter around one of the lines going to the growler winding, as shown in Fig. 5-42. Set the switch to the highest current range. Switch on the growler and then set the test unit to the appropriate current range. Rotate the rotor in the growler and take note of the current indication whenever the growler is energized. The bars and end rings in the rotor behave similarly to a shorted secondary of a transformer. The growler winding acts as the primary. A good rotor will produce approximately the same

125

REFRIGERATION AND AIR CONDITIONING TECHNOLOGY

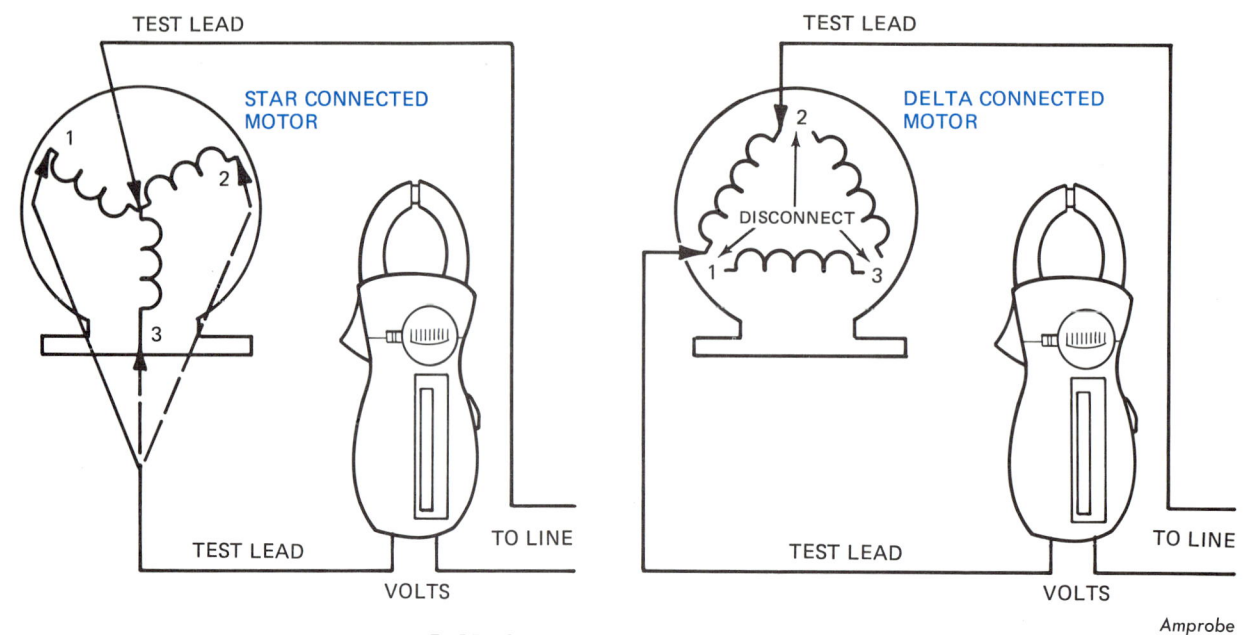

5-41. Locating an open in a motor.

current indications for all positions of the rotor. A defective rotor will exhibit a drop in the current reading when the open bars move into the growler field.

Testing the Centrifugal Switch in a Split-Phase Motor

A faulty centrifugal switch may not disconnect the start-winding at the proper time. To determine conclusively that the start-winding remains in the circuit, place the split-core ammeter around one of the start-winding leads. Set the instrument to the highest current range. Turn on the motor switch. Select the appropriate current range. Observe if there is any current in the start-winding circuit. A current indication signifies that the centrifugal switch did not open when the motor came up to speed. Fig. 5-43.

Test for Short Circuit Between Run- and Start-Windings

A short between run- and start-windings may be determined by using the ammeter and line voltage to check for continuity between the two separate circuits. Disconnect the run- and start-winding leads and connect the instrument as shown in Fig. 5-44. Set the meter on voltage. A full-line voltage reading will be obtained if the windings are shorted to one another.

Test for Capacitors

Defective capacitors are very often the cause of trouble in capacitor-type motors. Shorts, opens, grounds, and insufficient capacity in microfarads are con-

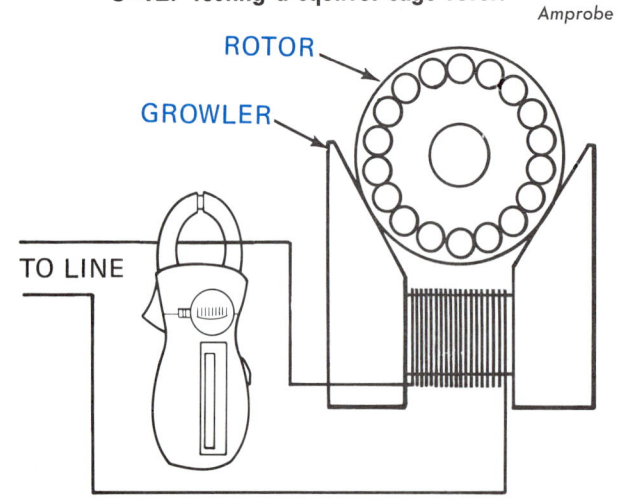

5-42. Testing a squirrel-cage rotor.

126

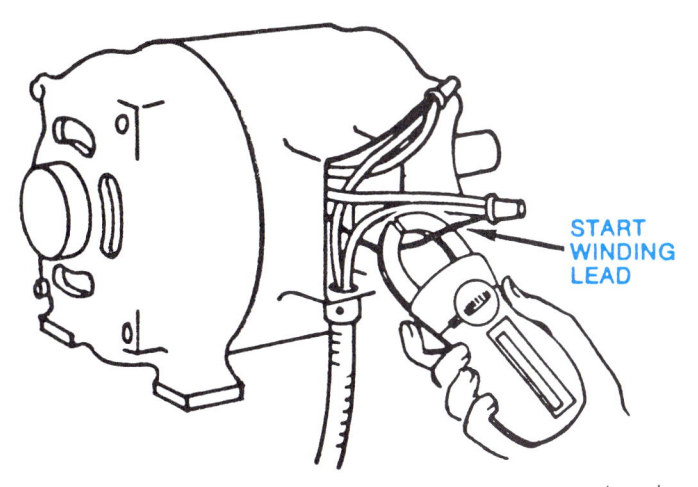

CHAPTER 5—ELECTRIC MOTORS AND CONTROLS

ing that is somewhat below line voltage. A negligible reading or a reading of no voltage will indicate that the capacitor is not grounded.

To measure the capacity of the capacitor, set the test unit's switch to the proper voltage range and read the line voltage indication. Then set to the appropriate current range and read the capacitor current indication. During the test, keep the capacitor on the line for a very short period of time, because motor starting electrolytic capacitors are rated for intermittent duty. Fig. 5-46. The capacity in microfarads is then computed by substituting the voltage and current readings in the following formula, assuming that a full 60-hertz line was used:

$$\text{Microfarads} = \frac{2650 \times \text{amperes}}{\text{volts}}$$

An open capacitor will be evident if there is no current indication in the test. A shorted capacitor is easily detected. It will blow the fuse when the line switch is turned on to measure line voltage.

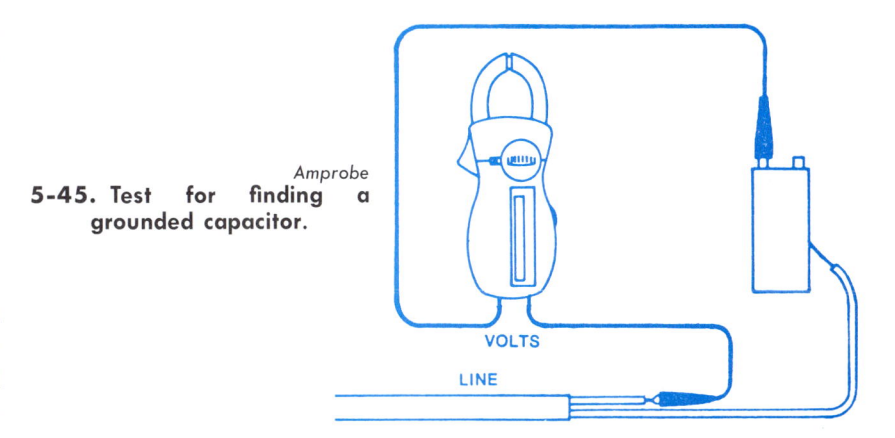

5-43. Testing a centrifugal switch on a motor.

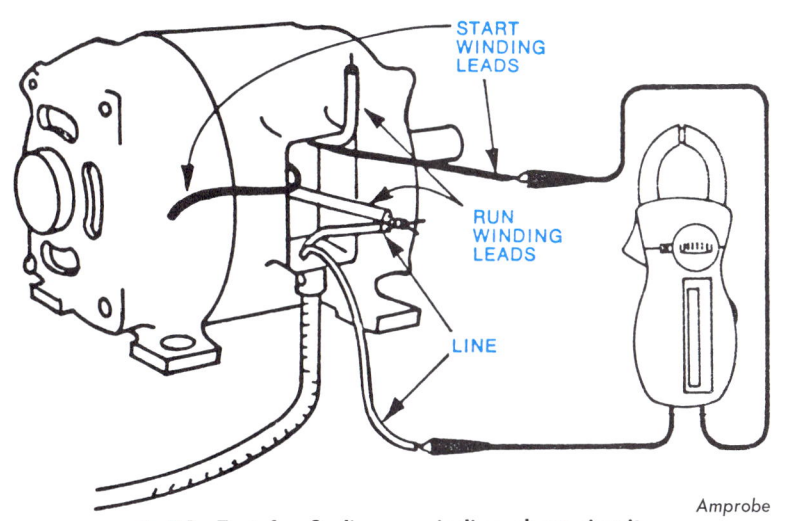

5-44. Test for finding a winding short circuit.

ditions for which capacitors should be tested to determine whether they are good.

To determine a grounded capacitor, set the instrument on the proper voltage range and connect the instrument and capacitor to the line as shown in Fig. 5-45. A full-line voltage indication on the meter signifies that the capacitor is grounded to the can. A high resistance ground will be evident by a voltage read-

5-45. Test for finding a grounded capacitor.

127

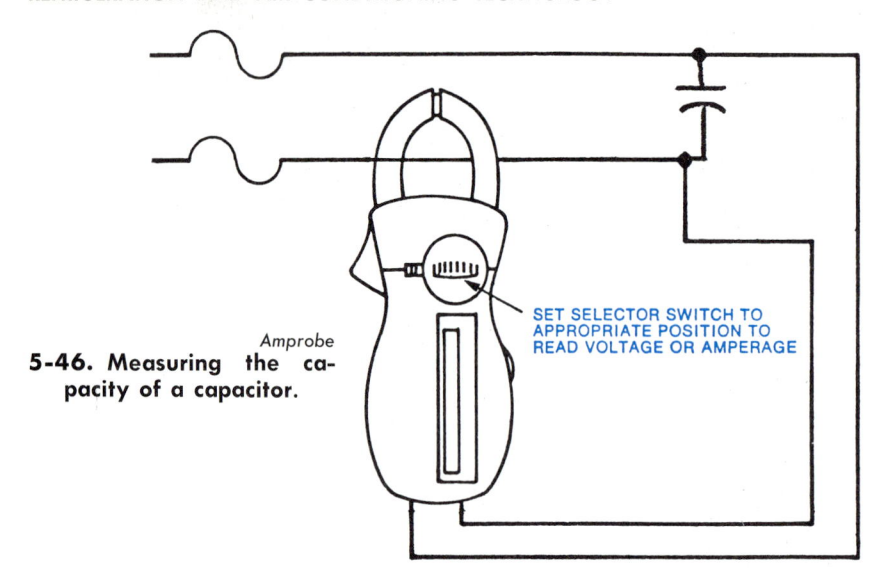

REFRIGERATION AND AIR CONDITIONING TECHNOLOGY

5-46. Measuring the capacity of a capacitor.

proper operation of motors, compressors, and other electrical equipment.

Some meggers use batteries. Others use a crank that turns a small coil of wire in a magnetic field. Turning the crank handle causes the coil of wire to generate an EMF. The EMF is usually of high voltage. Thus, the megger can shock you if you touch the lead ends when the handle is cranked. There is very low current, so there may be little actual damage caused by the electrical energy through your body. Needless to say, read the instructions and follow them closely. *Do not use a megger in an explosive atmosphere.*

Equipment under test with the megohmmeter may build up a capacitive charge from the testing. One model has a "press to read" button. When it is re-

USING THE MEGOHMMETER FOR TROUBLESHOOTING

The megohmmeter (sometimes called a *megger*) is a device that can be used to measure *millions* of ohms. Fig. 5-47. *Meg* means "million." The equipment usually uses high voltage to push a small amount of current through the insulator being measured. The insulation resistance is very important in the

5-47. Megohmmeters. (A) Megger or megohmmeter with a handcrank. (B) Megger or megohmmeter with a battery for power.

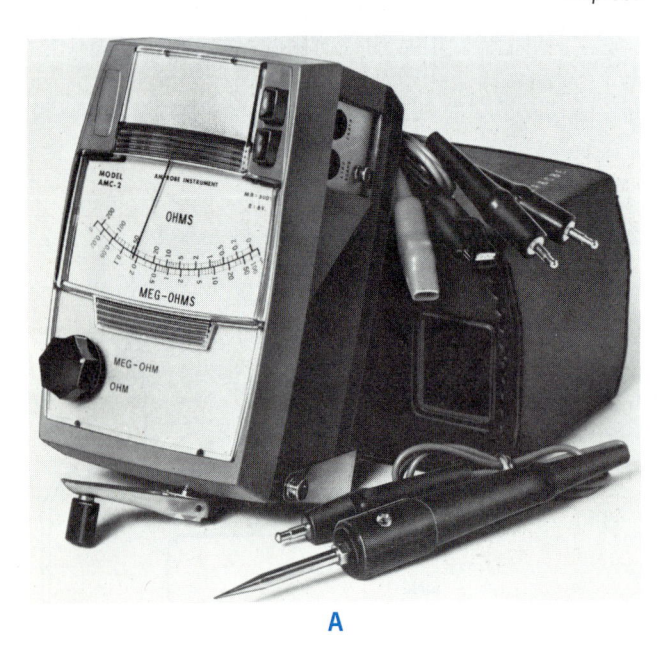

A 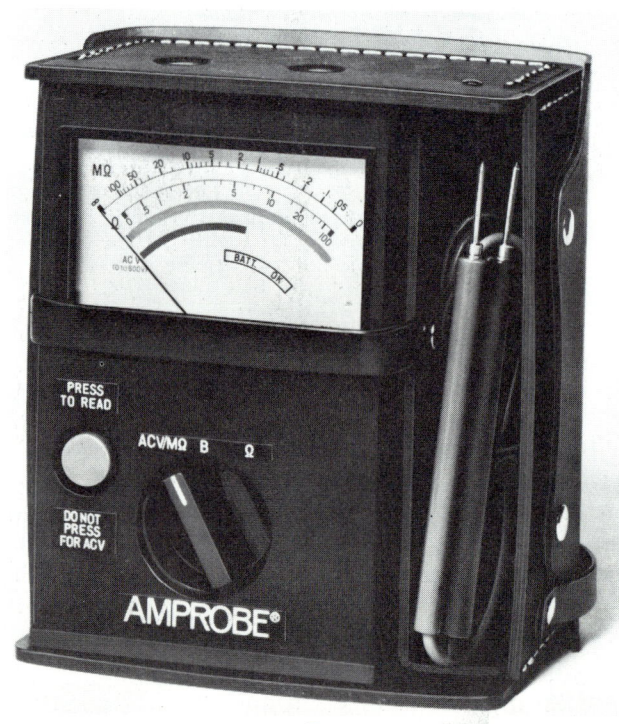   B

CHAPTER 5—ELECTRIC MOTORS AND CONTROLS

leased, it automatically discharges the capacitive charge. With other models you must wait a few minutes for the charge to dissipate or remove the test lead from the *earth* (*ground*) jack on the tester and touch to the equipment terminal that the other test lead, *line,* is connected to. Never use the megger on a *live* circuit. Since it has a self-contained power source, it is not necessary to draw current from the line.

There are two possible conducting or leakage paths in the insulation of all electrical apparatus—one through the insulating material and the other over its surface. By using the guard terminal, the surface leakage can be separated and a direct measurement made of the insulation itself. Fig. 5-48.

INSULATION RESISTANCE TESTING

The primary purpose of insulation is to keep electricity flowing in the desired path. The perfect insulation would have infinite resistance, which would prevent the flow of any current through the insulation to ground. However, there is no perfect insulation material. Thus, there is always some current flow. Good insulation is one that has and keeps a high resistance value to minimize the current flow.

Unless there is accidental damage, insulation failure is generally gradual, rather than sudden. This is because failure is generally the result of repeated heating and cooling, the related expansion and contraction, and dirt, physical abrasion, vibration, moisture, and chemicals.

When insulation starts to fail, its resistance decreases. This allows more current to flow through the insulation. If the resistance continues to decrease, the condition of the insulation may reach a point where it may permit through the insulation a current flow large enough to cause the blowing of a fuse, equipment damage, or fatal shock.

If you are responsible for the servicing, maintenance, or installation of electrical equipment, you must be concerned with insulation resistance.

An insulation resistance testing program helps reveal failing insulation before it causes a serious problem. Such a testing program consists of periodic insulation resistance tests on critical equipment and systems. The results are recorded on a control card or file for each piece of equipment or each test point in a system. Any trend that indicates a decreasing insulation resistance value is an indication that the insulation is failing and that corrective maintenance should be scheduled.

Measuring Insulation Resistance

Insulation resistance measurements are affected by a number of factors. *Temperature* and the *duration of the measurement* are two primary ones. *Humidity* may also affect readings. Thus, it is a good idea to make a note as to whether the air is dry or humid at the time of the measurement. You may find that insulation resistance readings are lower on humid days and higher on dry days. Wet or flooded equipment should be dried and cleaned as much as possible before measurements are taken. Lastly, dirt and other contaminants (corrosion, chemicals, etc.) can also affect readings. You should be certain that the contact points at which measurements are to be taken are reasonably clean.

The duration of the resistance measurement also affects the reading. If the insulation is good, the reading will continually increase as long as the megohmmeter is connected to the insulation. The most common megger measurement is taken at the end of a 60-second interval. This time period generally gives a satisfactory measurement of the insulation resistance.

A second test involves taking a reading after 30 seconds and 60 seconds. The 60-second reading divided by the 30-second reading is known as the *dielectric absorption ratio.* Comparing periodic ratios may prove more useful than comparing 1-minute readings. Generally speaking, a ratio of 1.25 is the bottom limit for borderline insulation. An extension of this test involves readings taken after 60 seconds and 10 minutes. The ratio of the 10-minute reading to the 60-second

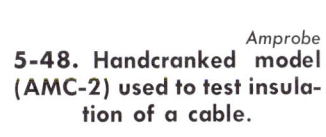

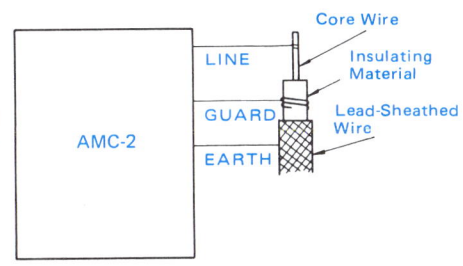

Amprobe
5-48. Handcranked model (AMC-2) used to test insulation of a cable.

129

REFRIGERATION AND AIR CONDITIONING TECHNOLOGY

reading is referred to as the *polarization index*. The resistance measurement taken at the end of 10 minutes should be considerably higher than that taken at 60 seconds. The measured insulation resistance of a dry winding in good condition should reach a relatively steady value in 10 minutes. If the winding is wet or dirty, the steady value will usually be reached in one or two minutes. The *index* is helpful in evaluating the winding dryness and fitness for over-potential testing.

As a guide, the recommended minimum value of the polarization index for alternating current and direct current rotating machines is 1.5 for 221° F. [105°C] (Class A) insulation systems and 2.0 for 266° F. [130°C] (Class B) insulation systems.

POWER TOOLS AND SMALL APPLIANCES

For double-insulated power tools, the megohmmeter lead shown connected to the housing would be connected to some metal part of the tool (such as the chuck or blade). Fig. 5-49. The switch of the power tool must be in the *on* position.

MOTORS

For testing (AC), disconnect the motor from the line by disconnecting the wires at the motor

5-49. Using a megohmmeter to check the insulation of a small hand drill.

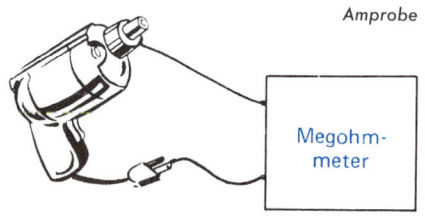

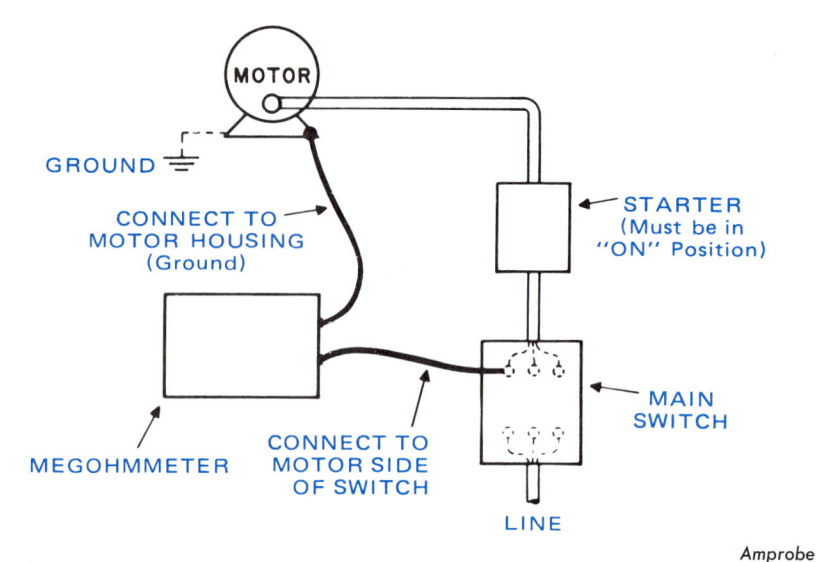

Amprobe

5-50. Using a megohmmeter to check the insulation of the windings of a motor.

Amprobe

terminals or by opening the main switch. If the main switch is used and the motor also has a starter, then the starter must be held in the *on* position. In the latter case, the measured resistance will include the resistances of the motor, wire, and all other components between the motor and the main switch. If a weakness is indicated, the motor and other components should be checked individually.

If the motor is disconnected at the motor terminals, connect one megohmmeter lead to the grounded motor housing. Connect the other lead to one of the motor leads. Fig. 5-50.

For testing (DC), disconnect the motor from the line. To test the brush rigging, field coils, and armature, connect one megohmmeter lead to the brush on the commutator. If the resistance measurement indicates a weakness, raise the brushes off the commutator and separately test the armature, field coils, and brush rigging. Do this by connecting one megohmmeter lead to each of them individually, leaving the other lead connected to the grounded motor housing.

CABLES

Disconnect the cable from the line. Also disconnect the opposite end to avoid errors due to leakage from other equipment. Check each conductor to ground and/or lead sheath by connecting one megohmmeter lead to each of the conductors in turn. Check insulation resistance between conductors by connecting megohmmeter leads to conductors in pairs. Fig. 5-51.

5-51. Using a megohmmeter to check the insulation qualities of wires between conductors.

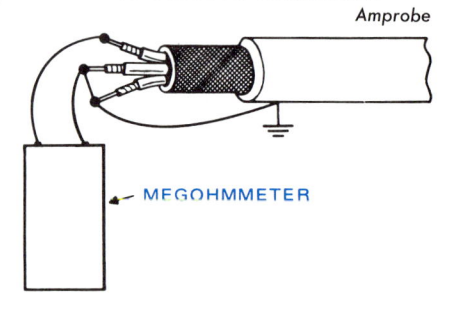

Amprobe

130

CHAPTER 5—ELECTRIC MOTORS AND CONTROLS

Table 5-A.
Moisture in Hermetic Compressor Systems

Megger Reading	Compressor Condition	Suggest Preventive Maintenance
100 megohms to infinity.	Good.	None necessary.
50 to 100 megohms.	Moisture present.	Change drier.
20 to 50 megohms.	Severe moisture and possible contaminated oil.	Change numerous driers. Change oil. Acid present.
0 to 20 megohms.	Severe contamination	Dump oil and entire refrigeration charge. Evacuate system. Install liquid and suction line driers. Recharge system with new oil and refrigerant.

HERMETIC COMPRESSOR SYSTEMS

Table 5-A may be used as a guide to determine the extent to which a system may be contaminated by moisture.

CIRCUIT BREAKERS AND SWITCHES

Circuit breakers and switches to be tested should be disconnected from the line. To test each terminal to ground, connect one megger lead to the frame or ground. Connect the other megger lead to each terminal of the circuit breaker or switch, one after the other. To test between terminals, connect the megger leads to pairs of terminals.

COILS AND RELAYS

Disconnect from the line relays and coils to be tested. To test the coil, connect one megger lead to one of the coil leads. The other megger lead goes to ground. Then connect the megger between one coil lead and the core.

To test a relay, connect one megger lead to the relay contact. The other megger lead goes to the coil. Then it goes to the core.

AC MOTOR CONTROL

Wound-rotor motors and AC commutator motors have only a limited application. The squirrel-cage induction motor is the most widely used motor. The control section of this chapter deals with such motors. The use of high voltages (2400 volts and higher) introduces requirements that are additional to those needed for 600-volt equipment. However, the basic principles are unchanged.

The motor, machine, and motor controller are interrelated and need to be considered as a package when choosing a specific device for a particular application. In general, five basic factors are considered when selecting a controller for a motor: the electrical service, motor, operating characteristics of the controller, environment, and National Electrical Code.

Motor Controller

A motor controller will perform some or all of the following functions: starting, stopping, overload protection, overcurrent protection, reversing, changing speed, jogging, plugging, sequence control, and pilot light indication. The controller can also provide the control for auxiliary equipment such as brakes, clutches, solenoids, heaters, and signals. A motor controller may be used to control a single motor or a group of motors.

The terms *starter* and *controller* mean practically the same thing. Strictly speaking, a starter is the simplest form of controller. It is capable of starting and stopping the motor and providing it with overload protection.

AC Squirrel-Cage Motor

The workhorse of industry is the AC squirrel-cage motor. Of the thousands of motors used today in general applications, the vast majority are of the squirrel-cage type. Squirrel-cage motors are simple in construction and operation.

The squirrel-cage motor gets its name because of its rotor construction. The rotor resembles a squirrel cage and has no wire winding. A number of terms need to be explained to understand motor control. *Full load current* (FLC) is the current required to produce full load torque at rated speed. *Locked rotor current* (LRC) is the inrush current when the motor is connected directly to the line. The locked rotor current can be from four to ten times the motor's full load current. The vast majority of motors have an LRC of about six times FLC. Therefore, this figure is generally used. The "six-times" value is expressed as 600% of FLC.

Motor speed depends on the number of poles in the motor's winding. On 60 hertz, a two-pole motor runs about 3450 rpm, a four-pole motor runs at 1725 rpm, and a six-pole motor runs at 1150 rpm. Motor nameplates are usually marked with actual full load speeds. However, frequently motors are referred to by their *synchronous speed.* Synchronous speeds are 3600 for the 3450-rpm, 1800 for the 1725-rpm, and 1200 for the 1150-rpm motor.

Torque is the "turning" or "twisting" force of the motor. It is usually measured in pound-feet. Except when the motor is accelerating to speed, the torque is related to the motor horsepower by the formula:

Torque (in pound-feet)
$$= \frac{hp \times 5252}{rpm}$$

The torque of a 25 hp motor running at 1725 rpm would be computed as follows:

$$Torque = \frac{25 \times 5252}{1725}$$

= approximately 76 pound-feet

If 90 pound-feet were required to drive a particular load, this motor would be overloaded and would draw a current in excess of *full load current.*

Temperature rise is the difference between the winding temperature of the motor when running and the ambient temperature. Current passing through the motor windings results in an increase in motor temperature. The temperature rise produced at full load is not harmful, provided the motor ambient temperature does not exceed 104° F. [40°C].

Higher temperature caused by increased current or higher ambient temperatures has a deteriorating effect on motor insulation and lubrication. One rule states that for each increase of 10° F. [5.5°C] above the rated temperature, motor life is cut by one half.

Duty rating is the rating of the motor for continuous or intermittent operation. Most motors have a continuous duty rating, permitting indefinite operation at rated load. Intermittent duty ratings are based on a fixed operating time (such as 5, 15, 30, or 60 minutes) after which the motor must be allowed to cool.

Motor service factor is given by the motor's manufacturer. It means that the motor can be allowed to develop more than its rated or nameplate hp without causing undue deterioration of the insulation. The service factor is a margin of safety. If, for example, a 10-hp motor has a service factor of 1.15, the motor can be allowed to develop 11.5 hp. The service factor depends on the motor design.

Jogging describes the repeated starting and stopping of a motor at frequent intervals for a short period of time. A motor would be jogged when a piece of driven equipment has to be positioned fairly closely. Thus, jogging might occur when positioning the table of a horizontal boring mill during setup or aligning any motor-driven device. If jogging is to occur more frequently than five times per minute, NEMA (National Electrical Manufacturer's Association) standards require that the starter be derated. For instance, a NEMA size-1 starter has a normal duty rating of 7½ hp at 230 V, polyphase. On jogging applications, this same starter has a maximum rating of 3 hp.

Plugging occurs when a motor running in one direction is momentarily reconnected to reverse the direction. It will be brought to rest very rapidly. If a motor is plugged more than five times per minute, derating of the controller is necessary. The contacts of the controller overheat. Plugging may be used only if the driven machine and its load will not be damaged by the reversal of the motor torque.

Enclosures

NEMA and other organizations have established standards of enclosure construction for control equipment. In general, equipment would be enclosed for one or more of the following reasons:

1. To prevent accidental contact with live parts.
2. To protect the control from harmful environmental conditions.
3. To prevent explosion or fires that might result from the electrical arc caused by the control.

Code

The National Electrical Code (NEC) deals with the installation of electrical equipment. It is primarily concerned with safety. It is adopted on a local basis, sometimes incorporating minor changes. NEC rules and provisions are enforced by governmental bodies exercising legal jurisdiction over electrical installations. The Code is used by insurance inspectors. Minimum safety standards are thus assured if the National Electrical Code is followed.

Protection of the Motor

Motors can be damaged, or their effective life reduced, when subjected to a continuous current only slightly higher than their full-load current rating times the service factor.

Damage to insulation and windings of the motor can also be sustained on extremely high currents of short duration. These occur when there are grounds and shorts.

All currents in excess of full load current can be classified as *overcurrents*. In general, a distinction is made based on the magnitude of the overcurrent and equipment to be protected. Overcurrent up to locked rotor current is usually the result of a mechanical overload on the motor. The National Electrical Code covers this in Article 430 (Part C).

Overcurrents due to short circuits or grounds are much higher than locked rotor currents. Equipment used to protect against damage due to this type of overcurrent must protect not only the motor, but also the branch circuit conductors and the motor controller.

The function of the overcurrent protective device is to protect the motor branch circuit conductors, control apparatus, and motor from short circuits or grounds. The protective devices commonly used to sense and clear overcurrents are thermal magnetic circuit breakers and fuses. The short circuit device shall be capable of carrying the starting current of the motor, but the device setting shall not exceed 250% of full load current, depending upon the code letter of the motor. Where the value is not sufficient to carry the starting current, it may be increased. However, it shall not exceed 400% of the motor full load current. The NEC (with a few exceptions) requires a means to disconnect the motor and controller from the line, in addition to an overcurrent protective device to clear short circuit faults.

Contactors, Starters, and Relays

If the condensing unit has a motor larger than $1\frac{1}{2}$ hp, it will have a starter or contactor. They are usually furnished with the unit.

Relays are a necessary part of many control and pilot light circuits. They are similar in design to contactors, but are generally lighter in construction so they carry smaller currents.

Magnetic contactors are normally used for starting polyphase motors, either squirrel cage or single phase. Contactors may be connected at any convenient point in the main circuit between the fuses and the motor. Small control wires may be run between the contactor and the point of control.

Protection of the motor against prolonged overload is accomplished by time-limit overload relays that are operative during the starting period and running period. Relay action is delayed long enough to take care of heavy starting currents and momentary overloads without tripping.

Motor Overload Protector

Motors for commercial condensing units are normally protected by a bimetallic switch operating on the thermo, or heating, principle. This is a built-in motor overload protector. It limits the motor winding temperature to a safe value. In its simplest form, the switch or motor protector consists essentially of a bimetal switch mechanism that is permanently mounted and connected in series with the motor circuit. Fig. 5-52.

When the motor becomes overloaded or stalled, excessive heat is generated in the motor winding due to the heavy current produced by this condition. The protector located inside the motor is controlled by the motor current passing through it and the motor temperature. The bimetal element is calibrated to open the motor circuit when the temperature, as a result of an excessive current, rises above a predetermined value. When the temperature decreases, the protector automatically resets and restores the motor circuit.

This device reduces service calls due to temporary overloads. The device stops the motor until it cools off and then allows it to start again when needed.

Servicing of motors with built-in overload devices must be handled with care. The compressor may be idle due to an overload. Hence, it will start as soon as the motor cools off. This could result in a serious mishap to the operator or repairperson. To avoid such difficulties, open the electrical circuit by pulling the line plug or switch prior to any repair or servicing operation.

Motor Winding Relays

A motor winding relay is usually incorporated in single-phase

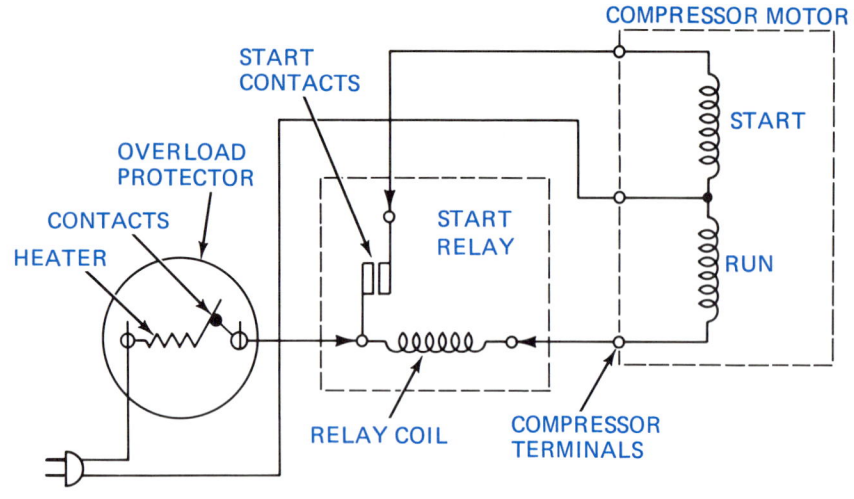

REFRIGERATION AND AIR CONDITIONING TECHNOLOGY

5-52. Circuit for a domestic refrigerator.

motor-compressor units. This relay is an electromagnetic device for making and breaking the electrical circuit to the start-winding. A set of normally closed contacts is in series with the motor start-winding. Fig. 5-52.

The electromagnetic coil is in series with the auxiliary winding of the motor. When the control contacts close, the motor start- and run-windings are energized. A fraction of a second later the motor comes up to speed and sufficient voltage is induced in the auxiliary winding to cause current to flow through the relay coil. The magnetic force is sufficient to attract the spring-loaded armature, which mechanically opens the relay starting contacts. With the starting contacts open, the start-winding is out of the circuit. The motor continues to run on only the run-winding. When the control contacts open, power to the motor is interrupted. This allows the relay armature to close the starting contacts. The motor is now ready to start a new cycle when the control contacts again close.

Solenoid Valves

Solenoid valves are used on multiple installations. They are electrically operated. A solenoid valve, when connected as in Fig. 5-53, remains open when current is supplied to it. It closes when the current is turned off. In general, solenoid valves are used to control the liquid refrigerant flow into the expansion valve, or the refrigerant gas flow from the evaporator when it or the fixture it is controlling reaches the desired temperature. The most common application of the solenoid valve is in the liquid line and operates with a thermostat. With this hookup, the thermostat may be set for the desired temperature in the fixture. When this temperature is reached, the thermostat will open the electrical circuit and shut off the current to the valve. The solenoid valve then closes and shuts off the refrigerant supply to the expansion valve. The condensing unit operation should be controlled by the low-pressure switch. In other applications, where the evaporator is to be in operation for only a few hours each day, a manually operated snap switch may be used to open and close the solenoid valve.

REFRIGERATION VALVE

The solenoid valve in Fig. 5-54 is operated with a normally closed status. A direct-acting metal ball and seat assure tight closing. The two-wire, class-W, coil is supplied standard for long life on low temperature service or sweating conditions. Current

5-53. Solenoid valve connected in the suction and liquid evaporator lines.

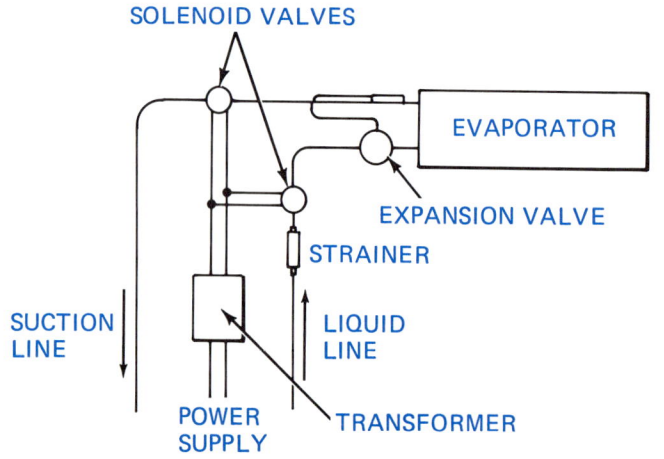

CHAPTER 5—ELECTRIC MOTORS AND CONTROLS

opened by energizing the coil and magnetically lifting the plunger and allowing full flow by the valve ball. De-energizing the coil permits the plunger and valve ball to return to the closed position.

The piloted piston solenoid valve is somewhat different. Fig. 5-55. It too, is normally closed. It can be used on all refrigerants except R-717.

When the solenoid is energized the plunger rises, lifting the pilot valve to allow pressure to bleed from above the piston. The pilot valve continues its rise and the piston follows due to the lower pressure effected above the piston. The piston is then held in a fully open position by the plunger and pilot stem to allow full flow through the valve with minimum pressure drop. When the solenoid is de-energized, the plunger drops and allows the pilot valve to seat. The pressure above the piston balances with that on the underside. The combined weight of the piston and plunger assembly causes the valve to return to the closed position.

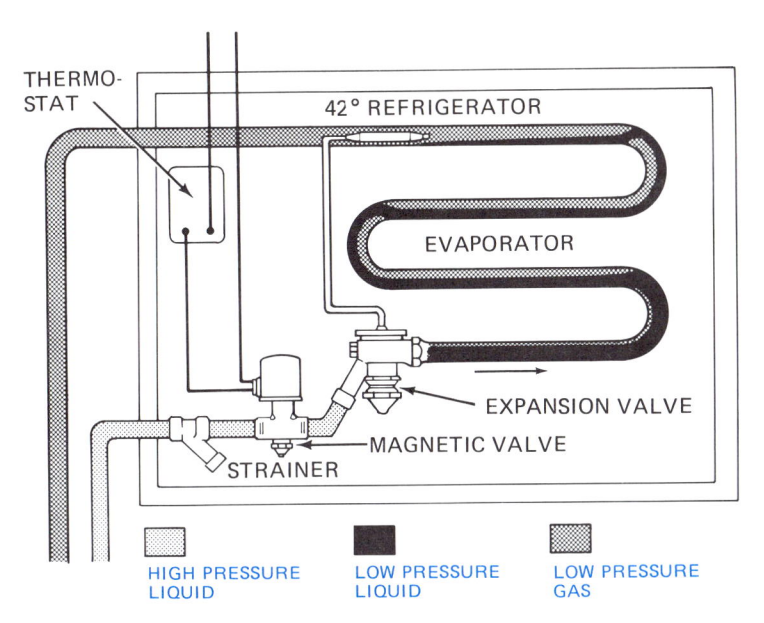

SCHEMATIC REFRIGERATION INSTALLATION

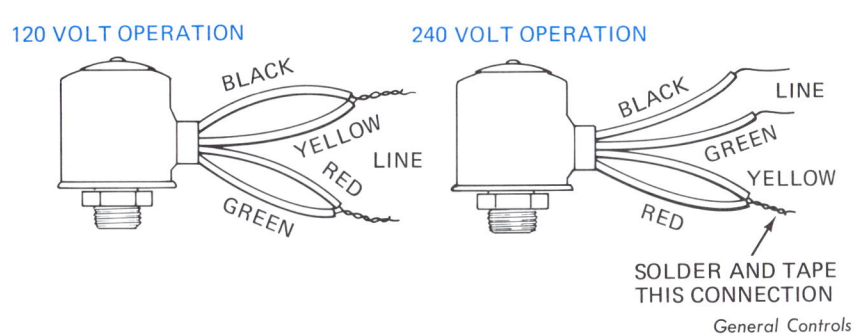

120 VOLT OPERATION   240 VOLT OPERATION

5-54. Solenoid valve locations.

General Controls

INSTALLATION

Install in a horizontal line with the solenoid upright. With the threaded type do not use the solenoid cover to turn the valve. Provide enough clearance for solenoid removal. On the solder type, remove the solenoid coil before installing the valve. Do not remove plunger tube. Wrap the valve with wet asbestos or a wet cloth while making up fittings. Improper handling may distort the cylinder and cause the piston to bind.

Table 5-B lists service suggestions for the solenoid valve.

failure or interruption will cause the valve to fail-safe in the closed position. The solenoid cover can be rotated 360° for easy installation. Explosion-proof models are available for use in hazardous areas.

APPLICATION

This solenoid valve is usable with all refrigerants except ammonia. Also it can be used for air, oil, water, detergents, butane or propane gas, and other non-corrosive liquids or gases.

A variety of temperature control installations can be accomplished with these valves. Such installations include bypass, defrosting, suction line, hot gas service, humidity control, alcohols, unloading, reverse cycle, chilled water, cooling tower, brine, and liquid line stop installations and ice makers.

OPERATION

The valves are held in the normally closed position by the weight of the plunger assembly and the fluid pressure on top of the valve ball. The valve is

135

REFRIGERATION AND AIR CONDITIONING TECHNOLOGY

Main Valve Closed Pilot Valve Open Main Valve Open

General Controls

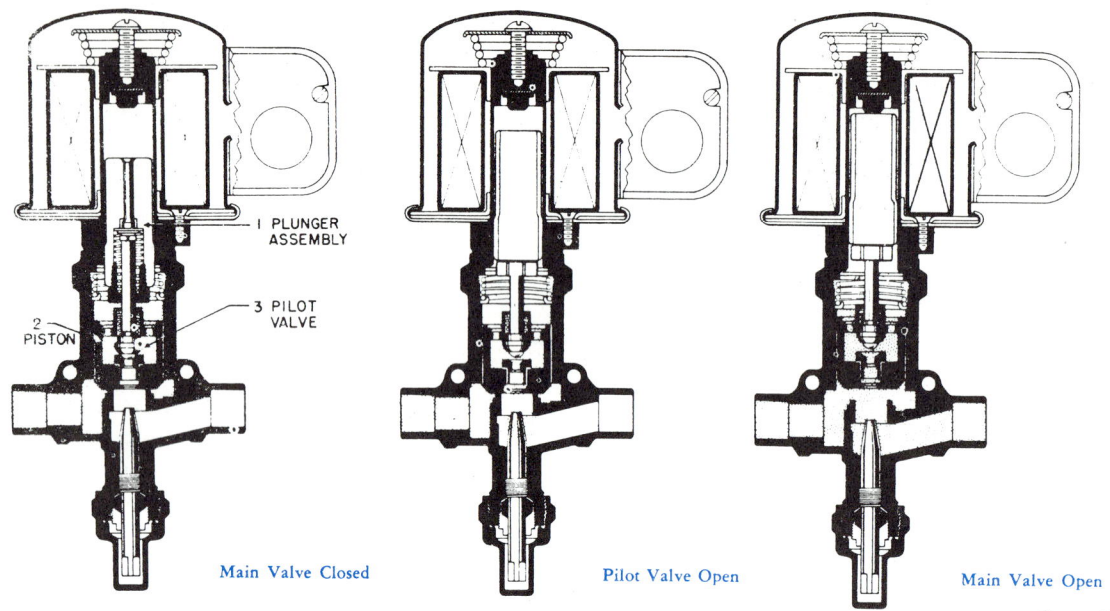

5-55. Operation of a solenoid valve.

TEMPERATURE CONTROLS

In modern condensing units, low-pressure control switches are being largely superseded by thermostatic control switches. A thermostatic control consists of three main parts—a bulb, a capillary tube, and a power element (switch). The bulb is attached to the evaporator in a manner that assures contact with the evaporator. It may contain a volatile liquid, such as a refrigerant. The bulb is connected to the power element by means of a small capillary tube. Fig. 5-56.

Operation of the thermostatic control switch is such that, as the evaporator temperature increases, the bulb temperature also increases. This raises the pressure of the thermostatic liquid vapor. This, in turn, causes the bellows to expand and actuate an electrical contact. The contact closes the motor circuit and the motor and compressor start operating. As the evaporator temperature decreases, the bulb becomes colder and the pressure decreases to the point where the bellows contracts sufficiently to open the electrical contacts controlling the motor circuit. In this manner, the condensing unit is entirely automatic. Thus, it is able to produce exactly the amount of refrigeration to meet any normal operating condition.

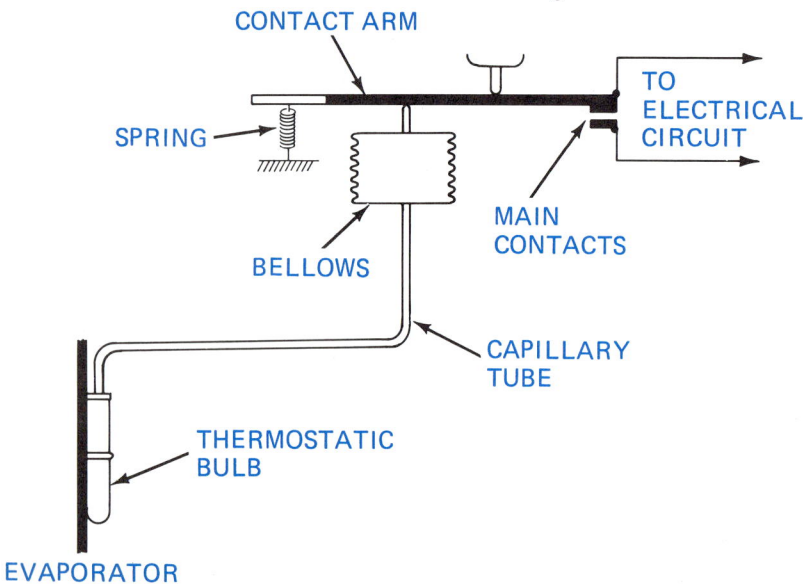

5-56. Thermostatic control switch using a bellows.

136

Table 5-B.
Service Suggestions

Trouble	Possible Cause	Remedy
Valve fails to open.	Timers, limit controls or other devices holding circuit open.	Check circuit for limit control operation, blown fuses, short circuit, and loose wiring.
	Solenoid coil shorted, burned out, or wrong voltage.	Replace with solenoid coil of correct voltage.
	Dirt, pipe compound, or other foreign matter restricting operation of piston or pilot valve.	Disassemble and clean internal parts with carbon tetrachloride. Install strainer ahead of valve.
Valve will not close.	Manual opening device holding valve open.	Turn manual opening stem counter-clockwise until stem backseats.
	Dirt, pipe compound, or other foreign matter restricting operation of piston or pilot valve.	Disassemble and clean internal parts with carbon tetrachloride. Install strainer ahead of valve.
	Damaged plunger tube preventing plunger operation.	Replace plunger tube.
Low leakage.	Foreign matter in valve interior or damaged seat or seat disc.	Clean valve interior with carbon tetrachloride. Check condition of seat and seat disc. Replace if necessary.

5-57. Working principles of a simple bimetallic thermostat.

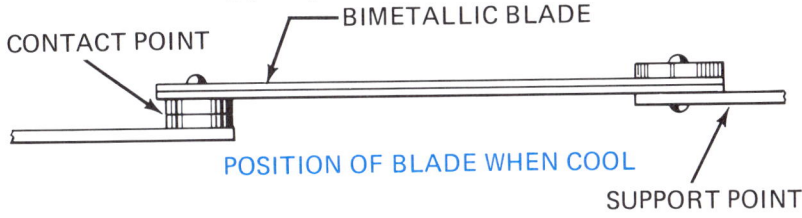

POSITION OF BLADE WHEN COOL

POSITION OF BLADE WHEN HEATED

An automatic temperature-control system is generally operated by making and breaking an electric circuit or by opening and closing a compressed-air line. When using the electric thermostat, the temperature is regulated by controlling the operation of an electric motor or valve. When using the compressed-air thermostat, temperature regulation is obtained by actuating a compressed-air operated motor or drive. Electrically-operated temperature-control systems are used generally by manufacturers for practically all installations. However, compressed-air temperature-control systems have applications in extremely large central and multiple installations in close temperature work. This is where a large amount of power is required for small control devices.

Bimetallic Thermostats

The bimetallic thermostat operates as a function of expansion or contraction of metals due to temperature changes. Bimetallic thermostats are designed for the control of heating and cooling in air-conditioning units, refrigeration storage rooms, greenhouses, fan coils, blast coils, and similar units.

The working principle of such a thermostat is shown in Fig. 5-57. As noted, two metals, each having a different coefficient of expansion, are welded together to form a bimetallic unit or blade. With the blade securely anchored at one end, a circuit is formed and the two contact points are closed to the passage of an electric current. Because an electric current provides heat in its passage through the bimetallic blade, the metals in the blade

REFRIGERATION AND AIR CONDITIONING TECHNOLOGY

5-58. Modern thermostat for both cooling and heating using a metallic strip's expansion ability to move a magnet close to a magnetic switch. (A) Typical thermostat, (B) interior of thermostat element, (C) typical sub-base showing switching and wiring terminal locations.

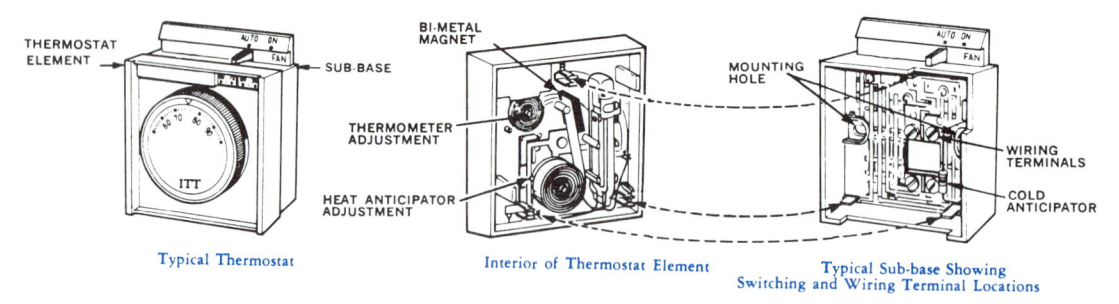

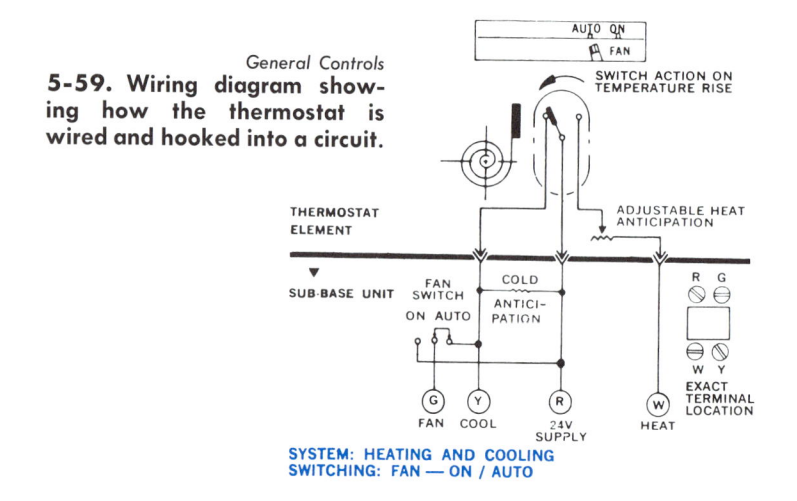

General Controls

5-59. Wiring diagram showing how the thermostat is wired and hooked into a circuit.

begin to expand, but at a different rate. The metals in the blade are so arranged that the one with a greater coefficient of expansion is placed at the bottom of the unit. After a certain time, the operating temperature is reached and the contact points become separated, thus disconnecting the appliance from its power source. After a short period, the contact blade will again become sufficiently cooled to cause the contact points to join, thus re-establishing the circuit and permitting the current again to actuate the circuit leading to the appliance. The foregoing cycle is repeated over and over again. In this way, the bimetallic thermostat prevents the temperature from rising too high or dropping too low.

Thermostat Construction and Wiring

Some thermostats are designed for use on both heating and cooling equipment. The thermostat shown in Fig. 5-58 is such a device. The basic thermostat element has a permanently sealed, magnetic SPDT switch. The thermostat element plugs into the sub-base and contains the heat anticipation, the magnetic switching, and a room temperature thermometer. The sub-base unit contains fixed cold anticipation and circuitry. This thermostat is for use with 24-volt equipment. In this case, the thermostatic element (bimetal) does not make direct contact with the electrical circuit. The expansion of the bimetal causes a magnet to move. This, in turn, causes a switch to close or open. Figure 5-59 illustrates the fact that the bimetal is not in the electrical circuit at all.

DEFROST CONTROLS

Automatic defrost is common in domestic refrigeration. It is accomplished in several ways. The control method used depends on the type of refrigeration system, the size and number of condensing units, and other factors.

Defrost Timer Operation

In small and medium-size domestic refrigerators, an automatic defrost-control clock may be set for a defrost cycle once every twenty-four hours or as often as deemed necessary. The defrosting is usually accomplished by providing one or more electric heaters. They are energized by the action of the electric

CHAPTER 5—ELECTRIC MOTORS AND CONTROLS

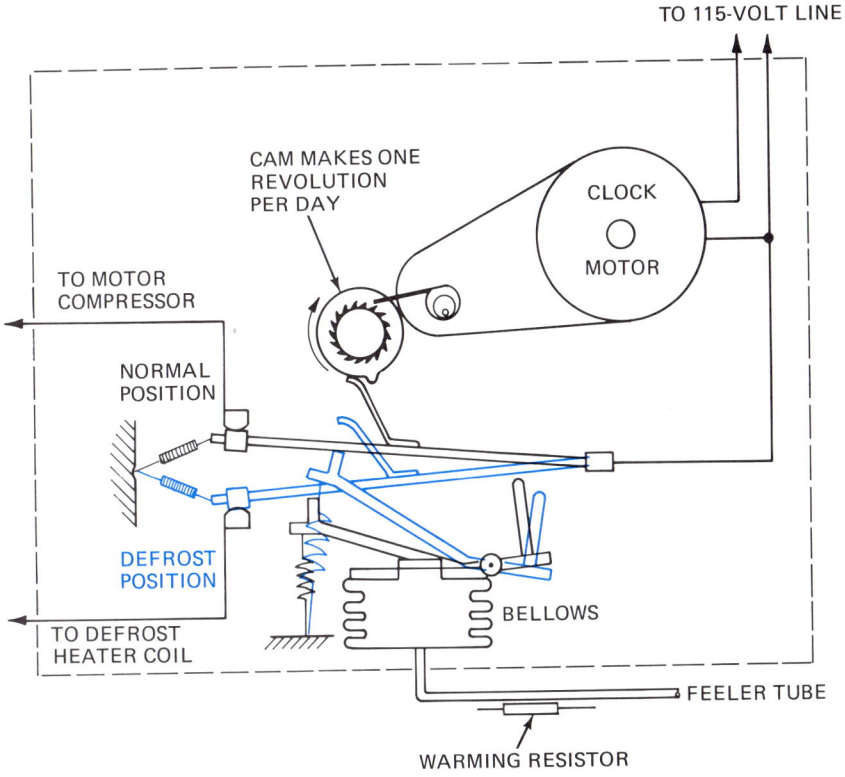

5-60. Twenty-four hour clock used to activate defrost-cycle.

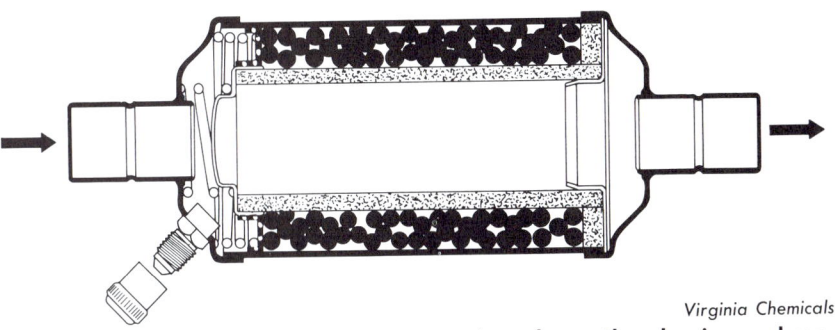

Virginia Chemicals

5-61. Low-side filter-drier. This cutaway view shows the alumina and molecular sieve desiccant for absorbing water and hydrochloric acid, organic acids, and oil breakdown products. Alumina seems to be more effective in removing organic acids than charcoal. The use of internal heat shields allows the drier to be installed with any of the low- or high-melting point brazing alloys, with no danger of burning the filter. A wet rag wrapped around the end of the drier is helpful in preventing excessive burning of the paint.

frost position by an electric clock. The switch arm is returned to the normal position by a power element that is responsive to changes in temperature. As the evaporator is warmed during a defrost period, the feeler tube of the defrost control is also warmed until it reaches the defrost cut-out point of approximately 45° F. [7°C]. The defrost-control bellows then forces the switch arm to snap from the defrost position to the normal position. Fig. 5-60. This starts the motor compressor.

Another common method of automatic defrosting is the so-called defrost cycle method. In this, the defrost cycle occurs during each compressor off-cycle. In a defrost system of this type, the defrost heaters are connected across the thermostat switch terminals. When the thermostat switch is closed, the heaters are shunted out of the circuit. When the thermostat opens, the heaters are energized, completing the circuit through the overload relay and compressor. Figure 5-62 (page 142) illustrates this type of defrosting. Note that the serpentine resistor is shorted out by the temperature control. This occurs when the temperature control is closed and the compressor motor is running normally.

Hot-Gas Defrosting

In any low-temperature room where the air is to be maintained below freezing, some adequate means for removing accumulated frost from the cooling surface should be provided. An improved hot-gas method of quick defrosting for direct-expansion low-temperature evaporators is now available. To apply this method, two or more evaporators

clock and provide the heating action necessary for complete defrosting.

The defrost controls, as usually employed, are essentially single-pole, double-throw (SPDT) switching devices in which the switch arm is moved to the de-

139

are needed in the system. This is because the hot gas required to defrost part of the system must be provided by the heat absorbed from the other cooling surface in a given system. Defrosting a plate bank with hot gas can be accomplished automatically by installing the proper controls. See Chapter 11.

MOTOR BURNOUT CLEANUP

The following cleanup methods are simple, rapid, and economical. They represent a drastic reduction in labor requirements over the obsolete flushing methods.

Procedure for Small Tonnage Systems

In systems up to 40 tons the refrigerant charge is relatively small. Motor burnout contaminants are not diluted to the extent that they are in large tonnage systems. As a result, there is a greater need to isolate the motor compressor from all harmful soluble and insoluble materials that might cause another burnout.

Driers should be installed in both the liquid and suction lines. Fig. 5-61 (page 139). The desiccant in the driers removes all harmful soluble chemicals that cause corrosion and attack motor winding insulation. Suction line filtration should be employed to prevent harmful solids above 5 microns [0.0002"] from returning to the compressor. Through abrasion, foreign particles such as casting dust, copper and aluminum dust, and flux contribute to motor burnouts and compressor damage.

The type of drier used in the suction line is of great importance. Throwaway-type liquid line driers, when used in the suction line, are usually too small. They may create a dangerously high pressure drop. This may cause overheating of the motor compressor and a repeat burnout.

Until recently, it has been necessary to use large replaceable cartridge-type driers for this purpose. Even these have a limited range. They are costly, heavy, and difficult to mount. Their filtering ability is very questionable. In addition, since the system must be opened to remove them, replacement of the liquid line drier and reevacuation of the system are needed. These operations add to the cost.

New low-side filter-driers eliminate these difficulties. Two essential components are combined into one. Thus, the filter-drier is both a suction line filter and a suction line drier. It is designed for permanent installation. The blended mixture of activated alumina and molecular sieves provides an enormous capacity for adsorbing moisture and other harmful soluble contaminants. It can adsorb inorganic and organic acids and oil breakdown materials.

Evidence indicates that these soluble contaminants are most easily removed in the suction line for the following reasons.

• Field experience shows success with soluble contaminants removal when properly-sized driers have been used in the suction line following hermetic motor burnouts.

• Modern drying materials have a substantially higher capacity for moisture and acids at lower temperatures. Suction line temperatures are normally from 20 to 60° F. [11 to 33°C] lower than liquid line temperatures, depending upon the application and ambient conditions.

• Since oil breakdown materials dissolve readily in oil, higher concentrations of contaminants are in contact with the desiccant. This results in conditions more favorable to maximum pickup.

• Liquid refrigerant does not compete with the desiccant to accept soluble contaminants. In the liquid line it does—by greatly diluting soluble materials. This greatly reduces contact time and, consequently, reduces the rate of pickup.

The low-side filter drier has an access valve on the inlet side for checking pressure drop and charge adjustment.

There is no method of cleanup after a burnout that does not carry some risk. No cleanup procedure will guarantee 100% success. The procedures that follow have been generally successful. They are practical at the field level and economical enough to be used by the equipment owner.

1. Discharge oil refrigerant mixture in liquid phase. If water-cooled, drain all water-containing areas first.

2. Remove burned-out compressor, taking care not to touch oil or sludge with bare hands.

3. Blow out coils and condenser with clean dry liquid refrigerant.

4. Install new motor compressor.

5. Install a moisture indicator and an oversize high-side filter-drier in the liquid line.

6. Install a low-side filter-drier in the suction line as close to the compressor as possible. If the system is larger than 20 actual tons, install two low-side filter-driers in parallel.

CHAPTER 5—ELECTRIC MOTORS AND CONTROLS

7. Triple evacuate to 500 microns, or as low as practical, and charge.

Optional: Check back in two weeks and perform an acid test on the oil. Use the acid test kit. If the oil is acidic or discolored, change the oil, both driers, and again evacuate. Another two-week checkup is desirable.

This method, due to line sizing and refrigerant cost, is applicable up to 40 tons. Consideration may also be given to saving the refrigerant if the charge is above 100 pounds.

Procedure for Large Tonnage Systems

In systems above 40 tons, the large refrigerant charge so dilutes the motor burnout contaminants that discarding the refrigerant is unnecessary. It cannot be justified from cost considerations. Oil breakdown materials and organic acids are more soluble in the oil than the refrigerant. They tend to concentrate in the oil. By repeated oil changes and drier changes, with the oil and drier extracting the contaminants, such systems can be cleaned up. The following procedure has been used over an extended period of time by many large contractors with successful cleanups from 40 tons to over 500 tons.

Due to design variations, the mechanics of carrying out the following procedures must be adapted to the system involved. The basic procedure is as follows.

1. If possible, wash out coil and condenser with clean refrigerant. In some designs, this is possible, but with others, completely impractical.

2. Reinstall the rebuilt compressor with a fresh, clean charge of oil.

3. Install the largest possible drier in the liquid phase of the system.

4. Operate 24 hours.

5. Change oil and drier or drier cores.

6. Operate 24 hours.

7. Change oil and drier or drier cores.

8. Operate 24 hours.

9. Change oil and drier or drier cores.

10. Triple evacuate to 500 microns or as low as practical, and charge.

11. Operate two weeks and check oil color. Perform an acid test on the oil. If it is neutral and the color normal, consider the job done. If the oil is acidic or discolored, repeat the above steps until neutrality is secure and the oil color is normal.

READING A SCHEMATIC

It is often difficult to read a schematic at first glance. Figure 5-62 shows the schematic for a home appliance. The voltage being used is 115 V. Follow the brown wires and see how they control the freezer light, cabinet light, and mullion heater. The brown wire on the right is spliced to an orange wire. This orange wire connects to one side of the freezer door switch, one side of the refrigerator door switch, and one side of the mullion heater. The brown wire from the left side of the schematic connects to two orange wires that attach to one side of the freezer light and one side of the cabinet light. The brown wire on the left connects to the other side of the mullion heater. There is a wire connecting the freezer light and the freezer door switch. Likewise, there is a wire connecting the cabinet light and the cabinet door switch.

Now, trace the schematic. Start at the top of the schematic at the 115-V lead.

Trace the left side first. The brown wire on the left side goes down to the orange wires that connect to the freezer light and cabinet light. The brown wire also connects to the mullion heater. Now, take the brown wire leading from the right side of the 115-V plug. It is spliced to the orange wire that connects to the door switch and the mullion heater. This means that the mullion heater is on when the plug is inserted into a power source. It also means that the freezer light does not come on until the door is open and the refrigerator door switch is closed. Likewise, the cabinet light does not come on until the door is open and the door switch is closed.

Referring again to Fig. 5-62, note the way in which the defrost controls are wired. Note in this case that the brown wire on the left side of the schematic—coming from the 115-V plug—has the temperature control inserted in series with the rest of the wiring and devices. Tracing from the left to right you will find that a black wire runs from the temperature control to the door switch. An ivory wire runs from the door switch to the freezer fan. Another ivory wire runs from the freezer fan to the defrost control (point 4). If the defrost control switch is up, it completes the path from point 3 to the brown wire that leads back to the 115-V plug. Thus, if the temperature control (refrigerator thermostat) and the freezer door switch are closed and the defrost control switch is

141

REFRIGERATION AND AIR CONDITIONING TECHNOLOGY

5-62. Schematic wiring diagram of a domestic refrigerator.

CHAPTER 5—ELECTRIC MOTORS AND CONTROLS

up, the circuit is complete for the freezer fan to operate.

Note that the defrost control is operated by a timer motor. The timer motor is in the circuit at all times when the temperature control switch is closed. This means the defrost control timer will operate and complete its cycle faster if the thermostat is closed. Therefore, the more the refrigerator compressor runs, the faster the defrost control advances to its predetermined point of operation.

To trace the defrost control's source of power, start at the 115-V plug. Trace from the left side through the temperature control and down the black wire to the defrost timer motor and through it to point 3, then to the brown wire from the other side of the power supply. The defrost timer motor is operating anytime that the temperature control (refrigerator thermostat) is closed.

The defrost solenoid in the circuit between the freezer fan and the timer motor is activated as follows. When the defrost control has its switch in the downward direction (from point 3 to point 2) the circuit is completed from the 115-V plug through the temperature control and defrost solenoid to point 2 on the defrost control and through the switch in the downward position to point 3. Point 3 is connected to the other side of the power line through the brown wire on the right side of the schematic. This completes the circuit for the defrost solenoid. As you can see, the defrost control must be in the downward position (connecting points 2 and 3) to complete the circuit and cause the defrosting cycle to begin. Note that the freezer fan motor is not in the circuit. Thus, the fan in the freezer is not running at this time.

The refrigerator motor is controlled as follows. Starting at the left side of the 115-V plug, trace the brown wire to the junction of the serpentine and temperature control. This temperature control switch shorts out the serpentine when the switch is closed. Thus, the serpentine is not in the circuit when the refrigerator is running. A black wire runs from the temperature control switch to the guardette (circuit breaker). A gray wire leads from the guardette to point L of the relay. From point L to point M on the relay is the relay's coil. This coil (point M to point R on the compressor motor) is in series with the run-winding of the compressor motor. Point R to point C of the compressor motor represents the run-winding of the compressor motor. Point C is *common* to start- and run-winding. Note the drawing of the compressor above the schematic. Here, the S, C, and R points are shown relative to their true location within the refrigerator. It can be seen that the temperature control and guardette must be closed for the run-winding to have a complete circuit to the power source lines.

The relay is in series with the run-winding. When the motor starts, the relay contacts are closed. Current through the contacts also completes its path to the common side of the power line (point C). Once the motor comes up to speed, the run-winding draws more current and causes the relay to be energized. Once energized, the relay opens the contact points and takes the start-winding out of the circuit. When the motor stops again (when the thermostat opens), the relay de-energizes and the contacts close. This means the relay is ready for the next starting sequence. If the relay contacts stick, the start-winding stays in the circuit and draws current. The guardette is brought into action and opens the circuit to protect the motor windings from overheating.

For the refrigerator fan motor to operate, it must have power. It runs when the temperature control and the guardette are closed. To trace the circuit for the fan motor, start at the left side of the 115-V plug. Trace the brown wire through the temperature control, the closed switch, and the guardette. From the #2 position on the guardette, a gray wire is connected to one side of the fan motor. The other side of the fan motor is connected by an orange wire to the brown wire leading to the other side of the 115-V plug. Thus, the temperature control switch and the guardette must be closed before the fan switch can run. Also, the fan motor runs whenever the compressor motor runs.

The serpentine heater is in the circuit whenever the temperature control is off or the refrigerator is not operating. It is a heating element wrapped around the evaporator coil. It prevents frost build-up beween defrosting cycles.

Look again at Fig. 5-62. See if you can more easily read the schematic.

143

REVIEW QUESTIONS

1. State the left-hand rule for current in a conductor.
2. State the right-hand rule for motors.
3. What is the main advantage of a DC series motor?
4. What is the difference between a single-phase motor and a three-phase motor?
5. How does the capacitor start motor differ from a split-phase motor?
6. What is the advantage of a three-phase motor over a single-phase motor?
7. What is a squirrel cage rotor?
8. What is a megger?
9. What is synchronous speed?
10. What is meant by the service factor of a motor?
11. What does NEMA stand for?
12. Describe the operation of a bimetallic thermostat.
13. How is automatic defrost accomplished in today's refrigerators?
14. Where are driers located in a refrigeration system?
15. What is a schematic?
16. What is a serpentine heater?

6

Refrigerants

Refrigerants are used in the process of refrigeration. Refrigeration is a process whereby heat is removed from a substance or a space.

A *refrigerant* is a substance that picks up latent heat when the substance evaporates from a liquid to a gas. This is done at a low temperature and pressure. A refrigerant expels latent heat when it condenses from a gas to a liquid at a high pressure and temperature. The refrigerant cools by absorbing heat in one place and discharging it in another area.

The desirable properties of a good refrigerant for commercial use are:
- Low boiling point.
- Safe and nontoxic.
- Easy to liquefy at moderate pressure and temperature.
- High latent heat value.
- Operation on a positive pressure.
- Not affected by moisture.
- Mixes well with oil.
- Noncorrosive to metal.

There are other qualities that all refrigerants have. These qualities are molecular weight, density, compression ratio, heat value, and temperature of compression. These qualities will vary with the refrigerants. The compressor displacement and compressor type or design will also influence the choice of refrigerant.

Table 6-A.
Characteristics of Typical Refrigerants

Name	Boiling Point in Degrees F.	Heat of Vaporization at Boiling Point Btu/lb. 1 At.
Sulfur dioxide	14.0	172.3
Methyl chloride	−10.6	177.8
Ethyl chloride	55.6	177.0
Ammonia	−28.0	554.7
Carbon dioxide	−110.5	116.0
Freezol (isobutane)	10.0	173.5
Freon 11	74.8	78.31
Freon 12	−21.7	71.04
Freon 13	−114.6	63.85
Freon 21	48.0	104.15
Freon 22	−41.4	100.45
Freon 113	117.6	63.12
Freon 114	38.4	58.53
Freon 115	−37.7	54.20
Freon 502	−50.1	76.46

CLASSIFICATIONS OF REFRIGERANTS

Refrigerants are classified according to their manner of absorption or extraction of heat from substances to be refrigerated. The classifications can be broken down into Class 1, Class 2, and Class 3.

Class 1 refrigerants are used in the standard compression type of refrigeration systems. Class 2 refrigerants are used as immediate cooling agents between Class 1 and the substance to be refrigerated. They do the same work for Class 3. Class 3 refrigerants are used in the standard absorption type of refrigerating systems.

- *Class 1.* This class includes those refrigerants that cool by absorption or extraction of heat from the substances to be refrigerated by the absorption of their latent heats. Table 6-A lists the characteristics of typical refrigerants.

- *Class 2.* The refrigerants in this class are those that cool substances by absorbing their sensible heats. They are air, calcium

145

chloride brine, sodium chloride (salt) brine, alcohol, and similar nonfreezing solutions.

• *Class 3.* This group consists of solutions that contain absorbed vapors of liquefiable agents or refrigerating media. These solutions function through their ability to carry the liquefiable vapors. The vapors produce a cooling effect by the absorption of their latent heat. An example is aqua ammonia, which is a solution composed of distilled water and pure ammonia.

COMMON REFRIGERANTS

Following are some of the more common refrigerants.

Sulfur Dioxide

Sulfur dioxide (SO_2) is a colorless gas or liquid. It is toxic, with a very pungent odor. When sulfur is burned in air, sulfur dioxide is formed. When sulfur dioxide combines with water it produces sulfuric and sulfurous acids. These acids are very corrosive to metal. They have an adverse effect on most materials. Sulfur dioxide is not considered a safe refrigerant. Sulfur dioxide is not considered safe when used in large quantities. As a refrigerant, sulfur dioxide operates on a vacuum to give the temperatures required. Moisture in the air will be drawn into the system when a leak occurs. This means the metal parts will eventually corrode, causing the compressor to seize.

Sulfur dioxide (SO_2) boils at 14° F. [−10°C] and has a heat of vaporization at boiling point (1 atmosphere) of 172.3 Btu/lb. It has a latent heat value of 166 Btu per pound.

To produce the same amount of refrigeration, sulfur dioxide requires about one-third more vapor than Freon® and methyl chloride. This means the condensing unit has to operate at a higher speed or the compressor cylinders must be larger. Since sulfur dioxide does not mix well with oil the suction line must be on a steady slant to the machine. Otherwise, the oil will trap out, constricting the suction line. This refrigerant is not feasible for use in some locations.

Methyl Chloride

Methyl chloride (CH_3Cl) has a boiling point of −10.6° F. [−23.3°C]. It also has heat of vaporization at boiling point (at 1 atmosphere) of 177.8 Btu/lb. It is a good refrigerant. However, because it will burn under some conditions some cities will not allow it to be used. It is easy to liquefy and has a comparatively high latent heat value. It does not corrode metal when in its dry state. However, in the presence of moisture it damages the compressor. A sticky black sludge is formed when excess moisture combines with the chemical. Methyl chloride mixes well with oil. It will operate on a positive pressure as low as −10° F. [−23°C]. The amount of vapor needed to cause discomfort in a person is in proportion to the following numbers:

Carbon dioxide	100
Methyl chloride	70
Ammonia	2
Sulfur dioxide	1

That means methyl chloride is 35 times safer than ammonia and 70 times safer than sulfur dioxide.

Methyl chloride is hard to detect with the nose or eyes. It does not produce irritating effects. Therefore, some manufacturers add a 1% amount of *acrolein* as a warning agent. Acrolein is a colorless liquid (C_3H_4O) with a pungent odor. It is produced by destructive distillation of fats.

Ammonia

Ammonia (NH_3) is used most frequently in large industrial plants. Freezers for packing houses usually employ ammonia as a refrigerant. It is a gas with a very noticeable odor. Even a small leak can be detected with the nose. Its boiling point at normal atmospheric pressure is −28° F. [−33°C]. Its freezing point is −107.86° F. [−77.7°C]. It is very soluble in water. Large refrigeration capacity is possible with small machines. It has high latent heat (555 Btu at 18° F. [−7.7°C]). It can be used with steel fittings. Water-cooled units are commonly used to cool down the refrigerant. High pressures are used in the lines (125 to 200 pounds per square inch). Anyone inside the refrigeration unit when it springs a leak is rapidly overcome by the fumes. Fresh air is necessary to reduce the toxic effects of ammonia fumes. Ammonia is combustible when combined with certain amounts of air (about one volume of ammonia to two volumes of air). It is even more combustible when combined with oxygen. It is very toxic. Heavy steel fittings are required since pressures of 125 to 200 pounds per square inch are common. The units must be water-cooled.

Carbon Dioxide

Carbon dioxide (CO_2) is a colorless gas at ordinary temperatures. It has a slight odor and an acid taste. Carbon dioxide is nonexplosive and nonflammable. It has a boiling point of 5° F. [−15°C]. A pressure of over 300 pounds per square inch is required to keep it from evaporation. To liquefy the gas, a condenser temperature of 80° F. [26.6°C] and a pressure of approximately 1000 pounds per square inch are needed. Its critical temperature is 87.8° F. [31°C]. It is harmless to breathe except in extremely large concentrations. The lack of oxygen can cause suffocation under certain conditions of carbon dioxide concentration.

Carbon dioxide is used aboard ships and in industrial installations. It is not used in household applications. The main advantage of using carbon dioxide for a refrigerant is that a small compressor can be used. The compressor is very small since a high pressure is required for the refrigerant. Carbon dioxide is, however, very inefficient, compared to other refrigerants. Thus, it is not used in household units.

Calcium Chloride

Calcium chloride ($CaCl_2$) is used only in commercial refrigeration plants. Calcium chloride is used as a simple carrying medium for refrigeration.

Brine systems are used in large installations where there is danger of leakage. They are used also where the temperature fluctuates in the space to be refrigerated. Brine is cooled down by the direct expansion of the refrigerant. It is then pumped through the material or space to be cooled. Here, it absorbs sensible heat.

Most modern plants operate with the brine at low temperature. This permits the use of less brine, less piping or smaller diameter pipe, and smaller pumps. It also lowers pumping costs. Instead of cooling a large volume of brine to a given temperature, the same number of refrigeration units are used to cool a smaller volume of brine to a lower temperature. This results in greater economy. The use of extremely low-freezing brine, such as calcium chloride, is desirable in the case of the shell-type cooler. Salt brine with a minimum possible freezing point of −6° F. [−20.9°C] may solidify under excess vacuum on the cold side of the refrigerating unit. This can cause considerable damage and loss of operating time. There are some cases, in which the cooler has been ruined.

Ethyl Chloride

Ethyl chloride (C_2H_5Cl) is not commonly used in domestic refrigeration units. It is similar to methyl chloride in many ways. It has a boiling point of 55.6° F. [13.1°C] at atmospheric pressure. Critical temperature is 360.5° F. [182.5°C] at a pressure of 784 pounds absolute. It is a colorless liquid or gas with a pungent ethereal odor and a sweetish taste. It is neutral toward all metals. This means that iron, copper, and even tin and lead can be used in the construction of the refrigeration unit. It does, however, soften all rubber compounds and gasket material. Thus, it is best to use only lead for gaskets.

Freon® Refrigerants*

The Freon® refrigerants have been one of the major factors responsible for the tremendous growth of the home refrigeration and air conditioning industries. The safe properties of these products have permitted their use under conditions where flammable or more toxic refrigerants would be hazardous to use. There is a Freon® refrigerant for every application—from home and industrial air conditioning to special low-temperature requirements.

The unusual combination of properties found in the Freon® compounds is the basis for the wide application and usefulness. Table 6-B presents a summary of the specific properties of some of the fluorinated products. Figure 6-1, on page 151, gives the absolute pressure and gage pressure of Freon® refrigerants at various temperatures.

MOLECULAR WEIGHTS

Compounds containing fluorine in place of hydrogen have higher molecular weights and often have unusually low boiling points. For example, methane (CH_4) with a molecular weight of 16 has a boiling point of −258.5° F. [−161.4°C] and is nonflammable. Freon® 14 (CF_4) has a molecular weight of 88 and a boiling point of −198.4° F. [−128°C] and is nonflammable. The effect is even more pronounced when chlorine is also present. Methylene chloride (CH_2Cl_2) has a molecular weight of 85 and boils at 105.2° F. [40.7°C] while Freon® 12 (CCl_2F_2, molecular weight 121) boils at −21.6° F. [−29.8°C]. It can be seen that Freon® compounds are high density materials with low boiling points, low viscosity, and low surface tension. Freon® includes products with boiling points covering a

---
*Freon® is DuPont's registered trademark for its fluorocarbon products.

Table 6-B.
Physical Properties of Freon* Products

	"FREON" 11	"FREON" 12	"FREON" 13	"FREON" 13B1	"FREON" 14
Chemical Formula	CCl_3F	CCl_2F_2	$CClF_3$	$CBrF_3$	CF_4
Molecular Weight	137.37	120.92	104.46	148.92	88.00
Boiling Point at 1 atm °C / °F	23.82 / 74.87	−29.79 / −21.62	−81.4 / −114.6	−57.75 / −71.95	−127.96 / −198.32
Freezing Point °C / °F	−111 / −168	−158 / −252	−181[1] / −294	−168 / −270	−184[2] / −299
Critical Temperature °C / °F	198.0 / 388.4	112.0 / 233.6	28.9 / 83.9	67.0 / 152.6	−45.67 / −50.2
Critical Pressure atm / lbs/sq in abs	43.5 / 639.5	40.6 / 596.9	38.2 / 561	39.1 / 575	36.96 / 543.2
Critical Volume cc/mol / cu ft/lb	247 / 0.0289	217 / 0.0287	181 / 0.0277	200 / 0.0215	141 / 0.0256
Critical Density g/cc / lbs/cu ft	0.554 / 34.6	0.588 / 34.8	0.578 / 36.1	0.745 / 46.5	0.626 / 39.06
Density, Liquid at 25°C (77° F.) g/cc / lbs/cu ft	1.476 / 92.14	1.311 / 81.84	1.298 / 81.05 @ −30°C (−22° F.)	1.538 / 96.01	1.317 / 82.21 @ −80°C (−112° F.)
Density, Sat'd Vapor at Boiling Point g/l / lbs/cu ft	5.86 / 0.367	6.33 / 0.395	7.01 / 0.438	8.71 / 0.544	7.62 / 0.476
Specific Heat, Liquid (Heat Capacity) at 25°C (77° F.) cal/(g)(°C) or Btu/(lb)(° F.)	0.208	0.232	0.247 @ −30°C (−22° F.)	0.208	0.294 @ −80°C (−112° F.)
Specific Heat, Vapor, at Const Pressure (1 atm) cal/(g)(°C) at 25°C (77° F.) or Btu/(lb)(° F.)	0.142 @ 38°C (100° F.)	0.145	0.158	0.112	0.169
Specific Heat Ratio at 25°C and 1 atm C_p/C_v	1.137 @ 38°C (100° F.)	1.137	1.145	1.144	1.159
Heat of Vaporization at Boiling Point cal/g / Btu/lb	43.10 / 77.51	39.47 / 71.04	35.47 / 63.85	28.38 / 51.08	32.49 / 58.48
Thermal Conductivity at 25°C (77° F.) Btu/(hr)(ft)(° F.) Liquid / Vapor (1 atm)	0.0506 / 0.00451	0.0405 / 0.00557	0.0378 / 0.00501 @ −30°C (−22° F.)	0.0234 / 0.00534	0.0361 / 0.00463 @ −80°C (−112° F.)
Viscosity[7] at 25°C (77° F.) Liquid centipoise / Vapor (1 atm) centipoise	0.415 / 0.0107	0.214 / 0.0123	0.170 / 0.0119 @ −30°C (−22° F.)	0.157 / 0.0154	0.23 / 0.0116 @ −80°C (−112° F.)
Surface Tension at 25°C (77° F.) dynes/cm	18	9	14 @ −73°C (−100° F.)	4	4 @ −73°C (−100° F.)
Refractive Index of Liquid at 25°C (77° F.)	1.374	1.287	1.199 @ −73°C (−100° F.)	1.238	1.151 @ −73°C (−100° F.)
Relative Dielectric Strength[8] at 1 atm and 25°C (77° F.) (nitrogen=1)	3.71	2.46	1.65	1.83	1.06
Dielectric Constant Liquid / Vapor (1 atm)[9a]	2.28 @ 29°C / 1.0036 @ 24°C[9b]	2.13 / 1.0032 @ 29°C (84° F.)	1.0024 @ 29°C (84° F.)		1.0012 @ 24.5°C (76° F.)
Solubility of "Freon" in Water at 1 atm and 25°C (77° F.) wt %	0.11	0.028	0.009	0.03	0.0015
Solubility of Water in "Freon" at 25°C (77° F.) wt %	0.011	0.009		0.0095 @ 21°C (70° F.)	
Toxicity	Group 5a[12]	Group 6[12]	probably Group 6[13]	Group 6[12]	probably Group 6[13]

Table 6-B. (Continued)
Physical Properties of Freon* Products

"FREON" 21	"FREON" 22	"FREON" 23	"FREON" 112	"FREON" 113	"FREON" 114	FC 114B2
CHCl$_2$F	CHClF$_2$	CHF$_3$	CCl$_2$F-CCl$_2$F	CCl$_2$F-CClF$_2$	CClF$_2$-CClF$_2$	CBrF$_2$-CBrF$_2$
102.93	86.47	70.01	203.84	187.38	170.93	259.85
8.92 48.06	−40.75 −41.36	−82.03 −115.66	92.8 199.0	47.57 117.63	3.77 38.78	47.26 117.06
−135 −211	−160 −256	−155.2 −247.4	26 79	−35 −31	−94 −137	−110.5 −166.8
178.5 353.3	96.0 204.8	25.9 78.6	278 532	214.1 417.4	145.7 294.3	214.5 418.1
51.0 750	49.12 721.9	47.7 701.4	34[3] 500	33.7 495	32.2 473.2	34.4 506.1
197 0.0307	165 0.0305	133 0.0305	370[3] 0.029	325 0.0278	293 0.0275	329 0.0203
0.522 32.6	0.525 32.76	0.525 32.78	0.55[3] 34	0.576 36.0	0.582 36.32	0.790 49.32
1.366 85.28	1.194 74.53	0.670 41.82	1.634 }[6] @ 30°C 102.1 } (86° F.)	1.565 97.69	1.456 90.91	2.163 135.0
4.57 0.285	4.72 0.295	4.66 0.291	7.02[5] 0.438	7.38 0.461	7.83 0.489	
0.256	0.300	0.345 @ −30°C (−22° F.)		0.218	0.243	0.166
0.140	0.157	0.176		0.161 @ 60°C (140° F.)	0.170	
1.175	1.184	1.191 @ 0 pressure		1.080 @ 60°C (140° F.)	1.084	
57.86 104.15	55.81 100.45	57.23 103.02	37 (est) 67	35.07 63.12	32.51 58.53	25 (est) 45 (est)
0.0592 0.00506	0.0507 0.00609	0.0569 } @ −30°C 0.0060 } (−22° F.)	0.040	0.0434 0.0044 (0.5 atm)	0.0372 0.0060	0.027
0.313 0.0114	0.198 0.0127	0.167 } @ −30°C 0.0118 } (−22° F.)	1.21[6]	0.68 0.010 (0.1 atm)	0.36 0.0112	0.72
18	8	15 @ −73°C (−100° F.)	23 @ 30°C (86° F.)	17.3	12	18
1.354	1.256	1.215 @ −73°C (−100° F.)	1.413	1.354	1.288	1.367
1.85	1.27	1.04	5 (est)	3.9 (0.44 atm)	3.34	4.02 (0.44 atm)
5.34 @ 28°C 1.0070 @ 30°C	6.11 @ 24°C 1.0071 @ 25.4°C	1.0073 @ 25°C[9b]	2.54 @ 25°C (77° F.)	2.41 @ 25°C (77° F.)	2.26 @ 25°C 1.0043 @ 26.8°C	2.34 @ 25°C (77° F.)
0.95	0.30	0.10	0.012 (Sat'n Pres)	0.017 (Sat'n Pres)	0.013	
0.13	0.13			0.011	0.009	
much less than Group 4, somewhat more than Group 5[12]	Group 5a[12]	probably Group 6[13]	probably less than Group 4, more than Group 5[13]	much less than Group 4, somewhat more than Group 5[12]	Group 6[12]	Group 5a[12]

149

REFRIGERATION AND AIR CONDITIONING TECHNOLOGY

Table 6-B. (Continued)
Physical Properties of Freon* Products

	"FREON" 115	"FREON" 116	"FREON" 500	"FREON" 502	"FREON" 503
Chemical Formula	CClF$_2$-CF$_3$	CF$_3$-CF$_3$	a	b	c
Molecular Weight	154.47	138.01	99.31	111.64	87.28
Boiling Point at 1 atm °C °F.	−39.1 −38.4	−78.2 −108.8	−33.5 −28.3	−45.42 −49.76	−87.9 −126.2
Freezing Point °C °F.	−106[10] −159	−100.6 −149.1	−159 −254		
Critical Temperature °C °F.	80.0 175.9	19.7[4] 67.5	105.5 221.9	82.2 179.9	19.5 67.1
Critical Pressure atm lbs/sq in abs	30.8 453	29.4[4] 432	43.67 641.9	40.2 591.0	43.0 632.2
Critical Volume cc/mol cu ft/lb	259 0.0269	225 0.0262	200.0 0.03226	199 0.02857	155 0.0284
Critical Density g/cc lbs/cu ft	0.596 37.2	0.612 38.21	0.4966 31.0	0.561 35.0	0.564 35.21
Density, Liquid g/cc at 25°C (77° F.) lbs/cu ft	1.291 80.60	1.587 [4] @ −73°C 99.08 (−100° F.)	1.156 72.16	1.217 75.95	1.233 @ −30°C 76.95 (−22° F.)[7]
Density, Sat'd Vapor g/l at Boiling Point lbs/cu ft	8.37 0.522	9.01[4] 0.562	5.278 0.3295	6.22 0.388	6.02 0.374
Specific Heat, Liquid (Heat Capacity) cal/(g)(°C) at 25°C (77° F.) or Btu/(lb)(° F.)	0.285	0.232 @ −73°C (−100° F.)[4]	0.258	0.293	0.287 @ −30°C (−22° F.)
Specific Heat, Vapor, at Const Pressure (1 atm) cal/(g)(°C) at 25°C (77° F.) or Btu/(lb)(° F.)	0.164	0.182[11] @ 0 pressure	0.175	0.164	0.16
Specific Heat Ratio at 25°C and 1 atm C$_p$/C$_v$	1.091	1.085 (est) @ 0 pressure	1.143	1.132	1.21 @ −34°C (−30° F.)
Heat of Vaporization cal/g at Boiling Point Btu/lb	30.11 54.20	27.97 50.35	48.04 86.47	41.21 74.18	42.86 77.15
Thermal Conductivity[7] at 25°C (77° F.) Btu/(hr) (ft) (° F.) Liquid Vapor (1 atm)	0.0302 0.00724	0.045 @ −73°C 0.0098 (−100° F.)	0.0432	0.0373 0.00670	0.0430 @ −30°C (−22° F.)[7]
Viscosity[7] at 25°C (77° F.) Liquid centipoise Vapor (1 atm) centipoise	0.193 0.0125	0.30 0.0148	0.192 0.0120	0.180 0.0126	0.144 @ −30°C (−22° F.)
Surface Tension at 25°C (77° F.) dynes/cm	5	16 @ −73°C (−100° F.)	8.4	5.9	6.1 @ −30°C (−22° F.)
Refractive Index of Liquid at 25°C (77° F.)	1.214	1.206 @ −73°C (−100° F.)	1.273	1.234	1.209 @ −30°C (−22° F.)
Relative Dielectric Strength[8] at 1 atm and 25°C (77° F.) (nitrogen=1)	2.54	2.02		1.3	
Dielectric Constant Liquid Vapor (1 atm)[9a]	1.0035 @ 27.4°C	1.0021 @ 23°C (73° F.)		6.11 @ 25°C 1.0035 (0.5 atm)	
Solubility of "Freon" in Water at 1 atm and 25°C (77° F.) wt %	0.006				0.042
Solubility of Water in "Freon" at 25°C (77° F.) wt %			0.056	0.056	
Toxicity	Group 6[12]	probably Group 6[13]	Group 5a	Group 5a[12]	probably Group 6[13]

*FREON is Du Pont's registered trademark for its fluorocarbon products.

a. CCl$_2$F$_2$/CH$_3$CHF$_2$ (73.8/26.2% by wt.)
b. CHClF$_2$/CClF$_2$CF$_3$ (48.8/51.2% by wt.)
c. CHF$_3$/CClF$_3$ (40/60% by wt.)

CHAPTER 6—REFRIGERATION

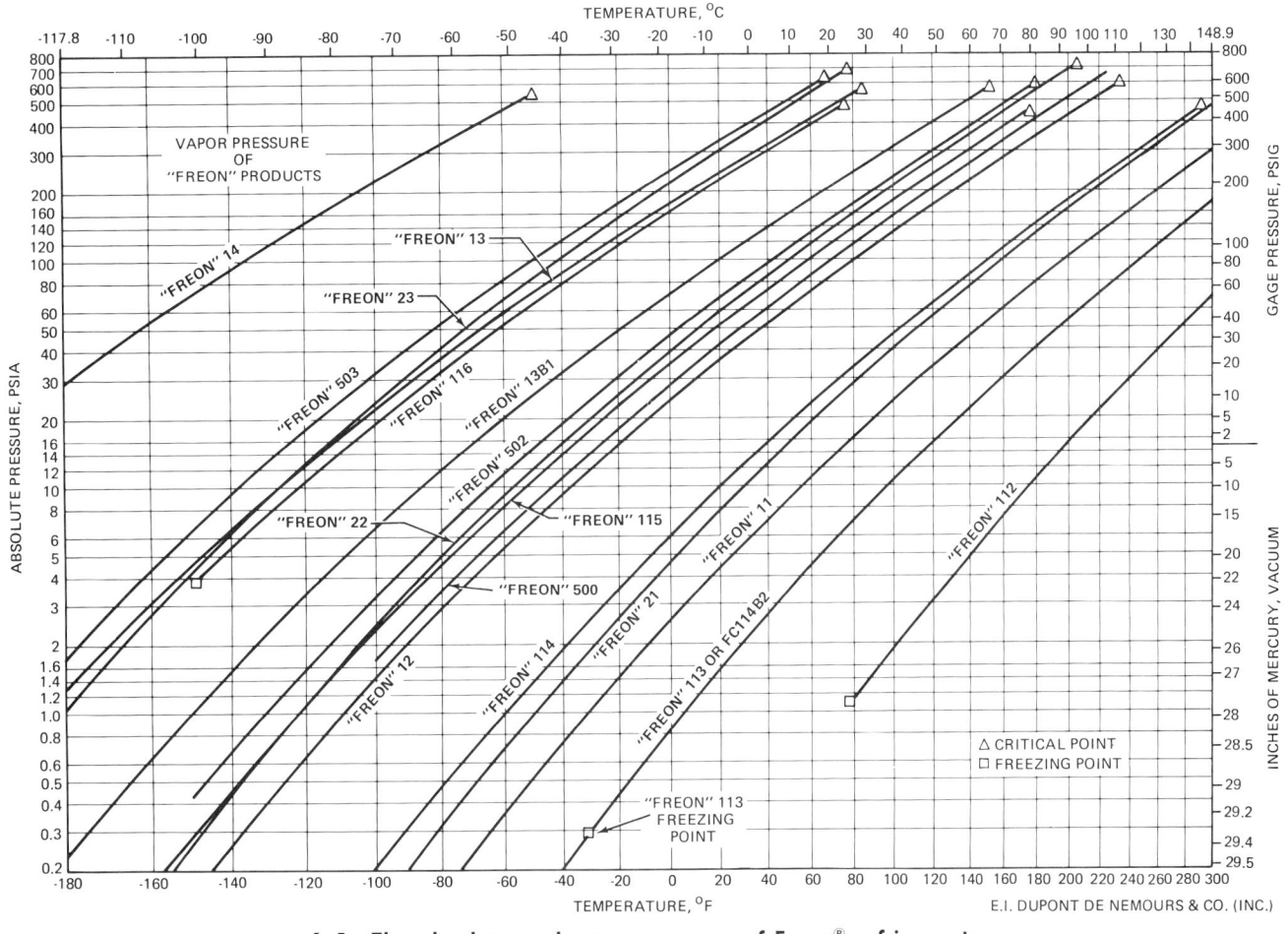

6-1. The absolute and gage pressures of Freon® refrigerants.

wide range of temperatures. Table 6-C.

The high molecular weight of the Freon® compounds also contributes to low vapor, specific heat values, and fairly low latent heats of vaporization. Tables of thermodynamic properties including enthalpy, entropy, pressure, density, and volume for the liquid and vapor are available from manufacturers.

Freon® compounds are poor conductors of electricity. In general, they have good dielectric properties.

FLAMMABILITY

None of the Freon® compounds are flammable or explosive. However, mixtures with flammable liquids or gases may be flammable and should be handled with caution. Partially halogenated compounds may also be flammable and must be individually examined.

TOXICITY

Toxicity means intoxicating or poisonous. One of the most important qualities of the Freon® fluorocarbon compounds is their low toxicity under normal conditions of handling and usage. However, the possibility of serious injury or death exists under unusual or uncontrolled exposures or in deliberate abuse by inhalation of concentrated vapors. The potential hazards of fluorocarbons are summarized in Table 6-D.

SKIN EFFECTS

Liquid fluorocarbons with boiling points below 32° F. [0°C] may freeze the skin, causing frostbite on contact. Suitable protective gloves and clothing give insulation protection. Eye protection should be used. In the event of frostbite, warm the affected area quickly to body temperature. Eyes should be flushed copiously with water. Hands may be held under armpits or im-

151

REFRIGERATION AND AIR CONDITIONING TECHNOLOGY

Table 6-C.
Fluoronated Products and Their Molecular Weight and Boiling Point

FREON PRODUCTS				
Product	Formula	Molecular Weight	Boiling Point °F.	Boiling Point °C
Freon 14	CF$_4$	88.0	−198.3	−128.0
Freon 503	CHF$_3$/CClF$_3$	87.3	−127.6	−88.7
Freon 23	CHF$_3$	70.0	−115.7	−82.0
Freon 13	CClF$_3$	104.5	−114.6	−81.4
Freon 116	CF$_3$–CF$_3$	138.0	−108.8	−78.2
Freon 13B1	CBrF$_3$	148.9	−72.0	−57.8
Freon 502	CHClF$_2$/CClF$_2$–CF$_3$	111.6	−49.8	−45.4
Freon 22	CHClF$_2$	86.5	−41.4	−40.8
Freon 115	CClF$_2$–CF$_3$	154.5	−37.7	−38.7
Freon 500	CCl$_2$F$_2$/CH$_3$CHF$_2$	99.3	−28.3	−33.5
Freon 12	CCl$_2$F$_2$	120.9	−21.6	−29.8
Freon 114	CClF$_2$–CClF$_2$	170.9	38.8	3.8
Freon 21	CHCl$_2$F	102.9	48.1	8.9
Freon 11	CCl$_3$F	137.4	74.9	23.8
Freon 113	CCl$_2$F–CClF$_2$	187.4	117.6	47.6
Freon 112	CCl$_2$F–CCl$_2$F	203.9	199.0	92.8

OTHER FLUORINATED COMPOUNDS				
Product	Formula	Molecular Weight	Boiling Point °F.	Boiling Point °C
FC 114B2	CBrF$_2$–CBrF$_2$	259.9	117.1	47.3
1,1-Difluoroethane*	CH$_3$–CHF$_2$	66.1	−13.0	−25.0
1,1,1-Chlorodifluoroethane**	CH$_3$–CClF$_2$	100.5	14.5	−9.7
Vinyl Fluoride	CH$_2$=CHF	46.0	−97.5	−72.0
Vinylidene Fluoride	CH$_2$=CF$_2$	64.0	−122.3	−85.7
Hexafluoroacetone	CF$_3$COCF$_3$	166.0	−18.4	−28.0
Hexafluoroisopropanol	(CF$_3$)$_2$CHOH	168.1	136.8	58.2

* Propellant or refrigerant 152a
** Propellant or refrigerant 142b
Copyright 1969 by E. I. du Pont de Nemours and Company, Wilmington, Delaware 19898

ORAL TOXICITY

Fluorocarbons are low in oral toxicity as judged by single-dose administration or repeated dosing over long periods of time.

However, direct contact of liquid fluorocarbons with lung tissue can result in chemical pneumonitis, pulmonary edema, and hemorrhage. Fluorocarbons 11 and 113, like many petroleum distillates, are fat solvents and can produce such an effect. If products containing these fluorocarbons were accidentally or purposely ingested, induction of vomiting would be contraindicted (medically wrong). In other words, *do NOT induce vomiting*.

CENTRAL NERVOUS SYSTEM (CNS) EFFECTS

Inhalation of concentrated fluorocarbon vapors can lead to CNS effects comparable to the effects of general anesthesia. The first symptom is a feeling of intoxication. This is followed by a loss of coordination and unconsciousness. Under severe conditions, death can result. If these symptoms are felt, the exposed individual should immediately go or be moved to fresh air. Medical attention should be sought promptly. *Individuals exposed to fluorocarbons should NOT be treated with adrenalin (epinephrine)*.

CARDIAC SENSITIZATION

Fluorocarbons can, in sufficient vapor concentration, produce cardiac sensitization. This is a sensitization of the heart to adrenaline brought about by exposure to high concentrations of organic vapors. Under severe exposure, cardiac arrhythmias

mersed in warm water. Get medical attention immediately.

Fluorocarbons with boiling points at or above ambient temperature tend to dissolve protective fat from the skin. This leads to skin dryness and irritation, particularly after prolonged or repeated contact. Such contact should be avoided by using rubber gloves or plastic gloves. Eye protection and face shields should be used if splashing is possible. If irritation occurs following contact, seek medical attention.

CHAPTER 6—REFRIGERANTS

Table 6-D.
Potential Hazards of Fluorocarbons

Condition	Potential Hazard	Safeguard
Vapors may decompose in flames or in contact with hot surfaces.	Inhalation of toxic decomposition products.	Good ventilation. Toxic decomposition products serve as warning agents. Avoid misuse.
Vapors are 4 to 5 times heavier than air. High concentrations may tend to accumulate in low places.	Inhalation of concentrated vapors can be fatal.	Forced-air ventilation at the level of vapor concentration. Individual breathing devices with air supply.
Deliberate inhalation to produce intoxication.	Can be fatal.	Lifelines when entering tanks or other confined areas. Do not administer epinephrine or other similar drugs.
Some fluorocarbon liquids tend to remove natural oils from the skin.	Irritation of dry, sensitive skin.	Gloves and protective clothing.
Lower boiling liquids may be splashed on skin.	Freezing.	Gloves and protective clothing.
Liquids may be splashed into eyes.	Lower boiling liquids may cause freezing. Higher boiling liquids may cause temporary irritation and if other chemicals are dissolved, may cause serious damage.	Wear eye protection. Get medical attention. Flush eyes for several minutes with running water.
Contact with highly reactive metals.	Violent explosion may occur.	Test the proposed system and take appropriate safety precautions.

DuPont

may result from sensitization of the heart to the body's own levels of adrenaline. This is particularly so under conditions of emotional or physical stress, fright, panic, etc. Such cardiac arrhythmias may result in ventricular fibrillation and death. Exposed individuals should immediately go or be removed to fresh air. There, the hazard of cardiac effects will rapidly decrease. Prompt medical attention and observation should be provided following accidental exposures. A worker adversely affected by fluorocarbon vapors should NOT be treated with adrenalin (epinephrine) or similar heart stimulants since these would increase the risk of cardiac arrhythmias.

THERMAL DECOMPOSITION

Fluorocarbons decompose when exposed directly to high temperatures. Flames and electrical resistance heaters, for example, will chemically decompose fluorocarbon vapors. Products of this decomposition in air include halogens and halogen acids (hydrochloric, hydrofluoric, and hydrobromic), as well as other irritating compounds. Although much more toxic than the parent fluorocarbon, these decomposition products tend to irritate the nose, eyes, and upper respiratory system. This provides a warning of their presence. The practical hazard is relatively slight. It is difficult for a person to remain voluntarily in the presence of decomposition products at concentrations where physiological damage occurs.

When such irritating decomposition products are detected, the area should be evacuated and ventilated. The source of the problem should be corrected.

Applications of Freon® Refrigerants

There is a Freon® refrigerant for every application from home and industrial air conditioning to special low-temperature requirements. Following are a few of the Freon® refrigerants.

• *Freon® 11* (CCl_3F) has a boiling point of 74.9° F. [23.8°C] and is widely used in centrifugal compressors for industrial and commercial air-conditioning systems. It is also used for industrial process water and brine cooling. Its low viscosity and freezing point have also led to its use as a low-temperature brine.

• *Freon® 12* (CCl_2F_2) has a boiling point of −21.6° F. [−29.8°C] and is the most widely known and used of the Freon® refrigerants. It is used principally in household and commercial refrigeration and air conditioning. It is used for refrigerators, frozen food locker plants, water coolers, room and window air-conditioning units and similar equipment. It is generally used in reciprocating compressors ranging in size from fractional to 800 horsepower. It is also used in the smaller rotary-type compressors. Fig. 6-2.

• *Freon® 13* ($CClF_3$) has a boiling point of −114.6° F. [−81.4°C] and is used in low-temperature specialty applications using reciprocating compressors and generally in cascade with Freon® 12, Freon® 22, or Freon® 522.

• *Freon® 22* ($CHClF_2$) has a boiling point of −41.4° F. [−40.8°C] and is used in all types of household and commercial refrigeration and air-conditioning applications with reciprocating compressors. The outstanding thermodynamic properties of Freon® 22 permit the use of smaller equipment than is possible with similar refrigerants. This makes it especially attractive for uses where size is a problem. Fig. 6-3.

• *Freon® 113* ($CCl_2F \cdot CClF_2$) has a boiling point of 117.6° F. [47.6°C]. It is used in commercial and industrial air conditioning and process water and brine cooling with centrifugal compression. It is especially useful in small tonnage applications.

• *Freon® 114* ($CClF_2 \cdot CClF_2$) has a boiling point of 38.8° F.

6-2. Freon can be purchased in a number of sizes. (A) R-12 can be bought in 14-ounce cans for automotive air conditioners or for adding to systems. (B) The 30-lb. containers are for those who use a great deal of refrigerant.
Virginia Chemicals

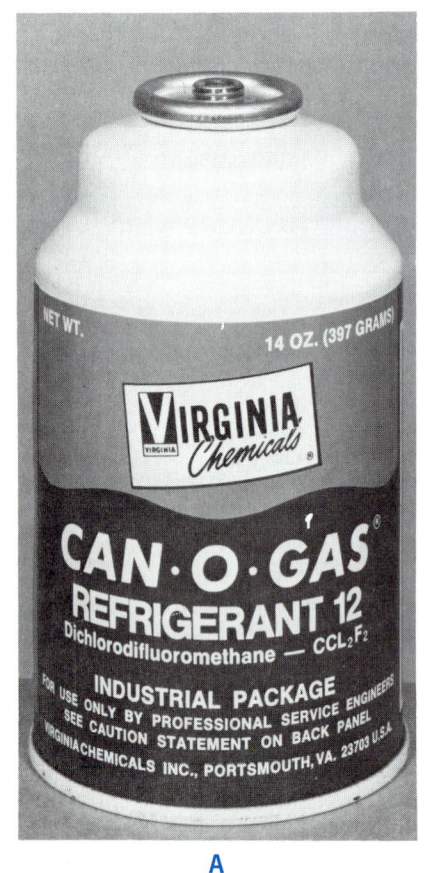

A   B

CHAPTER 6—REFRIGERANTS

6-3. Freon® 22 is marketed in containers of various sizes, such as 1-lb., 2-lb., and 15-lb. cans.
Virginia Chemicals

posed of 73.8 percent Freon® 12 (CCl_2F_2) and 26.2 percent CH_3CHF_2. It boils at $-28.3°$ F. [$-33.5°C$]. It is used in home and commercial air conditioning in small and medium-size equipment and in some refrigeration applications.

- *Freon® 502.* This is an azeotropic mixture also. It consists of 48.8 percent of Freon® 22 and 51.2 percent of Freon® 115, by weight. It boils at $-49.8°$ F. [$-45.4°C$]. With Freon® 502, refrigeration capacity is greater than with Freon® 22. Note the pressure differences on the pressure gage in Fig. 6-4. Discharge temperatures are comparable to those found with Freon® 12. It is finding new applications in low- and medium-temperature cabinets for the display and storage of foodstuffs, in food freezing, and in heat pumps.

- *Freon® 503.* This is an azeotropic mixture of CHF_3 and $CClF_3$. The weight ratio is 40 percent CHF_3 and 60 percent $CClF_3$. The boiling point of this mixture is $-127.6°$ F. [$-88.7°C$]. It is used in low-temperature cascade systems.

- *Freon® 13B1.* $CBrF_3$ boils at $-72°$ F. [$-57.8°C$]. It serves the temperature range between Freon® 502 and Freon® 13. Specific applications are under development.

REACTION OF FREON® TO VARIOUS MATERIALS FOUND IN REFRIGERATION SYSTEMS

Metals

Most of the commonly used construction metals—such as steel, cast iron, brass, copper, tin, lead, and aluminum—can be used satisfactorily with the Freon® compounds under normal conditions of use. At high temperatures some of the metals may act as catalysts for the breakdown of the compound. The tendency of metals to promote thermal decomposition of the Freon® compounds is in the following general order. Those metals that *least* promote thermal decomposition are listed first.

Inconel®.
Stainless steel.
Nickel.
1340 steel.
Aluminum.
Copper.
Bronze.
Brass.
Silver.

The above order is only approximate. Exceptions may be found for individual Freon® compounds or for special conditions of use.

Magnesium alloys and aluminum containing more than 2% magnesium are not recommended for use in systems

6-4. Pressure gage for R-12, R-22, and R-502.
Marsh

[3.8°C]. It is used in small refrigeration systems with rotary-type compressors. It is used in large industrial process cooling and air-conditioning systems using multistage centrifugal compressors.

- *Freon® 500* (CCl_2F_2) is an azeotropic mixture. *Azeotropic* means that a mixture is liquid, maintains a constant boiling point, and produces a vapor of the same composition as the mixture with CH_3CHF_2. It is com-

155

containing Freon® compounds where water may be present.

Zinc is not recommended for use with Freon® 113. Experience with zinc and other Freon® compounds has been limited and no unusual reactivity has been observed. However, it is more chemically reactive than other common construction metals. Thus, it would seem wise to avoid its use with the Freon® compounds unless adequate testing is carried out.

Some metals may be questionable for use in applications requiring contact with Freon® compounds for long periods of time or unusual conditions of exposure. These metals, however, can be cleaned safely with Freon® solvents. Cleaning applications are usually for short exposures at moderate temperatures.

Most halocarbons may react violently with highly reactive materials, such as sodium, potassium, and barium in their free metallic form. Materials become more reactive when finely ground or powdered. In this state, magnesium and aluminum may react with fluorocarbons, especially at higher temperatures. Highly reactive materials should not be brought into contact with fluorocarbons until a careful study is made and appropriate safety precautions are taken.

Plastics

A brief summary of the effect of Freon® compounds on various plastic materials follows. However, compatibility should be tested for specific applications. Differences in polymer structure and molecular weight, plasticizers, temperature, and pressure may alter the resistance of the plastic toward the Freon® compound.

- *Teflon - TFE - fluorocarbon resin.* No swelling observed when submerged in Freon® liquids, but some diffusion found with Freon® 12 and Freon® 22.
- *Polychlorotrifluoroethylene.* Slight swelling, but generally suitable for use with Freon® compounds.
- *Polyvinyl alcohol.* Not affected by the Freon® compounds, but very sensitive to water. Used especially in tubing with an outer protective coating.
- *Vinyl.* Resistance to the Freon® compounds depends on vinyl type and plasticizer. Considerable variation is found. Samples should be tested before use.
- *Orlon—acrylic fiber.* Generally suitable for use with the Freon® compounds.
- *Nylon.* Generally suitable for use with Freon® compounds, but may tend to become brittle at high temperatures in the presence of air or water. Tests at 250° F. [121°C] with Freon® 12 and Freon® 22 showed the presence of water or alcohol to be undesirable. Adequate testing should be carried out.
- *Polyethylene.* May be suitable for some applications at room temperatures. However, it should be thoroughly tested since greatly different results have been found with different samples.
- *Lucite®—acrylic resin (methacrylate polymers).* Dissolved by Freon® 22. However, it is generally suitable for use with Freon® 12 and Freon® 114 for short exposure. On long exposure, it tends to crack, craze, and become cloudy. Use with Freon® 113 may be questionable. It probably should not be used with Freon® 11.

Cast *Lucite®* acrylic resin is much more resistant to the effect of solvents than extruded resin. It can probably be used with most of the Freon® compounds.

- *Polystyrene.* Considerable variation found in individual samples. However, it is generally not suited for use with Freon® compounds. Some applications might be all right with Freon® 114.
- *Phenolic resins.* Usually not affected by the Freon® compounds. However, composition of resins of this type may be quite different. Samples should be tested before use.
- *Epoxy resins.* Resistant to most solvents and entirely suitable for use with the Freon® compounds.
- *Cellulose acetate or nitrate.* Suitable for use with Freon® compounds.
- *Delrin-acetal resin.* Suitable for use with Freon® compounds under most conditions.
- *Elastomers.* Considerable variation is found in the effect of the Freon® compounds on elastomers. The effect depends on the particular compound and elastomer type. In nearly all cases a satisfactory combination can be found. In some instances the presence of other materials, such as oils, may give unexpected results. Thus, preliminary testing of the system involved is recommended.

REFRIGERANT PROPERTIES

Refrigerants can be characterized by a number of properties. These properties are pressure, temperature, volume, density, and enthalpy. Also, flammabil-

CHAPTER 6—REFRIGERANTS

Table 6-E.
Operating Pressures

Refrigerant	Evaporating Pressure (PSIG) at 5° F.	Condensing Pressure (PSIG) at 86° F.
R–11	24.0 in. Hg	3.6
R–12	11.8	93.2
R–22	28.3	159.8
R–717	19.6	154.5
R–718	29.7	28.6

ity, ability to mix with oil, moisture reaction, odor, toxicity, leakage tendency, and leakage detection are important properties that characterize refrigerants.

Freon® refrigerants R-11, R-12, R-22, plus ammonia and water will be used to show their properties in relationship to the abovementioned categories. Freon® R-11, R-12, and R-22 are common Freon® refrigerants. The number assigned to ammonia is R-717, while water has the number R-718.

Pressure

The pressure of a refrigeration system is important. It determines how sturdy the equipment must be to hold the refrigerant. The refrigerant must be compressed and sent to various parts of the system under pressure. The main concern is keeping the pressure as low as possible. The ideal low-side pressure or evaporating pressure should be as near atmospheric pressure (14.7 pounds per square inch) as possible. This keeps down the price of the equipment. It also puts positive pressure on the system at all points. By having a small pressure, it is possible to prevent air and moisture from entering the system. In the case of a vacuum or a low pressure, it is possible for a leak to suck in air and moisture. Note the five refrigerants and their pressures in Table 6-E.

Freon® R-11 is used in very large systems because it requires more refrigerant than others—even though it has the best pressure characteristics of the group. Several factors must be considered before a suitable refrigerant is found. There is no ideal refrigerant for all applications.

Table 6-F.
Boiling Temperature

Refrigerant	Temperature (° F.)
R–11	74.7
R–12	–21.6
R–22	–41.4
R–717 (ammonia)	–28.0
R–718 (water)	212.0

Temperature

Temperature is important in selecting a refrigerant for a particular job. The boiling temperature is that point at which a liquid is vaporized upon the addition of heat. This, of course, depends upon the refrigerant and the absolute pressure at the surface of the liquid and vapor. Note that in Table 6-F, R-22 has the lowest boiling temperature. Water (R-718) has the highest boiling temperature. Atmospheric pressure is 14.7 pounds per square inch.

Once again, there is no ideal atmospheric boiling temperature for a refrigerant. However, temperature-pressure relationships are important in choosing a refrigerant for a particular job.

Volume

Specific volume is defined as the definite weight of a material. Usually expressed in terms of cubic feet per pound, the volume is the reciprocal of density.

The specific volume of a refrigerant is the number of cubic feet of gas that is formed when one pound of the refrigerant is vaporized. This is an important factor to be considered when choosing the size of refrigeration system components. Compare the specific volumes (at 5° F.) of

Table 6-G.
Specific Volumes at 5° F.

Refrigerant	Liquid Volume (cubic feet/lb.)	Vapor Volume (cubic feet/lb.)
R–11	0.010	12.27
R–12	0.011	1.49
R–22	0.012	1.25
R–717	0.024	8.15
R–718 (water)	0.016	12 444.40

REFRIGERATION AND AIR CONDITIONING TECHNOLOGY

Table 6-H.
Liquid Density at 86° F.

Refrigerant	Liquid Density (lbs./cu. ft.)
R-11	91.4
R-12	80.7
R-22	73.4
R-717	37.2
R-718	62.4

Table 6-I.
Enthalpy (Btu/lb. at 5° F. [−15°C])

Refrigerant	Liquid Enthalpy	+	Latent Heat of Vaporization	=	Vapor Enthalpy
R-11	8.88	+	84.00	=	92.88
R-12	9.32	+	60.47	=	78.79
R-22	11.97	+	93.59	=	105.56
R-717	48.30	+	565.00	=	613.30
R-718 (at 40° F.)	8.05	+	1071.30	=	1079.35

the five refrigerants we have chosen. Freon® R-12 and R-22 (the most often used refrigerants) have the lowest specific volumes as vapors. Refer to Table 6-G.

Density

Density is defined as the mass or weight per unit of volume. In the case of a refrigerant, it is the weight in terms of volume given in pounds per cubic foot (lb./cu. ft.). Note in Table 6-H that the density of R-11 is the greatest. The density of R-717 (ammonia) is the least.

Enthalpy

Enthalpy is the *total* heat in a refrigerant. The sensible heat *plus* the latent heat makes up the total heat. Latent heat is the amount of heat required to change the refrigerant from a liquid to a gas. The latent heat of vaporization is a measure of the heat per pound that the refrigerant can absorb from an area to be cooled. It is, therefore, a measure of the cooling potential of the refrigerant circulated through a refrigeration system. Table 6-I. Latent heat is expressed in Btu per pound.

Flammability

Of the five refrigerants mentioned so far, the only one that is flammable is ammonia. None of the Freon® compounds are flammable or explosive. However, mixtures with flammable liquids or gases may be flammable and should be handled with caution. Partially halogenated compounds may also be flammable and must be individually examined. If the refrigerant is used around fire, its flammability should be carefully considered. Some city codes specify which refrigerants cannot be used within city limits.

Capability of Mixing with Oil

Some refrigerants mix well with oil. Others, such as ammonia and water, do not. The ability to mix with oil has advantages and disadvantages. If the refrigerant mixes easily, parts of the system can be lubricated easily by the refrigerant and its oil mixture. The refrigerant will bring the oil back to the compressor and moving parts for lubrication.

There is a disadvantage to the mixing of refrigerant and oil. If it is easily mixed, the refrigerant can mix with the oil during the off cycle and then carry off the oil once the unit begins to operate again. This means that the oil needed for lubrication is drawn off with the refrigerant. This can cause damage to the compressor and moving parts. With this condition, there is foaming in the compressor crankcase and loss of lubrication. In some cases, the compressor is burned out. Procedures for cleaning up a burned-out motor will be given later.

Moisture and Refrigerants

Moisture should be kept out of refrigeration systems. It can corrode parts of the system. Whenever low temperatures are produced, the water or moisture can freeze. If freezing of the metering device occurs, then refrigerant flow is restricted or cut off. The system will have a low efficiency or none at all. The degree of efficiency will depend upon the amount of icing or the part affected by the frozen moisture.

All refrigerants will absorb water to some degree. Those that absorb very little water permit free water to collect and freeze at low-temperature points. Those that absorb a high amount of moisture will form corrosive acids and corrode the system. Some systems will allow water to be absorbed and frozen. This causes corrosion.

Hydrolysis is the reaction of a material, such as Freon® 12 or methyl chloride, with water. Acid materials are formed. The hydrolysis rate for the Freon®

CHAPTER 6—REFRIGERANTS

Table 6-J.
Hydrolysis Rate in Water
Grams/Litre of Water/Year

Compound	1 atm Pressure 86° F. Water Alone	1 atm Pressure 86° F. With Steel	Saturation Pressure 122° F. With Steel
CH_3Cl	*	*	110
CH_2Cl_2	*	*	55
Freon 113	<0.005	ca. 50**	40
Freon 11	<0.005	ca. 10**	28
Freon 12	<0.005	0.8	10
Freon 21	<0.01	5.2	9
Freon 114	<0.005	1.4	3
Freon 22	<0.01	0.1	*
Freon 502	<0.01†	<0.1†	

* Not measured †Estimated ** Observed rates vary

Table 6-K.
Molecular Weights of Selected Refrigerants

Refrigerant	Molecular Weight
R-11	137.4
R-12	120.9
R-22	86.5
R-717 (ammonia)	17.0
R-718 (water)	18.0

compounds as a group is low compared with other halogenated compounds. Within the Freon® group, however, there is considerable variation. Temperature, pressure, and the presence of other materials also greatly affect the rate. Typical hydrolysis rates for the Freon® compounds and other halogenated compounds are given in Table 6-J.

With water alone at atmospheric pressure, the rate is too low to be determined by the analytical method used. When catalyzed by the presence of steel, the hydrolysis rates are detectable but still quite low. At saturation pressures and a higher temperature, the rates are further increased.

Under neutral or acidic conditions, the presence of hydrogen in the molecule has little effect on the hydrolytic stability. However, under alkaline conditions compounds containing hydrogen, such as Freon® 22 and Freon® 21, tend to be hydrolyzed more rapidly.

Odor

The five refrigerants are characterized by their distinct odor or the absence of it. Freon® R-11, R-12, and R-22 have a slight odor. Ammonia (R-717) has a very acrid odor and can be detected even in small amounts. Water (R-718), of course, has no odor.

A slight odor is needed in a refrigerant so that its leakage can be detected. A strong odor may make it impossible to service equipment. Special gas masks may be needed. Some refrigerated materials may be ruined if the odor is too strong. About the only time that an odor is preferred in a refrigerant is when a toxic material is used for a refrigerant. A refrigerant that may be very inflammable should have an odor so that its leakage can be detected easily to prevent fire or explosions.

Toxicity

Toxicity is the characteristic of a material that makes it intoxicating or poisonous. Some refrigerants can be very toxic to humans. Others may not be toxic at all. The halogen refrigerants (R-11, R-12, and R-22) are harmless in their normal condition or state. However, they form a highly toxic gas when an open flame is used around them.

Water, of course, is not toxic. However, ammonia can be toxic if present in sufficient quantities. Make sure the manufacturer's recommended procedures for handling are followed when handling refrigerants.

Tendency to Leak

The size of the molecule makes a difference in the tendency of a refrigerant to leak. The greater the molecular weight, the larger the hole must be for the refrigerant to escape. A check of the molecular weight of a refrigerant will indicate the problem it may present to a sealed refrigeration system. Table 6-K shows that R-11 has the least tendency to leak, whereas ammonia is more likely to leak.

DETECTING LEAKS

There are several tests used to

check for leaks in a closed refrigeration system. Most of them are simple. Following are some useful procedures:

• Hold the joint or suspected leakage point under water and watch for bubbles.

• Coat the area suspected of leakage with a strong solution of soap. If a leak is present, soap bubbles will be produced.

Sulfur Dioxide

To detect sulfur dioxide *leaks,* an ammonia swab may be used. The swab is made by soaking a sponge or cloth—tied onto a stick or piece of wire—in aqua ammonia. Household ammonia may also be used. A dense white smoke forms when the ammonia comes in contact with the sulfur dioxide. The usual soap bubble or oil test may be used when no ammonia is available.

• If ammonia is used, check for leakage in the following ways:
   a. Burn a sulfur stick in the area of the leak. If there is a leak, a dense white smoke will be produced. The stronger the leak, the denser the white smoke.
   b. Hold a wet litmus paper close to the suspected leak area. If there is a leak, the ammonia will cause the litmus paper to change color.

• Refrigerants that are halogenated hydrocarbons (Freon® compounds) can be checked for leakage with a halide leak test. This involves holding a torch or flame close to the leak area. If there is a refrigerant leak, the flame will turn green. In every instance, the room should be well ventilated when the torch test is made. There is presently

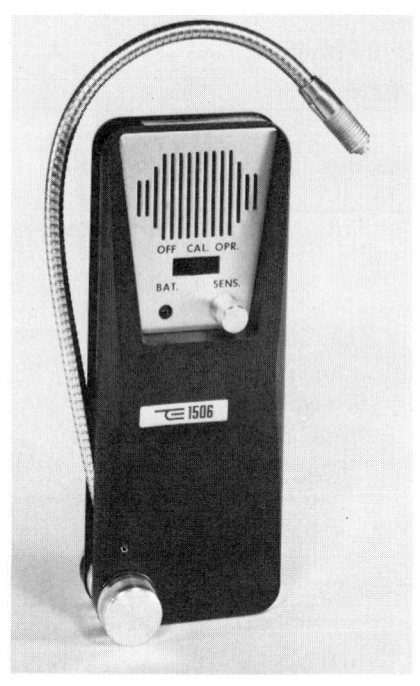

Thermal Engineering
6-5. A hand-held electronic leak detector.

available an electronic detector for such refrigerant leaks. The detector gives off a series of rapid clicks if the refrigerant is present. The higher the concentration of the refrigerant, the more rapid the clicks. Fig. 6-5.

Carbon Dioxide

Leaks can be detected with a soap solution if there is internal pressure on the part to be tested. When carbon dioxide is present in the condenser water, the water will turn yellow with the addition of bromothymol blue.

Ammonia

Leaks are detected (in small amounts of ammonia) when a lit sulfur candle is used. The candle will give off a very thick, white smoke when it contacts the ammonia leak. The use of phenolphthalein paper is also considered a good test. The smallest trace of ammonia will cause the moistened paper strip to turn pink. A large amount of ammonia will cause the phenolphthalein paper to turn a vivid scarlet.

Methyl Chloride

Leaks are detected by a leak-detecting halide torch. Fig. 6-6. Some torches use alcohol for fuel and produce a colorless flame. When a methyl chloride leak is detected, the flame turns *green.* A brilliant blue flame is produced

6-6. A halide gas leak detector.
Turner

when large or stronger concentrations are present. In every instance, the room should be well ventilated when the torch test is made. The combustion of the refrigerant and the flame produces harmful chemicals. If a safe atmosphere is not present, the soap bubble test or oil test should be used to check for leaks.

As mentioned, methyl chloride is hard to detect with the nose or eyes. It does not produce irritating effects. Therefore, some manufacturers add a 1% amount of *acrolein* as a warning agent. Acrolein is a colorless liquid (C_3H_4O) with a pungent odor.

REVIEW QUESTIONS

1. List five desirable properties of a good refrigerant for commercial use.
2. How are refrigerants classified?
3. Where is calcium chloride used in refrigeration systems?
4. What type of refrigerant is used in home refrigerators?
5. Are Freon® compounds flammable?
6. What does toxicity mean?
7. What is a fluorocarbon?
8. What is a halocarbon?
9. Why is the pressure of a refrigeration system important?
10. Define specific volume.
11. How are refrigerant leaks detected?

7
Compressors

REFRIGERATION COMPRESSORS

Refrigeration compressors can be classified according to the following:
- Number of cylinders.
- Method of compression.
- Type of drive.
- Location of the driving force or motor.

The method of compression may be *reciprocating, centrifugal,* or *rotary.* The location of the power source also classifies compressors. Independent compressors are belt driven. Semihermetic compressors have direct drive, with the motor and compressor in separate housings. The hermetic compressor has direct drive, with the motor and compressor in the same housing.

Reciprocating units have a piston in a cylinder. The piston acts as a pump to increase the pressure of the refrigerant from the low side to the high side of the system. A reciprocating compressor can have twelve or more cylinders. Fig. 7-1.

Two methods of capacity control are generally applied to the reciprocating refrigeration compressor on commercial air conditioning systems. Both methods involve mechanical means of unloading cylinders by holding open the suction valve.

The most common method of capacity control uses an internal multiple-step valve. This applies compressor oil pressure or high-side refrigerant pressure to a bellows or piston that actuates the unloader.

The second method of capacity control uses an external solenoid valve for each cylinder. The solenoid valve allows compressor oil or high-side refrigerant to pass to the unloader.

A centrifugal compressor is basically a fan or blower that builds refrigerant pressure by forcing the gas through a funnel-shaped opening at high speed. Fig. 7-2.

Capacity is usually controlled by opening and closing vanes. These vanes regulate the amount of refrigerant gas allowed to enter the fan or turbine. Fig. 7-3. When the vanes restrict the flow of refrigerant, the turbine cannot do its full amount of work on the refrigerant. Thus, its capacity is limited. Most centrifugal machines can be limited to 10 to 25% of full capacity by this method. Some will operate at almost zero capacity, however. Another, though less common, method is to control the speed of the motor that is turning the turbine.

CONDENSERS

A condenser must take the superheated vapor from the compressor, cool it to its condensing temperature, and then condense it. This action is opposite to that of an evaporator. Generally, two types of condensers

Trane
7-1. Reciprocating compressor.

CHAPTER 7—COMPRESSORS

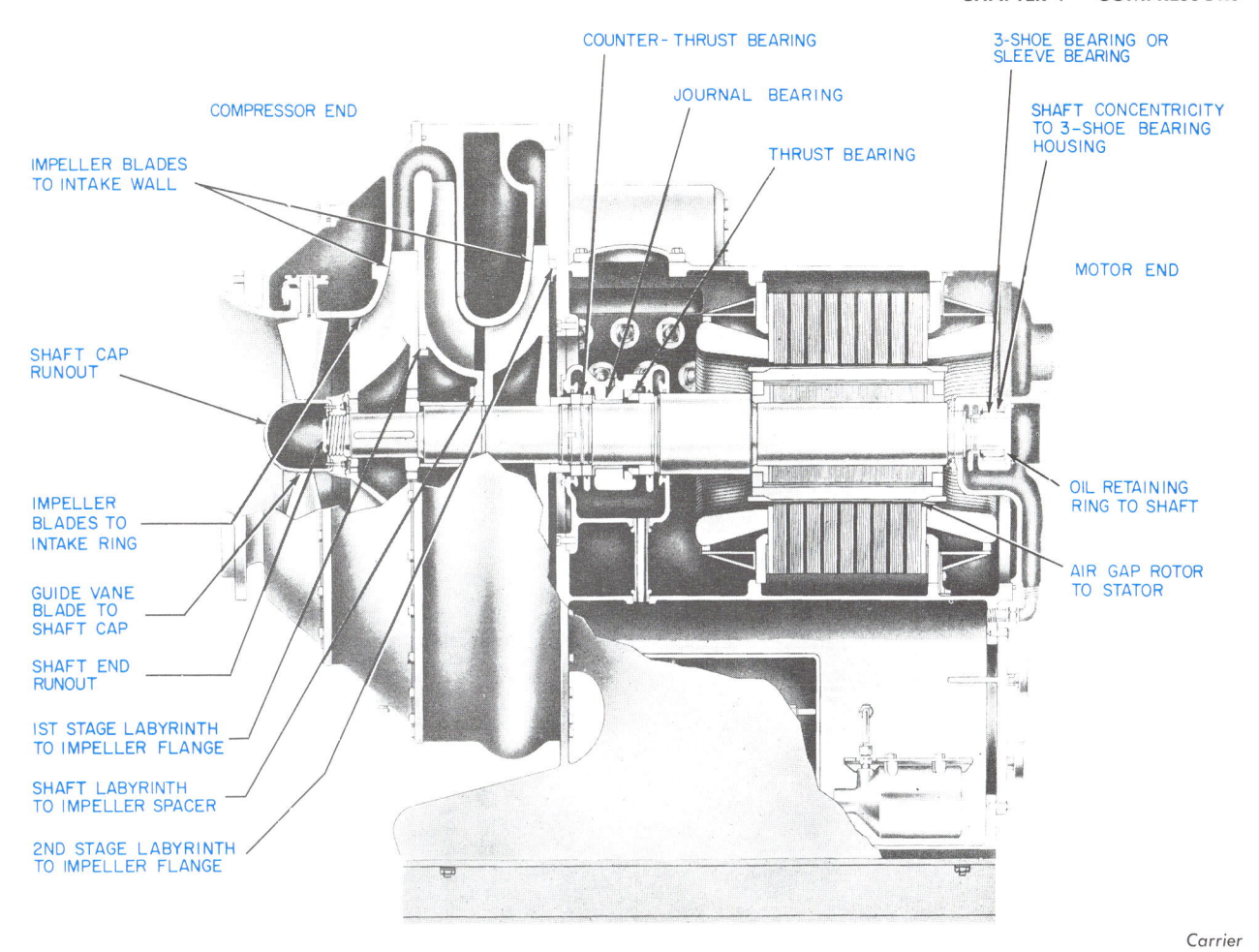

7-2. Centrifugal compressor.

Carrier

are used—air-cooled and water-cooled.

Air-Cooled Condensers

Air-cooled condensers are usually of the fin-and-tube type, with the refrigerant inside the tubes and air flowing in direct contact over the outside. Usually, a fan forces the air over the coil. This increases its cooling capabilities. Figs. 7-4 and 7-5.

Water-Cooled Condensers

In the water-cooled condenser, the refrigerant is cooled with water within pipes. Fig. 7-6. The tubing containing water is placed inside a pipe or housing containing the warm refrigerant. The heat is then transferred from the refrigerant through the tubing to the water. Water-cooled condensers are more efficient than air-cooled condensers. However, they must be supplied with large quantities of water. This water must be either discharged or reclaimed by cooling it to a temperature that makes it reusable.

Reclaiming is usually accomplished by a cooling tower. Figs. 7-7 and 7-8. The tower chills the water by spraying it into a closed chamber. Air is forced over the spray. Cooling towers may be equipped with fans to force the air over the sprayed water.

Another device used to cool refrigerant is an evaporative condenser. Fig. 7-9. Here, the gas-filled condenser is placed in an enclosure. Water is sprayed on it and air forced over it to cool the condenser by evaporation.

HERMETIC COMPRESSORS

A hermetic compressor is a

163

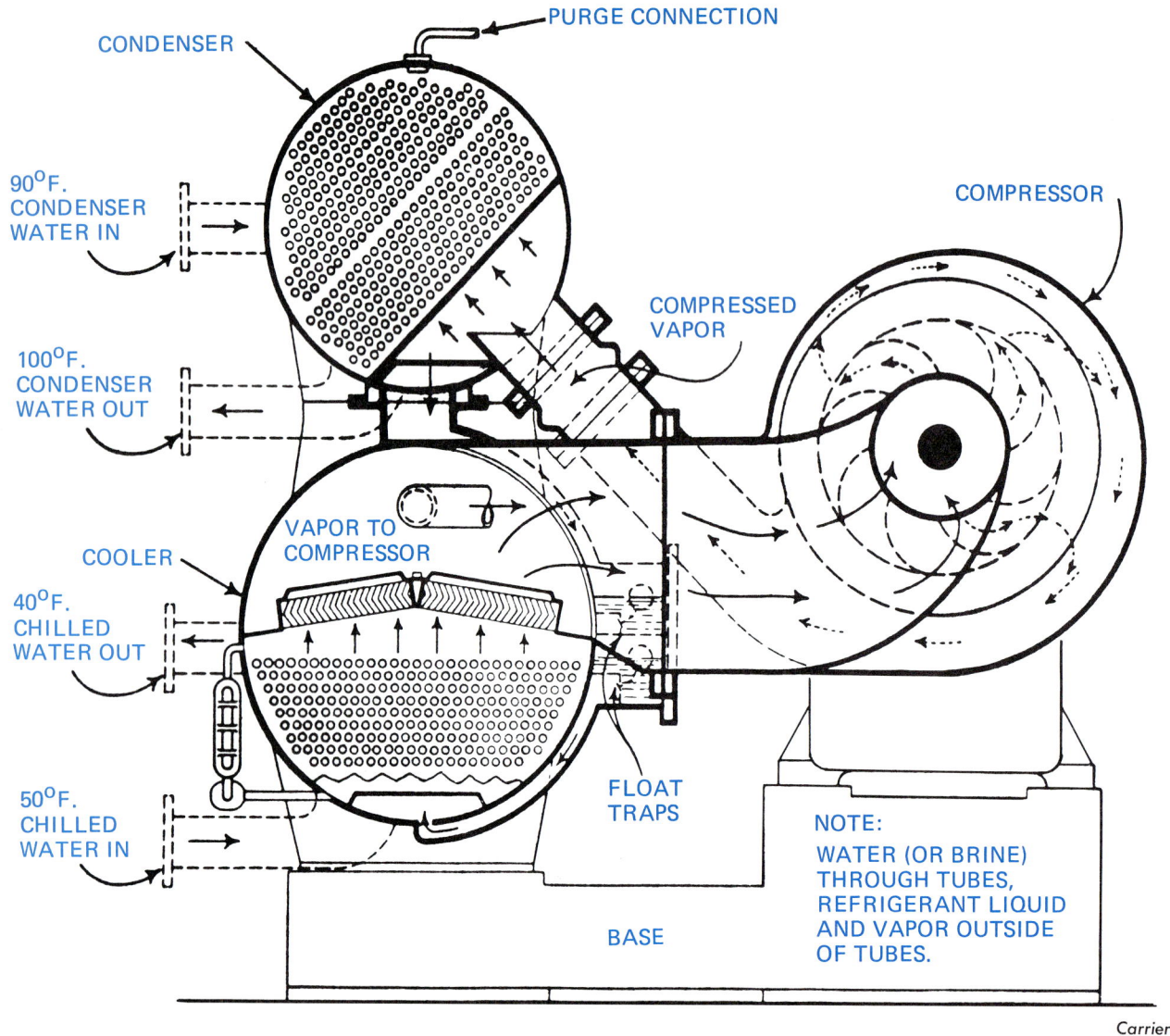

REFRIGERATION AND AIR CONDITIONING TECHNOLOGY

7-3. Centrifugal compressor system.

Carrier

7-4. Schematic of air-cooled condenser.

Johnson

7-5. Air-cooled condenser.

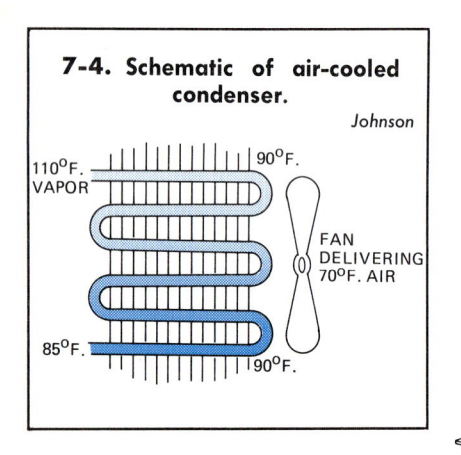

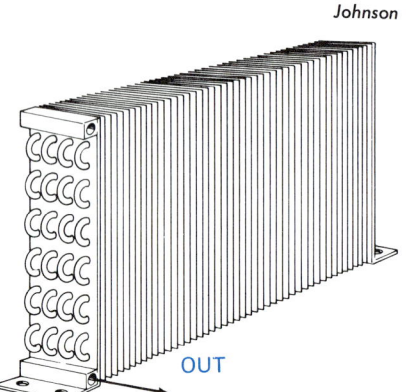

Johnson

direct-connected motor compressor assembly enclosed within a steel housing. It is designed to pump low-pressure refrigerant gas to a higher pressure.

A hermetic container is one that is tightly sealed so no gas or liquid can enter or escape. The container is usually sealed by welding.

Tecumseh hermetic compressors have a low-pressure shell, or housing. This means that the in-

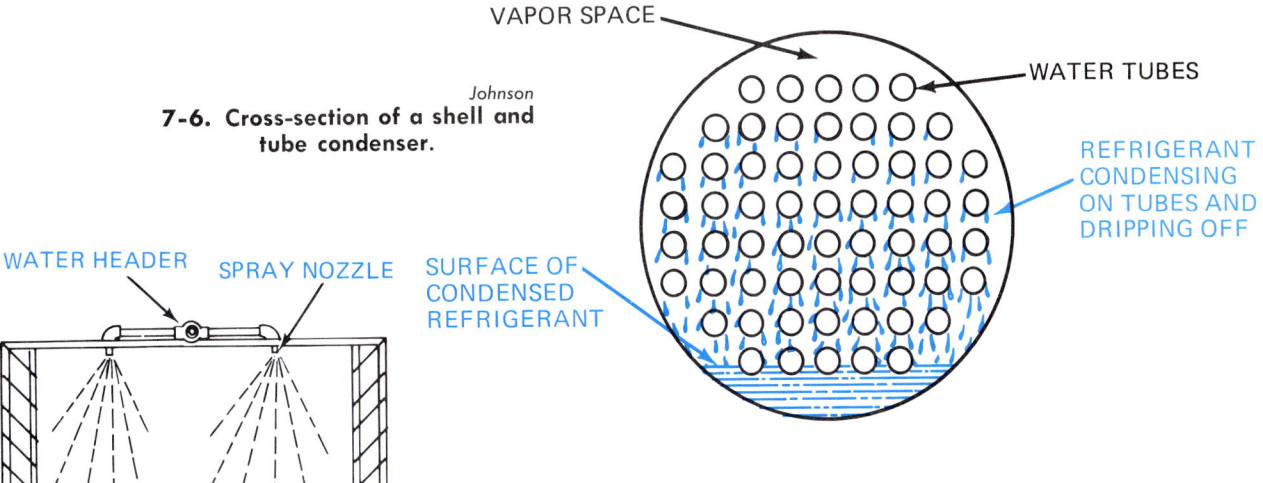

CHAPTER 7—COMPRESSORS

7-6. Cross-section of a shell and tube condenser.

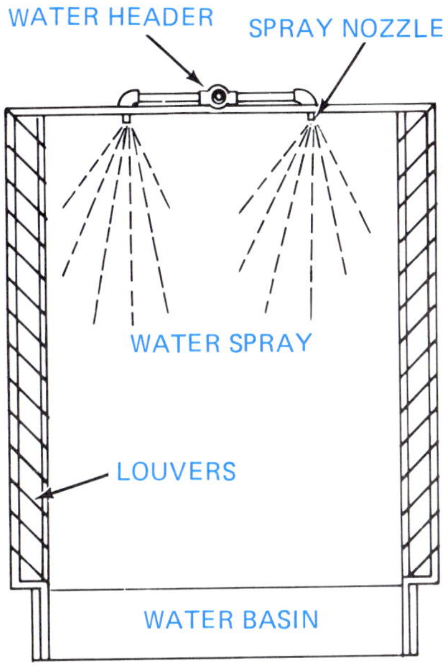

7-7. Spray-type cooling tower.

7-8. Deck-type cooling tower.

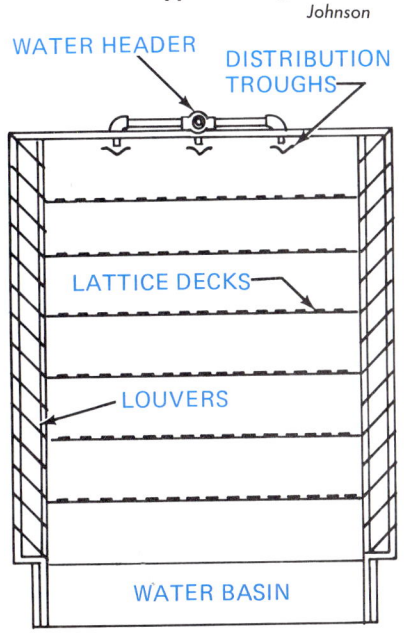

7-9. Evaporative condenser.

terior of the compressor housing is subjected only to suction pressure. It is not subjected to the discharge pressure created by the piston stroke. This point is emphasized to stress the hazard of introducing high-pressure gas into the compressor shell at pressures above 150 psig.

The major internal parts of a hermetic compressor are shown in Fig. 7-10.

The suction is drawn into the compressor shell, then to and through the electric motor that provides power to the crankshaft. The crankshaft revolves in its bearings, driving the piston or pistons in the cylinder or cylinders. The crankshaft is designed to carry oil from the oil pump in the bottom of the compressor to all bearing surfaces. Refrigerant gas surrounds the compressor crankcase and the motor as it is drawn through the compressor shell and into the cylinder or cylinders, through the suction muf-

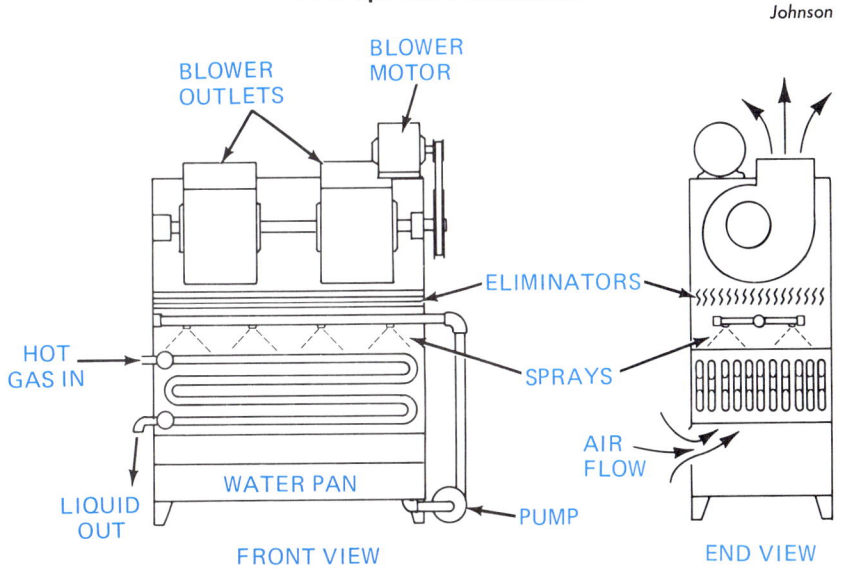

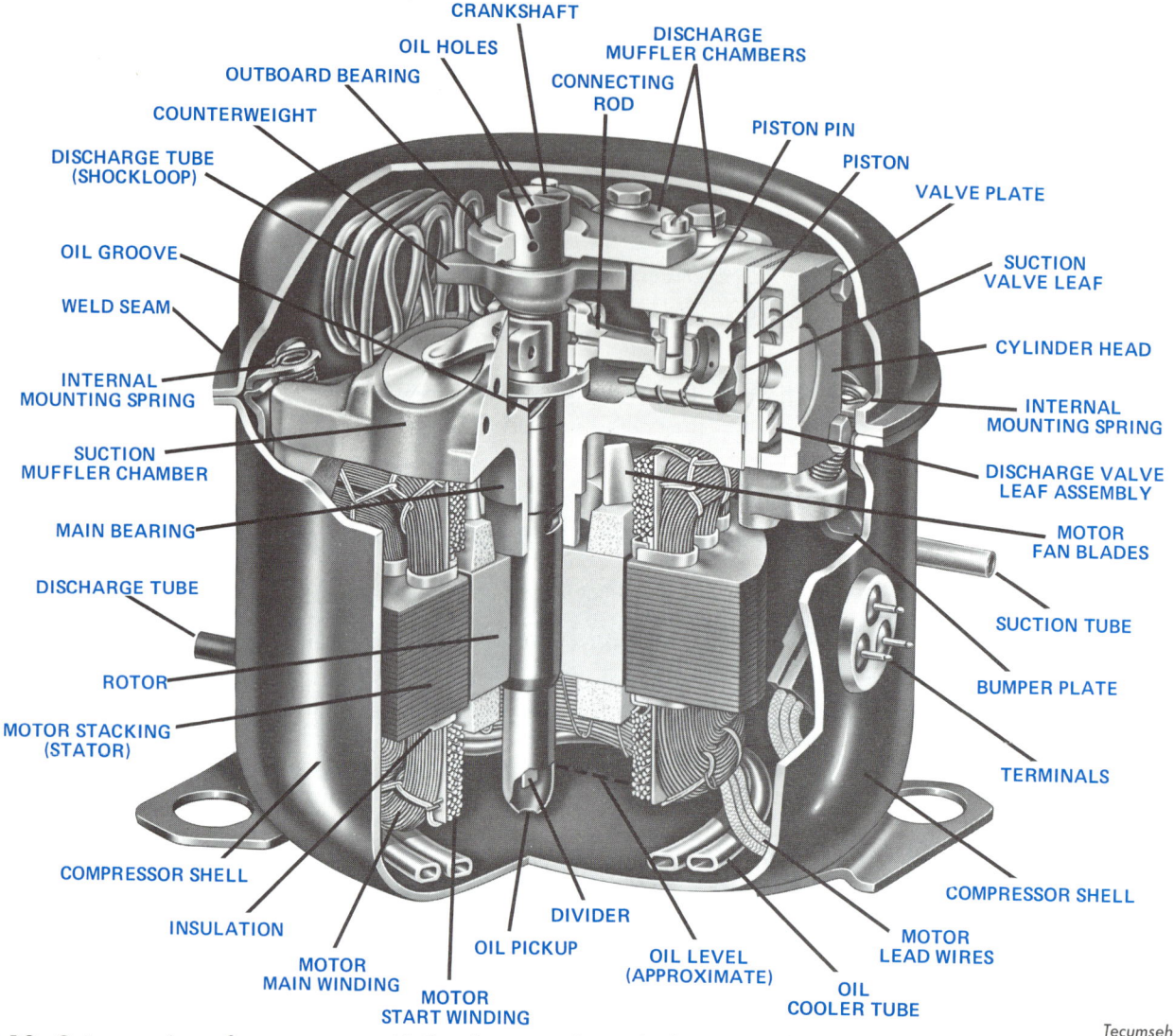

REFRIGERATION AND AIR CONDITIONING TECHNOLOGY

7-10. Cutaway view of a compressor. Notice the motor is on the bottom and the compressor piston is on the top. *Tecumseh*

fler and suction valves. As the gas is compressed by the moving piston, it is released through the discharge valves, discharge muffler, and compressor discharge tube.

Compressor Types

Hermetic compressors have different functions. Some are used for home refrigeration. Some are used to produce air conditioning. Others are used in home or commercial freezers. Hermetic compressors are also used for food display cases.

One of the largest manufacturers of hermetic compressors is Tecumseh. Their plants are located at Tecumseh, Michigan, and Marion, Ohio. The serial plate on the compressor tells several things about the compressor. Fig. 7-11. Also, notice that several manufacturers make the motor for the compressor. Delco, Emerson, General Electric, Wagner, Ranco, A.O. Smith, Aichi, and Westinghouse make electric motors for Tecumseh compressors. Newer models are rated in hertz (Hz), instead of cycles.

The following information is only a brief outline of some of the points to be remembered in servicing.

The *pancake* models are designated with a P as the first letter of the serial number. They are made with $1/20$- to $1/3$-horsepower

CHAPTER 7—COMPRESSORS

TYPICAL SERIAL PLATE 1953-1958

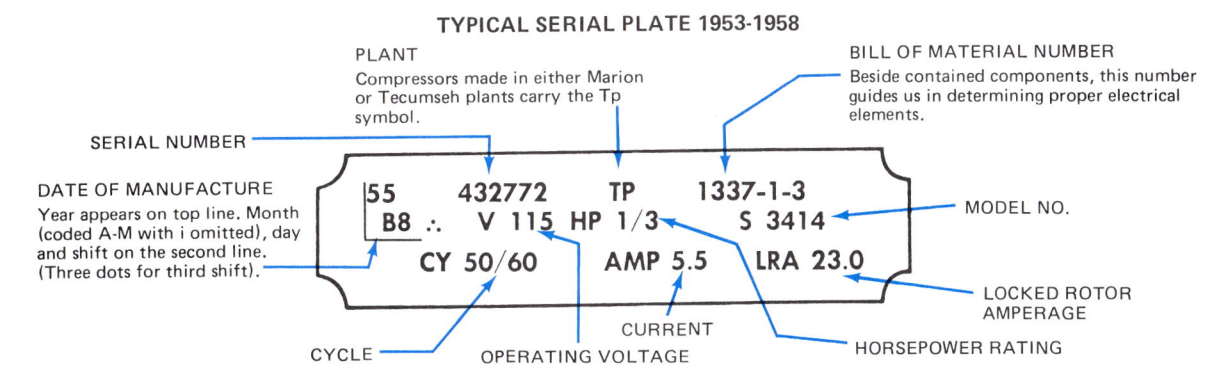

PLANT — Compressors made in either Marion or Tecumseh plants carry the Tp symbol.

BILL OF MATERIAL NUMBER — Beside contained components, this number guides us in determining proper electrical elements.

SERIAL NUMBER

DATE OF MANUFACTURE — Year appears on top line. Month (coded A-M with i omitted), day and shift on the second line. (Three dots for third shift).

```
55      432772    TP      1337-1-3
B8 ∴    V 115     HP 1/3  S 3414
        CY 50/60  AMP 5.5 LRA 23.0
```

MODEL NO.

LOCKED ROTOR AMPERAGE

HORSEPOWER RATING

CURRENT

OPERATING VOLTAGE

CYCLE

TYPICAL SERIAL PLATE 1958-1964 (AUGUST)

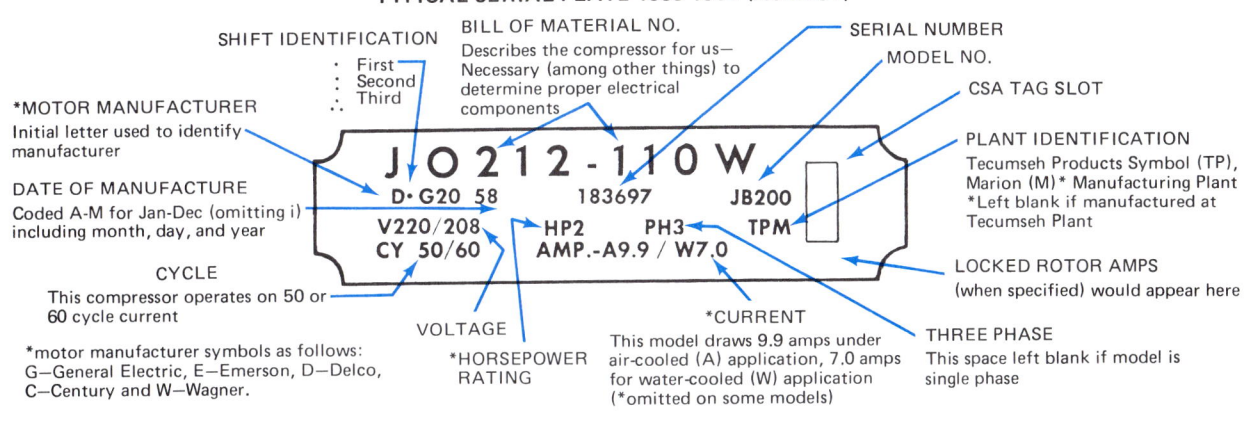

SHIFT IDENTIFICATION
- · First
- : Second
- ∴ Third

BILL OF MATERIAL NO. — Describes the compressor for us— Necessary (among other things) to determine proper electrical components

SERIAL NUMBER

MODEL NO.

*MOTOR MANUFACTURER — Initial letter used to identify manufacturer

DATE OF MANUFACTURE — Coded A-M for Jan-Dec (omitting i) including month, day, and year

```
        J O 2 1 2 - 1 1 0 W
        D·G20 58   183697    JB200
        V220/208   HP2  PH3  TPM
        CY 50/60   AMP.-A9.9 / W7.0
```

CSA TAG SLOT

PLANT IDENTIFICATION — Tecumseh Products Symbol (TP), Marion (M)* Manufacturing Plant *Left blank if manufactured at Tecumseh Plant

LOCKED ROTOR AMPS (when specified) would appear here

THREE PHASE — This space left blank if model is single phase

CYCLE — This compressor operates on 50 or 60 cycle current

*motor manufacturer symbols as follows:
G—General Electric, E—Emerson, D—Delco, C—Century and W—Wagner.

VOLTAGE

*HORSEPOWER RATING

*CURRENT — This model draws 9.9 amps under air-cooled (A) application, 7.0 amps for water-cooled (W) application (*omitted on some models)

TYPICAL CURRENT SERIAL PLATE

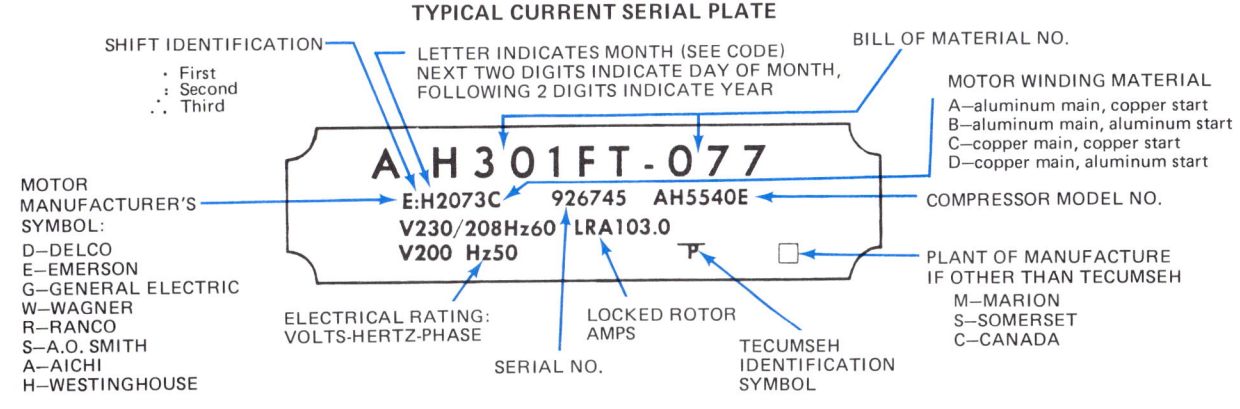

SHIFT IDENTIFICATION
- · First
- : Second
- ∴ Third

LETTER INDICATES MONTH (SEE CODE) NEXT TWO DIGITS INDICATE DAY OF MONTH, FOLLOWING 2 DIGITS INDICATE YEAR

BILL OF MATERIAL NO.

MOTOR WINDING MATERIAL
A—aluminum main, copper start
B—aluminum main, aluminum start
C—copper main, copper start
D—copper main, aluminum start

```
        A H 3 0 1 F T - 0 7 7
        E:H2073C   926745    AH5540E
        V230/208Hz60  LRA103.0
        V200 Hz50              TP
```

COMPRESSOR MODEL NO.

MOTOR MANUFACTURER'S SYMBOL:
D—DELCO
E—EMERSON
G—GENERAL ELECTRIC
W—WAGNER
R—RANCO
S—A.O. SMITH
A—AICHI
H—WESTINGHOUSE

PLANT OF MANUFACTURE IF OTHER THAN TECUMSEH
M—MARION
S—SOMERSET
C—CANADA

ELECTRICAL RATING: VOLTS-HERTZ-PHASE

LOCKED ROTOR AMPS

SERIAL NO.

TECUMSEH IDENTIFICATION SYMBOL

DATE OF MANUFACTURE

The date of manufacture is determined by a code on the serial plate or unit nameplate. This code is as follows:
Starting in January 1940 the date designation on all hermetic compressors was simplified to one letter and one figure. The months are lettered as follows:

| January—A | March—C | May—E | July—G | September—J | November—L |
| February—B | April—D | June—F | August—H | October—K | December—M |

Preceding this letter is a numeral indicating the year this compressor was built. For example, 1A would indicate the compressor was built January 1941, 7C would indicate the compressor was built March 1947. This system will hold for compressors manufactured from 1940 through 1949.

For compressors manufactured from 1950 to 1952, the year precedes the letter designating the month. For example, 51L is a compressor manufactured in November 1951. From 1953 to 1958 the year is the first numeral and the month and day are on the second line.

From 1958 on, the second line reading from left to right is: Letter indicating motor, dots to identify shift, letter for month, 2 digits for day and 2 digits for year of manufacture.

7-11. Serial plate information on Tecumseh's hermetic compressors.

167

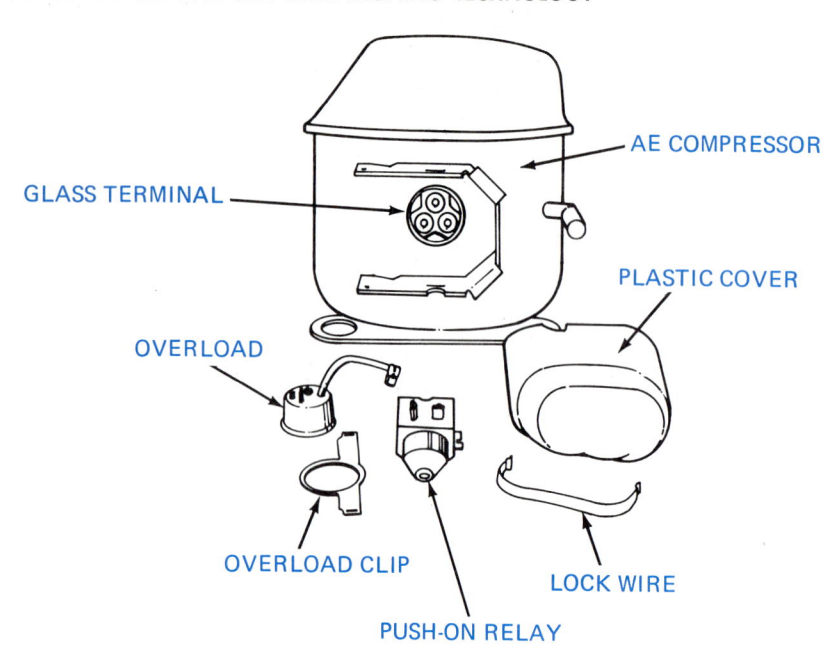

REFRIGERATION AND AIR CONDITIONING TECHNOLOGY

7-12. An AE compressor showing the glass terminal, overload, overload clip, push-on relay, plastic cover, and lock wire.

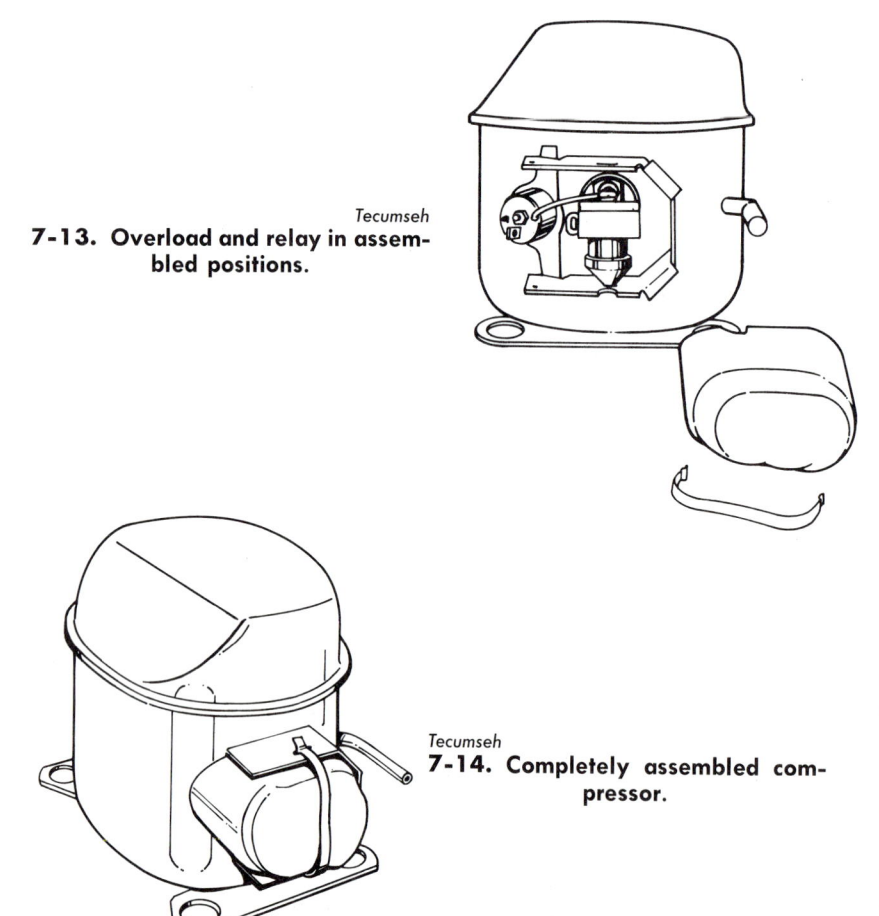

7-13. Overload and relay in assembled positions.

7-14. Completely assembled compressor.

motors. All of them use an oil charge of 22 ounces and R-12 as a refrigerant. They have a temperature range of 20° F. [−6°C] to 55° F. [13°C]. The smaller horsepower models are used where −30° F. [−34°C] to 10° F. [−12°C] is required.

The T and AT compressor models have $\frac{1}{6}$-, $\frac{1}{5}$-, $\frac{1}{4}$-, and $\frac{1}{3}$-horsepower motors. All of these models use R-12 refrigerant. The smaller sizes use a 38-ounce oil charge, while the larger horsepower models use 32 ounces. They have temperature ranges of −30° F. [−34°C] to 10° F. [−12°C] and 20° F. [−6°C] to 55° F. [13°C].

The AE compressors are used for household refrigerators, freezers, dehumidifiers, vending machines, and water coolers. Fig. 7-12. They are made in $\frac{1}{20}$-, $\frac{1}{12}$-, $\frac{1}{8}$-, $\frac{1}{6}$-, $\frac{1}{5}$-, and $\frac{1}{4}$-horsepower units. The oil charge may be 10, 16, 20, or 23 ounces. This AE compressor model line uses R-12 and, in some cases, R-22 as a refrigerant.

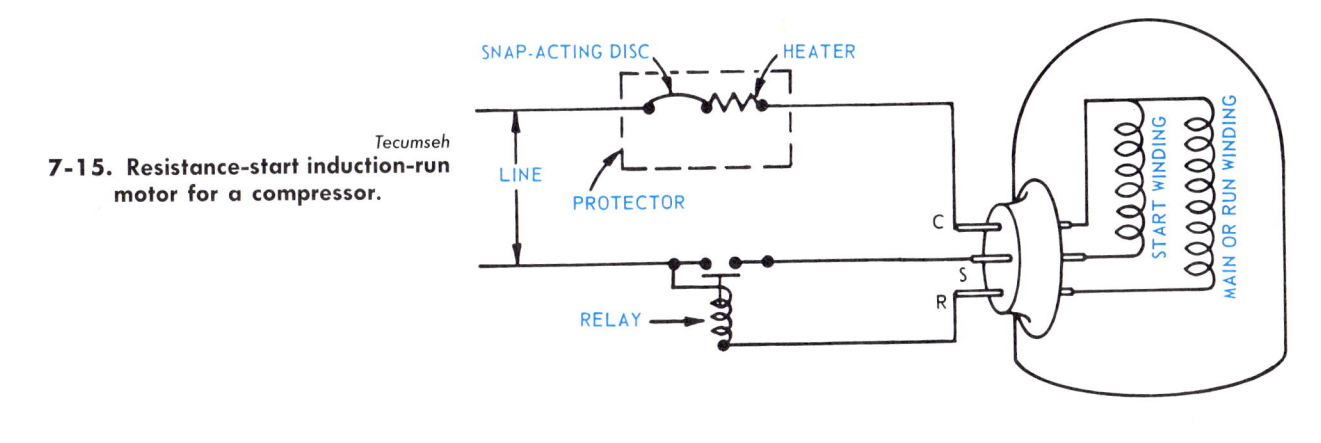

CHAPTER 7—COMPRESSORS

Tecumseh
7-15. Resistance-start induction-run motor for a compressor.

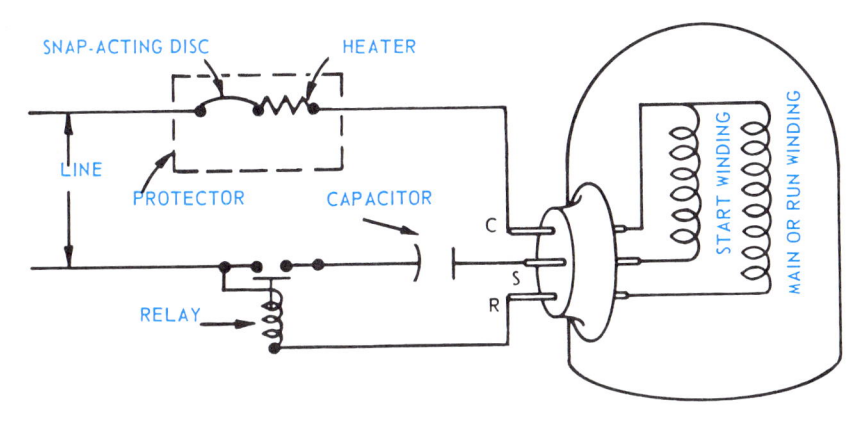

Tecumseh
7-16. Capacitor-start induction-run motor for a compressor.

Figure 7-13 shows the overload relay in its proper location with the cover removed to indicate the proper positioning. Figure 7-14 shows all parts assembled under the cover. The cover is secured to the fence with a bale strap.

This type of compressor may have a resistance-start induction-run motor. Fig. 7-15. It may have a capacitor-start induction-run motor. Fig. 7-16. These motor types will be discussed later.

Model AK compressors are rated in Btu per hour. They have a 7,000 to 12,000 Btu rating range. All model AK compressors are used for air conditioning units. The refrigerant used is R-22. A 17-ounce charge of oil is used on all models.

The AB compressors also are used for air conditioning units. However, they are larger, starting with the 19,000 Btu/hr rating and extending up to 24,000 Btu, or a 2-ton limit. Keep in mind that 12,000 Btu equal one ton. Refrigerant 22 is used with a 36-ounce charge of oil.

The AU and AR compressors are made in $\frac{1}{2}$-, $\frac{3}{4}$-, 1-, and $1\frac{1}{4}$-horsepower sizes. They are used primarily for air conditioning units. Most of the models use R-22, except for a few models that use R-12. A 30-ounce charge of oil is standard, except in one of the $\frac{1}{2}$-horsepower models. Because of such exceptions, you must refer to the manufacturer's specifications chart to obtain the information relative to a specific model number within a series.

The ISM (internal spring mount) series of compressors ranges in size from $\frac{1}{8}$ horsepower to 1 horsepower. Their temperature range is from −30° F. [−34°C] to 10° F. [−12°C] and from 20° F. [−6°C] to 55° F. [13°C]. The oil charge is either 40 or 45 ounces, depending upon the particular model.

The AH compressors are de-

169

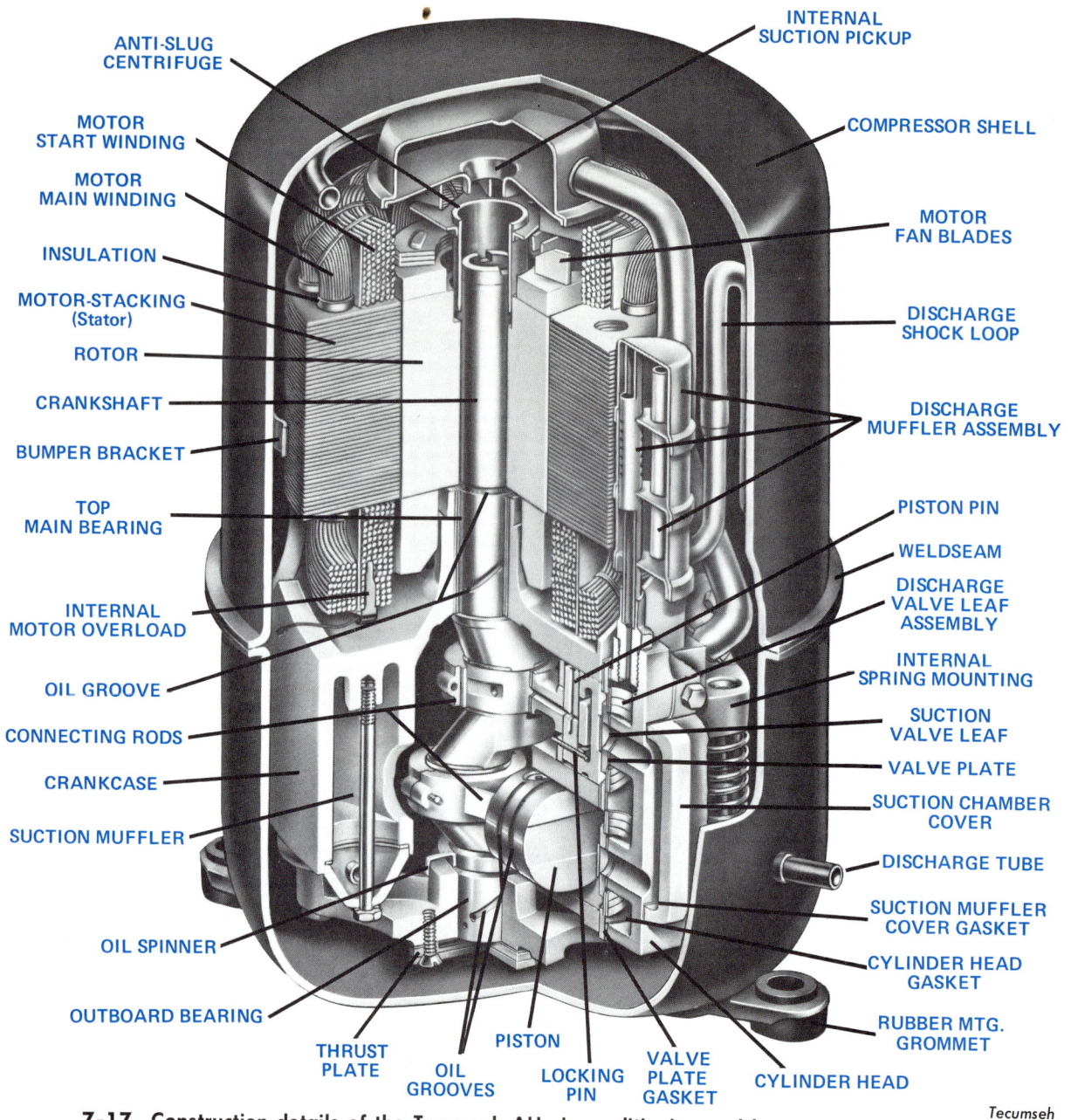

REFRIGERATION AND AIR CONDITIONING TECHNOLOGY

7-17. Construction details of the Tecumseh AH air conditioning and heat pump compressors. *Tecumseh*

signed for residential and commercial air conditioning and heat pump applications. Fig. 7-17. They can be obtained with either three- or four-point mountings. Fig. 7-18. The internal line-break motor protector is used. It is located precisely in the center of the heat sink portion of the motor windings. Thus, it detects excessive motor winding temperature and safely protects the compressor from excessive heat and/or current draw. Fig. 7-19.

The snap-on terminal cover assembly is shown in Fig. 7-20. It is designed for assembly without tools. The molded fiberglass terminal cover is secured by a simple bale strap.

This AH compressor series has

170

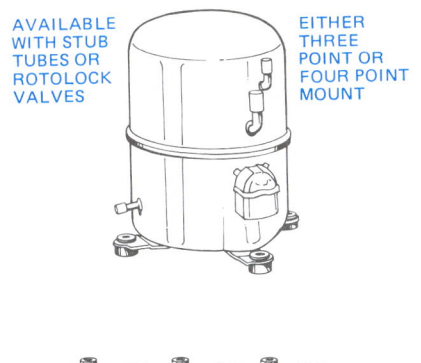

CHAPTER 7—COMPRESSORS

AVAILABLE WITH STUB TUBES OR ROTOLOCK VALVES

EITHER THREE POINT OR FOUR POINT MOUNT

MOUNTING GROMMETS & SPACERS

Tecumseh

7-18. External view of the Tecumseh AH air conditioning and heat pump compressor with its grommets and spacers.

a run capacitor in the circuit shown in Fig. 7-21. This compressor is designed for single-phase operation. Figure 7-22 shows the terminal box with the position of the terminals and the ways in which they are connected for run, start, and common.

The AH compressors are rated in Btu/hr. They range from 3500 to 40,000 Btu/hr. These models use 45 ounces of oil for the charge. They are used as air conditioning units and for almost any other temperature range applications. They use R-12 or R-22 for refrigerant.

The B and C model compres-

FIBERGLAS COVER
COVER GASKET
BALE STRAP

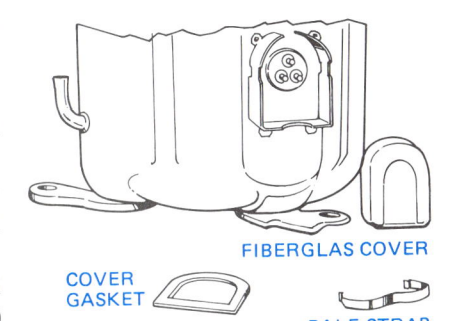

Tecumseh

7-20. Snap-on terminal cover assembly.

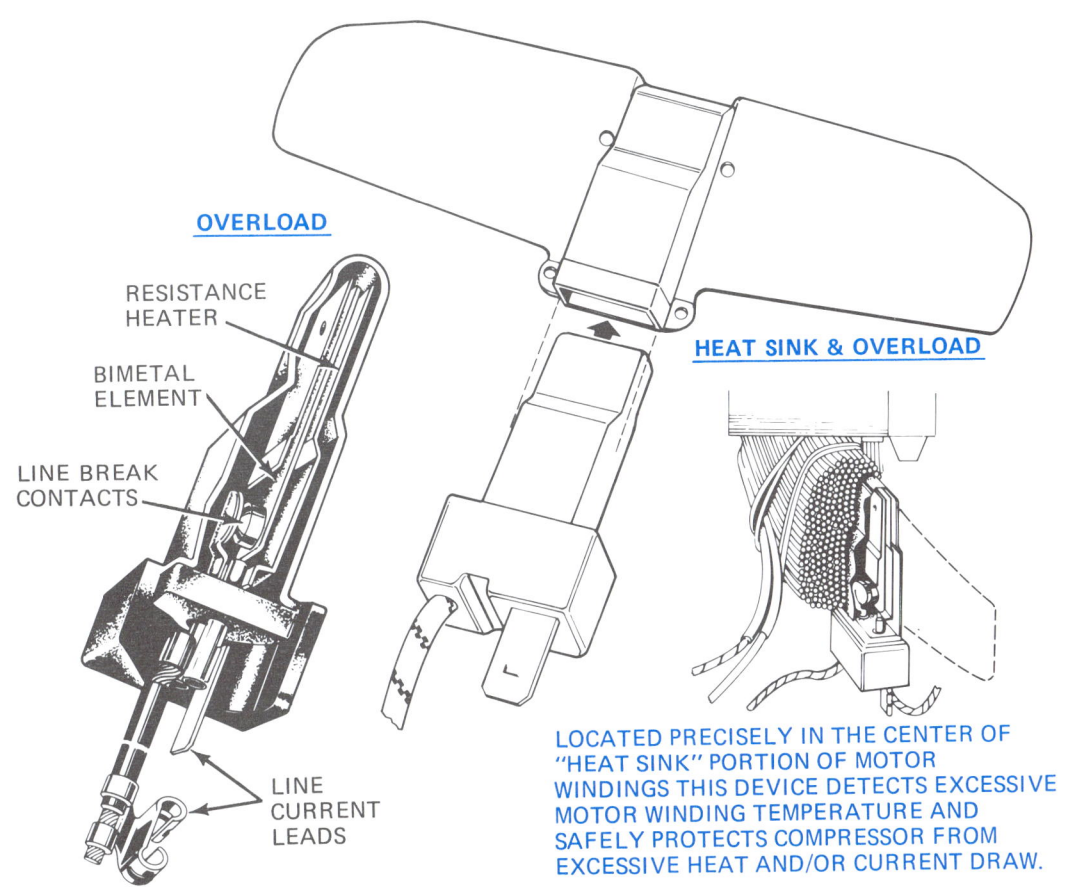

sors are available in ⅓-, ½-, ¾-, 1-, 1½-, 1¾-, and 2-horsepower units. All of them use a 45-ounce

OVERLOAD
- RESISTANCE HEATER
- BIMETAL ELEMENT
- LINE BREAK CONTACTS
- LINE CURRENT LEADS

HEAT SINK & OVERLOAD

LOCATED PRECISELY IN THE CENTER OF "HEAT SINK" PORTION OF MOTOR WINDINGS THIS DEVICE DETECTS EXCESSIVE MOTOR WINDING TEMPERATURE AND SAFELY PROTECTS COMPRESSOR FROM EXCESSIVE HEAT AND/OR CURRENT DRAW.

Tecumseh

7-19. Internal line-break motor protector.

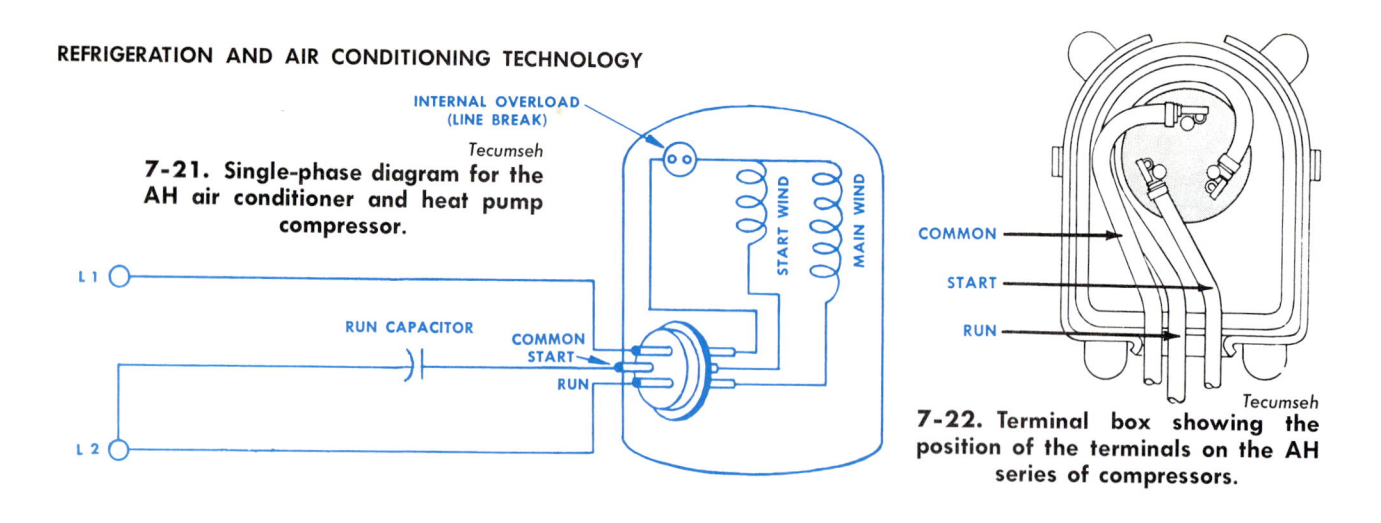

REFRIGERATION AND AIR CONDITIONING TECHNOLOGY

7-21. Single-phase diagram for the AH air conditioner and heat pump compressor.

7-22. Terminal box showing the position of the terminals on the AH series of compressors.

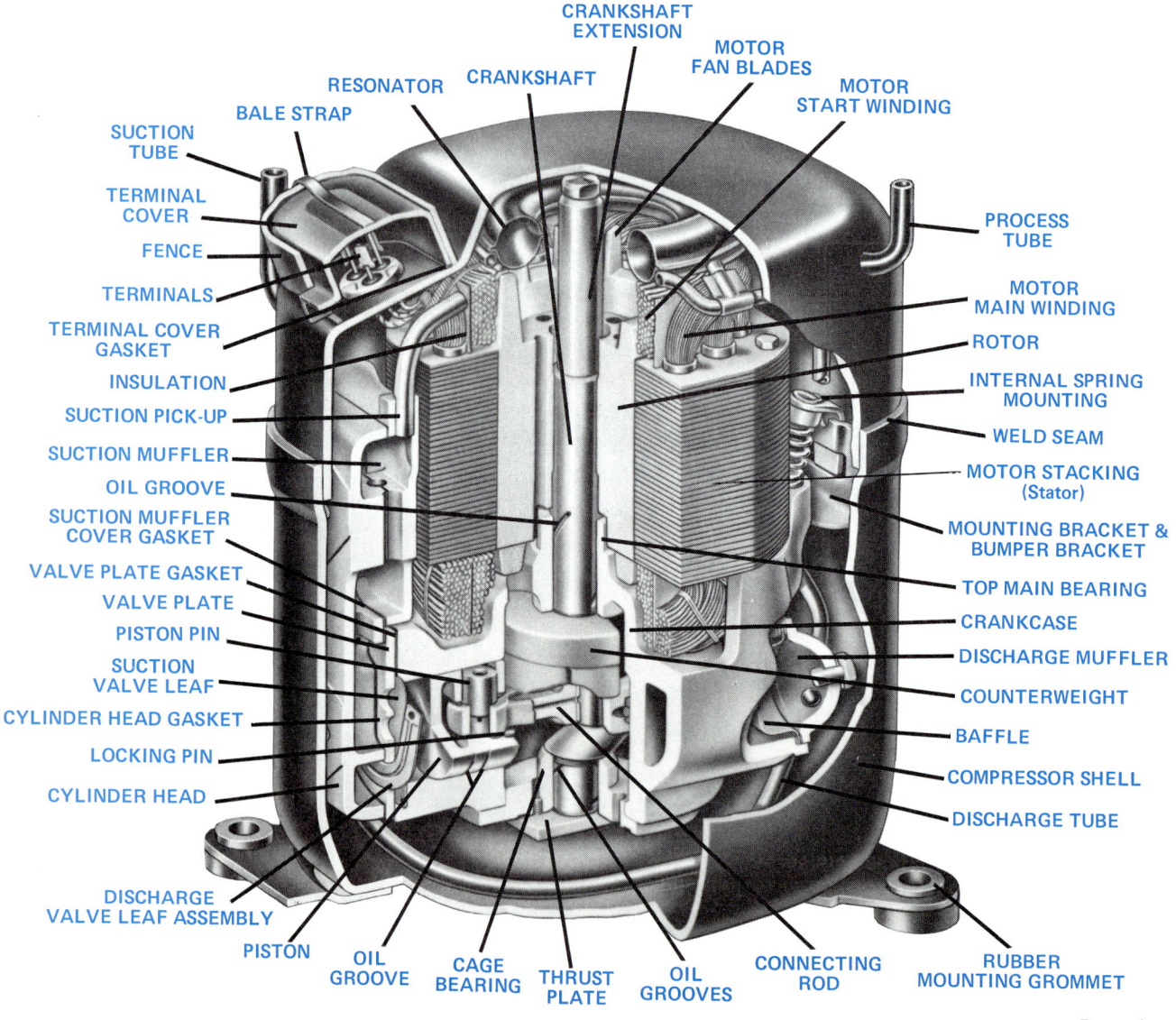

7-23. Cutaway view of the AJ series of air conditioning compressors.

172

CHAPTER 7—COMPRESSORS

THREE POINT OR FOUR POINT MOUNTING

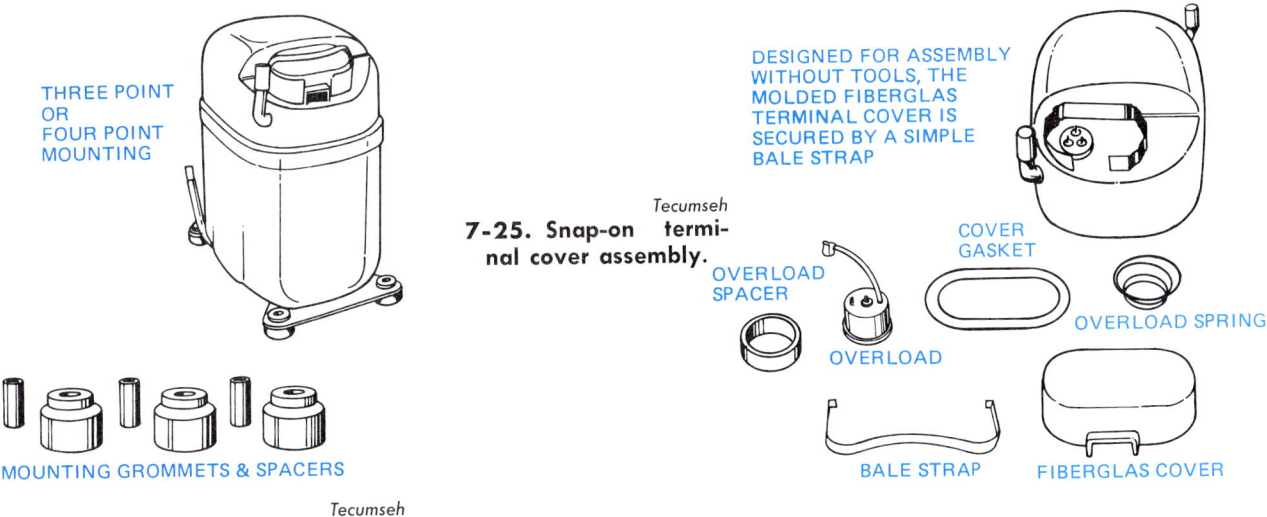

MOUNTING GROMMETS & SPACERS

Tecumseh
7-24. External view of the AJ compressor.

DESIGNED FOR ASSEMBLY WITHOUT TOOLS, THE MOLDED FIBERGLAS TERMINAL COVER IS SECURED BY A SIMPLE BALE STRAP

Tecumseh
7-25. Snap-on terminal cover assembly.

OVERLOAD SPACER — OVERLOAD — COVER GASKET — OVERLOAD SPRING — BALE STRAP — FIBERGLAS COVER

oil charge. They have a wide variety of temperature and air conditioning applications. The model number may be preceded by a B or a C to indicate this series of compressors.

The AJ series of air conditioning compressors ranges in size from 1100 to 19,500 Btu. Fig. 7-23. An oil charge of 26 or 30 ounces is standard, depending upon the model. They are mounted on three or four points.

Fig. 7-24. A snap-on terminal cover allows quick access to the connections under the cover. Fig. 7-25. This particular model has an anti-slug feature that is standard on all AJM12 and larger models. Fig. 7-26. (An anti-slug feature keeps the liquid refrigerant moving.)

This type of compressor relies upon the permanent split-capacitor motor. In this instance, the need for both start and run capacitor is not presented. The starting relay and the starting capacitor are eliminated in this arrangement. Fig. 7-27. With the PSC motor, the running capacitor acts as both a starting and running capacitor. It is never disconnected. Both motor windings are always engaged while the compressor is starting and running.

PSC motors provide good running performance and adequate starting torque for low line voltage conditions. They reduce po-

7-26. An anti-slug feature is standard on all AJ1M12 models and on larger models of the AJ series.
Tecumseh

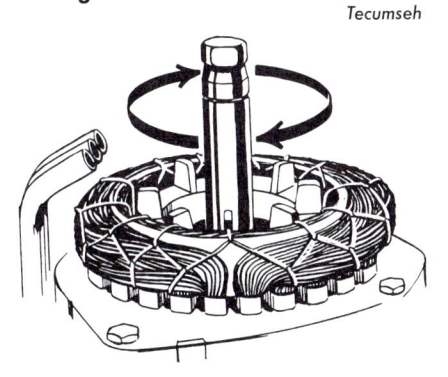

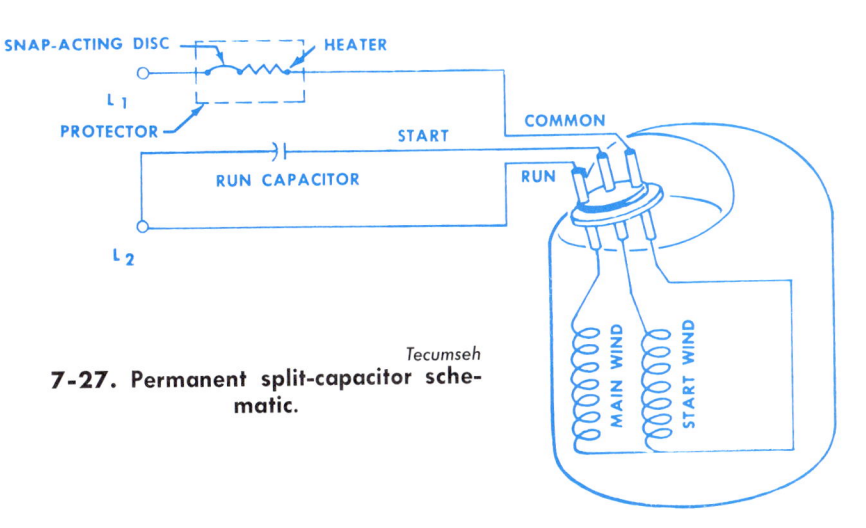

Tecumseh
7-27. Permanent split-capacitor schematic.

173

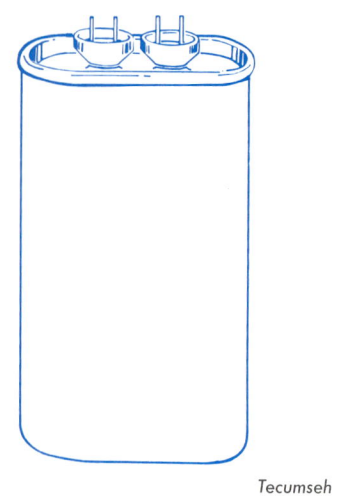

REFRIGERATION AND AIR CONDITIONING TECHNOLOGY

7-28. Running capacitor. *Tecumseh*

tential motor trouble since the electrical circuit is simplified. Figure 7-28 shows a running capacitor designed for continuous duty. It increases the motor efficiency while improving power and reducing current drain from the line.

Do not operate a compressor without its designated running

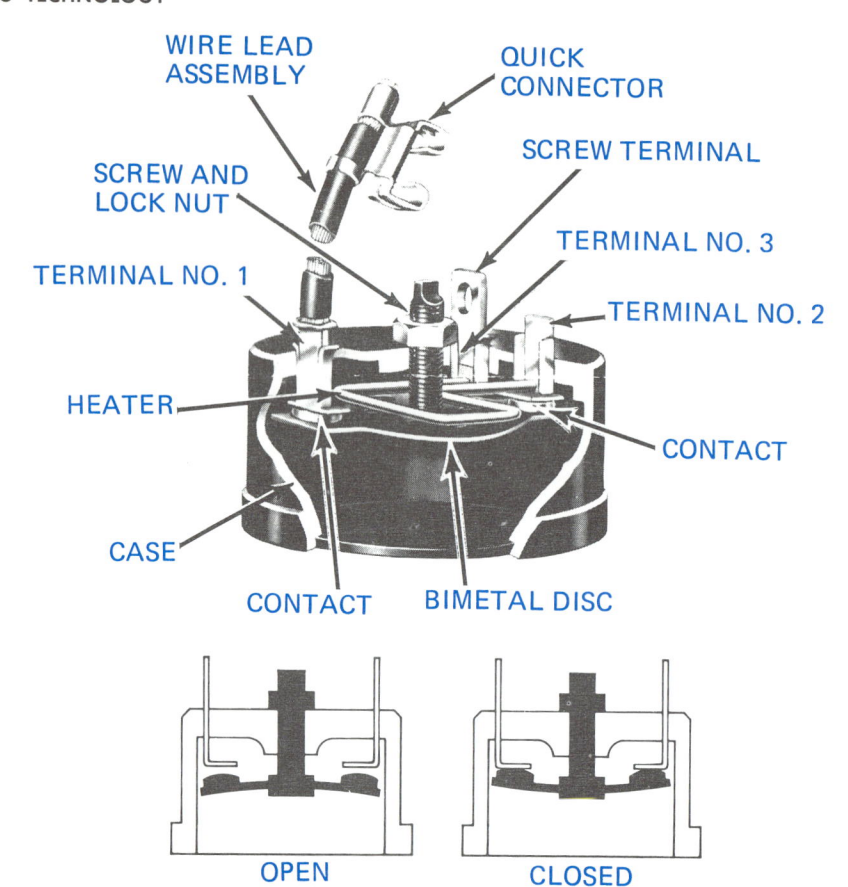

7-30. External line-break overload. *Tecumseh*

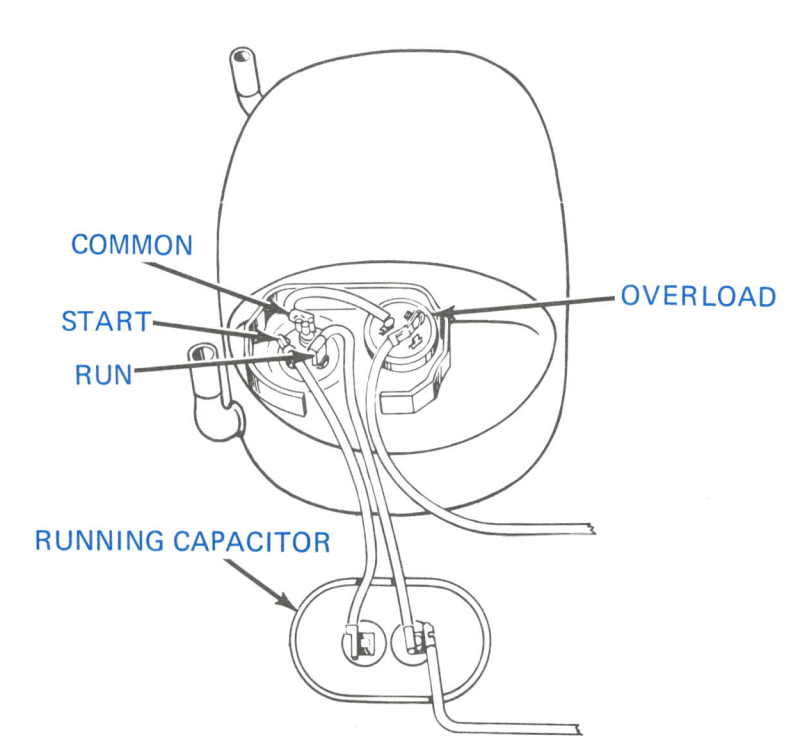

7-29. A PSC motor hookup. *Tecumseh*

174

CHAPTER 7—COMPRESSORS

capacitor. Otherwise, an overload results in the loss of starting and running performance. Adequate motor overload protection is not available either. A run capacitor in the circuit causes the motor to have some rather unique characteristics. Such motors have better pull-out characteristics when a sudden load is applied.

Figure 7-29 shows how this particular series of compressors is wired for using the capacitor in the run and start circuit. Note the overload is an external line breaker. This motor overload device is firmly attached to the compressor housing. It quickly senses any unusual temperature rise or excess current draw. The bimetal disc reacts to either excess temperature and/or excess current draw. It flexes downward, thereby disconnecting the compressor from the power source. Fig. 7-30.

The CL compressor series is designed for residential and commercial air conditioning and heat pumps. These compressors are made in 2½-, 3-, 3½-, 4-, and 5-horsepower sizes. They can be operated on three-phase or single-phase. Fig. 7-31. Since this is one of the larger compressors, it has two cylinders and pistons. It needs a good protection system for the motor. This one has an internal thermostat to interrupt the control circuit to the motor contactor. The contactor then disconnects the compressor from the power source. Figure 7-32 shows the location of the internal thermostat.

There is a supplementary overload in the compressor terminal box so it can be reached for service. Fig. 7-33, found on page 178. A locked rotor or another condition producing excessive current draw causes the bimetal disc to flex upward. This opens the pilot circuit to the motor contactor. The contactor then disconnects the compressor from the power source. Single-phase power re-

7-31a. Single-phase hookup for the CL air conditioning and heat pump compressors.

Tecumseh

175

REFRIGERATION AND AIR CONDITIONING TECHNOLOGY

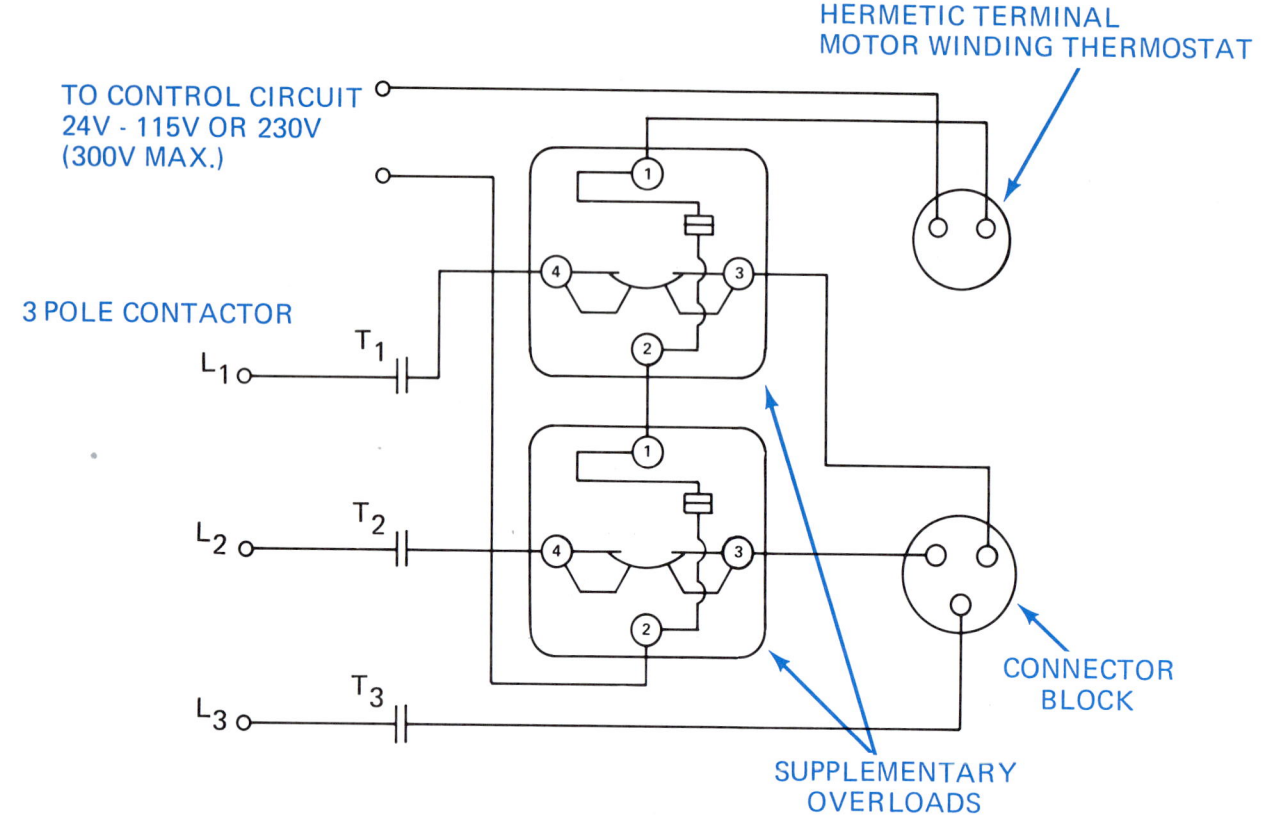

7-31b. Three-phase hook-up for the CL air conditioning and heat pump compressors.

quires one supplementary overload. Fig. 7-34. Three-phase power requires two supplementary overloads. Fig. 7-35. This CL line of compressors uses R-22 and R-12 refrigerants. They also use an oil charge of either 45 or 55 ounces. In some cases, when the units are interconnected, they use 65 ounces.

The H, J, and PJ compressors vary from $\frac{3}{4}$ to 3 horsepower. They have a wide temperature range. All of these models use a 55-ounce oil charge. Both R-22 and R-12 refrigerants are used.

The F and PF compressors are 2-, 3-, 4-, and 5-horsepower units. They use either 115 or 165 ounces of oil. They, too, are used for a number of temperature ranges.

HERMETIC COMPRESSOR MOTOR TYPES

There are four general types of single-phase motors. Each has distinctly different characteristics. Compressor motors are designed for specific requirements regarding starting torque and running efficiency. These are two of the reasons why different types of motors are required to meet the various demands.

Resistance Start-Induction Run

The resistance start-induction run (RSIR) motor is used on many small hermetic compressors through $\frac{1}{3}$ horsepower. The motor has low starting torque. It must be applied to completely self-equalizing capillary tube systems such as household refrigerators, freezers, small water coolers, and dehumidifiers. This motor has a high resistance start winding that is not designed to remain in the circuit after the motor has come up to speed. A current relay is necessary to disconnect the start winding as the motor comes up to design speed. Fig. 7-36.

Capacitor Start-Induction Run

The capacitor start-induction

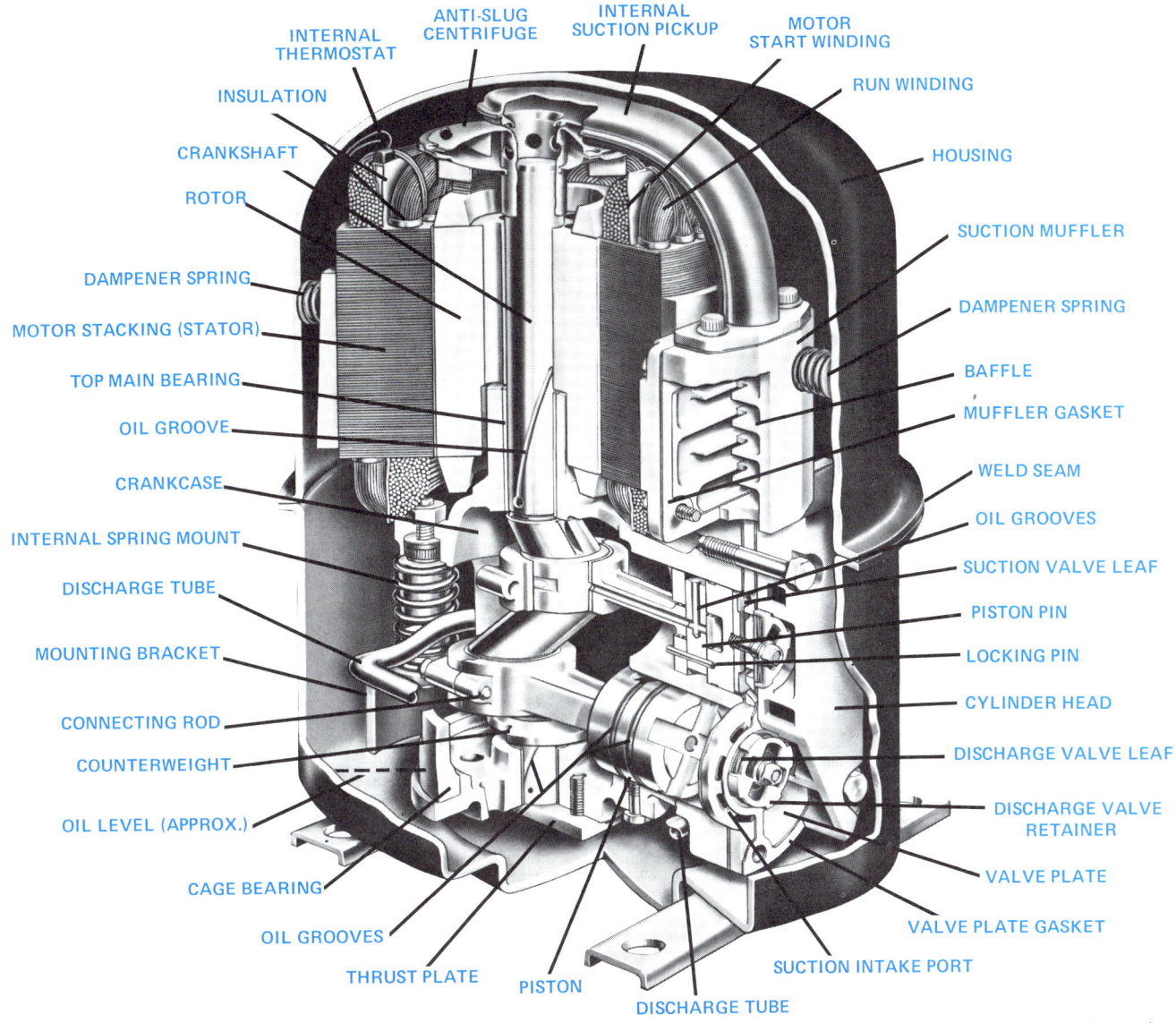

7-32. Construction details of the CL compressor.

run (CSIR) motor is similar to the RSIR. However, a start capacitor is included in series with the start winding to produce a higher starting torque. This motor is commonly used on commercial refrigeration systems with a rating through 3/4 horsepower. Fig. 7-37.

Capacitor Start and Run

The capacitor start and run (CSR) motor arrangement uses a start capacitor and a run capacitor in parallel with each other and in series with the motor start winding. This motor has high starting torque and runs efficiently. It is used on many refrigeration and air conditioning applications through 5 horsepower. A potential relay removes the start capacitor from the circuit after the motor is up to speed. Potential relays must be accurately matched to the compressor. Fig. 7-38. Efficient operation depends on this.

177

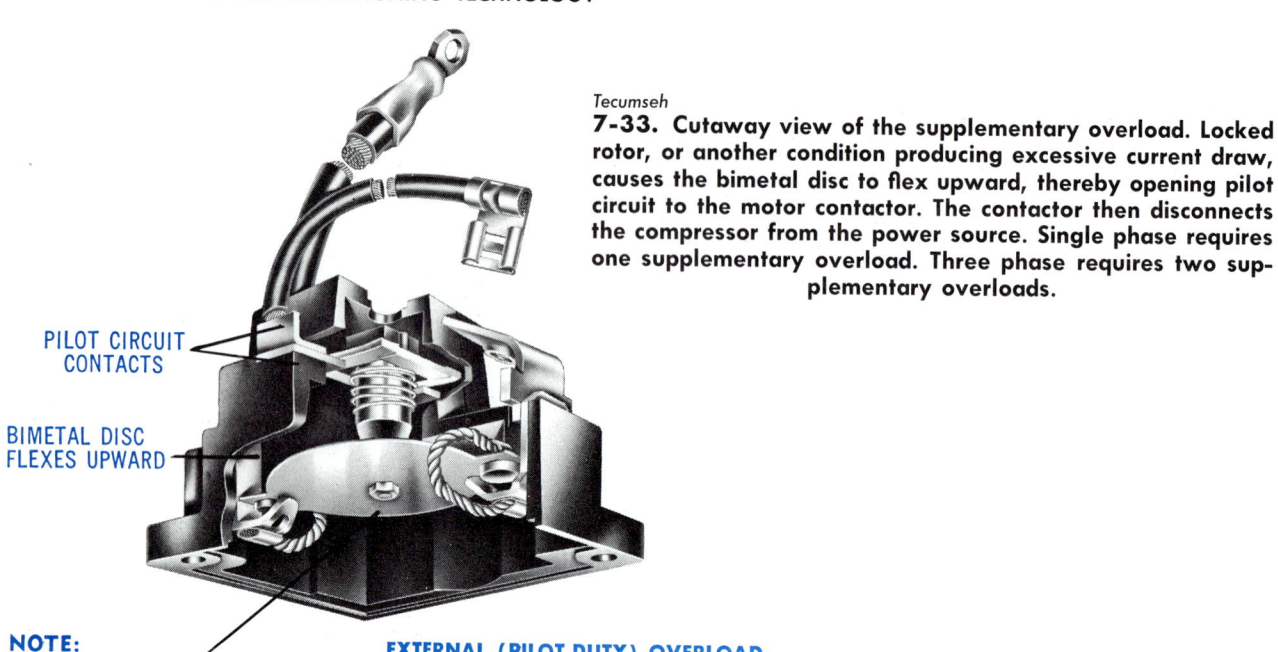

REFRIGERATION AND AIR CONDITIONING TECHNOLOGY

Tecumseh

7-33. Cutaway view of the supplementary overload. Locked rotor, or another condition producing excessive current draw, causes the bimetal disc to flex upward, thereby opening pilot circuit to the motor contactor. The contactor then disconnects the compressor from the power source. Single phase requires one supplementary overload. Three phase requires two supplementary overloads.

PILOT CIRCUIT CONTACTS

BIMETAL DISC FLEXES UPWARD

NOTE: BIMETAL DISC DOES NOT BREAK LINE CURRENT

EXTERNAL (PILOT DUTY) OVERLOAD SENSING MOTOR CURRENT

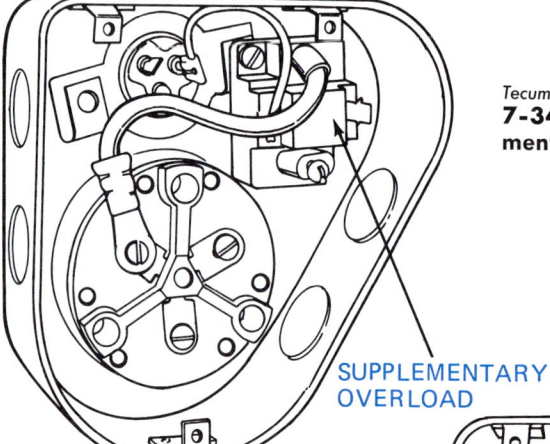

Tecumseh

7-34. Location of the supplementary overload on the CL compressor series.

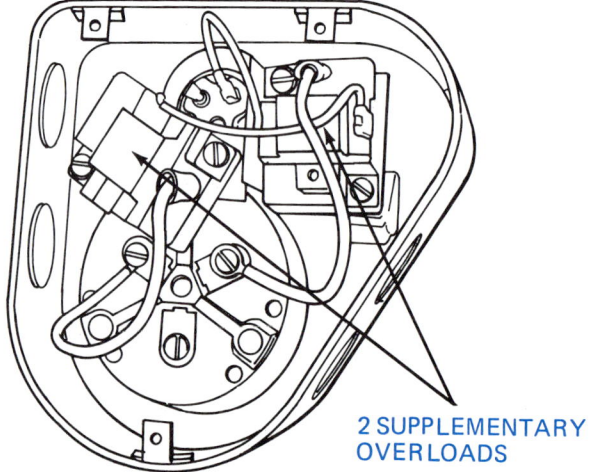

SUPPLEMENTARY OVERLOAD

Tecumseh

7-35. Location of two supplementary overloads in the terminal box makes it applicable for three-phase power connections.

2 SUPPLEMENTARY OVERLOADS

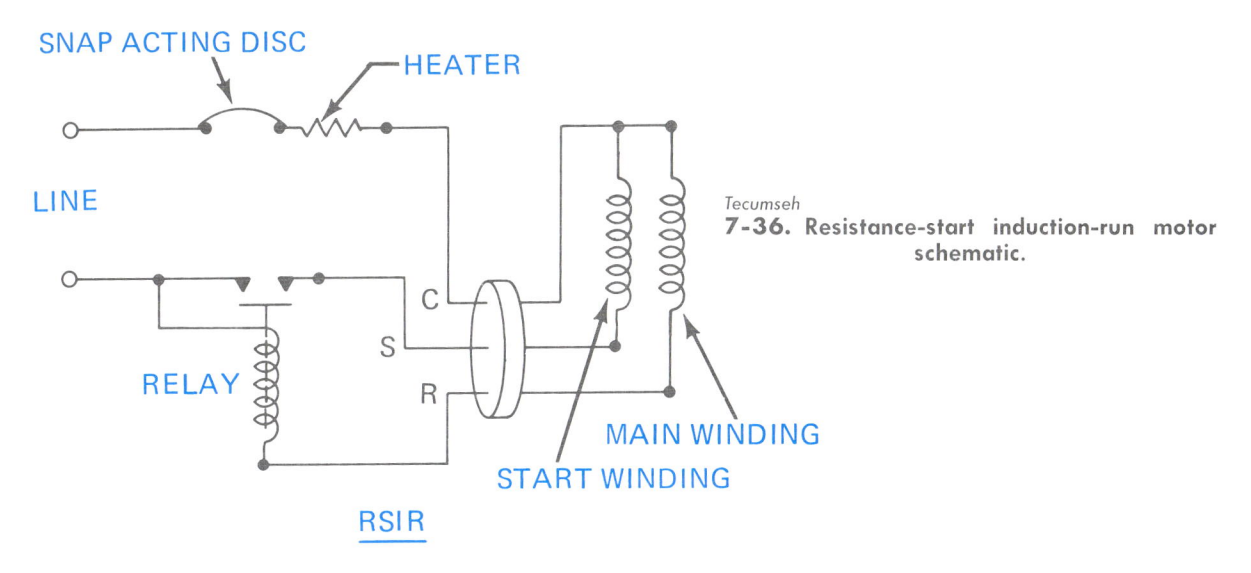

7-36. Resistance-start induction-run motor schematic.

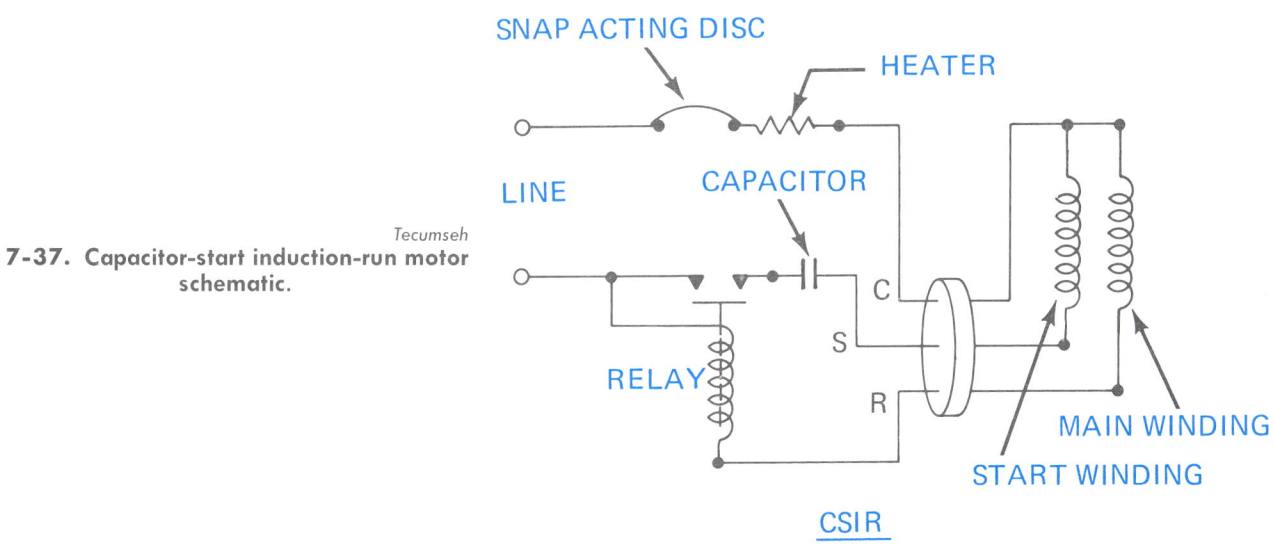

7-37. Capacitor-start induction-run motor schematic.

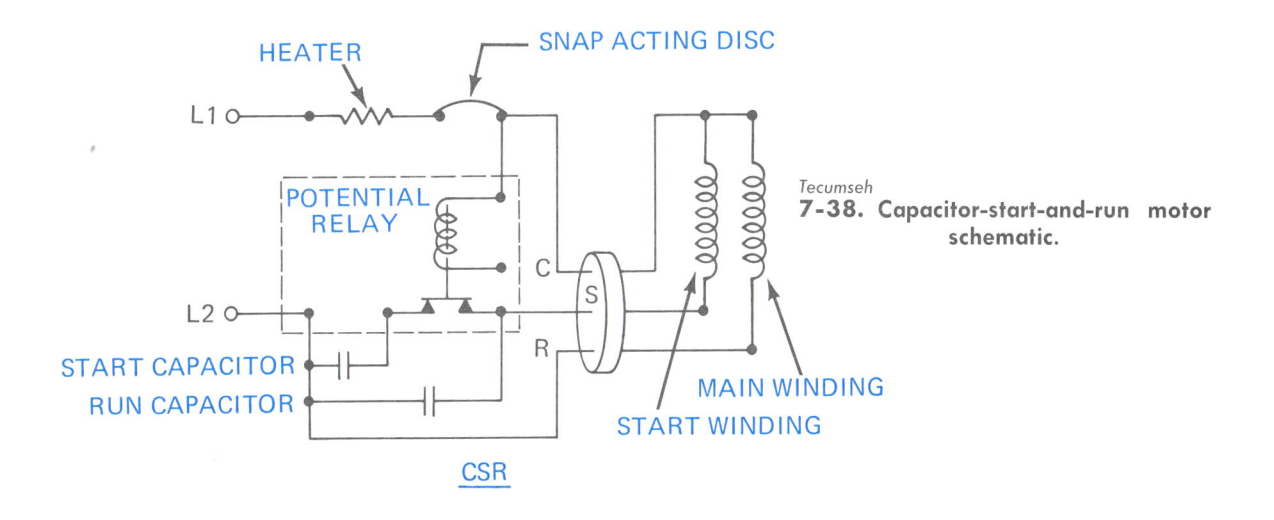

7-38. Capacitor-start-and-run motor schematic.

179

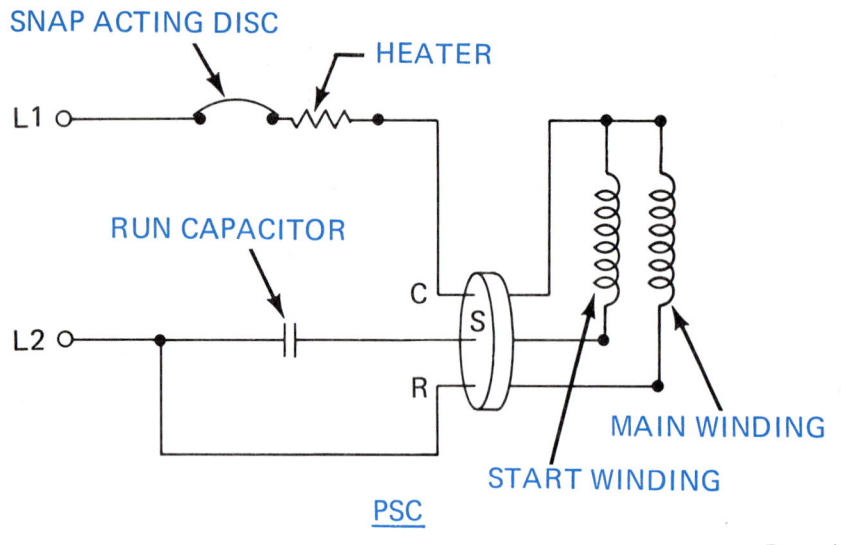

REFRIGERATION AND AIR CONDITIONING TECHNOLOGY

7-39. Permanent split-capacitor motor schematic.

Permanent Split Capacitor

The permanent split capacitor (PSC) has a run capacitor in series with the start winding. Both run capacitor and start winding remain in the circuit during start and after the motor is up to speed. Motor torque is sufficient for capillary and other self-equalizing systems. No start capacitor or relay is necessary. The PSC motor is basically an air conditioning compressor motor. It is very common through 3 horsepower. It is also available in 4- and 5-horsepower sizes. Fig. 7-39.

COMPRESSOR MOTOR RELAYS

A hermetic compressor motor relay is an automatic switching device designed to disconnect the motor start winding after the motor has attained a running speed.

There are two types of motor relays used in the refrigeration and air conditioning compressors—the *current-type* relay and the *potential-type* relay.

Current-Type Relay

The current-type relay is generally used with small refrigeration compressors up to $\frac{3}{4}$ horsepower. When power is applied to the compressor motor, the relay solenoid coil attracts the relay armature upward. This causes bridging contact and stationary contact to engage. Fig. 7-40. This energizes the motor start winding. When the compressor motor attains running speed, the motor main winding current is such that the relay solenoid coil de-energizes. This allows the relay contacts to drop open. This disconnects the motor start winding.

The relay must be mounted in true vertical position so that the armature and bridging contact will drop free when the relay solenoid is de-energized.

Potential-Type Relay

This relay is generally used with large commercial and air conditioning compressors. The motors may be capacitor start-capacitor run types up to 5 horsepower. Relay contacts are normally closed. The relay coil is wired across the start winding. It senses voltage change. Start winding voltage increases with motor speed. As the voltage increases to the specific pick-up value, the armature pulls up, opening the relay contacts and de-energizing the start winding. After switching, there is still suf-

7-40. Current-type relay. This is generally used with small refrigeration compressors up to $\frac{3}{4}$ horsepower.

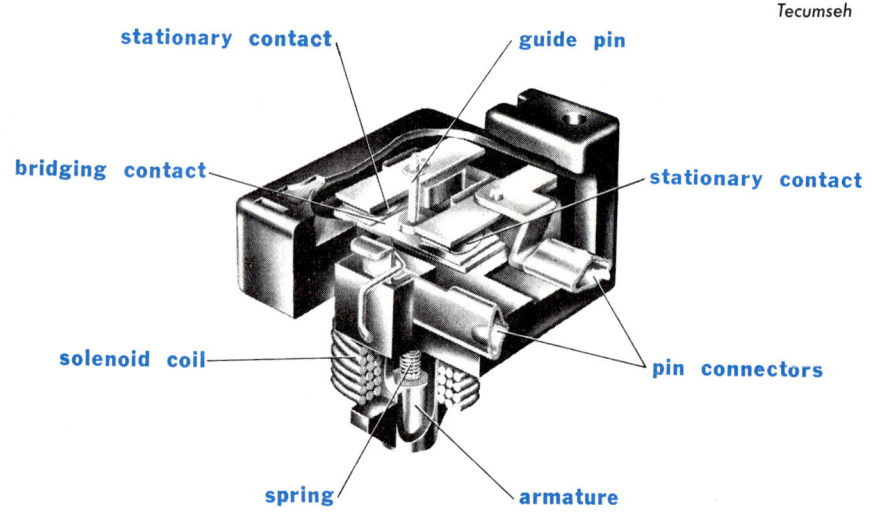

CHAPTER 7—COMPRESSORS

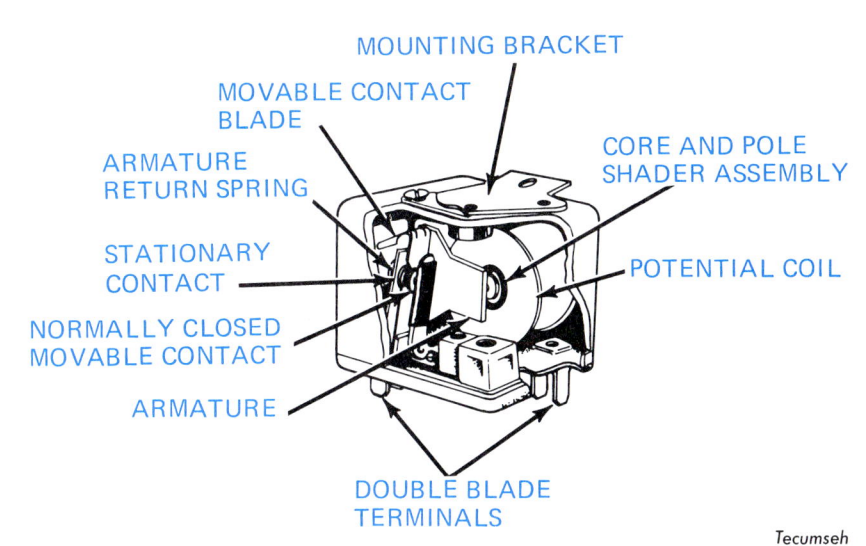

7-41. Potential-type relay. This is generally used with large commercial and air conditioning compressors (capacitor-start, capacitor-run) to 5 horsepower.

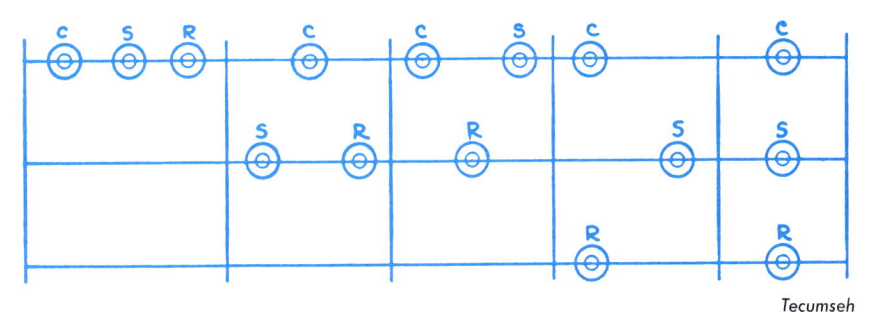

Tecumseh

7-42. Identification of compressor terminals.

Tecumseh

ficient voltage induced in the start winding to keep the relay coil energized and the relay starting contacts open. When power is shut off to the motor, the voltage drops to zero. The coil is de-energized and the start contacts reset. Fig. 7-41.

Many of these relays are extremely position sensitive. When changing a compressor relay, care should be taken to install the replacement in the same position as the original. Never select a replacement relay solely by horsepower or other generalized rating. Select the correct relay from the parts guide book furnished by the manufacturer.

COMPRESSOR TERMINALS

For the compressor motor to run properly it must have the power correctly connected to its terminals outside of the hermetic shell. There are several different types of terminals used on the various models of Tecumseh compressors.

Tecumseh terminals are *always* thought of in the order of common, start, and run. Read the terminals in the same way you would read the sentences on a book's page. Start at the top left-hand corner and read across the first line from left to right. Then, read the second line from left to right. In some cases three lines must be "read" to complete the identification process. Figure 7-42 shows the different arrangements of terminals. All Tecumseh compressors, except one model, follow one of these patterns. The exception is the old twin-cylinder, internal-mount compressor built at Marion. This was a 90° piston model designated with an "H" at the beginning of the model number (that is, HA100). The terminals were reversed on the H models and read run, start, and common. Fig. 7-43. These compressors were replaced by the J-model series in 1955. All J models follow the usual pattern for common, start, and run.

Built-up Terminals

Some built-up terminals have screw- and nut-type terminals for the attaching of wires. Fig. 7-44.

RUN START COMMON

Tecumseh

7-43. Built-up terminals. These are on the obsolete twin-cylinder internal-mount H models.

7-44. Built-up terminals. These are on all external-mount B and C twin-cylinder models and on F, PF, and CF four-cylinder external-mount models.

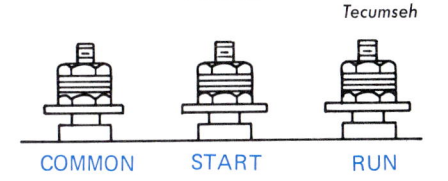

Tecumseh

COMMON START RUN

181

REFRIGERATION AND AIR CONDITIONING TECHNOLOGY

7-45. Built-up terminals on pancake compressors manufactured before 1952. *Tecumseh*

7-46. Built-up terminals on S and C single-cylinder ISM models. *Tecumseh*

7-47. Built-up terminals on twin-cylinder internal-mount J and PJ models. *Tecumseh*

7-48. Glass quick-connect terminals. These are used on S and C single-cylinder ISM models, as well as on AK and CL models. Note the location of the internal thermostat terminals found on the CL models. *Tecumseh*

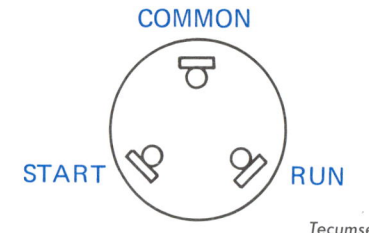

7-49. Glass quick-connect terminals. This arrangement is found on AU, AE, AB, AJ, and AH models. *Tecumseh*

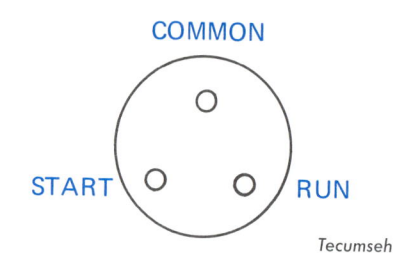

7-50. Glass terminals. The pancake-type compressors P, R, AP, AR, and T, as well as AT models, use this terminal configuration. *Tecumseh*

Others may have different arrangements. The pancake compressors built in 1953 and after have glass terminals that look something like those shown in Fig. 7-45. The terminal arrangement for S and C single-cylinder ISM (Internal Spring Mount) models resembles that shown in Fig. 7-46. Models J and PJ with twin-cylinder internal mount have a different terminal arrangement. It looks like that shown in Fig. 7-47.

Glass Quick-Connect Terminals

Figure 7-48 shows the quick-connect terminals used on S and C single-cylinder ISM models. The AK and CL models also use this type of arrangement. Many of the CL models have the internal thermostat terminals located closeby.

Quick-connect glass terminals are also used on AU and AR air conditioning models. The AE air conditioning models also use glass quick connects. Models AB, AJ, and AH also use glass quick connects, but notice how their arrangement of common, start, and run varies from that shown in Fig. 7-48. Figure 7-49 shows how the AU, AR, AE, AB, AJ, and AH models terminate.

Glass terminals are also used on pancake-type compressors with P, R, AP, and AR designations. Fig. 7-50. The T and AT models, as well as the AE refrigeration models, also use the glass terminals, but without the quick connect.

Keep in mind that you should never solder any wire or wire termination to a compressor terminal. Heat applied to a terminal is liable to crack the glass terminal base or loosen the built-up terminals. This will, in turn, cause a refrigerant leak at the compressor.

MOTOR MOUNTS

To dampen vibration, hold the compressor while in shipment, and cushion horizontal thrust when the compressor starts or stops, some type of mounting is necessary. Several different arrangements are used. However, each of them uses a base plate. Also, some space is allowed between the rubber grommet and the washer on the nut. The rubber grommet absorbs most of the vibration. Fig. 7-51. In Fig. 7-52, you can see the use of a spring to prevent damage to the rubber grommet. This is used for the heavier compressors. There are usually three, but sometimes four, of these rubber motor mounts on each compressor model. One of the greatest uses of this type of mount is to make sure that vibrations are not transferred to other parts of the refrigeration system or passed on to the pipes. There, they would weaken the soldered joints.

CHAPTER 7—COMPRESSORS

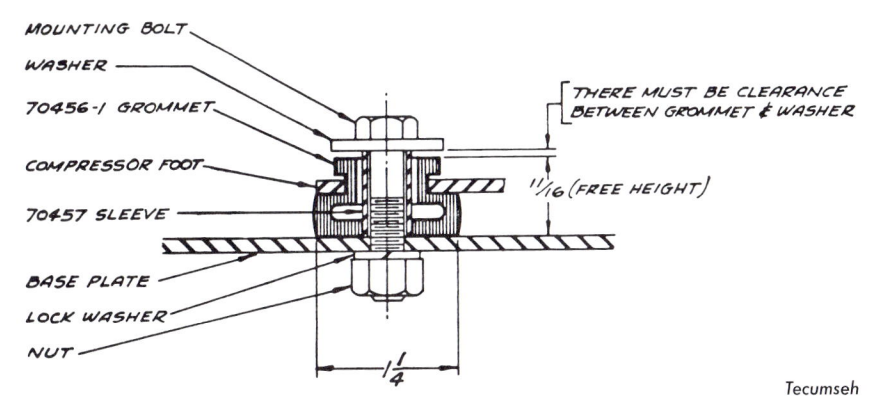

7-51. Mounting grommet assembly.

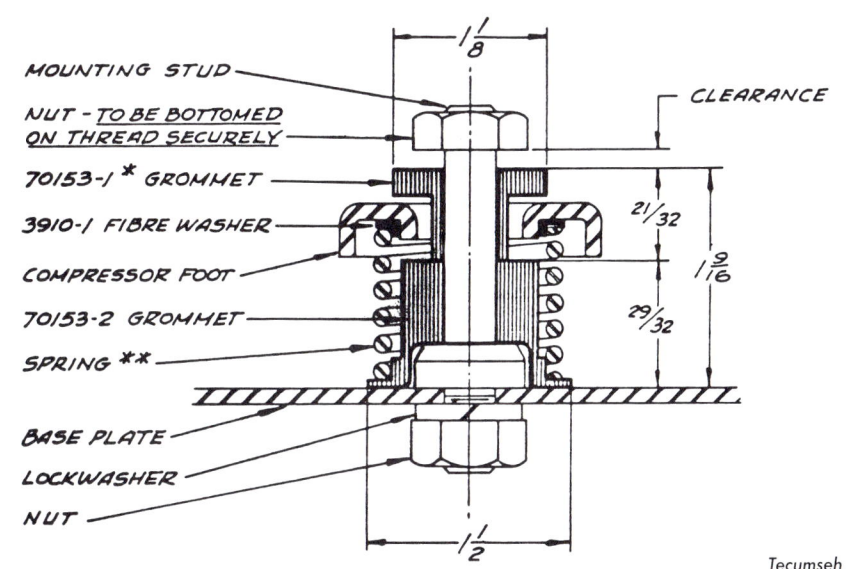

7-52. Mounting spring and grommet assembly.

CRANKCASE HEATERS

Most compressors have crankcase heaters. This is because most air conditioning and commercial systems are started up with a large part of the system refrigerant charge in the compressor. This is especially when the unit has been idle for some time or when the compressor is being started for the first time. On start-up, the refrigerant boils off, taking the oil charge with it. This means the compressor is forced to run for as long as three or four minutes until the oil charge circulates through the system and returns to the crankcase. Obviously, this shortens the service life of the compressor.

The solution is to charge the system so that little or no refrigerant collects in the crankcase and to operate the crankcase heater at least twelve hours before start-up or after a prolonged down time.

Two types of crankcase heaters are in common use on compressors. The wraparound type is usually referred to as the "belly band." The other type is the run capacitance off-cycle heat method.

The wraparound heater should be strapped to the housing below the oil level and in close contact with the housing. A good heater will maintain the oil at least 10° F. [5°C] above the temperature of any other system component. When the compressor is stopped it will maintain it at or above a minimum temperature of 80° F. [27°C].

In the run capacitance off-cycle heat method, single-phase compressors are stopped by opening only one leg (L_1). Thus, the other leg to the power supply (L_2) of the run capacitor remains "hot." A trickle current through the start windings results, thereby warming the motor windings. Thus, the oil is warmed on the "off-cycle."

Make sure you pull the switch that disconnects the whole unit from the power source before working on such a system.

Capacitance crankcase heat systems can be recognized by one or more of the following:

• Contactor or thermostat breaks only one leg to the compressor and condenser fan.

• Equipment carries a notice that power is on at the compressor when it is not running and that the main breaker should be opened before servicing.

• Run capacitor is sometimes split (it has three terminals) so that only part of the capacitance is used for off-cycle heating.

CAUTION: Make sure you use an exact replacement when changing such dual-purpose run capacitors. The capacitor must

183

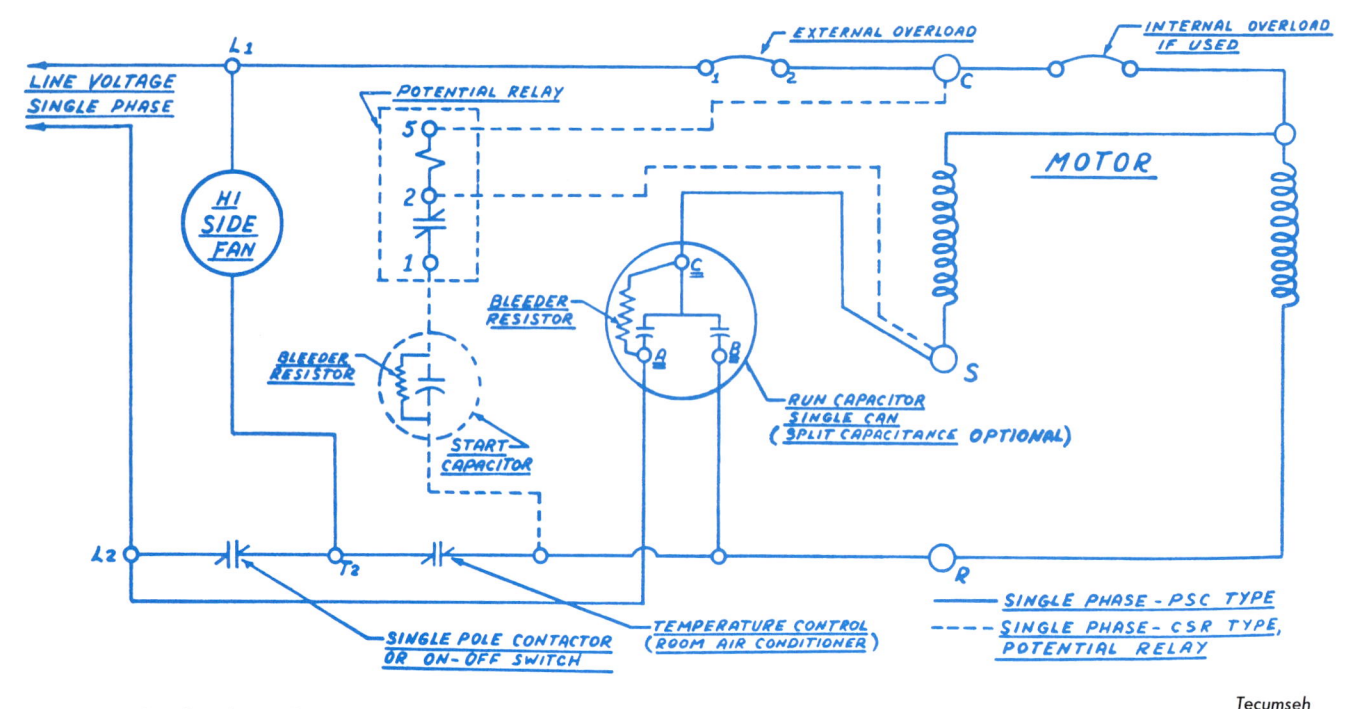

REFRIGERATION AND AIR CONDITIONING TECHNOLOGY

7-53. Single-phase CSR- or PSC-type compressor motor hookup with internal or external line-break overloads. *Tecumseh*

be fused and carry a bleed resistor across the terminals.

The basic wiring diagram for a PSC compressor with a run capacitance off-cycle heat is shown in Fig. 7-53.

ELECTRICAL SYSTEMS FOR COMPRESSOR MOTORS

Most of the problems associated with hermetic compressors are electrical. Most of the malfunctions are in the current relay, potential relay, circuit breaker, or loose connections. In most cases, internal parts of the compressor housing can be checked with an ohmmeter.

Normal Starting Torque Motors (RSIR) with a Current-Type Relay

Normal starting torque motors (RSIR) with a current-type relay mounted on the compressor terminals require several tests that must be performed in the listed sequence. Figure 7-54 shows a two-terminal external overload device in series with the start and run windings. The fan motor runs from point 1 on the current relay to point 3 on the overload device. L_2 has the relay coil inserted in series with the run winding. When the winding draws current, the solenoid is energized. This is done by the initial surge of current through the run winding. When the relay energizes with sufficient current, it closes the contacts (points 1 and S) and places the start winding in the circuit. The start winding stays in the circuit until the relay de-energizes. When the motor comes up to about 75 percent of its run speed, the relay de-energizes since the current through the run winding drops off. This change in current makes it a very sensitive circuit. The sensing relay must be in good operating condition. Otherwise, it will not energize or de-energize at the proper times.

The starting contacts on the current-type relay are normally open, as shown in Fig. 7-54. The electrical system on this type of compressor system can be checked by using a voltmeter to check line voltage. Then, use an ohmmeter to check continuity. That means the power must be off. Make sure the circuit breaker is off at the main power supply for this unit. If a fan is used (as shown in the dotted lines of Fig. 7-54) make sure one lead is disconnected from the line. Next, check continuity across the following:

1. Check continuity across L_1 and point 3 of overload. There is

CHAPTER 7—COMPRESSORS

no continuity. Close control contacts by hand. If there is still no continuity, replace the control.

2. Check continuity across No. 3 and No. 1 on overload. If there is no continuity, the protector may be tripped. Wait 10 minutes and check again. If there is still no continuity, the protector is defective. Replace it.

3. Pull the relay off the compressor terminals. *Keep it in an upright position!*

4. Check continuity across relay terminal 1 (or L) and S. If there is continuity, relay contacts are closed, when they should be open. Replace relay.

5. Check continuity across No. 1 and M. If there is no continuity, replace the relay. The solenoid is open.

6. Check continuity across compressor terminals C and R. If there is no continuity, there is an open run winding. Replace the compressor.

7. Check continuity across compressor terminals C and R. If there is no continuity, there is an open start winding. Replace the compressor.

8. Check continuity across compressor terminal C and the shell of the compressor. There is continuity. This means the motor is grounded. Replace the compressor.

9. Check the winding resistance values against those published by the manufacturer of the particular model.

If all the tests prove satisfactory, there is no capillary restriction, and the unit still fails to operate properly, change the relay. The new relay will eliminate any electrical problems, such as improper pickup or dropout, that cannot be determined by the tests listed above. If a good relay fails to correct the difficulty, the compressor is inoperative due to internal defects. It must be replaced.

High Starting Torque Motors (CSIR) with a Current-Type Relay

High starting torque motors (CSIR) with a current-type relay mounted on the compressor terminals can be easily checked for proper operation. Remember from the previous type that the current-type relay normally has its contacts open.

Use a voltmeter first to check the power source. Use an ohmmeter to check continuity. Make sure the power is off and the fan motor circuit is open. The electri-

7-54. Normal starting motors (RSIR) with current relay mounted on the compressor terminals.

Tecumseh

185

REFRIGERATION AND AIR CONDITIONING TECHNOLOGY

cal system on this type of hermetic system can be checked as follows. Fig. 7-55.

1. Check continuity across L₁ and 3. If there is no continuity, close the control contacts. If there is still no continuity, replace the control.

2. Check continuity across No. 3 and No. 1 on the overload. If there is no continuity, the protector may be tripped. Wait for 10 minutes and check again. If there is still no continuity, the protector is defective. Replace it.

3. Pull relay off compressor terminals. *Keep it upright!*

4. Check continuity across relay terminal 1 and S. If there is continuity, the relay contacts are closed when they should be open. Replace the relay.

5. Check continuity across relay terminals 2 and M. If there is no continuity, replace the relay.

6. Check continuity across compressor terminals C and R. If there is no continuity, there is an open run winding. Replace the compressor.

7. Check continuity across compressor C and S. If there is no continuity, there is an open start winding. Replace the compressor.

8. Check continuity across C and shell of the compressor. If there is continuity, there is a grounded motor. Replace the compressor.

9. Check the winding resistance against the values given in the manufacturer's resistance tables.

10. Check continuity across relay terminals 1 and 2. Place the meter on the R × 1 scale. If there is continuity, there is a shorted capacitor. Replace the start capacitor. Place the meter on the R × 100,000 scale. If there is no needle deflection, there is an open capacitor. Replace the start capacitor.

If all the tests prove satisfactory, there is no capillary restriction, and the unit still fails to operate properly, change the relay. The new relay will eliminate electrical problems such as improper pickup and dropout. These cannot be determined with the tests listed above. If a good relay fails to correct the difficulty, the compressor is inoperative due to internal defects. It must be replaced.

High Starting Torque Motors (CSIR) with a Two-Terminal External Overload and a Remote-Mounted Potential Relay

High starting torque motors (CSIR) with a two-terminal external overload and a remote-mounted potential relay represent another type that must be checked. These are used in compressors for light air conditioning units and also for commercial and residential refrigeration units.

In this type of motor the starting contacts on the potential-type relay are normally closed. The

7-55. High starting torque motors (CSIR) with current relay mounted on the compressor terminals.

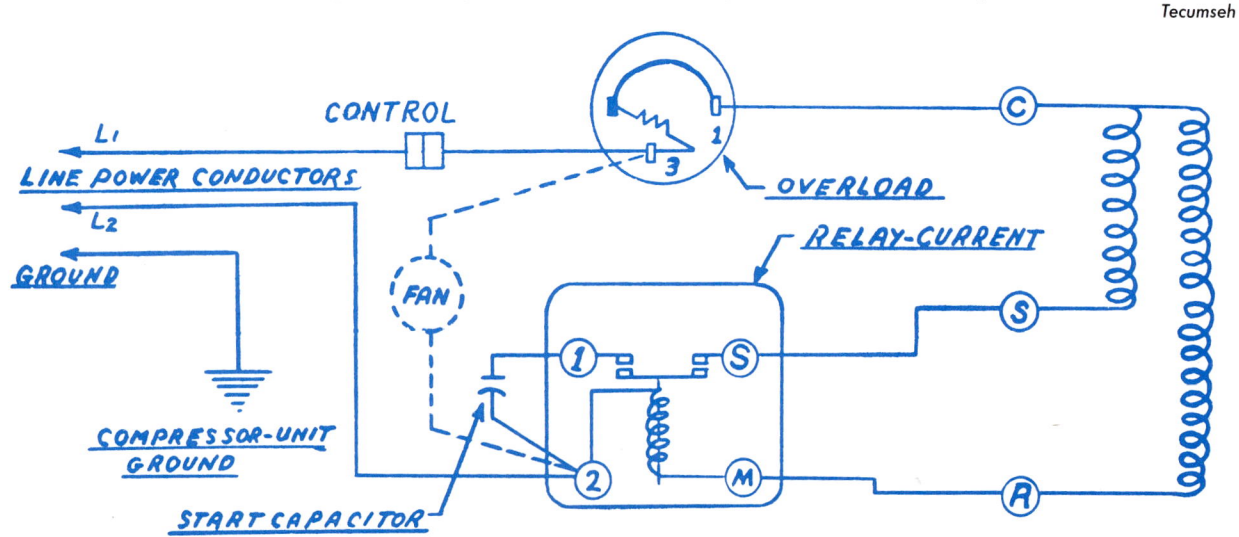

Tecumseh

CHAPTER 7—COMPRESSORS

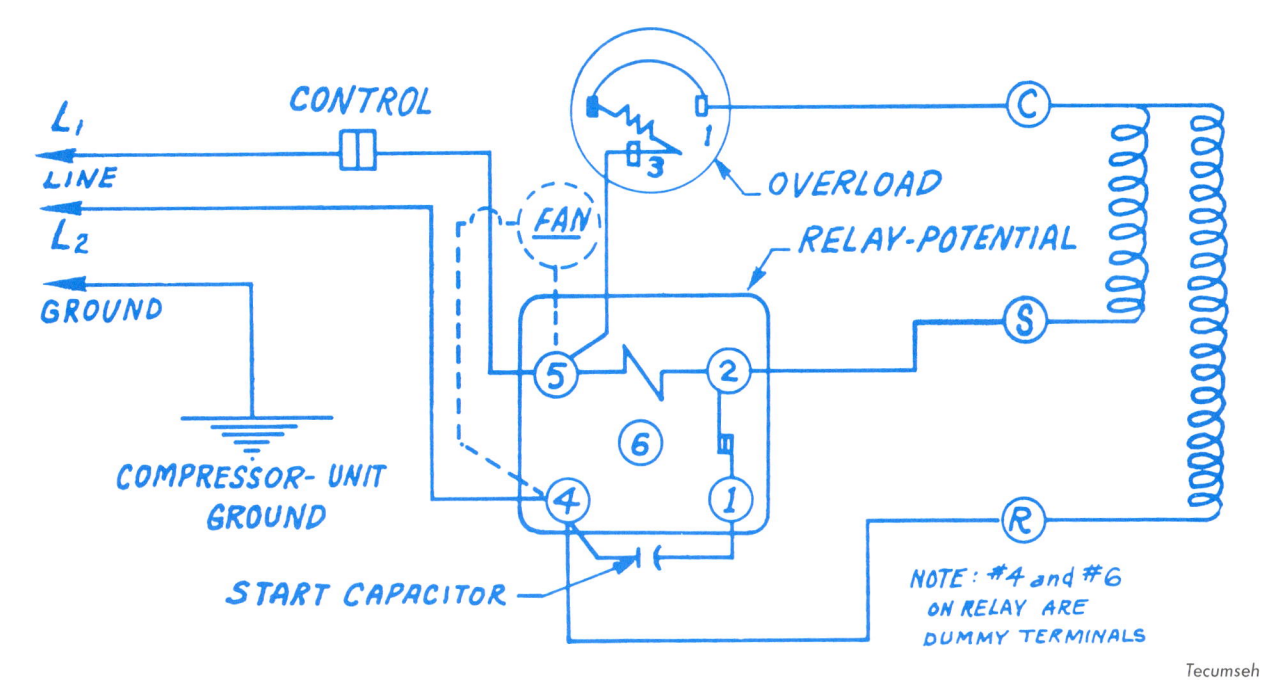

7-56. High starting torque motors (CSIR) with two-terminal external overload and potential relay mounted remote.

Tecumseh

electrical system on this type of hermetic system can be seen in Fig. 7-56. Use a voltmeter to check the power source. Then use an ohmmeter, with the power turned off, to check continuity. Make sure leads 2S and 4R are disconnected. Open the fan circuit, if there is one.

Now, using the ohmmeter, check continuity across the following:

1. Check continuity across L_1 and 3. If there is no continuity, close the control contacts. If there is still no continuity, replace the control.

2. Check continuity across No. 3 and No. 1 on the overload. If there is no continuity, the protector may be tripped. Wait 10 minutes and try again. If there is still no continuity, the protector is defective. Replace the protector.

3. Check continuity across No. 3 on the overload and No. 5 on the relay. If there is no continuity, check the leads between No. 3 on the protector and No. 5 on the relay.

4. Check continuity across No. 1 on the overload and C on the compressor. If there is no continuity, check the leads between No. 1 on the overload and C on the compressor.

5. Check continuity across C and S on the compressor. If there is no continuity, an open start winding is indicated. Replace the compressor.

6. Check continuity across C and R on the compressor. If there is no continuity, there is an open in the run winding of the compressor. Replace the compressor.

7. Check continuity across No. 5 on the relay and No. 2 on the relay. If there is no continuity, the solenoid's coil is open. The relay is defective. Replace the relay.

8. Check continuity across No. 2 and No. 1 on the relay. If there is no continuity, the contacts are open when they should be closed. Replace the relay.

9. Check continuity across No. 1 on the relay and No. 4 on the relay with the meter on the $R \times 1$ scale. If there is continuity, the capacitor is shorted. Replace the start capacitor. No needle deflection on the meter when it is on the $R \times 100{,}000$ scale means the capacitor is open. Replace the capacitor.

10. Check between C and the shell of the compressor. If there is continuity, there is a short. The motor is grounded. Replace the compressor.

11. Check the motor winding resistances against the manufacturer's specification sheet.

12. Check continuity between

187

REFRIGERATION AND AIR CONDITIONING TECHNOLOGY

leads 2S and 4R and reconnect the unit.

If all the tests prove satisfactory and the unit still does not operate properly, change the relay. The new relay will eliminate any electrical problems, such as improper pickup and dropout, that cannot be determined with the checks just performed. If a good relay fails to correct the difficulty, the compressor is inoperative due to internal defects. It must be replaced.

High Starting Torque Motors (CSR) with Three-Terminal Overloads and Remote-Mounted Relays

High starting torque motors (CSR) with three-terminal overloads and remote-mounted potential relays are another type of motor used in the hermetic compressor systems. Fig. 7-57.

Starting contacts on the potential type of relay are normally closed. The electrical system on this type of compressor can be checked by using the voltmeter to check the power source. Then, use the ohmmeter to check continuity across the following locations. First, disconnect the leads so that no external wiring connects terminals 5-C, S-2 on the relay, and R-2 on the overload. Using the ohmmeter, check the following:

1. Check continuity across the control contacts—L$_1$ and C on the compressor. The control contacts must be closed. If they are open, replace the compressor.

2. Check continuity across No. 5 on the relay and No. 2 on the relay. No continuity indicates an open potential coil. Replace the relay.

3. Check continuity across

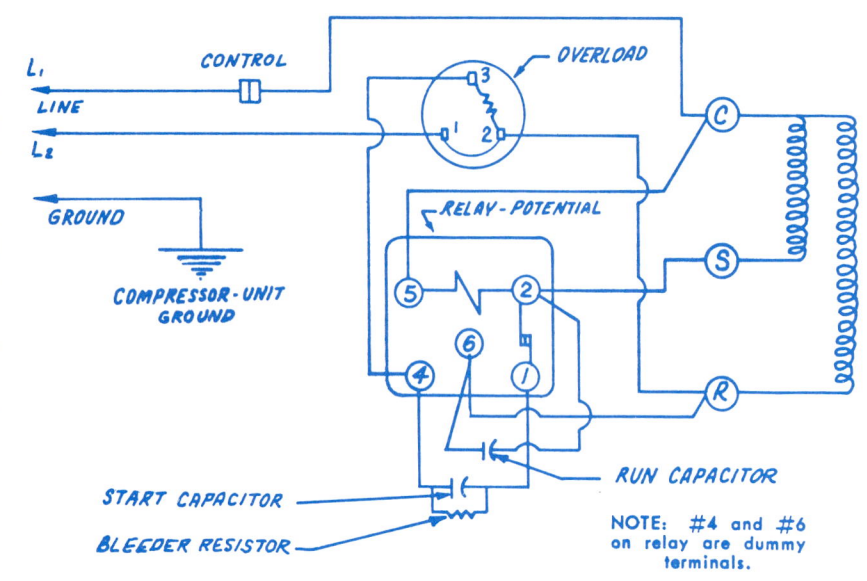

7-57. High starting torque motors (CSR) with a three-terminal external overload and potential relay mounted remote.

No. 2 and No. 1 on the relay. No continuity indicates an open contact situation. Replace the relay.

4. Check continuity across terminals C and S on the compressor. No continuity indicates an open start winding. Replace the compressor.

5. Check continuity across terminals C and R on the compressor. No continuity indicates an open run winding. Replace the compressor.

6. Check continuity across No. 6 and No. 2 on the relay with the meter on the R × 1 scale. Continuity shows a shorted capacitor. Replace the run capacitor. Set the meter on the R × 100,000 scale. If there is no needle deflection, the capacitor is open. Replace the run capacitor.

7. Check continuity across No. 1 on the relay and No. 3 on the overload. Check as in Step 6 above.

8. Check continuity across No. 1 and No. 3 on the overload.

No continuity indicates the overload is open and should be replaced. However, it should have been given at least 10 minutes to replace itself properly.

9. Check continuity across C terminal on the compressor and the other ohmmeter lead to the shell of the compressor. Continuity indicates the motor has become grounded to the shell. Replace the compressor.

10. Check the resistance of the motor windings against the values given in the manufacturer's resistance tables.

11. Check continuity of the leads removed above and reconnect terminals 5 to C, S to 2 on the relay, and R and 2 on the overload.

If the tests prove satisfactory and the unit still does not operate properly, replace the relay. The new relay will eliminate any electrical problems, such as improper pickup and dropout, which cannot be determined

CHAPTER 7—COMPRESSORS

with the checks just performed. If a good relay fails to correct the difficulty, the compressor is inoperative due to internal defects. It must be replaced.

PSC Motor with a Two-Terminal External Overload and Run Capacitor

Another type of motor used on compressors is the PSC. Fig. 7-58. It has a two-terminal external overload and a run capacitor. It does not have a start capacitor or relay.

Use a voltmeter to check the source voltage. Then, using an ohmmeter, perform the following checks. Disconnect the run capacitor from terminals S and R before starting the tests.

1. L_1 and No. 3 on the overload show no continuity. Close the control contacts. If there is still no continuity, replace the control.

2. C and S terminals on the compressor show no continuity. This means the start winding is open. Replace the compressor.

3. C and R terminals on the compressor show no continuity. This means the run winding is open. A replacement compressor is needed to correct the problem.

4. C and 1 on the overload show no continuity. A defective lead from C to 1 is the probable cause.

5. No. 1 and No. 3 on the overload indicate no continuity. The protector may be tripped. Wait 10 minutes before checking again. If there is still no continuity, the protector is defective. Replace the overload protector.

6. C and the shell of the compressor show continuity. The motor is shorted to the shell or ground. Replace the compressor.

7. Check the motor windings against the manufacturer's tables.

8. Check across the run capacitor with the meter on the R × 1 scale. If it shows continuity, the capacitor is shorted and must be replaced. Set the meter on R × 100,000 scale. No needle deflection indicates that the capacitor is open and needs to be replaced.

9. Reconnect the capacitor to the circuit at terminals S and R. The marked terminal should go to R.

If the above PSC tests reveal no difficulties, but the compressor does not operate properly, add the proper relay and start capacitor to provide additional starting torque. Figure 7-59 gives the proper wiring for a field-installed relay and capacitor. If the unit still fails to operate, the

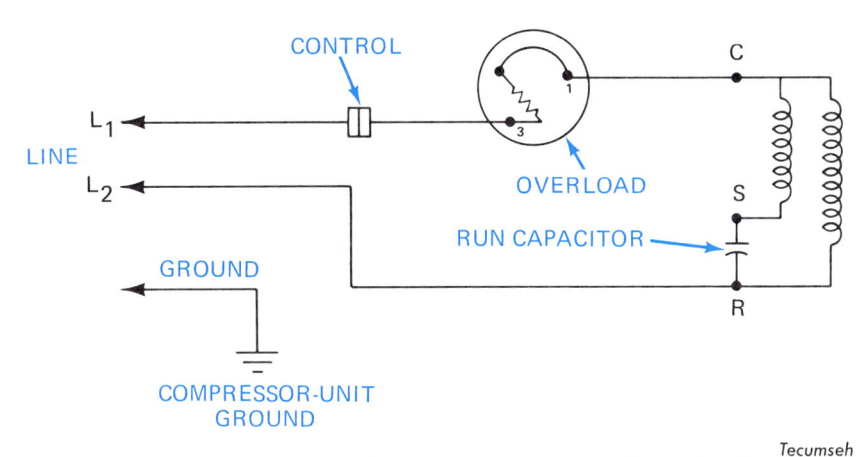

Tecumseh

7-58. PSC motors with two-terminal external overload.

7-59. PSC motors with two-terminal external overload with start components field installed.

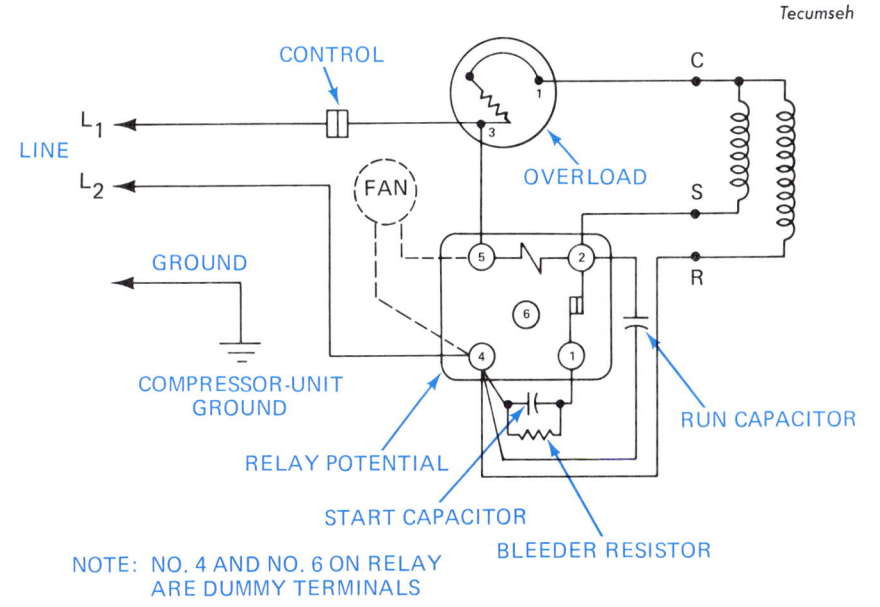

Tecumseh

NOTE: NO. 4 AND NO. 6 ON RELAY ARE DUMMY TERMINALS

189

REFRIGERATION AND AIR CONDITIONING TECHNOLOGY

compressor is inoperative due to internal defects. It must be replaced.

PSC Motor with an Internal Overload (Linebreaker)

Those PSC motors with an internal overload (linebreaker) are somewhat different from those just checked. Thus, the testing sequence varies somewhat. This compressor has an internal line break overload and a run capacitor. It does not have a start capacitor or relay. Fig. 7-60.

1. Use a voltmeter to check the power source. Check the voltage at compressor terminals C and R. If there is no voltage, the control circuit is open.

2. Unplug the unit and check continuity across the thermostat and/or contactor. Check the contactor holding coil.

3. If the line voltage is present between terminals C and R and the compressor does not operate, unplug the unit and disconnect the run capacitor from S and R.

Note: The compressor shell *must* be at 130° F. [54°C] or less for the following checks. This temperature can be read by a Tempstik. A less reliable guide is that the hand can remain in contact with the compressor shell without discomfort at a temperature of 130° F. [54°C] or less.

4. Using the ohmmeter, check the following:

a. Check continuity between R and S. If there is continuity, it can be assumed that both windings are intact. If there is no continuity, it can be assumed that one or both of the windings are open and the compressor should be replaced.

b. Check continuity between R and C. If there is no continuity, the internal overload is tripped. Wait for it to cool off and close. It sometimes takes more than an hour.

c. There is continuity between R and S, but no continuity between R and C (or S and C). If the motor is cool enough (below 130° F. [54°C]) to have closed the overload, then it can be assumed that the overload is defective. The compressor should be replaced.

d. Check continuity between the S terminal and the compressor shell and between the R terminal and the compressor shell. If there is continuity in either or both cases, the motor is grounded. The compressor should be replaced.

e. Check the motor winding resistance against the values given in the manufacturer's charts.

f. Check across the run capacitor with the meter on the R × 1 scale. If there is continuity, the capacitor is shorted and should be replaced.

g. Check across the run capacitor with the meter on the R × 100,000 scale. If there is no needle deflection on the meter, the capacitor is open and should be replaced.

5. Reconnect the run capacitor into the circuit at S and R.

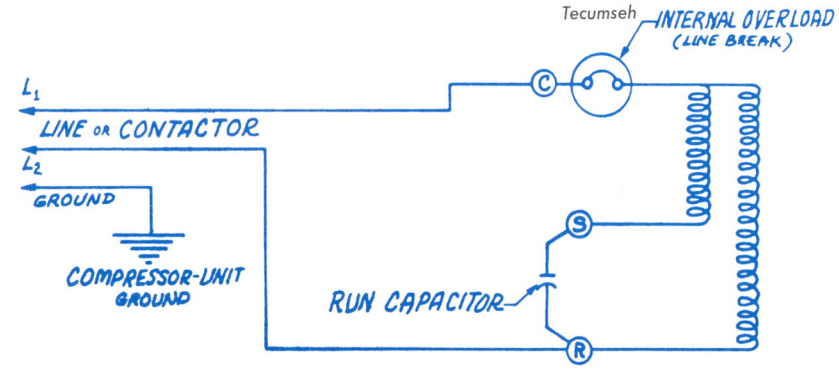

CSR or PSC Motor with the Start Components and an Internal Overload or Linebreaker

The next combination is the CSR or PSC motor with the start components and an internal overload or linebreaker. The run capacitor, start capacitor, and potential relay are the major components outside the compressor. Fig. 7-61.

1. Using the voltmeter, check the power source. Check voltage at the compressor terminals C and R. If there is no voltage, the control circuit is open.

2. Unplug the unit and check continuity across the thermostat and/or contactor. Check the contactor holding coil.

3. If the line voltage is present between terminals C and R and the compressor does not operate, unplug the unit and disconnect the connections to the compressor terminals.

Note: The compressor shell must be at 130° F. [54°C] or less

7-60. PSC motors with internal overload or linebreaker.

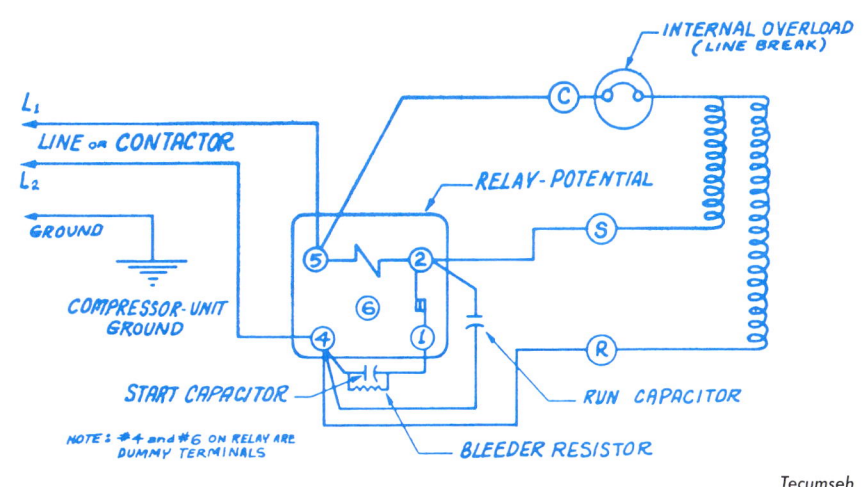

7-61. A CSR or PSC motor with start components and internal overload (linebreak). *Tecumseh*

for the following checks. This temperature can be read by a Tempstik. A less reliable guide is that at this temperature the hand can remain in contact with the compressor shell without discomfort.

4. Using the ohmmeter, check the following:

a. Check continuity between R and S. If there is continuity, it can be assumed that both windings are intact. If there is no continuity, it can be assumed that one or both of the windings are open. The compressor should be replaced.

b. Check continuity between R and C. If there is no continuity, the internal overload is tripped. Wait for it to cool off and close. It sometimes takes more than an hour.

c. There is continuity between R and S, but no continuity between R and C (or S and C). If the motor is cool enough (130° F. [54°C]) to have closed the overload, then it can be assumed that the overload is defective. The compressor must be replaced.

d. Check continuity between the S terminal and the compressor shell and between R and the compressor shell. If there is continuity in either or both cases, the motor is grounded and the compressor should be replaced.

e. Check the motor winding resistance against the values given in the manufacturer's tables for the specific model being tested.

f. Check continuity across the run capacitor with the meter on the R × 1 scale. If there is continuity, the capacitor is shorted and should be replaced.

g. Check continuity across the run capacitor with the meter on the R × 100,000 scale. If there is no needle deflection, the capacitor is open and should be replaced.

5. Check continuity across 5 and 2 on the relay. No continuity indicates an open potential coil. Replace the relay. The electrolytic capacitor used for the start usually has its contents on the outside of the compressor housing. If the coil did not energize properly, it leaves the start capacitor in the circuit too long (only a few seconds). This means the capacitor will get too hot. When this happens, the capacitor will spew its contents outside the container.

6. Check continuity across No. 2 and No. 1 on the relay. No continuity shows an open contacts condition. Replace the relay.

7. Check continuity across No. 4 and No. 1 on the relay with the meter on the R × 1 scale. Continuity indicates a shorted capacitor. Replace. With the meter on the R × 100,000 scale, if there is no needle deflection, the start capacitor is open. Replace the capacitor.

If all of the tests prove satisfactory and the unit still fails to operate properly, change the relay. If a new relay does not solve the problem, then it is fairly safe to assume that the compressor is defective and should be replaced.

Compressors with Internal Thermostat, Run Capacitor, and Supplementary Overload

Some compressors have an internal thermostat, a run capacitor, and a supplementary overload. However, they do not have a start capacitor or relay. The schematic for such a compressor is shown in Fig. 7-62.

The supplementary overload has normally closed contacts connected in series with the normally closed contacts of the internal thermostat in the motor. Operation of either of these devices will open the control circuit to drop out the contactor. Make

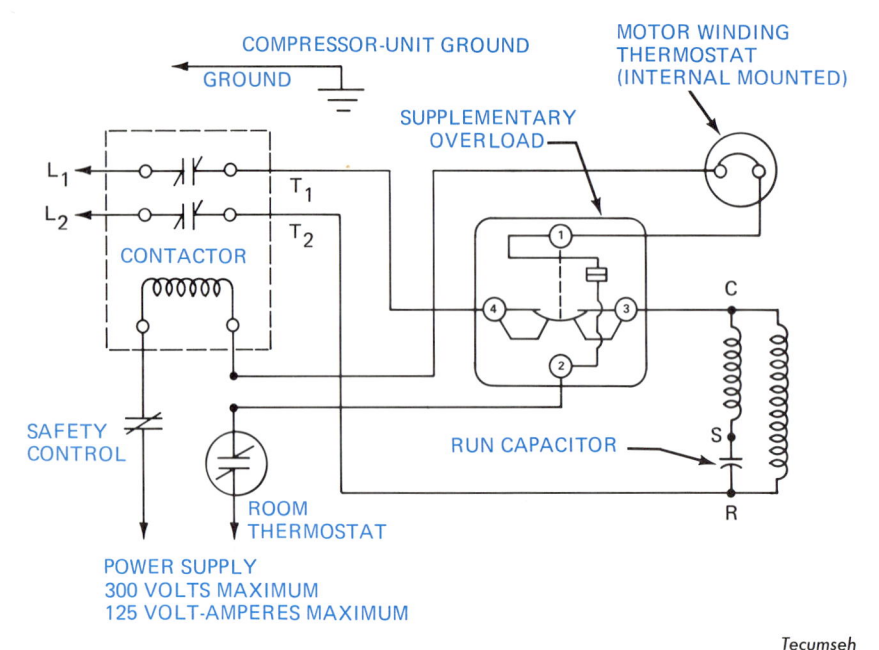

REFRIGERATION AND AIR CONDITIONING TECHNOLOGY

7-62. PSC motors with internal thermostat and supplementary external overload.
Tecumseh

sure the control thermostat and the system safety controls are closed. Using a voltmeter, check the power source at L_1, L_2, and the control circuit power supply. If the contactor is not energized, the contactor holding coil is defective or the control circuit is open in either the supplementary overload or the motor thermostat. Unplug the unit and disconnect the run capacitor from terminals S and R.

Using the ohmmeter, check the continuity across the following:

1. Check continuity across No. 3 and No. 4 on the overload. No continuity means the supplementary overload is defective. Replace it.

2. Check continuity across No. 1 and No. 2 on the overload. No continuity can mean the overload may be tripped. Wait 10 minutes. Test again. If there is still no continuity, the overload is defective. Replace overload.

3. Check continuity across the internal (motor winding) thermostat terminals at the compressor. Check Fig. 7-48 for the location of the internal thermostat terminals. If there is no continuity, the internal thermostat may be tripped. Wait for it to cool off and close. It sometimes takes an hour. If the compressor is cool to the touch (below 130° F. [54°C]) and there is still no continuity, internal thermostat circuitry is open and the compressor must be replaced.

4. Check continuity across terminals C and S on the compressor. No continuity indicates an open start winding. Replace the compressor.

5. Check continuity across terminals C and R. No continuity indicates an open run winding. Replace the compressor.

6. Check continuity across terminal C and the shell of the compressor. Continuity shows a grounded compressor. Replace the compressor.

7. Check the motor winding resistance with the chart given by the manufacturer.

8. Check continuity across the run capacitor with a meter on the $R \times 1$ scale. If there is continuity, the capacitor is shorted. Replace it. Place the meter on the $R \times 100,000$ scale. If there is no needle deflection, the capacitor is open. Replace the capacitor.

9. Reconnect the capacitor to the circuit at terminals S and R.

CSR or PSC Motor with Start Components, Internal Thermostat, and Supplementary External Overload

Another arrangement for single-phase compressors is the CSR or PSC motor with start components, internal thermostat, and supplementary external overload. Fig. 7-63.

This type of compressor is equipped with an internal thermostat, run capacitor, start capacitor, potential relay, and supplemental overload.

The supplemental overload has normally closed contacts connected in series with the normally closed contacts of the internal thermostat located in the motor. Operation of either of these devices will open the control circuit to drop out the contactor. Figure 7-64 shows the details of the internal thermostat.

Make sure the control thermostat and system safety controls are closed. Using the voltmeter, check the power source at L_1 and L_2. Also check the control circuit power supply with the voltmeter.

192

CHAPTER 7—COMPRESSORS

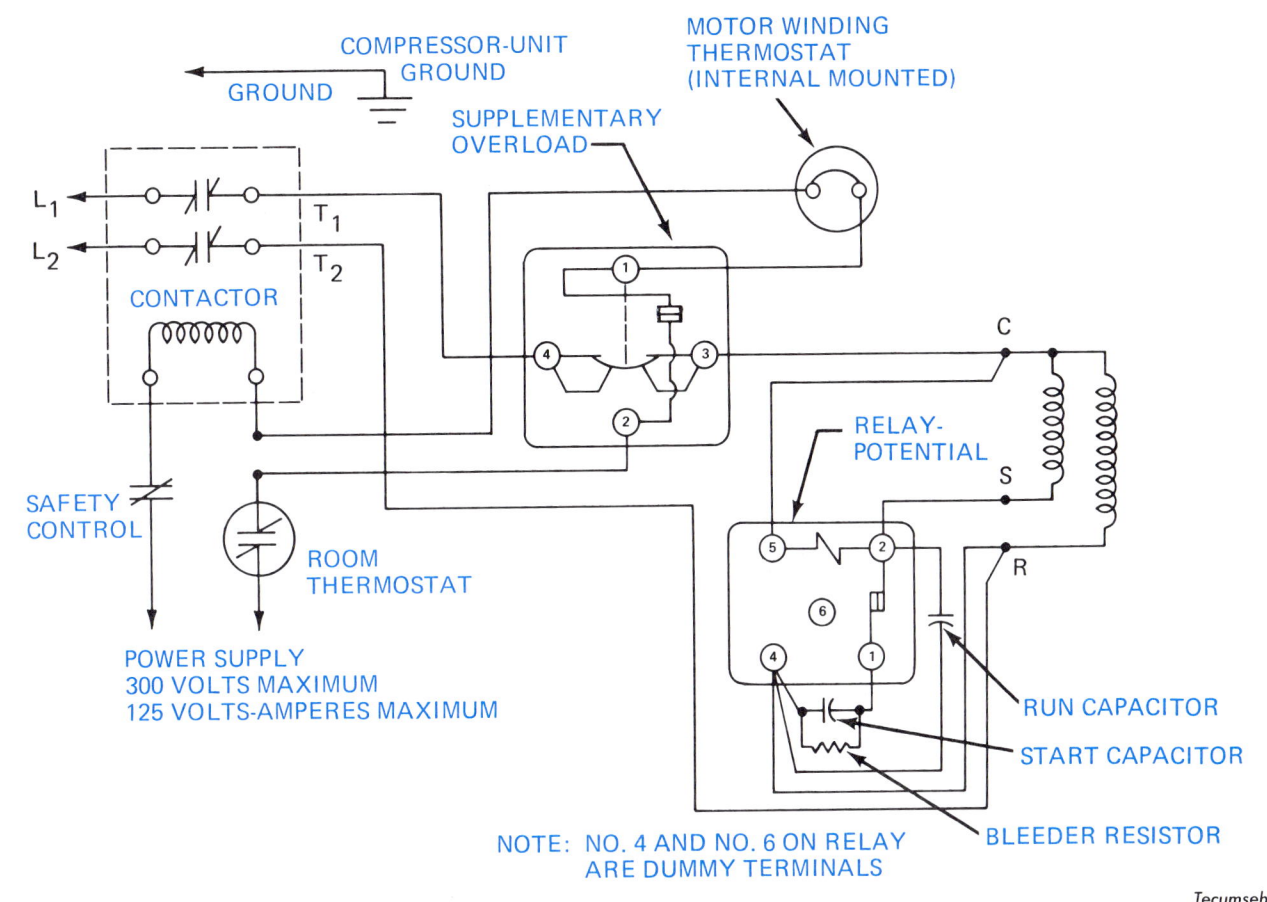

7-63. A CSR or PSC motor with start components and internal thermostat and supplementary external overload.

7-64. Internal thermostat embedded in the motor winding.
Tecumseh

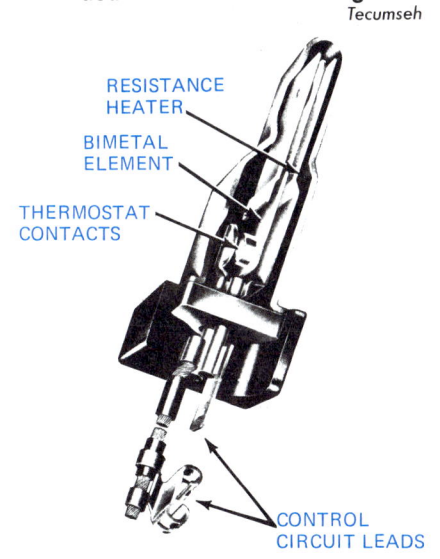

If the contactor is not energized, the contactor holding coil is defective or the control circuit is open in either the supplemental overload or the motor thermostat.

Unplug the unit and disconnect the connections to the compressor terminals.

Use the ohmmeter and check for continuity across the following:

1. With the control circuit power supply off, check the continuity of the contactor holding coil.

2. Check continuity across No. 4 and No. 3 of the supplemental overload. No continuity means the overload is defective. Replace the overload.

3. Check continuity across No. 1 and No. 2 of the overload. No continuity means the overload may be tripped. Wait at least 10 minutes and test again. If there is still no continuity, the overload is defective. Replace the defective overload.

4. Check continuity across the internal thermostat terminals at the compressor. See Fig. 7-48 for the location of the terminals of the internal thermostat. If there is no continuity, the internal thermostat may be tripped. Wait for it to cool off and close. It sometimes takes more than one

193

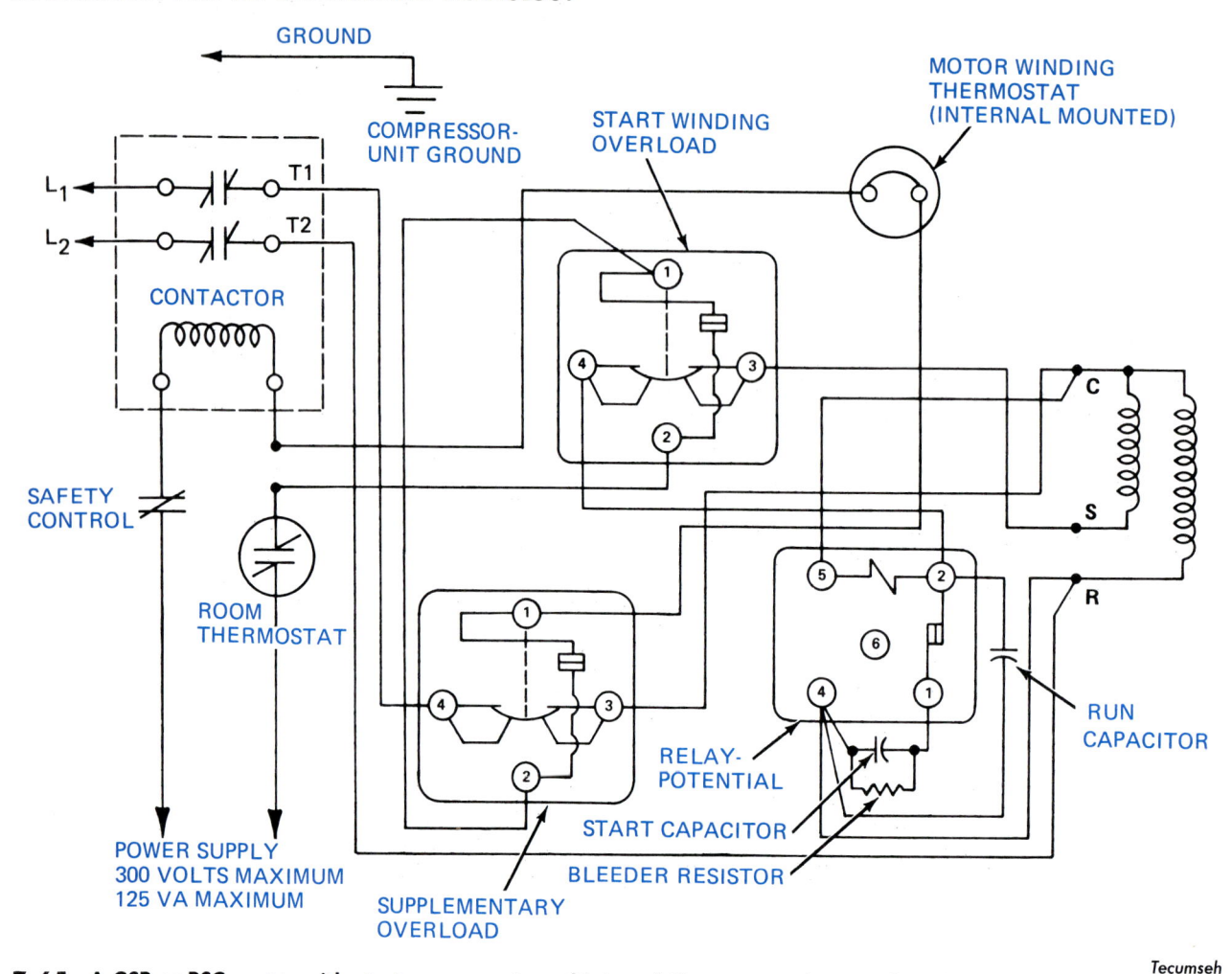

REFRIGERATION AND AIR CONDITIONING TECHNOLOGY

7-65. A CSR or PSC motor with start components and internal thermostat plus supplemental external overload and start winding overload. Note: #4 and #6 on the relay are dummy terminals.

Tecumseh

hour to cool. If the compressor is cool to the touch (below 130° F. [54°C]), and there is still no continuity, the internal thermostat circuitry is open and the compressor must be replaced.

5. Check for continuity across terminals R and S on the compressor. If there is no continuity, one or both of the windings are open. Replace the compressor.

6. Check for continuity across terminal S and the compressor shell. Check for continuity across terminal R and the compressor shell. If there is continuity in either or both cases, the motor is grounded and the compressor should be replaced.

7. Check the motor winding resistances with the chart furnished by the compressor manufacturer.

8. Check for continuity across terminals 5 and 2 on the relay. No continuity indicates an open potential coil. Replace the relay.

9. Check for continuity across terminals 2 and 1 on the relay. No continuity indicates open contacts. Replace the relay.

10. Check for continuity across terminals 4 and 1 on the relay with the meter on the R × 1 scale. If continuity is read, it indicates a shorted capacitor. Replace the capacitor. Repeat with the meter on the R × 100,000 scale. No needle deflection indicates the start capacitor is open. Replace the start capacitor.

11. Discharge the run capacitor by placing a screwdriver across the terminals. Remove the leads from the run capacitor. With the meter set on the R × 1 scale, continuity across the capacitor terminals indicates a

CHAPTER 7—COMPRESSORS

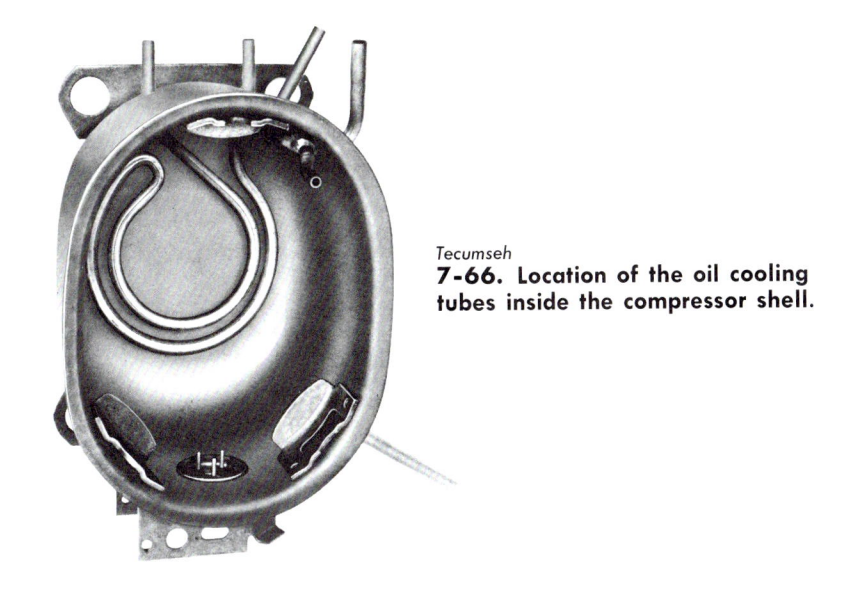

Tecumseh
7-66. Location of the oil cooling tubes inside the compressor shell.

shorted capacitor. Replace the capacitor. Repeat the same test with the meter set on the R × 100,000 scale. No needle deflection indicates an open capacitor. Replace the run capacitor.

If all the tests prove satisfactory and the unit still fails to operate properly, change the relay. The new relay will eliminate electrical problems such as improper pickup or dropout, which cannot be determined by the above tests. If a good relay fails to correct the difficulty, the compressor is inoperative due to internal defects. It must be replaced.

One other arrangement for a compressor using single-phase current is a CSR or PSC motor with start components, internal thermostat, supplemental external overload, and start winding overload. Fig. 7-65.

The diagnosis is identical to that described for the previous type of motor circuit. However, there is an additional start winding overload in the control circuit in series with the internal thermostat and supplemental overload.

Check the start winding overload in the same way the supplemental overload is checked.

COMPRESSOR CONNECTIONS AND TUBES

Tecumseh, like other compressor manufacturers, made compressors for many manufacturers of refrigerators, air conditioning systems, and coolers. Because of this, the same compressor model may be found in the field in many suction and discharge variations. Each variation depends upon the specific application for which the compressor was designed.

Suction connections can usually be identified as the stub tube with the largest diameter in the housing. If two stubs have the same outside diameter, then the one with the heavier wall will be the suction connection. If both of the largest stub tubes have the same outside diameter and wall thickness, then either can be used as the suction connection. However, the one farthest from the terminals is preferred.

The stub tube not chosen for the suction connection may be used for processing the system.

Compressor connections can usually be easily identified. However, occasionally some question arises concerning oil cooler tubes and process tubes.

Oil cooler tubes are found only in low temperature refrigeration models. These tubes connect to a coil or hairpin bend within the compressor oil sump. Fig. 7-66. This coil or hairpin bend is not open inside the compressor. Its only function is to cool the compressor sump oil. The oil cooler tubes are generally connected to an individually separated tubing circuit in the air-cooled condenser.

Process Tubes

Process tubes are installed in compressor housings at the factory as an aid in factory dehydration and charging. These can be used in place of the suction tube if they are of the same diameter and wall thickness as the suction tube.

Standard discharge tubing arrangements for Tecumseh hermetic compressors are shown in Fig. 7-67. Discharge tubes are generally in the same position within any model family. Suction and process tube positions may vary.

Other Manufacturers of Compressors

Besides Tecumseh, there are other manufacturers of compressors for the air conditioning and refrigeration trade. One is Americold Compressor Corporation. Two of the models made by

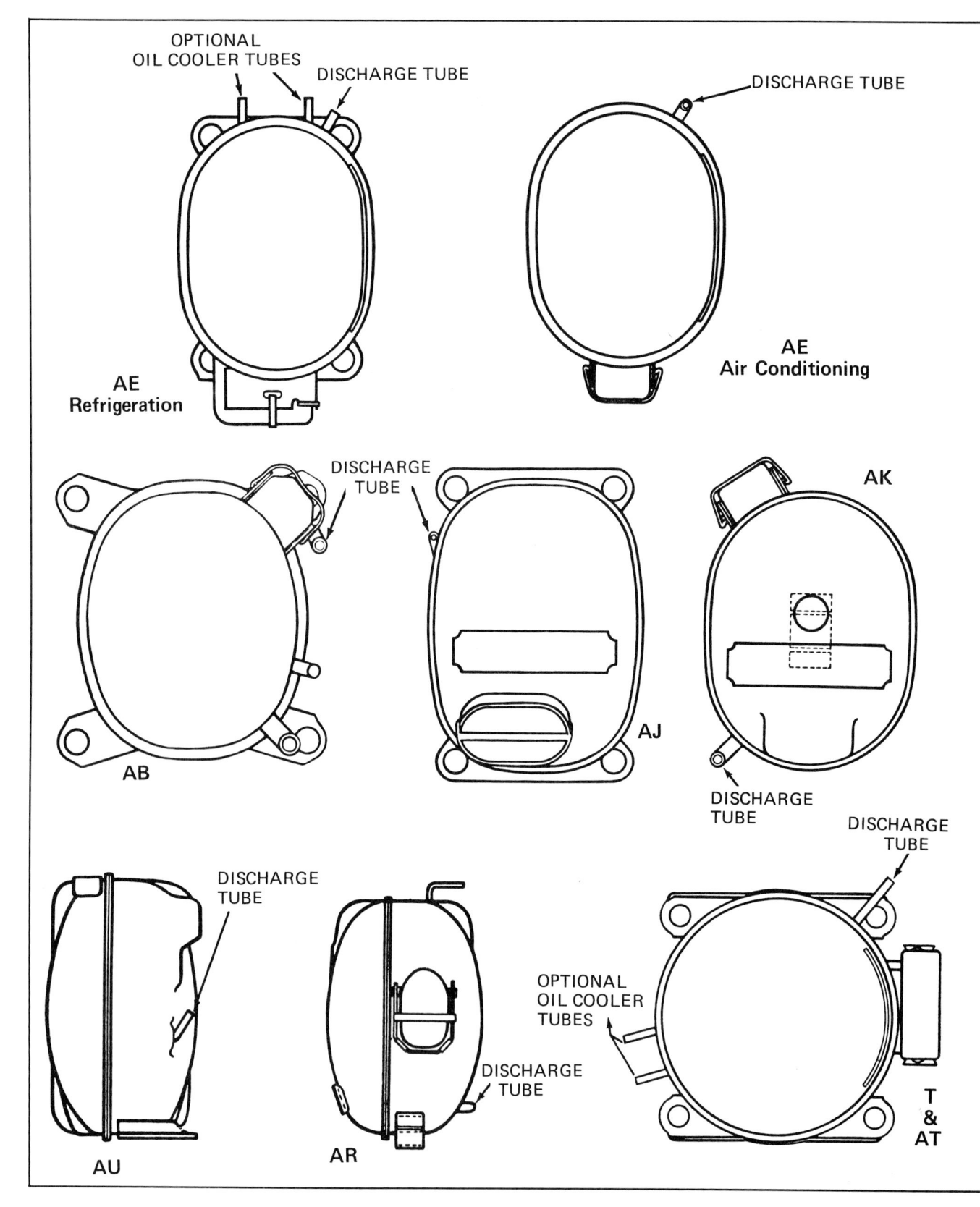

REFRIGERATION AND AIR CONDITIONING TECHNOLOGY

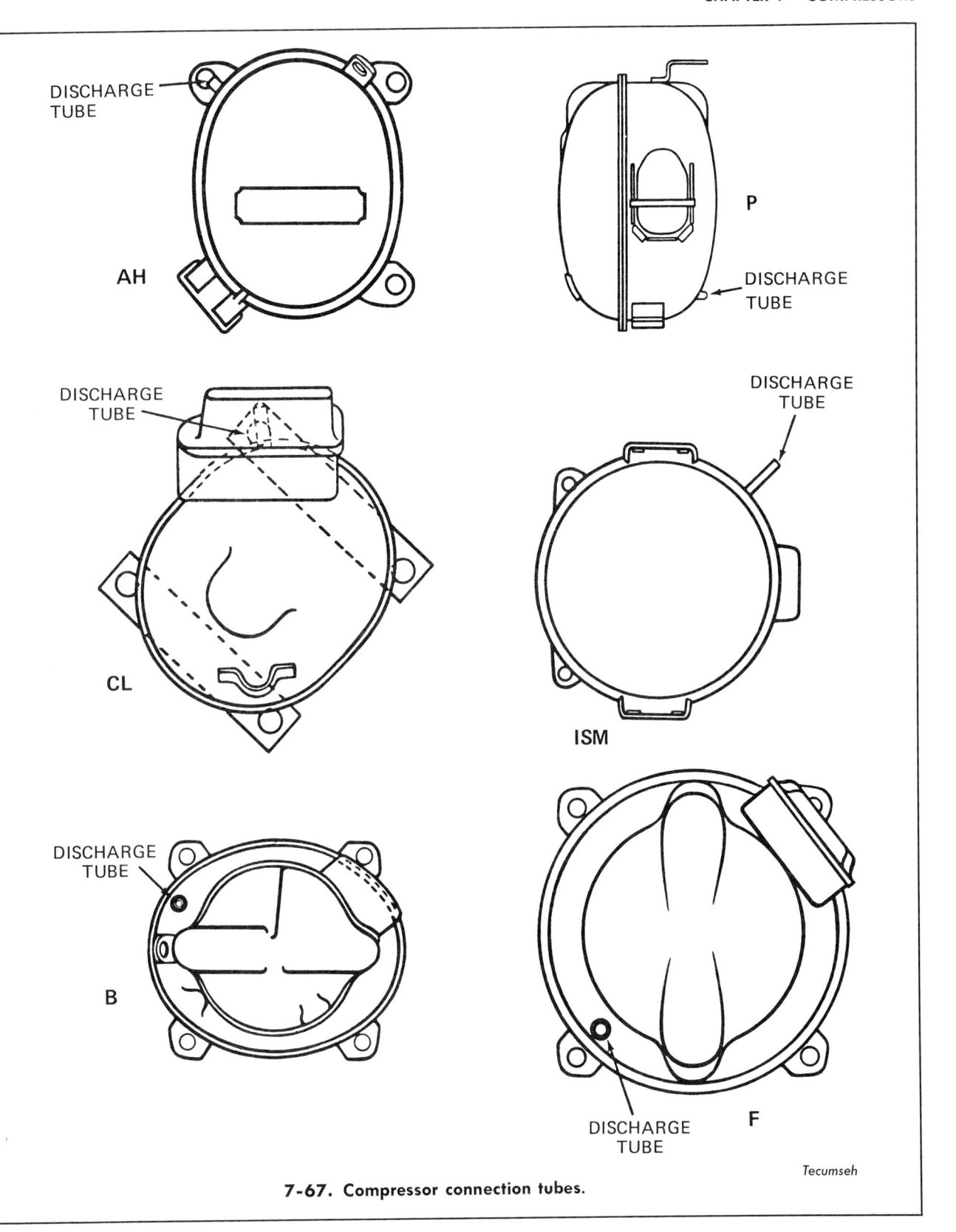

CHAPTER 7—COMPRESSORS

7-67. Compressor connection tubes.

Tecumseh

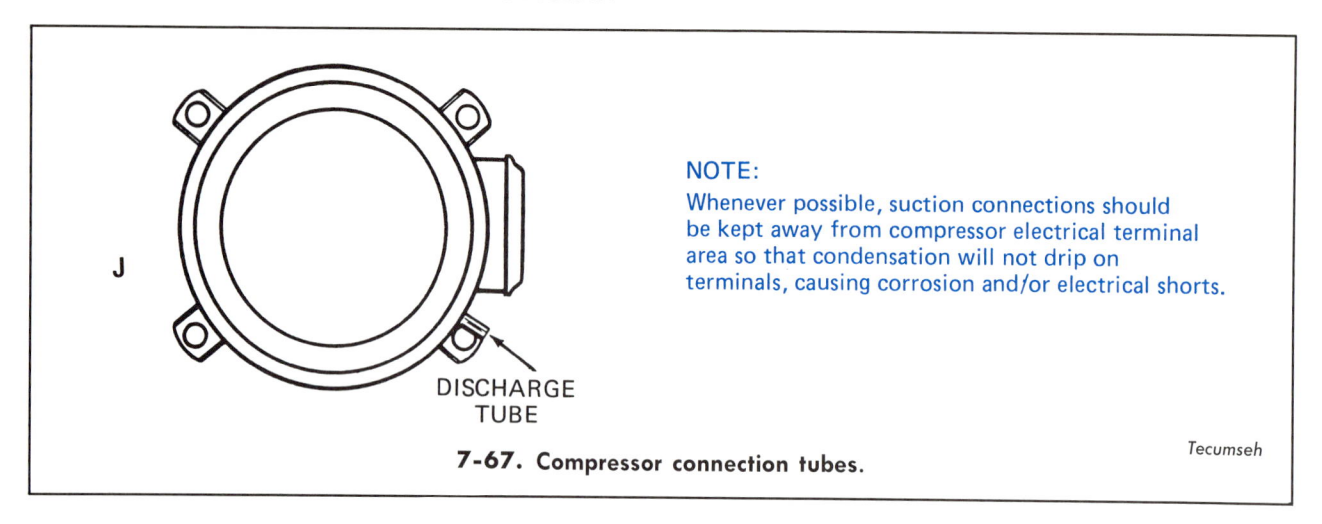

REFRIGERATION AND AIR CONDITIONING TECHNOLOGY

NOTE:
Whenever possible, suction connections should be kept away from compressor electrical terminal area so that condensation will not drip on terminals, causing corrosion and/or electrical shorts.

7-67. Compressor connection tubes.

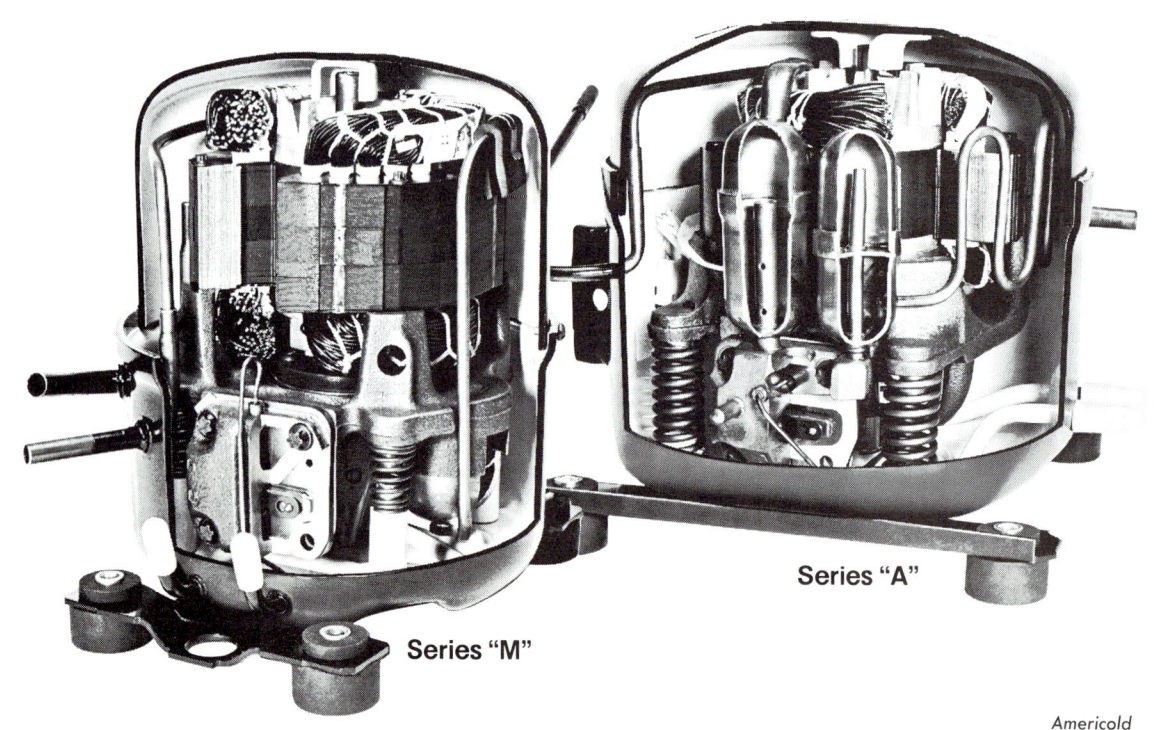

Tecumseh

7-68. Series M and series A compressors made by Americold.

Americold

Americold are the M series and the A series. Fig. 7-68. Both use the same overload relay and current relay connections. Fig. 7-69. All of these models use R-12 as the refrigerant. They are made in sizes ranging from $1/10$ through $1/4$ horsepower. They weigh 21 to 25 pounds. Figure 7-70 shows the location of the suction and discharge stubs as well as the process tube.

ROTARY COMPRESSORS

The rotary compressor is made in two different configurations—the *stationary blade* rotary compressor and the *rotating blade* rotary compressor. The sta-

198

CHAPTER 7—COMPRESSORS

A - OVERLOAD PROTECTOR
B - CURRENT RELAY

Americold

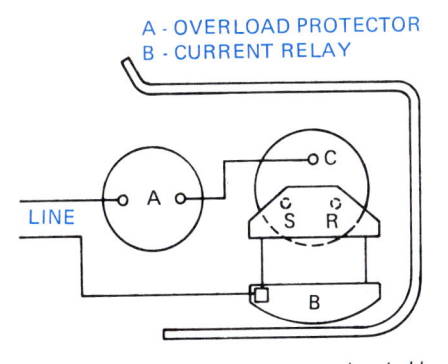

7-69. Location of the terminals for the compressors and electrical connections on Americold compressors.

Figure 7-72 shows how the rotation of the off-center cam compresses the gas refrigerant in the cylinder of the rotary compressor. The cam is rotated by an electric motor. As the cam spins it carries the ring with it. The ring rolls on its outer rim around the wall of the cylinder.

To be brought into the chamber, the gas must have a pathway. Note that in Fig. 7-73 the vapor comes in from the freezer and goes out to the condenser

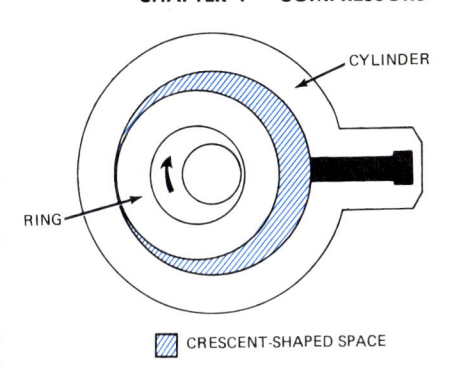

General Motors

7-72. Operation of a rotary compressor.

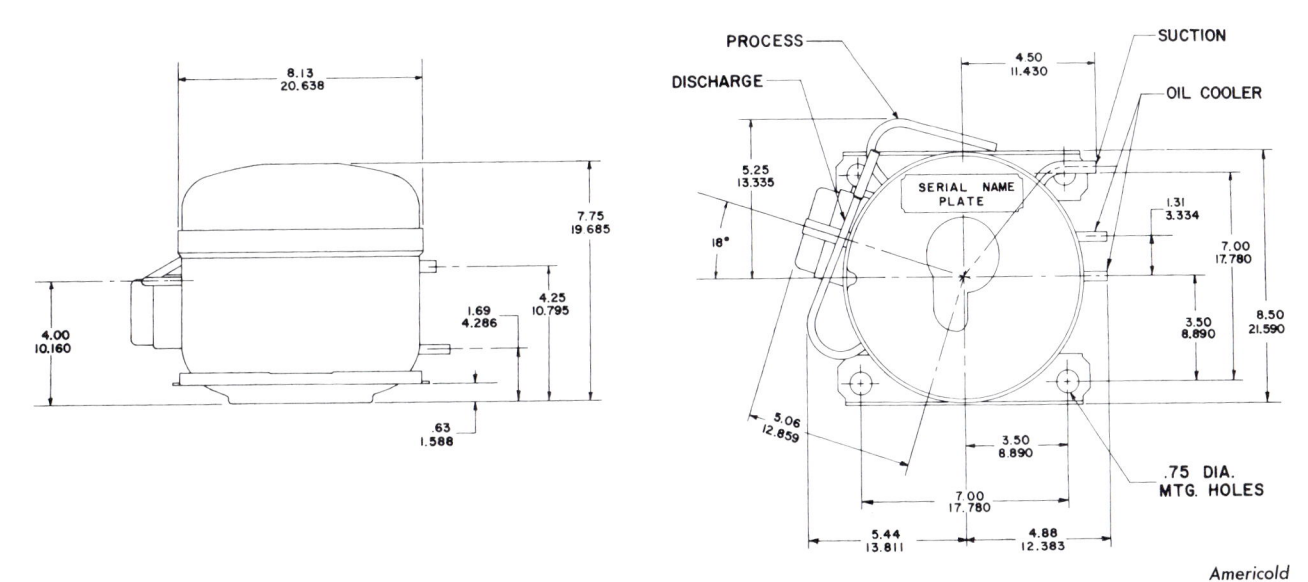

Americold

7-70. Location of process, discharge, suction, and oil cooler stubs on Americold compressors.

tionary blade rotary compressor is the type that has just been described. Both of these compressors have problems regarding lubrication. This problem has been partly solved.

Stationary Blade Rotary Compressors

The only moving parts in a stationary blade rotary compressor are a steel ring, an eccentric or cam, and a sliding barrier. Fig. 7-71.

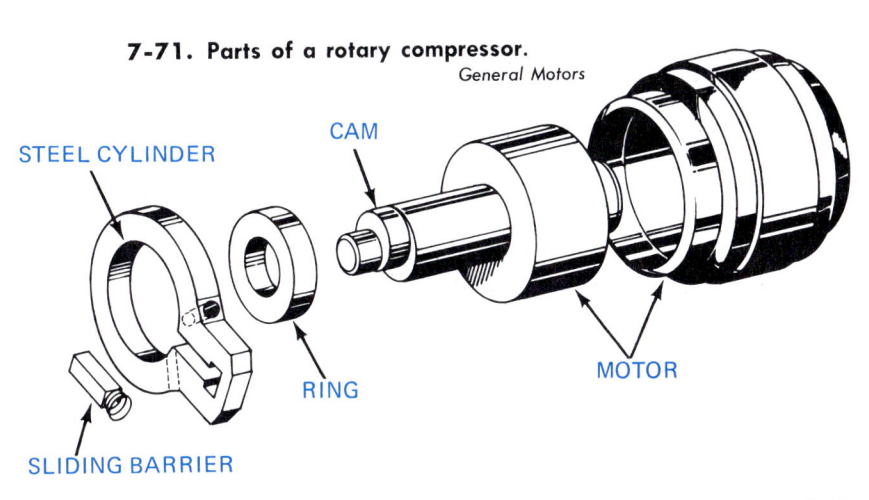

7-71. Parts of a rotary compressor.
General Motors

199

REFRIGERATION AND AIR CONDITIONING TECHNOLOGY

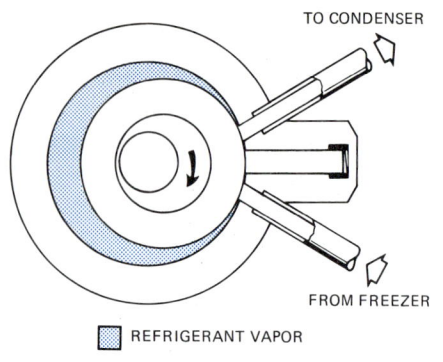

7-73. Beginning of the compression phase in a rotary compressor. *General Motors*

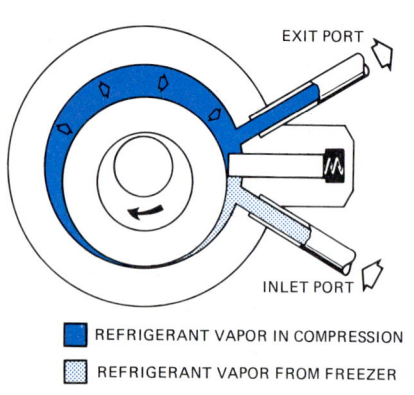

7-74. Beginning of the intake phase in a rotary compressor. *General Motors*

it compresses the gas and passes it on to the condenser. Fig. 7-75. The finish of the compression portion of the stroke or operation is shown in Fig. 7-76. The ring rotates around the cylinder wall. It is held in place by the spring tension of the barrier's spring and the pressure of the cam being driven by the electric motor. This type of compressor is not used as much as the reciprocating hermetic type of compressor.

Rotating Blade Rotary Compressors

The rotating blade rotary compressor has its roller centered on a shaft that is eccentric to the center of the cylinder. Two spring-loaded roller blades are mounted 180° apart. They sweep the sides of the cylinder. The roller is mounted so that it touches the cylinder at a point between the intake and the discharge ports. The roller rotates. In rotating, it pulls the vapor into the cylinder through the intake port. Here, the vapor is trapped in the space between the cylinder wall, the blade, and the point of contact between the roller and the cylinder. As the next blade passes the contact point, the vapor is compressed. The space for the vapor becomes smaller and smaller as the blade rotates. Once the vapor has reached the pressure determined by the compressor manufacturer, it exits through the discharge port to the condenser.

On this type of rotating blade rotary compressor the seals on the blades present a particular problem. There also are lubrica-

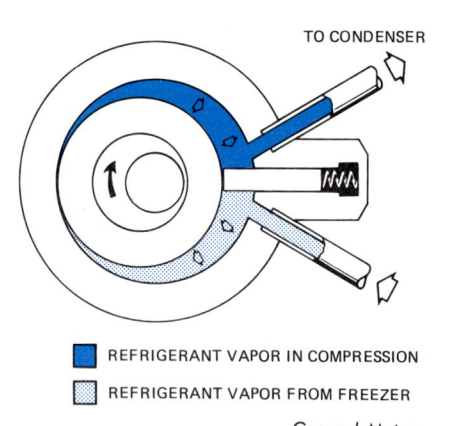

7-75. Compression and intake phases half completed in a rotary compressor. *General Motors*

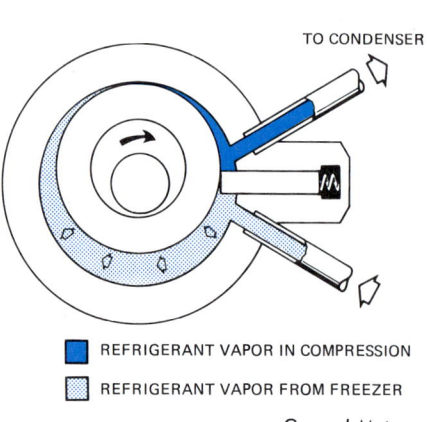

7-76. Finish of the compression phase of the rotary compressor. *General Motors*

tion problems. However, a number of rotary compressors are still in operation in home refrigerators.

Some manufacturers make rotary blade compressors for commercial applications. They are used primarily with ammonia. Thus, there is no copper or copper alloy tubing or parts. Most of the ammonia tubing and working metal is stainless steel.

through holes that have been drilled in the compressor frame. Note that the gas is compressed by an offset rotating ring. Figure 7-74 shows how the refrigerant vapor in the compressor is brought from the freezer. Then, the exit port is opening. When the compressor starts to draw in the vapor from the freezer the barrier is held against the ring by a spring. This barrier separates the intake and exhaust ports. As the ring rolls around the cylinder

200

REVIEW QUESTIONS

1. List three classifications of refrigeration compressors.
2. List three types of hermetic compressor motors.
3. What are two types of motor relays used in refrigeration and air conditioning compressors?
4. Describe a glass quick-connect terminal.
5. What is the purpose of the rubber grommet used to hold a compressor in place?
6. What are the two types of crankcase heaters?
7. What does RSIR stand for in motor terminology?
8. What does CSIR stand for in motor terminology?
9. What is a stub tube?
10. Why are oil cooler tubes needed?
11. What is a process tube?
12. What are the two types of rotary compressors?

8

Condensers and Cooling Towers

The condenser is a heat transfer device. It is used to remove heat from hot refrigerant vapor. Using some method of cooling, the condenser changes the vapor to a liquid. There are three basic methods of cooling the condenser's hot gases. The method used to cool the refrigerant and return it to the liquid state serves to categorize the two types of condensers. Thus, there are two types of condensers: air cooled and water cooled. Cooling towers are also used to cool the refrigerant.

Most commercial or residential home air-conditioning units are air cooled. Water is also used to cool the refrigerant. This is usually done where there is an adequate supply of fairly clean water. Industrial applications rely upon water to cool the condenser gases. The evaporative process is also used to return the condenser gases to the liquid state. Cooling towers use the evaporative process.

CONDENSERS

Air-Cooled Condensers

Figure 8-1 illustrates the refrigeration process within an air-cooled condenser. Figure 8-2 shows some of the various types of compressors and condensers mounted as a unit. These units may be located outside the cooled space. Such a location makes it possible to exhaust the heated air from the cooled space. Note that the condenser has a large-bladed fan that pushes air through the condenser fins. The fins are attached to coils of copper or aluminum tubing. The tubing houses the liquid and the gaseous vapors. When the blown air contacts the fins, it cools them. The heat from the compressed gas in the tubing is thus transferred to the cooler fin.

Heat given up by the refrigerant vapor to the condensing medium includes both the heat absorbed in the evaporator and the heat of compression. Thus, the condenser always has a load that is the sum of these two heats. This means the compressor must handle more heat than that generated by the evaporator. The quantity of heat (in Btu) given off by the condenser is rated in heat per minute per ton of evaporator capacity. These condensers are rated at various suction and condensing temperatures.

The larger the condenser area exposed to the moving airstream, the lower will be the temperature of the refrigerant when it leaves the condenser. The temperature of the air leaving the vicinity of the condenser will vary with the load inside the area being cooled. If the evaporator picks up the additional heat and transfers it to the condenser, then the condenser must transmit this heat to the air passing over the surface of the fins. The temperature rise in the condensing medium passing through the condenser is directly proportional to the condenser load. It is inversely proportional to the quantity and specific heat of the condensing medium.

To exhaust the heat without causing the area being cooled to heat up again, it is common practice to locate the condenser outside of the area being conditioned. For example, for an air-conditioned building, the condenser is located on the rooftop or on an outside slab at grade level. Fig. 8-3.

Some condensers are cooled by natural air flow. This is the case in domestic refrigerators. Such natural convection condensers use either plate surface or finned tubing. Fig. 8-4.

Air-cooled condensers that use fans are classified as chassis-

CHAPTER 8—CONDENSERS AND COOLING TOWERS

PRODUCES HEATED OUTDOOR AIR   PRODUCES COOLED INDOOR AIR

LIQUID

COOLING OF COMPRESSED GASEOUS REFRIGERANT CHANGES IT TO A LIQUID

CONDENSER   EVAPORATOR

HEATING OF LIQUID REFRIGERANT CHANGES IT TO A GAS

COMPRESSED GAS   VAPOR

OUTDOOR AIR FAN   INDOOR AIR FAN

COMPRESSOR

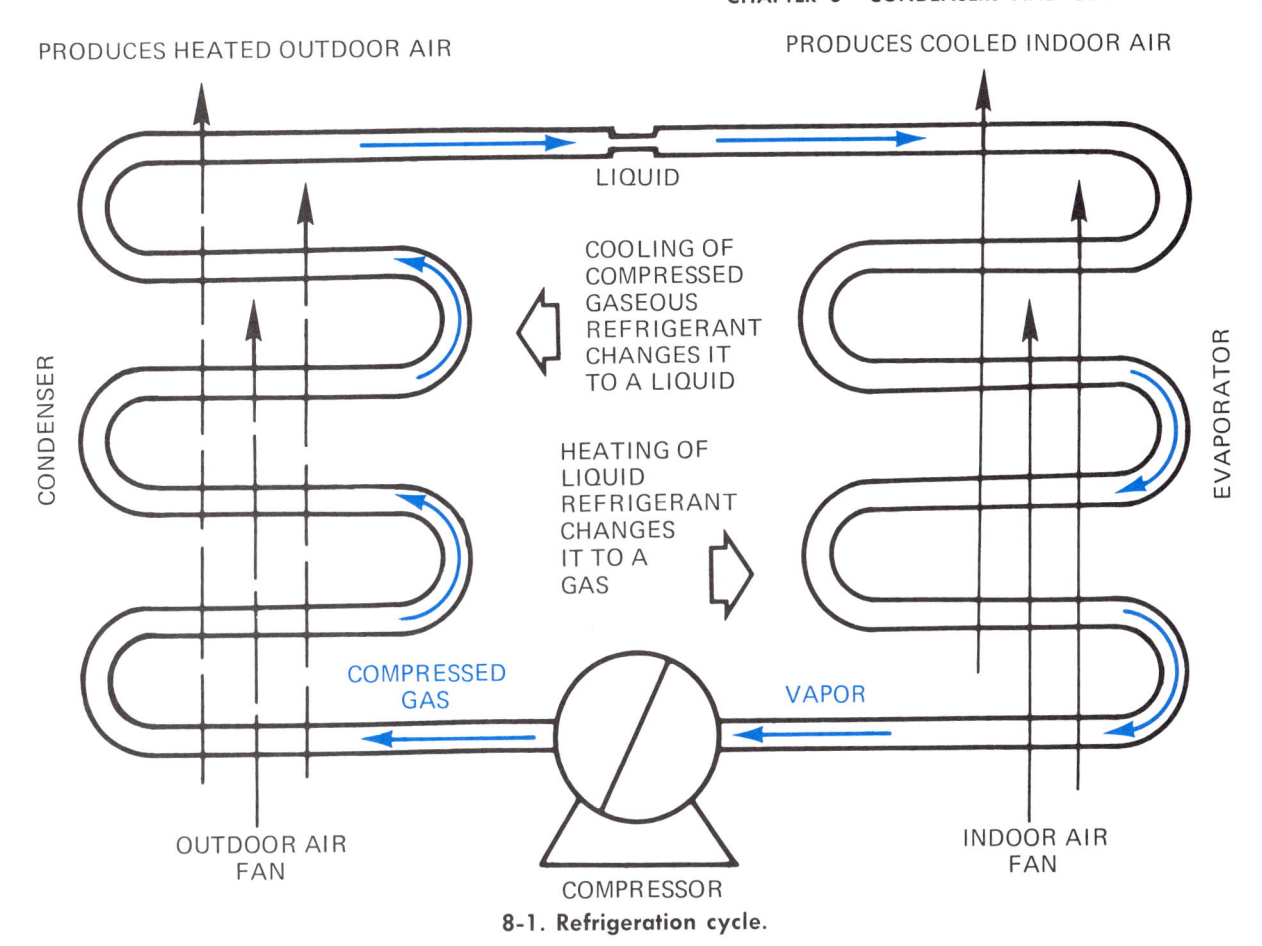

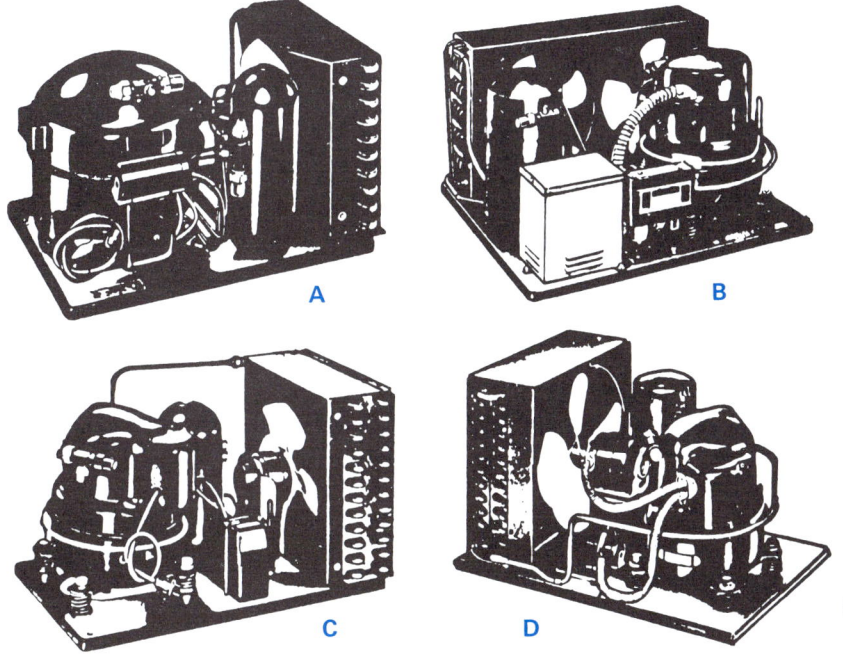

8-1. Refrigeration cycle.

A   B

C   D

mounted and remote. The chassis-mounted type is shown in Fig. 8-2. Here, the compressor, fan, and condenser are mounted as one unit. The remote type is shown in Fig. 8-3. Remote air-cooled condensers can be obtained in sizes that range from 1 ton to 100 tons. The chassis-mounted type is usually limited to 1 ton or less.

Water-Cooled Condensers

Water is used to cool condensers. One method is to cool con-

Tecumseh
8-2. A condenser, fan, and compressor, self-contained, in one unit.

203

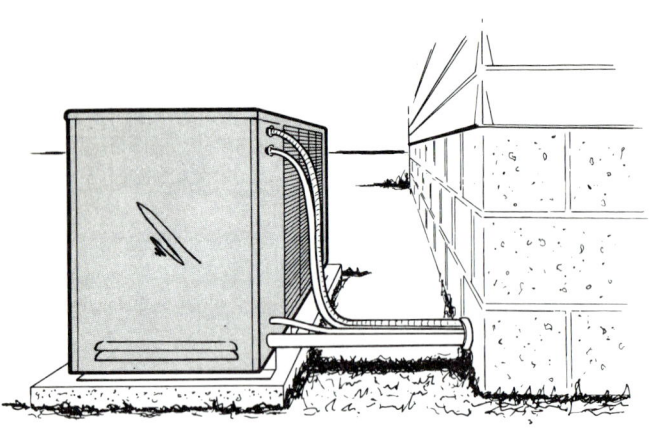

REFRIGERATION AND AIR CONDITIONING TECHNOLOGY

Unit on slab at grade level

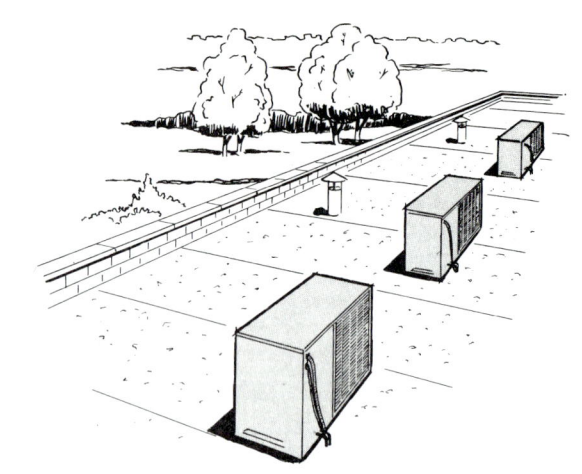

Multiple units on rooftop

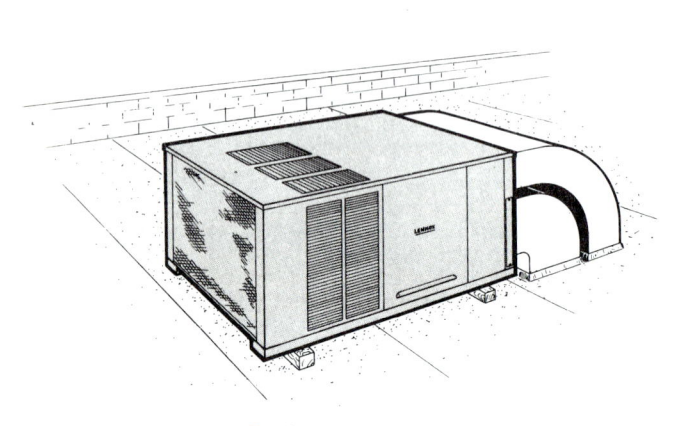

Rooftop installation

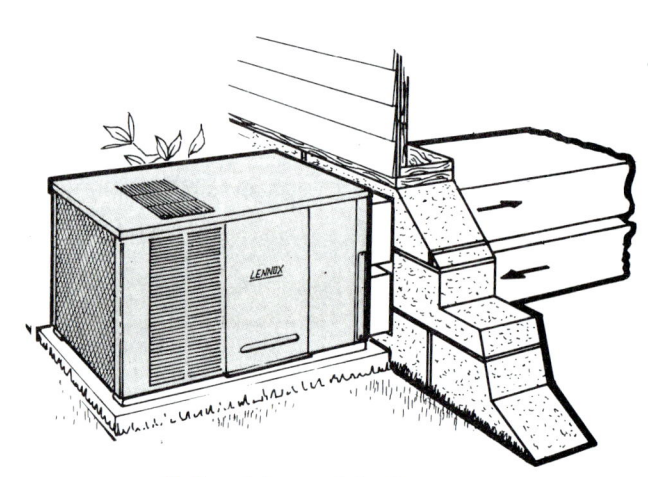

Unit on slab at grade level

Lennox

8-3. Condensers mounted on rooftops and at grade level.

densers with water from the city water supply and then exhaust the water into the sewer after it has been used to cool the refrigerant. This method can be expensive and, in some instances, is not allowed by law. When there is a sewer problem, a limited sewer treatment plant capacity, or drought, it is impractical to use this cooling method.

The use of recirculation to cool the water for re-use is more practical. However, in recirculation, the power required to pump the water to the cooling location is part of the expense of operating the unit.

There are three types of water-cooled condensers. They are the double-tube, the shell-and-coil, and the shell-and-tube types.

The double-tube type consists of two tubes, one inside the other. Fig. 8-5. Water is piped through the inner tube. Refrigerant is piped through the tube that encloses the inner tube. The refrigerant flows in the opposite direction than the water. Fig. 8-6.

This type of coaxial water-cooled condenser is designed for use with refrigeration and air conditioning condensing units where space is limited. These condensers can be mounted vertically, horizontally, or at any angle.

They can be used with cooling towers also. They perform at peak heat of rejection with water

204

CHAPTER 8—CONDENSERS AND COOLING TOWERS

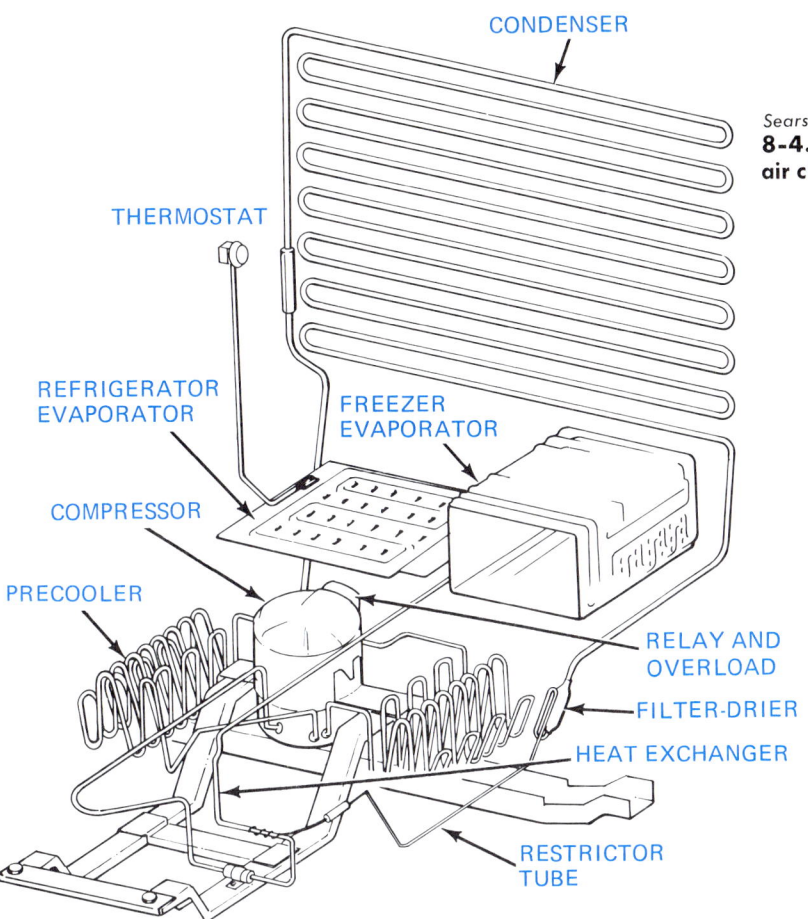

8-4. Flat, coil-type condenser with natural air circulation. This is used in refrigerators in the home.

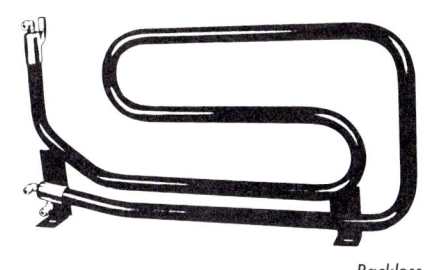

8-5. Coaxial water-cooled condenser. It is used with refrigeration and air-conditioning units where space is limited.

pressure drop of not more than five pounds per square inch, utilizing flow rates of three gallons per minute per ton.

The typical counter-flow path shows the refrigerant going in a 105° F. [41°C] and the water going in at 85° F. [30°C] and leaving at 95° F. [35°C]. Fig. 8-7.

The counter-swirl design shown in Fig. 8-6 gives heat transfer performance of superior quality.

The tube construction provides for excellent mechanical stability. The water flow path is turbulent. This provides a scrubbing action that maintains cleaner surfaces. The construction method shown also has very high system pressure resistance.

The water-cooled condenser shown in Fig. 8-5 can be obtained in a number of combinations. Some of these combina-

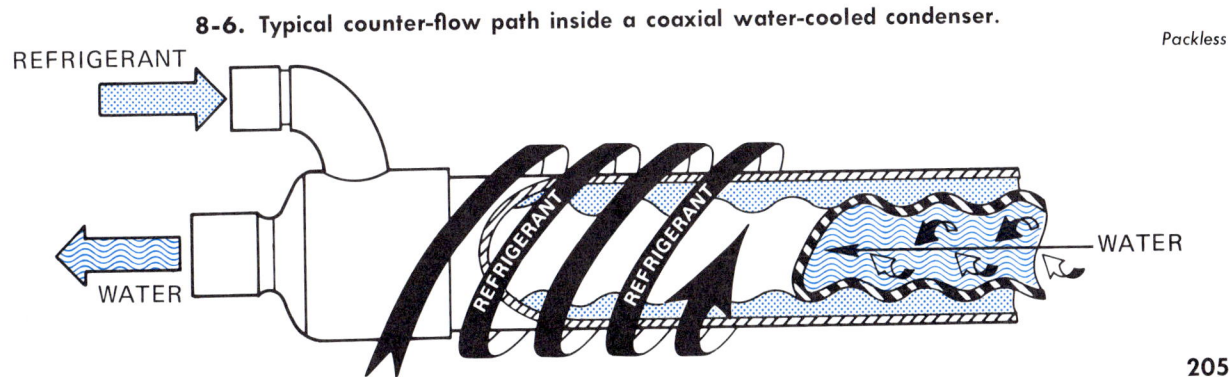

8-6. Typical counter-flow path inside a coaxial water-cooled condenser.

205

REFRIGERATION AND AIR CONDITIONING TECHNOLOGY

Packless
8-7. Water and refrigerant temperatures in a counter-flow, water-cooled condenser.

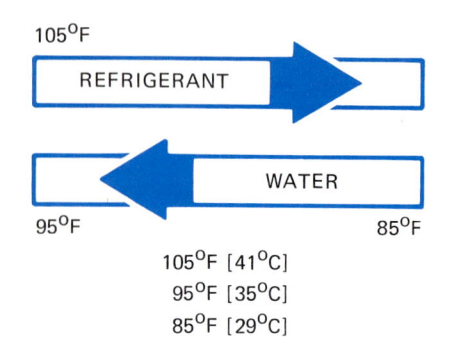

105°F [41°C]
95°F [35°C]
85°F [29°C]

Table 8-A.
Some Possible Metal Combinations in Water-Cooled Condensers

Shell Metal	Tubing Metal
Steel	Copper
Copper	Copper
Steel	Cupronickel
Copper	Cupronickel
Steel	Stainless Steel
Stainless Steel	Stainless Steel

tions are listed in Table 8-A. Copper tubing is suggested for use with fresh water and with cooling towers. The use of cupronickel is suggested when salt water is used for cooling purposes.

Convolutions in the water tube result in a spinning, swirling water flow that inhibits the accumulation of deposits on the inside of the tube. This contributes to the antifouling characteristics in this type of condenser. Figure 8-8 shows the various types of construction for the condenser.

This type of condenser may be added as a booster to standard air-cooled units.

Figure 8-9 shows some of the configurations of this type of condenser. The spiral, the helix, and the trombone are shown here. Note the input for the water and the input for the refrigerant. The condensers can be further cooled by using a cooling tower to furnish water to contact the outside tube. Also, a water tower can be used to cool the water sent through the inside tube for cooling purposes. This type of condenser is usable where refrigeration or air conditioning requirements are $\frac{1}{3}$ ton to 3 tons.

The shell-and-coil condenser are made by placing a bare tube or a finned tube inside a steel shell. Fig. 8-10. Water circulates

8-9. The three configurations of coaxial water-cooled condensers.
Packless

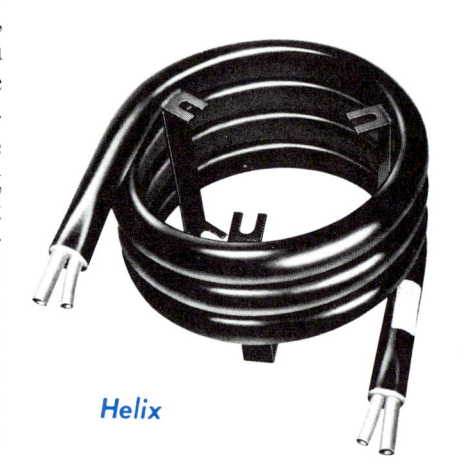

Spiral

Helix

8-8. Different types of tubing fabrication inside the coaxial-type water-cooled condenser.

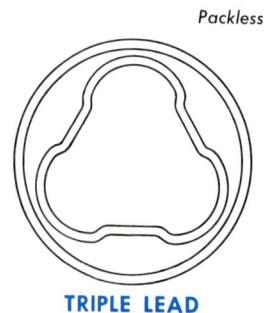

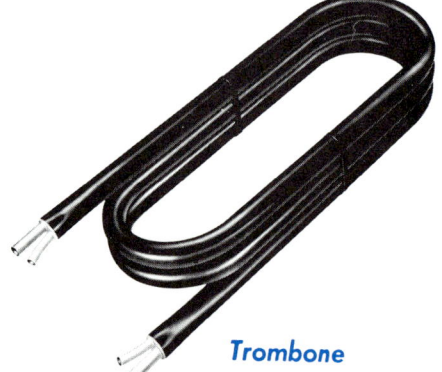

Packless

SINGLE LEAD

DOUBLE LEAD

TRIPLE LEAD

Trombone

CHAPTER 8—CONDENSERS AND COOLING TOWERS

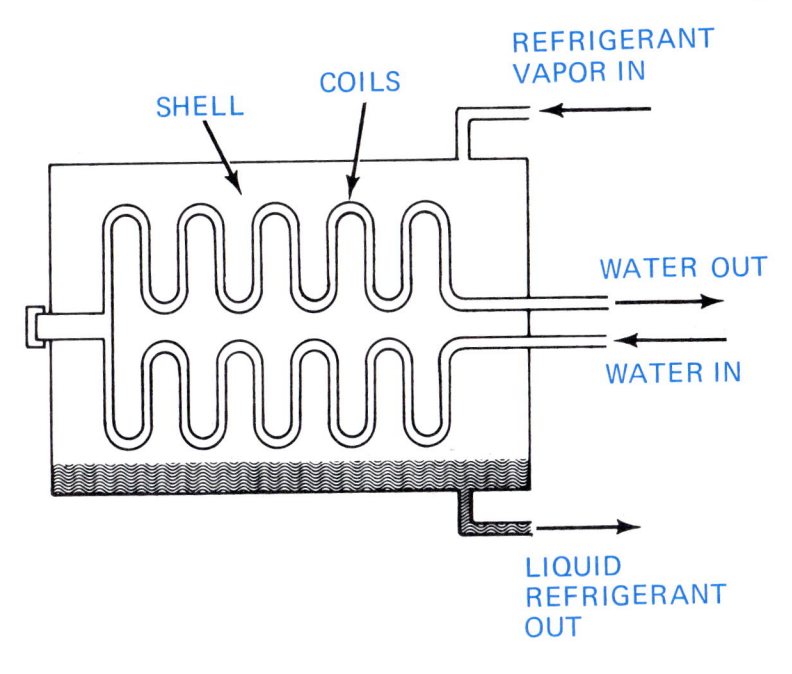

8-10. The shell-and-coil condenser. This is a series type of coil arrangement inside the shell, which the refrigerant enters as a vapor and leaves as a liquid.

through the coils. Refrigerant vapor is injected into the shell. The hot vapor contacts the cooler tubes and condenses. The condensed vapor drains from the coils and drops to the bottom of the tank or shell. From there it is recirculated through the refrigerated area by way of the evaporator. In most cases, the unit is cleaned by placing chemicals into the water. The chemicals have a tendency to remove the deposits that build up on the tubing walls.

CHILLERS

A chiller is part of a condenser. Chillers are used to cool water or brine solutions. The cooled (chilled) water or brine is then fed through pipes to evaporators. This cools the area in which the evaporators are located. This type of cooling, using chilled water or brine, can be used in large air conditioning units. It can also be used for industrial processes where cooling is required for a particular operation.

Figure 8-11 illustrates such an operation. Note how the compressor sits atop the condenser. Chillers are the answer to requirements of 200 tons to 1600 tons of refrigeration. They are used for process cooling, comfort air conditioning, and nuclear power plant cooling. In some cases, they are used to provide ice for ice skating rinks. The arrows in Fig. 8-11 indicate the refrigerant flow and the water or brine flow through the large pipes. Figure 8-12 shows the machine in a cutaway view. The following explanation of the various cycles will provide a better understanding of the operation of this type of equipment.

Refrigeration Cycle

The machine compressor continuously draws large quantities of refrigerant vapor from the cooler, at a rate determined by the size of the guide vane opening. This compressor suction reduces the pressure within the cooler, allowing the liquid refrigerant to boil vigorously at a fairly low temperature (typically 30 to 35° F. [−1 to 2°C]).

Liquid refrigerant obtains the energy needed for the change to vapor by removing heat from the water in the cooler tubes. The cold water can then be used in the air conditioning process.

After removing heat from the water, the refrigerant vapor enters the first stage of the compressor. There, it is compressed and flows into the second stage of the compressor. Here it is mixed with flash-economizer gas and further compressed.

Compression raises the refrigerant temperature above that of the water flowing through the condenser tubes. When the warm (typically 100 to 105° F. [38 to 41°C]) refrigerant vapor contacts the condenser tubes, the relatively cool condensing water (typically 85 to 95° F. [29 to

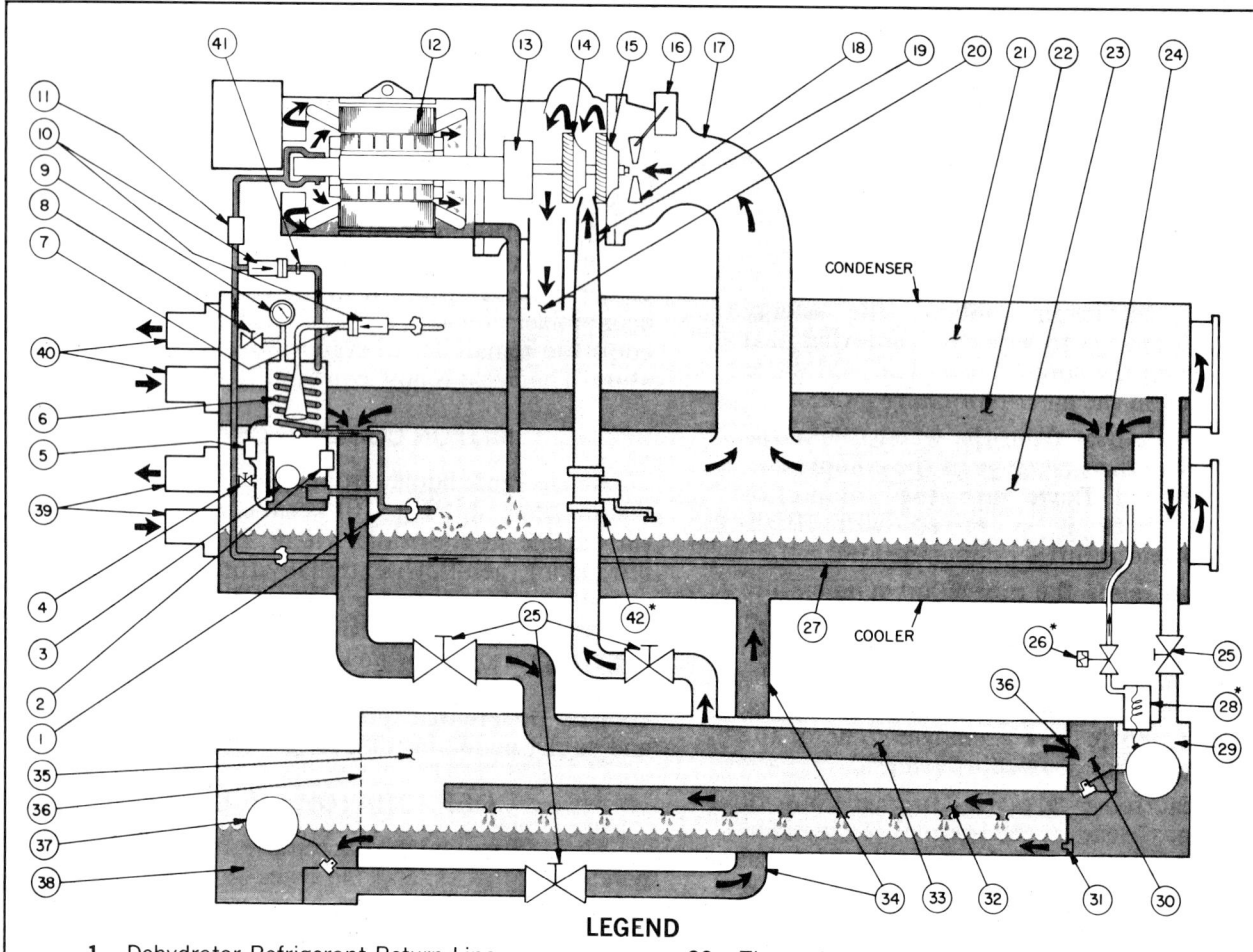

REFRIGERATION AND AIR CONDITIONING TECHNOLOGY

LEGEND

1 – Dehydrator Refrigerant Return Line
2 – Liquid Level Sight Glass
3 – Dehydrator Float Valve
4 – Water Drain Valve
5 – Water Sight Glass
6 – Dehydrator Condensing Coil
7 – Dehydrator Refrigerant Sampling Line and 1/16-in. Orifice
8 – Dehydrator Air Relief Valve
9 – Dehydrator Pressure Gage
10 – Refrigerant Strainers
11 – Refrigerant Strainer or Filter
12 – Compressor Motor
13 – Transmission
14 – Second-Stage Impeller
15 – First-Stage Impeller
16 – Guide Vane Actuator
17 – Compressor Suction Elbow
18 – Variable Guide Vanes
19 – Flash Economizer Gas Line
20 – Compressor Discharge
21 – Condenser
22 – Thermal Economizer
23 – Cooler
24 – Sump
25 – Isolation Valves (4)
26 – Refrigerant Feed Control Solenoid Valve*
27 – Motor Cooling and Dehydrator Supply Line
28 – Refrigerant Feed Control*
29 – High-Side Float Chamber
30 – High-Side Valve Chamber
31 – Refrigerant Orifice and Screen
32 – Flash Economizer Spray Pipe
33 – Condenser Refrigerant Drain Line
34 – Refrigerant Supply Line to Cooler
35 – Utility Vessel
36 – Refrigerant Screen (2)
37 – Low-Side Float Valve
38 – Low-Side Float Chamber
39 – Chilled Water (Brine) Connections
40 – Condenser Water Connections
41 – Orifice, 1/8-in.
42 – Economizer Gas Damper Valve*

*On some machines.

Carrier

8-11. The chiller, compressor, condenser, and cooler are combined in one unit.

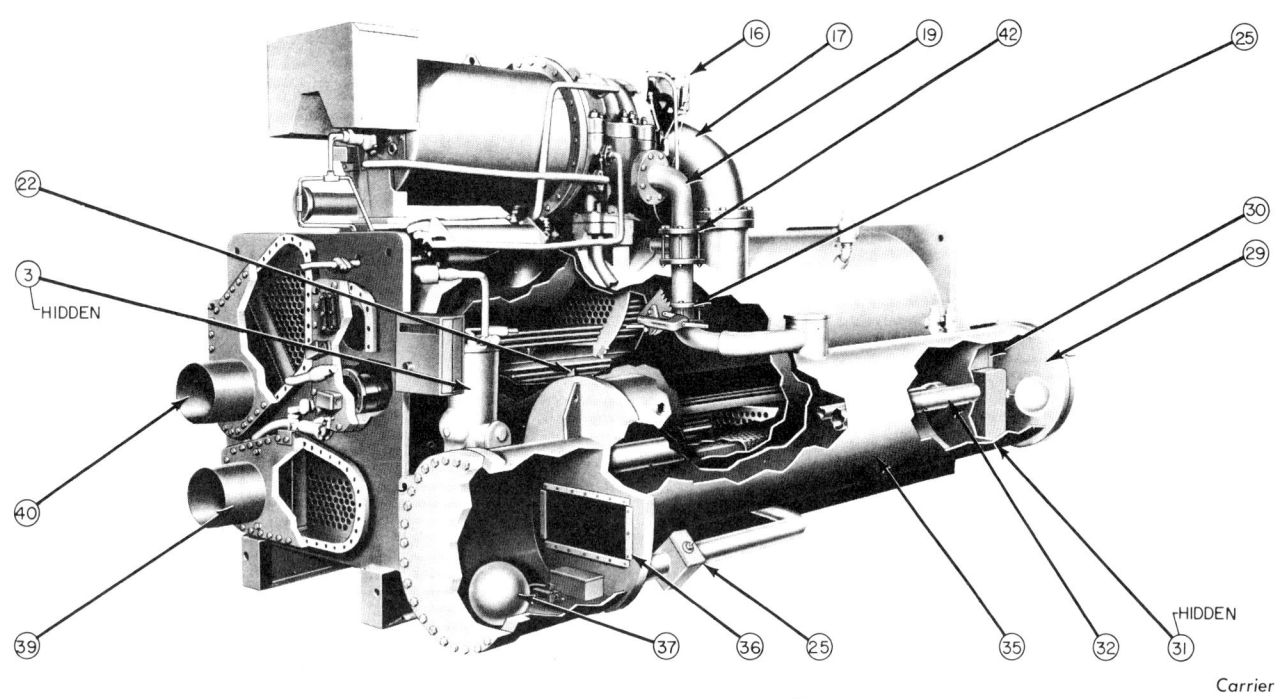

8-12. Cutaway view of a chiller.

35°C]) removes some of the heat and the vapor condenses into a liquid.

Further heat removal occurs in the group of condenser tubes that form the thermal economizer. Here, the condensed liquid refrigerant is subcooled by contact with the coolest condenser tubes. These are the tubes that contain the entering water.

The subcooled liquid refrigerant drains into a high-side valve chamber. This chamber maintains the proper fluid level in the thermal economizer and meters the refrigerant liquid into a flash economizer chamber. Pressure in this chamber is intermediate between condenser and cooler pressures. At this lower pressure, some of the liquid refrigerant flashes to gas, cooling the remaining liquid. The flash gas, having absorbed heat, is returned directly to the compressor's second stage. Here, it is mixed with gas already compressed by the first stage impeller. Since the flash gas must pass through only half the compression cycle to reach condenser pressure, there is a savings in power.

The cooled liquid refrigerant in the economizer is metered through the low-side valve chamber into the cooler. Because pressure in the cooler is lower than economizer pressure, some of the liquid flashes and cools the remainder to evaporator (cooler) temperature. The cycle is now complete.

Motor Cooling Cycle

Refrigerant liquid from a sump in the condenser (No. 24 in Fig. 8-11) is subcooled by passage through a line in the cooler (No. 27 in Fig. 8-11). The refrigerant then flows externally through a strainer and variable orifice (No. 11 in Fig. 8-11) and enters the compressor motor end. Here it sprays and cools the compressor rotor and stator. It then collects in the base of the motor casing. Here, it drains into the cooler. Differential pressure between the condenser and cooler maintains the refrigerant flow.

Dehydrator Cycle

The dehydrator removes water and noncondensable gases. It indicates any water leakage into the refrigerant. See No. 6 in Fig. 8-11.

This system includes a refrigerant condensing coil and chamber, water drain valve, purging valve, pressure gage, refrigerant float valve, and refrigerant piping.

A dehydrator sampling line continuously picks up refrigerant vapor and contaminants, if any, from the condenser. Vapor is

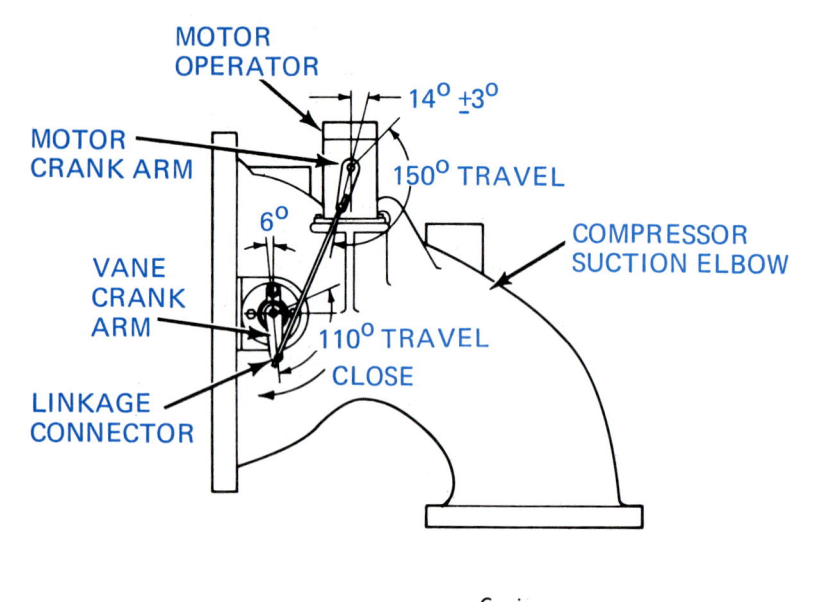

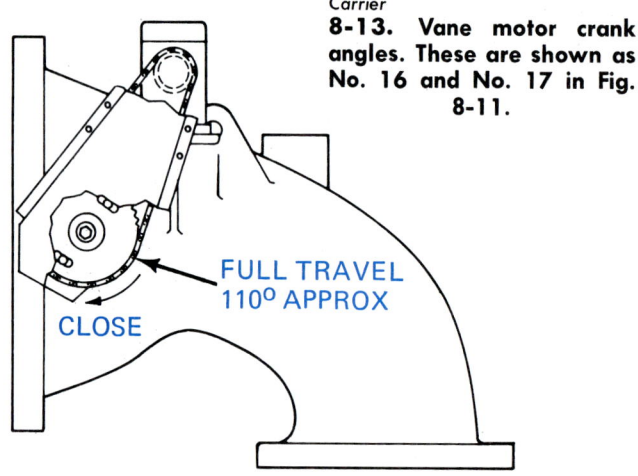

REFRIGERATION AND AIR CONDITIONING TECHNOLOGY

8-13. Vane motor crank angles. These are shown as No. 16 and No. 17 in Fig. 8-11. *Carrier*

condensed into a liquid by the dehydrator condensing coil. Water, if present, separates and floats on the refrigerant liquid. The water level can be observed through a sight glass.

Water may be withdrawn manually at the water drain valve. Air and other noncondensable gases collect in the upper portion of the dehydrator condensing chamber. The dehydrator gage indicates the presence of air or other gases through a rise in pressure. These gases may be manually vented through the purging valve.

A float valve maintains the refrigerant liquid level and pressure difference necessary for the refrigerant condensing action. Purified refrigerant is returned to the cooler from the dehydrator float chamber.

Lubrication Cycle

The oil pump and oil reservoir are contained within the unishell. Oil is pumped through an oil filter-cooler that removes heat and foreign particles. A portion of the oil is then fed to the compressor motor-end bearings and seal. The remaining oil lubricates the compressor transmission, compressor thrust and journal bearings, and seal. Oil is then returned to the reservoir to complete the cycle.

Controls

The cooling capacity of the machine is automatically adjusted to match the cooling load by changes in the position of the compressor inlet guide vanes. Fig. 8-13.

A temperature-sensing device in the circuit of the chilled water leaving the machine cooler continuously transmits signals to a solid-state module in the machine control center. The module, in turn, transmits the amplified and modulated temperature signals to an automatic guide vane actuator.

A drop in the temperature of the chilled water leaving the circuit causes the guide vanes to move towards the closed position. This reduces the rate of refrigerant evaporation and vapor flow into the compressor. Machine capacity decreases. A rise in chilled water temperature opens the vanes. More refrigerant vapor moves through the compressor and the capacity increases.

The modulation of the temperature signals in the control center allows precise control of guide vane response, regardless of the system load.

SOLID-STATE CAPACITY CONTROL

In addition to amplifying and modulating the signals from chilled water sensor to vane actu-

CHAPTER 8—CONDENSERS AND COOLING TOWERS

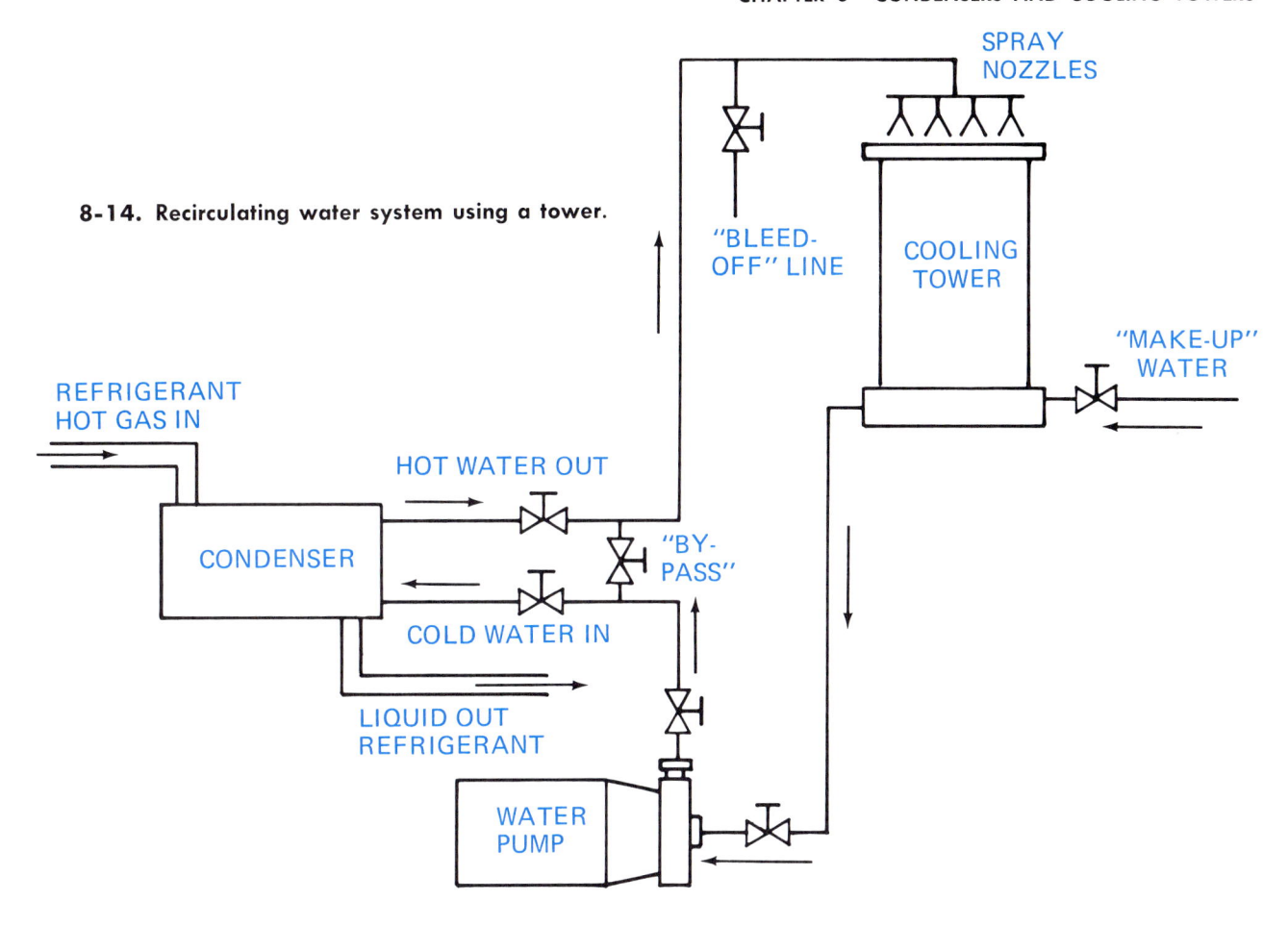

8-14. Recirculating water system using a tower.

ator, the solid-state module in the control center provides a means for preventing the compressor from exceeding full load amperes. It also provides a means for limiting motor current down to 40% of full load amperes to reduce electrical demand rates.

A throttle adjustment screw eliminates guide vane hunting. A manual capacity control knob allows the operator to open, close, or hold the guide vane position when desired.

COOLING TOWERS

Cooling towers are used to conserve or recover water. In one design the hot water from the condenser is pumped to the tower. There, it is sprayed into the tower basin. The temperature of the water decreases as it gives up heat to the air circulating through the tower. Some of the towers are rather large, since they work with condensers yielding 1600 tons of cooling capacity. Fig. 8-14.

Most of the cooling that takes place in the tower results from the evaporation of part of the water as it falls through the tower.

The lower the wet bulb temperature of the incoming air, the more efficient the air is in decreasing the temperature of the water being fed into the tower.

The following factors influence the efficiency of the cooling tower.

1. Mean difference between vapor pressure of the air and pressure in the tower water.
2. Length of exposure time and amount of water surface exposed to air.
3. Velocity of air through the tower.
4. Direction of air flow relative to the exposed water surface (parallel, transverse, or counter).

Theoretically, the lowest temperature to which the water can be cooled is the temperature of the air (wet bulb) entering the tower. However, in practical terms, it is impossible to reach

REFRIGERATION AND AIR CONDITIONING TECHNOLOGY

the temperature of the air. In most instances, the temperature of the water leaving the tower will be no lower than 7 to 10° F. [4 to 6°C] above the air temperature.

The range of the tower is the temperature of the water going into the tower and the temperature of the water coming out of the tower. This range should be matched to the operation of the condenser for maximum efficiency.

Cooling Systems Terms

The following terms apply to cooling tower systems.

Cooling range is the number of degrees Fahrenheit through which the water is cooled in the tower. It is the difference between the temperature of the hot water entering the tower and the temperature of the cold water leaving the tower.

Approach is the difference in degrees Fahrenheit between the temperature of the cold water leaving the cooling tower and the wet-bulb temperature of the surrounding air.

Heat load is the amount of heat "thrown away" by the cooling tower in Btu per hour (or per minute). It is equal to the pounds of water circulated multiplied by the cooling range.

Cooling tower pump head is the pressure required to lift the returning hot water from a point level with the base of the tower to the top of the tower and force it through the distribution system.

Drift is the small amount of water lost in the form of fine droplets retained by the circulating air. It is independent of, and in addition to, evaporation loss.

Bleed-off is the continuous or intermittent wasting of a small fraction of circulating water to prevent the build-up and concentration of scale-forming chemicals in the water.

Make-up is the water required to replace the water that is lost by evaporation, drift and bleed-off.

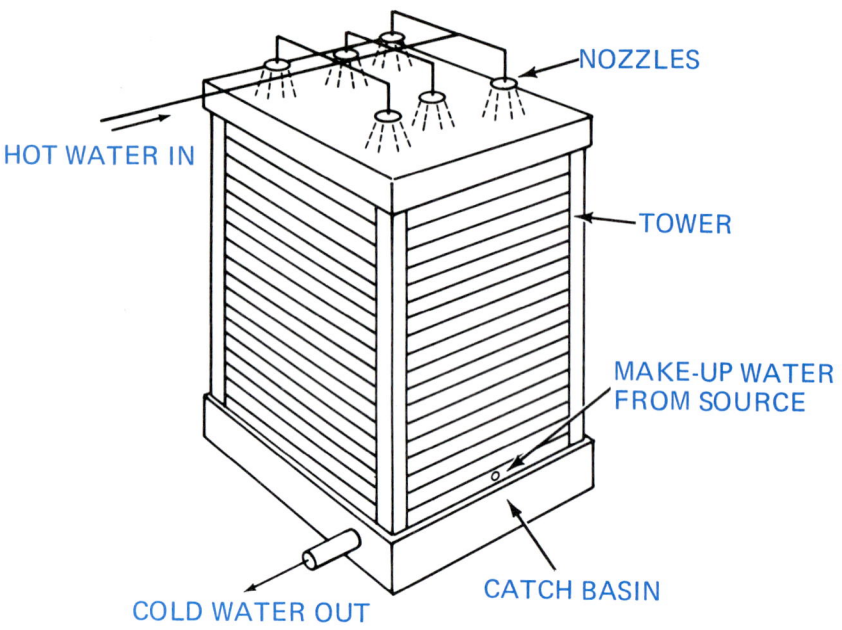

8-15. Natural-draft cooling tower.

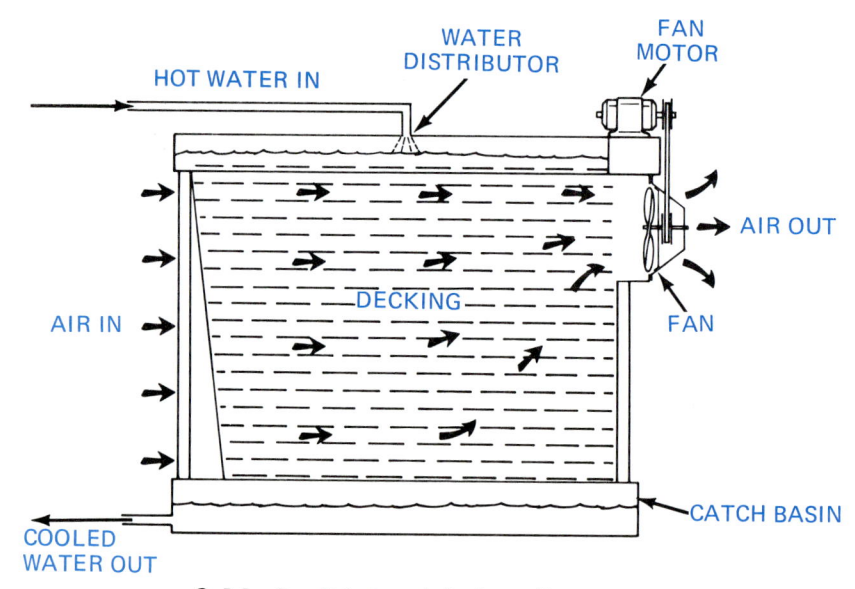

8-16. Small induced-draft cooling tower.

Design of Cooling Towers

Classified by the air circulation method used, there are two types of cooling towers. They are either natural-draft or mechanical-draft towers. Figure 8-15 shows the operation of the natural-draft

212

CHAPTER 8—CONDENSERS AND COOLING TOWERS

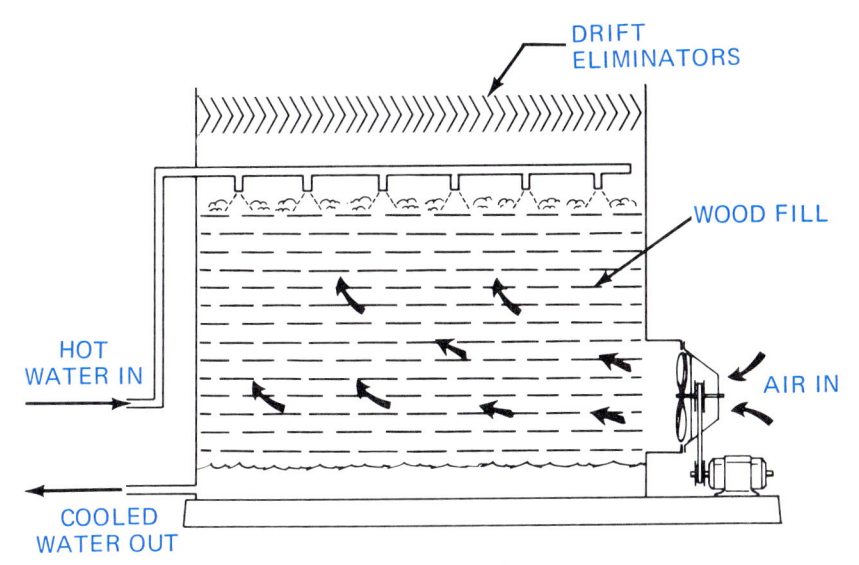

8-17. Forced-draft cooling tower.

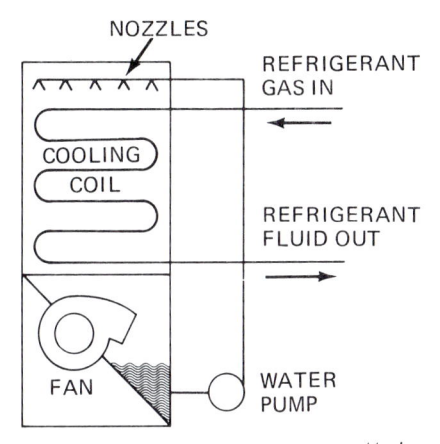

8-18. Evaporative cooler has no fill deck. The water cools process fluid directly. *Marley*

cooling tower. Figure 8-16 shows the operation of the mechanical-draft cooling tower. The forced-draft cooling tower shown in Fig. 8-17 is just one example of the mechanical-draft designs available today.

Cooling tower ratings are given in tons. This is based on heat transfer capacity of 250 Btu per minute per ton. The normal wind velocity taken into consideration for tower design is 3 miles per hour. The wet bulb temperature is usually 80° F. [27°C] for design purposes. The usual flow of water over the tower is 4 gallons per minute for each ton of cooling desired. Several charts are available with current design technology. Manufacturers supply the specifications for their towers. However, there are some important points to remember when use of a tower is being considered:

1. In tons of cooling, the tower should be rated at the same capacity as the condenser.
2. The wet bulb temperature must be known.
3. The temperature of the water leaving the tower should be known. This would be the temperature of the water entering the condenser.

Towers present some maintenance problems. These stem primarily from the water used in the cooling system. Chemicals are employed to control the growth of bacteria and other substances. Scale in the pipes and on parts of the tower also must be controlled. Chemicals are used for each of these controls. This problem will be discussed in Chapter 9.

EVAPORATIVE CONDENSERS

The evaporative condenser is a condenser and a cooling tower combined. Figure 8-18 illustrates how the nozzles spray water over the cooling coil to cool the fluid or gas in the pipes. This is a very good water conservation tower. In the future, this system will probably become more popular. The closed-circuit cooler should see increased use because of dwindling water supplies and more expensive treatment problems. The function of this cooler is to process the fluid in the pipes. This is a sealed contamination-free system. Instead of allowing the water to drop onto slats or other deflectors, this unit sprays the water directly onto the cooling coil.

NEW DEVELOPMENTS

All-metal towers with housing, fans, fill, piping, and structural members made of galvanized or stainless steel are now being built. Some local building codes are becoming more restrictive with respect to fire safety. Low maintenance is another factor in the use of all-metal towers.

Engineers are beginning to specify towers less subject to deterioration due to environmental conditions. Thus, all-steel or all-metal towers are called for. Already, galvanized steel towers have made inroads into the air conditioning and refrigeration market. Stainless-steel towers are being specified in New York City, northern New Jersey, and Los Angeles. This is due primarily to a polluted atmosphere,

213

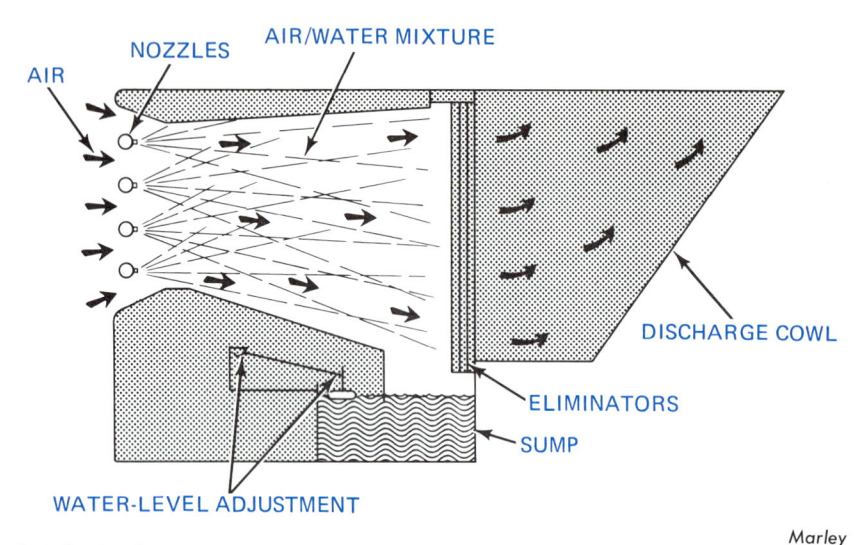

8-19. Cooling tower with natural draft properties. There are no moving parts in the cooling tower. *Marley*

which can lead to early deterioration of nonmetallic towers and, in some cases, metals.

Figure 8-19 shows a no-fans design for a cooling tower. Large quantities of air are drawn into the tower by cooling water as it is injected through spray nozzles at one end of a venturi plenum. No fans are needed. Effective mixing of air and water in the plenum permits evaporative heat transfer to take place without the fill required in conventional towers.

The cooled water falls into the sump and is pumped through a cooling-water circuit to return for another cycle. The name applied to this design is Baltimore Air-coil. In 1981, towers rated at 10 to 640 tons with 30 to 1920 gallons per minute were standard. The nozzle clogging problem has been minimized by using pre-strainers in the high-pressure flow. There are no moving parts in the tower. This results in very low maintenance costs.

Air-cooled condensers are reaching 1000 tons in capacity. Air coolers and air condensers are quite attractive for use in refineries and natural gas compressor stations. They are also used for cooling in industry, as well as for commercial air conditioning purposes. Figure 8-19 shows how the air-cooled condensers are used in a circuit system that is completely closed. These are very popular where there is little or no water supply.

REVIEW QUESTIONS

1. What is the purpose of a condenser in a refrigeration system?
2. List the three basic methods for cooling hot gases.
3. How does a chiller serve as a cooling system?
4. Describe the dehydrator cycle in a chiller operation.
5. What is the purpose of the solid-state module?
6. Why are cooling towers necessary?
7. How are cooling towers rated?
8. Describe the term "make-up water."
9. Why are stainless-steel towers needed?

9

Cooling Water Problems

Three-fourths of the earth's surface is covered with water. The earth is blanketed with water vapor, which is an indispensable part of the atmosphere. Heat from the sun shining on oceans, rivers, and lakes evaporates some water into the atmosphere. Warm, moisture-laden air rises and cools. The cooling vapor condenses to form clouds. Wind currents carry clouds over land masses where the precipitation may occur in the form of rain, snow, or sleet. Because of the sun and upper air currents, this process is repeated again and again. Pure water has no taste and no odor. Pure water, however, is actually a rarity.

All water found in oceans, rivers, lakes, streams, and wells contains various amounts of minerals picked up from the earth. Even rainwater is not completely pure. As rain falls to earth, it washes from the air various gases and solids such as oxygen, carbon dioxide, industrial gases, dust, and even bacteria. Some of this water sinks into the earth and collects in wells or forms underground streams. The remainder runs over the ground and finds its way back into various surface water supplies.

Water is often referred to as the *universal solvent*. Water that runs over and through the earth, mixes with many minerals. Some of these mineral solids are dissolved or disintegrated by water.

Pure water and sanitary water are the same as far as municipalities are concerned. "Pure" in this case means that the water is free from excessive quantities of germs and will not cause disease. Mineral salts or other substances in water do not have to be removed by water treatment plants unless they affect sanitary conditions. Mineral salts are objectionable in water used for many other purposes. These uses include generating power, heating buildings, processing materials, and manufacturing. *Water fit for human consumption is not necessarily acceptable for use in boilers or cooling equipment.*

Water is used in many types of cooling systems. Heat removal is the main use of water in air conditioning or refrigeration equipment. Typical uses include once-through condensers, open recirculating cooling systems employing cooling towers, evaporative condensers, chilled water systems, and air washers. In evaporative condensers, once-through systems and cooling towers, water removes heat from refrigerant and then is either wasted or cooled by partial evaporation in air. Knowledge of impurities in water used in any of these systems aids in predicting possible problems and methods of preventing them.

FOULING, SCALING, AND CORROSION

Fouling reduces water flow and heat transfer. Fouling can be caused by the collection of loose debris over pump suction screens in sumps, growth of algae in sunlit areas, and slime in shade or dark sections of water systems. Material can clog pipes, or other parts of a system, after it has broken loose and been carried into the system by the water stream.

Scaling also reduces water flow and heat transfer. Scaling is caused by the depositing of dissolved minerals on equipment surfaces. This is particularly so in hot areas, where heat transfer is most important.

Corrosion is caused by impurities in the water. In addition to reducing water flow and heat transfer, it also damages equipment. Eventually, corrosion will reduce operational efficiency. It may lead to expensive repairs or even equipment replacement.

Impurities have at least five confirmed sources. One is the earth's atmosphere. Water falling through the air, whether it be natural precipitation or water showering through a cooling tower, picks up dust, as well as

215

oxygen and carbon dioxide. Similarly, synthetic atmospheric gases and dust affect the purity of water. Heavily industrialized areas are susceptible to such impurities being introduced into their water systems.

Decaying plant life is a source of water impurity. Decaying plants produce carbon dioxide. Other products of vegetable decay cause bad odor and taste. The by-products of plant decay provide a nutrient for slime growth.

These three sources of impurities contaminate water with material that makes it possible for water to pick up more impurities from a fourth source, minerals. Minerals found in the soil beneath the earth's surface are probably the major source of impurities in water. Many minerals are present in subsurface soil. They are more soluble in the presence of the impurities from the first three sources, mentioned above.

Industrial and municipal wastes are a fifth major source of water impurities. Municipal waste affects bacterial count. Therefore, it is of interest to health officials, but is not of primary concern from a scale or corrosion standpoint. Industrial waste, however, can add greatly to the corrosive nature of water. It can indirectly cause a higher than normal mineral content.

Fouling is caused by the correction or generation of finely divided material that has the appearance of mud or silt. This sludge is normally composed of dirt and trash from the air. Silt is introduced with make-up water. Leaves and dust are blown in by wind and washed from the air by rain. This debris settles in sumps or other parts of cooling systems. Plant growth also causes fouling. Bacterial or algae spores in water will result in the formation of large masses of algae and slime. These may clog system water pipes and filters. Fouling is also caused by paper, bottles, and other trash.

Prevention of Scaling

There are three ways to prevent scaling. The first is to eliminate or reduce the hardness minerals from the feed water. Control of factors that cause hardness salts to become less soluble is important. Hardness minerals are defined as water-soluble compounds of calcium and magnesium. Most calcium and magnesium compounds are much less soluble than are corresponding sodium compounds. By replacing the calcium and magnesium portion of these minerals with sodium, solubility of the sulfates and carbonates is improved to such a degree that scaling no longer is a problem. This is the function of the water softener.

The second method of preventing scale is by controlling water conditions that affect the solubility of scale-forming minerals. The five factors that affect the rate of scale formation are:
- Temperature.
- TDS (total dissolved solids).
- Hardness.
- Alkalinity.
- pH.

These factors can, to some extent, be regulated by proper design and operation of water-cooled equipment. Proper temperature levels are maintained by ensuring a good water flow rate and adequate cooling in the tower. Water flow in recirculating systems should be approximately 3 gallons per minute per ton. Lower flow levels allow the water to remain in contact with hot surfaces of the condenser for a longer time and pick up more heat. Temperature drop across the tower should be 8 to 10° F. [4.5 to 5.5°C] for a compression refrigeration system, and 18 to 20° F. [10 to 11°C] in most absorption systems. This cooling effect, due to evaporation, is dependent on tower characteristics and uncontrollable atmospheric conditions.

Air flow through the tower and the degree of water breakup are two factors that determine the amount of evaporation that will occur. Since heat energy is required for evaporation, the amount of water that is changed into vapor and lost from the system determines the amount of heat. That is, the number of Btu to be dissipated is the heat factor. One pound of water, at cooling tower temperatures, requires 1050 Btu to be converted from liquid to vapor. Therefore, the greater the weight of water evaporated from the system, the greater the cooling effect or temperature drop across the tower.

Total dissolved solids, hardness, and alkalinity are affected by three interrelated factors. These are evaporation, make-up, and bleed or blowdown rates. Water, when it evaporates, leaves the system in a pure state, leaving behind all dissolved matter. Water volume of evaporative cooling systems is held at a relatively constant figure through the use of float valves.

Fresh make-up water brings with it dissolved material. This is added to that already left behind by the evaporated water. Theo-

CHAPTER 9—COOLING WATER PROBLEMS

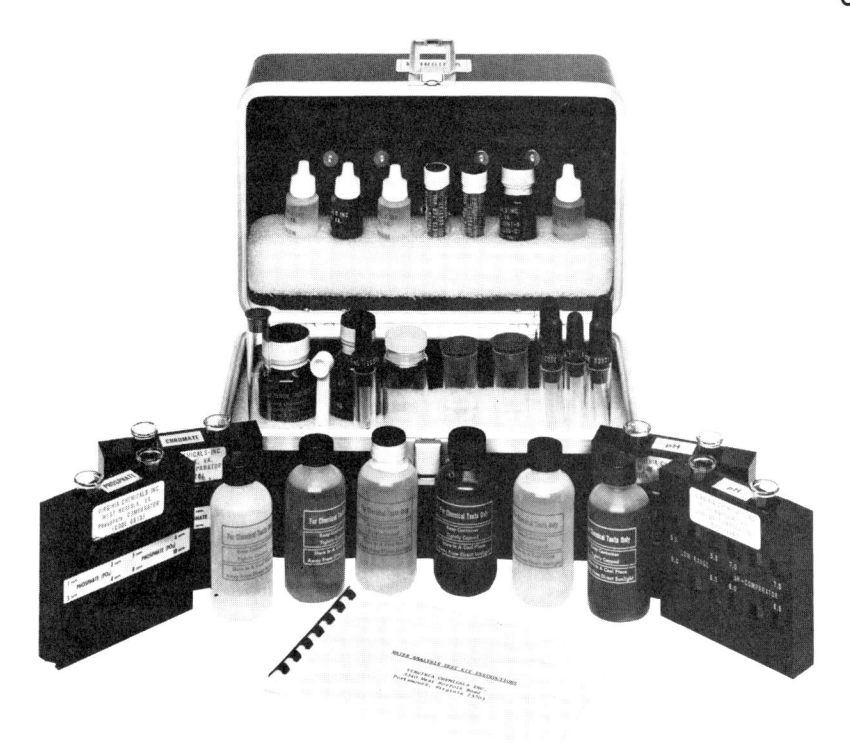

Virginia Chemicals

9-1. This compact field test kit provides equipment for testing pH, phosphate, chromate, total hardness, calcium hardness, alkalinity, and chloride.

by cleaning systems in a given area over an extended period of time. In this way, the pretreatment procedure and the amount of scale remover required to remove the type of scale most often found in this area become common knowledge. Unless radical changes in feed water quality occur, the type of scale encountered remains fairly constant. Experience is further developed through the use of the two other methods. Figure 9-1 shows a water analysis kit.

FIELD TESTING

Field tests, which are quite simple to perform, determine the reactivity of scale with the cleaning solution. A small sample of cleaning solution is prepared by adding 1 tablespoon of liquid scale remover, or 1 teaspoon of solid scale remover, to $\frac{1}{2}$ pint of water. A small piece of scale is then dropped into the cleaning solution. The reaction rate usually will determine the type of scale. The reaction between scale remover and carbonate scale results in vigorous bubbling. The scale eventually dissolves or disintegrates. However, if the scale sample is of hard or flinty composition, and little or no bubbling in the acid solution is observed, heat should be applied. Sulfate scale will dissolve at 140° F. [60°C]. The small scale sample should be consumed in about an hour. If the scale sample contains a high percentage of silica, little or no reaction will be observed. Iron scale is easily identified by appearance. Testing with a clean solution usually is not required.

Since this identification procedure is quite elementary, and combinations of all types of scale are often encountered, it is obvi-

retically, assuming that all the water leaves the system by evaporation and the system volume stays constant, the concentration of dissolved material will continue to increase indefinitely. For this reason, a bleed or blowdown is used. There is a limit to the amount of any material that can be dissolved in water. When this limit is reached, the introduction of additional material will cause either sludge or scale to form. Controlling the rate at which dissolved material is removed controls the degree to which this material is concentrated in circulating water.

SCALE IDENTIFICATION

Scale removal depends on the chemical reaction between scale and the cleaning chemical. Scale identification is important. Of the four scales most commonly found, only carbonate is highly reactive with cleaning chemicals generally regarded as safe for use in cooling equipment. The other scales require a pretreatment that renders them more reactive. This pretreatment depends on the type of scale to be removed. Attempting to remove a problem scale without proper pretreatment can waste time and money.

Scale identification can be accomplished in one of three ways: experience, field tests, and laboratory analysis. With the exception of iron scale, which is orange, it is very difficult, if not impossible, to identify scale by appearance. Experience is gained

ous that more precise methods may be required. Such methods are most easily carried out in a laboratory. Many chemical manufacturers provide this service. Scale samples that cannot be identified in the field may be mailed to these laboratories. Here a complete breakdown and analysis of the problem scale will be performed. Detailed cleaning recommendations will be given to the sender.

Most scales are predominately carbonate, but they may also contain varying amounts of sulfate, iron, or silica. Thus, the quantities of scale remover required for cleaning should be calculated specifically for the type of scale present. The presence of sulfate, iron, or silica also affects other cleaning procedures.

CORROSION

There are four basic causes of corrosion: corrosive acids, oxygen, galvanic action, and biological organisms.

Corrosive Acids

Aggressive or strong acids, such as sulfurous, sulfuric, hydrochloric, and nitric, are found in most industrial areas. These acids are formed when certain industrial waste gases are washed out of the atmosphere by water showering through a cooling tower. The presence of any of these acids will cause a drop in circulating water pH. Water and carbon dioxide are found everywhere. When carbon dioxide is dissolved in water, carbonic acid is formed. This acid is less aggressive than the acids already mentioned. Because it is always present, however, serious damage to equipment can result from the corrosive effects of this acid.

Corrosion by Oxygen

Corrosion by oxygen is another problem. Water that is sprayed into the air picks up oxygen. This oxygen then is carried into the system. Oxygen reacts with any iron in equipment. It forms iron oxide, which is a porous material. Flaking or blistering of oxidized metal allows corrosion of the freshly exposed metal. Blistering also restricts water flow and reduces heat transfer. Reaction rates between oxygen and iron increase rapidly as temperatures increase. Thus, the most severe corrosion takes place in hot areas of equipment with iron parts.

Oxygen also affects copper and zinc. Zinc is the outer coating of galvanized material. Here, damage is much less severe because oxidation of zinc and copper forms an inert metal oxide. This sets up a protective film between the metal and the attacking oxygen.

Galvanic Action

Galvanic corrosion is the third cause of corrosion. Galvanic corrosion is basically a reaction between two different metals in electrical contact. This reaction is both electrical and chemical in nature. The following three conditions are necessary to produce galvanic action.

1. Two dissimilar metals possessing different electrochemical properties must be present.
2. An electrolyte, a solution through which an electrical current can flow, must be present.
3. An electron path to connect these two metals is also required.

Many different metals are used to fabricate air conditioning and refrigeration systems. Copper and iron are two dissimilar met-

Table 9-A.
Galvanic Series

Anodic (Corroded End)
Magnesium
Magnesium Alloy
Zinc
Aluminum
Mild Steels
Alloy Steels
Wrought Iron
Cast Iron
Soft Solders
Lead
Tin
Brass
Copper
Bronze
Copper-Nickel Alloys
Nickel
Silver
Gold
Platinum
Cathodic (Protected End)

als. Add a solution containing ions, and an electrolyte is produced. Unless the two metals are placed in contact, no galvanic action will take place. A coupling is made when two dissimilar metals, such as iron and copper, are brought into contact with one another. This sets up an electrical path or a path for electron movement. This allows electrons to pass from the copper to the iron. As current leaves the iron and reenters the solution to return to the copper, corrosion of the iron takes place. Copper-iron connections are common in cooling systems.

Greater separation of metals in the galvanic series results in their increased tendency to corrode. For example, if platinum is joined with magnesium, with a proper electrolyte, then platinum would be protected and magnesium would corrode. Since they are so far apart on the scale, the corrosion would be rapid. See Table 9-A. If iron and copper are joined, we can tell by their relative positions in the series that iron would corrode, but to a lesser degree than the magnesium mentioned in the previous example. Nevertheless, corrosion would be extensive enough to be very damaging. However, if copper and silver are joined together, then the copper would corrode. Consequently, the degree of corrosion is determined by the relative positions of the two metals in the galvanic series.

Improperly grounded electrical equipment or poor insulation can also initiate or accelerate galvanic action. Stray electrical currents cause a similar type of corrosion, usually referred to as *electrolytic corrosion*. This generally results in the formation of deep pits in metal surfaces.

Biological Organisms

Another cause of corrosion is biological organisms. These are algae, slime, and fungi. Slimes thrive in complete absence of light. Some slimes cling to pipes and will actually digest iron. This localized attack results in the formation of small pits which, over a period of time, will expand to form holes.

Other slimes live on mineral impurities, especially sulfates, in water. When doing so, they give off hydrogen sulfide gas. The gas forms weak hydrosulfuric acid. (Do not confuse this with strong sulfuric acid.) This acid slowly, but steadily, deteriorates pipes and other metal parts of the system. Slime and algae release oxygen into the water. Small oxygen bubbles form and cling to pipes. This oxygen may act in the same manner as a dissimilar metal and cause corrosion by galvanic action. This type of corrosion is commonly referred to as *oxygen cell corrosion*.

ALGAE

Algae are a very primitive form of plant life. They are found almost everywhere in the world. The giant Pacific kelp are algae. Pond scums and the green matter that grows in cooling towers are also algae. Live algae range in color from yellow, red, and green to brown and grey. Like bacterial slime, they need a wet or moist environment and prefer a temperature between 40 and 80° F. [4 and 27°C]. Given these conditions, they will find mineral nourishment for growth in virtually any water supply.

SLIME

Slime is caused by bacteria. Slime bacteria can grow and reproduce at temperatures from well below freezing (32° F. [0°C]) to the temperature of boiling water (212° F. [100°C]). However, they prefer temperatures between 40 and 80° F. [4 and 27°C]. They usually grow in dark places. Some types of slime also grow when exposed to light in cooling towers. The exposure of the dark-growing organisms to daylight will not necessarily stop their growth. The only condition essential to slime propogation is a wet or moist environment.

FUNGI

Fungi are a third biological form of corrosion. Fungi attack and destroy the cellulose fibers of wood. They cause what is known as brown rot or white rot. If fungal decay proceeds unchecked, serious structural damage will occur in a tower.

CONTROL OF ALGAE, SLIME, AND FUNGI

It is essential that a cooling system be kept free of biological growths as well as scale. Fortunately, several effective chemicals are available for controlling algae and slime. Modern algaecides and slimicides fall into three basic groups: the chlorinated phenols (pentachlorophenates), quarternary ammonium compounds, and various organometallic compounds.

A broad range of slime and algae control agents is required to meet the various conditions that exist in water-cooled equipment. Product selection is dependent on the following:

- The biological organism present.
- The extent of the infestation.
- The resistance of the existing growths to chemical treatment.
- The type and specific location of the equipment to be treated.

There is considerable difference of opinion in the trade as to how often algaecides should be added and whether "slug" or continuous feeding is the better method.

In treating heavy biological growths, remember that when these organisms die they break loose and circulate through the system. Large masses can easily block screens, strainers, and condenser tubes. Some provision should be made for preventing them from blocking internal parts of the system. The best way

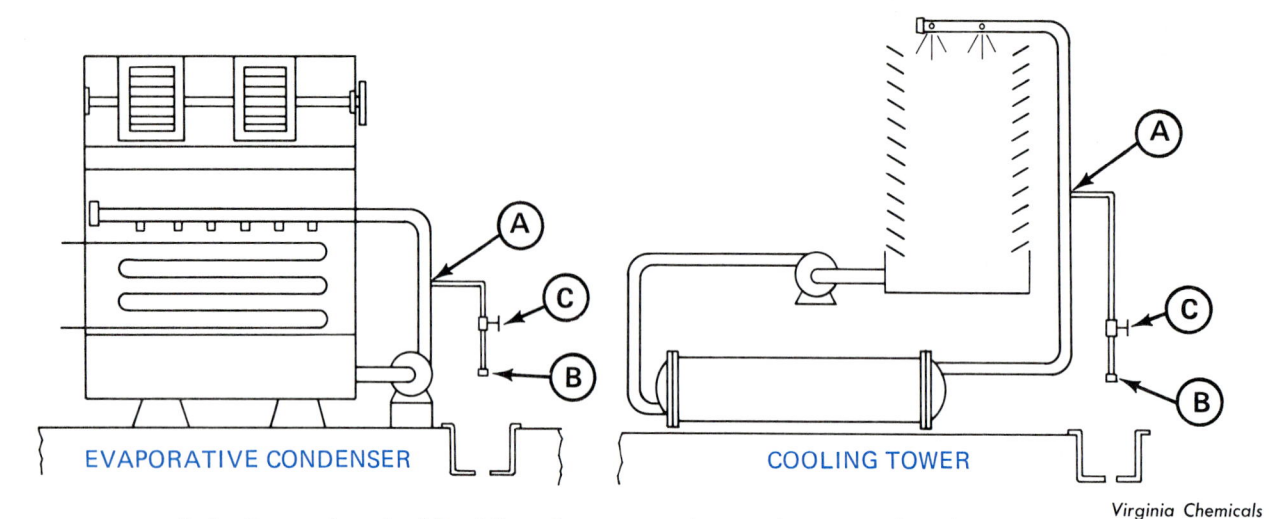

REFRIGERATION AND AIR CONDITIONING TECHNOLOGY

9-2. Connections for bleed lines for evaporative condensers and cooling towers.

Virginia Chemicals

to do this is to remove the thick, heavy growths *before* adding treatment. The day after treatment is completed, thoroughly drain and flush the system and clean all strainers.

THE PROBLEM OF SCALE

Air conditioning or refrigeration is basically the controlled removal of heat from a specific area. The refrigerant that carries heat from the cooled space must be cooled before it can be reused. Cooling and condensation of refrigerant require the use of a cooling medium that, in many systems, is water.

There are two types of water-cooled systems. The first type uses once-through operation. The water picks up heat and is then discarded or wasted. In effect, this is 100%, or total, bleed. Little, if any, mineral concentration occurs. The scale that forms is due to the break down of bicarbonates by heat. These form carbonates, which are less soluble at high temperatures than at low. Such scale can be prevented through use of a treatment chemical.

The other type of water-cooled system is the type in which heat is removed from water by partial evaporation. The water is then recirculated. Water volume lost by evaporation is replaced. This type of system is more economical from the standpoint of water use. However, the concentration of dissolved minerals leads to conditions which, if not controlled and chemically treated, may result in heavy scale formation.

Evaporative Systems

One method of operating evaporative recirculating systems involves 100% evaporation of the water with no bleed. This, of course, causes excessive mineral concentration. Without a bleed on the system, water conditions will soon exceed the capability of any treatment chemical. A sec-

9-4. Calculating the amount of water in a round tank.

Virginia Chemicals

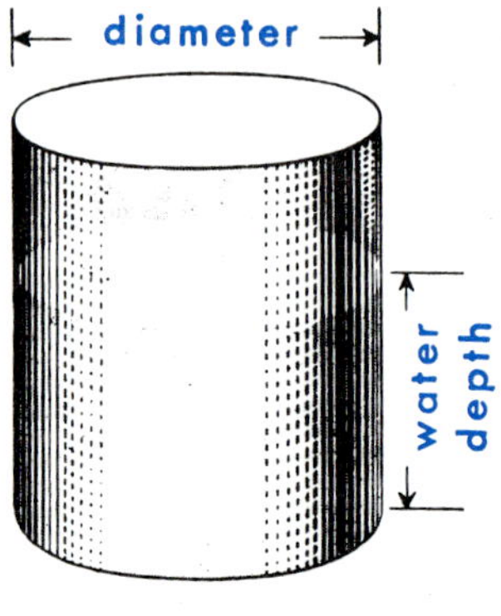

Virginia Chemicals
9-3. The method of calculating the amount of water in a rectangular tank or sump.

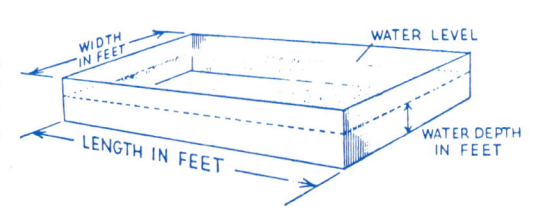

220

CHAPTER 9—COOLING WATER PROBLEMS

ond method of operation employs a high bleed rate without chemical treatment. Scale will form and water is wasted. The third method is the reuse of water, with a bleed to control concentration of scale-forming minerals. Thus, by the addition of minimum amounts of chemical treatment, good water economy can be realized. This last approach is the most logical and least expensive. Figure 9-2 shows how connections for bleed lines are made on evaporative condensers and cooling towers.

Scale Formation

Scale is formed as a direct result of mineral insolubility. This in turn, is a direct function of temperature, hardness, alkalinity, pH, and total dissolved solids. Generally speaking, as these factors increase, solubility or stability of scale-forming minerals decreases. Unlike most minerals, scale-forming salts are less soluble at high temperatures. For this reason, scale forms most rapidly on heat exchanger surfaces.

HOW TO CLEAN COOLING TOWERS AND EVAPORATIVE CONDENSERS

To clean cooling towers and evaporative condensers, first determine the amount of water in the system. This is done by:

Determining the Amount of Water in the Sump

1. Measure the length, width, and water depth in feet. See Fig. 9-3.

Use the following formula— length × width × water depth × 7.5 = gallons of water in the sump. Example: A sump is 5 feet long and 4 feet wide, with a water depth of 6 inches. Complete the equation: $5 \times 4 \times 0.5 \times 7.5 = 75$ gallons of water in the sump.

Determining the Amount of Water in the Tank

1. Measure the diameter of the tank and the depth of the water in feet. See Fig. 9-4.

2. Use the following formula: diameter^2 × water depth × 6 = gallons of water in tank. Example: A tank has a diameter of 3 feet and the water is 3 feet deep. Complete the equation: $3^2 \times 3 \times 6$, or $9 \times 3 \times 6 = 162$ gallons of water in the tank.

Total Water Volume

The two formulas above will give you the water volume in either the tank or sump. Each is figured separately since they are both part of the system's circulating water supply. There is also water in the connecting lines. These lines must be measured for total footage. Once you find the pipe footage connecting the system you can figure its volume of water too. Simply take 10% of the water volume in the sump for each 50 feet of pipe run. This is added to the water in the sump and the water in the tank to find the *total* system water volume.

For example, a system has 75 gallons of water in the sump and 162 gallons of water in the tank. The system has 160 feet of pipe.

75 gallons + 162 gallons = 237 gallons

160 feet ÷ 50 feet = 3.2

75 gallons in sump ÷ 10 = 7.5 gallons for every 50 feet in the pipes

7.5 gallons × 3.2 = 24 gallons in the total *pipe* system

237 gallons (in *tank* and *sump*) + 24 gallons (in *pipes*) = 261 gallons in the total system

This is the amount of water that must be treated to keep the system operating properly.

Now that you have determined the volume of water in the system, you can calculate the amount of chemicals needed.

1. Drain the sump. Flush out, or remove manually, all loose sludge and dirt. This is important because they waste the chemicals.

2. Close the bleed line and refill the sump with fresh water to the lowest level at which the circulating pump will operate. See Fig. 9-5.

3. Calculate the total gallons of water in the system. Next, while the water is circulating, add starting amounts of either chemical slowly, as follows. Note, these amounts are for *hot water systems*. For cold water systems, see page 222.

• Solid scale remover: 5 pounds per 10 gallons of water.

• Regular liquid scale remover: 1 gallon per 15 gallons of water.

• Concentrated liquid scale remover: 1 gallon per 20 gallons of water.

Refer to Fig. 9-5. The scale removers can be introduced at the water tower distribution plate (A), the sump (B), or the water tank (C). Convenience is the keyword here. The preferred addition point is directly into the pump suction area.

4. When using liquid scale remover, add 1 ampule of antifoam reagent per gallon of chemical. This will usually prevent excessive foaming if added before the scale remover. Extra antifoam is available in 1-pint bottles. When using the solid scale remover, stir the crystals in a plastic pail or drum until completely dissolved. Then pour

221

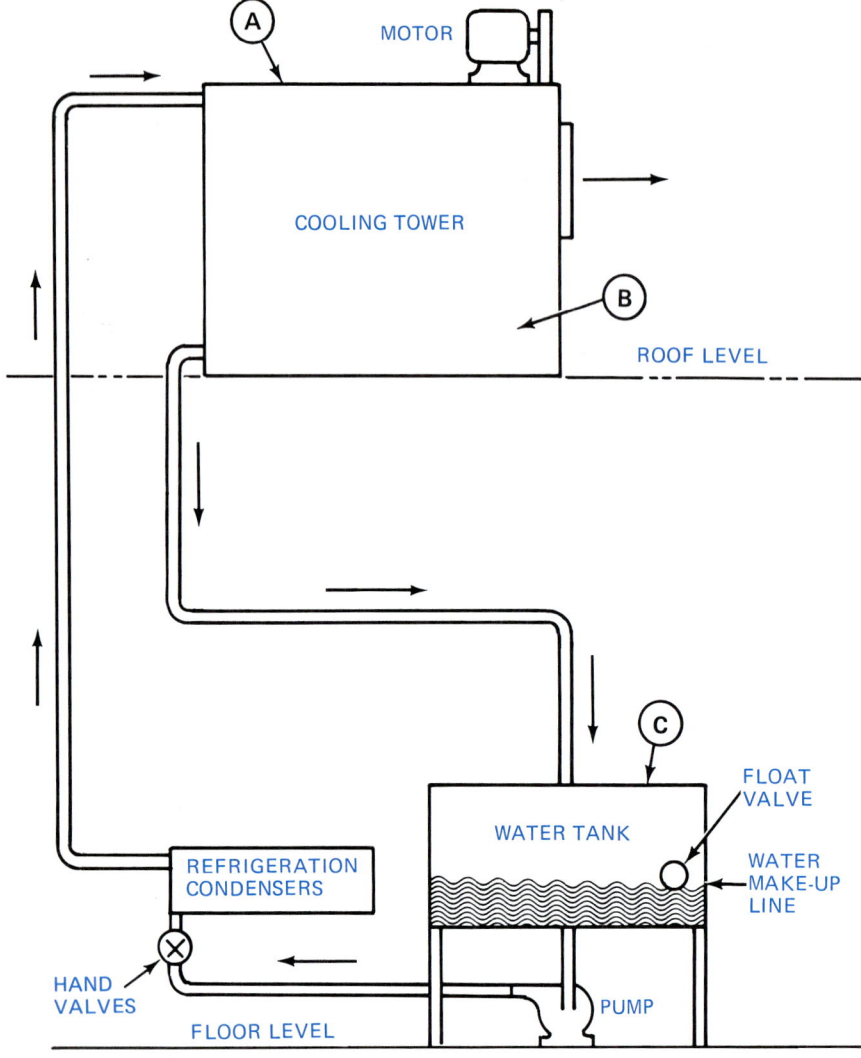

REFRIGERATION AND AIR CONDITIONING TECHNOLOGY

9-5. Forced- and natural-draft towers. (A) Water tower distribution plate, (B) sump, (C) water tank.
Virginia Chemicals

Drain or pump this strong solution into the system. Repeat this procedure until the required weight of the chemical has been added in concentrated solution. Then, add fresh water to fill the system. The treatment should be repeated once each year for best results.

If make-up water is needed during the year, be sure to treat this water also at the rate of 6 pounds per 100 gallons of water added.

Chilled Water Systems

For chilled water systems, follow the instructions just outlined for hot water systems. However, *use only 3 pounds* of circulating water treatment solution for each 100 gallons of water in the system. This treatment solution should be compatible with antifreeze solutions.

9-6. Preparing crystals in a drum or tank.
Virginia Chemicals

slowly as a liquid. Loose crystals, if allowed to fall to the bottom of the sump, will not dissolve without much stirring. If not dissolved, they might damage the bottom of the sump.

Figure 9-6 shows how to prepare the crystals in the drum. Use a 55-gallon drum. Install a drain or spigot about 6 to 8" from the bottom. Set the drum in an upright position. Fill the drum with fresh water within 6" from the top, preferably warm water at about 80° F. [27°C]. Since the fine particles of the water treatment crystals are quite irritating to the nose and eyes, immerse each plastic bag in the water. Cut the bag below the surface of the water. See Fig. 9-6. Stir the crystals until dissolved. About 6 to 8 pounds of crystals will dissolve in each gallon of water.

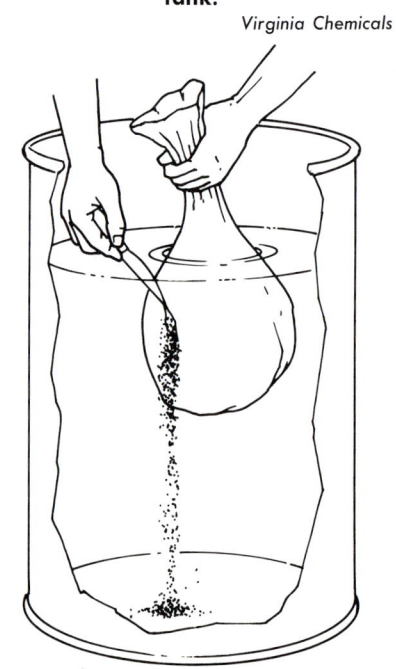

CHAPTER 9—COOLING WATER PROBLEMS

CLOSED SYSTEM INSTALLATIONS

AROUND CIRCULATING PUMP
In recirculating systems

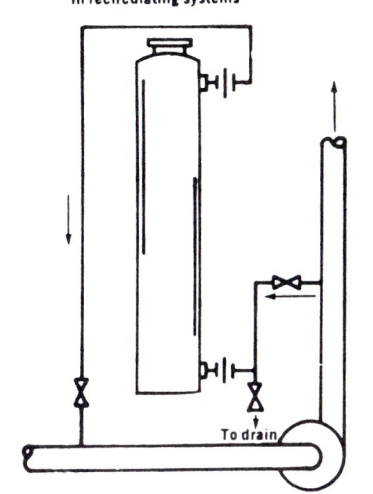

BY-PASS ARRANGEMENT
Non-recirculating, once through systems

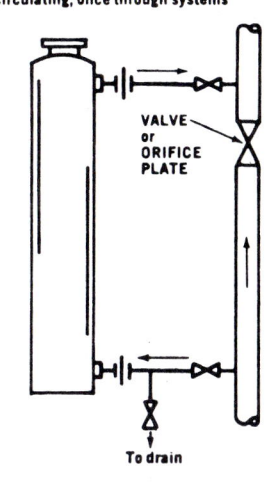

Virginia Chemicals

9-7. Feed through a bypass feeder.

Feed through a Bypass Feeder

For easy feeding of initial and repeat doses of water treatment solution, install a crystal feeder in a bypass line. The crystals will dissolve as water flows through the feeder. See Fig. 9-7.

Install the feeder in either a by-pass or in-line arrangement, depending upon the application. Place the feeder on a solid, level floor or foundation. Connection with standard pipe unions is recommended.

Connecting pipe threads should be carefully cut and cleaned to remove all burrs or metal fragments. Apply a good grade of pipe dope. Use the dope liberally.

Always install valves in the inlet and the outlet lines. Install a drain line with the valve in the bottom (inlet) line.

Before opening the feeder, always close the inlet and the outlet valves. Open the drain valve to relieve the pressure and drain as much water as necessary. When adding crystals or chemicals molded in shapes (balls, briquettes, etc.), it is advisable to have the feeder about one-half filled with water.

If stirring in the feeder is necessary, use only a soft wood. Stir gently to avoid damaging the epoxy lining.

Fill the feeder to the level above the outlet line. Coat the top opening, the gasket, and the locking grooves of the cap with petroleum jelly or a heavier lubricant.

Open the outlet valve fully. Then slowly open the inlet valve. If throttled flow is desired for control of treatment feed rate, throttle with inlet valve only.

Normal Operation

Once the chemicals have been properly introduced, operate the system in the normal manner. Check the scale remover strength in the sump by observing the color of the solution when using the solid scale remover or using test papers. When Virginia Chemicals removers are used, a green solution indicates a very strong cleaner. A blue solution indicates normal cleaning strength. A purple solution indicates more cleaner is needed. If necessary, dip a sample of the sump solution in a glass to aid color check.

If Virginia Chemicals scale remover is used in either solid or liquid form, use test papers to check for proper mixture and solution strength. Red test paper indicates there is enough cleaner. Inspection of the evaporative condenser tubes or lowering of head pressure to normal will indicate when the unit is clean. With shell and tube condensers, inspection of the inside of the water outlet pipe of the condenser will indicate the amount of scale in the unit.

After scale removal is completed, drain the spent solution to the sewer. Thoroughly rinse out the system with at least two fillings of water. Do not drain spent solutions to lawns or near valuable plants. The solution will cause plant damage, just as will any other strong salt solution. Do not drain to a septic tank. Refill the sump with fresh water and resume normal operation.

HOW TO CLEAN SHELL (TUBE OR COIL) CONDENSERS

Isolate the condenser to be cleaned from the cooling tower system by an appropriate valve arrangement or by disconnecting the condenser piping. Pump in at the lowest point of the condenser. Venting the high points with tubing returning to the solution drum is necessary in some

223

REFRIGERATION AND AIR CONDITIONING TECHNOLOGY

units to assure complete liquid filling of the water side.

As shown in Fig. 9-8, start circulating from a plastic pail or drum the minimum volume of water necessary to maintain circulation. After adding antifoam reagent or *solid* scale remover, slowly add *liquid* scale remover until the test strips indicate the proper strength for cleaning. Test frequently and observe the sputtering in the foam caused by carbon dioxide in the return line. Add scale remover as necessary to maintain strength. Never add more than 1 pound of solid scale remover per gallon of solution. Most condensers with moderate amounts of carbonate scale can be cleaned in about one hour. Circulation for 30 to 40 minutes without having to add cleaner to maintain cleaning strength usually indicates that action has stopped and that the condenser is clean.

Empty and flush the condenser after cleaning with at least two complete fillings of water. Reconnect the condenser in the line.

If a condenser is completely clogged with scale, it is sometimes possible to open a passageway for the cleaning solution by using the standpipe method, as shown in Fig. 9-9. Enough scale remover-liquid is mixed with an equal volume of water to fill the two vertical pipes to a level slightly above the condenser. Some foaming will result from the action of the cleaner solution on the scale. Thus, some protective measures should be taken to prevent foam from injuring surrounding objects. The antifoam reagent supplied with each package will help control this nuisance. When the cleaning operation has been completed, drain the spent solution to the sewer. Rinse the condenser with at least two fillings of fresh water.

SAFETY

Scale remover contains acid and can cause skin irritation. Avoid contact with your eyes, skin, and clothing. In case of contact, flush the skin or eyes with plenty of water for at least 15 minutes. If the eyes are affected, get medical attention. Keep scale remover and other chemicals out of the reach of children.

Do not drain the spent solution to the roof or to a septic tank. Always drain the spent solution to the storm sewer when possible.

Safety is all-important. All chemicals, especially acids, should be treated with great respect and handled with care. Rubber gloves, acid-proof coveralls, and safety goggles should be worn when working with chemicals. Cleaning a system through the tower, although easier and faster than some of the other methods, presents one unique hazard. Wind drift, even with the tower fan off, is a definite possibility.

Wind drift will carry tiny droplets of acid that can burn eyes and skin. These acid droplets will also damage automobile finishes and buildings. Should cleaning solution contact any part of the person, it should be washed off immediately with soap and water.

Forethought and reasonable precaution will prevent grief and expense.

9-8. Cleaning a shell (tube or coil) condenser.
Virginia Chemicals

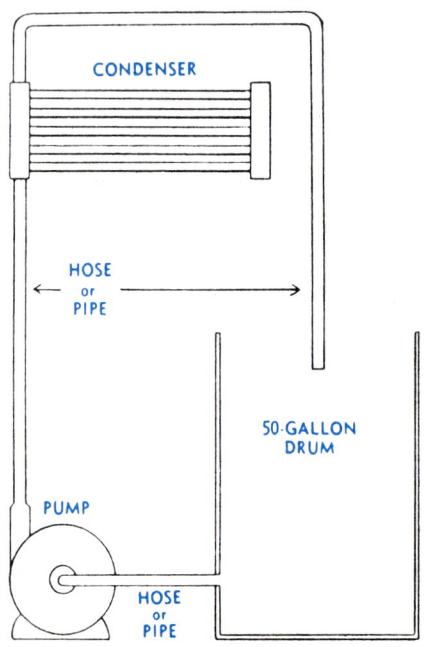

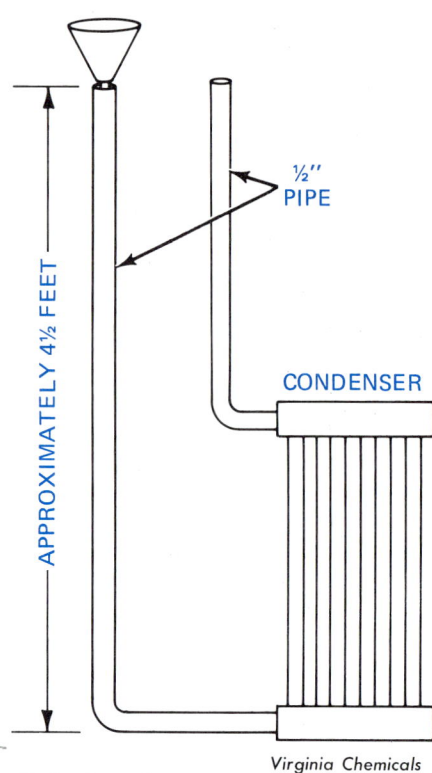

9-9. Opening a passageway for cleaning solution by using the standpipe method.
Virginia Chemicals

SOLVENTS AND DETERGENTS

There are several uses for solvents and detergents in the ordinary maintenance schedule of air-cooled fin coil condensers, evaporator coils, permanent-type air filters, and fan blades. In most instances, a high-pressure spray washer is used to clean the equipment with detergent. Then a high-pressure spray rinse is used to clean the unit being scrubbed. The pump is usually rated at 2 gallons per minute at 500 pounds per square inch of pressure. The main function is to remove dirt and grease from fans and cooling surfaces. It takes about 10 to 15 minutes for the cleaning solution to do its job. It is then rinsed with clean water.

Permanent-type filters are cleaned by using the dipping method. Prepare a cleaning solution—one part detergent to one part water. Use this solution as a bath in which the filters may be immersed briefly. After dipping, set the filter aside for 10 to 15 minutes. Flush with a stream of water. If water is not available, good results may be obtained by brisk agitation in a tank filled with fresh water.

REVIEW QUESTIONS

1. Why is rainwater not pure?
2. Why is water that is fit for human consumption not usable in boilers and cooling equipment?
3. How can scaling be prevented?
4. List some aggressive or strong acids.
5. In what way does oxygen cause corrosion?
6. What is meant by galvanic action?
7. What is algae?
8. What is slime?
9. What damage does fungi cause?
10. How are shell condensers cleaned?
11. What safety precautions should be taken when using scale remover?
12. How does wind drift affect tower operation?
13. How is detergent used to clean condensers and evaporator coils?
14. How are permanent-type filters cleaned?

10

Evaporators

The evaporator removes heat from the space being cooled. As the air is cooled, it condenses water vapor. This must be drained. If the water condensing on the evaporator coil freezes—when the temperature is below 32° F. [0°C]—the refrigerator or freezer must work harder. Frozen water or ice acts as an insulator. It reduces the efficiency of the evaporator. When evaporators are operated below 32° F. they must be defrosted periodically. This eliminates frost buildup on the coils or the evaporator plates.

There are several types of evaporators. The *coiled evaporator* is used in warehouses for refrigerating large areas. The *fin evaporator* is used in the air conditioning system that is part of the furnace in a house. Fig. 10-1. The finned evaporator has a fan that blows air over its thin metal surfaces. *Plate evaporators* use flat surfaces for their cooling surface. Fig. 10-2. They are commonly used in freezers. If the object to be cooled or frozen is placed directly in contact with the evaporator plate, the cold is transferred more efficiently.

Figure 10-3 shows a home refrigerator cooling system. Note the evaporator.

COILED EVAPORATOR

Evaporator coils on air condi-

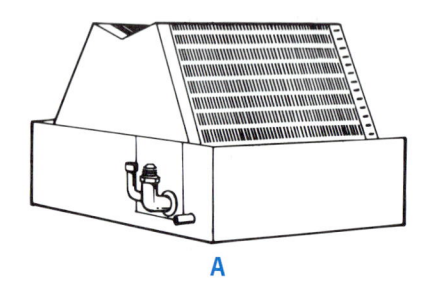

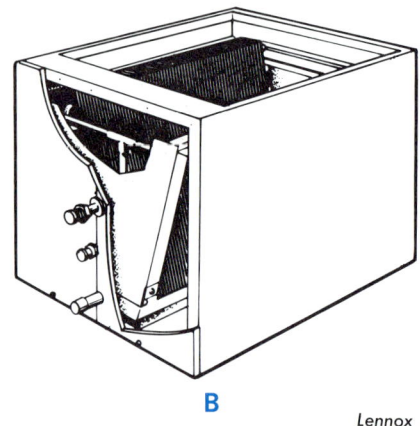

A

B
Lennox

10-1. Evaporators. (A) Evaporator used in home air conditioning system where the unit is placed in the "bonnet" of the hot air furnace. (B) Slanted evaporator used in home air conditioning.

tioning units fall into two categories: finned tube coil and shell-and-tube chiller.

Finned Tube Coil

The finned tube coil is placed in the airstream of the unit. Refrigerant vaporizes in it. The refrigerant in the tubes and the air flowing around the fins attached to the tubes draw heat from the air. This is commonly referred to as a direct expansion cooling system. Fig. 10-4.

Shell-and-Tube Chiller

Shell-and-tube units are used to chill water for air-cooling purposes. Usually, the refrigerant is in tubes mounted inside a tank or shell containing the water or liquid to be cooled. The refrigerant in the tubes draws the heat through the tube wall and from the liquid as it flows around the tubes in the shell. This system can be reversed. Thus, the water would be in the tubes and the

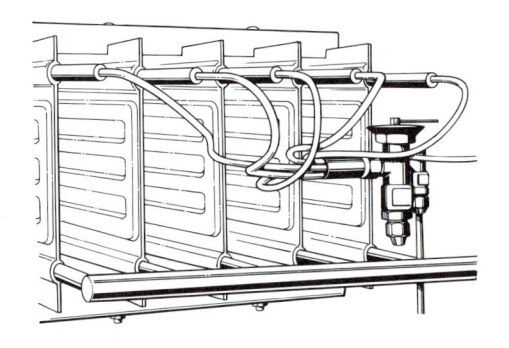

Sporlan
10-2. Plate evaporator.

226

CHAPTER 10—EVAPORATORS

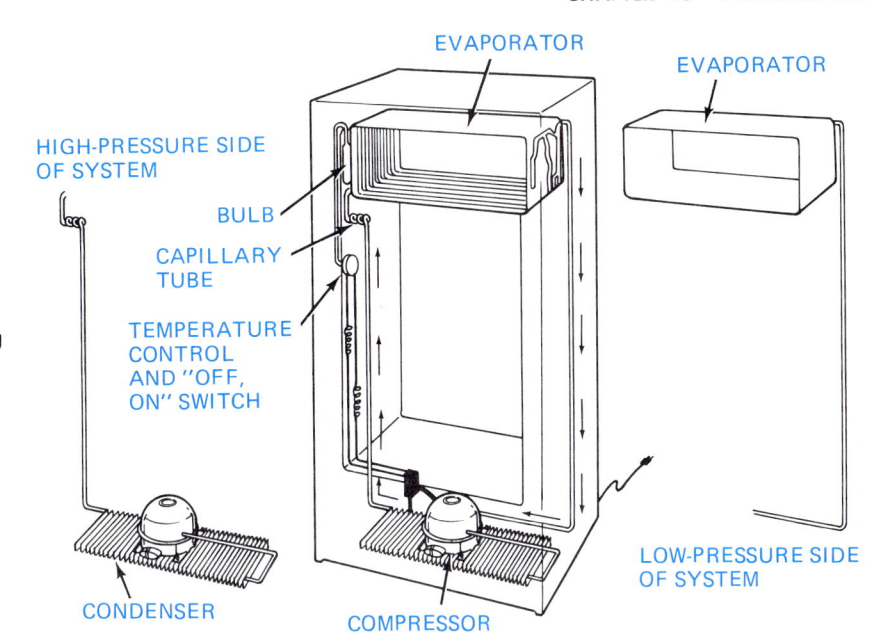

10-3. A home refrigerator cooling system.

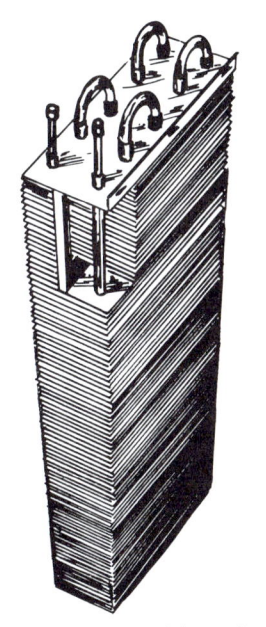

Johnson Controls
10-4. Finned coil evaporator.

refrigerant would be in the tank. As the gas passes through the tank over the tubes, it would draw the heat from the water in the tubes. Fig. 10-5.

Figure 10-5 shows how R-12 is used in a standard vapor-compression refrigeration cycle. System water for air conditioning and other uses is cooled as it flows through the evaporator tubes. Heat is transferred from the water to the low-temperature, low-pressure refrigerant. The heat removed from the water causes the refrigerant to evaporate. The refrigerant vapor is drawn into the first stage of the compressor at a rate controlled by the size of the guide vane opening. The first stage of the compressor raises the temperature and pressure of the vapor. This vapor, plus vapor from the flash economizer, flows into the second stage of the compressor. There, the saturation temperature of the refrigerant is raised above that of the condenser water.

This vapor mixture is discharged directly into the condenser. There, relatively cool condenser water removes heat from the vapor, causing it to condense again to liquid. The heated water leaves the system, returning to a cooling tower or other heat-rejection device.

A thermal economizer in the bottom section of the condenser brings warm condensed refrigerant into contact with the inlet water tubes. These are the coldest water tubes. They may hold water with a temperature as low as 55° F. [13°C]. This subcools the refrigerant so that when it moves on in the cycle, it has greater cooling potential. This improves cycle efficiency and reduces power per ton requirements. The liquefied refrigerant leaves the condenser through a plate-type control. It flows into the flash economizer or utility vessel. Here, the normal flashing of part of the refrigerant into vapor cools the remaining refrigerant. This flash vapor is diverted directly to the second stage of the compressor. Thus, it does not need to be pumped through the

227

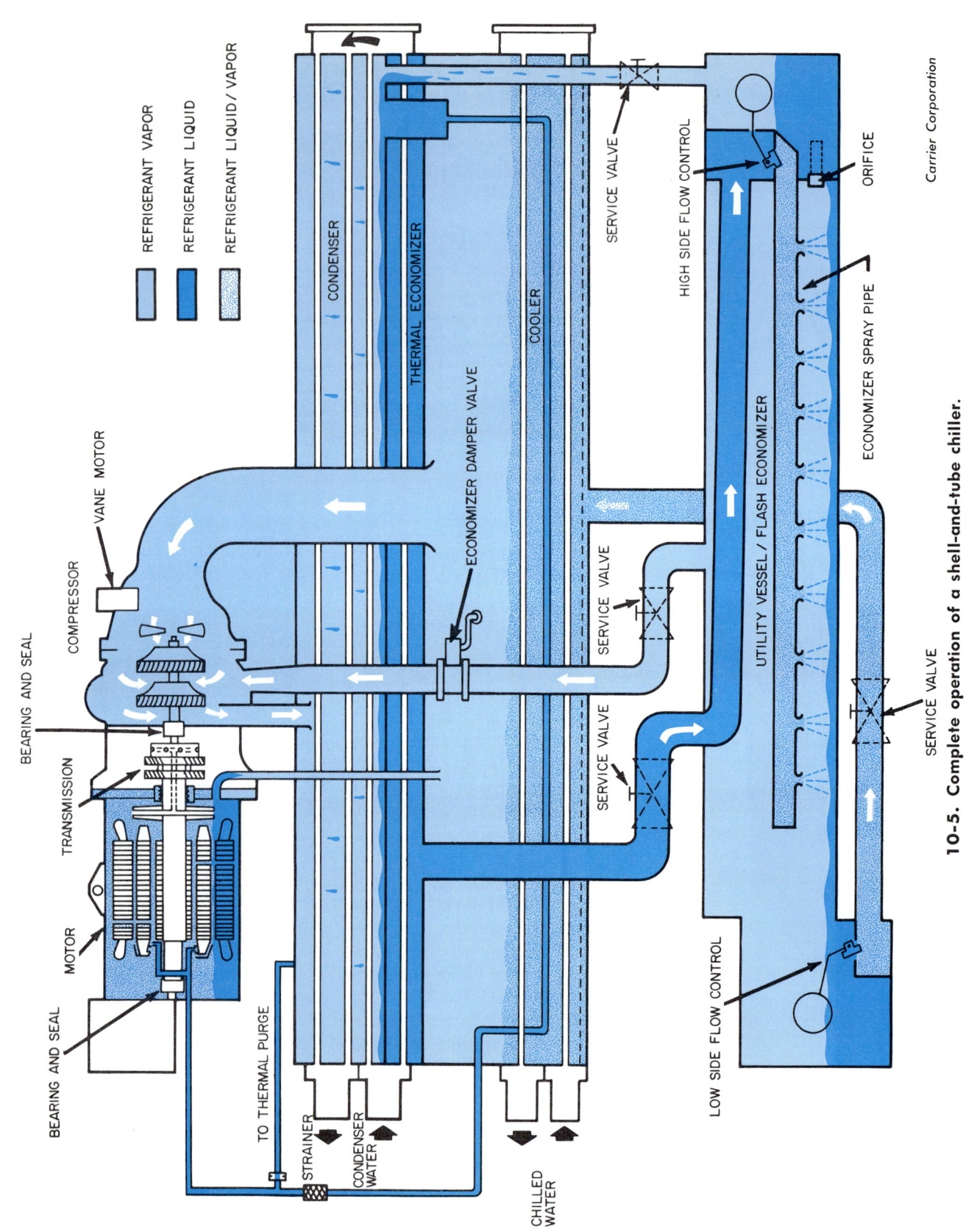

10-5. Complete operation of a shell-and-tube chiller.

CHAPTER 10—EVAPORATORS

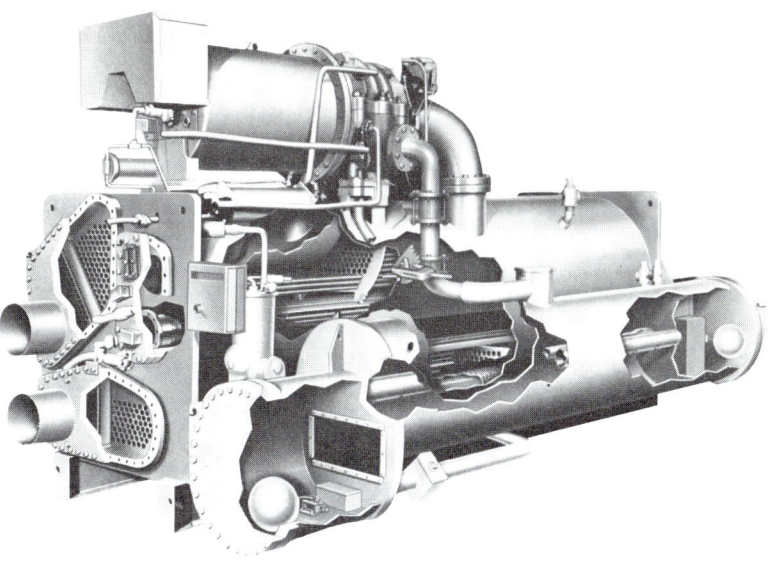

Carrier Corporation
10-6. Cutaway view of the chiller portion of the shell-and-tube chiller shown in Fig. 10-5.

full compression cycle. The net effect of the flash economizer is energy savings and lower operating costs. A second plate-type control meters the flow of liquid refrigerant from the utility vessel back to the cooler, where the cycle begins again. Fig. 10-6.

APPLICATION OF CONTROLS FOR HOT-GAS DEFROST OF AMMONIA EVAPORATORS

To defrost ammonia evaporators, it is sometimes necessary to check the plumbing arrangement and the valves used to accomplish the task. To enable hot gas defrost systems to operate successfully, several factors must be considered. There must be an adequate supply of hot gas. The gas should be at a minimum of 100 psig. The defrost cycle should be accurately timed. Condensate removal or storage must be provided. An automatic suction accumulator or heat reservoir should be used to protect compressors from liquid refrigerant slugs if surge drums or other evaporators are not adequate to handle the excess gas and condensates. See Fig. 10-7.

Controls must be used to direct and regulate the pressure and flow of ammonia and hot gas during refrigeration and defrost cycles.

Direct-Expansion Systems

Figure 10-7 shows a high temperature system (above 32° F. [0°C]) with no drip pan defrost. During the normal cooling cycle, controlled by a thermostat, the room temperature may rise above the high setting of the thermostat. This indicates a need for refrigeration. The liquid solenoid (valve A), pilot solenoid (valve B), and the dual-pressure regulator (valve D) open, allowing refrigerant to flow. When solenoid (valve D) is energized, the regulator is controlled by the low-pressure adjusting bonnet. The regulator maintains the predetermined suction pressure in the evaporator.

When the room temperature reaches the low setting on the thermostat, there is no longer need for refrigeration. At this time, solenoid valve A and solenoid valve D close and remain closed until further refrigeration is required.

The hot gas solenoid (valve C) remains closed during the normal refrigeration cycle. When the three-position selector switch is turned to DEFROST, liquid solenoid valve A and valve D with a built-in pilot solenoid close. This allows valve D to operate as a defrost pressure regulator on the high setting. The hot gas solenoid (valve C) opens to allow hot gas to enter the evaporator. When the defrost is complete, the system is switched back to the normal cooling cycle.

The system may be made completely automatic by replacing the manual switch with an electric time clock. Table 10-A shows the valve sizes needed for this system.

Valves Used in Direct-Expansion Systems

The pilot solenoid valve (B) is a $\frac{1}{8}''$ ported solenoid valve that

REFRIGERATION AND AIR CONDITIONING TECHNOLOGY

is direct-operated and suitable as a liquid, suction, hot gas, or pilot valve at pressures to 300 pounds.

Solenoid valve A is a one-piston, pilot-operated valve suitable for suction, liquid, or gas lines at pressures to 300 pounds. It is available with a $9/16''$ or $3/4''$ port.

Solenoid valve C is a rugged, pilot-operated, two-piston valve with spring return for positive closing under the most adverse conditions. It is used for compressor unloader, suction, liquid, and hot gas applications.

The dual-pressure regulator valve (D) is designed to operate at two predetermined pressures without resetting or adjustment. By merely opening and closing a pilot solenoid, it is capable of maintaining either the low- or high-pressure setting.

Figure 10-8 shows a pilot light assembly. It is placed on valves when it is essential to know their condition for troubleshooting procedures.

A low-temperature defrost system with water being used to defrost the drain pan is shown in Fig. 10-9.

Cooling Cycle

During the normal cooling cycle controlled by a thermostat, as room temperature rises above the high setting on the thermostat there is a need for refrigera-

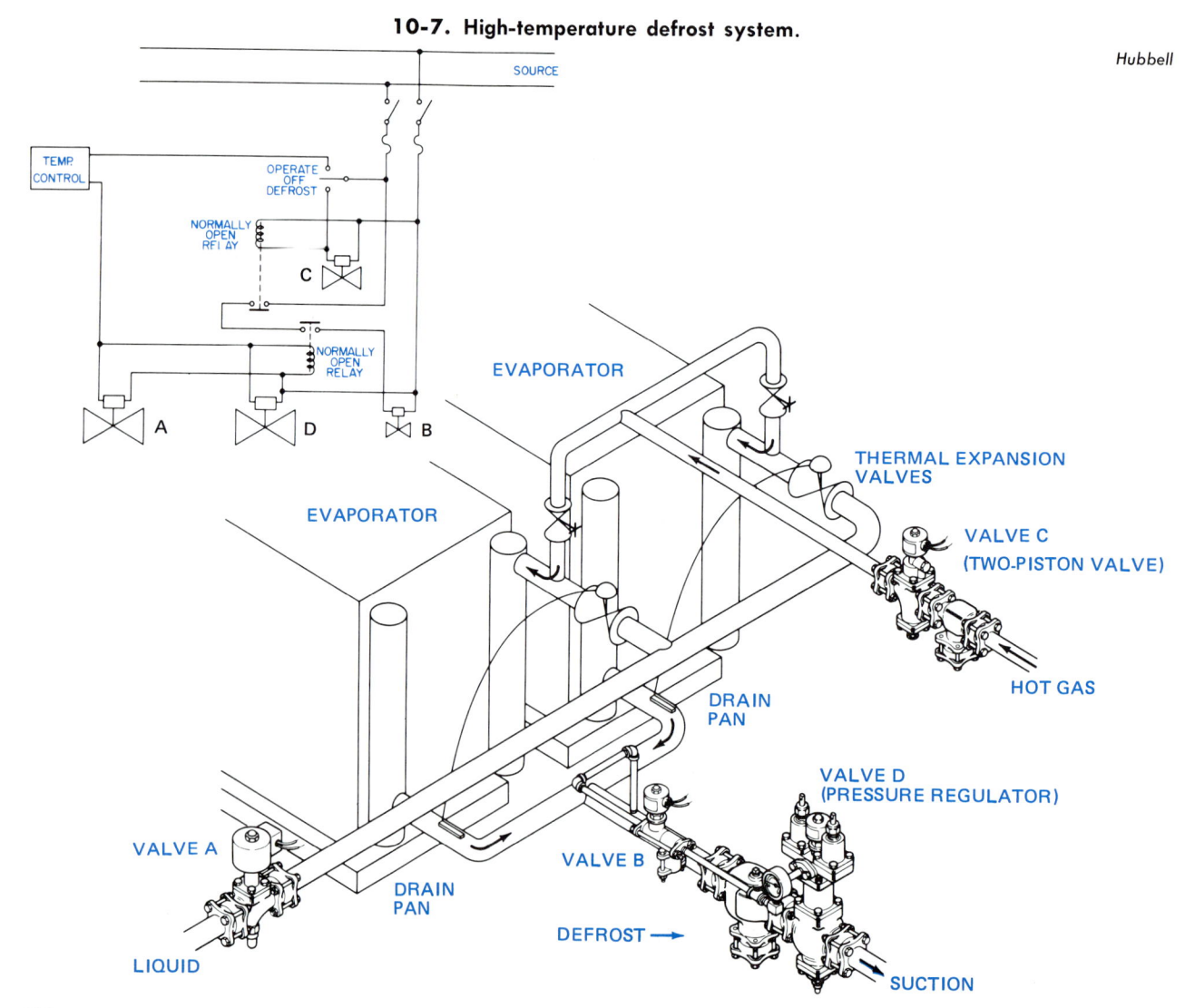

10-7. High-temperature defrost system.

Hubbell

CHAPTER 10—EVAPORATORS

Table 10-A.
Valve Sizing for High Temperature System

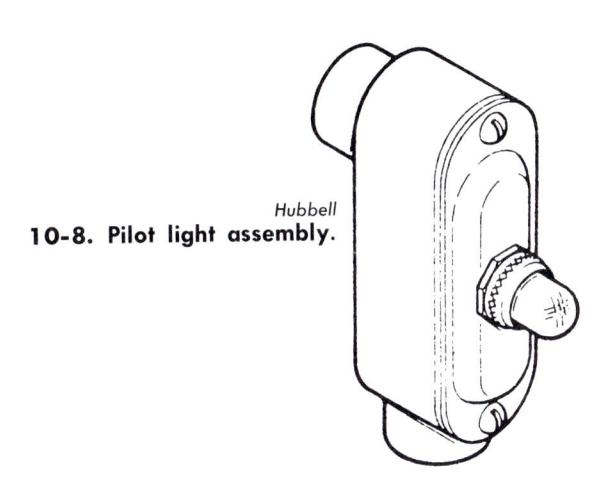

Tons Refrigerant	Liquid Solenoid	Hot Gas Solenoid	Pilot Solenoid	Dual-pressure Regulator
3	1/2″	1/2″	1/4″	3/4″
5	1/2″	1/2″	1/4″	3/4″
7	1/2″	1/2″	1/4″	1″
10	1/2″	1/2″	1/4″	1 1/4″
12	1/2″	1/2″	1/4″	1 1/4″
15	1/2″	1/2″	1/4″	1 1/2″
20	1/2″	1/2″	1/4″	1 1/2″
25	1/2″	1/2″	1/4″	2″
30	1/2″	1/2″	1/4″	2″
35	1/2″	1/2″	1/4″	2″
40	3/4″	3/4″	1/4″	2″
45	3/4″	3/4″	1/4″	2 1/2″
50	3/4″	3/4″	1/4″	2 1/2″

10-8. Pilot light assembly. *Hubbell*

tion. Liquid solenoid (A) and the built-in pilot valve (D) open, allowing refrigerant to flow. The opening of the built-in pilot allows the pressure to bypass the sensing chamber of valve D. This forces it to remain wide open with resultant minimum pressure drop through the valve.

When the room temperature drops to the low setting on the thermostat, there is no longer need for refrigeration. Solenoid valve A and pilot valve D close.

They remain closed until refrigeration is again required. Hot gas valve C and defrost-water solenoid valve E remain closed during the cooling cycle.

Defrost Cycle

When the three-position selector switch is turned to DEFROST, solenoid valve A and pilot solenoid valve D close as hot gas valve C and evaporator pilot valve B open. This allows hot gas to enter the evaporator. Valve D now acts as a back pressure regulator, maintaining a predetermined pressure above the freezing point. After a regulated delay, preferably toward the end of the defrost cycle, the time delay relay allows the water solenoid to open. This causes water to spray over the evaporator, melting ice that may be lodged between coils and flushing the drain pan.

When the evaporator is defrosted, the system is returned to the cooling cycle by turning the three-position selector switch. The hot gas solenoid (valve C) and built-in pilot (valve E) close as the liquid solenoid (valve A) opens.

This system can be made completely automatic by replacing the manual selector with an electric time clock.

Table 10-B shows some of the valve sizing for the low-temperature system.

DIRECT EXPANSION WITH TOP HOT-GAS FEED

In the evaporator shown in

REFRIGERATION AND AIR CONDITIONING TECHNOLOGY

Fig. 10-10, when the defrost cycle is initiated, the hot gas is introduced through the hot gas solenoid valve to the manifold. It then passes through the balancing globe valve and the pan coil to a check valve that prevents liquid crossover. From the check valve, hot gas is directed to the top of the evaporator. Here, it forces the refrigerant and accumulated oil from the relief regulator (valve A). This regulator has been de-energized to convert it to a relief regulator set at about

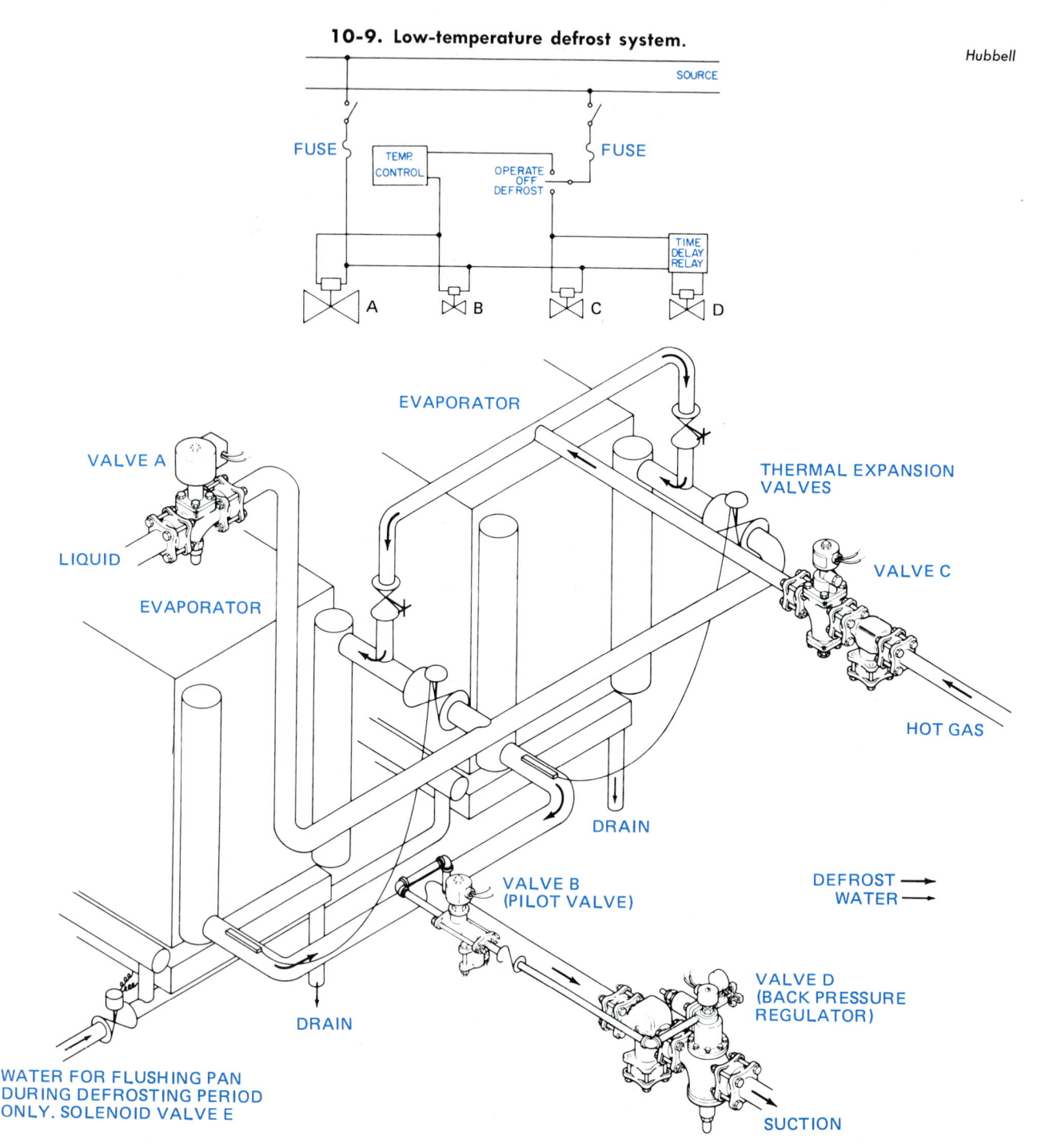

10-9. Low-temperature defrost system.

Hubbell

CHAPTER 10—EVAPORATORS

Table 10-B.
Valve Sizing for Low Temperature System

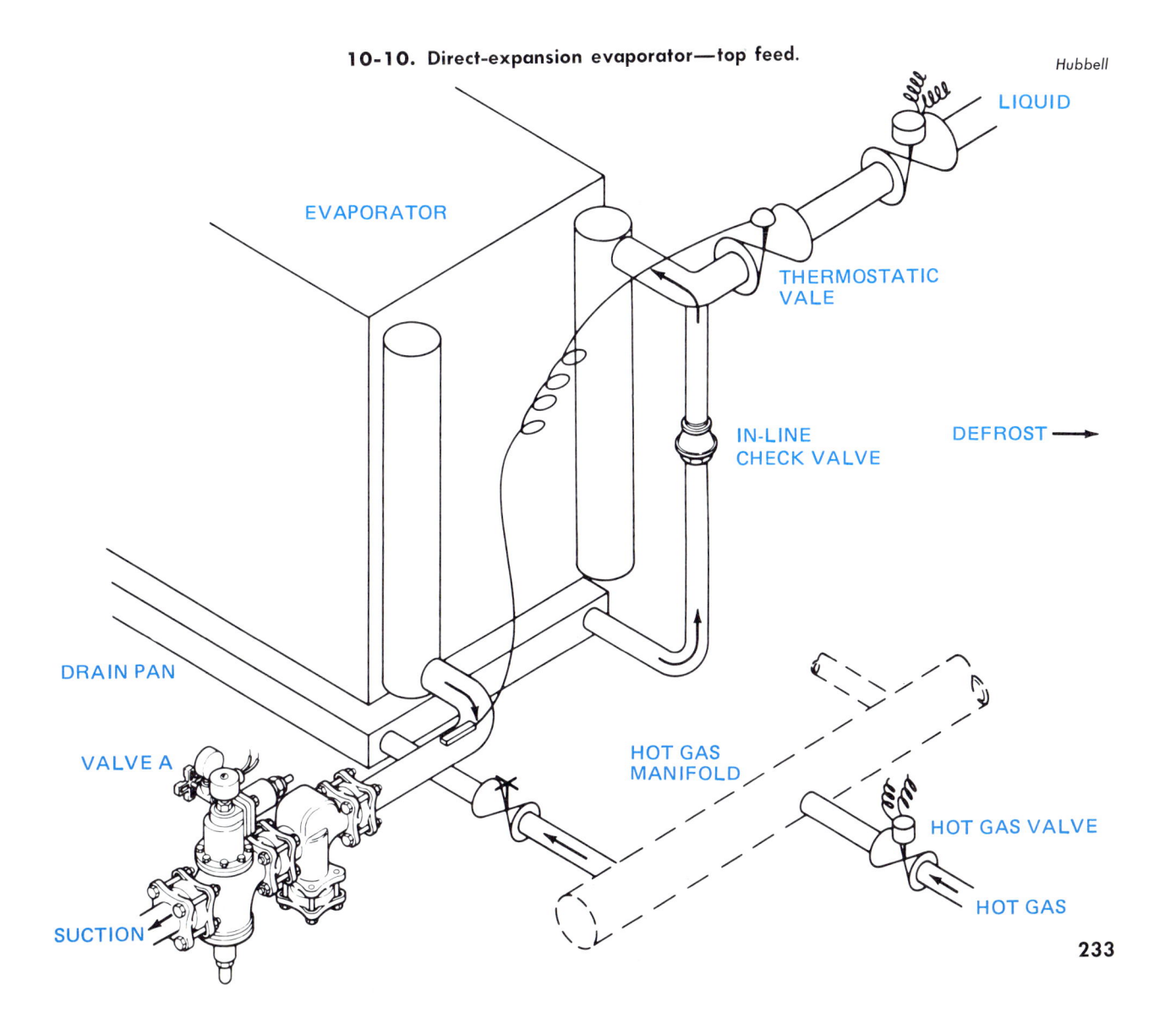

Tons Refrigerant	Liquid Solenoid	Hot Gas Solenoid	Back Pressure Regulator	Pilot Solenoid	Defrost Water
3	1/2″	1/2″	1″	1/4″	1/2″
5	1/2″	1/2″	1 1/4″	1/4″	3/4″
7	1/2″	1/2″	1 1/4″	1/4″	3/4″
10	1/2″	1/2″	1 1/2″	1/4″	1″
12	1/2″	1/2″	1 1/2″	1/4″	1″
15	1/2″	1/2″	2″	1/4″	1 1/4″
20	1/2″	1/2″	2″	1/4″	1 1/4″
25	1/2″	1/2″	2 1/2″	1/4″	1 1/2″
30	1/2″	1/2″	2 1/2″	1/4″	1 1/2″
35	1/2″	1/2″	3″	1/4″	2″
40	3/4″	3/4″	3″	1/4″	2″
45	3/4″	3/4″	3″	1/4″	2″
50	3/4″	3/4″	3″	1/4″	2″

10-10. Direct-expansion evaporator—top feed.

Hubbell

REFRIGERATION AND AIR CONDITIONING TECHNOLOGY

70 psig. It meters defrost condensate to the suction line and accumulator.

DIRECT EXPANSION WITH BOTTOM HOT-GAS FEED

Compare the systems shown in Fig. 10-10 and Fig. 10-11. In the system shown in Fig. 10-11, the defrost hot gas is introduced into the bottom of the evaporator through the drain pan. The system operates similarly to that shown in Fig. 10-10. However, most of the liquid refrigerant is retained in the evaporators as defrost proceeds from the bottom to the top.

FLOODED LIQUID SYSTEMS

Figure 10-12 shows a flood-gas

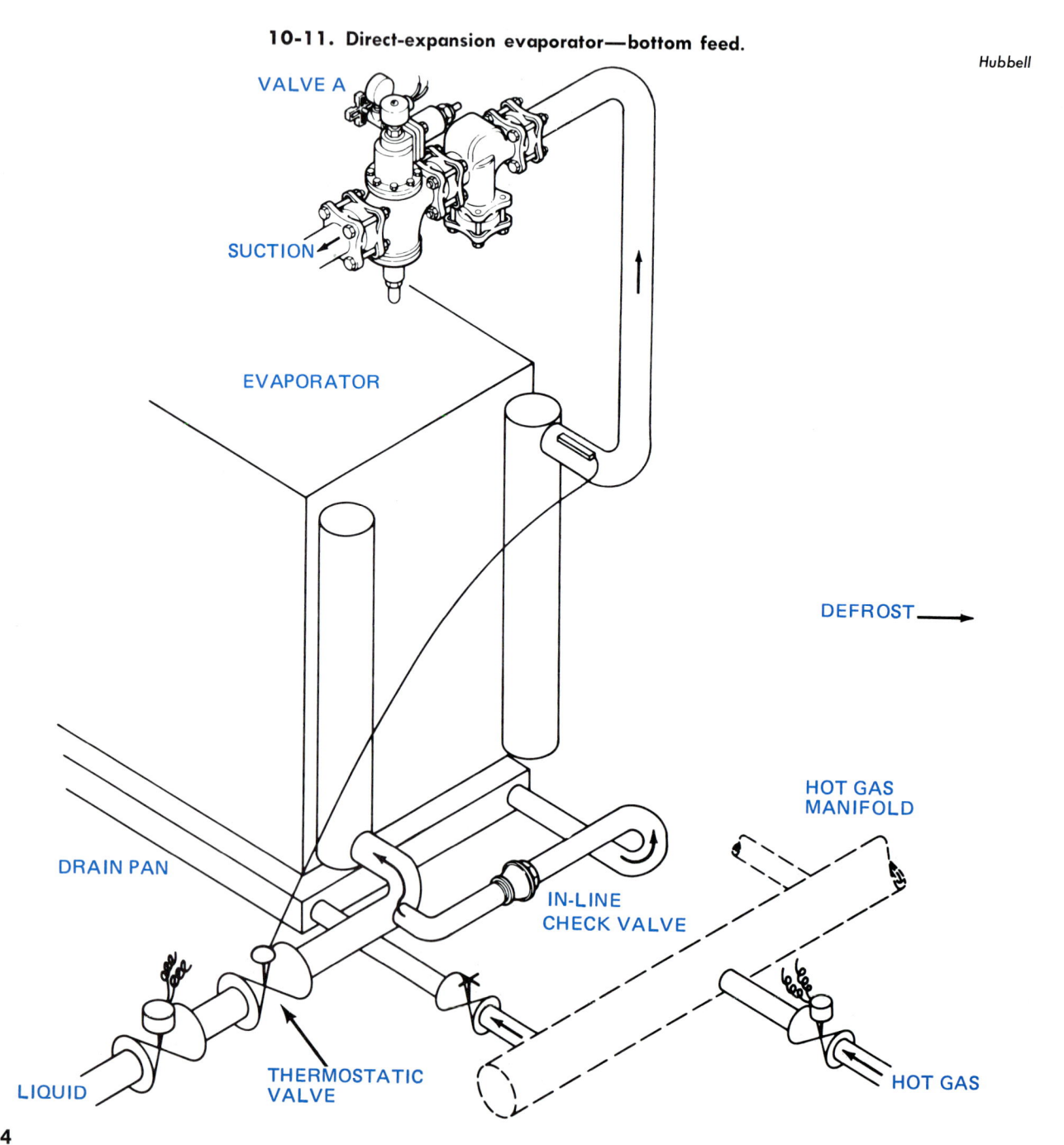

10-11. Direct-expansion evaporator—bottom feed.

Hubbell

CHAPTER 10—EVAPORATORS

and liquid leg shutoff (top hot-gas feed) system. Here, the gas-powered valve is used in both ends of the evaporator. It is a gas-powered check valve. At defrost, the normally closed type-A pilot solenoid is energized. Hot gas pressure closes the gas-powered check valves. Hot gas flows through the solenoid, globe valves, pan coil, and in-line check valve into the top of the evaporator. Here, it purges the evaporator of fluids. The evaporator is discharged at the metered rate through valve B, which has been de-energized and acts as a regulator during defrost.

At the end of the defrost cycle, excess pressure will bleed from the relief line at a safe rate through the energized valve B. The gas-powered valves will not open the evaporator to the surge drum until the gas pressure is nearly down to the system pressure.

FLOODED-GAS LEG SHUTOFF (BOTTOM HOT-GAS FEED)

The system shown in Fig. 10-13 is similar to that shown in Fig. 10-12. However, the liquid leg of the evaporator dumps directly into the surge drum without a relief valve. In this system, valve C is a defrost regulator. It is placed in the suction line, where it is normally open. During defrost, valve C is de-energized, converting to a defrost regulator. In such a system, it is recommended that a large-capacity surge drum or valve A be used as a bypass valve. This will bleed defrost pressure gradually around valve C into the suction line. Note how the in-line check

10-12. Gas and liquid leg shutoff—top feed.

Hubbell

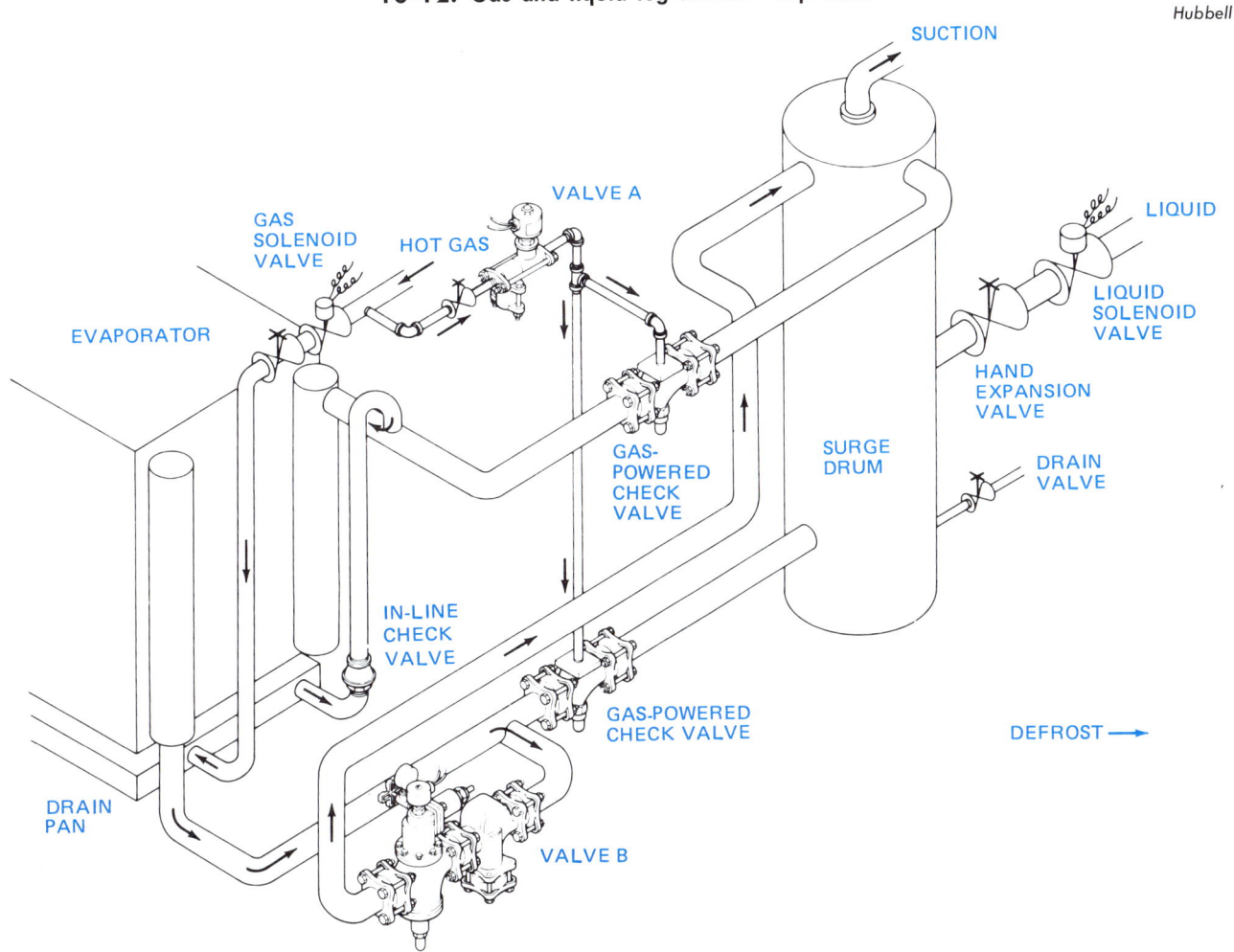

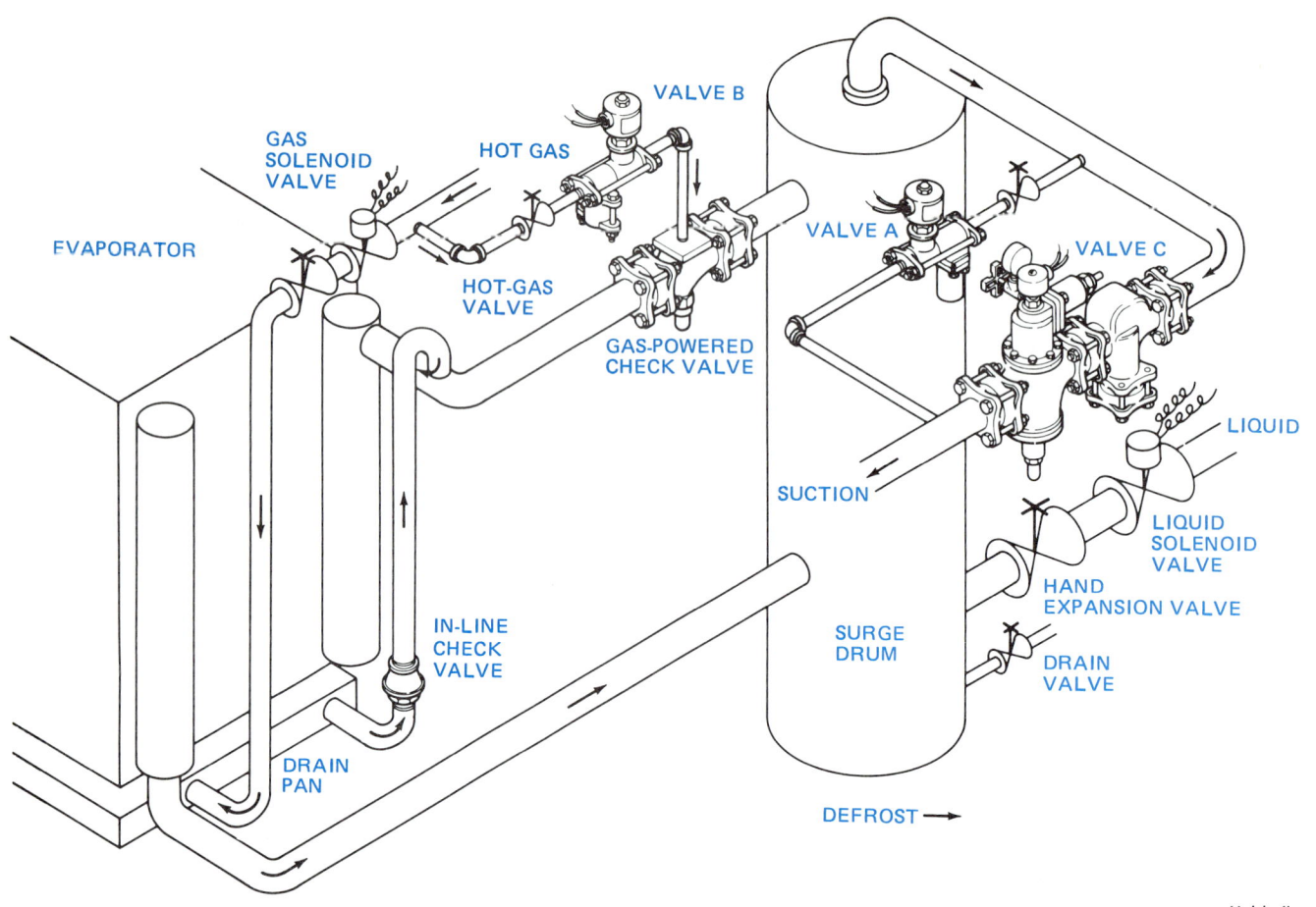

REFRIGERATION AND AIR CONDITIONING TECHNOLOGY

10-13. Gas leg shutoff—bottom feed.

Hubbell

valve is used to prevent cross flow.

FLOODED CEILING EVAPORATOR—LIQUID LEG SHUTOFF (BOTTOM HOT-GAS FEED)

Figure 10-14 illustrates a flooded ceiling evaporator. Upon initiation of the defrost sequence, the hot gas solenoid (Number 1) is opened. Gas flows to gas-powered check valve, isolating the bottom of the surge tank from the evaporator. The hot gas flows through the pan coil and the in-line check valve into the evaporator. Excess gas pressure is dumped into the surge tank. It will bleed through valve A. During defrost, this valve has been de-energized to perform as a relief regulator set at approximately 70 psig.

FLOODED CEILING EVAPORATOR—LIQUID LEG SHUTOFF (TOP HOT-GAS FEED)

Figure 10-15 shows a multiple flooded-evaporator system using input and output headers to connect the various evaporators to the surge drum. Note that, upon defrost, the fluid and condensate are purged from the evaporator and surge drum into the remote accumulator through the relief regulator, which is a reseating safety valve. This is usually set at about 70 psig. The accumulator must be sized to accept the refrigerant, plus hot gas condensate.

FLOODED CEILING BLOWER (TOP HOT-GAS FEED)

Figure 10-16 shows a modification of the system shown in

236

CHAPTER 10—EVAPORATORS

Fig. 10-15. In the system shown in Fig. 10-16, top-fed hot defrost gas forces the evaporator fluid directly to the bottom of the large surge drum. The defrost regulator (valve A), which is normally open, is de-energized during the defrost to act as a relief regulator. To minimize heating of the ammonia that accumulates in the surge drum during defrost, a thermostat bulb should be used to sense the temperature rise in the bottom header. This thermostat can be used to terminate the defrost cycle. Once again, the gas-powered check valve isolates the evaporator from the surge drum until the gas pressure is shut off.

FLOODED CEILING BLOWER (HOT-GAS FEED THROUGH SURGE DRUM)

Figure 10-17, on page 240, shows a simple defrost arrangement. It is a setup for the refrigeration system shown in Fig. 10-16. However, both the evaporator and surge drum are emptied during the defrost, necessitating the use of an ample suction accumulator to protect the compressor. In this system, the pilot solenoid valve in conjunction with the reverse-acting pressure regulator limits the system pressure. This permits the use of a simple solenoid valve and globe valve for rate control in the relief line.

FLOODED FLOOR-TYPE BLOWER (GAS AND LIQUID LEG SHUTOFF)

Figure 10-18, on page 241, illustrates a flooded floor unit suitable for operation down to $-70°$ F. $[-57°C]$.

The gas-pressure-powered valve used in this circuit has a solenoid pilot operator. This pro-

10-14. Ceiling evaporator, liquid leg shutoff—bottom feed.

Hubbell

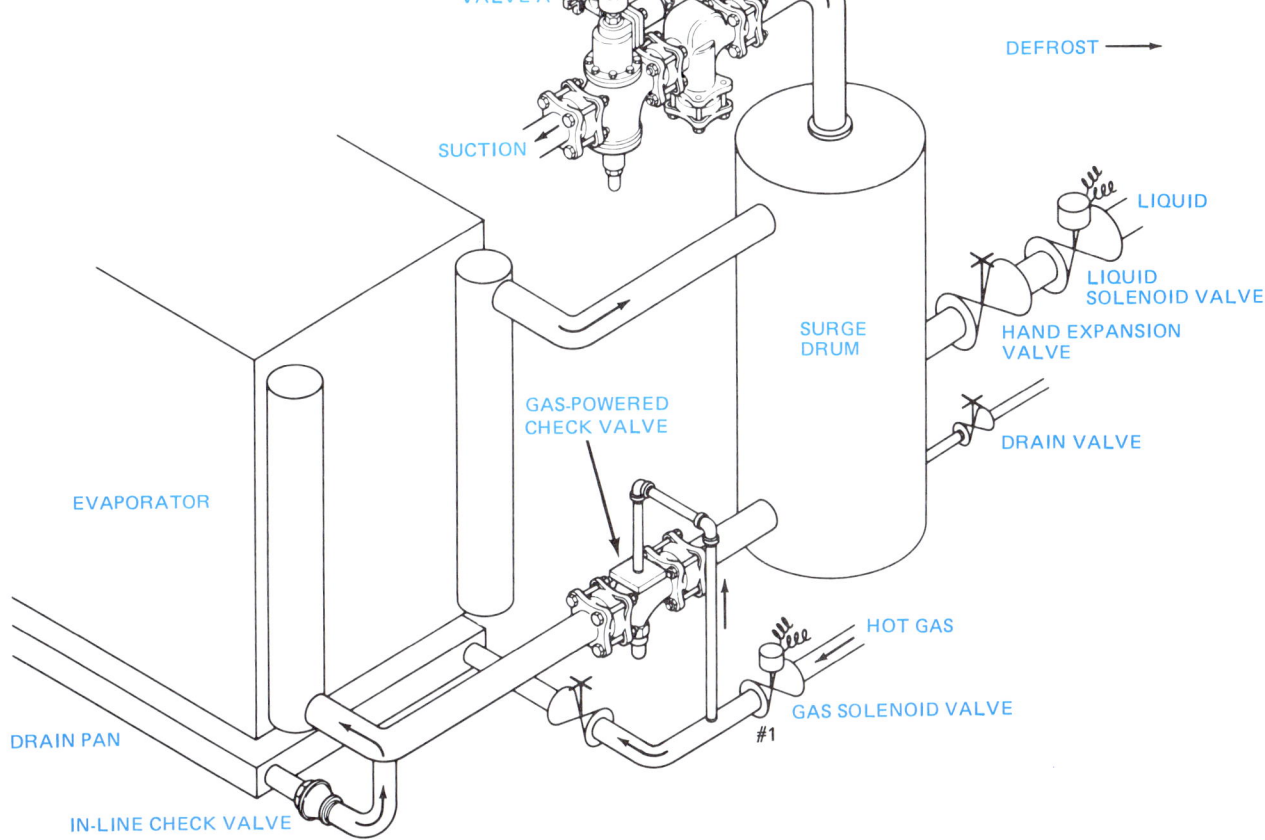

237

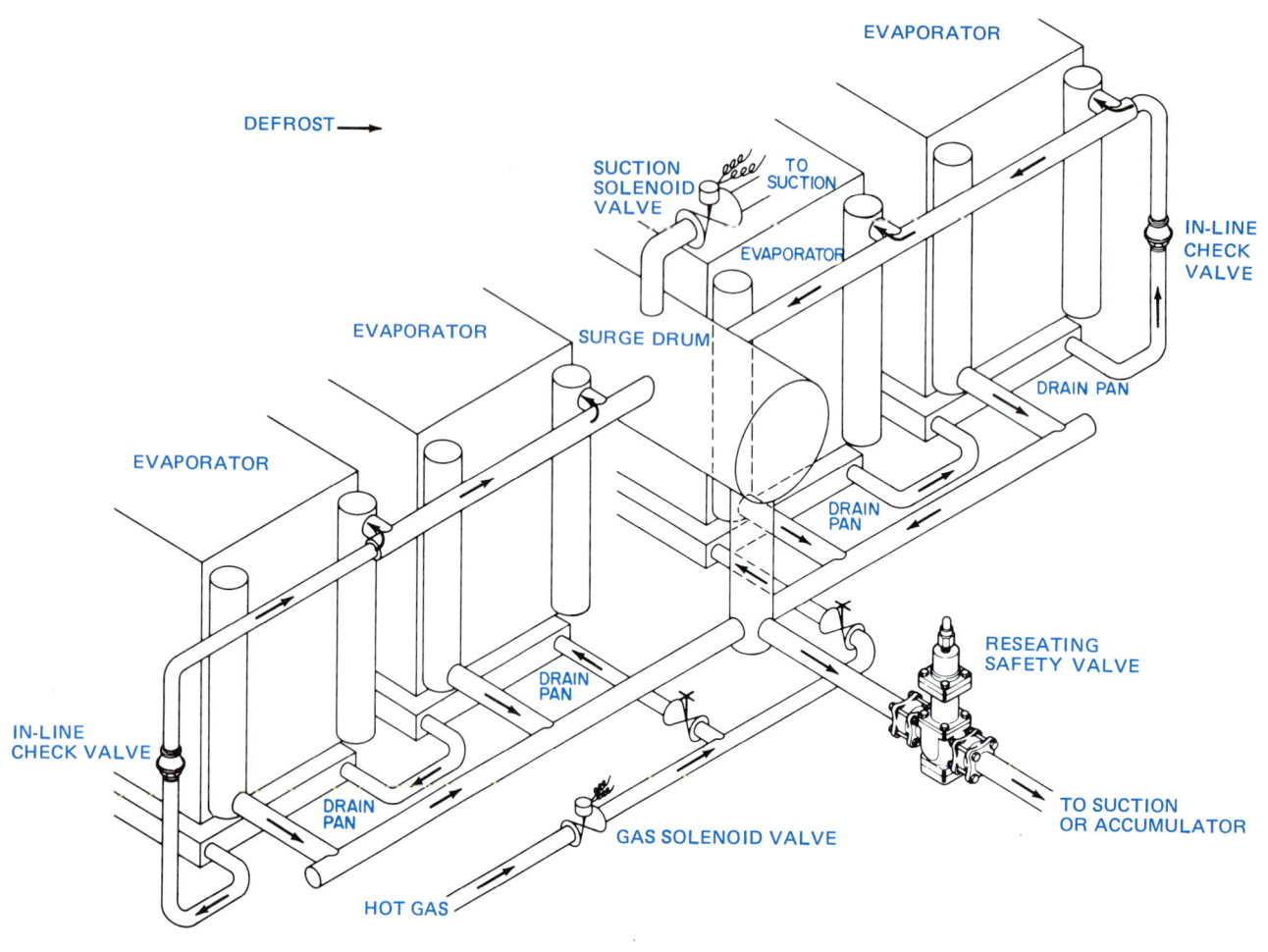

REFRIGERATION AND AIR CONDITIONING TECHNOLOGY

10-15. Ceiling evaporator, liquid leg shutoff—top feed.

Hubbell

vides positive action with gas or liquid loads at high or low temperatures and pressures.

To defrost a group of evaporators without affecting the temperatures of the common surge drum, the gas-powered valve is used in each end of the evaporator. A reseating safety valve is a relief regulator. It controls the defrost pressure to the relief line accumulator. A check valve prevents back flow into the relief line. The in-line check valve prevents crossover between adjacent evaporators.

At high temperatures (above −25° F. [−31°C]), use of the gas-powered check valve in place of the gas-powered solenoid valve is recommended.

FLOODED FLOOR-TYPE BLOWER (GAS LEG SHUTOFF)

The system shown in Fig. 10-19, on page 242, is similar to that shown in Fig. 10-18. However, a single gas-type, pressure-powered valve is used.

Overpressure at the surge drum is relieved by valve B, a defrost-relief regulator. This is normally wide open. It becomes a regulating valve when its solenoid is de-energized during defrost.

Defrost gas flows through the hot gas solenoid when energized. It then flows through the globe

CHAPTER 10—EVAPORATORS

10-16. Ceiling blower—top feed.

Hubbell

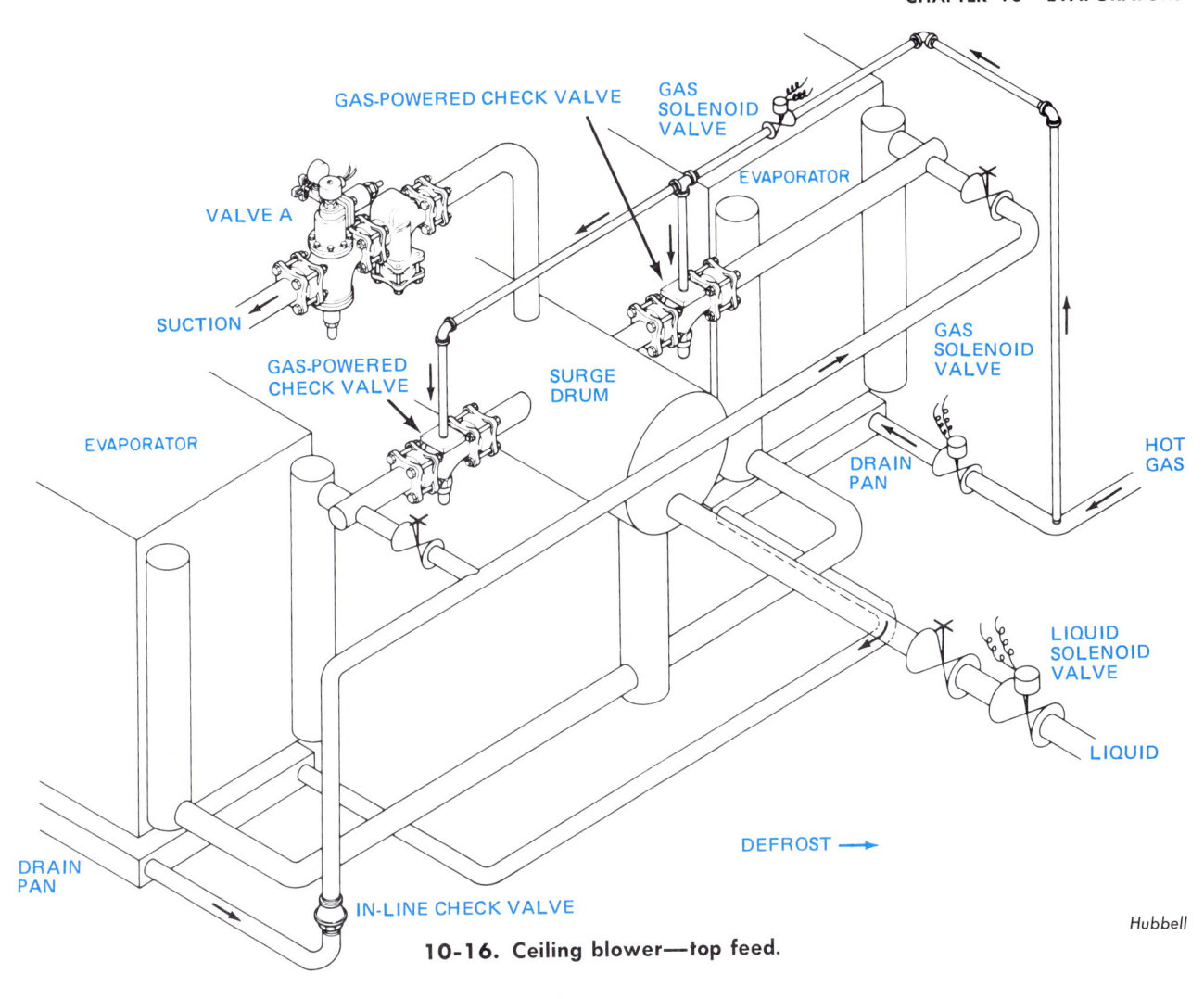

valve and the in-line check valve to force the evaporator fluid into the surge drum.

An optional hot-gas thermostat bulb may be used to sense heating of the bottom of the evaporator. Thus, it can act as a backup for the timed defrost cycle.

LIQUID RECIRCULATING SYSTEMS

Liquid refrigerant recirculating systems are frequently fed by liquid flow upward through their evaporators. These systems are called bottom fed. This is accomplished by either mechanical or gas displacement recirculators during the refrigerant cycle. Fig. 10-20 on page 243.

In some systems, more than a single evaporator is fed from the same recirculator, as shown in Fig. 10-20. Then, a proper distribution of liquid between evaporators must be maintained to achieve efficient operation of each evaporator. This balance is usually accomplished by the insertion of adjustable globe valves or orifices into the liquid feeder lines. Similarly, adjustment of the globe valves or insertion of orifices is also often used properly to distribute hot gas during the defrost cycle.

Equalizing orifices or globe valves are not used if the hot gas

239

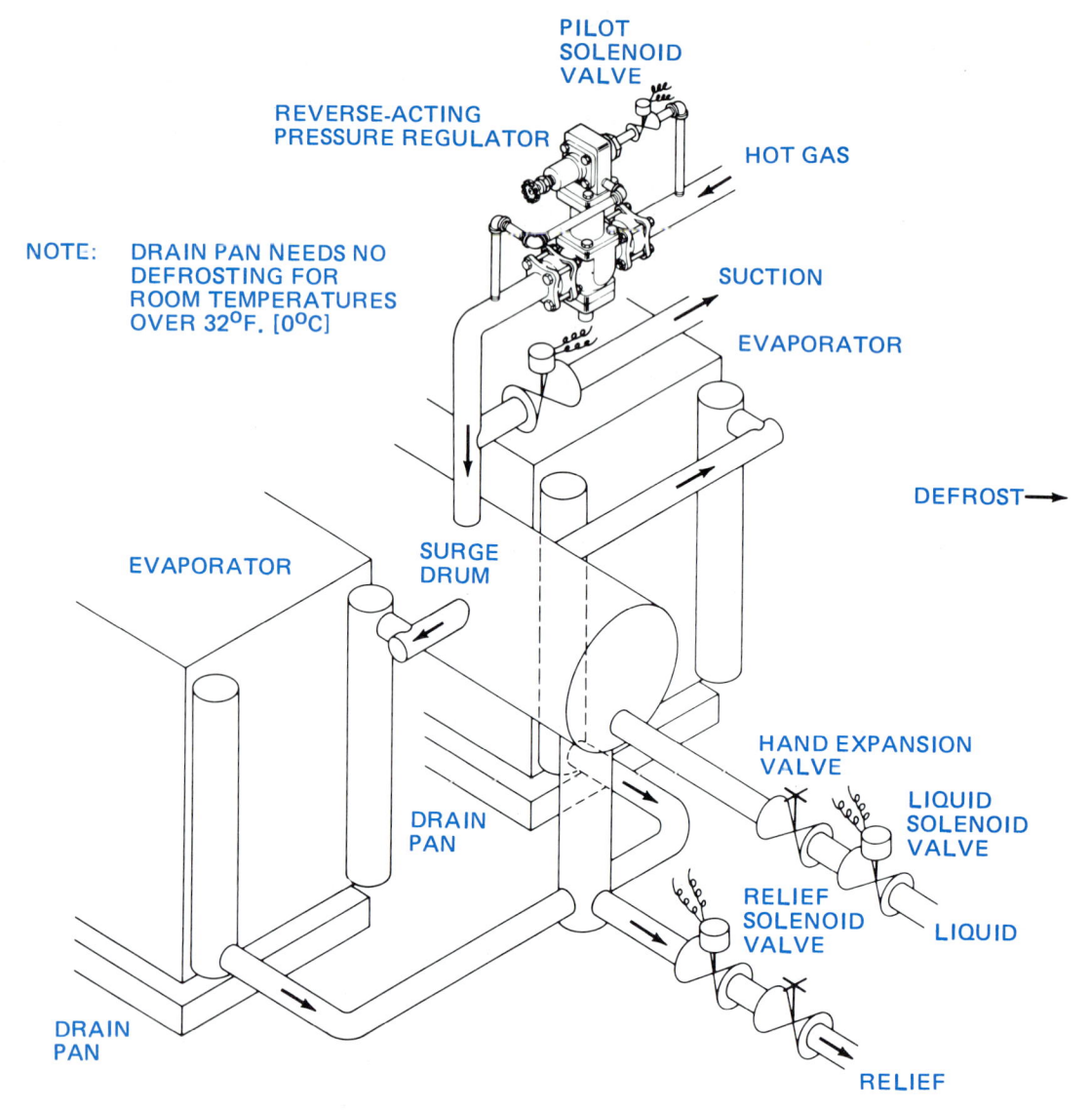

REFRIGERATION AND AIR CONDITIONING TECHNOLOGY

10-17. Ceiling blower—feed through surge drum.

Hubbell

used for defrosting is fed to the bottom of the evaporators as shown in Fig. 10-20. In such cases, most of the hot gas could flow through the circuits nearest the hot gas supply line. The same would also happen in circuits where both vertical and horizontal headers are used, as in Fig. 10-21. The more remote circuits could remain full of cold liquid. Consequently, they would not defrost.

Supplying hot gas to the top of the evaporator forces liquid refrigerant down through the evaporator and out through a reseating safety valve relief regulator into the suction line return to the accumulator. Fig. 10-21 on page 244. Reseating safety valve relief regulators are usually set to relieve at 60 to 80 psig to provide rapid defrost.

The use of check valves is important in flooded liquid recirculating systems fed by mechanical

CHAPTER 10—EVAPORATORS

gas displacement liquid recirculators. The check valves are used where the pressure of the hot gas used for defrost is higher than the system pressure. The reseating safety check valve must be used to stop this gas at high pressure from flowing back into the liquid supply line.

FLOODED RECIRCULATOR (BOTTOM HOT-GAS FEED)

The multiple system shown in Fig. 10-20 shows the check valve mounted in each of the liquid refrigerant branch lines. A single solenoid valve is used in the main refrigerant line. The defrost gas is bottom fed.

FLOODED RECIRCULATOR (TOP HOT-GAS FEED)

The system illustrated in Fig. 10-21 shows a check valve mounted directly at the outlet of each of the liquid solenoid valves. The defrost gas is top fed. This system permits selective defrosting of each evaporator. A single accumulator is used to protect the compressor during defrost, as well as to accumulate both liquid refrigerant and defrost condensate. This protection is accomplished by using a differential pressure regulator valve in an evaporator bypass circuit.

The differential pressure regulator valve will open sufficiently to relieve excess pressure across the compressor inlet. The pressure will discharge as this excess pressure differential occurs. When the pressure differential is less than the regulator valve set-

10-18. Floor blower—gas and liquid leg shutoff.

Hubbell

REFRIGERATION AND AIR CONDITIONING TECHNOLOGY

10-19. Floor blower—gas leg shutoff.

Hubbell

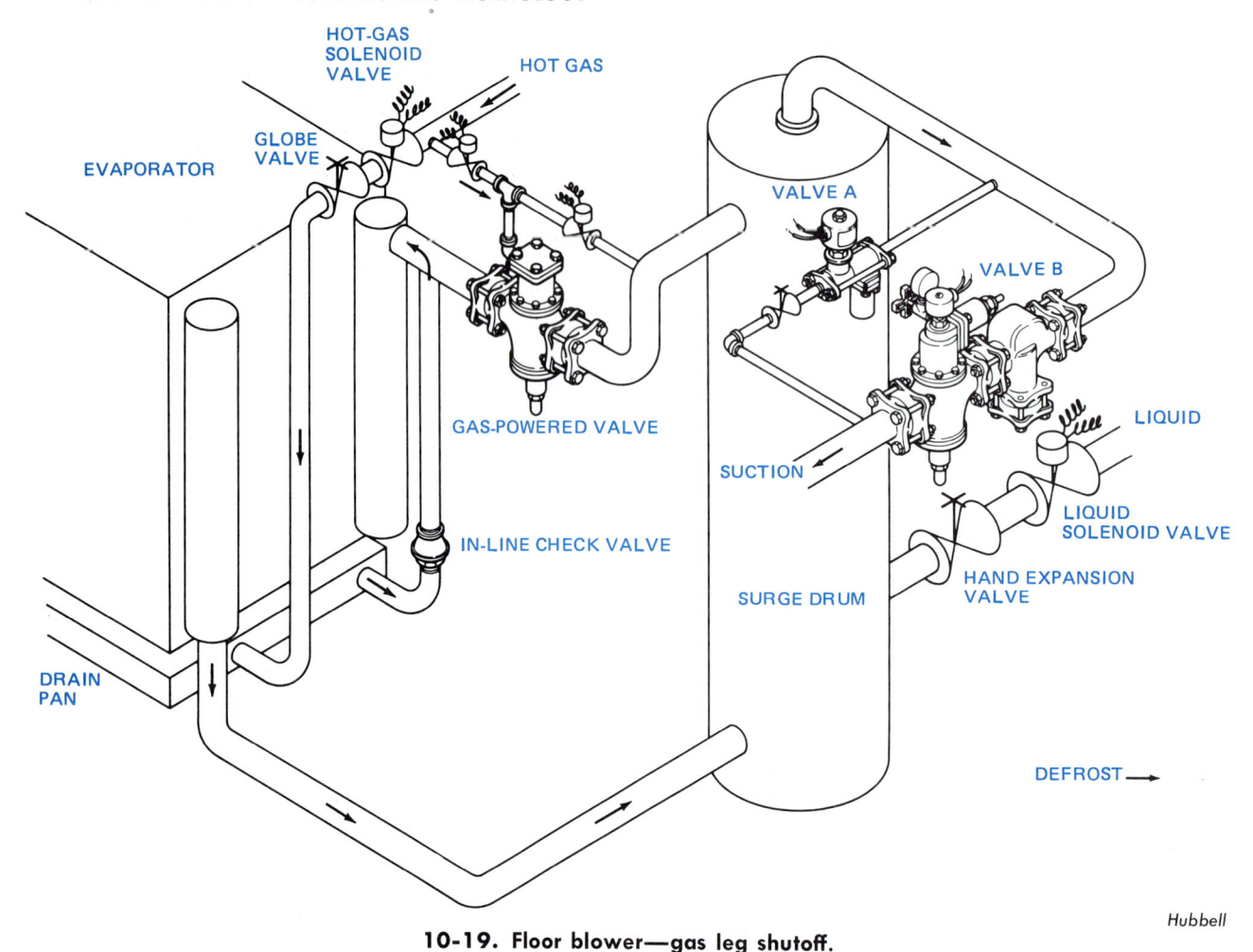

ting, the regulator will be tightly closed.

LOW-TEMPERATURE CEILING BLOWER

The low-temperature liquid recirculating system illustrated in Fig. 10-22 uses several controls. During the cooling cycle, #1 pilot valve is opened and #2 pilot valve is closed, holding the gas-powered solenoid valve wide open. This allows flow of liquid through the energized liquid solenoid valve from the recirculator and then through the circuit of the unit. The in-line check valve installed between the drain pan coil header and suction line prevents drainage of liquid into the drain pan coil.

For defrost, the liquid solenoid valve is closed. The #1 pilot solenoid is de-energized. The #2 solenoid is opened, closing the gas-powered solenoid valve tightly. The hot gas solenoid is energized. This allows distribution of the hot gas through the drain pan coils, the in-line check valve, the top of the suction header, and the coil. The gas comes out the bottom of the liquid header.

Check valve A prevents the flow of the high pressure gas into the liquid line. Therefore, the gas is relieved through the safety valve relief regulator (B). This is set to maintain pressure in the evaporator to promote rapid, efficient defrost.

YEAR-ROUND-AUTOMATIC CONSTANT LIQUID PRESSURE CONTROL SYSTEM

The constant liquid control system is a means of increasing

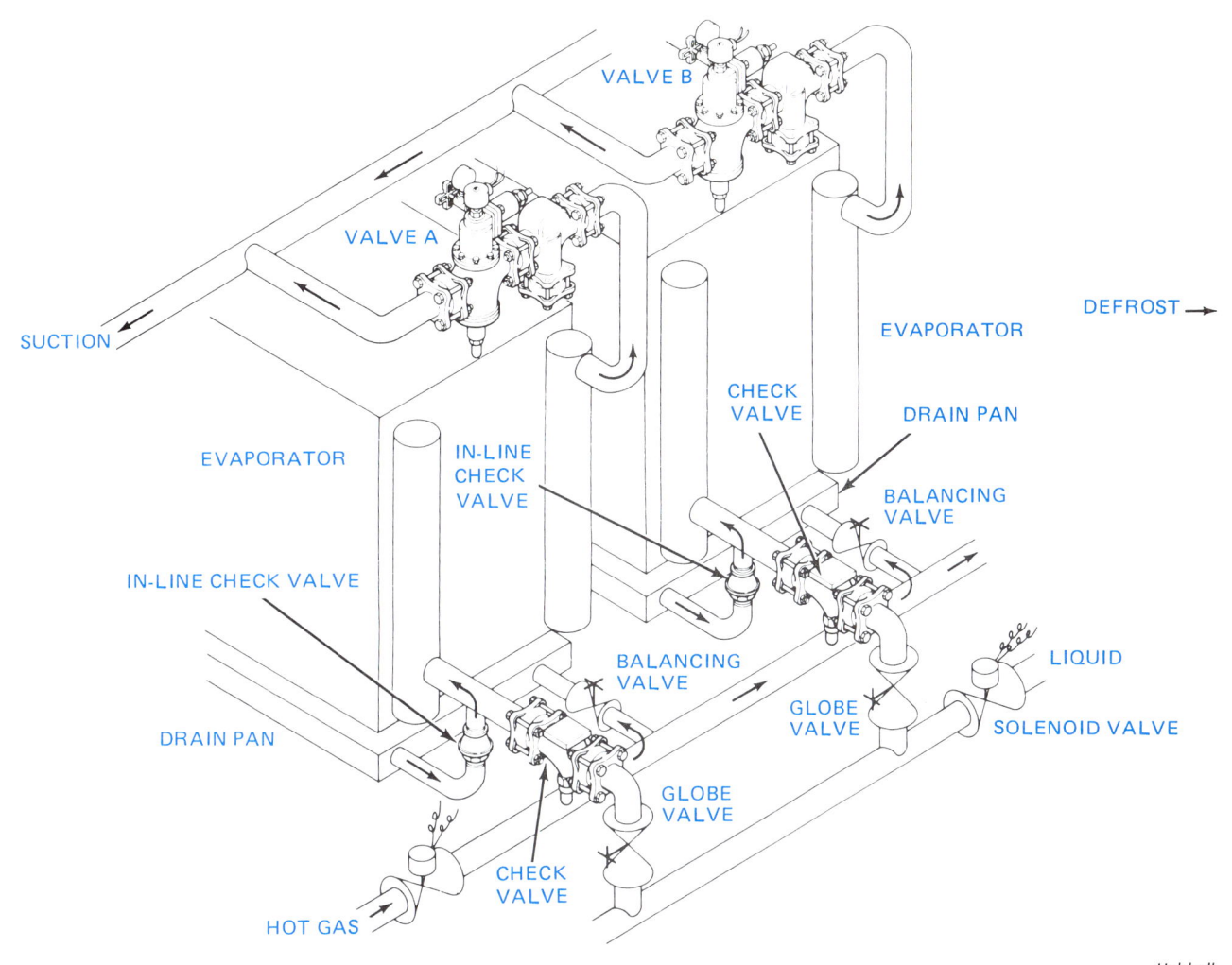

10-20. Flooded recirculator—bottom feed.

the efficiency of a refrigeration system that utilizes air-cooled, atmospheric, or evaporative condensers. Fig. 10-23, page 246.

This is accomplished by automatically maintaining a constant liquid pressure throughout the year to assure efficient operation. Constant liquid pressure on thermal expansion valves, float controls, and other expansion devices results in efficient low side operation. Hot gas defrosting, liquid recirculation, or other refrigerant control systems require constant liquid pressure for successful operation. Liquid pressure is reduced by cold weather and extremely low wet-bulb temperatures with low refrigeration loads.

To compensate for a decrease in liquid measure, it is necessary automatically to throttle the discharge to a predetermined point and regulate the flow of discharge pressure to the liquid line coming from the condenser and going to the receiver. Thus, predetermined pressure is applied to the top of the liquid in the receiver. The constant liquid pressure control does this. In addition, when the compressor "start and stop" is controlled by pressure-stats, the pressure-operated hot-gas flow control valve is a tight closing stop valve during stop periods. This permits efficient "start and stop" operation of the compressor by pressure control of the low side.

243

REFRIGERATION AND AIR CONDITIONING TECHNOLOGY

The three valves in the system shown in Fig. 10-23 are the reverse-acting pressure regulator, the pressure-operated hot-gas flow-control valve, and the relief check valve. The function of the control system is to maintain a constant liquid pressure (A). This is accomplished by the reverse-acting pressure regulator valve, which is a modulating-type valve. It maintains a constant predetermined pressure on the downstream side of the regulator.

To maintain a constant pressure (A) it is necessary to maintain a discharge pressure (B) approximately 5 psi above (A). This is accomplished by the hot-gas control valve, which will maintain a constant pressure (B) on the upstream or inlet side of the regulator. Due to the design of the regulator, a constant supply of gas will be available at a predetermined pressure to supply the pressure regulator to maintain pressure A. Excess hot gas is not required to maintain A fill flow into the condenser.

The relief check valve prevents pressure A from causing backflow into the condenser. When the compressor shuts down, the hot-gas flow-control valve closes tightly and shuts off the discharge line. This prevents gas from flowing into the condenser.

10-21. Flooded recirculator—top feed.

Hubbell

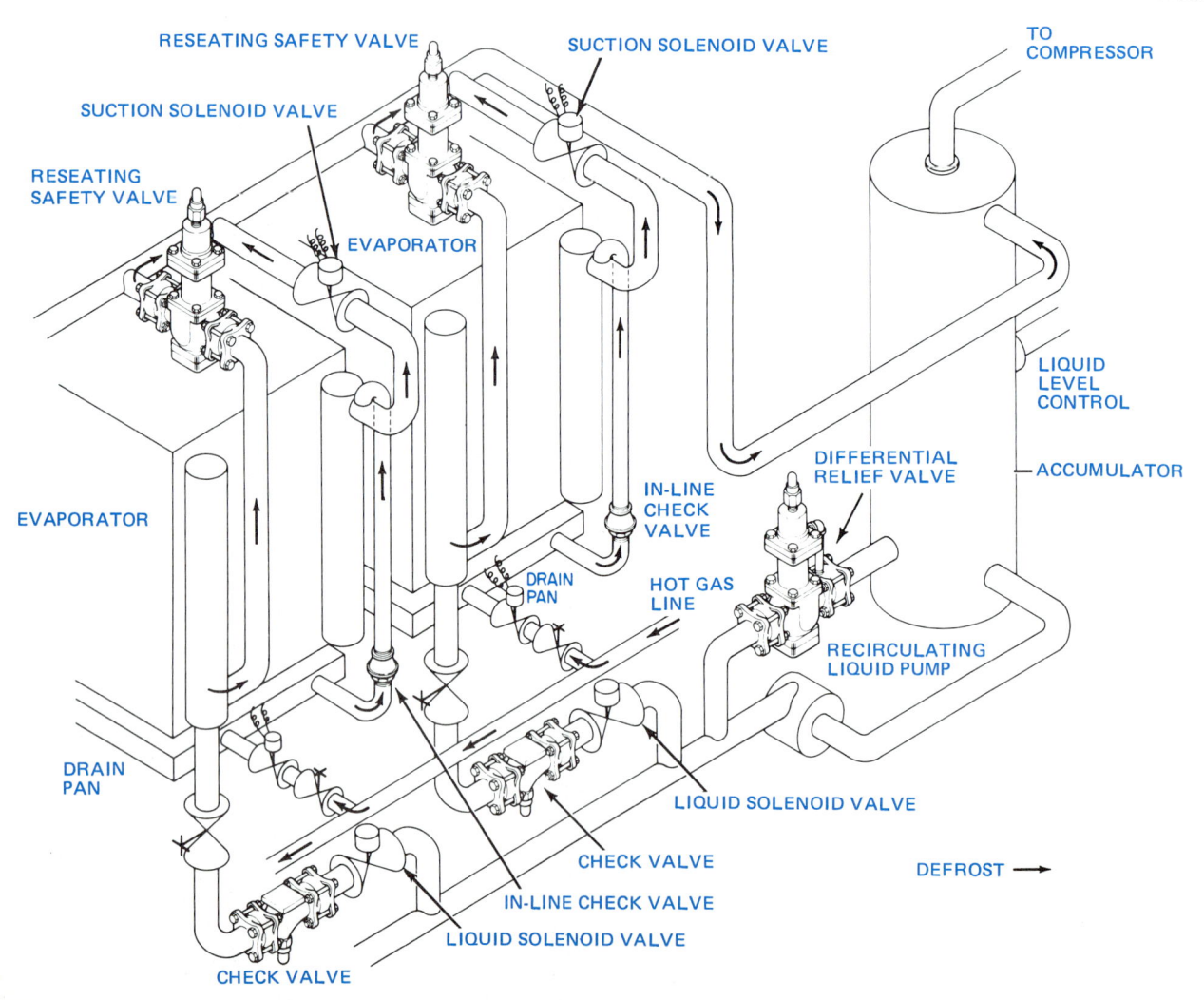

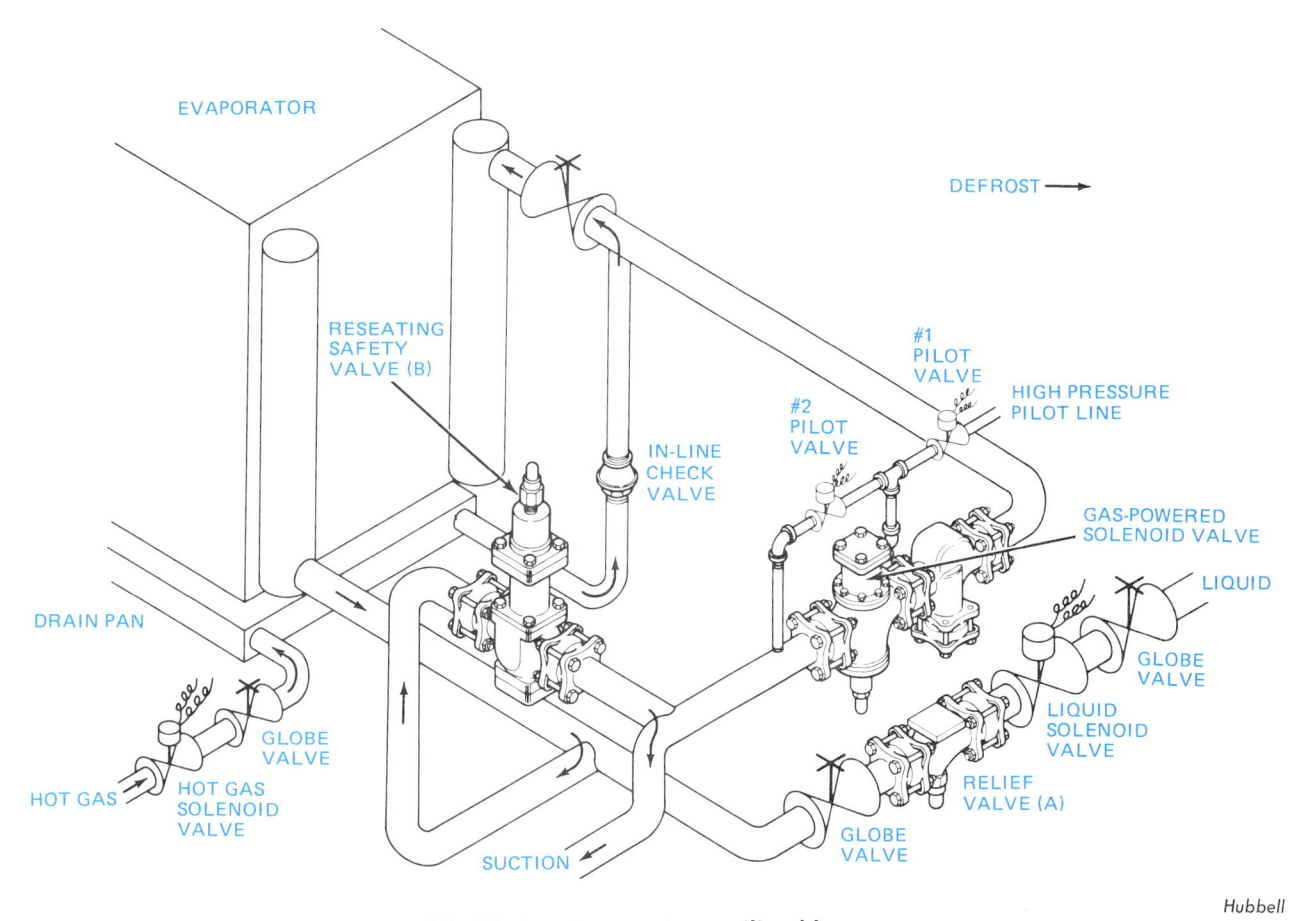

CHAPTER 10—EVAPORATORS

10-22. Low-temperature ceiling blower.

Hubbell

The check valve actually prevents the back flow of liquid into the condenser. Thus, liquid cannot back up into the condenser in extremely cold weather. Sufficient low-side pressure will be maintained to start the compressor when refrigeration is required.

DUAL-PRESSURE REGULATOR

A dual-pressure regulator is shown in Fig. 10-24. It is used on a shell-and-tube cooler. The dual-pressure regulator is particularly adaptable for the control of shell-and-tube brine or water coolers, which at intervals may be subjected to increased loads. Such an arrangement is shown in Fig. 10-24.

The high-pressure diaphragm is set at a suction pressure suitable for the normal load. The low-pressure diaphragm is set for a refrigerant temperature low enough to take care of any intermittent additional loads on the cooler.

In this case, the transfer between low and high pressure is effected by a thermostat. The remote bulb of the thermostat is located in the water or brine line leaving the cooler. A temperature increase at this bulb indicating an increase in load will cause the thermostat to open the electric pilot and transfer control of the cooler to the low temperature diaphragm. Upon removal of the excess load, the thermostat will cause the electric pilot to close the low pressure port. The cooler is then automatically transferred to the normal pressure for which the high pressure diaphragm is set. The diaphragms may be set at any two evaporator pressures at which it is desirable to operate. Any electric switching device responsive to load change may be used to change from one evaporator pressure to the other.

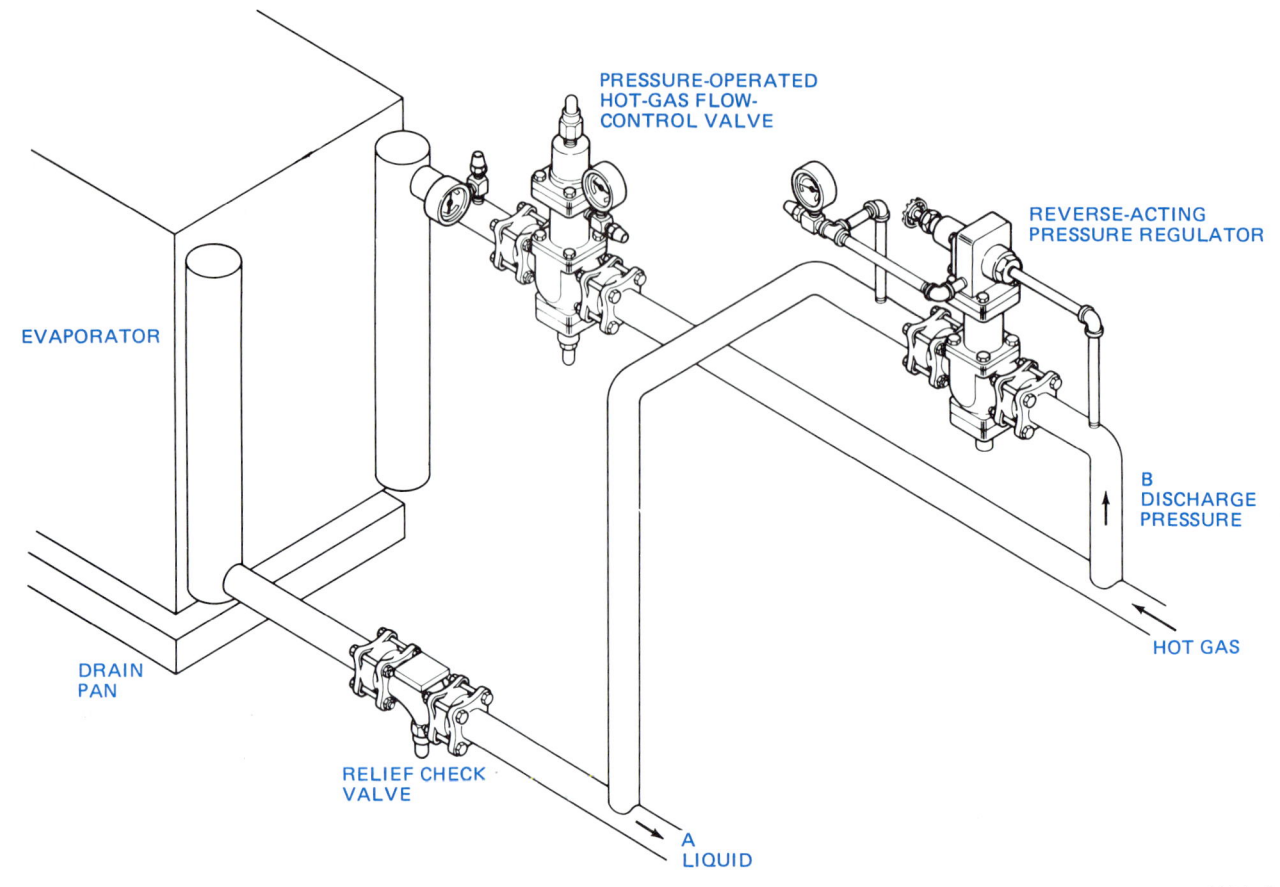

REFRIGERATION AND AIR CONDITIONING TECHNOLOGY

10-23. Year-round automatic control system.

Hubbell

VALVES AND CONTROLS FOR HOT-GAS DEFROST OF AMMONIA-TYPE EVAPORATORS

The following valves and controls are used in the hot-gas defrost systems of ammonia-type evaporators:

• *Hot-gas or pilot solenoid valve.* The valve is a 1/8" ported solenoid valve. It is a direct-operated valve suitable as a liquid, suction, hot gas, or pilot valve at pressures to 300 pounds.

• *Suction, liquid, or gas solenoid valve.* The suction solenoid valve is a one-piston, pilot-operated valve suitable for suction, liquid, or gas lines at pressures to 300 pounds. It is available with a 9/16" or 3/4" port.

• *Pilot-operated solenoid valve.* The valve is a one-piston, pilot-operated solenoid valve used as a positive stop valve for applications above −30° F. [−34°C] on gas or liquid.

• *Pilot-operated two-piston valve.* The solenoid valve is a rugged, pilot-operated, two-piston valve with spring return for positive closing under the most adverse conditions. It is used for compressor unloader, suction, liquid, and hot gas applications.

• *Gas-powered solenoid valve.* The gas-powered solenoid valve is a power-piston type of valve that uses high pressure to force the valve open through the control of pilot valves. Because of the high power available to open these valves, heavy springs may be used to close the valves positively at temperatures down to −90° F. [−68°C].

• *Dual-pressure regulator valve.* The dual-pressure regulator

246

CHAPTER 10—EVAPORATORS

valve is designed to operate at two predetermined pressures without resetting or adjustment. By merely opening and closing a pilot solenoid, either the low or high pressure setting is maintained.

● *Reseating safety valve.* The reseating safety valve is generally used as a relief regulator to maintain a predetermined system pressure. The pressure maintained by the valve is adjustable manually.

● *Back-pressure regulator arranged for full capacity.* The back-pressure regulator is normally used where pressure control of the evaporator is not required—as in a direct expansion system. A pilot solenoid is energized, allowing pressure to bypass the sensing chamber of the regulator holding the valve wide open. De-energizing the pilot valve allows the valve to revert to its function as a back-pressure regulator maintaining a preset pressure upstream of the valve. The valve performs both as a suction solenoid and as a relief regulator.

● *Differential relief valve.* The differential relief valve is a modulating regulator for liquid or gas use. It will maintain a constant preset pressure differential between the upstream and downstream side of a regulator.

● *Reverse-acting pressure regulator.* The reverse-acting pressure regulator is used to maintain a constant predetermined pressure downstream of the valve. When complete shutoff of the

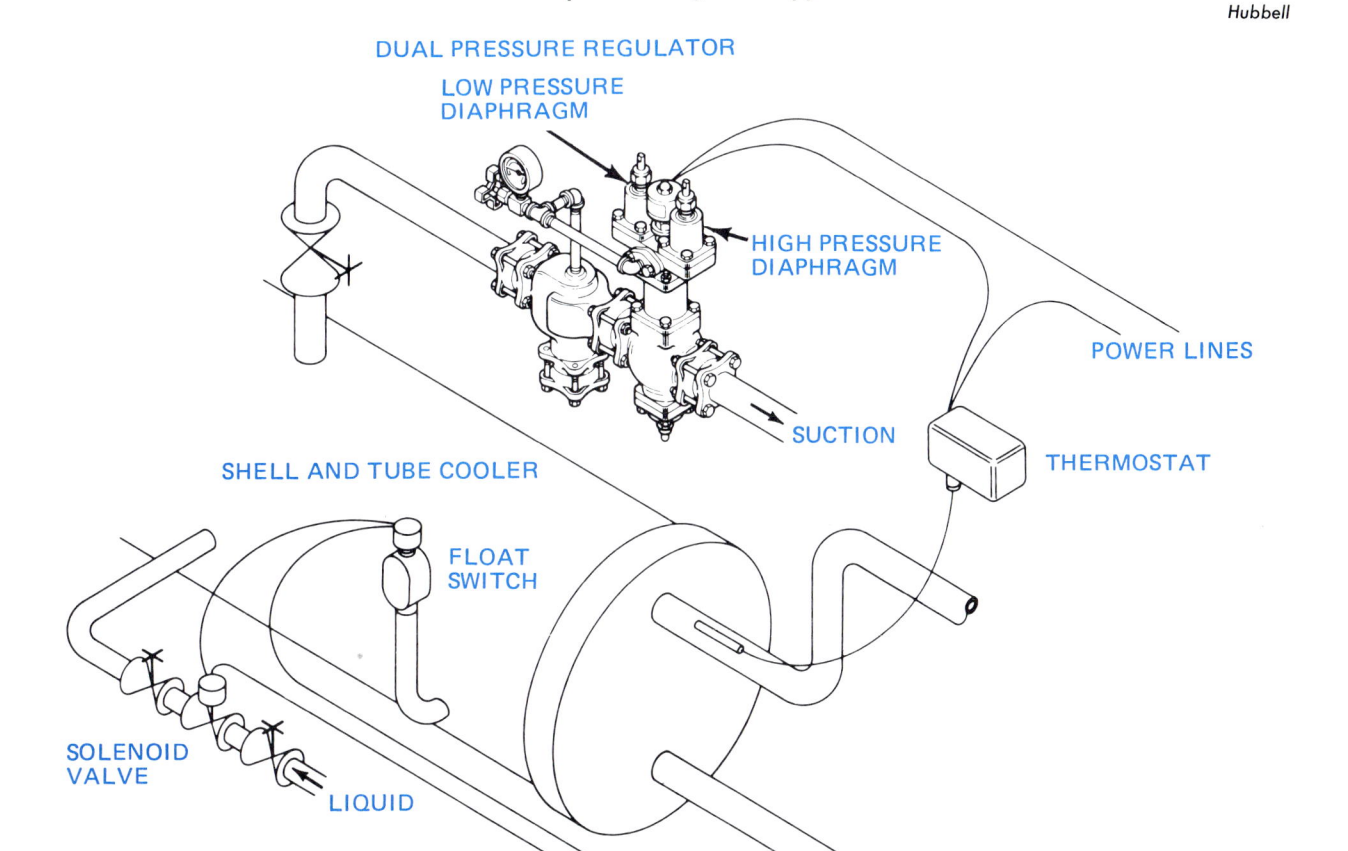

10-24. Dual-pressure regulator application.

Hubbell

247

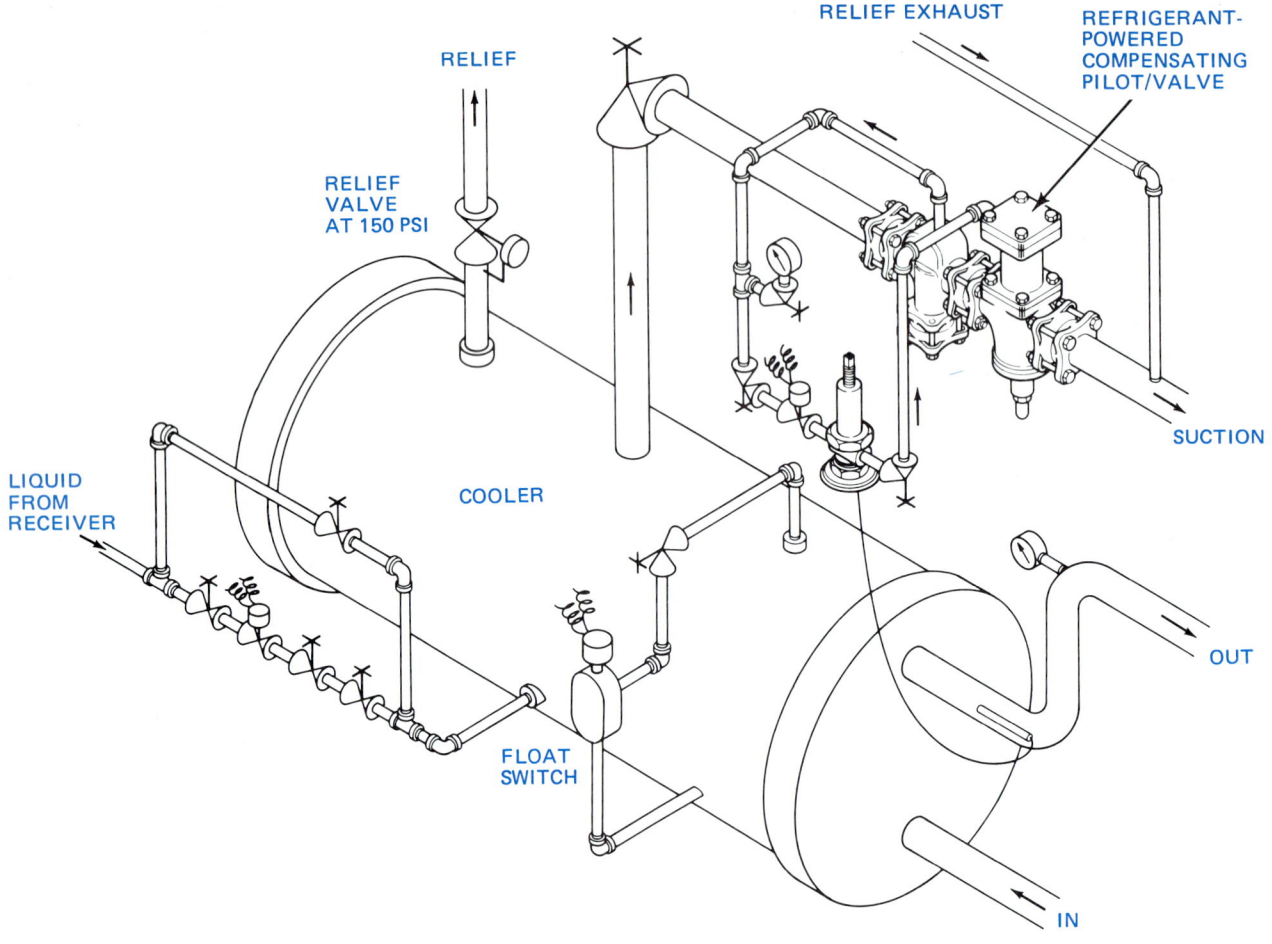

REFRIGERATION AND AIR CONDITIONING TECHNOLOGY

10-25. Thermal-compensating back-pressure regulator.

regulator is required, a pilot valve is installed in the upstream feeder line. When the solenoid valve is closed, the regulator closes tightly. When the solenoid valve is open, the regulator is free to operate as the pressure demands. With the solenoid installed as described above, this becomes a combination reverse-acting regulator and stop valve.

● *Gas-powered check valve.* The gas-powered check valve is held in a normally open position by a strong spring. Gas pressure applied at the top of the valve closes the valve positively against the high system pressures. A manual opening stem is standard.

● *Check valve.* The check valve is a spring-loaded positive check valve with manual opening stem. It is used to prevent backup of relatively high pressure into lower pressure lines.

● *In-line check valve.* The in-line check valve is used in multiple-branch liquid lines fed by a single solenoid valve. This check valve prevents circulation between evaporators during refrigeration.

The in-line check valve is also used between drain pans and evaporators to prevent frosting of the drain pan during refrigeration.

These valves and controls are necessary. They cause defrosting

248

operations to take place in large evaporators used in commercial jobs. Some manufacturing operations also call for large-capacity refrigeration equipment.

BACK-PRESSURE REGULATOR APPLICATIONS OF CONTROLS

In a refrigeration system designed to maintain a predetermined temperature at full load, any decrease in load would tend to lower below full load temperature the temperature of the medium being cooled.

To maintain constant temperatures in applications having varying loads, means must be provided to change refrigerant temperature to meet varying load requirements.

Refrigerant temperature is a function of evaporator pressure. Thus, the most direct means of changing refrigerant temperature to meet varying load requirements is to vary the system pressure. This variation of system pressure is accomplished by adjusting the setting of a back pressure regulator.

A number of back pressure valve controls are available. Some of them are shown in the following:

Refrigerant-Powered Compensating-Type Pilot Valve

The upper portion of the valve head is similar to a standard pressure regulating head. On the lower portion of the head another diaphragm is connected to the main diaphragm by a push rod. As the thermal bulb warms, the liquid in it expands, pushing up on the rod and opening the regulator. Because this is accomplished by an outside power source, the pressure drop through the head is reduced considerably. The valve head will function in connection with the regulator on a $\frac{1}{2}$- to $\frac{3}{4}$-pound overall pressure drop. The point at which the modulation or compensation takes place may be adjusted by turning the adjusting stem. By turning the stem in, the product temperature is increased. By turning the stem out, the product temperature is decreased. The back pressure valve will remain wide open, taking advantage of the line suction pressure until the product being cooled approaches the temperature at which modulation is to begin. The valve head will hold the temperature of the product to within $\pm\frac{1}{2}°$ F. [0.28°C] of the desired temperature. In the case of failure of the thermal element, the valve head can be used as a straight back-pressure valve by readjusting it to the predetermined suction pressure at which you desire the system to operate. Fig. 10-25.

Air-Compensating Back-Pressure Regulator

A standard regulator is reset by manually turning the adjusting stem, which increases the spring pressure on top of the diaphragm. In an air-compensated regulator, a change of pressure on top of the diaphragm is accomplished by introducing air pressure into the airtight bonnet over the diaphragm. As this air pressure is increased, the setting of the regulator will be increased. This will produce like changes of evaporator pressure and refrigerant temperature. The variations in air pressure are produced by the temperature changes of the thermostatic remote bulb placed in the stream of the medium being cooled as it leaves the evaporator.

Temperature changes in the medium being cooled over the remote bulb of the thermostat will cause the thermostat to produce air pressures in the regulator bonnet within a range of 0 to 15 pounds. This will cause the regulator to change the evaporator suction pressure in a like amount. A more definite understanding of this operation is obtained by assuming certain working conditions for the purpose of illustration. In cases where a larger range of modulation is required, a three-to-one air relay may be installed. This will permit a 45-pound range of modulation. Fig. 10-26.

Electric-Compensating Back-Pressure Regulator

A standard regulator is reset manually by turning the adjusting stem, usually found at the top of the regulator. In an electrically compensated regulator, turning the stem to obtain different refrigerant pressures and temperatures in the evaporator is accomplished by a small electric motor. This motor rotates the adjusting stem in accordance with temperature variations in a thermostatic bulb placed in the medium being cooled as it leaves the evaporator. The adjusting stem, spring, and controlling diaphragm have been separated from their positions at the top of the regulator. They have been placed in a small remote unit mounted on a com-

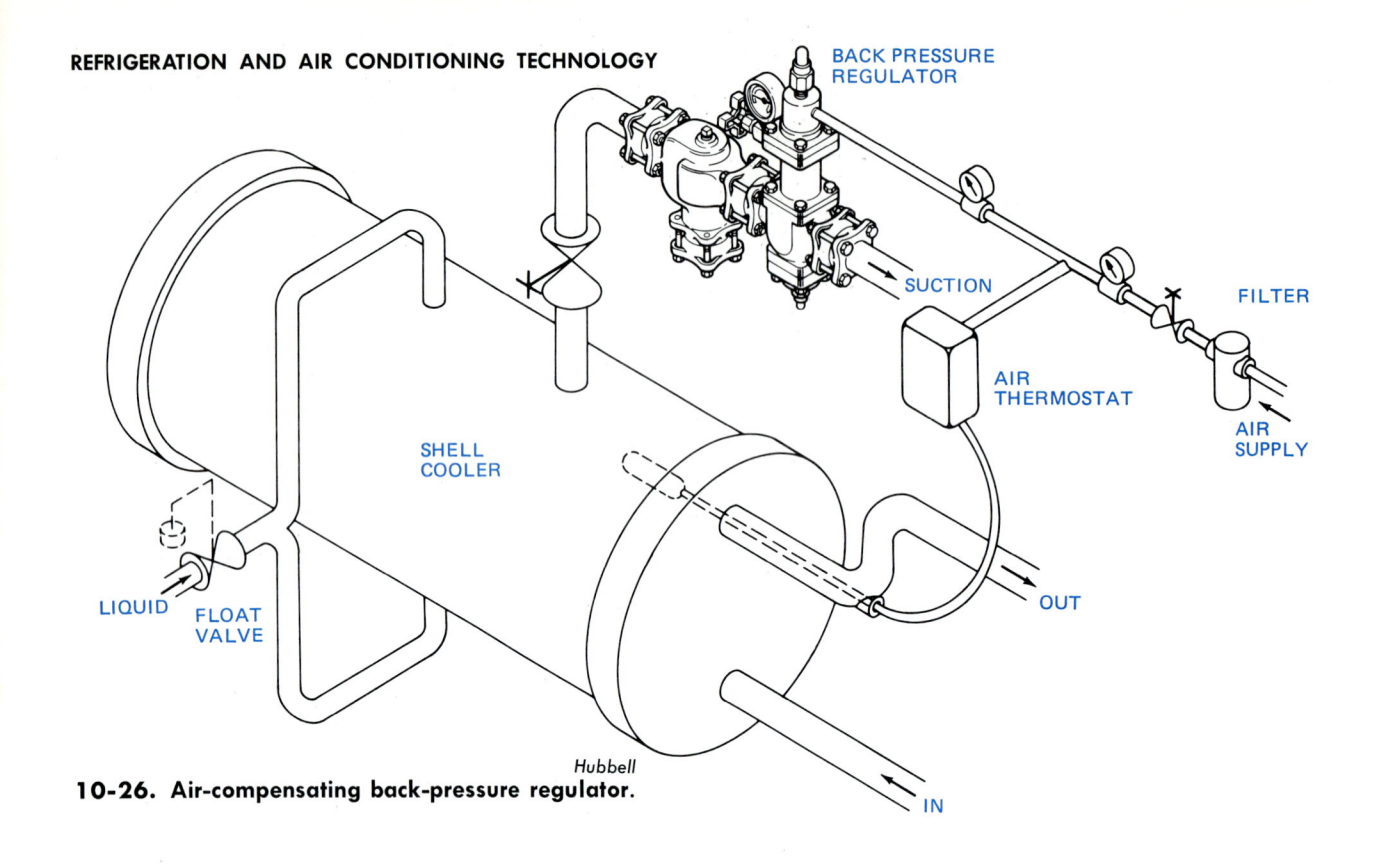

REFRIGERATION AND AIR CONDITIONING TECHNOLOGY

10-26. Air-compensating back-pressure regulator.

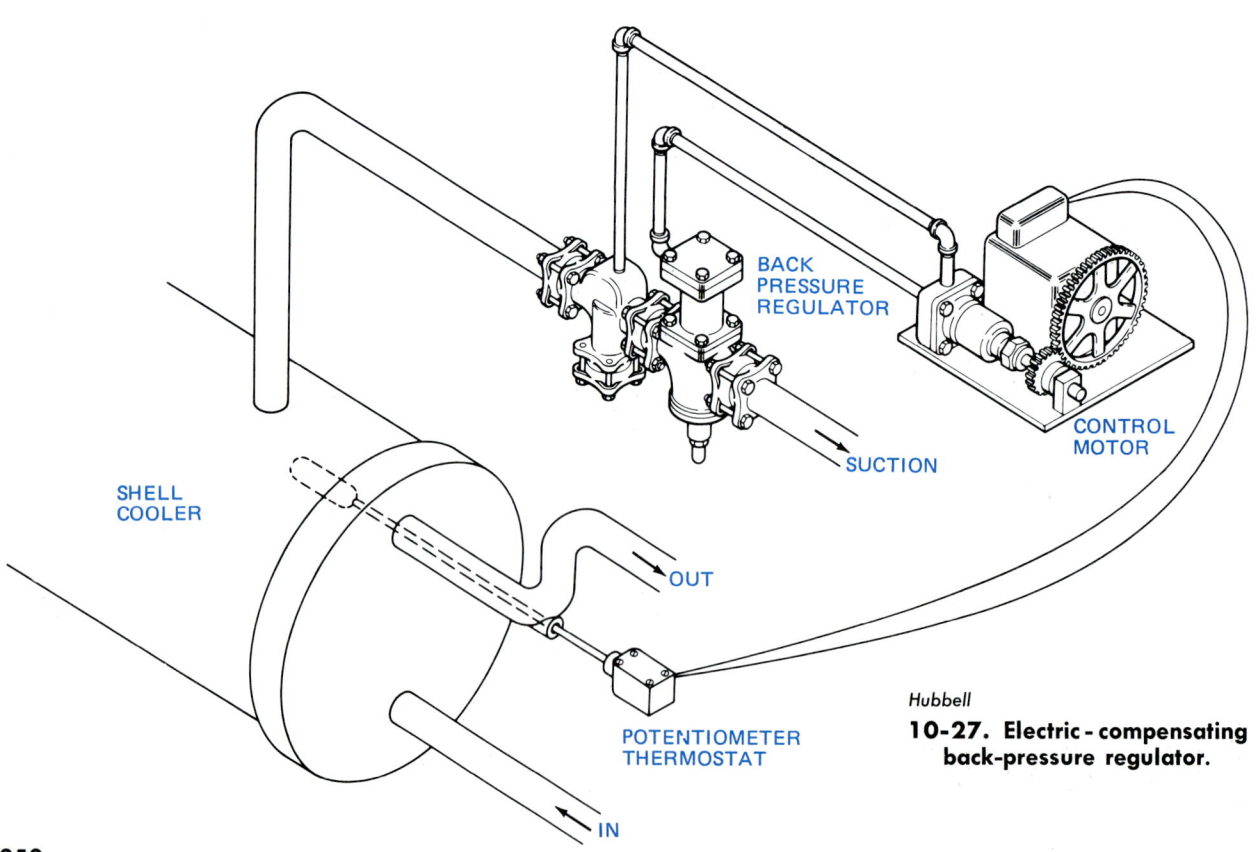

10-27. Electric-compensating back-pressure regulator.

250

CHAPTER 10—EVAPORATORS

Table 10-C.
Troubleshooting a Differential Pressure Relief Regulator

Symptom	Probable Cause	*Remedy
Erratic Operation. No adjustment. Regulator remains open.	Damaged pilot seat bead and/or diaphragms.	Replace.
	Dirt binding power or disc pistons.	Clean, repair and/or replace damaged items.
	Dirt lodged in seat disc or pilot seat bead area.	
	Tubing sensing downstream pressure blocked.	Remove obstruction.
	Manual opening stem holding disc piston open.	Turn opening stem to automatic position.
Short cycling, hunting, or chattering.	Regulator too large for load conditions.	Install properly sized metered orifice control.
	Power piston bleed hole enlarged.	Replace or contact factory for sizing.
	O.D. of power piston worn, creating excessive clearance.	Replace piston.
Excessive pressure drop.	Regulator too small for load.	Replace with correctly sized regulator.
	Passage to sensing chamber blocked.	Remove obstructions.
	Strainer blocked.	Clean strainer—replace screen if damaged.
No adjustment over 90 psig.	Range spring rated at 2 to 90 psig.	Order range kit rated at 75–300 psig.
No adjustment under 2 psig.	Range spring rated at 2 to 90 psig.	Order range kit rated at 25″ vacuum to 50 psig.

* IF REPAIR REQUIRES METAL REMOVAL—REPLACE PART.

mon base with the motor and gear drive. This compensating unit may be located in any convenient place within 20′ of the main regulator. The unit is connected to it by two small pipelines. These convey the pressure changes set up by the control diaphragm.

The total arc of rotation of the motor and the large gear on the motor acting through the smaller pinion on the adjusting stem of the diaphragm unit will rotate the stem about two turns. This is sufficient to cause the regulator to vary the evaporator pressure through a total range of about 13 pounds. Fig. 10-27.

VALVE TROUBLESHOOTING

Most of the problems in an evaporator system occur in the valves that make the defrost system operate properly. Every valve has its own particular problems. A differential pressure relief regulator valve is shown in Fig. 10-28.

A listing of its component parts should help you see the areas where trouble may occur. Table 10-C lists possible causes and remedies.

The valve difficulties and remedies listed in Table 10-C are for one particular type of valve. Manufacturers issue troubleshooting tables such as the one shown. These should be consulted when troubleshooting the valves of the evaporator system.

251

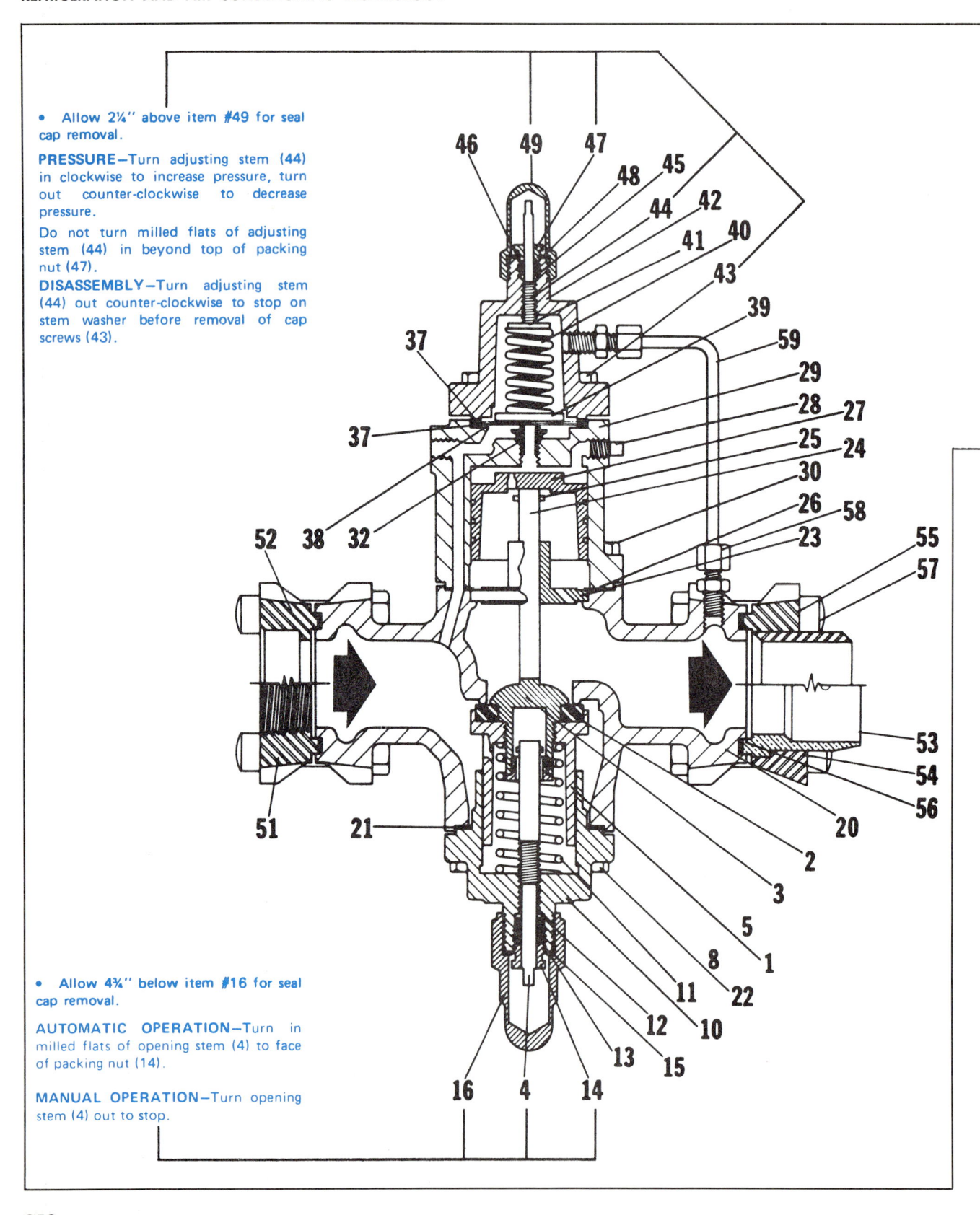

REFRIGERATION AND AIR CONDITIONING TECHNOLOGY

- Allow 2¼" above item #49 for seal cap removal.

PRESSURE—Turn adjusting stem (44) in clockwise to increase pressure, turn out counter-clockwise to decrease pressure.

Do not turn milled flats of adjusting stem (44) in beyond top of packing nut (47).

DISASSEMBLY—Turn adjusting stem (44) out counter-clockwise to stop on stem washer before removal of cap screws (43).

- Allow 4¾" below item #16 for seal cap removal.

AUTOMATIC OPERATION—Turn in milled flats of opening stem (4) to face of packing nut (14).

MANUAL OPERATION—Turn opening stem (4) out to stop.

252

1	DISC PISTON	23	GUIDE PLATE	43	CAP SCREW
2	SEAT DISC	24	PUSH ROD	44	PRESSURE ADJUSTING STEM
3	SEAT DISC RETAINER	25	ROLL PIN	45	PRESSURE ADJ. STEM WASHER
4	OPENING STEM	26	CYLINDER GASKET	46	PACKING
5	ROLL PIN		**CYLINDER ASSEMBLY**	47	PACKING NUT
8	STEM RETAINING NUT	27	POWER PISTON	48	SEAL CAP GASKET
10	BOTTOM CAP	28	PIPE PLUG	49	SEAL CAP
11	DISC PISTON SPRING	29	CYLINDER	50	NAME PLATE (NOT SHOWN)
12	OPENING STEM WASHER	30	CAP SCREW	51	THREADED FLANGE
13	PACKING		**BONNET ASSEMBLY**	52	SOCKET WELD FLANGE
14	PACKING NUT	32	PILOT SEAT BEAD	53	O.D.S. FITTING
15	SEAL CAP GASKET	37	DIAPHRAGM GASKET	54	O.D.S. FLANGE
16	SEAL CAP	38	DIAPHRAGM	55	WELD NECK FLANGE
	BODY ASSEMBLY	39	ADJUSTING SPRING PLATE	56	GASKET
20	BODY (SQUARE)	40	ADJUSTING SPRING	57	BOLT & NUT
21	BOTTOM CAP GASKET	41	ADJUSTING SPRING GUIDE	58	4 × 4 MALE CONNECTOR
22	CAP SCREW	42	BONNET	59	.250 DIA. TUBING

Hubbell

10-28. Differential pressure relief regulators automatically maintain a pre-set differential between the upstream (inlet) and the downstream (outlet) side of the control valve.

REVIEW QUESTIONS

1. What is the purpose of an evaporator?
2. What is the purpose of a shell-and-tube chiller?
3. Describe the cooling cycle.
4. Describe the defrost cycle.
5. What is meant by direct expansion with top hot-gas feed?
6. What is a flooded ceiling evaporator?
7. How are liquid refrigerant recirculating systems fed?
8. Where are equalizing orifices or globe valves used?
9. Why is the constant liquid control system used in refrigeration systems?
10. What is a dual-pressure regulator?

11
Refrigerant Flow Control

METERING DEVICES

Metering devices divide the high side from the low side of the refrigeration system. Acting as a pressure control, metering devices allow the correct amount of refrigerant to pass into the evaporator.

Hand Expansion Valve

Of the several types of metering devices, the hand expansion valve is the simplest. Fig. 11-1. Used only on manually controlled installations, the hand expansion valve is merely a needle valve with a fine adjustment stem. When the machine is shut down, the hand expansion valve must be closed to isolate the liquid line.

Automatic Expansion Valve

The automatic expansion valve controls liquid flow by responding to the suction pressure of the unit acting on its diaphragm or bellows. Fig. 11-2. When the valve opens, liquid refrigerant passes into the evaporator. The resulting increase in pressure in the evaporator closes the valve.

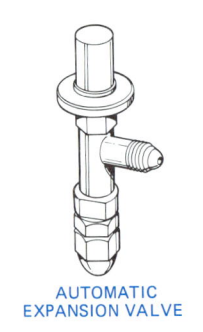

Meanwhile, the compressor is pulling gas away from the coils, reducing the pressure. This pressure reduction allows the expansion valve to open again. In operation, the valve never quite closes. The needle floats just off

Mueller Brass
11-1. Hand expansion valve.

Mueller Brass
11-2. Automatic expansion valve.

the seat and opens wide when the unit calls for refrigeration. When the machine is shut down, the pressure building up in the coils closes the expansion valve until the unit starts up.

Thermostatic Expansion Valve

The thermostatic expansion valve, used primarily in commercial refrigeration and in air con-

Sporlan Valve
11-3. A thermostatic expansion valve.

ditioning, is a refinement of the automatic expansion valve. Fig. 11-3. A bellows or diaphragm responds to pressure from a remote bulb charged with a substance similar to the refrigerant in the system. The bulb is attached to the suction line near the evaporator outlet. It is connected to the expansion valve by a capillary tube.

In operation, the thermostatic expansion valve keeps the frost line of the unit at the desired location by reacting to the superheat of the suction gas. Superheat cannot be present until all liquid refrigerant in the evaporator has been vaporized. Thus, it is possible to obtain a range of evaporator temperatures by adjusting the superheat control of the thermostatic expansion valve.

The prime importance of this type of metering device is its ability to prevent the flood-back of slugs of liquid through the suction line to the compressor. If this liquid returns to the compressor, it could damage it. The compressor is designed to pump vapors, not liquids.

Capillary Tubing

Small-bore capillary tubing is used as a metering device. It is used on everything from the household refrigerator to the heat pump. Essentially, it is a carefully measured length of very small diameter tubing. It creates a predetermined pressure drop in the system. The capillary has no moving parts.

Because a capillary tube cannot stop the flow of refrigerant when the condensing unit stops, such a refrigeration unit will always equalize high side and low side pressures on the off-cycle. For this reason, it is important that the refrigerant charge be of such a quantity that it can be held on the low side of the system without damage to the compressor. In a charge of several pounds, this "critical charge" of refrigerant may have to be carefully weighed.

An accumulator, or enlarged chamber, is frequently provided on a capillary tube system to prevent slugs of liquid refrigerant from being carried into the suction line.

Float Valve

A float valve, either high side or low side, can serve as a metering device. The high-side float, located in the liquid line, allows the liquid to flow into the low side when a sufficient amount of refrigerant has been condensed to move the float ball. No liquid remains in the receiver. A charge of refrigerant just sufficient to fill the coils is put into the system on installation. This type of float, formerly used extensively, is now limited to use in certain types of industrial and commercial systems.

The low-side float valve keeps the liquid level constant in the evaporator. It is used in flooded-type evaporators where the medium being cooled flows through tubes in a bath of refrigerant. The low-side float is more critical in operation than the high-side float and must be manufactured more precisely. A malfunction will cause the evaporator to fill during shutdown. This condition will result in serious pounding and probable compressor trouble on start-up.

Needle valves, either diaphragm or packed type, may be used as hand expansion valves. As such, they are usually installed in a bypass line around an automatic or thermostatic expansion valve. They are placed in operation when the normal control is out of order or is removed for repairs.

FITTINGS AND HARDWARE

Modern refrigerants can escape through the most minute openings. Since porosity in a fitting could create such an opening, it is mandatory that porosity be eliminated from fittings and accessories that are to be used with refrigerants. Fig. 11-4.

One way to eliminate porosity in fittings is to either forge or draw them from brass rod. This creates a final grain structure that prevents the seepage of refrigerant due to porosity. The threads on fittings must be machined with some degree of accuracy to prevent leaks. Solder-type fittings should be made of wrought copper, brass rod, or brass forgings. Fig. 11-5. This eliminates the possibility of leaks due to porosity of the metal. The tube is not weakened by the cutting of threads, as is the case with iron pipe. A soldered joint allows the use of a much lighter wall tube with complete safety and with significant cost savings.

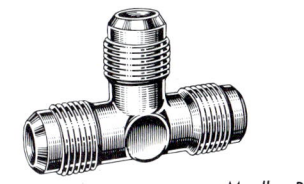

Mueller Brass
11-4. 45° flare fitting.

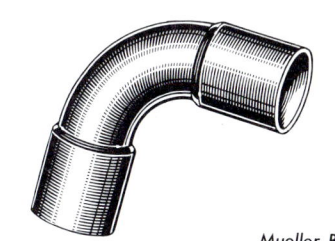

Mueller Brass
11-5. Wrought copper, solder-to-solder fitting.

One advantage of copper pipe over iron is the elimination of scale and corrosion. In service, a light coating of copper oxide forms on the outside of the copper tube. This coating prevents chemical attack. There is no "rusting out" of copper tube.

Copper Tubing

For flare fitting applications, seamless soft copper tube is recommended. Fig. 11-6. This tube is furnished with sealed ends. It is supplied in 50' lengths in sizes from $\frac{1}{8}''$ through $\frac{3}{4}''$ OD for flaring and through $1\frac{3}{8}''$ OD for soldering.

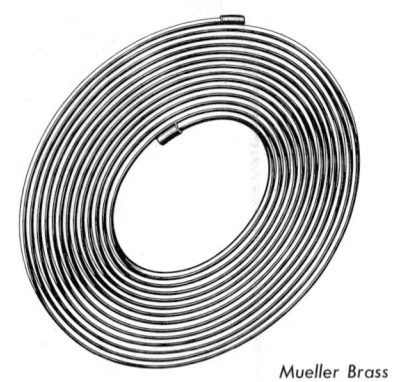

REFRIGERATION AND AIR CONDITIONING TECHNOLOGY

11-6. Refrigeration service tube in a 50-foot coil.
Mueller Brass

The chief demand for this tube is in sizes from $\frac{1}{4}''$ through $\frac{5}{8}''$ OD. Sizes smaller than $\frac{1}{4}''$ are seldom used in commercial refrigeration. To uncoil the tube without kinks, hold one free end against the floor or on a bench and uncoil along the floor or bench. The tube may be cut to length with a hacksaw or tube cutter. In either case, deburr the end before flaring. Bending is readily accomplished with either external or internal bending springs or lever-type bending tools.

ACR (air-conditioning, refrigeration) tube is frequently used. It is cleaned, degreased, dried, and end-sealed at the factory. This assures the user that he or she is installing a clean, trouble-free tube. Some tubing is available with an inert gas (nitrogen). Fig. 11-7. The nitrogenized ACR tube is purged, charged with clean, dry nitrogen, and then sealed with reusable plugs. After cutting the tube, the remaining length can easily be re-plugged. The remaining nitrogen limits excess oxides during succeeding brazing operations. It comes in 20′ lengths. Type-L hard tube has from $\frac{3}{8}''$ through $3\frac{1}{8}''$ OD. Type-K tube is also available.

Where tubing will be exposed inside food compartments, tinned copper is recommended. Type-L, hard-temper copper tube is recommended for field installations using solder-type fittings. Type-M is sufficiently strong for any pressures of the commonly used refrigerants. However, it is used chiefly in manufactured assemblies where external damage to the tube is not as likely as in field installations. For maximum protection against possible external damage to refrigerant lines, a few cities require the use of type-K copper tube.

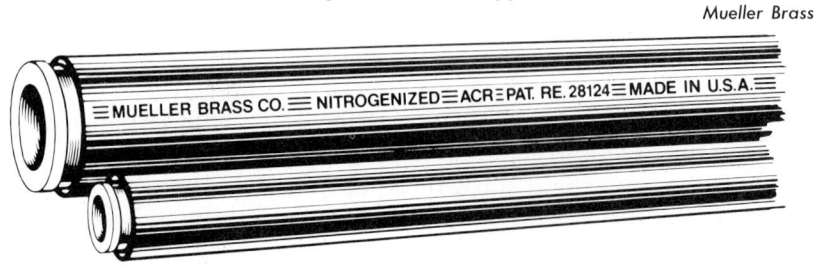

Line

Correct line sizes are essential to obtaining maximum efficiency from refrigeration equipment. In supermarkets, for example, the long lines running under the floor from the display cases to the machine room at the rear of the store must be fully engineered. Otherwise, problems of oil return, slugging, or erratic refrigeration are quite likely. Table 11-A lists refrigerant line sizes.

When available, the manufacturer's recommendations must be followed regarding step-sizing, risers, traps, and the like.

Available information on Refrigerant 502 claims performance at temperatures below 5° F. [−15°C] when compared with Refrigerant 22.

Solder

Each solder is designed for a certain job. For instance, 50-50 solder, which consists of 50% tin and 50% lead, will not function well in some instances. In fact, 50-50 solder will deteriorate in some refrigerated food storage compartments where normally wet refrigerant lines and high carbon dioxide content are present. For this reason, No. 95 solder is recommended. It has 95% tin and 5% antimony. Number 122 solder (45% silver brazing alloy) is used for joints in refrigerant lines where 50-50 solder may deteriorate.

Suction Line P-Traps

For years, the P-trap was made by forming two or more fittings. It has now become available in one piece. Fig. 11-8. The newer one-piece P-trap promotes efficient oil migration in refrigeration systems. This is increasingly important today. Many large food markets place their compressors and condensers on balconies or mezzanines. Such remote condensing units are likely to have long horizontal suction lines or vertical risers exceeding 3′ in height in the suction line. In such cases, the oil concentration in the circulating refrigerant may be expected to be above 0.6%. A low vapor velocity may be en-

11-7. Nitrogenized ACR copper tube.
Mueller Brass

Table 11-A.
Sizes of Refrigerant Lines

Btu Per Hour	REFRIGERANT 12 Liquid Line	REFRIGERANT 12 Suction Line 5°F. [−15°C]	REFRIGERANT 12 Suction Line 40°F. [4.4°C]	REFRIGERANT 22 Liquid Line	REFRIGERANT 22 Suction Line 5°F. [−15°C]	REFRIGERANT 22 Suction Line 40°F. [4.4°C]	REFRIGERANT 40 Liquid Line	REFRIGERANT 40 Suction Line 5°F. [−15°C]	REFRIGERANT 40 Suction Line 40°F. [4.4°C]	REFRIGERANT 502 Liquid Line	REFRIGERANT 502 Suction Line 5°F. [−15°C]	REFRIGERANT 502 Suction Line 40°F. [4.4°C]
3,000	1/4	1/2	1/2	1/4	1/2	1/2	1/4	1/2	1/2	1/4	1/2	1/2
6,000	3/8	5/8	5/8	3/8	5/8	5/8	1/4	1/2	1/2	3/8	5/8	5/8
9,000	3/8	7/8	5/8	3/8	7/8	5/8	3/8	5/8	5/8	3/8	7/8	5/8
12,000	3/8	1 1/8	7/8	3/8	7/8	7/8	3/8	7/8	7/8	3/8	7/8	7/8
15,000	3/8	1 1/8	7/8	3/8	1 1/8	7/8	3/8	7/8	7/8	3/8	1 1/8	7/8
18,000	3/8	1 1/8	7/8	3/8	1 1/8	7/8	3/8	1 1/8	7/8	3/8	1 1/8	7/8
21,000	1/2	1 1/8	1 1/8	1/2	1 1/8	1 1/8	3/8	1 1/8	7/8	1/2	1 1/8	1 1/8
24,000	1/2	1 3/8	1 1/8	1/2	1 1/8	1 1/8	1/2	1 1/8	7/8	1/2	1 1/8	1 1/8
30,000	5/8	1 3/8	1 1/8	1/2	1 3/8	1 1/8	1/2	1 1/8	1 1/8	5/8	1 3/8	1 1/8
36,000	5/8	1 3/8	1 1/8	5/8	1 3/8	1 1/8	1/2	1 3/8	1 1/8	5/8	1 3/8	1 1/8
42,000	5/8	1 5/8	1 3/8	5/8	1 3/8	1 3/8	1/2	1 3/8	1 1/8	5/8	1 3/8	1 3/8
48,000	5/8	1 5/8	1 3/8	5/8	1 5/8	1 3/8	1/2	1 3/8	1 1/8	5/8	1 5/8	1 3/8
54,000	5/8	1 5/8	1 3/8	5/8	1 5/8	1 3/8	5/8	1 3/8	1 1/8	5/8	1 5/8	1 3/8
60,000	7/8	1 5/8	1 3/8	5/8	1 5/8	1 3/8	5/8	1 5/8	1 3/8	7/8	1 5/8	1 3/8
72,000	7/8	2 1/8	1 5/8	7/8	1 5/8	1 3/8	5/8	1 5/8	1 3/8	7/8	2 1/8	1 5/8
96,000	7/8	2 1/8	1 5/8	7/8	2 1/8	1 5/8	7/8	2 1/8	1 5/8	7/8	2 1/8	1 5/8
108,000	7/8	2 5/8	2 1/8	7/8	2 1/8	1 5/8	7/8	2 1/8	1 5/8	7/8	2 1/8	1 5/8
120,000	7/8	2 5/8	2 1/8	7/8	2 1/8	1 5/8	7/8	2 1/8	1 5/8	7/8	2 1/8	1 5/8
150,000	1 1/8	2 5/8	2 1/8	7/8	2 1/8	2 1/8	7/8	2 1/8	2 1/8	1 1/8	2 1/8	2 1/8
180,000	1 1/8	2 5/8	2 1/8	1 1/8	2 5/8	2 1/8	7/8	2 5/8	2 1/8	1 1/8	2 5/8	2 1/8
210,000	1 1/8	3 1/8	2 1/8	1 1/8	2 5/8	2 1/8	7/8	2 5/8	2 1/8	1 1/8	2 5/8	2 1/8
240,000	1 3/8	3 1/8	2 5/8	1 3/8	2 5/8	2 1/8	7/8	2 5/8	2 1/8	1 3/8	2 5/8	2 1/8
300,000	1 3/8	3 1/8	2 5/8	1 3/8	3 1/8	2 5/8	1 1/8	2 5/8	2 1/8	1 3/8	3 1/8	2 5/8
360,000	1 3/8	3 5/8	2 5/8	1 1/8	3 1/8	2 5/8	1 1/8	3 1/8	2 5/8	1 3/8	3 1/8	2 5/8
420,000	1 5/8	3 5/8	3 1/8	1 3/8	3 1/8	2 5/8	1 1/8	3 1/8	2 5/8	1 5/8	3 1/8	2 5/8
480,000	1 5/8	4 1/8	3 1/8	1 5/8	3 5/8	3 1/8	1 1/8	3 1/8	2 5/8	1 5/8	3 5/8	3 1/8
540,000	1 5/8	4 1/8	3 1/8	1 5/8	3 5/8	3 1/8	1 3/8	3 5/8	3 1/8	1 5/8	3 5/8	3 1/8
600,000	1 5/8	4 1/8	3 1/8	1 5/8	4 1/8	3 1/8	1 3/8	3 5/8	3 1/8	1 5/8	4 1/8	3 1/8

To convert Btu per hour to tons of refrigeration—divide by 12,000
Suction temperature, condensing medium, compressor design and many other factors determine horsepower required for a ton of refrigerating capacity. Consult ASHRAE Handbook.

Mueller Brass

countered. This results in unsatisfactory oil return to the compressor. Tests have proven that

11-8. Suction line P-trap.
Mueller Brass

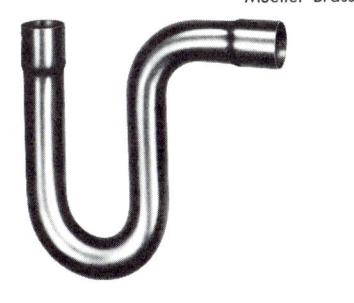

with a P-trap installed, vapor velocity can fall as low as 160' per minute and satisfactory oil return can still be achieved. The P-trap drains the oil from the horizontal runs approaching the risers. This oil, in turn, migrates up through the riser to the compressor in one of three different forms: as a rippling oil film, as a mist, or as a transparent colloidal dispersion in the vaporized refrigerant. The method of oil migration depends upon the vapor velocity in the suction line.

Compressor Valves

There are three types of compressor valves: adjustable, double port, and single port. Open and semihermetic compressors are usually fitted with compressor service valves, one each at the suction and discharge ports. The service valve has no operating function. Nevertheless, it is indispensable when service is to be performed on any part of the refrigeration system. Fig. 11-9.

Compressor service valves are back-seating. They are con-

REFRIGERATION AND AIR CONDITIONING TECHNOLOGY

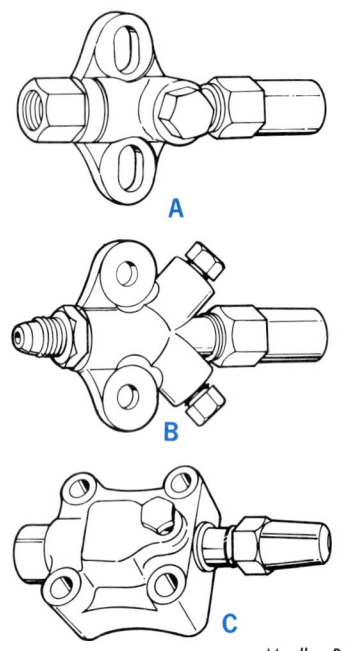

11-9. Compressor valves. (A) Adjustable compressor valve, (B) double-port compressor valve, (C) single-port compressor valve.

11-10. Packless line valves.

Mueller Brass

TYPE I

TYPE III

1. Handwheel screw
2. Metal handwheel
3. Bronze operating screw stem
4. Nylon operating screw gasket
5. Bronze operating screw nose
6. Neoprene "O" ring
7. Forged brass bonnet
8. Spring steel bearing washer
9. Diaphragms
10. Stainless steel spring
11. Stem assembly
12. Nylon seat disc
13. Forged brass body
14. Spring support ring
15. Identification tag

structed so that the stem forms a seal against a seat whether the stem is full-forward or full-backward. Valve packing is depended upon only when the stem is in the intermediate position. In one style of construction, the front seat, including the one connection, is threaded. Silver is brazed into the body after the stem has been assembled. When the valve is full-open (normal position when the unit is running), the gage and charging port plug or cap may be removed without loss of refrigerant. A charging line or pressure gage may be attached to this side port. It is also possible to repack the valve without interruption of service.

Line Valves

Line valves are essential components of refrigerant systems.

Fig. 11-10. Installed in key locations, line valves make it possible to isolate any portion of a system or, in a multiple hookup, to separate one system from the rest. Local codes frequently specify the location of line valves in commercial and industrial refrigeration and air conditioning systems.

There are two types of line valves—packed and packless. They must be designed to prevent refrigerant leakage. Since refrigerants are difficult to retain, packed valves are usually equipped with seal caps. Some seal caps are designed to be removed and used as wrenches for operating the valves.

In large packed valves, such as that shown in Fig. 11-11, O-rings are used as seals between the bonnets and valve bodies. They are available in either straight-through ($7/8''$ to $4 1/8''$) or angle-type ($7/8''$ through $3 1/8''$) construction.

The packless design is often preferred for smaller valves. The packless-type valve is used to good advantage on charging boards. The valves contain triple diaphragms, one of phosphor bronze and two of stainless steel. These valves must be frost-proof. They must be designed for use where condensation is likely to occur. During the off-cycle, condensation may seep down the stem of a non-frost-proof valve into the bonnet. There, it will alternately freeze and thaw. The eventual buildup of ice against the diaphragms may close the valve. Another factor that should be considered is whether or not the valve has back-seating. This prevents all pressure pulsations while the valve is open. The back-seating should allow in-

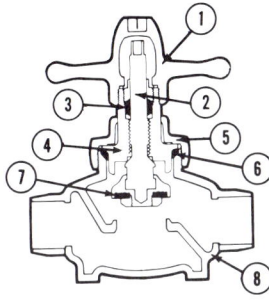

CHAPTER 11—REFRIGERANT FLOW CONTROL

11-12. Filter-drier. *Mueller Brass*

11-13. Suction line filter-drier. *Mueller Brass*

1. Cast bronze wing-seal cap
2. Bronze stem
3. Molded stem packing
4. Forged brass bonnet in sizes over 1⅛
5. Forged brass union collar
6. Neoprene "O" ring
7. Nylon seat disc
8. Cast bronze body

Mueller Brass
11-11. Packed line valve.

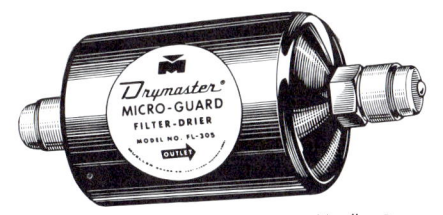

spection of the diaphragms without shutting down the system.

DRIERS, LINE STRAINERS, AND FILTERS

Driers

Most authorities agree that moisture is the most detrimental material in a refrigeration system. A unit can stand only minute amounts of water. For this reason, most refrigeration and air conditioning systems, both field and factory-assembled, contain driers. Fig. 11-12.

MOISTURE

Moisture or water is always present in refrigeration systems. Acceptable limits vary from one unit to another and from one refrigerant to another. Moisture is harmful even if freeze-ups do not occur. Moisture is an important factor in the formation of acids, sludge, and corrosion. To be safe, keep the moisture level as low as possible.

Moisture will react with today's halogen-type refrigerants to form harmful hydrochloric and hydrofluoric acids within the system. To minimize the possibility of freeze-up or corrosion, the following maximum safe limits of moisture should be observed:

Refrigerant 12   15 parts per million (15 ppm)

Refrigerant 22   60 parts per million (60 ppm)

Refrigerant 502   30 parts per million (30 ppm)

If the moisture exceeds these figures, corrosion is possible. Also, excess water may freeze at the metering device if the system operates below 32° F. [0°C]. Freeze-ups do not occur in air conditioning systems where evaporator temperatures are normally above 40° F. [4.4°C].

A drier charged with a moisture-removing substance and installed in the liquid line is the most practical way to remove moisture. With a drier of the proper size, excess water is stored in the drier. Here, it can neither react with the refrigerant nor travel through the system.

Many materials have been tried as desiccants, or drying agents. Today, the desiccant materials most commonly used are silica gel, activated alumina, calcium sulfate, and the zeolite-type materials, such as molecular sieves and micro traps.

The total drier design considers not only drying and filtering, but also maintaining maximum refrigerant flow. Filter-driers must allow free flow of refrigerant. They must also prevent fine particles of the adsorbent or other foreign matter from passing through to the metering device, usually located downstream from the drier. Fig. 11-13.

DIRT

Dirt, sludge, flux, and metallic particles are frequently found in refrigeration systems. Numerous metallic contaminants—cast iron dust, rust, and scale, plus steel, copper, and brass filings—can damage cylinder walls and bearings. They can plug capillary tubes and thermostatic expansion valve screens. These contaminants are catalytic and contribute to decomposition of the refrigerant-oil mixture at high temperatures.

ACIDS

By themselves, Refrigerants 12

and 22 are very stable—even when heated to a high temperature. However, under some conditions, reactions occur that can result in the formation of acids. For example, at elevated temperatures, Refrigerant 12 will react with the oil to form hydrochloric and hydrofluoric acids. These acids are usually present as a gas in the system and are highly corrosive. Where an "acid acceptor" such as electrical insulation paper is present, Refrigerant 22 will decompose at high temperatures to form hydrochloric acid. The reaction of refrigerants with water may cause hydrolysis and the formation of hydrochloric and hydrofluoric acids. In ordinary usage this reaction is negligible. However, in a very wet system operating at abnormally high temperatures, some hydrolysis may occur.

All of these reactions are increased by elevated temperature and are catalytic in effect. They result in the formation of corrosive compounds.

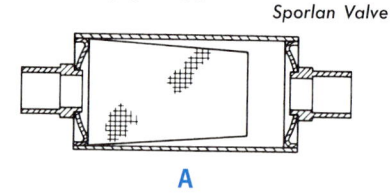

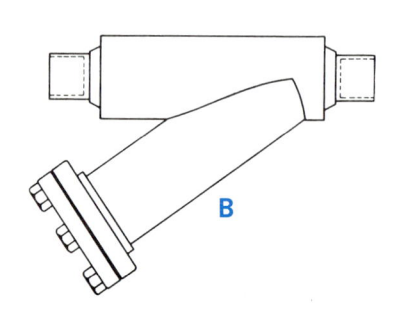

Another source of acidity in refrigeration systems is the organic acid formed from oil breakdown. Appreciable amounts of organic acid are found in the majority of oil samples analyzed in the laboratory. These acids will also corrode the metals in a system. Therefore, they must be removed.

Acid may be neutralized by the introduction of an alkali, but the chemical combination of the two creates further hazards. They release additional moisture and form a salt. Both of these are detrimental to the system.

SLUDGE AND VARNISH

The utmost care may be taken in the design and fabrication of a system. Nonetheless, in operation, unusually high discharge temperatures will cause the oil to break down and form sludge and varnish.

Temperatures may vary in different makes of compressors and under different operating conditions. Temperatures of 265° F. [130°C] are not unusual at the discharge valve under normal operation. Temperatures well above 300° F. [150°C] frequently occur under unusual conditions. Common causes of high temperatures in refrigeration systems are dirty condensers, noncondensible gases in the condenser, high compression ratio, high superheat of suction gas returned to the compressor, and fan failure on forced convection condensers.

In addition to high discharge temperatures, certain catalytic metals contribute to oil-refrigerant mixture breakdown. The most significant of these is iron. It is used in all systems and is an active catalyst. Copper is a catalyst also, but its action is slower. However, the end result is the same. The reaction causes sludges and other corrosive materials that will hinder the normal operation of compressor valves and control devices. In addition, air in a system will also accelerate oil deterioration.

Line Strainers and Filters

It is impossible to keep all foreign matter out of factory or field-assembled refrigeration systems. Core sand from the compressor casting, brazing oxides in piping or tubing, chips from cutting or flaring, sawdust, and dirt are found in most refrigeration systems. This is especially so with field-assembled installations. Tubing, for example, may have been exposed to air-carried dirt for several days. Cleanliness is difficult to maintain in the field. All tubing to be used for refrigeration applications should be protected by capping or sealing. It should be recapped or sealed after each use.

Moving through the system with the flowing refrigerant, particles of foreign matter may score critical moving parts or clog orifices. To prevent such damage, strainers or filters are frequently installed in the system. Fig. 11-14. A filter drier placed in the liquid line ahead of the metering device is the normal precaution. However, on multiple installations, it is usual to install a strainer upstream of each metering device just ahead of key valves and controls. To protect the compressor, most engineers also specify filters for the suction line.

The simplest strainer consists of a set of metal screens of the proper mesh. Fig. 11-15. Adding felt pads or asbestos cloth creates

11-14. Strainers. (A) Noncleanable strainer, (B) Y-type line strainer.
Sporlan Valve

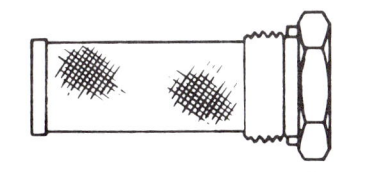

CHAPTER 11—REFRIGERANT FLOW CONTROL

11-15. Strainer-filter, metal screen. *Mueller Brass removable*

a very effective filter, rather than a mere strainer. A cellulose fiber core as used in Fig. 11-13 is also very effective for this purpose.

Suction line strainers and filters are designed with sufficient flow capacity to prevent excessive pressure drop. Since determining the need for cleaning or replacement of a suction line filter is related to pressure drop, some designs are offered with pressure taps. These permit pressure gage installation to determine the degree of pressure drop.

Strainers are made in several designs. They are also supplied in a wide range of sizes. Screen areas are large. In cartridge-type strainers, provision is made for removal of the screens or filters. Fig. 11-16.

LIQUID INDICATORS

Liquid indicators are inserted in a refrigerant line to indicate the amount of refrigerant in a system. Fig. 11-17. Proper operation of a refrigeration system depends upon there being the correct amount of refrigerant in the unit. Looking through the window of a liquid indicator is the simplest way to determine whether there is a refrigerant shortage. A shortage of refrigerant may be due to a leak in the system or to failure to charge enough refrigerant into a unit after field service.

Liquid indicators normally disclose a shortage of refrigerant by the appearance of bubbles. Some use a special assembly that shows by the appearance of the word FULL that there is sufficient refrigerant at that point in the system.

Liquid indicators are manufactured in single-port, double-port, and straight-through types. In the single- and double-port indicators, an internal compression bushing seals the glass firmly against the body. Assem-

11-16. A strainer and its replacement core-type filter. *Sporlan Valve*

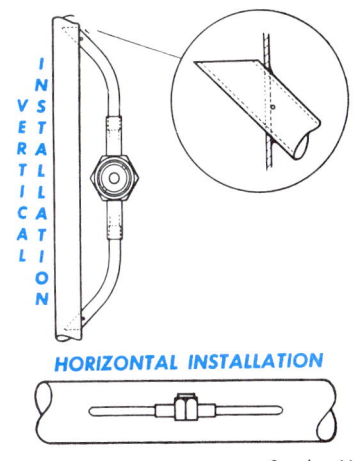

11-17. Liquid line indicator installation. *Sporlan Valve*

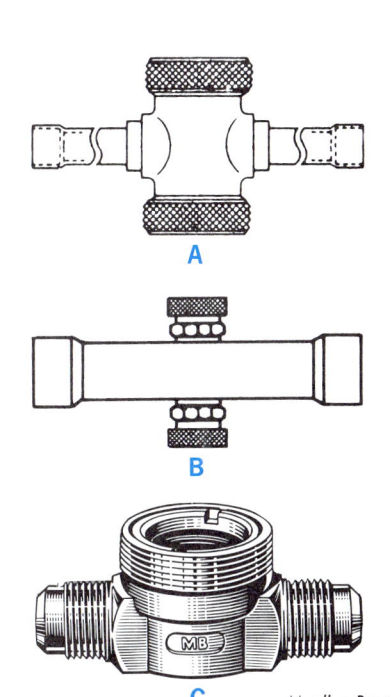

11-18. Liquid indicators. (A) and (B) Double port, (C) single port. *Mueller Brass*

blies are furnished with a protective dust cap or seal cap. To observe the liquid stream, it is necessary to remove the cap. Fig. 11-18.

261

REFRIGERATION AND AIR CONDITIONING TECHNOLOGY

Table 11-B.
Moisture Content (In Parts Per Million)

Unit Shows	Liquid Line Temp. (°F.)	Refrigerants 11 & 12 75°	Refrigerants 11 & 12 **100°**	Refrigerants 11 & 12 125°	Refrigerant 22 75°	Refrigerant 22 **100°**	Refrigerant 22 125°	Refrigerant 500 75°	Refrigerant 500 **100°**	Refrigerant 500 125°	Refrigerants 502, 113, & 114 75°	Refrigerants 502, 113, & 114 **100°**	Refrigerants 502, 113, & 114 125°
Green DRY		Below 5	Below **10**	Below 20	Below 30	Below **45**	Below 60	Below 40	Below **60**	Below 100	Below 10	Below **20**	Below 30
Chartreuse CAUTION		5–15	**10–30**	20–50	30–90	**45–130**	60–180	40–90	**60–150**	100–230	10–45	**20–65**	30–110
Yellow WET		Above 15	Above **30**	Above 50	Above 90	Above **130**	Above 180	Above 90	Above **150**	Above 230	Above 45	Above **65**	Above 110

Sporlan Valve

BOLD figures are for the average design conditions of refrigerant liquid lines operating at 100° F. Since the actual temperature is not critical, a satisfactory estimate can be made by comparing it to body temperature. If it feels cool to the touch, use 75° F. If it feels warm, use 125° F. column figures.

The unit calibration information given above is based on detailed experimental data for Refrigerants 12, 22, 500, 502, and 113. The calibration information on other refrigerants and solvents was obtained from a comparison of their properties with 12, 22, 500, 502, and 113. For the less common liquids, the following moisture calibration is suggested:

Refrigerant 13.........use......"12" calibration
Refrigerant 21.........use......"22" calibration
Trichloroethylene.......use......"22" calibration
Perchloroethylene..........use......"113" calibration
Carbon Tetrachloride.......use.........."12" calibration
Propane or Butane.........use......."500" calibration

AIR TEST—Recent tests on AIR show that the unit changes color in the range of 0.5% to 2.0%. In ordinary air lines this means that the unit will change color at dew points in the range of −40 to −60° F.

A relatively new addition to the function of the liquid indicator is that of moisture detection. Special materials used in the ports of liquid-moisture indicators change color to indicate excessive moisture in the system.

Indicators with solder-type ends but without extended ends are normally furnished disassembled so the heating required for soldering will not damage glass or gaskets. In addition, single- and double-port types are supplied assembled with extended ends. These make it possible to solder without damaging the indicator, as long as normal precautions are observed.

Construction

The indicator is a porous filter paper impregnated with a chemical salt that is sensitive to moisture. The salt changes color according to the moisture content (relative saturation) of the refrigerant. The indicator changes color below moisture levels generally accepted as a safe operation range. This device is not suitable for use with ammonia or sulphur dioxide. However, it does have a full application with Refrigerants 11, 12, 22, 113, 114, 500, and 502. See Table 11-B.

The indicator should be installed after the filter-drier and ahead of the expansion device. Prior to installation, the indicator will be yellow, indicating a wet condition. This is a normal situation, since the air in contact with the element is above 0.5% relative humidity. This does not affect the operation or calibration of the indicator. As soon as it is installed in a system, the indicator element will begin to change according to the moisture content in the refrigerant. The action of the indicator element is completely reversible. The element will change color as often as the moisture content of the system varies. Some change may take place rapidly at the start-up of a new system or after replacement of a drier on existing installations. However, the equipment should be operated for about 12 hours to allow the system to reach equilibrium before deciding if the drier needs to be changed.

Installation

Indicators with $\frac{1}{4}''$ through $1\frac{1}{8}''$ ODF (outside diameter flanged) connections should not be disassembled in the field for brazing or any other purpose.

262

The long fittings on sweat models are copper-plated steel and do not conduct heat as readily as copper fittings.

On indicators with 1 3/8", 1 5/8", and 2 1/8" ODF connections, the indicator cartridge must be removed from the brass saddle fitting before brazing the indicator in the main liquid line. It is shipped hand tight for easy removal.

Bypass Installations

On systems having liquid lines larger than 2 1/8" OD, the indicator should be installed in a bypass line. During the operating cycle, this will provide sufficient flow to obtain a satisfactory reading for both moisture and liquid indication.

Best results will be obtained if the bypass line is parallel to the main liquid line and the take-off and return tubes project *into* the main liquid line at a 45° angle. Preformed 1/4" and 3/8" tubing is available. It can be used with either flare or sweat-type indicators.

Excess Oil and the Indicator

When a system is circulating an excessive amount of oil, the indicator may become saturated. This causes the indicator to appear brown or translucent and lose its ability to change color. However, this does not damage the indicator. Let the indicator unit remain in the line. The circulating refrigerant will remove the excess oil and the indicator element will return to its proper color.

Alcohol

Do not install the color-changing indicator in a system that has methyl alcohol or a similar liquid dehydrating agent. Remove the alcohol by using a filter and then install the indicator. Otherwise, the alcohol will damage the color indicator.

Leak Detectors

Dye-type visual leak detectors will also mask the color-changing indicator. Here again, use a filter to remove all leak detector color from the system before installing the indicator.

Liquid Water

On occasion, it is possible for large quantities of water to enter a refrigeration system. An example would be a broken tube in a water-cooled condenser. If the free water contacts the indicator element, the element will be damaged. All moisture indicators are made of a chemical salt. These salts must be soluble in water to change color. If excessive water is present, the salts will dissolve. Permanent damage to the indicator will result. The indicator may remain yellow, or even turn white.

Hermetic Motor Burnouts

After a hermetic motor burnout, install a filter to remove the acid and sludge contamination. When the system has operated for 48 hours, replace the filter. At the same time, install the color indicator for moisture.

The acid formed by the burnout may damage the indicator element of the color-changing unit. Thus, it should be installed only after the greater percentage of contaminants has been removed.

Hardware and Fittings

In assembling a unit in the factory or the field, strict standards of quality must be observed. Cleanliness is very important. The cleanliness of a part can determine the efficiency of a piece of equipment. Figure 11-19 illustrates some of the hardware and fittings.

THERMOSTATIC EXPANSION VALVE (TEV)

Several different valves are used to control the flow of refrigerants. All refrigerants are relatively expensive. They will leak through fittings and tubing capable of retaining water at high pressures. A leak results in the loss of expensive refrigerant and in possible product loss, such as of frozen food. For this reason, all refrigerant lines and fittings must be absolutely seepage-proof.

Proper fittings and controls also have a bearing on the efficiency and capacity of a refrigerating machine. The capacity of a condensing unit depends, among other things, upon the suction pressure at which the unit operates. Normally, the higher the suction pressure, the greater the efficiency of the compressor.

Suction pressure at the compressor is governed by the design of the evaporator. The desired temperature in the medium being cooled and the pressure drop in the suction line from the evaporator to the compressor also govern the design pressure. This pressure drop in the suction line can be kept to a minimum by use of ample line sizes, fittings, and accessories designed to eliminate restrictions. Pressure drop is also a factor in liquid lines between the receiver and the metering device. Excessive pressure drop will result in "flashing," or par-

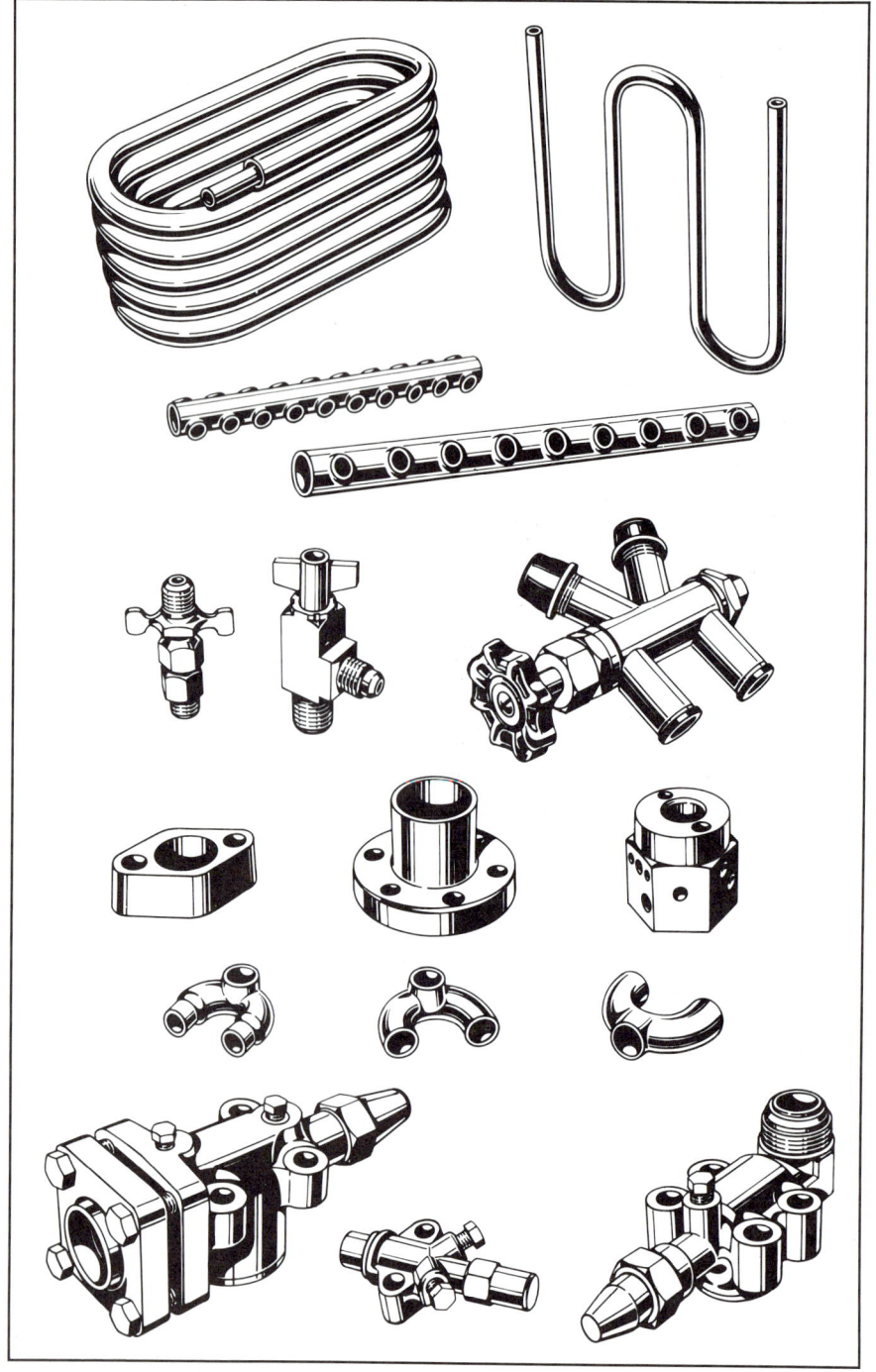

REFRIGERATION AND AIR CONDITIONING TECHNOLOGY

Serpentine, coiled or formed tubes

Single and double row headers with multiple openings

Check, relief, cylinder and by-pass valves

Special machined flanges and fittings

Return bends with side or top outlets

Cast iron and brass (¼" thru 4⅛") compressor valves

Mueller Brass

11-19. Hardware and fittings for refrigeration and air conditioning installation.

tial vaporization, of the liquid refrigerant before it reaches the metering device. The metering device is designed to handle liquid. It will not function properly if fed a mixture of vapor and liquid. Here, valves play an important role in controlling and metering the flow of liquid in the system.

The thermostatic expansion valve uses the fluctuations of the

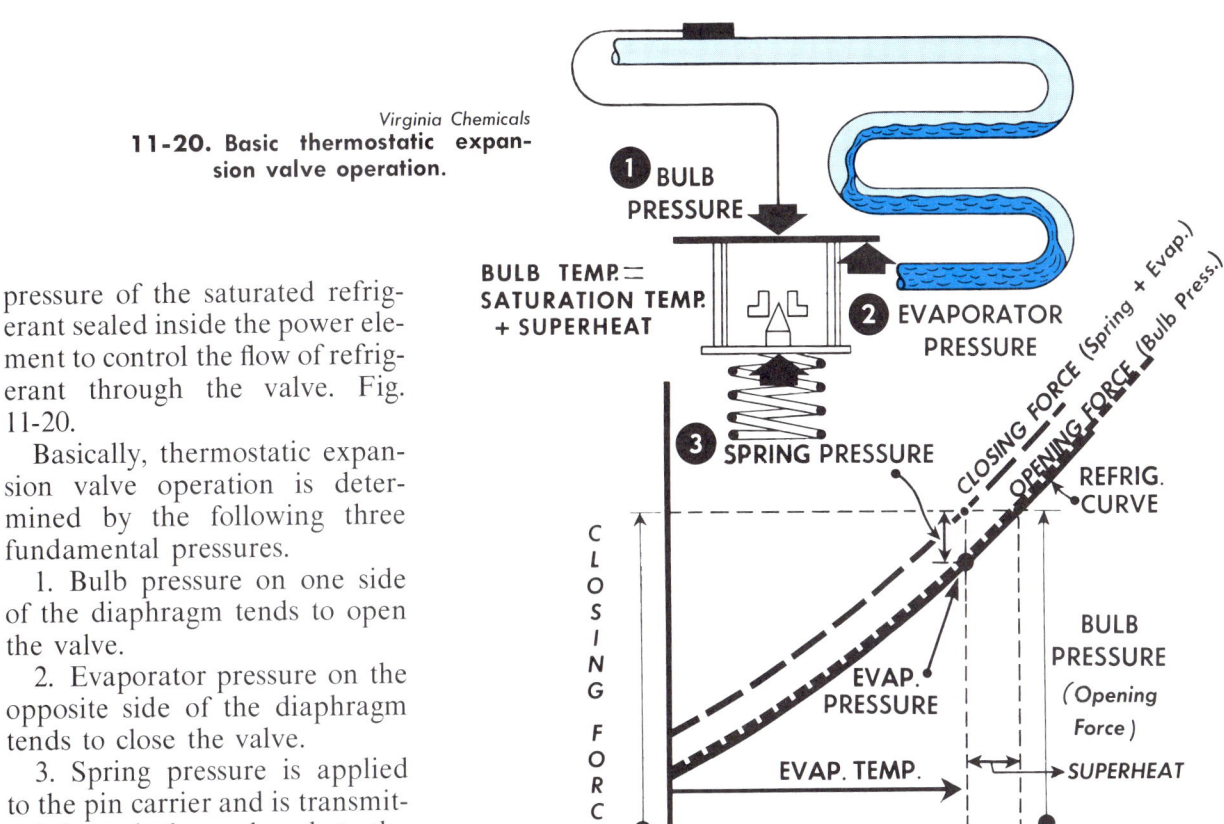

CHAPTER 11—REFRIGERANT FLOW CONTROL

11-20. Basic thermostatic expansion valve operation.
Virginia Chemicals

pressure of the saturated refrigerant sealed inside the power element to control the flow of refrigerant through the valve. Fig. 11-20.

Basically, thermostatic expansion valve operation is determined by the following three fundamental pressures.

1. Bulb pressure on one side of the diaphragm tends to open the valve.
2. Evaporator pressure on the opposite side of the diaphragm tends to close the valve.
3. Spring pressure is applied to the pin carrier and is transmitted through the push rods to the evaporator side of the diaphragm. This assists in closing the valve.

When the valve is modulating, bulb pressure is balanced by the evaporator pressure and spring pressure. When the same refrigerant is used in the thermostatic element and refrigeration system, each will exert the same pressure if their temperatures are identical. After evaporation of the liquid refrigerant in the evaporator, the suction gas is superheated. Its temperature will increase. However, the evaporator pressure, neglecting pressure drop, is unchanged. This warmer vapor flowing through the suction line increases the bulb temperature. Since the bulb contains both vapor and liquid refrigerant, its temperature and pressure increase. This higher bulb pressure acting on the top (bulb side) of the diaphragm is greater than the opposing evaporator pressure and spring pressure, which causes the valve pin to be moved away from the seat. The valve is opened until the spring pressure—combined with the evaporator pressure—is sufficient to balance the bulb pressure. Fig. 11-20.

If the valve does not feed enough refrigerant, the evaporator pressure drops or the bulb temperature is increased by the warmer vapor leaving the evaporator (or both). The valve then opens. This admits more refrigerant until the three pressures are again in balance. Conversely, if the valve feeds too much refrigerant, the bulb temperature is decreased, or the evaporator pressure increases (or both). The spring pressure tends to close the valve until the three pressures are in balance.

With an increase in evaporator load, the liquid refrigerant evaporates at a faster rate and increases the evaporator pressure. The higher evaporator pressure results in a higher evaporator temperature and a correspondingly higher bulb temperature. The additional evaporator pressure (temperature) acts on the bottom of the diaphragm. The additional bulb pressure (temperature) acts on the top of the diaphragm. Thus, the two pressure increases on the diaphragm cancel each other. The valve eas-

265

REFRIGERATION AND AIR CONDITIONING TECHNOLOGY

11-21. Installation of TEV with the compressor above the evaporator.
Virginia Chemicals

Table 11-C.
Vertical Lift and Pressure Drop

Refrig-erant	Vertical Lift—Feet				
	20	40	60	80	100
	Static Pressure Loss—psi				
12	11	22	33	44	55
22	10	20	30	40	50
500	10	19	29	39	49
502	10	21	31	41	52
717 (Ammonia)	5	10	15	20	25

Refrigerant	Average Pressure Drop Across Distributor
12	25 psi
22	35 psi
500	25 psi
502	35 psi
717 (Ammonia)	40 psi

Sporlan Valve

ily adjusts to the new load condition with a negligible chance in superheat.

Valve Location

Thermostatic expansion valves may be mounted in any position. However, they should be installed as close to the evaporator inlet as possible. If a refrigerant distributor is used, mount the distributor directly to the valve outlet for best performance. If a hand valve is located on the outlet side of the thermostatic expansion valve (TEV), it should have a full-sized port. No restrictions should appear between the thermostatic expansion valve and the evaporator, except a refrigerant distributor if one is used.

When the evaporator and thermostatic expansion valve are located above the receiver, there is a static pressure loss in the liquid line. This is due to the weight of the column of liquid refrigerant. This weight may be interpreted in terms of pressure loss in pounds per square inch. Table 11-C. If the vertical lift is great enough, vapor, or *flash gas*, will form in the liquid line. This greatly reduces the capacity of the thermostatic expansion valve.

When an appreciable vertical lift is unavoidable, precautions should be taken to prevent the accompanying pressure loss from producing liquid line vapor. This can be accomplished by providing enough subcooling to the liquid refrigerant, either in the condenser or after the liquid leaves the receiver. Subcooling is found by subtracting the actual liquid temperature from the condensing temperature (corresponding to the condensing pressure). The amount of subcooling necessary to prevent vapor formation in the liquid line is usually available in a table. Table 11-D. CAUTION: *Ammonia valves should never be permitted to operate with vapor in the liquid line. This causes severe pin and seat erosion. It also will drastically reduce the life of the valve.*

Bulb Location

The location of the bulb is extremely important. In some cases, it determines the success or failure of the refrigerating plant. For satisfactory expansion valve control, good *thermal contact* between the bulb and suction line is essential. The bulb should be securely fastened with two bulb straps to a clean, straight section of the suction line.

Application of the bulb to a horizontal run of suction line is preferred. If a vertical installation cannot be avoided, the bulb should be mounted so that the capillary tubing comes out at the top. On suction lines 7/8" OD and larger, the surface temperature may vary slightly around the circumference of the line. On these lines, it is generally recommended that the bulb be installed at a point midway on the side of the horizontal line and

CHAPTER 11—REFRIGERANT FLOW CONTROL

Table 11-D.
Pressure Loss and Required Subcooling for 100° F. & 130° F. Condensing of Refrigerants

Refrigerant	100° F. [37.8°C] Condensing Pressure Loss—psi					
	5	10	20	30	40	50
	Required Subcooling—°F.					
12	3	6	12	18	25	33
22	2	4	8	11	15	19
500	3	5	10	15	21	27
502	2	3	7	10	14	18
717 (Ammonia)	2	4	7	10	14	17

Refrigerant	130° F. [54.4°C] Condensing Pressure Loss—psi					
	5	10	20	30	40	50
	Required Subcooling—°F.					
12	3	5	9	14	18	23
22	2	4	6	9	12	14
500	2	4	8	11	15	19
502	1	3	5	8	11	13
717 (Ammonia)	2	3	5	7	10	12

Sporlan Valve

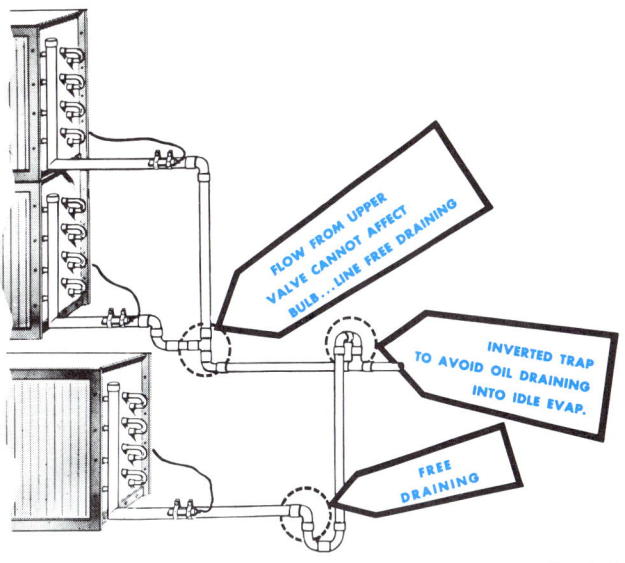

Virginia Chemicals

11-22. Installation of the TEV with multiple evaporators, above and below main suction line.

parallel to the direction of flow. On smaller lines the bulb may be mounted at any point around the circumference. However, locating the bulb on the bottom of the line is not recommended, since an oil-refrigerant mixture is generally present at that point. Certain conditions peculiar to a particular system may require a different bulb location than that normally recommended. In these cases, the proper bulb location may be determined by trial.

Accepted principles of good suction line piping should be followed to provide a bulb location that will give the best possible valve control. Never locate the bulb in a trap or pocket in the suction line. Liquid refrigerant or a mixture of liquid refrigerant and oil boiling out of the trap will falsely influence the temperature of the bulb and result in poor valve control.

Recommended suction line piping includes a horizontal line leaving the evaporator to which the thermostatic expansion valve bulb is attached. This line is pitched slightly downward. When a vertical riser follows, a short trap is placed immediately ahead of the vertical line. Fig. 11-21. The trap will collect any liquid refrigerant or oil passing through the suction line and prevent it from influencing the bulb temperature.

On multiple evaporator installations the piping should be arranged so that the flow from any valve cannot affect the bulb of another. Approved piping practices, including the proper use of traps, insure individual control for each valve without the influence of refrigerant and oil flow from other evaporators. Fig. 11-22.

For recommended suction line piping when the evaporator is located above the compressor, see Fig. 11-23. The vertical riser extending to the height of the evaporator prevents refrigerant from draining by gravity into the compressor during the off-cycle. When a pump-down control is used, the suction line may turn down without a trap.

On commercial and low temperature applications, the bulb should be the same as the evaporator temperature during the off-cycle. This will insure tight closing of the valve when the compressor stops. If bulb insulation is used on lines operating

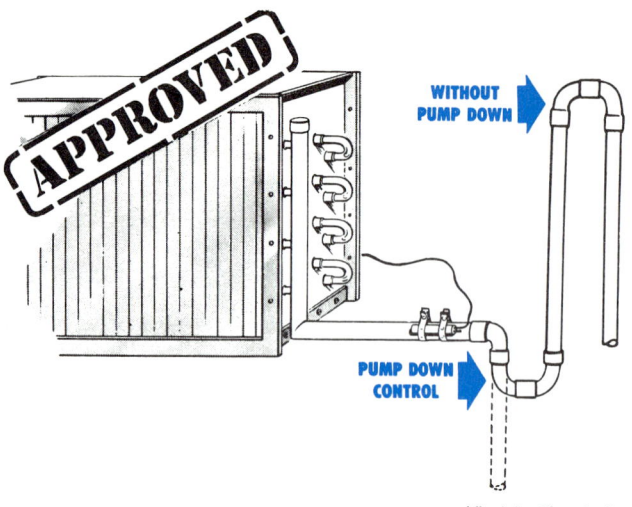

REFRIGERATION AND AIR CONDITIONING TECHNOLOGY

11-23. Installation of the TEV with the compressor below the evaporator.

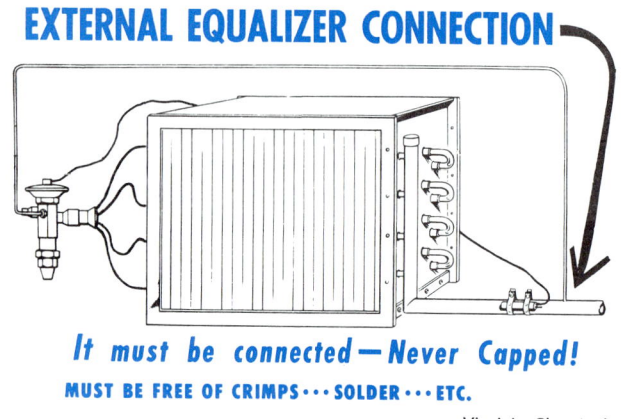

11-24. External equalizer connection.

below 32° F. [0°C], use non-water absorbing insulation to prevent water from freezing around the bulb.

On brine tanks and water coolers the bulb should be below the liquid surface. Here, it will be at the same temperature as the evaporator during the off-cycle. A solenoid valve must be used ahead of the thermostatic expansion valve.

Some air conditioning applications have thermostatic expansion valves equipped with charged elements. Here, the bulb may be located inside or outside the cooled space or duct. The valve body should not be located in the airstream leaving the evaporator. Avoid locating the bulb in the return airstream unless the bulb is well insulated.

External Equalizer

As the evaporating temperature drops, the maximum pressure drop that can be tolerated between the valve outlet and the bulb location without serious capacity loss for an internally equalized valve also decreases. This is shown in Table 11-D. There are, of course, applications that may satisfactorily employ the internal equalizer when higher pressure drop is present. This should usually be verified by laboratory tests. The general recommendations given in Table 11-D are suitable for most field-installed systems. Use the external equalizer when pressure drop between the outlet and bulb locations exceeds values shown in Table 11-D. When the expansion valve is equipped with an external equalizer, it must be connected. Never cap an external equalizer. The valve may flood, starve, or regulate erratically. There is no operational disadvantage in using an external equalizer, even if the evaporator has a low pressure drop. Note: The external equalizer *must* be used on evaporators that use a pressure-drop type refrigerant distributor. Fig. 11-24. Generally, the external equalizer connection is in the suction line immediately downstream of the bulb. Fig. 11-24. However, equipment manufacturers sometimes select other locations that are compatible with their specific design requirements.

Field Service

The thermostatic expansion valve is erroneously considered by some to be a complex device. As a result, many valves are needlessly replaced when the cause of the system malfunction is not immediately recognized.

Actually, the thermostatic expansion valve performs only one very simple function. It keeps the evaporator supplied with enough refrigerant to satisfy all load con-

CHAPTER 11—REFRIGERANT FLOW CONTROL

ditions. It is not a temperature control, suction pressure control, a control to vary the compressor running time, or a humidity control.

The effectiveness of the valve's performance is easily determined by measuring the superheat. Fig. 11-25. Observing the frost on the suction line or considering only the suction pressure may be misleading. Checking the superheat is the first step in a simple and systematic analysis of thermostatic expansion valve performance.

If insufficient refrigerant is being fed to the evaporator the superheat will be high. If too much refrigerant is being fed to the evaporator the superheat will be low.

Although these symptoms may be attributed to improper thermostatic expansion valve control, more frequently the origin of the trouble lies elsewhere.

CRANKCASE PRESSURE REGULATING VALVES

Crankcase pressure regulating valves are designed to prevent overloading of the compressor motor. They limit the crankcase pressure during and after a defrost cycle or after a normal shutdown period. When properly installed in the suction line, these valves automatically throttle the vapor flow from the evaporator until the compressor can handle the load. They are available in the range of 0 to 60 psig.

Operation of the Valve

Crankcase pressure regulating valves (CROS) are sometimes called suction pressure regulating valves. They are sensitive only to their outlet pressure. This would be the compressor crankcase or suction pressure. To indicate this trait, the designation describes the operation: *C*lose on *R*ise of *O*utlet pressure, or CRO. As shown in Fig. 11-26, the inlet pressure is exerted on the underside of the bellows and on top of the seat disc. Since the effective area of the bellows is equal to the area of the port, the inlet pressure cancels out and does not affect valve operation. The valve outlet pressure acting on the bottom of the disc exerts a force in the closing direction. This force is opposed by the adjustable spring force. These are the operating forces of the CRO. The CRO's pressure setting is determined by the spring force. Thus, by increasing the spring force, the valve setting or the pressure at which the valve will close is increased.

As long as the valve outlet pressure is greater than the valve pressure setting, the valve will remain closed. As the outlet pressure is reduced, the valve will open and pass refrigerant vapor into the compressor. Further reduction of the outlet pressure will allow the valve to open to its rated position, where the rated pressure drop will exist across the valve port. An increase in the outlet pressure will cause the valve to throttle until the pressure setting is reached.

The operation of a valve of this type is improved by an anti-chatter device built into the valve.

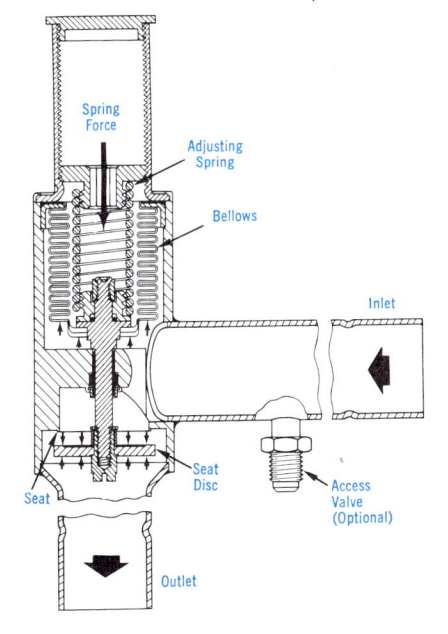

11-25. How to figure superheat.

11-26. Crankcase pressure regulating valve.
Sporlan Valve

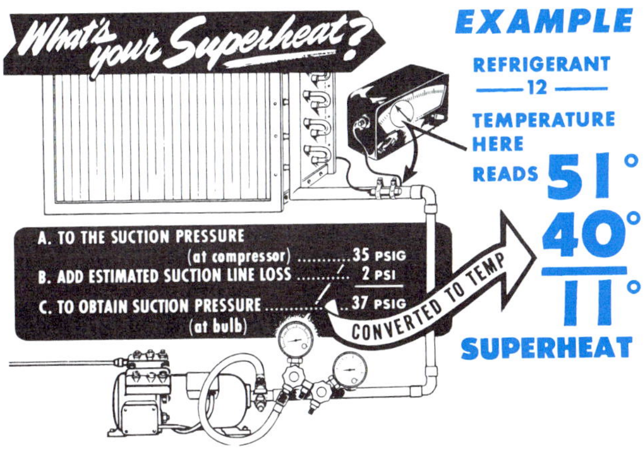

REFRIGERATION AND AIR CONDITIONING TECHNOLOGY

Without this device, the CRO would be susceptible to compressor pulsations that greatly reduce the life of a bellows. This feature allows the CRO to function at low load conditions without any chattering or other operation difficulties.

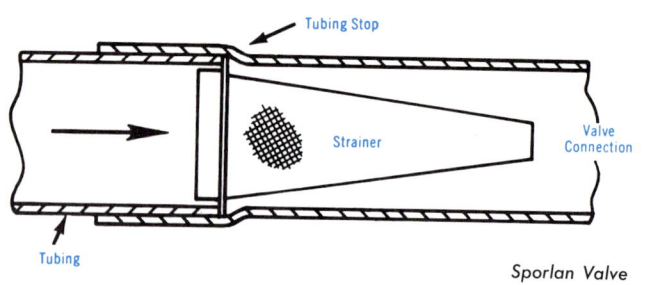

Valve Location

As Fig. 11-27 indicates, the CRO valve is applied in the suction line between the evaporator and the compressor. Normally, the CRO is installed downstream of any other controls or accessories. However, on some applications it may be advisable or necessary to locate other system components, such as an accumulator, downstream of the CRO. This is satisfactory as long as the CRO valve is applied only as a crankcase pressure regulating valve. CRO valves are designed for application in the suction line only. They should not be applied in hot gas bypass lines or any other refrigerant line of a system.

Strainer

Just as with any refrigerant flow control device, the need for an inlet strainer is a function of system cleanliness and proper installation procedures. Fig. 11-28. When the strainer is used, the tubing is inserted in the valve connection up to the tubing stop. Thus, the strainer has been locked in place. Moisture and particles too small for the inlet strainer are harmful to the system and must be removed. Therefore, it is recommended that a filter-drier be installed according to the application recommendations.

11-28. Strainer for cleanliness.

11-27. CRO valve applied in the suction line between the evaporator and the compressor.

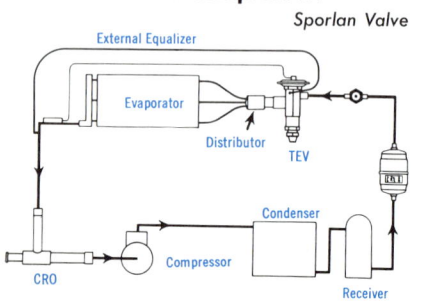

Brazing Procedures

When installing CROs with solder connections, the internal parts must be protected by wrapping the valve with a *wet* cloth to keep the body temperature below 250° F. [121°C]. The tip of the torch should be large enough to avoid prolonged heating of the connections. Overheating can also be minimized by directing the flame away from the valve body.

Test and Operating Pressures

Excessive leak testing or operating pressures may damage these valves by reducing the life of the bellows. For leak detection, an inert gas such as nitrogen or CO_2 may be added to an idle system to supplement the refrigerant pressure.

CAUTION: *Inert gas must be added to the system carefully. Use a pressure regulator. Unregulated gas pressure can seriously damage the system and endanger human life. Never use oxygen or explosive gases. The valves will withstand 200 to 300 psig. However, check the manufacturer's recommendations first.*

Adjusting the Pressure

The standard setting by the factory for CROs in the 0/60 psig range is 30 psig. Since these valves are adjustable, the setting may be altered to suit the specific system requirements. CROs should be adjusted at start-up when the pressure in the evaporator is above the desired setting. The final valve setting should be below the maximum suction pressure recommended by the compressor or unit manufacturer.

The main purpose of the CRO is to prevent the compressor motor from overloading due to high suction pressure. Thus, it is important to arrive at the correct pressure setting. The best way to see if the motor is overloaded is to check the current draw at start-up or after a defrost cycle. If overloading is evident, a suction gage should be put on the compressor. The CRO setting may be too high and may have to be adjusted. If the compressor is overloaded and the CRO valve is to be reset, the following procedure should be followed.

The unit should be shut off long enough for the system pressure to equalize. Observe the suction pressure as the unit is started, since this is the pressure the valve is controlling. If the setting is to be decreased, slowly adjust the valve in a counterclockwise direction approximately one-quarter turn for each 1 psi pressure change required. After a few moments of operation, the unit should be cycled off and the system pressure allowed to equalize again. Observe the suction pressure (valve setting) as the unit is started up. If the setting is still too high, the adjustment should be repeated. The proper size hex wrench is used to adjust these valves. A clockwise rotation increases the valve setting, while a counterclockwise rotation decreases the setting.

When CROs are installed in parallel, each should be adjusted the same amount. If one valve has been adjusted more than the other, best performance will occur if both are adjusted all the way in before resetting them an equal amount.

Service

Since CRO valves are hermetic and cannot be disassembled for inspection and cleaning, they are usually replaced if inoperative. If a CRO fails to open, close properly, or will not adjust, solder or other foreign material is probably lodged in the port. It is sometimes possible to dislodge these materials by turning the adjustment nut all the way in with the system running. If the CRO develops a refrigerant leak around the spring housing, it probably has been overheated during installation or the bellows has failed due to severe compressor pulsations. In either case, the valve must be replaced.

EVAPORATOR PRESSURE REGULATING VALVES

Evaporator pressure regulating valves offer an efficient means of balancing the system capacity and the load requirements during periods of low loads. They also are able to maintain different evaporator conditions on multitemperature systems. The main function of this valve is to prevent the evaporator pressure from falling below a predetermined value at which the valve has been set.

Control of evaporator pressure by cycling the compressor with a thermostat or some other method is quite adequate on most refrigeration systems. Control of the evaporator pressure also controls the saturation temperature. As the load drops off, the evaporating pressure starts to decrease and the system performance falls off. These valves automatically throttle the vapor flow from the evaporator. This maintains the desired minimum evaporator pressure. As the load increases, the evaporating pressure will increase above the valve setting and the valve will open further.

Operation

For any pressure sensitive valve to modulate to a more closed or open position, a change in the operating pressure is required. The unit change in the valve stroke for a given change in the operating pressure is called the *valve gradient*. Every valve has a specific gradient designed into it for the best possible operation. Valve sensitivity and the valve's capacity rating are functions of the valve gradient. Thus, a relatively sensitive valve is needed when a great change in the evaporating temperature cannot be tolerated. Therefore, the valves have nominal ratings based on an 8 psi evaporator pressure change, rather than a full stroke.

Evaporator pressure regulating valves respond only to variations in their inlet pressure (evaporator pressure). Thus, the designation for evaporator pressure regulating valves is ORI (*O*pens on *R*ise of *I*nlet pressure). Fig. 11-29.

Pressure at the outlet is exerted on the underside of the bellows and on top of the seat disc. The effective area of the bellows is equal to the area of the port. Thus, the outlet pressure cancels out and the inlet pressure acting on the bottom of the seat disc opposes the adjustable spring force. These two forces are the operating forces of the ORIT. (The *T* added to the valve designation indicates an access valve

11-29. Evaporator pressure regulating valve.
Sporlan Valve

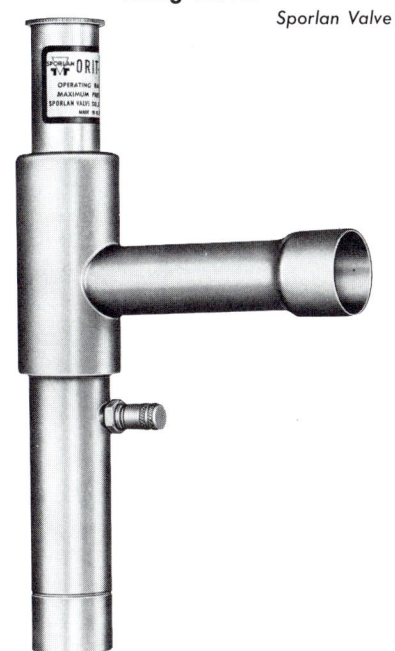

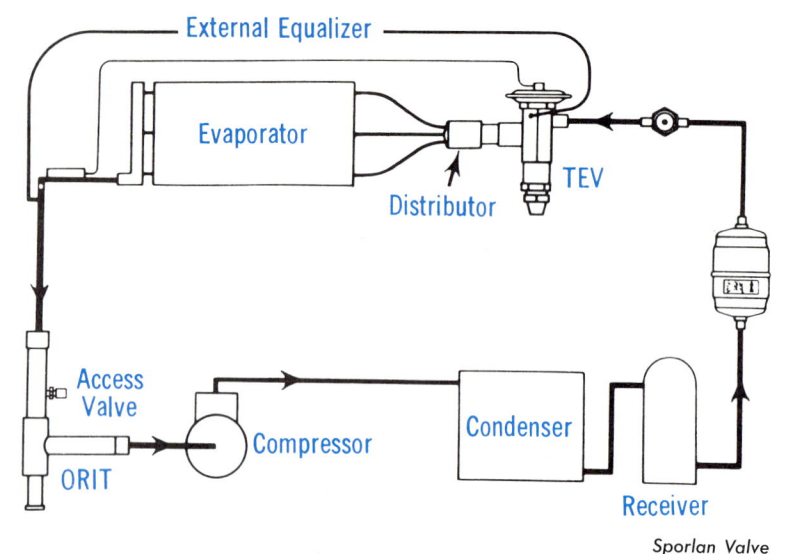

REFRIGERATION AND AIR CONDITIONING TECHNOLOGY

11-30. Valve location in a single evaporator system.
Sporlan Valve

Type of System

The proper application of the evaporator pressure regulating valve involves the consideration of several system factors.

One type of system is a single evaporator type, such as a water chiller. Here, the valve is used to prevent freeze-up at light loads. Fig. 11-30.

Another type of system is a multitemperature refrigeration system with evaporators operating at different temperatures. Fig. 11-31. A valve may be required on one or more of the evaporators to maintain pressures higher than that of the common suction line. For example, if evaporator A in Fig. 11-31 is designed for 35° F. [1.7°C] (72.6 psig on Refrigerant 502), evaporator B for 32° F. [0°C] on the same refrigerant (68.2 psig), and other evaporators for 25° F. [−3.9°C] (58.7 psig), the valves (ORIT) are used to maintain a pressure of 72.6 psig in evaporator A and 68.2 psig in evaporator B. However, some multitemperature systems may require an ORIT on each evaporator, depending on the type of product being refrigerated.

on the inlet connection.) When the evaporator load changes, the ORIT opens or closes in response to the change in evaporator pressure. An increase in inlet pressure above the valve setting tends to open the valves. If the load drops, less refrigerant is boiled off in the evaporator and the evaporator pressure will decrease. The decrease in evaporator pressure tends to move the ORIT to a more closed position. This, in turn, keeps the evaporator pressure up. The result is that the evaporator pressure changes as the load changes. The operation of a valve of this type is improved by an anti-chatter device built into the valve. Without this device, the ORIT would be susceptible to compressor pulsations that can reduce the life of a bellows. This anti-chatter feature allows the ORIT to function at low load conditions without chattering or other operating difficulties.

11-31. Multitemperature refrigeration system.
Sporlan Valve

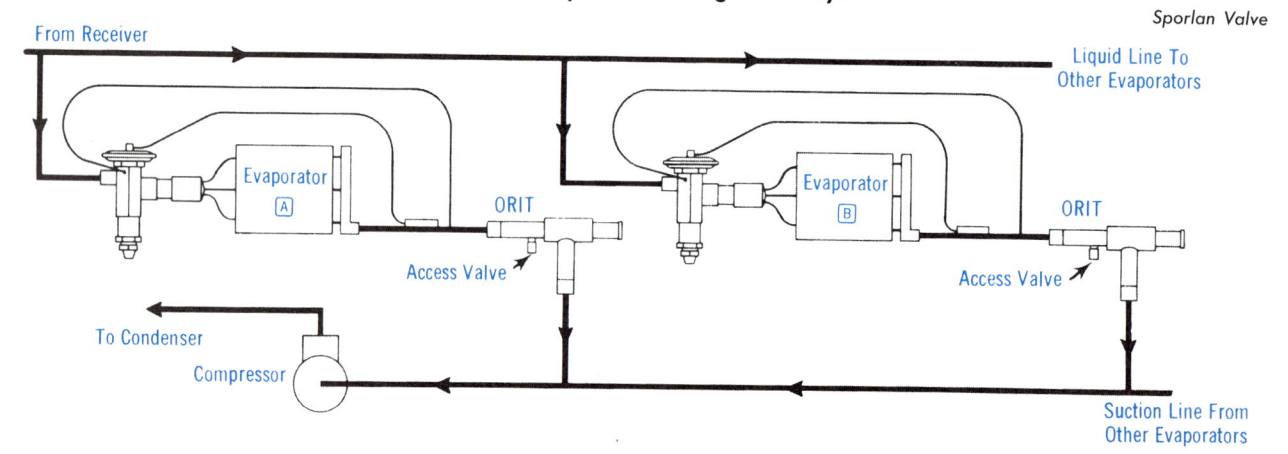

CHAPTER 11—REFRIGERANT FLOW CONTROL

Valve Location

ORITs must be installed upstream of any other suction line controls or accessories. These valves may be installed in the position most suited to the application. However, these valves should be located so that they do not act as an oil trap or so that solder cannot run into the internal parts during brazing in the suction line. Since these valves are hermetic, they cannot be disassembled to remove solder trapped in the internal parts.

Installation of a filter-drier and a strainer may be worth the expense to keep the system clean and operational. Brazing procedures are the same as for other valves of this type. The valve core of the access valve is shipped in an envelope attached to the access valve. If the access valve connection is to be used as a reusable pressure tap to check the valve setting, the ORIT must be brazed in *before* the core is installed. This protects the synthetic material of the core. If the access valve is to be used as a permanent pressure tap, the core and access valve cap may be discarded.

Test and Operating Pressures

As with other pressure valves, it is possible to introduce nitrogen or CO_2 in an idle system to check for correct pressure settings.

The usual precautions for working with gases apply here. The standard factory setting for the 0/50 psig range is 30 psig. For the 30/100 psig range, it is 60 psig. Since these valves are adjustable, the setting may be altered to suit the system.

The main purpose of an ORIT valve is to keep the evaporator pressure above some given point at minimum load conditions. The valves are selected on the basis of the pressure drop at full load conditions. Nevertheless, they should be adjusted to maintain the minimum allowable evaporator pressure under the actual minimum load conditions.

These valves can be adjusted by removing the cap and turning the adjustment screw with a hex wrench of the proper size. A clockwise rotation increases the valve setting, while a counter-clockwise rotation decreases the setting. To obtain the desired setting, a pressure gage should be utilized on the inlet side of the valve. Thus, the effects of any adjustments can be observed.

When these valves are installed in parallel, each should be adjusted the same amount. If one valve has been adjusted more than the other, the best performance will occur if both are adjusted all the way in before resetting them an equal amount.

Service

Since these valves are hermetic and cannot be disassembled for inspection and cleaning, they usually must be replaced if found defective or inoperative. It is possible sometimes to adjust the valve until the obstruction is dislodged. This usually works best when the system is running. If it leaks around the spring housing, it will have to be replaced. The bellows have been permanently damaged.

HEAD PRESSURE CONTROL VALVES

Design of air conditioning and refrigeration systems using air-cooled condensing units involves two main problems that must be solved if the system is to be operated reliably and economically. These problems are high ambient and low ambient operation. If the condensing unit is properly sized, it will operate satisfactorily during extreme ambient temperatures. However, most units will be required to operate at ambient temperatures below their design dry bulb temperature during most of the year. Thus, the solution to low ambient operation is more complex.

Without good head pressure control during low ambient operation, the system can have running-cycle and off-cycle problems. Two running-cycle problems are of prime concern:

1. The pressure differential across the thermostatic expansion valve port affects the rate of refrigerant flow. Thus, low head pressure generally causes insufficient refrigerant to be fed to the evaporator.

2. Any system using hot gas for defrost or compressor capacity control must have a normal head pressure to operate properly. In either case, failure to have sufficient head pressure will result in low suction pressure and/or iced evaporator coils.

The primary off-cycle problem is the possible inability to get the system on-the-line if the refrigerant has migrated to the condenser. The evaporator pressure may not build up to the cut-in point of the low pressure control. The compressor cannot start, even though refrigeration is required. Even if the evaporator pressure builds up to the cut-in setting, insufficient flow through the TEV will cause a low suction pressure, which results in compressor cycling.

273

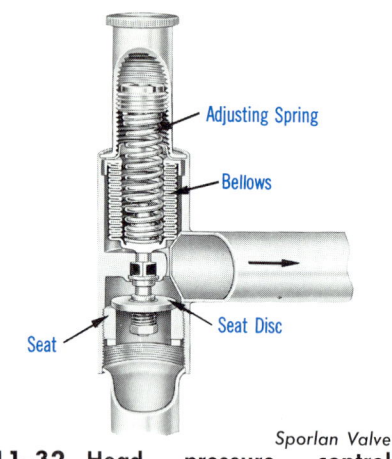

REFRIGERATION AND AIR CONDITIONING TECHNOLOGY

11-32. Head pressure control valve. *Sporlan Valve*

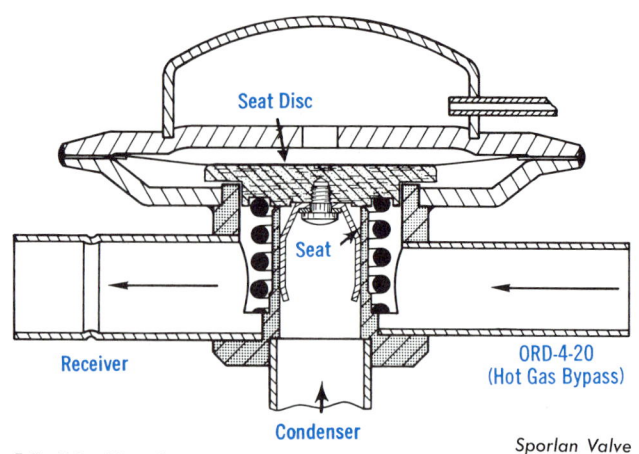

11-33. Head pressure control valve that opens on rise of outlet pressure (ORO). *Sporlan Valve*

There are nonadjustable and adjustable methods of head pressure control by valves. Each method uses two valves designed specifically for this type of application. Low ambient conditions are encountered during fall-winter-spring operation on air-cooled systems, with the resultant drop in condensing pressure. Then, the valve's purpose is to hold back enough of the condensed liquid refrigerant to make part of the condenser surface inactive. This reduction of active condensing surface raises condensing pressure and sufficient liquid line pressure for normal system operation.

Operation

The ORI head pressure control valve is an inlet pressure regulating valve. It responds to changes in condensing pressure only. The valve designation stands for *O*pens on *R*ise of *I*nlet pressure. As shown in Fig. 11-32, the outlet pressure is exerted on the underside of the bellows and on top of the seat disc. Since the effective area of the bellows is equal to the area of the port, the outlet pressure cancels out. The inlet pressure acting on the bottom of the seat disc opposes the adjusting spring force. These two forces are the operating forces of the ORI.

When the outdoor ambient temperature changes, the ORI opens or closes in response to the change in condensing pressure. An increase in inlet pressure above the valve setting tends to open the valve. If the ambient temperature drops, the condenser capacity is increased and the condensing pressure drops off. This causes the ORI to start to close or assume a throttling position.

ORO Valve Operation

The ORO head pressure control valve is an *outlet* pressure regulating valve that responds to changes in receiver pressure. The valve designation stands for *O*pens on *R*ise of *O*utlet pressure. Fig. 11-33. The inlet and outlet pressures are exerted on the underside of the seat disc in an opening direction. Since the area of the port is small in relationship to the diaphragm area, the inlet pressure has little direct effect on the operation of the valve. The outlet or receiver pressure is the control pressure. The force on top of the diaphragm that opposes the control pressure is due to the air charge in the element. These two forces are the operating forces of the ORO.

When the outdoor ambient temperature changes, the condensing pressure changes. This causes the receiver pressure to fluctuate accordingly. As the receiver pressure decreases, the ORO throttles the flow of liquid from the condenser. As the receiver pressure increases, the valve modulates in an opening direction to maintain a nearly constant pressure in the receiver. Since the ambient temperature of the element affects the valve pressure setting, the control pressure may change slightly when the ambient temperature changes. However, the valve and element temperature remain fairly constant.

ORD Valve Operation

The ORD valve is a pressure *differential* valve. It responds to changes in the pressure difference across the valve. Fig. 11-34. The valve designation stands for *O*pens on *R*ise of *D*ifferential pressure. Therefore, the ORD is dependent on some other control valve or action for its operation. In this respect, it is used with either the ORI or ORO for head pressure control.

As either the ORI or ORO valve starts to throttle the flow of liquid refrigerant from the condenser, a pressure differential is created across the ORD. When the differential reaches 20 psi, the ORD starts to open and bypasses hot gas to the liquid drain line. As the differential increases, the ORD opens further until its full stroke is reached at a differential of 30 psi. Due to its function in the control of head pressure, the full stroke can be utilized in selecting the ORD. While the capacity of the ORD increases as the pressure differential increases, the rating point at 30 psi is considered a satisfactory maximum value.

The standard pressure setting for the ORD is 20 psig. For systems where the condenser pressure drop is higher than 10 or 12 psi, an ORD with a higher setting can be ordered.

Head pressure control can be improved with an arrangement such as that shown in Fig. 11-35. In this operation, a constant receiver pressure is maintained for normal system operation. The ORI is adjustable over a nominal range of 100 to 225 psig. Thus, the desired pressure can be maintained for all of the commonly used refrigerants—12, 22, and 502.

The ORI is located in the liquid drain line between the condenser and the receiver. The ORD is located in a hot gas line bypassing the condenser. During periods of low ambient temperature, the condensing pressure falls until it approaches the setting of the ORI valve. The ORI then throttles, restricting the flow of liquid from the condenser. This causes refrigerant to back up in the condenser, thus reducing the active condenser surface. This raises the condensing pressure. Since it is really receiver pressure that needs to be maintained, the bypass line with the ORD is required.

The ORD opens after the ORI has offered enough restriction to cause the differential between condensing pressure and receiver pressure to exceed 20 psi. The hot gas flowing through the ORD heats up the cold liquid being passed through the ORI. Thus, the liquid reaches the receiver warm and with sufficient pressure to assure proper expansion valve operation. As long as sufficient refrigerant charge is in the system, the two valves modulate the flow automatically to maintain proper receiver pressure regardless of outside ambient temperature.

Installation

To insure proper performance, head pressure control valves

11-34. Head pressure control valve that opens on rise of differential across the valve (ORD).
Sporlan Valve

11-35. Adjustable ORI/ORD system.
Sporlan Valve

must be selected and applied correctly. These valves can be installed in either horizontal or vertical lines. If possible, the valves should be oriented so solder cannot run into the internal parts during brazing. Care should be taken to install the valves with the flow in the proper direction. The ORI and ORO valves *cannot* be installed in the discharge line for any reason.

In most cases the valves are located at the condensing unit. When the condenser is remote from the compressor, the usual location is near the compressor. In all cases it is important that some precautions be taken in mounting the valves. While the heaviest valve is approximately 2.5 pounds [1.14 kilograms] in weight, it is suggested that they be adequately supported to prevent excessive stress on the connections. Since discharge lines are a possible source of vibrations that result from discharge gas pulses and inertia forces associated with the moving parts, fatigue in tubing, fittings, and connections may result. Pulsations are best handled by placing a good muffler as close to the compressor as possible.

Vibrations from moving parts of the compressor are best isolated by flexible loops or coils (discharge lines 1/2" or smaller) or flexible metal hoses for larger lines. For best results, the hoses should be installed as close to the compressor shut-off valves as possible. The hoses should be mounted horizontally and parallel to the crankshaft or vertically. The hoses should *never* be mounted horizontally and 90° from the crankshaft. A rigid brace should be placed on the outlet end of the hose. This brace will prevent vibrations beyond the hose.

Brazing Procedures

Any of the commonly used brazing alloys for high side usage are satisfactory. It is very important that the internal parts be protected by wrapping the valve with a wet cloth to keep the body temperature below 250° F. [121°C]. Also, when using high temperature solders, the torch tip should be large enough to avoid prolonged heating of the copper connections. Always direct the flame away from the valve body.

Test and Operating Pressures

Excessive leak testing or operating pressures may damage these valves and reduce the life of the operating members. For leak detection, an inert dry gas such as nitrogen or CO_2 can be added to an idle system to supplement the refrigerant pressure.

Remove the cap and adjust the adjustment screw with the proper wrench. Check the manufacturer's recommended pressures before making adjustments.

Refrigerant and charging procedures require that enough refrigerant be available for flooding the condenser at the lowest expected ambient temperature. There must still be enough charge in the system for proper operation.

A shortage of refrigerant will cause hot gas to enter the liquid line and the expansion valve. Refrigeration will cease.

The receiver must have sufficient capacity to hold at least all of the excess liquid refrigerant in the system. This is because such refrigerant will be returned to the receiver when high ambient conditions prevail. If the receiver is too small, liquid refrigerant will be held back in the condenser during high ambient conditions. Excessively high discharge pressures will be experienced.

CAUTION: *All receivers must utilize a pressure relief valve or device according to the applicable standards or codes (ARI Standard 495).*

Follow the manufacturer's recommendations for charging the system. Procedures may vary with different valve manufacturers.

Service

There are several possible causes for system malfunction with "refrigerant side" head pressure control. These may be difficult to isolate from each other. As with any form of system troubleshooting, it is necessary to know the existing operating temperatures and pressures before system problems can be determined. Once the malfunction is established, it is easier to pinpoint the cause and then take suitable action. Table 11-E lists the most common malfunctions, the possible causes, and the remedies.

Nonadjustable ORO/ORD System Operation

The nonadjustable ORO head pressure control valve and the ORD pressure differential valve offer the most economical system of refrigerant side-head pressure control. Just as the ORI/ORD system simplified this type of control, the ORO/ORD system offers the capability of locating the condenser and receiver on the *same* elevation. See Fig. 11-36. By making these two valves available either separately or

CHAPTER 11—REFRIGERANT FLOW CONTROL

Table 11-E.
Troubleshooting Head Pressure Control Valves

MALFUNCTION—LOW HEAD PRESSURE	
Possible Cause	Remedy
1. Insufficient refrigerant charge to adequately flood condenser.	1. Add charge.
2. Low pressure setting on ORI.	2. Increase setting.
3. ORI fails to close due to foreign material in valve.	3. Turn adjustment out so material passes through valve. If unsuccessful, replace ORI.
4. ORI fails to adjust properly.	4. See 3 above.
5. Wrong setting on ORO (e.g., 100 psig on Refrigerant 22 or 502 system).	5. Replace ORO with valve with correct setting.

6. ORO fails to close due to:
  a. Foreign material in valve.
 	
  b. Loss of air charge in element.
	6. See below:
	  a. Cause ORO to open by raising condensing/receiver pressure above valve setting by cycling condenser fan. If foreign material does not pass through valve, replace ORO.
	  b. Replace ORO.

7. ORD fails to open (on ORI/ORD system only) due to:
  a. Less than 20 psi pressure drop across ORD.
  b. Internal parts damaged by overheating when installed.
	7. See below:
	  a. Check ORI causes/remedies above: 2, 3, or 4.
	  b. Replace ORD.

8. Refrigerant leak at adjustment housing of ORI.	8. Replace ORI.

MALFUNCTION—HIGH HEAD PRESSURE	
Possible Cause	Remedy
1. Dirty condenser coil.	1. Clean coil.
2. Air on condenser blocked off.	2. Clear area around unit.
3. Too much refrigerant charge.	3. Remove charge until proper head pressure is maintained.
4. Undersized receiver.	4. Check receiver capacity against refrigerant required to maintain desired head pressure.
5. Non-condensibles (air) in system.	5. Purge from system.
6. High pressure setting on ORI.	6. Decrease setting.
7. ORI or ORO restricted due to inlet strainer being plugged.	7. Open inlet connection to clean strainer.
8. ORI fails to adjust properly or to open due to foreign material in valve.	8. Turn adjustment out so material passes through valve. If unsuccessful, replace ORI.
9. Wrong setting on ORO (e.g., 180 psig on Refrigerant 12 system).	9. Replace ORO with valve with correct setting.
10. ORD fails to open due to internal parts being damaged by overheating when installed (only when used with ORO).	10. Replace ORD.

11. ORD bypassing hot gas when not required due to:
  a. Internal parts damaged by overheating when installed.
  b. Pressure drop across condenser coil, ORI or ORO, and connecting piping above 14 psi.
	11. See below:
	  a. Replace ORD.
	  b. Reduce pressure drop (e.g., use larger ORI **or** ORI or ORO valves in parallel) or order ORD-4 with higher setting.

Sporlan Valve

brazed together, there is added flexibility in the piping layout.

The operation of the ORO/ORD system is such that a nearly constant receiver pressure is maintained for normal operation. As the temperature of the ORO element decreases, the pressure setting decreases ac-

Sporlan Valve
11-36. The ORO is located in the liquid drain line between the condenser and the receiver.

277

cordingly. However, by running the bypassed hot gas through the ORO the element temperature is adequately maintained so the ORO/ORD system functions well to ambient temperatures of −40° F. [−40°C] and below. This third connection on the ORO also eliminates the need for a tee connection in the liquid drain line.

Note that in Fig. 11-36 the ORO is located in the liquid drain line between the condenser and the receiver, while the ORD is located in a hot gas line bypassing the condenser. Other than the fact that the ORO operates in response to its outlet pressure (receiver pressure), the ORO/ORD operates in the same basic manner as the ORI/ORD system previously explained.

DISCHARGE BYPASS VALVES

On many air conditioning and refrigeration systems it is desirable to limit the minimum evaporating pressure. This is so especially during periods of low load either to prevent coil icing or to avoid operating the compressor at a lower suction pressure than it was designed for. Various methods of operation have been designed to achieve the result—integral cylinder unloading, gas engines with variable speed control, or multiple smaller systems. Compressor cylinder unloading is used extensively on larger systems. However, it is too costly on small equipment, usually 10 hp or below. Cycling the compressor with a low pressure cutout control has had widespread usage, but is being reevaluated for three reasons:

1. On-off control on air conditioning systems is uncomfortable and does a poor job of humidity control.

2. Compressor cycling reduces equipment life.

3. In most cases, compressor cycling is uneconomical because of peak load demand charges.

One solution to the problem is to bypass a portion of the hot discharge gas directly into the low side. This is done by the modulating control valve—commonly called a *Discharge Bypass Valve* (DBV). This valve, which opens on a decrease in suction pressure, can be set to maintain automatically a desired minimum evaporating pressure, regardless of the decrease in evaporator load.

Operation

Discharge bypass valves (DBV) respond to changes in downstream or suction pressure. Fig. 11-37. When the evaporating pressure is above the valve setting, the valve remains closed. As the suction pressure drops below the valve setting, the valve responds and begins to open. As with all mod-

11-37. Discharge bypass valve.

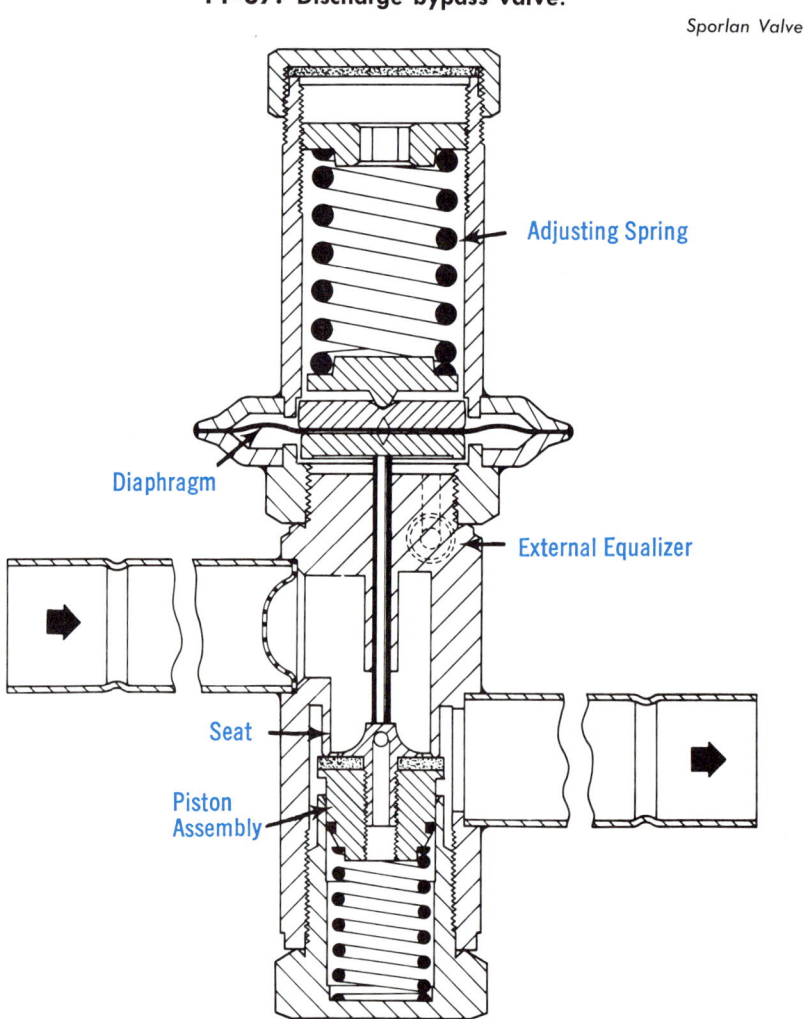

Sporlan Valve

278

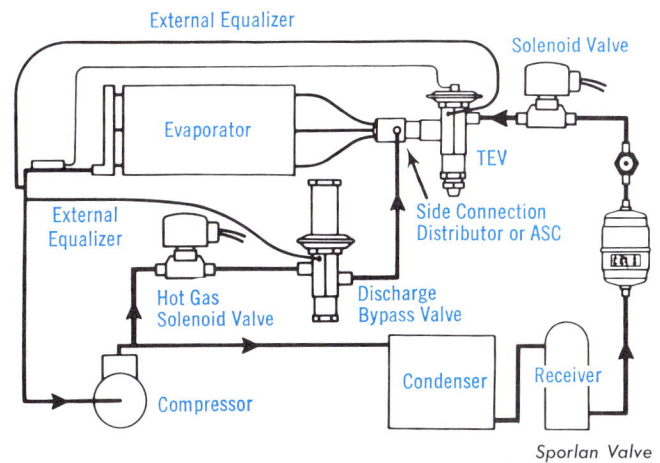

11-38. Connection arrangement for a discharge bypass valve. *Sporlan Valve*

ulating-type valves, the size of the opening is proportional to the change in the variable being controlled. In this case, the variable is the suction pressure. As the suction pressure drops, the valve opens further until the limit of the valve stroke is reached. However, on normal applications there is not sufficient pressure change to open these valves to the limit of their stroke. The amount of pressure change available from the point at which it is desired to have the valve closed to the point at which it is to be open varies widely with the refrigerant used and the evaporating temperature. For this reason, DBVs are rated on the basis of allowable evaporator temperature change from closed position to rated opening. A 6° F. [3.3°C] change is considered normal for most applications and is the basis of capacity ratings.

Application

DBVs provide an economical method of compressor capacity control in place of cylinder unloaders or of handling unloading requirements below the last step of cylinder unloading.

On air conditioning systems, the minimum allowable evaporating temperature that will avoid coil icing depends on evaporator design. The amount of air passing over the coil also determines the allowable evaporator minimum temperature. The refrigerant temperature may be below 32° F. [0°C]. However, coil icing will not usually occur with high air velocities, since the external surface temperature of the tube will be above 32° F. [0°C]. For most air conditioning systems the minimum evaporating temperature should be 26 to 28° F. [−3.3 to −2.2°C]. DBVs are set in the factory. They start to open at an evaporating pressure equivalent to 32° F. [0°C] saturation temperature. Therefore, evaporating temperature of 26° F. [−3.3°C] is their rated capacity. However, since they are adjustable, these valves can be set to open at a higher evaporating temperature.

On refrigeration systems, discharge bypass valves are used to prevent the suction pressure from going below the minimum value recommended by the compressor manufacturer. A typical application would be a low temperature compressor designed for operation at a minimum evaporating temperature on Refrigerant 22 of −40° F. [−40°C]. The required evaporating temperature at normal load conditions is −30° F. [−34°C]. A discharge bypass valve would be selected that would start to open at the pressure equivalent to −34° F. [−36°C] and bypass enough hot gas at −40° F. [−40°C] to prevent a further decrease in suction pressure. Valve settings are according to manufacturer's recommendations.

The discharge bypass valve is applied in a branch line off the discharge line as close to the compressor as possible. The bypassed vapor can enter the low side at one of the following locations:

1. To evaporator inlet with distributor.
2. To evaporator inlet without distributor.
3. To suction line.

Figure 11-38 shows the bypass to evaporator inlet with a distributor. The primary advantage of this method is that the system thermostatic expansion valve will respond to the increased superheat of the vapor leaving the evaporator and will provide the liquid required for desuperheating. The evaporator also serves as an excellent mixing chamber for the bypassed hot gas and the liquid-vapor mixture from the expansion valve. This ensures that dry vapor reaches the compressor suction. Oil return from the evaporator is also improved,

REFRIGERATION AND AIR CONDITIONING TECHNOLOGY

since the velocity in the evaporator is kept high by the hot gas.

Externally Equalized Bypass Valves

The primary function of the DBV is to maintain suction pressure. Thus, the compressor suction pressure is the control pressure. It must be exerted on the underside of the valve diaphragm. When the DBV is applied as shown in Fig. 11-38, where there is an appreciable pressure drop between the valve outlet and the compressor suction, the externally equalized valve must be used. This is true because when the valve opens, a sudden rise in pressure occurs at the valve outlet. This creates a false control pressure, which would cause the internally equalized valve to close.

Many refrigeration systems and water chillers do not use refrigerant distributors but may require some method of compressor capacity control. This type of application provides the advantages discussed previously.

Bypass to Evaporator Inlet without Distributor

On many applications, it may be necessary to bypass directly into the suction line. This is generally true of systems with multi-evaporators or remote condensing units. It may also be true for existing systems where it is easier to connect to the suction line than the evaporator inlet. The latter situation involves systems fed by TEVs or capillary tubes. When hot gas is bypassed, temperature starts to increase. This can cause breakdown of the oil and refrigerant, possibly resulting in a compressor burnout. On close-coupled systems, this can be eliminated by locating the main expansion valve bulb downstream of the bypass connection, as shown in Fig. 11-39.

Installation

Bypass valves can be installed in horizontal or vertical lines, whichever best suits the application and permits easy accessibility to the valves. However, consideration should be given to locating these valves so they do not act as oil traps. Also solder must not run into the internal parts during brazing.

The discharge bypass valve should always be installed at the condensing unit, rather than at the evaporator section. This will insure the rated bypass capacity of the discharge bypass valve. It will also eliminate the possibility of hot gas condensing in the bypass line. This is especially true on remote systems.

When externally equalized lines are used, the equalizer con-

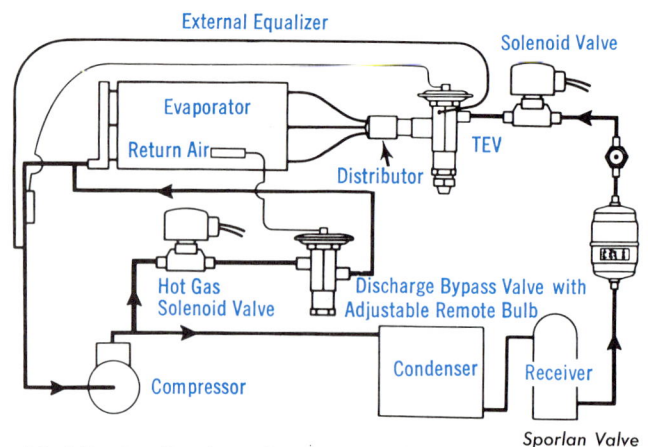

11-39. Application of a hot gas bypass to an existing system with only minor piping changes.
Sporlan Valve

11-40. Externally equalized discharge bypass valve.
Sporlan Valve

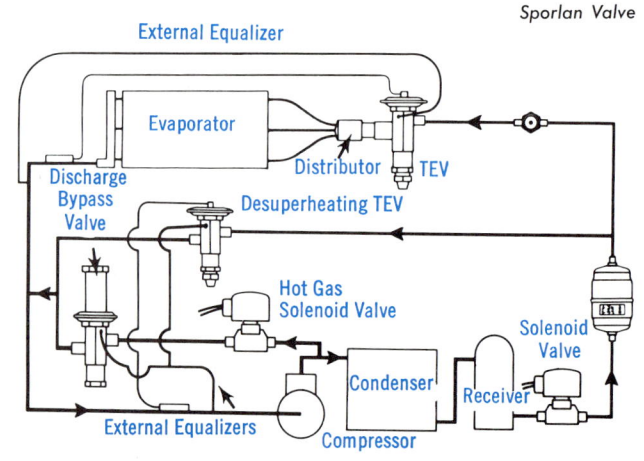

280

nection must be connected to the suction line where it will sense the desired operating pressure. Fig. 11-40.

Since the discharge bypass valve is applied in a bypass line between the discharge line and the low side of a system, the valve is subjected to compressor vibrations. Unless the valve, connecting fittings, and tubing are properly isolated from the vibrations, fatigue failures may occur. While the heaviest valve weighs only 3.5 pounds [1.6 kilograms], it should be adequately supported to prevent excessive stress on the connections.

If the *remote-bulb* type bypass valve is used, the bulb must be located in a fairly constant ambient temperature because the element-bulb assembly is air charged. These valves are set at the factory in an 80° F. [27°C] ambient temperature. Thus, any appreciable variation from this temperature will cause the pressure setting to vary from the factory setting. For a nonadjustable valve, the remote bulb may be located in an ambient of 80° F. ± 10° F. [27°C ± 5.5°C]. The adjustable remote bulb model can be adjusted to operate in a temperature of 80° F. ± 30° F. [27°C ± 16.7°C]. On many units the manufacturer will have altered the pressure setting to compensate for an ambient temperature appreciably different than 80° F. [27°C]. Therefore, on some units it may be necessary to consult with the equipment manufacturer for the proper opening pressure setting of the bypass valve.

There are numerous places on a system where the remote bulb can be located. Two possible locations are the return airstream and a structural member of the unit, if it is located in a conditioned space. Other locations, where the temperature is fairly constant but different than 80° F. [27°C] are also available. These include the return water line on a chiller, the compressor suction line, or the main liquid line. As previously mentioned, the setting may have been altered.

A bulb strap with bolts and nuts is usually supplied with each remote-bulb type DBV. This strap is for use in fastening the bulb in place.

Special Considerations

If a DBV is applied on a system with an evaporator pressure regulating valve (ORIT or other type), the DBV may bypass into either the evaporator inlet or the suction line. The bypass will depend on the specific system. Valve function and the best piping method to protect the compressor should be deciding factors. If the DBV is required on a system with a crankcase pressure regulating valve (CRO or other type), the bypass valve can bypass to the low side of the evaporator inlet or the suction line without difficulties. The only decision necessary is whether an internally or externally equalized valve is required. This depends on where the hot gas enters the low side. The pressure setting of the DBV must be lower than the CRO setting for each valve to function properly.

The hot gas solenoid valve is to be located upstream of the bypass valve. If the solenoid valve is installed downstream of the DBV, the oil and/or liquid refrigerant may be trapped between the two valves. Depending on the ambient temperature surrounding the valves and piping, this could be dangerous.

If the hot gas solenoid valve is required for pumpdown control, it should be wired in parallel with the liquid solenoid valve so it can be de-energized by a thermostat.

The hot gas solenoid is sometimes used for protection against high superheat conditions because the compressor does not have an integral temperature protection device. If this is done, the solenoid valve is wired in series with a bimetal thermostat fastened to the discharge line close to the compressor.

Testing and Operating Pressures

Excessive leak testing or operating pressures may damage these valves and reduce the life of the operating members. Since a high side test pressure differential of approximately 350 psig or higher will force the DBV open, the maximum allowable test pressures for DBV are the same as for the high and low side of the system. If greater high side test pressures than those given in the manufacturer's specifications are to be encountered, some method of isolating the DBV from these high pressures must be found.

Valve setting and adjustment must be done according to the manufacturer's recommendations. Proper instrumentation must be used to determine exactly when these valves are open.

Hot Gas

Hot gas may be required for other system functions besides bypass capacity control. Hot gas may be needed for defrost and head pressure control. Normally,

Table 11-F.
Troubleshooting Discharge Bypass Valves

Valve Type*	Malfunction	Cause	Remedy
FULLY ADJUSTABLE MODELS—ADR TYPE			
ADRS-2 ADRSE-2 ADRP-3 ADRPE-3	Failure to open	1. Dirt or foreign material in valve.	1. Disassemble valve and clean.
	Failure to close	1. Dirt or foreign material in valve. 2. Diaphragm failure. 3. Equalizer passageway plugged. 4. External equalizer not connected or equalizer line pinched shut.	1. Disassemble valve and clean. 2. Replace element only. 3. Disassemble valve and clean. 4. Connect or replace equalizer line.
ADRH-6 ADRHE-6	Failure to open	1. Dirt or foreign material in valve. 2. Equalizer passageway plugged. 3. External equalizer not connected or equalizer line pinched shut.	1. Disassemble valve and clean. 2. Disassemble valve and clean. 3. Connect or replace equalizer line.
	Failure to close	1. Dirt or foreign material in valve. 2. Diaphragm failure.	1. Disassemble valve and clean. 2. Replace element only.
"LIMITED" ADJUSTABLE MODELS—DR-AR TYPE			
DRP-3-AR DRPE-3-AR	Failure to open	1. Dirt or foreign material in valve. 2. Diaphragm failure. 3. Air charge in element lost.	1. Disassemble valve and clean. 2. Replace element only. 3. Replace element only.
	Failure to close	1. Dirt or foreign material in valve. 2. Equalizer passageway plugged. 3. External equalizer not connected or equalizer line pinched shut.	1. Disassemble valve and clean. 2. Disassemble valve and clean. 3. Connect or replace equalizer line.
DRH-6-AR DRHE-6-AR	Failure to open	1. Dirt or foreign material in valve. 2. Diaphragm failure. 3. Equalizer passageway plugged. 4. External equalizer not connected or equalizer line pinched shut. 5. Air charge in element lost.	1. Disassemble valve and clean. 2. Replace element only. 3. Disassemble valve and clean. 4. Connect or replace equalizer line. 5. Replace element only.
	Failure to close	1. Dirt or foreign material in valve.	1. Disassemble valve and clean.
NON-ADJUSTABLE MODELS—REMOTE BULB and DOME TYPE			
DRS-2 DRSE-2 DRP-3 DRPE-3	Failure to open	1. Dirt or foreign material in valve. 2. Diaphragm failure. 3. Air charge in element lost.	1. Disassemble valve and clean. 2. Replace element only. 3. Replace element only.
	Failure to close	1. Dirt or foreign material in valve. 2. Equalizer passageway plugged. 3. External equalizer not connected or equalizer line pinched shut.	1. Disassemble valve and clean. 2. Disassemble valve and clean. 3. Connect or replace equalizer line.
DRH-6 DRHE-6	Failure to open	1. Dirt or foreign material in valve. 2. Diaphragm failure. 3. Equalizer passageway plugged. 4. External equalizer not connected or equalizer line pinched shut. 5. Air charge in element lost.	1. Disassemble valve and clean. 2. Replace element only. 3. Disassemble valve and clean. 4. Connect or replace equalizer line. 5. Replace element only.
	Failure to close	1. Dirt or foreign material in valve.	1. Disassemble valve and clean.

*The model numbers are for Sporlan valves.

Sporlan Valve

CHAPTER 11—REFRIGERANT FLOW CONTROL

these functions will not interfere with each other. However, compressor cycling on low suction pressure may be experienced on system start-up when the discharge bypass valve is operating and other functions require the hot gas also. For example, the head pressure control requires hot gas to pressurize the receiver and liquid line to get the thermostatic expansion valve operating properly. In this case, the discharge bypass valve should be prevented from functioning by keeping the hot gas solenoid valve closed until adequate liquid line or suction pressure is obtained.

Malfunctions

There are several reasons for system malfunctions. Possible causes of trouble when hot gas bypass for capacity control is used are listed in Table 11-F.

Valves are coded by the manufacturer. The part numbers given in Table 11-F are those of the Sporlan Valve Company. Note that each letter and number has a meaning. The coded part numbers in Fig. 11-41 are given as examples. Similar codes are used by other valve manufacturers. To be informed of such codes, you will need the manufacturers' bulletins. A good file of such bulletins will enable you quickly to identify the various valve problems.

LEVEL CONTROL VALVES

Capillary tubes and float valves are used to control the refrigerant in a system.

Capillary Tubes

Capillary tubes are used to control pressure and temperature in a refrigeration unit. They are most commonly used in domestic refrigeration, milk coolers, ice-cream cabinets, and smaller units. Commercial refrigeration units use other devices. The capillary tube consists of a tube with a very small diameter. The length of the tube depends on the size of the unit to be served, the refrigerant used, and other physical considerations. To effect the necessary heat exchange, this tube is usually soldered to the suction line between the condenser and the evaporator. The capillary tube acts as a constant throttle or restrictor on the refrigerant. Its length and diameter offer sufficient frictional resistance to the flow of refrigerant to build up the head pressure needed to condense the gas.

If the condenser and evaporator were simply connected by a large tube, the pressure would rapidly adjust itself to the same value in both of them. A small diameter water pipe will hold back water, allowing a pressure to be built up behind the water column, but with a small rate of flow. Similarly, the small diameter capillary tube holds back the liquid refrigerant. This enables a high pressure to be built up in the condenser during the operation of the compressor. At the same time, this permits the refrigerant to flow slowly into the evaporator. Fig. 11-42. A filter-drier is usually inserted between the condenser and the capillary tube. This is necessary because the line or tube is so small that it is easily clogged.

Capillary tubes may be cleaned and unplugged by the method suggested in Chapter 2. Replacement should be performed in the shop after discharging the unit. In replacing the capillary tube, make sure that the same length of tube is used. The bore or inner diameter should be exactly the

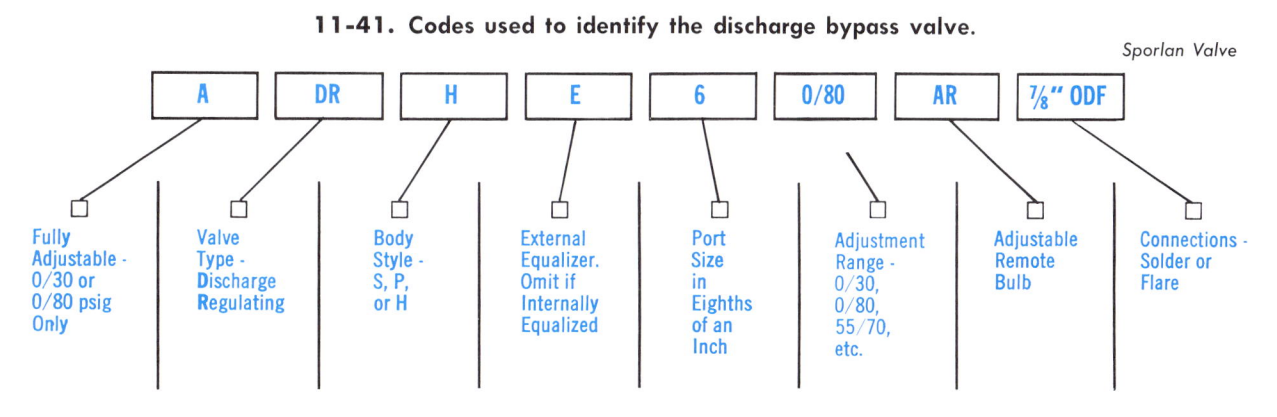

11-41. Codes used to identify the discharge bypass valve.

Sporlan Valve

A	DR	H	E	6	0/80	AR	7/8" ODF
Fully Adjustable - 0/30 or 0/80 psig Only	Valve Type - Discharge Regulating	Body Style - S, P, or H	External Equalizer. Omit if Internally Equalized	Port Size in Eighths of an Inch	Adjustment Range - 0/30, 0/80, 55/70, etc.	Adjustable Remote Bulb	Connections - Solder or Flare

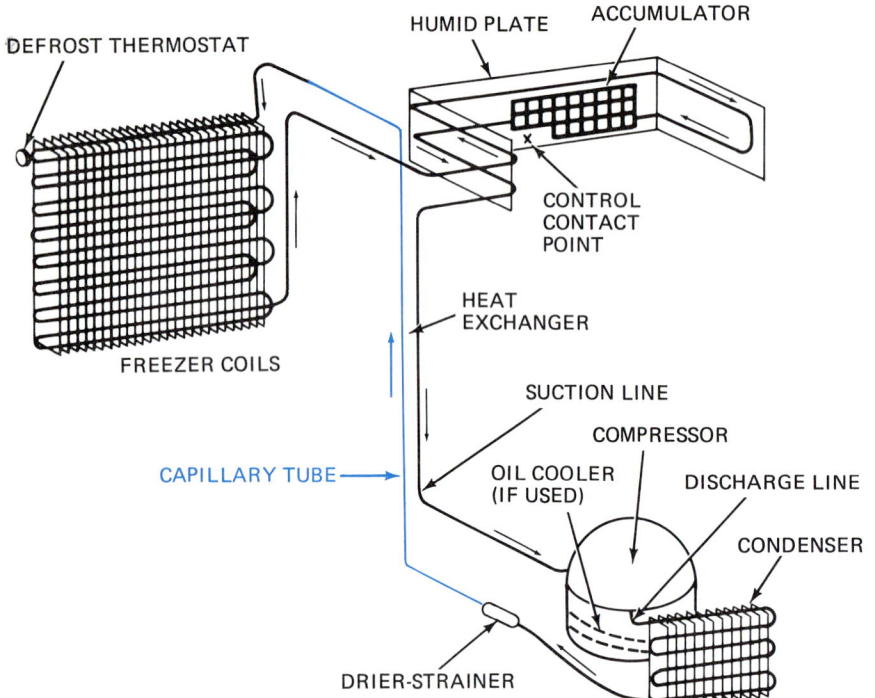

REFRIGERATION AND AIR CONDITIONING TECHNOLOGY

11-42. Refrigerant flow with a capillary tube in the line.

for handling the larger quantities of liquid.

Installation

The following precautions must be observed before installation of a float valve:

1. Most float controls are designed for a maximum differential pressure of 200 pounds.
2. If the pressure will exceed 190 psi, there are stems and orifices of special size available for low temperature use.
3. In any application, keep the bottom equalizing line above the bottom of the evaporator to avoid oil logging.
4. Make sure there are no traps in the equalizing line.
5. The stems of a globe valve must be in a horizontal plane.
6. Refrigerant flow must be same as the old tube. It is easy to check with the proper tool. This tool is described in Chapter 2.

Float Valve

A hollow float is sometimes used to control the level of refrigerant. Fig. 11-43. The float is fastened to a lever arm. The arm is pivoted at a given point and connected to a needle that seats at the valve opening. If there is no liquid in the evaporator, the ball lever arm rests on a stop and the needle is not seated, thus leaving the valve open. Once liquid refrigerant under pressure from the compressor enters the float chamber, the float rises with the liquid level until, at a predetermined level, the needle closes the needle-valve opening.

In Freon®-12 plants of large size, multiple ports are provided

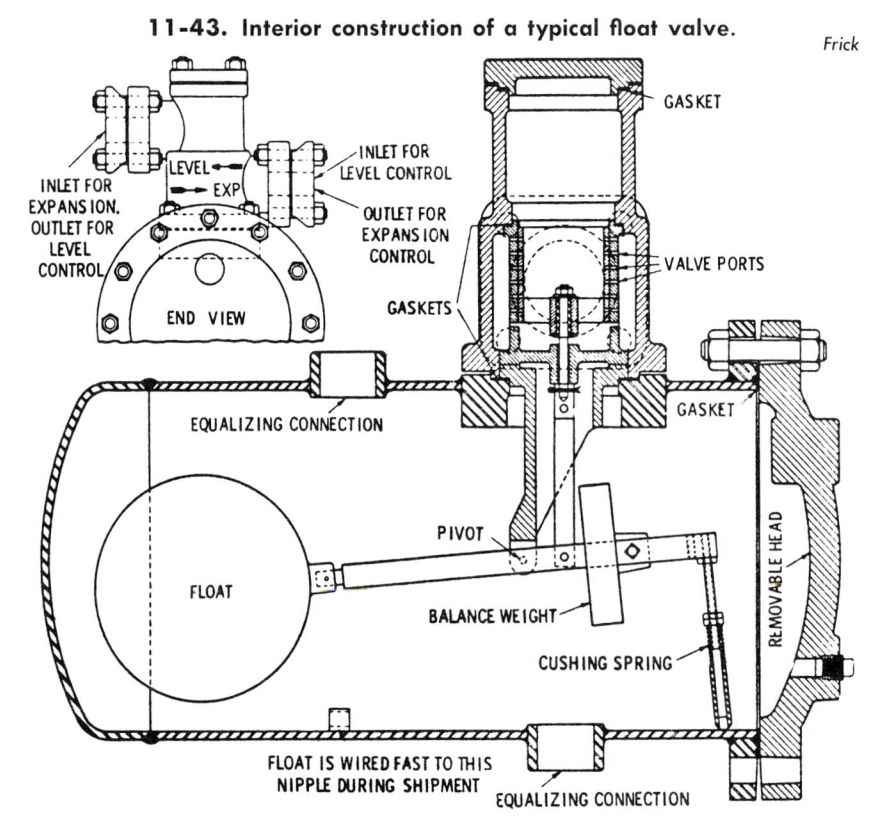

11-43. Interior construction of a typical float valve.
Frick

CHAPTER 11—REFRIGERANT FLOW CONTROL

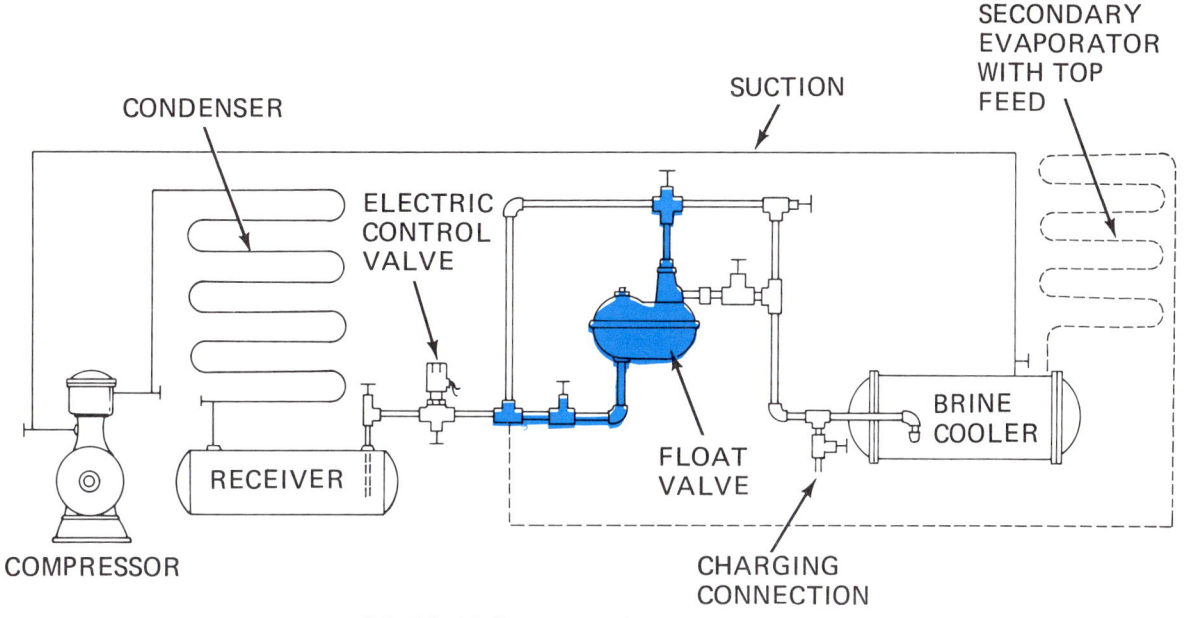

11-44. High-pressure float control system.

kept to less than 100' per minute where a bottom float equalizing connection is made to the header or accumulator return. That means the header and accumulator pipe must be properly sized.

7. Accumulators of a small diameter with a velocity of over 50 fpm are not suitable for accurate float application. However, the float may control within wider limits with higher velocities. The top equalizing connection must be connected to a point of practically zero gas velocity.

8. In automatic plants, always provide a solenoid valve in the liquid line ahead of the float control. This solenoid valve is to close either when the temperatures are satisfactory or when the compressor stops.

Figure 11-44 illustrates the connections for a high-pressure float control. There have been new developments in the control of liquid level since the early days of refrigeration.

LEVEL-MASTER CONTROL

The level-master control is a positive liquid level control device suitable for application to all flooded evaporators. Fig. 11-45.

The level-master control is a standard thermostatic expansion valve with a level-master element. The combination provides a simple, economical, and highly effective liquid level control. The bulb of the conventional thermostatic element has been modified to an insert type of bulb that incorporates a low-wattage heater. A 15-watt heater is supplied as standard. For applications below −60° F. [−51°C] evaporating temperature, a special 25-watt heater is needed.

The insert bulb is installed in the accumulator or surge drum at the point of the desired liquid level. As the level at the insert bulb drops, the electrically added heat increases the pressure within the thermostatic element and opens the valve. As the liquid level at the bulb rises, the electrical input is balanced by the heat transfer from the bulb to the liquid refrigerant. The level-master control either modulates or eventually shuts off. The evaporator pressure and spring assist in providing a positive closure.

Installation

The level-master control is applicable to any system that has been specifically designed for flooded operation.

The valve is usually connected to feed into the surge drum above the liquid level. It can feed into the liquid leg or coil header. The insert bulb can be installed directly in the shell, surge drum, or liquid leg on new or existing installations. Existing float systems can be easily converted

285

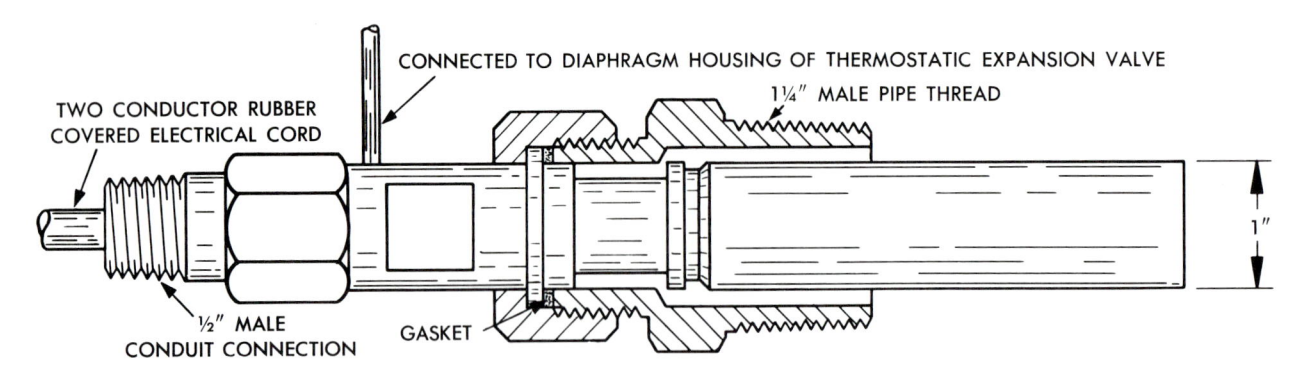

REFRIGERATION AND AIR CONDITIONING TECHNOLOGY

LEVEL MASTER ELEMENT WITH ½" MALE CONDUIT CONNECTION

Sporlan Valve

11-45. Level-master control.

by installing the level-master control insert bulb in the float chamber.

Electrical Connections

The heater is provided with a two-wire neoprene-covered cord 2' in length. It runs through a moisture-proof grommet and a ½" male conduit connection affixed to the insert bulb assembly. Fig. 11-46.

The heater circuit must be interrupted when refrigeration is not required. Wire the heater in parallel with the holding coil of the compressor line starter or solenoid valve—not in series.

Hand Valves

On some installations, the valve is isolated from the surge drum by a hand valve. A 2- to 3-pound pressure drop from the valve outlet to the bulb location is likely. For such installations, an externally equalized valve is recommended.

Oil Return

All reciprocating compressors will allow some oil to pass into the discharge line along with the discharge gas. Mechanical oil separators are used extensively. However, they are never completely effective. The untrapped oil passes through the condenser, liquid line, expansion device, and into the evaporator.

In a properly designed direct expansion system, the refrigerant velocity in the evaporator tubes and the suction line is sufficiently high to insure a continuous return of oil to the compressor

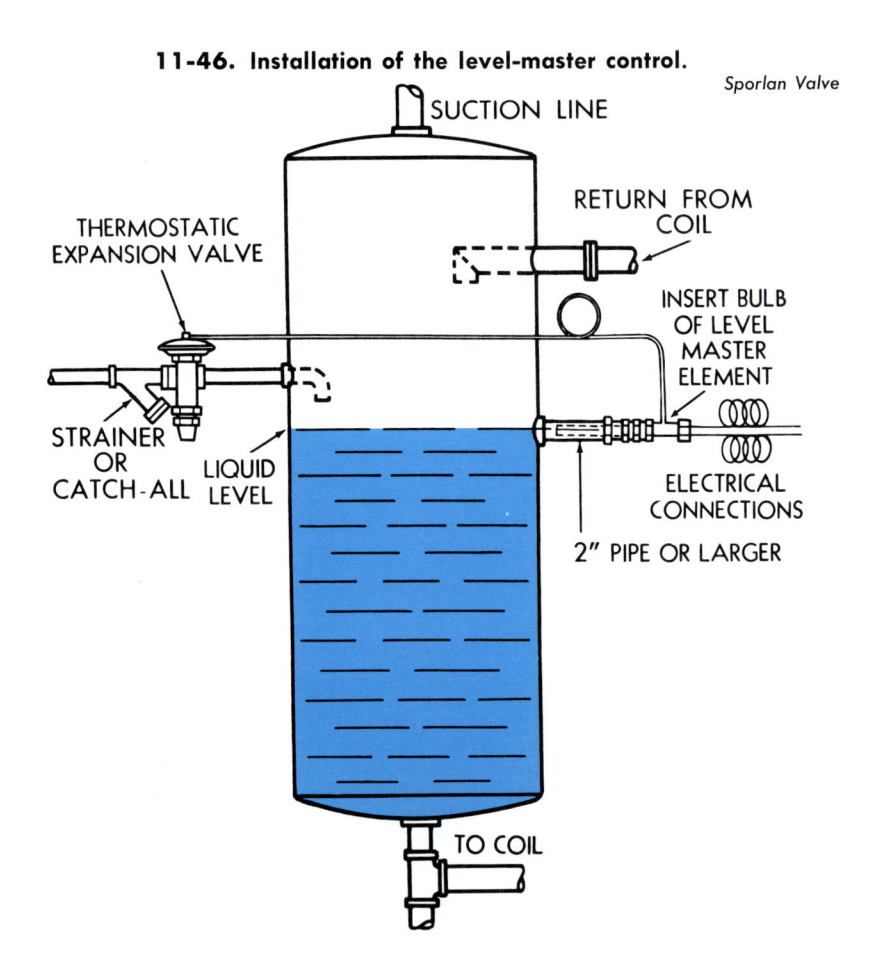

11-46. Installation of the level-master control.

Sporlan Valve

CHAPTER 11—REFRIGERANT FLOW CONTROL

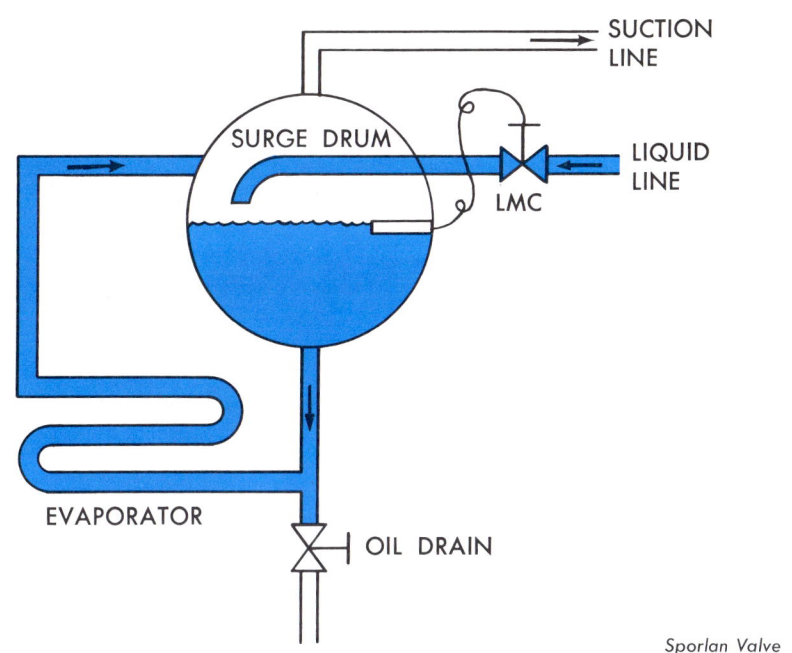

11-47. Location of LMC in liquid line.
Sporlan Valve

crankcase. However, this is not characteristic of flooded systems. Here, the surge drum is designed for a relatively low vapor velocity. This prevents entrainment of liquid refrigerant droplets and consequent carry-over into the suction line. This design also prevents the return of any oil from the low side in the normal manner.

If oil is allowed to concentrate at the insert bulb location of the level-master control, overfeeding with possible floodback can occur. The tendency to overfeed is due to the fact that the oil does not convey the heat from the low wattage heater element away from the bulb as rapidly as does pure liquid refrigerant. The bulb pressure is higher than normal and the valve remains in the open or partially open position.

Oil and Ammonia Systems

For all practical purposes, liquid ammonia and oil are *immiscible* (not capable of being mixed). Since the density of oil is greater than that of ammonia, it will fall to the bottom of any vessel containing such a mixture if the mixture is relatively placid. Therefore, the removal of oil from an ammonia system is a comparatively simple task. Generally, on systems equipped with a surge drum, the liquid leg is extended downward below the point where the liquid is fed off to the evaporator. A drain valve is provided to allow periodic manual draining. Fig. 11-47.

For flooded chillers that do not use a surge drum, a sump with a drain valve is usually provided at the bottom of the chiller shell. These methods are quite satisfactory, except possibly on some low-temperature systems. Here, the drain leg or sump generally must be warmed prior to attempting to draw off the oil. The trapped oil becomes quite viscous at lower temperatures.

If oil is not drained from a flooded ammonia system, a reduction in the evaporator heat transfer rate can occur due to an increase in the refrigerant film resistance. Difficulty in maintaining the proper liquid level with any type of flooded control can also be expected.

With a float valve, you can expect the liquid level in the evaporator to increase with high concentration of oil in a remote float chamber. If a level-master control is used with the insert bulb installed in a remote chamber, oil concentration at the bulb can cause overfeeding with possible floodback. The lower or liquid balance line must be free of traps and be free-draining into the surge drum or chiller, as shown in Fig. 11-48. The oil drain leg or sump must be located at the lowest point in the low side.

Oil and Halocarbon Systems

With halocarbon systems (Refrigerants 12, 22, 502, etc.) the oil and refrigerant are *miscible* (capable of being mixed) under certain conditions. Oil is quite soluble in liquid Refrigerant 12 and partially so in liquid Refrigerant 22 and 502. For example, for a 5% (by weight) solution of a typical napthenic (a petroleum-based oil) oil in liquid refrigerant, the oil will remain in solution down to about $-75°$ F. $[-59°C]$ for Refrigerant 12, down to about $0°$ F. $[-18°C]$ for Refrigerant 22, and down to about $20°$ F. $[-7°C]$ for Refrigerant 502. Depending upon the type of oil and the percentage of oil present, these figures can vary. However, based on the foregoing, we can

287

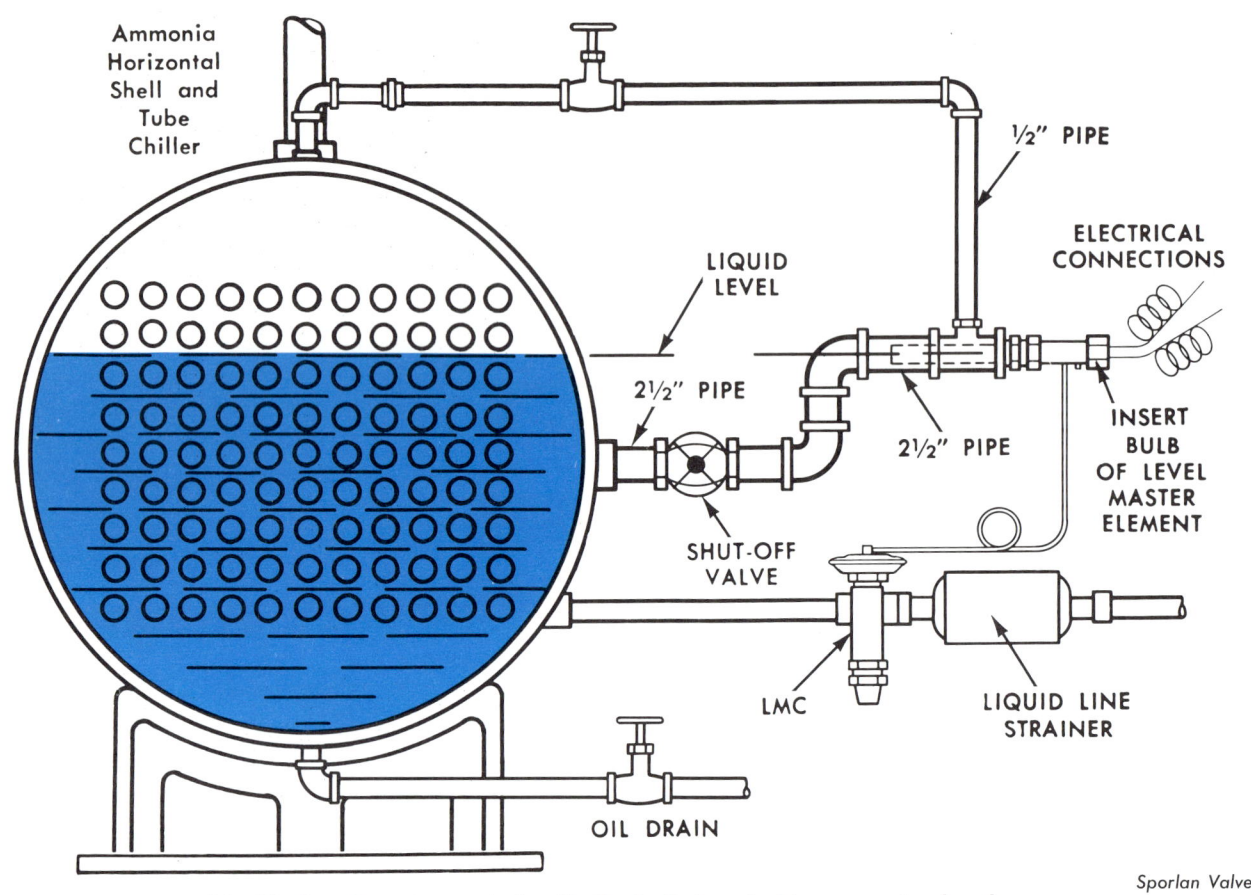

11-48. Level-master control with the bulb inserted in a remote chamber.

Sporlan Valve

assume that for the majority of Refrigerant-12 systems the oil and refrigerant are completely miscible at all temperatures normally encountered. However, at temperatures below 0° F. [−18°C] with Refrigerant 22 and a 5% oil concentration and temperatures below 20° F. [−7°C] with Refrigerant 502 and a 5% oil concentration, a liquid phase separation occurs. An oil-rich solution will appear at the top and a refrigerant-rich solution will lay at the bottom of any relatively placid remote bulb chamber.

Oil in a halocarbon-flooded evaporator can produce many results. Oil as a contaminant will raise the boiling point of the liquid refrigerant. For example, with Refrigerant 12, the boiling point increases approximately 1° F. [0.56°C] for each 5% of oil (by weight) in solution. As in an ammonia system, oil can foul the heat transfer surface with a consequent loss in system capacity. Oil can produce foaming and possible carry-over of liquid into the suction line. Oil can also affect the liquid level control.

With a float valve you can normally expect the liquid level in the evaporator to decrease with increasing concentrations of oil in the float chamber. This is due to the difference in density between the lighter oil in the chamber and the lower balance leg and the heavier refrigerant/oil mixture in the evaporator. A lower column of dense mixture in the evaporator will balance a higher column of oil in the remote chamber and piping. This is similar to a "U" tube manometer with a different fluid in each leg.

With the level-master control, the heat transfer rate at the bulb is decreased, producing overfeeding and possible floodback. What can be done? First of all, the oil concentration must be kept as low as possible in the evaporator, surge drum, and remote insert bulb chamber (if one is used). With Refrigerant 12, since the oil/refrigerant mixture is

CHAPTER 11—REFRIGERANT FLOW CONTROL

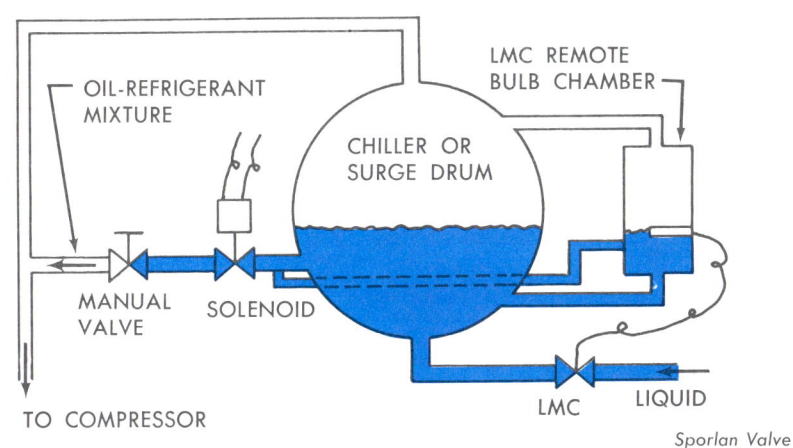

11-49. Direct drain of oil to the suction line is one of three ways to recover oil in flooded systems. Heat from the environment or a liquid-suction heat exchanger is required to vaporize the liquid refrigerant so drained. Vapor velocity carries oil back to the compressor.

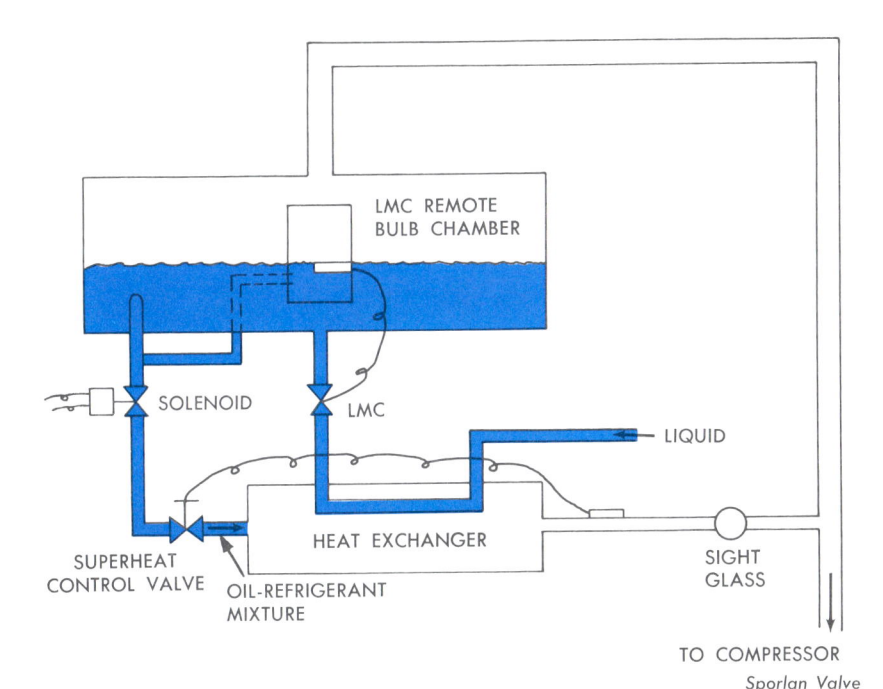

Sporlan Valve

11-50. Oil return by draining oil-refrigerant mixture through a heat exchanger is shown here. Heat in incoming liquid vaporizes refrigerant to prevent return of liquid to the compressor. Liquid feed is controlled by a thermostatic or hand expansion valve.

Sporlan Valve

- Direct drain into the suction line.
- Drain through a high-pressure, liquid-warmed heat exchanger.
- Drain through a heat exchanger with the heat supplied by an electric heater.

Draining directly into the suction line, as shown in Fig. 11-49, is the simplest method. However, the hazard of possible floodback to the compressor remains.

Draining through a heat exchanger, as indicated in Fig. 11-50, is a popular method. The liquid refrigerant floodback problems are minimized by using the warm liquid to vaporize the liquid refrigerant in the oil/refrigerant mixture.

The use of a heat exchanger with an insert electric heater, as shown in Fig. 11-51, is a variation of the preceding method.

In all of the return arrangements discussed, a solenoid valve should be installed in the drain line and arranged to close when the compressor is not in operation. Otherwise, liquid refrigerant could drain from the low side into the compressor crankcase during the off cycle.

If the insert bulb is installed directly into the surge drum or chiller, oil return is necessary only from this point. However, the insert bulb is sometimes located in a remote chamber that is tied to the surge drum or chiller with liquid and gas balance lines. Then oil return should be made from both locations, as shown in Figs. 11-49, 11-50, and 11-52.

Conclusions

The problem of returning oil from a flooded system is not highly complex. There are undoubtedly other methods in use today that

homogenous, it can be drained from almost any location in the chiller, surge drum, or remote chamber that is below the liquid level. With Refrigerants 22 and 502, the drain must be located at or slightly below the surface of the liquid, since the oil-rich layer is at the top. There are many types of oil return devices:

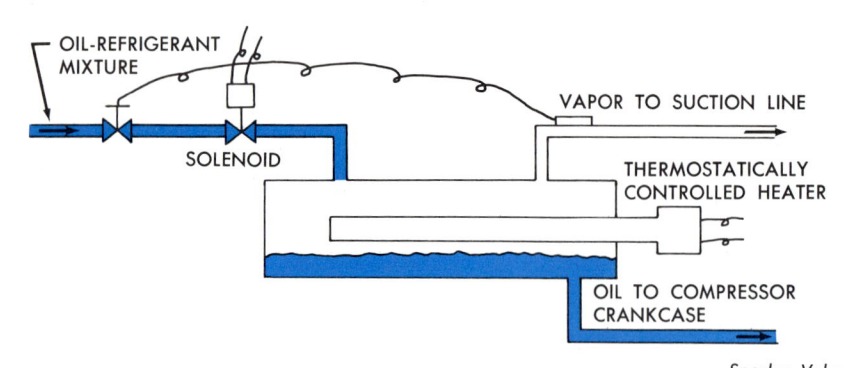

REFRIGERATION AND AIR CONDITIONING TECHNOLOGY

11-51. Electric heater may also be added to separate oil and refrigerant. This system is similar to that shown in Fig. 10-49, except that the heat required for vaporization is added electrically.

Sporlan Valve

are comparable to those outlined here. Regardless of how it is accomplished, oil return must be provided for proper operation of any flooded system. This is necessary not only with the level-master control, but also with a float or other type of level control device.

OTHER TYPES OF VALVES

Service Valves on Sealed Units

Hermetic refrigeration systems, also called sealed units, normally have no service valves on the compressor. Instead, a charging plug or valve may be mounted on the compressor. A special tool is needed to operate the charging device, which varies on different makes.

A service engineer needs the correct valve-operating device for a unit. Thus, a kit is made that contains adapters and wrench ends to fit many makes of sealed units.

Essentially, the device is a body with a union connection and provisions for charging line and pressure gage connections. The stem may be turned or pushed in or out of the body as required.

Figure 11-53 shows a line piercing valve. They are used for charging, testing, or purging those hermetic units not provided with a charging plug or valve. These valves may be permanently attached to the line without danger of refrigerant loss.

Water Valves

Manually operated valves are installed on water circuits associated with refrigeration systems—either on cooling towers or in secondary brine circuits. They are installed for convenience in servicing and for flexibility in operating conditions. These valves make it possible to re-circuit, bypass, or shut off water flow as desired. Fig. 11-54.

These manually operated shut-off or flow-control valves are available in a wide variety of styles and sizes. Valve stems and body seats are accurately machined to close tolerances, ensuring easy and positive shutoff. They are made of nonporous cast bronze.

There are three main types: stop valves, globe valves, and gate valves. Fig. 11-54.

Check Valves

Some refrigeration systems are designed in which the refrigerant liquid or vapor flows to several components, but must never flow back through a given line. A check valve is needed in such installations. As its name implies, a check valve checks or prevents

11-52. Level-master control inserted in remote chamber.

Sporlan Valve

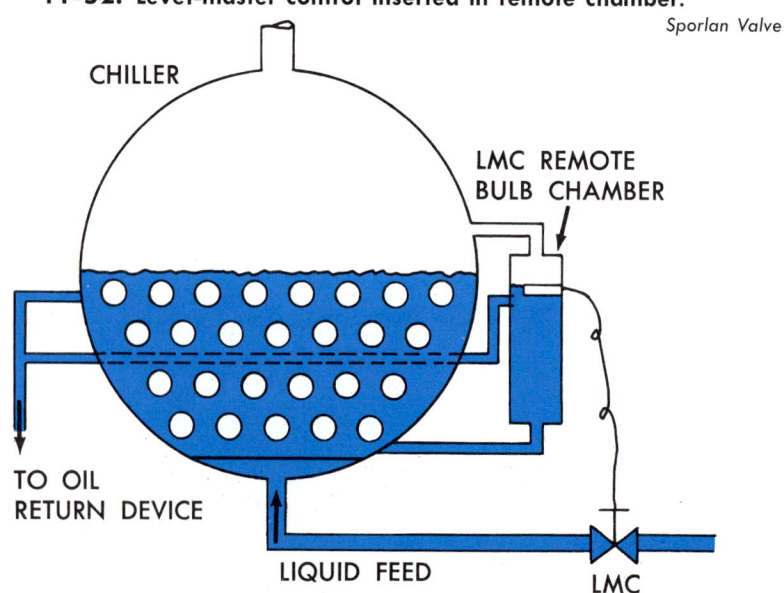

CHAPTER 11—REFRIGERANT FLOW CONTROL

11-53. Line piercing valve. *Mueller Brass*

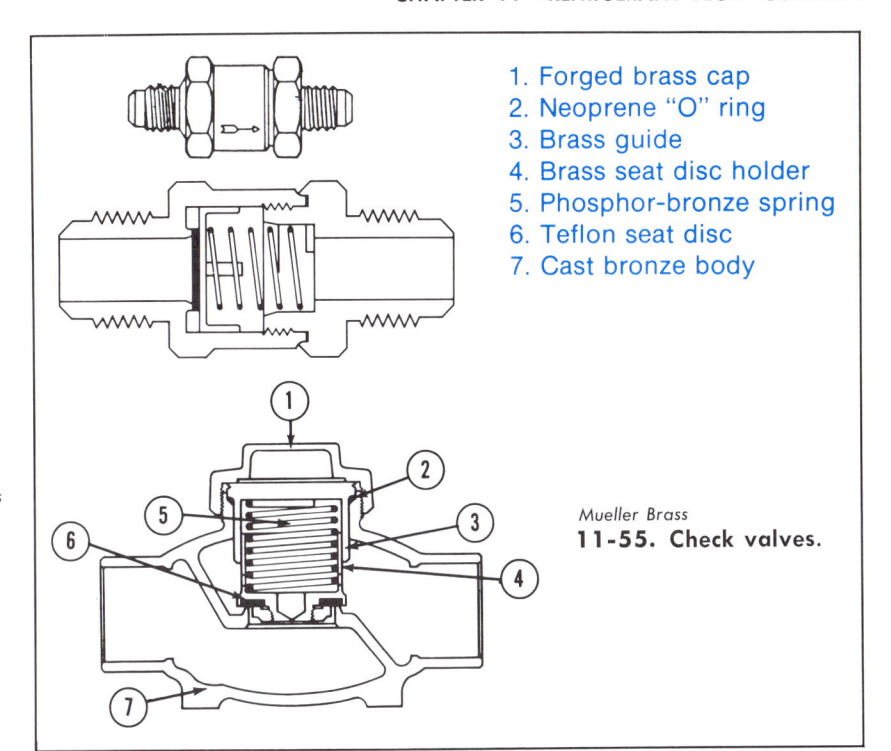

1. Forged brass cap
2. Neoprene "O" ring
3. Brass guide
4. Brass seat disc holder
5. Phosphor-bronze spring
6. Teflon seat disc
7. Cast bronze body

11-55. Check valves. *Mueller Brass*

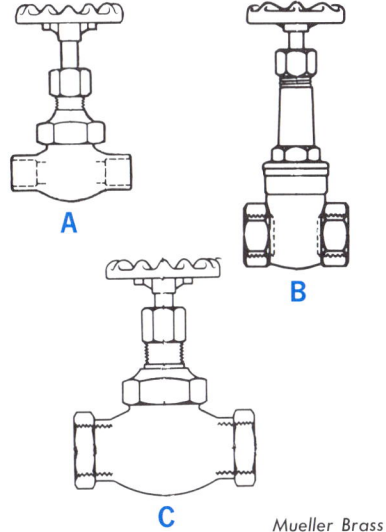

11-54. Water valves. (A) Stop valve, (B) gate valve, (C) globe valve. *Mueller Brass*

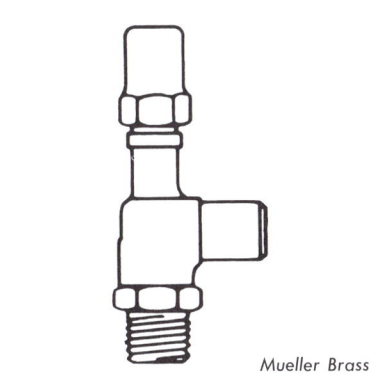

11-56. Receiver angle valve. *Mueller Brass*

the flow of refrigerant in one direction, while allowing free flow in the other direction. For example, two evaporators might be controlled by a single condensing system. In this case, a check valve should be placed in the line from the lower temperature evaporator to prevent the suction gas from the higher temperature evaporator from entering the lower temperature evaporator. Fig. 11-55.

Check valves are designed to eliminate chattering and to give maximum refrigerant flow when the unit is operating. If the spring tension is sufficient to overcome the weight of the valve disc, the check valve may be mounted in any position.

Receiver Valves

Receivers may be fitted with two valves—an inlet valve and an outlet valve. The outlet valve may have the inlet in the form of an ordinary connection, such as an elbow. An inlet valve permits closing the receiver should a leak develop between the compressor and the receiver. The receiver outlet valve is important when the system is "pumped down," when for reasons of service all the refrigerant is conveyed to the receiver for temporary storage. Fig. 11-56.

ACCUMULATORS

Accumulators have been used for years on original equipment. More recently they have been field installed. The significance with respect to accumulator and system performance has never been clarified. Engineers have been forced to evaluate each model in terms of the system on which it is to be applied.

Application in the field has

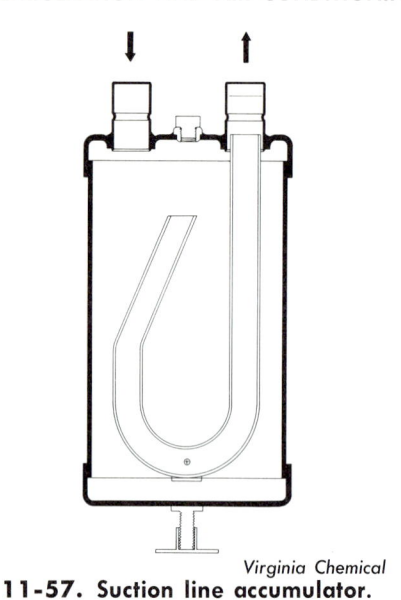

REFRIGERATION AND AIR CONDITIONING TECHNOLOGY

11-57. Suction line accumulator. *Virginia Chemical*

been primarily based on choosing a model with fittings that will accommodate the suction line and be large enough to hold about half of the refrigerant charge.

There is no standard rating system for accumulators. The accuracy of rating data becomes a function of the type of equipment used to determine the ratings. Some data is now available to serve as a guide to those checking the use of an accumulator.

Purpose

The purpose of an accumulator is to prevent compressor damage due to slugging of refrigerant and oil. They provide a positive oil return at all rated conditions. They are designed to operate at −40° F. [−40°C] evaporator temperature. Pressure drop is low across them. They act as a suction muffler. They can take suction gas temperatures as low as 10° F. [−12.2°C] at the accumulator. Most of them can withstand a working pressure of 300 psi and have fusible relief devices.

Compressors are designed to compress vapors, not liquids. Many systems, especially low-temperature systems, are subject to the return of excessive quantities of liquid refrigerant. This returned refrigerant dilutes the oil and washes out bearings. In some cases, it causes complete loss of oil in the crankcase. This results in broken valve reeds, and damage to pistons, rods, crankshafts, and other moving parts. The accumulator acts as a reservoir to hold temporarily the excess oil-refrigerant mixture and return it at a rate the compressor can safely handle. Figure 11-57 shows the interior view.

RATING DATA

Refrigerant Holding Capacity

The refrigerant holding capacity of the accumulator is based on an average condition of 65% fill under running conditions. It is obvious that directly on start-up or after long off-cycles the amount held may fluctuate from empty to nearly full.

Minimum Evaporator Temperature and Minimum Temperature of Suction Gas at the Accumulator

The oil-refrigerant mixture in the suction line has been studied over the range of −50° F. to +40° F. [−46°C to +4°C].

The value of −40° F. [−40°C] was chosen as a minimum evaporator temperature because it appears adequate for commercial refrigeration. Yet, it is conservative enough to provide a margin of safety. More important is the requirement that the temperature of the suction gas at the accumulator be 10° F. [−12°C] or higher. Particularly with refrigerants such as Freon® 502 in the low temperature range up to 0° F. [−17.8°C], the oil and refrigerant separate into two layers, with the upper layer being the oil-rich layer. At these low temperatures, the oil-rich layer can become so viscous that it will not flow. When the refrigerant below the heavy oil layer leaves the accumulator, the very thick oil settles over the oil return port and stops all oil return. This condition will occur regardless of accumulator design. If temperatures below 10° F. [−12°C] at the accumulator are to be used, auxiliary heat must be added to keep the oil fluid.

Maximum recommended actual tonnage is based on pressure drop through the accumulator equivalent to an effect of 1° F. [0.56°C] on evaporator temperature.

Minimum recommended actual tonnage is based on the minimum flow through the accumulator necessary to insure positive oil return.

For operating conditions outside the manufacturer's published ratings, contact the manufacturer for recommendations.

INSTALLATION OF THE ACCUMULATOR

Locate the accumulator as close to the compressor as possible. In systems employing reverse cycle, the accumulator must be installed *between* the reversing valve and the compressor. Proper inlet (from the evaporator) and outlet (to the compressor) must be observed. The accumulator must be installed vertically.

Proper sizing of an accumulator may not necessarily result in the accumulator connections matching the suction line size. This new technology must replace the dangerous and outmoded practice of matching the accumulator connections to the suction line size. To accommodate mismatches, bushing down may be required.

The accumulator should not be installed in a bypass line or in suction lines that experience other than total refrigerant flow.

When installing an accumulator with solder connections, direct the torch away from the top access plug to prevent possible damage to the O-ring seal. When installing a model equipped with a fusible plug, a dummy plug should be inserted in place of the fusible plug until all brazing or soldering is complete.

REVIEW QUESTIONS

1. What is capillary tubing?
2. Why do some cities require the use of K-type copper tubing?
3. What is the composition of No. 95 solder?
4. Name two types of line valves.
5. What is the most detrimental material in a refrigeration system?
6. What happens to halogen-type refrigerants when they combine with water?
7. What happens to Refrigerants 12 and 22 when heated to a high temperature?
8. How are sludge and varnish formed in a refrigeration system?
9. What happens to a liquid indicator if the system has too much oil?
10. What does TEV stand for?
11. What is flash gas?
12. What is the purpose of the crankcase pressure regulating valve?
13. What does CRO stand for?
14. Where are ORITs installed?
15. What is the purpose of an ORO head pressure valve?
16. What is the function of a discharge bypass valve?
17. How is a remote bulb used in a bypass valve?
18. List two possible locations for the mounting of the remote bulb.
19. How are capillary tubes used in a refrigeration unit?
20. What is an accumulator?
21. What is the refrigerant holding capacity of the accumulator?

12

Refrigerators

Basically, a refrigerator is an insulated cabinet with cold surfaces inside to absorb heat. This heat absorption cools the inside of the cabinet. An icebox was a refrigerator because it was an insulated cabinet holding a block of ice. The ice remained at a temperature of 32° F. [0°C]. This cooled the air in the refrigerator to about 45° F. [7.2°C]. The cool air, in time, cooled the foods in the cabinet. However, the ice eventually melted in an icebox. Thus, the ice had to be replaced and the water removed. The electric refrigerating system overcame this great disadvantage.

Figure 12-1 shows the schematic diagram of the electric refrigerator. Note that a coil of tubing called an *evaporator* replaces the ice. The evaporator contains a liquid refrigerant that evaporates. The evaporator produces the cold surface inside the cabinet. In the process of evaporating to a gas, the liquid cools the coils. The effect is much the same as when alcohol on the skin cools the skin as it evaporates.

The compressor pumps the refrigerant gas from the evaporator through a tube called the *suction line*. The compressor then compresses the gas and pumps it into the condenser. The process of compressing heats the gas. (Have you noticed that a tire pump gets hot when pumping air into a tire?)

Since the air is heated above room temperature, the air in the room cools the gas. This causes it to condense into a liquid. An example of this is the condensation of hot steam into a liquid on the bathroom mirror. The condenser processes the gas into a liquid.

From the condenser, the liquid refrigerant passes through a smaller tube called a *capillary tube*. The condensed gas is now liquid and does not occupy as much space in its new form. The capillary tube slowly meters the liquid refrigerant into the evaporator. There, it evaporates again to produce a cooling effect. Then the process repeats itself.

Refer to Fig. 12-2, which shows a single-door refrigerator. The evaporator consists of a coil of aluminum tubing wrapped

12-1. Schematic diagram of a refrigerator.

General Electric

294

around and brazed to an aluminum box. It can maintain a temperature of 10° F. [−12.2°C], which is considerably colder than ice. Therefore, it can also maintain a lower temperature in the fresh food section, namely 40° F. [4.4°C], instead of 45° F. [7.2°C].

The refrigerant evaporates in the coils. The gas is pumped out through the suction line by the compressor. The compressor compresses the gas and pumps it into the condenser on the back of the cabinet. There it condenses into a liquid. The liquid refrigerant is metered slowly through the capillary tube back into the evaporator. At this point the refrigerating process begins all over again.

Note that, in Fig. 12-2, the hot capillary tube is soldered to the cool suction line. This increases efficiency by reducing the temperature of the liquid refrigerant before it enters the evaporator.

This single-door refrigerator is a great improvement over the icebox. However, it is not without disadvantages. For instance, the evaporator temperature cannot be lowered below 10° F. [−12.2°C] without running the risk of freezing the fresh food. Also, the evaporator must be defrosted manually every week or so.

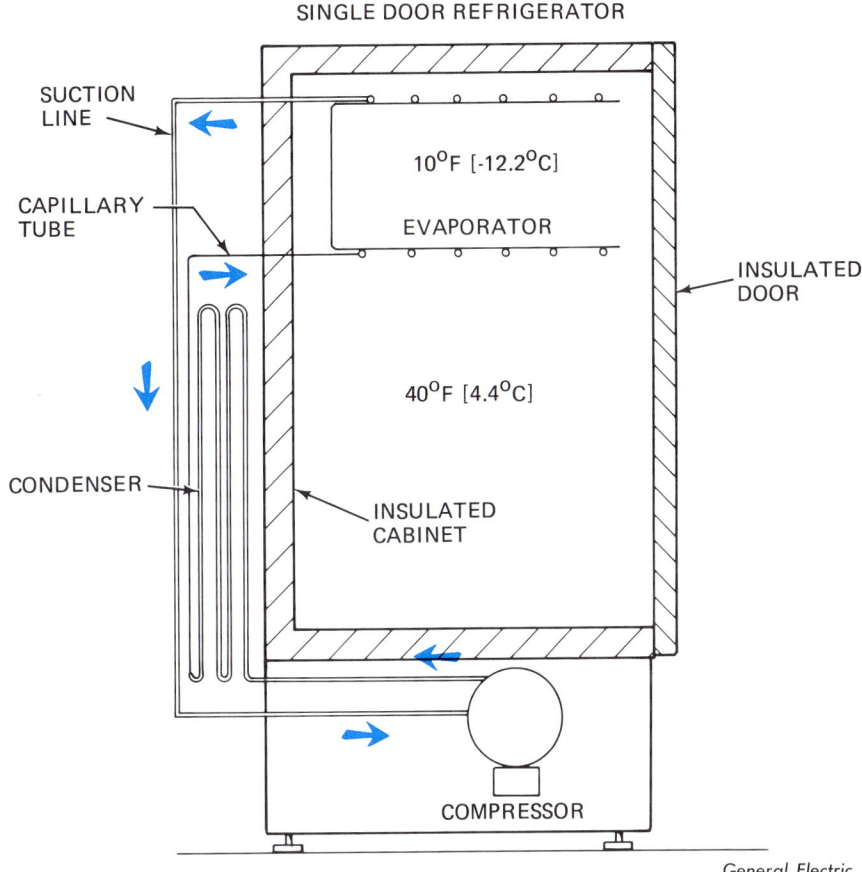

FEATURES OF THE SINGLE-DOOR REFRIGERATOR

A single-door manual-defrost refrigerator has several distinctive features.

Cabinet Styling and Construction

Single-door cabinets have a one-piece, pre-painted steel wrapper consisting of top, bottom, and sides with an interlocking, snap-in galvanized back panel.

The cabinet wrapper is then "pop" riveted to the cabinet base, making the base a permanent part of the cabinet. All cabinet seams are internally treated with special sealing materials, which act as vapor barriers.

Insulation and Inner Liners

The single-door model is foam insulated, forming a slim, solid, three-ply wall of single-unit construction.

The one-piece compartment (inner wall) liner is tough, corrosion-proof "ABS" material. It is vacuum formed. Both outer and inner surfaces are welded together with a core of urethane foamed insulation. They provide a solid, seamless compartment. The ABS inner wall is not removable.

Evaporator Door

The evaporator door has an automatic door closer. The closer is a coil spring wound over the lower hinge pin. This spring winds (or tightens) when the door is opened. It unwinds when the door closes.

To remove, open the evaporator door, remove the nylon spacer between the upper hinge and evaporator door. Lift up the

12-2. Single-door refrigerator.
General Electric

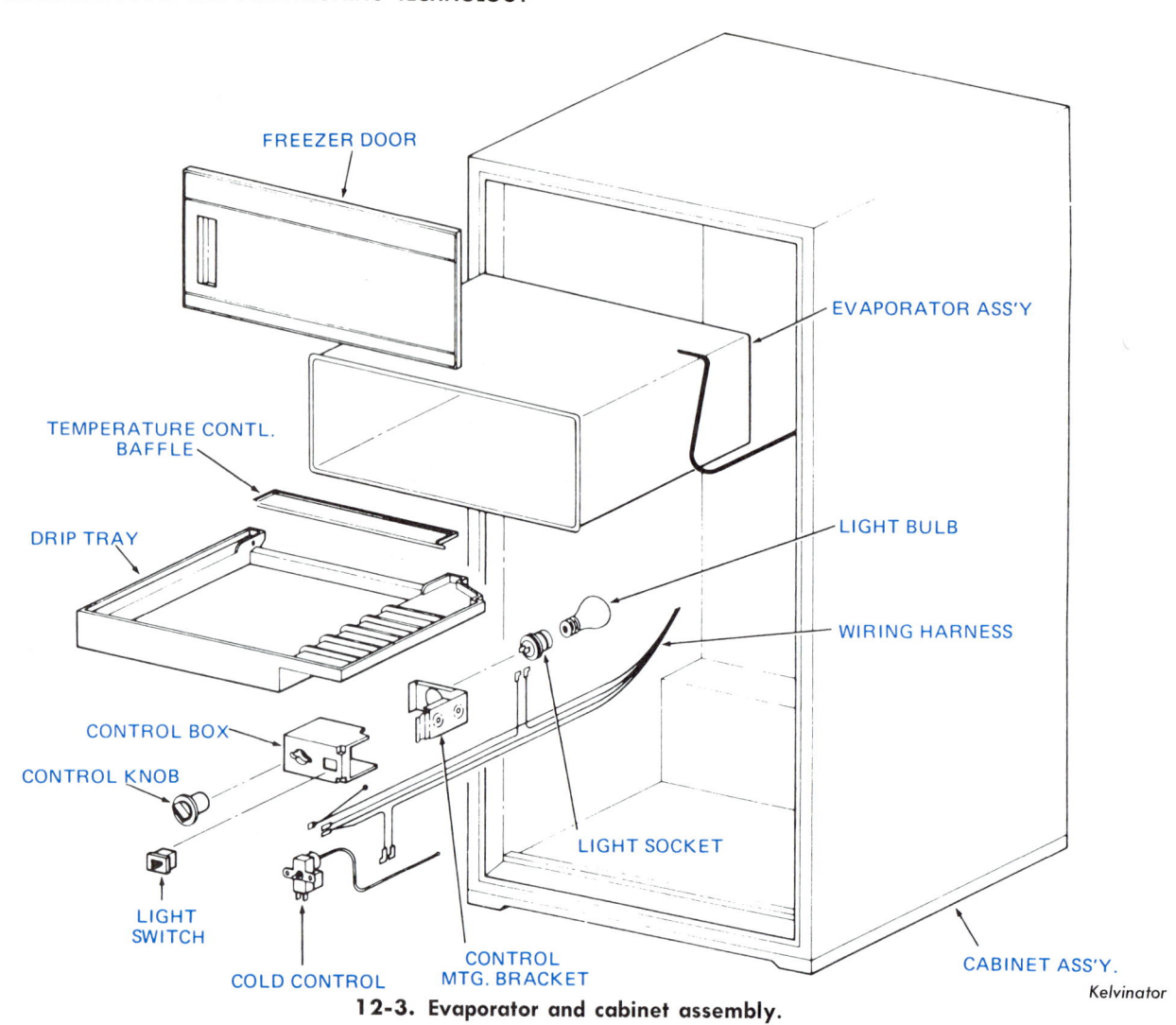

REFRIGERATION AND AIR CONDITIONING TECHNOLOGY

12-3. Evaporator and cabinet assembly.

Kelvinator

bottom of the door, releasing the lower hinge and coil spring. Pull the bottom of the door forward slightly. Then, pull down to release the hinge. Figure 12-3 shows the refrigerator cabinet.

Drip Tray Assembly

The refrigerator food and freezer compartments are separated by a removable drip tray assembly. The drip tray has a two-position temperature control baffle located at the rear. See Fig. 12-3.

The purpose of the temperature control baffle is to give additional control of temperatures in the freezer or food compartment. When colder temperatures are desired in the freezer compartment, close the control baffle. The open position should be used when colder temperatures are desired in the food compartment. The baffle should be in a closed position during defrosting. Being closed during defrosting, it will deflect ice and water into the drip tray.

Refrigerator Door

The outer panel is made from one piece of heavy-gage, deep-drawn, cold-rolled steel. A wide flange at the periphery is formed integral with the outer panel. The wide flange serves as an attachment point for the door inner panel and door gasket. The door inner panel is vacuum formed from ABS material.

Door Alignment and Adjustment

The door can be readily ad-

CHAPTER 12—REFRIGERATORS

justed in the field. If the door does not fit properly, level the cabinet. Then check for hinge bind. Hinge bind will result in excessive compression of the door gasket on the hinge side. This will cause the gasket to roll, preventing the gasket from sealing against the cabinet flange. To correct, add a shim under the bottom hinge and adjust the top hinge forward.

Door Gasket

The door gasket is the magnetic type. Full-length magnets are located along the four sides of the gasket in the front section of the balloon portion, as shown in Fig. 12-4. The balloon section projects beyond the edge of the gasket on the handle side.

To replace the gasket, remove the screws that secure the gasket and door inner panel to the door outer panel. Remove the gasket and door inner panel as an assembly. Remove the gasket from the door inner panel.

Install new gasket on the door inner panel. Assemble the door inner panel and the gasket to the door outer panel. Then adjust the door properly. Details on door adjustment are given later in this chapter.

Door Handles

Door handles are secured to the edge of the door outer panel with two screws that thread into the door panel. Door handles can be replaced without disassembling the door.

Nameplate

Nameplates are attached to the door outer panel with double-face tape. To replace a nameplate, place a protective pad on the surface of the door next to the nameplate. Force the blade of the putty knife or suitable tool between the nameplate and the door. Be careful not to damage the door and the nameplate.

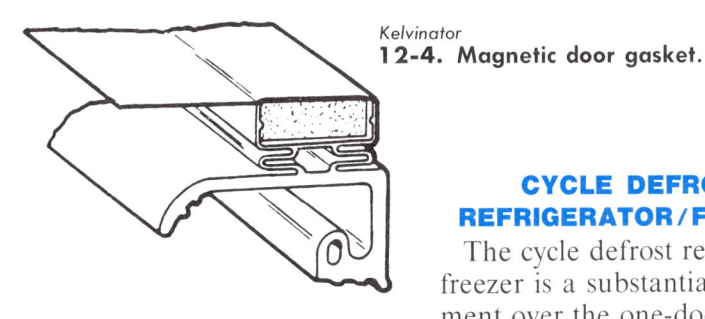

Kelvinator
12-4. Magnetic door gasket.

CYCLE DEFROST REFRIGERATOR/FREEZER

The cycle defrost refrigerator/freezer is a substantial improvement over the one-door refrigerator. It can hold a temperature of 0° F. [−17.8°C] or lower in the freezer without freezing the fresh food in the refrigerator section. The two compartments are isolated and insulated from each

12-5. Cycle defrost refrigerator/freezer.
General Electric
CYCLE DEFROST REFRIGERATOR-FREEZER

297

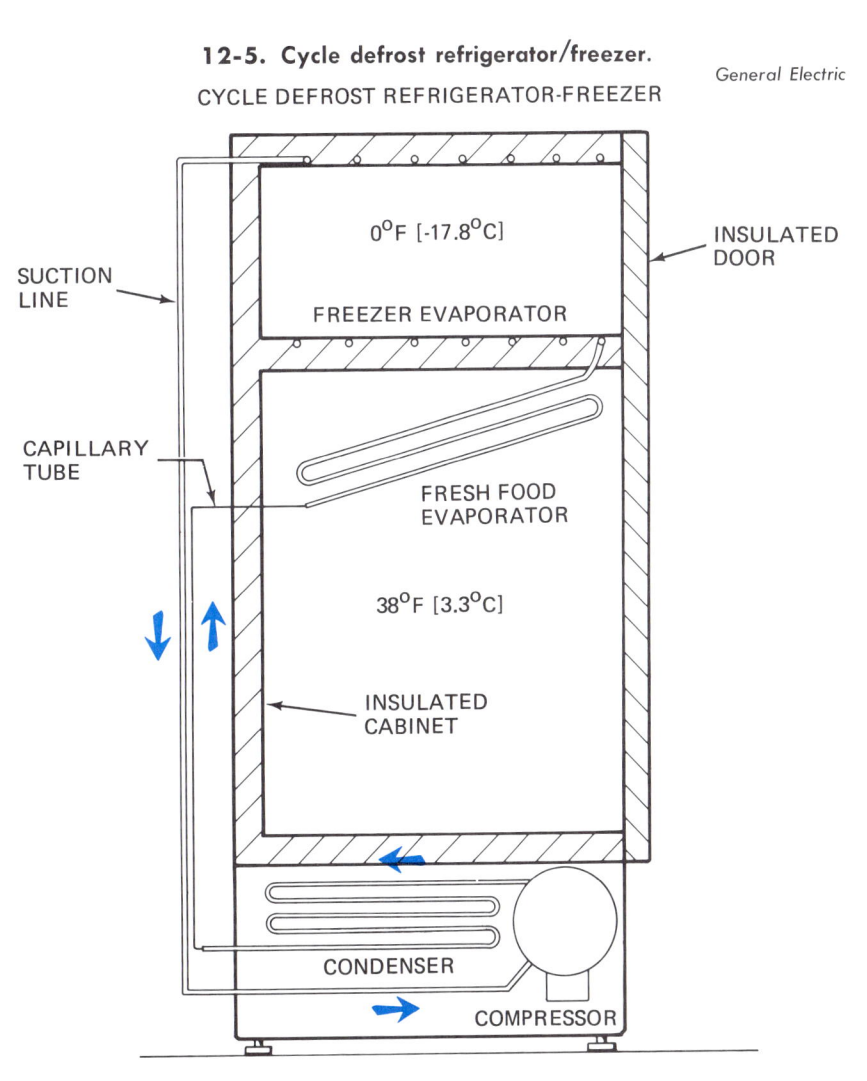

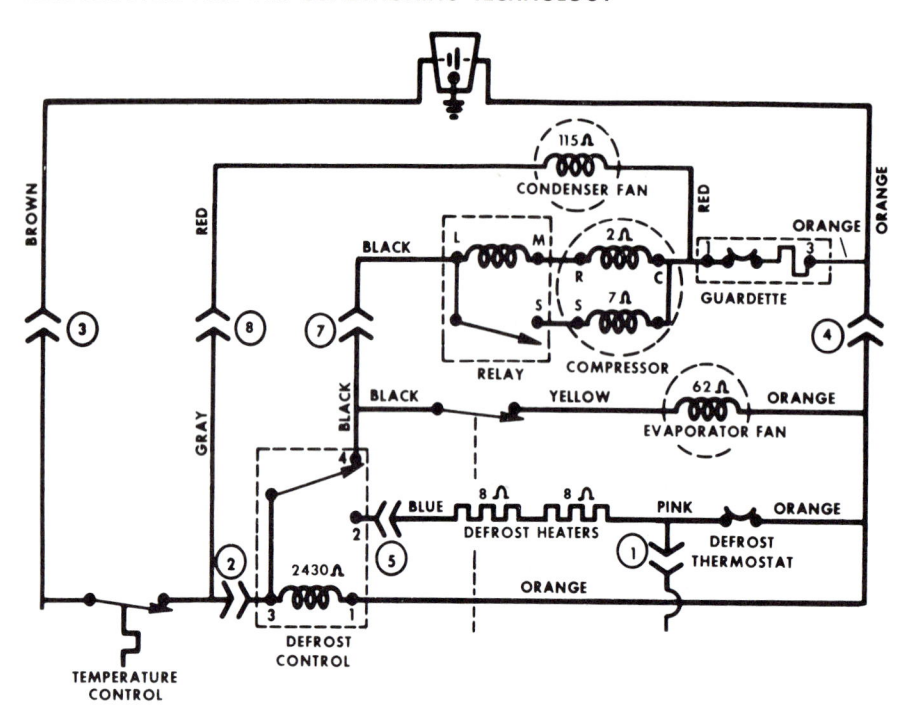

REFRIGERATION AND AIR CONDITIONING TECHNOLOGY

General Electric
12-6. Cycle defrost refrigerator-freezer electrical schematic.

other. Fig. 12-5. This freezer requires manual defrosting only every three or four months because its door is opened less frequently than the single-door refrigerator.

In the cycle defrost refrigerator/freezer, there are two evaporators. There is an evaporator for the freezer cabinet and an evaporator for the refrigerator section. They are in series with each other. The fresh food evaporator defrosts automatically every time the machine turns off. This is called a *cycle*. Cycle defrost derives its name from this on-off cycle.

The thermostat is placed inside the refrigerated cabinet to cause the compressor to turn on and off according to the temperature inside the cabinet. Figure 12-6 is an electrical diagram of the cycle defrost refrigerator/freezer. Figure 12-7 shows the cabinet.

In cycle defrosting, the thermostat or temperature control is attached to the fresh food evaporator. It is a special design that has a "constant on" characteristic. When you set it for a colder temperature, it lowers the temperature. At that time, it cuts off the compressor. It does not affect the temperature at the time it turns on the compressor. This cut-on temperature is fixed at about 37° F. [2.7°C] and cannot be changed. At 37° F. [2.7°C], all of the frost on the fresh food evaporator is melted. Therefore, whenever the compressor is off, it stays off until the fresh food evaporator has automatically defrosted. No heaters are needed because the fresh food evaporator is exposed to the 38° F. [3.3°C] air in the refrigerator section. Thus, with a cycle defrost refrigerator, the refrigerator section defrosts automatically every cycle. The freezer must be defrosted manually every three or four months. The freezer provides temperatures of 0° F. [−17.8°C] for long-term frozen food storage.

NO-FROST TOP-MOUNT REFRIGERATOR/FREEZER

The no-frost top-mount freezer also provides temperatures of 0° F. [−17.8°C]. However, the freezer never requires defrosting. Fig. 12-8. This type of refrigerator/freezer has the same elements as other types of refrigerator/freezers. However, it has only one evaporator. This evaporator is located in the partition between the freezer and the refrigerator.

This evaporator produces cold air, which is blown by a fan to the freezer on top and the refrigerator on the bottom. Fig. 12-9. There is a defrost heater mounted in the center of the evaporator. A timer measures the com-

298

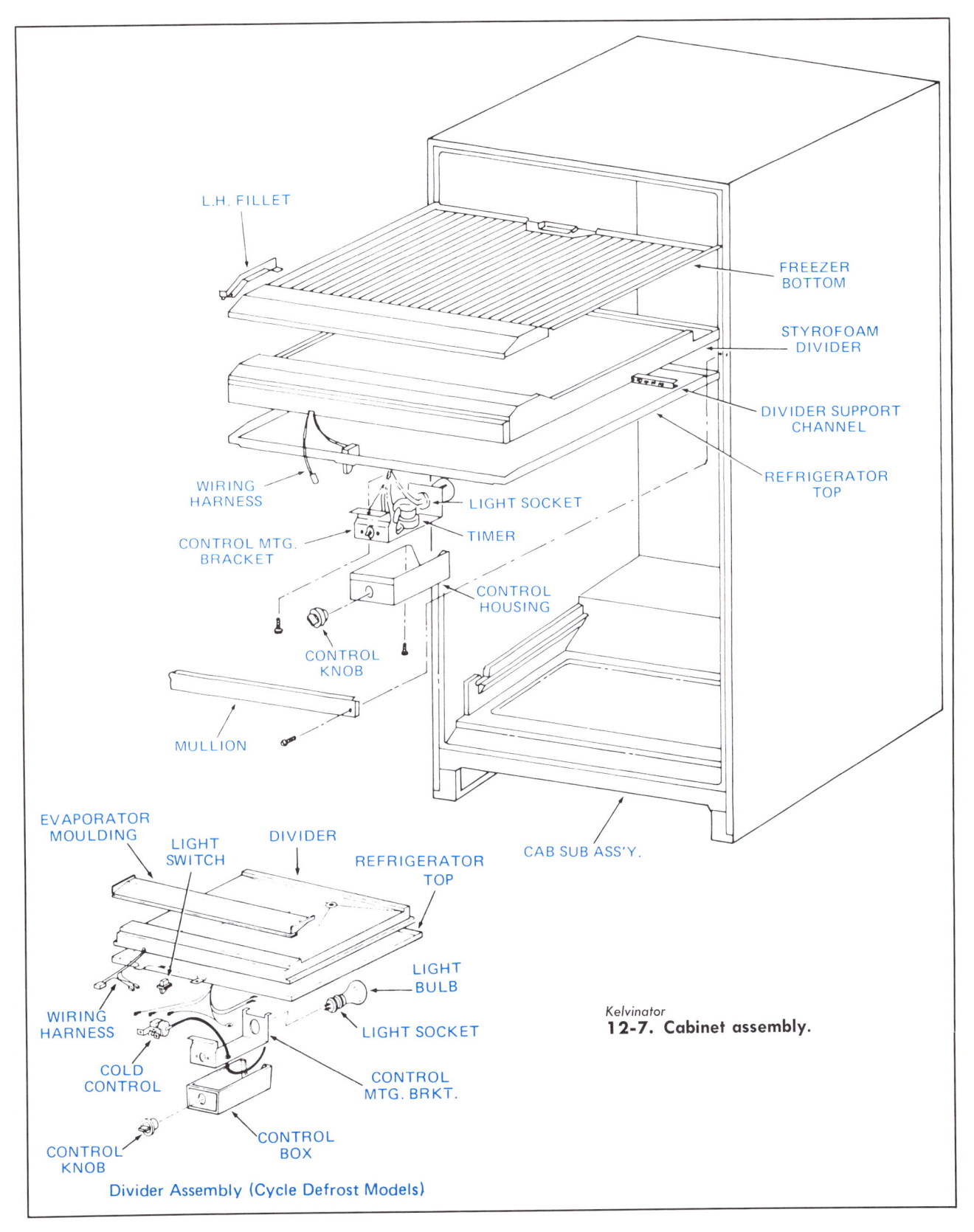

CHAPTER 12—REFRIGERATORS

Kelvinator
12-7. Cabinet assembly.

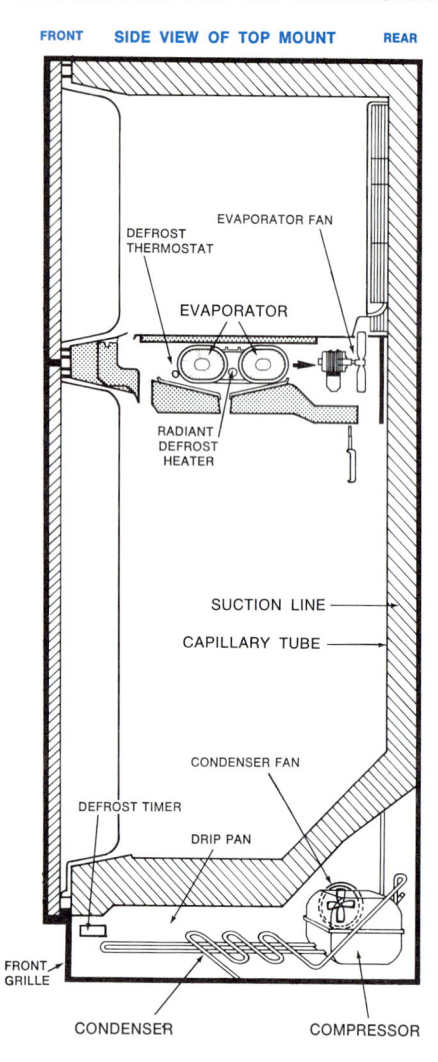

REFRIGERATION AND AIR CONDITIONING TECHNOLOGY

General Electric
12-8. No-frost top-mount refrigerator/freezer.

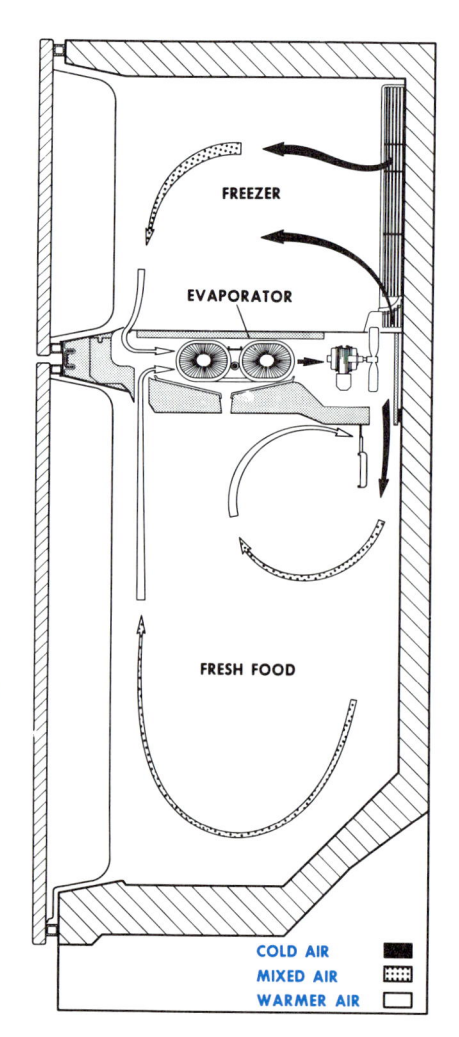

General Electric
12-9. Air circulation within the no-frost top-mount refrigerator/freezer. Placement of parts on a top-mount freezer, side view.

pressor running time. When the compressor has run for six hours, it automatically turns on the defrost heater. This heater melts the frost from the evaporator.

A thermostat turns off the heater when the evaporator is warm enough to have removed all of the frost. During the defrost cycle, neither the compressor nor the evaporator fan will run.

The compressor running time is affected by the door openings. Frost accumulation is also affected by door openings. The compressor running time then is directly related to the formation of frost. Thus, the timer controls the defrost cycle. The refrigerator will defrost less frequently with light usage. If a straight-time timer were used it would defrost every six hours whether it was needed or not.

Figure 12-10 shows the Kelvinator no-frost model divider assembly. Note difference between this and the cycle defrost model.

NO-FROST SIDE-BY-SIDE REFRIGERATOR/FREEZER

The no-frost side-by-side has exactly the same elements as the

CHAPTER 12—REFRIGERATORS

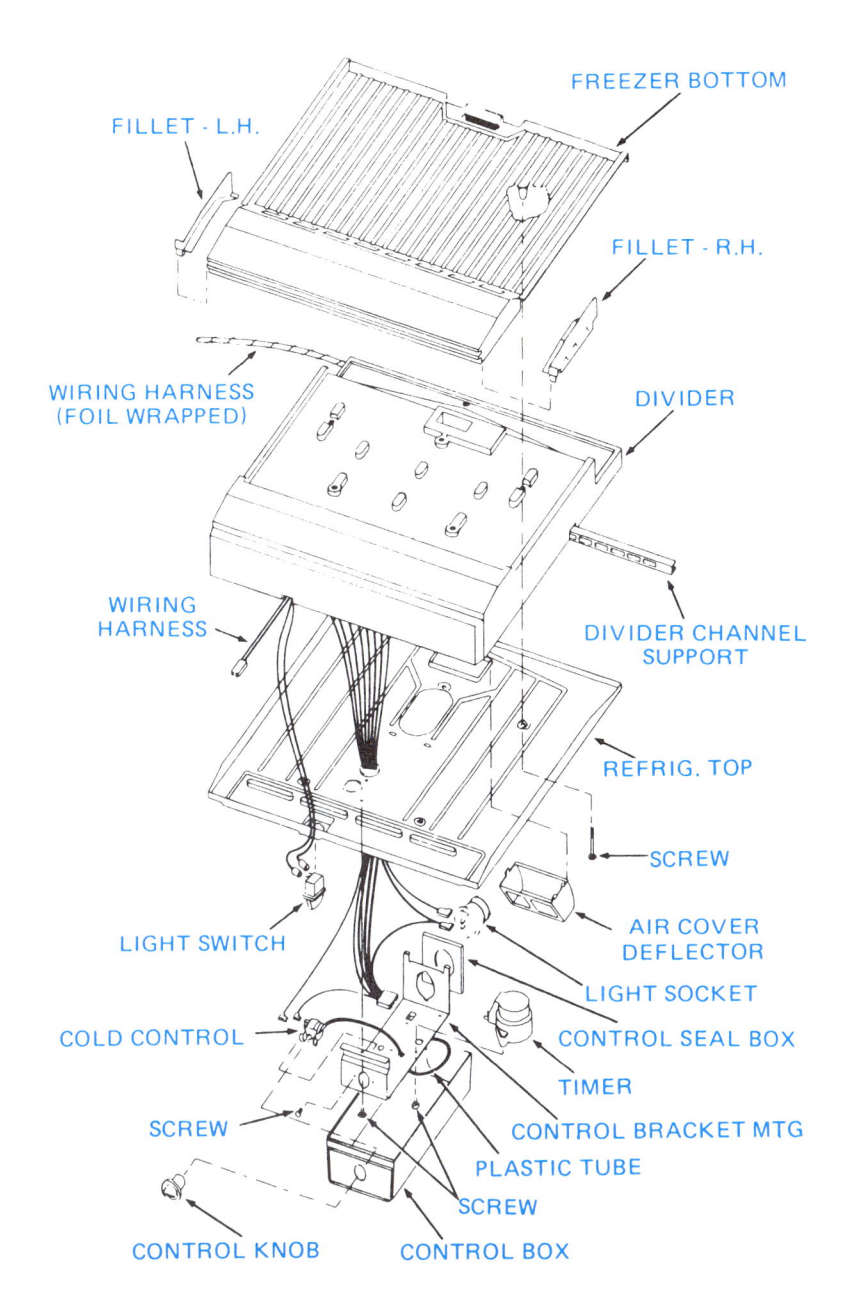

Kelvinator
12-10. Divider assembly for no-frost models.

top-mount no-frost. The difference is that the evaporator is mounted vertically on the back wall of the freezer. Figure 12-11 is an end view showing the cross-section of the freezer compartment. It shows the cold air in the freezer circulating from top to bottom. Remember cold air is heavy. Warm air is light and rises.

Figure 12-12 clearly points out the recognizable feature of a side-by-side refrigerator/freezer. The refrigerator cold air is blown from the evaporator through a hole in the partition at the top. It is returned through a hole at the bottom next to the meat drawer. Figure 12-13, on page 304, is a schematic of one side-by-side refrigerator/freezer.

ICEMAKER

Some refrigerators come with icemakers already installed. Others come with the option. This means they may be added after the customer has purchased the refrigerator. The complete icemaker kit comes with instructions on how to install it. A representative kit is shown in Fig. 12-14 on page 305.

301

REFRIGERATION AND AIR CONDITIONING TECHNOLOGY

12-11. No-frost side-by-side refrigerator, end view. *General Electric*

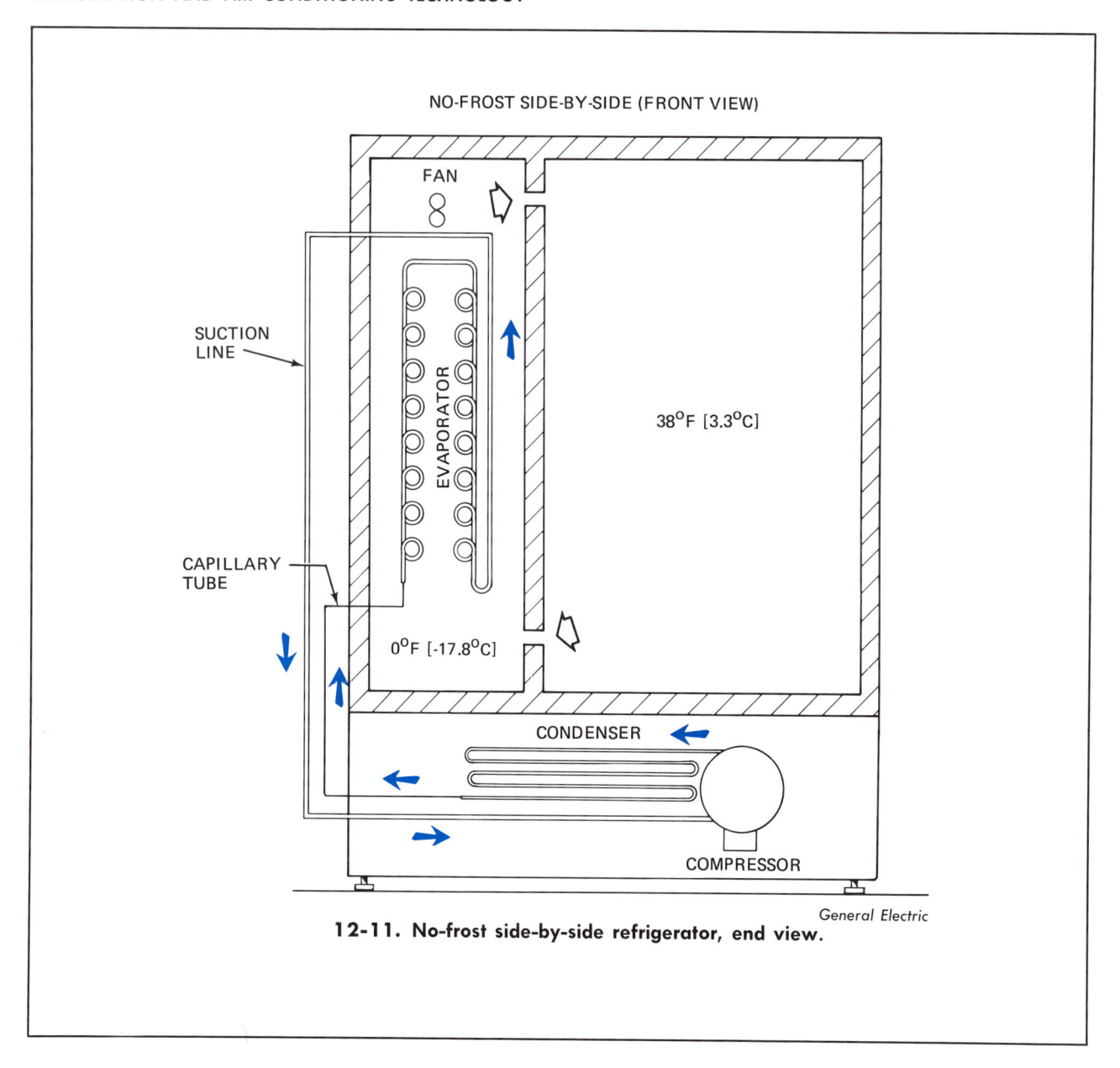

Figure 12-15 shows how the different icemaker models can be identified. Figure 12-16, on page 306, shows an example of a label that indicates what model of icemaker can be used with a refrigerator. Figure 12-17, on page 306, shows a complete unit with ice in the container.

These icemakers operate with a thermostat with limits of 26 to 16° F. [−3.3 to −8.9°C]. The mold heater unit consumes 120 watts. The motor has a speed of 1 rpm. The ejection cycle is two to four minutes per cycle. The ice cube rate is 0.2 pounds per cycle. In twenty-four hours, the average is 4.5 to 5.5 pounds of ice made and ready for use.

On some exterior ice service refrigerators (with a photo-sensing type icemaker) these additional items should be checked if the icemaker is inoperative:

• Point 1 in Fig. 12-18, found on page 306. The ice bucket

302

CHAPTER 12—REFRIGERATORS

NO-FROST SIDE-BY-SIDE (END VIEW)

12-12. No-frost side-by-side, front view.

General Electric

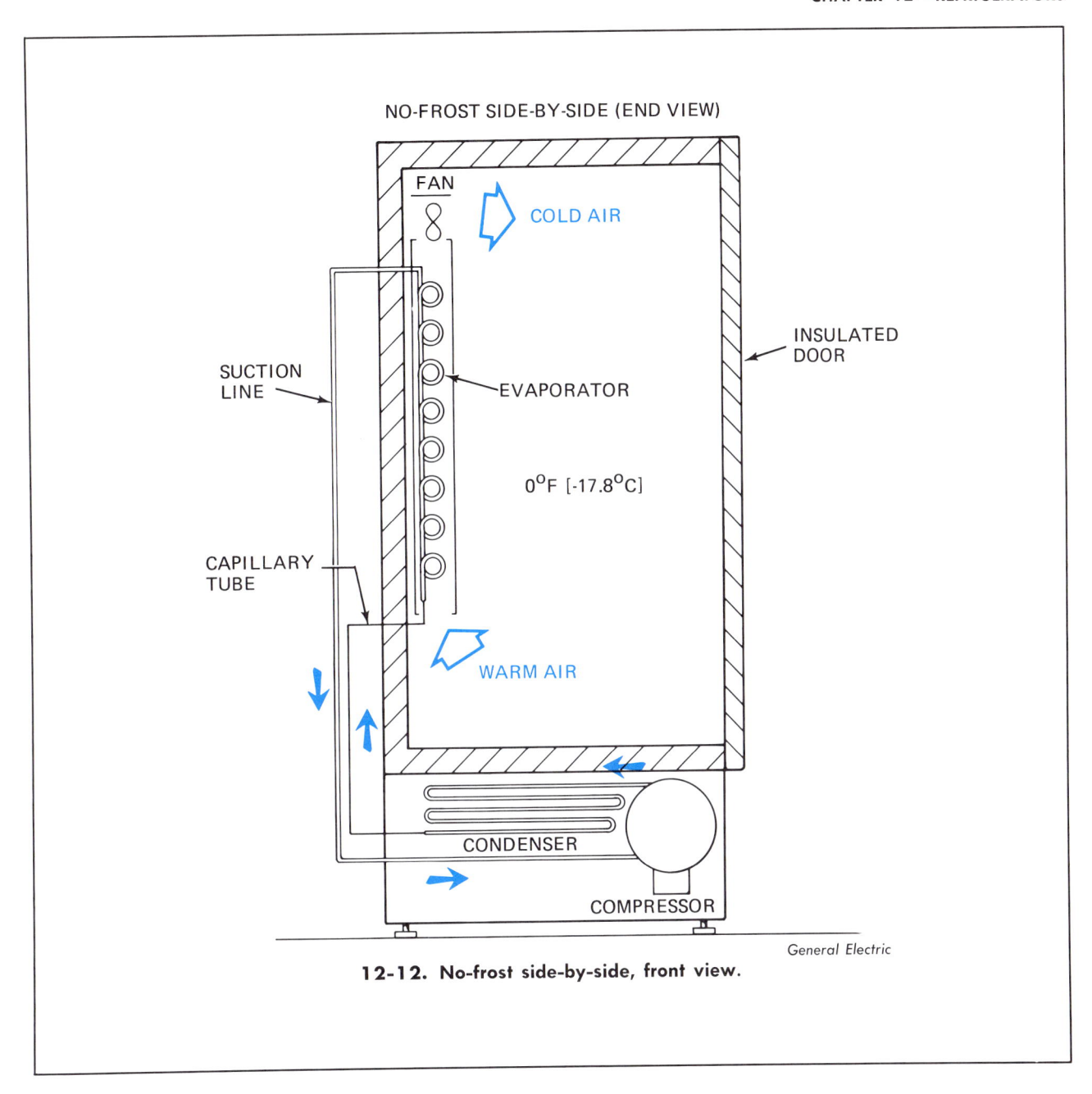

should be in place and properly positioned.

• Point 2 in Fig. 12-18. Inspect the opening in the photo switch plate for obstructions.

• Point 3 in Fig. 12-18. Examine the light source for foreign matter on the lens, for a burned-out bulb, for a darkened bulb, or a missing lens.

• Point 4 in Fig. 12-18. Determine if the ice service door switch has the plunger depressed sufficiently when the door is closed. Be sure plunger actuator arm is positioned properly.

The Kelvinator refrigerator has a slightly different icemaker. It is discussed here to give you some knowledge of the components of icemakers.

303

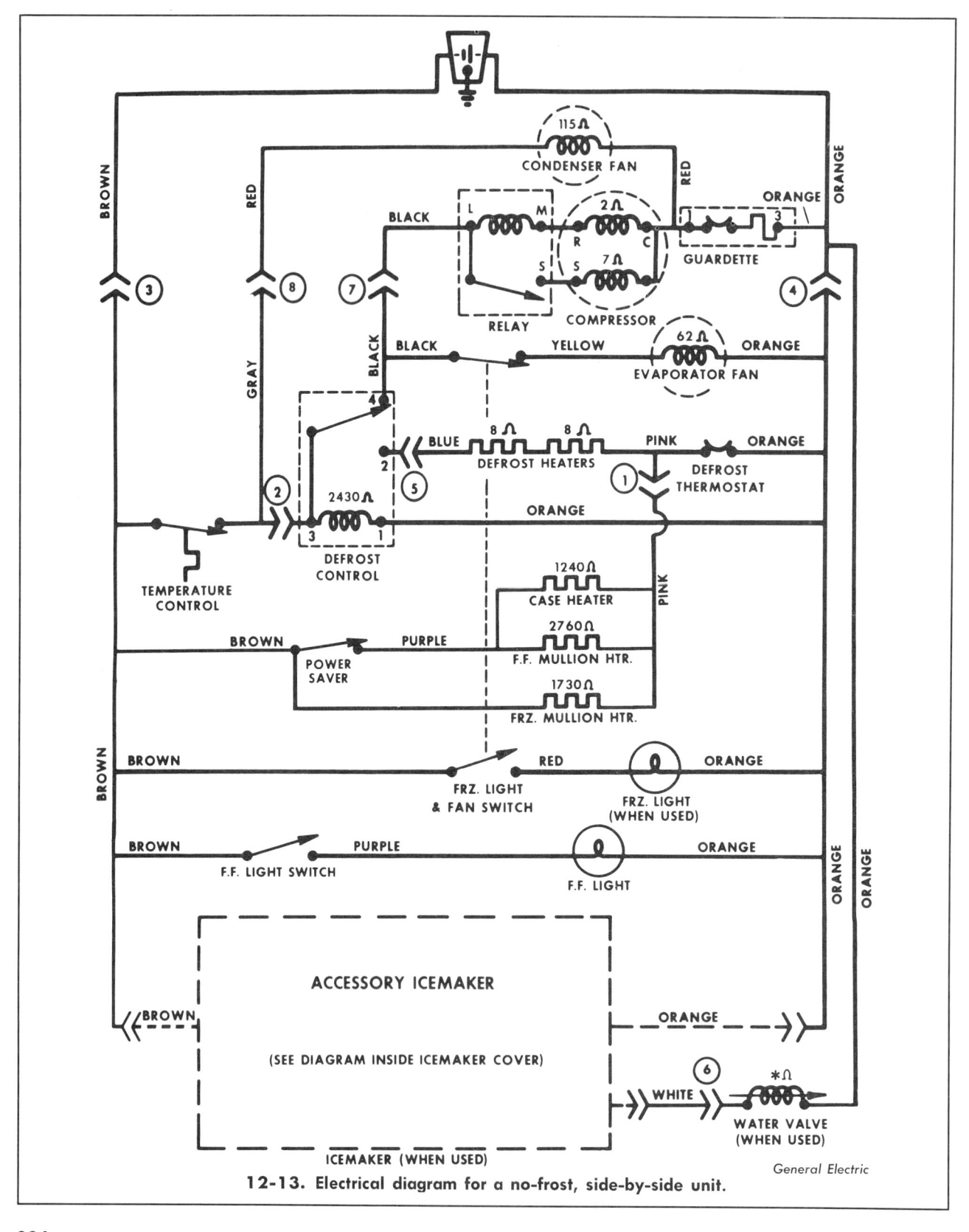

REFRIGERATION AND AIR CONDITIONING TECHNOLOGY

12-13. Electrical diagram for a no-frost, side-by-side unit.

General Electric

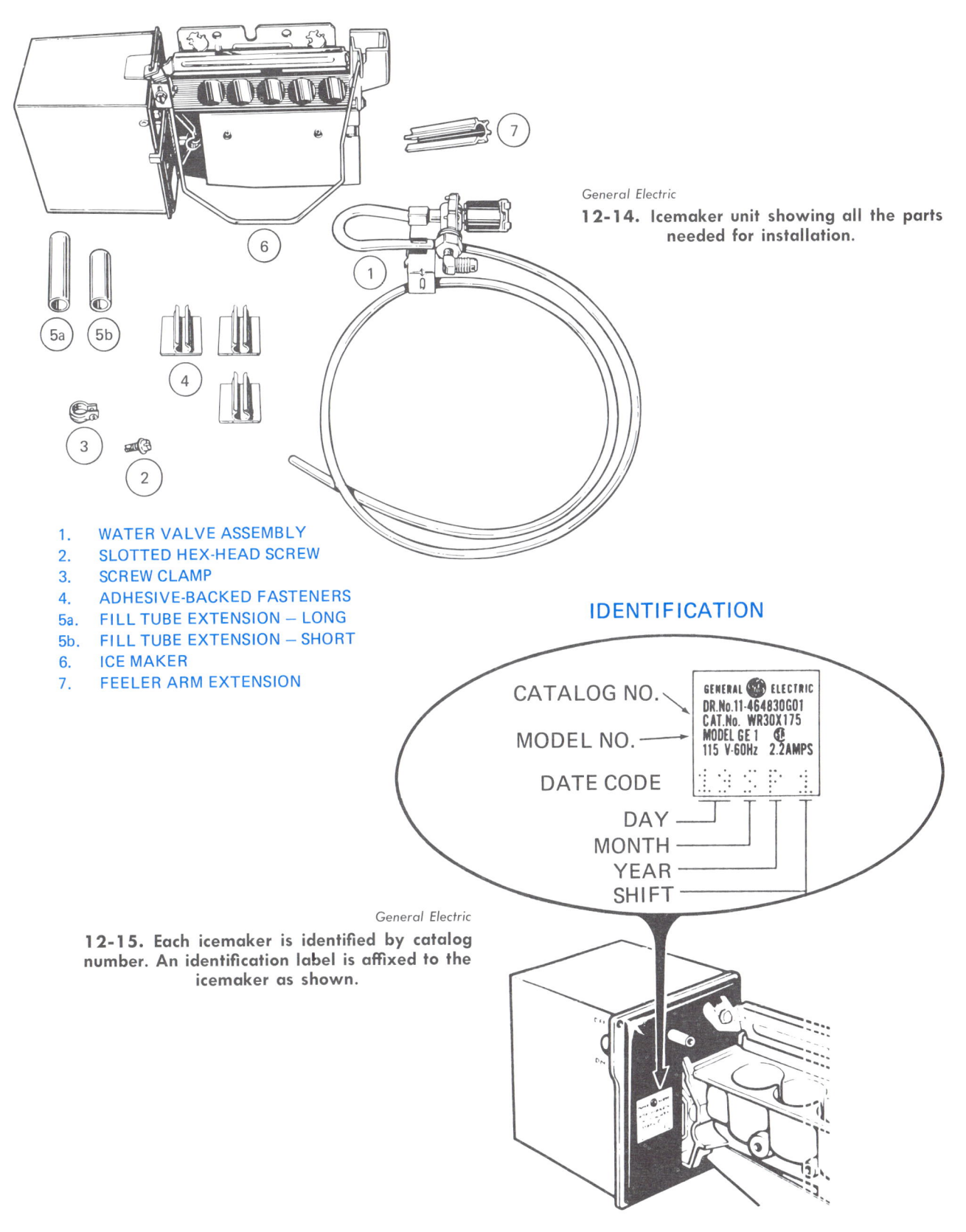

CHAPTER 12—REFRIGERATORS

General Electric
12-14. Icemaker unit showing all the parts needed for installation.

1. WATER VALVE ASSEMBLY
2. SLOTTED HEX-HEAD SCREW
3. SCREW CLAMP
4. ADHESIVE-BACKED FASTENERS
5a. FILL TUBE EXTENSION — LONG
5b. FILL TUBE EXTENSION — SHORT
6. ICE MAKER
7. FEELER ARM EXTENSION

IDENTIFICATION

CATALOG NO.
MODEL NO.
DATE CODE
DAY
MONTH
YEAR
SHIFT

General Electric
12-15. Each icemaker is identified by catalog number. An identification label is affixed to the icemaker as shown.

305

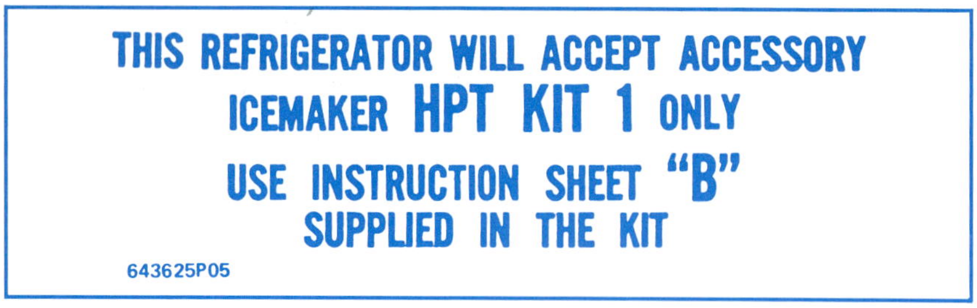

REFRIGERATION AND AIR CONDITIONING TECHNOLOGY

> **THIS REFRIGERATOR WILL ACCEPT ACCESSORY ICEMAKER HPT KIT 1 ONLY**
>
> **USE INSTRUCTION SHEET "B" SUPPLIED IN THE KIT**
>
> 643625P05

General Electric
12-16. The label on the back of the refrigerator tells the service technician which kit to use for a refrigerator.

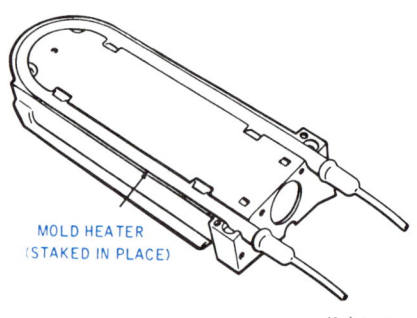

Kelvinator
12-20. Mold heater (staked in place).

MOLD HEATER (STAKED IN PLACE)

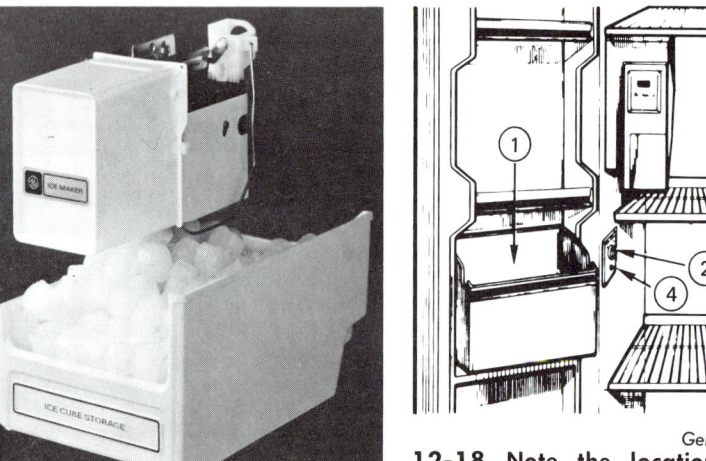

General Electric
12-17. Complete icemaker.

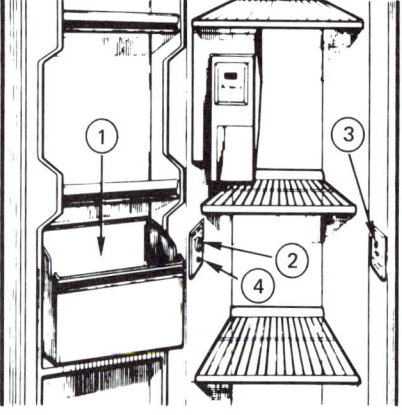

General Electric
12-18. Note the location of the photo switch on this icemaker.

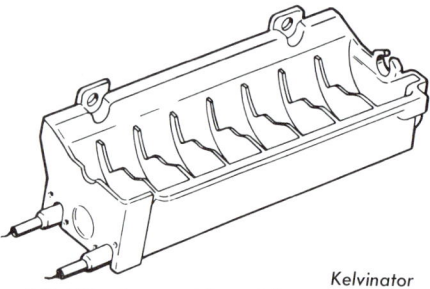

Kelvinator
12-19. Ice mold and ice stripper.

The ice mold of the Kelvinator is an aluminum die-casting. Fig. 12-19. The front of the device is bonded to the icemaker thermostat. The ice mold has a semicircular interior partitioned into compartments of equal size. Water enters at the rear of the mold through a fill trough. An opening in each partition permits all compartments to fill. A film of silicone grease on the top edge of the mold prevents siphoning of water by capillary action.

Mold Heater

A mold heater, rated at 165 watts and covered with an aluminum sheath, is embedded in a grooved section on the underside of the mold. When the mold heater is energized, heat melts the ice contact surface within the mold. This allows the ice pieces to be harvested. See Fig. 12-20.

The mold heater is wired in series with the ice maker thermostat. This acts as a safety device.

The original heater is staked in place, but can be removed for replacement. The replacement element is secured to the mold by four flat-head retaining screws, which thread into holes in the mold adjacent to the heater. To get good thermal contact, Alumilastic® is used between the heat and the mold.

Ice Stripper

An ice stripper is attached to the mold. It prevents ice pieces from falling back into the mold. It also serves as a decorative side cover. See Fig. 12-19.

Ice Ejector

The ejector blades are molded from Delrin®, a plastic material. They extend from a central shaft that turns in nylon bearings at the front and rear. Each blade sweeps an ice section from the mold. The drive end of the ice ejector is D-shaped. Silicone grease is used to lubricate the bearing surfaces. Fig. 12-21.

Water Valve Assembly

The water valve is solenoid operated. When it energizes, it releases water from the supply line into the ice mold. The amount of water released is directly proportional to the length

CHAPTER 12—REFRIGERATORS

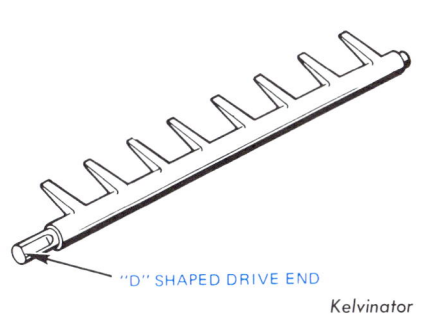

12-21. Ice ejector.

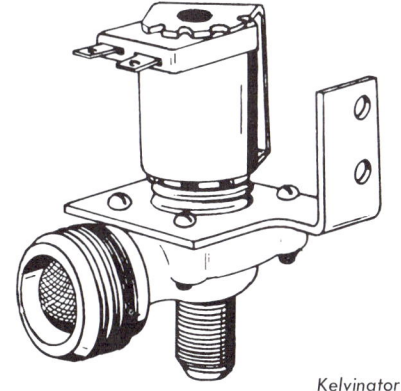

12-22. Water valve for icemaker.

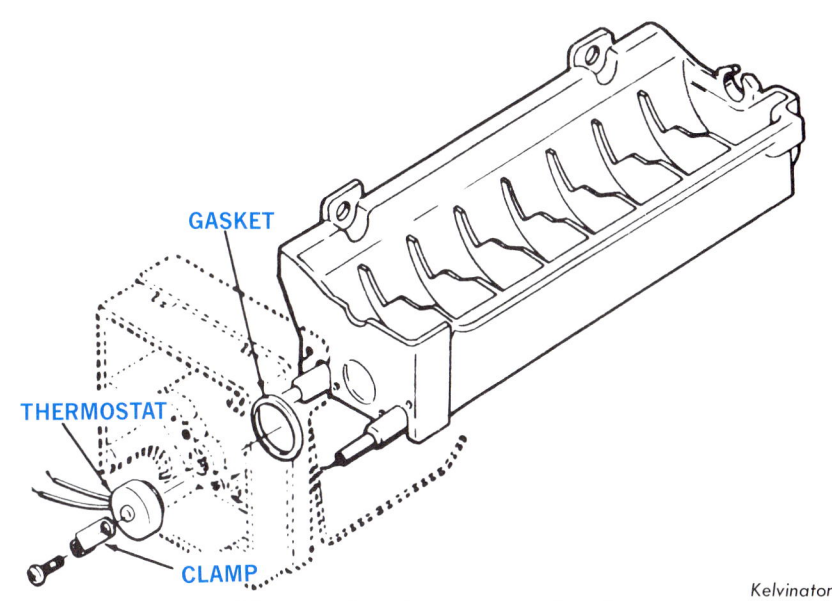

12-23. Icemaker thermostat and gasket.

of time the water valve switch is energized. A flow washer inside the water valve maintains a constant rate of water flow over a supply line pressure range of 15 to 100 psi. It will not compensate for pressures below 15 psi, or greater than 100 psi. A 60-mesh screen, placed ahead of the flow washer, filters out foreign materials. See Fig. 12-22.

The solenoid coil draws 10 to 15 watts at 105 volts. The coil is wired in series with the mold heater. It is connected directly across the 120-volt line. The voltage drop across the heater is 15 volts. This leaves a voltage drop of 105 volts across the solenoid.

Thermostat

The thermostat is a single-pole, single-throw, bimetallic, disc-type thermal switch. It automatically starts the harvest cycle when the ice is frozen. The thermostat closes at a temperature of 18° ± 6° F. [−7.8°C ± 3.3°C]. It resets at a temperature of 50° ± 6° F. [10°C ± 3.3°C]. Wired in series with the mold heater, the icemaker thermostat acts as a safety device against overheating in case of mechanical failure. An Alumilastic® bond is provided where the thermostat is mounted against the mold. A gasket prevents water from leaking into the support housing. Fig. 12-23.

Signal Arm and Linkage

The signal arm is cam-driven and operates a switch to control the quantity of ice produced. In the harvest cycle, the arm is raised and lowered during each of the two revolutions of the timing cam. If the signal arm comes to rest on top of the ice in the storage container during either revolution, the switch will remain open. It will stop the icemaker at the end of that revolution. When sufficient ice is removed from the container to allow the arm to lower, ice production will resume. To stop the icemaker manually, raise the signal arm until it locks in the upper position. Operation is resumed when the arm is manually lowered. Fig. 12-24.

Timing Switches

The three timing switches used are of the single-pole, double-throw type. They are identical except for function, and can be used interchangeably.

The holding switch assures completion of a revolution once the icemaker operation has started. The water valve switch opens the water valve during the fill cycle. It is the only adjustable component in the icemaker. The shut-off switch stops the icemaker operation when the storage container is full of ice.

Timing Cam and Coupler

Three separate cams are combined in one molded Delrin®

307

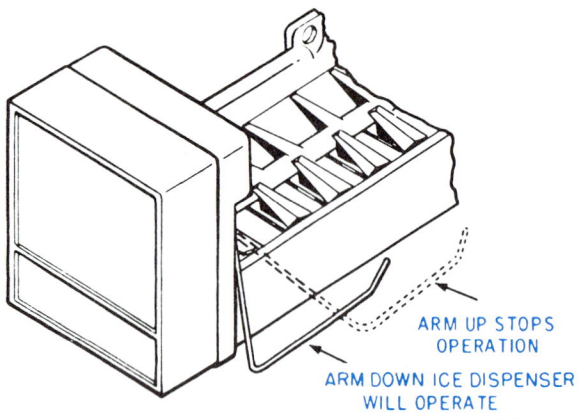

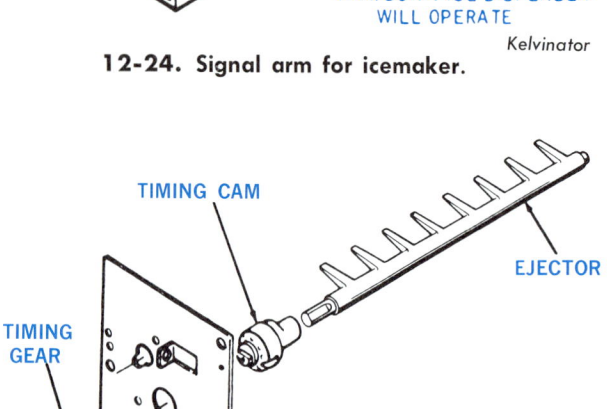

REFRIGERATION AND AIR CONDITIONING TECHNOLOGY

12-24. Signal arm for icemaker.

12-25. Timing gear and cam for an icemaker.

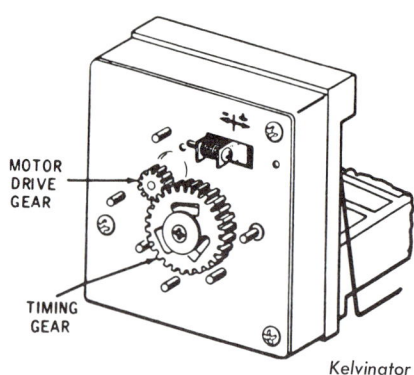

12-26. Motor drive and timing gears for the icemaker.

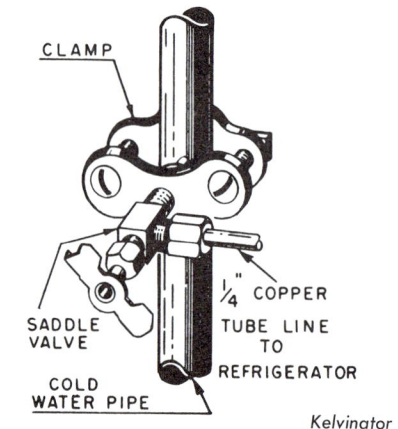

12-27. Saddle valve installation.

part. One end is attached to a large timing gear. The other end is coupled to the ejector.

The inner cam operates the shut-off switch lever arm. The center cam operates the holding switch. The outer cam operates the water valve switch.

Timing Gear

The large molded plastic gear is driven by the motor. It, in turn, rotates the cam and ejector. A D-shaped hole in the gear fits over the timing cam hub. Spacer tabs on the backside of the gear prevent the gear from binding on the mounting plate. Fig. 12-25.

Motor

A low-wattage stall-type motor drives the timing gear. This gear turns the timing cam and ejector blades approximately one revolution every 3 minutes ($\frac{1}{3}$ r/min). Fig. 12-26.

Fill Trough

The fill trough is molded nylon. It supports the inlet tube and directs the water into the mold. It also forms a bearing for one end of the ejector blades.

Wiring

A four-prong plug connects the icemaker wiring to the cabinet wiring harness. The icemaker assembly is wired across the line and will harvest during the refrigeration or defrost cycles.

Installing Water Supply Line to an Icemaker

Make sure that the installation complies with the local code. All parts normally required, except the $\frac{1}{4}''$ copper tubing, are supplied by the refrigerator manufacturer, providing the local code permits the use of a saddle valve.

To connect the water supply to the icemaker, proceed as follows:

1. Turn off the main water supply.

CHAPTER 12—REFRIGERATORS

Test Cycling the Icemaker

It may be necessary to test-cycle an icemaker to check its operation. This can be done on the repair bench or while mounted in the refrigerator.

If the icemaker is in an operating refrigerator, take precaution against the formation of condensate by allowing the cold metal components to warm up before removing the front cover. This can be expedited by cycling the assembly with the cover in place and the water supply valve closed.

To cycle the icemaker manually, slowly turn the ejector blades clockwise until the holding switch circuit is completed to the motor. When the motor starts, all components except the icemaker thermostat should perform normally to complete the cycle. Then, remove the front cover by prying it loose with the blade of a screwdriver at the bottom of the support housing. If further test cycling is necessary, place the blade of a screwdriver in the slot located in the motor drive gear. Turn counterclockwise until the holding switch circuit is completed to the motor. Fig. 12-29. NOTE: This is the procedure for test cycling only this particular model of Kelvinator icemaker. Other models and icemakers made by other manufacturers will require different methods.

Water Valve Switch and Water Fill Volume

The amount of water fill is directly proportional to the length of time terminals C-NC of the water valve switch are closed. This occurs when the switch plunger drops into a cavity formed in the cam. Different

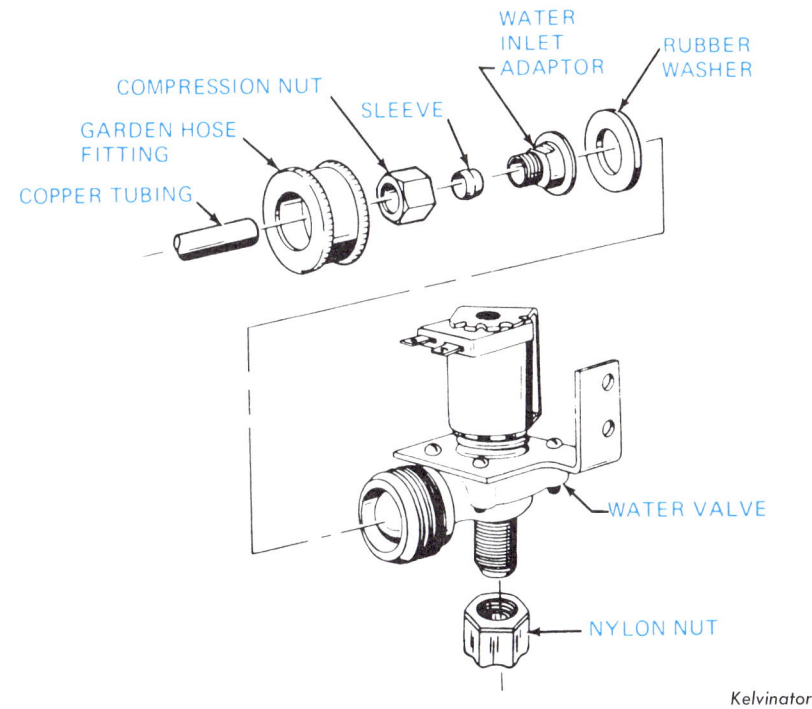

12-28. Water valve connector.

2. Select a vertical cold water line ($\frac{1}{2}''$ or $\frac{3}{4}''$ pipe) as close to the refrigerator as possible. Drill a $\frac{1}{4}''$ hole in this water supply line.

NOTE: If a vertical cold water line is not available, a horizontal line may be used, providing the hole is drilled in the *side* of the pipe and not in the bottom. A hole drilled in the bottom of the pipe might become clogged with scale and foreign material. This would stop water flow.

3. Install the saddle valve. See Fig. 12-27. Make sure the gasket fits tightly around the hole. Tighten the clamp screws evenly and just enough to provide a watertight seal.

4. Connect $\frac{1}{4}''$ copper tubing to the saddle valve and flush water through the tubing. It is advisable to provide a surplus loop of tubing behind the refrigerator. If this is done, the refrigerator may be moved away from the wall for servicing.

5. In some installations, it may be desirable to install a $\frac{1}{4}''$ shut-off valve in the $\frac{1}{4}''$ copper tubing line at a convenient location behind the refrigerator. A $\frac{1}{4}''$ shut-off valve is not usually included with the refrigerator. It can be obtained at a hardware store.

6. Remove the protection cap from the water valve on the back of the refrigerator cabinet. Slip the garden hose fitting and compression nut and sleeve onto the $\frac{1}{4}''$ copper tube. See Fig. 12-28. Assemble the valve inlet adapter on the end of the $\frac{1}{4}''$ copper tube and tighten the compression nut. Insert the rubber washer into the garden hose fitting and assemble to the water valve on the back of the refrigerator cabinet.

7. Turn on main water supply and check all connections for leaks.

309

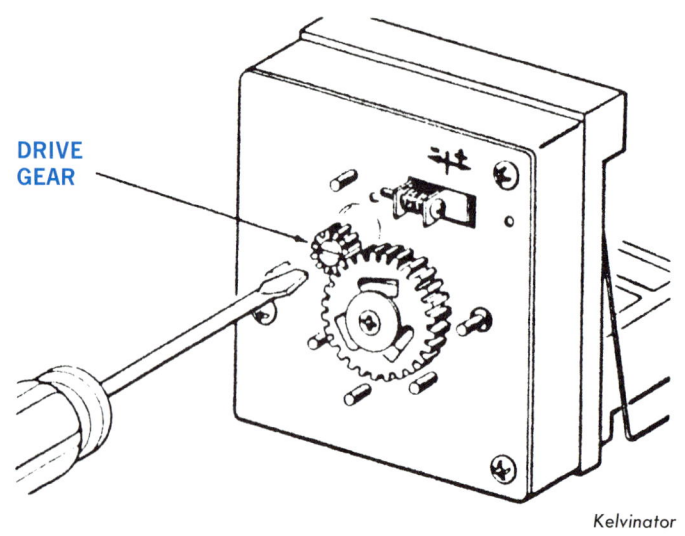

REFRIGERATION AND AIR CONDITIONING TECHNOLOGY

12-29. Test-cycling an icemaker. *Kelvinator*

water valves have different flow rates. For this reason, when a water valve is replaced, the water valve switch must be adjusted. Figure 12-30 shows the water-fill adjusting screw on one model of icemaker.

The correct water fill volume on this model is 145 cubic centimetres, or about 5 ounces. To measure, test-cycle the icemaker and collect the water. Measure in a container calibrated in cubic centimetres or ounces. The fill volume is adjusted by increasing or decreasing the length of time the water valve switch remains closed.

To adjust the valve switch, first determine how much more or less water is needed. The adjusting screw is calibrated so that one complete revolution changes the water fill about 18 cubic centimetres. Turning the screw clockwise decreases the fill, while turning it counterclockwise increases the fill.

For example, an icemaker is test-cycled. The water fill sample is 158 cubic centimetres. Subtracting 145 cubic centimetres from 158 cubic centimetres, it is found that the adjustment needed is 13 cubic centimetres. Since one turn of the adjusting screw changes the fill 18 centimetres, a three-quarters turn clockwise would reduce the fill about 13 cubic centimetres, or the desired amount.

Icemaker Parts Replacement

First, disconnect the appliance service cord from the power supply. If the refrigerator is operating and cold, allow the icemaker to warm to room temperature before removing the front cover. This prevents moisture from condensing on the metal components.

To replace the front cover, proceed as follows:

1. Be sure the icemaker is at room temperature before removing the cover.
2. Place the blade of a screwdriver in the slot at the bottom of the mold support. Pry the cover loose.
3. When installing the cover, be sure the retaining tabs inside the cover are located on top and bottom. Then snap the cover into place.

To replace the fill trough and bearings, proceed as follows:

1. Push the retaining tab back, away from the mold.
2. Rotate counterclockwise until the trough is clear.
3. Pull from the back to detach the bearing from the mold and ejector blades.
4. Replace in reverse order.

To replace the ejector blades, proceed as follows:

1. Turn the blades to the twelve-o'clock position.
2. Remove the fill trough and the bearing.
3. Force back and up to disengage from the motor coupling.
4. Replace in reverse order, noting that the blades lock in the motor coupling at the twelve-o'clock position.
5. Lubricate the bearing ends at the ejector with silicone grease.

To replace the ice stripper, proceed as follows:

1. Remove the icemaker from the refrigerator.
2. Remove the retaining screw at the back of the mold.
3. Force the ice stripper back to disengage it from the front of the mold support housing.
4. Replace in reverse order.

To replace the motor and the switch mounting plate, proceed as follows:

1. Remove the front cover.
2. Remove the three mounting-plate-to-support-housing attaching screws.
3. Carefully remove the mounting plate. Disengage the end of the signal arm. Note the relative position of the signal arm spring.

CHAPTER 12—REFRIGERATORS

4. Transfer the motor, switches, cam, gear, and water-fill adjusting spring and arm to the replacement mounting plate.

5. Install the mounting plate on the support housing. Be sure the wiring harness is properly positioned and the signal arm spring is in place.

6. Check the water fill cycle and adjust as required.

To replace the motor, proceed as follows:

1. Remove the front cover.
2. Remove the three mounting-plate-to-support-housing attaching screws.
3. Disconnect the motor leads.
4. Remove the two motor mounting screws.
5. Replace in reverse order.

To replace the water valve switch, proceed as follows:

1. Remove the front cover.
2. Remove the three mounting-plate-to-support-housing attaching screws.
3. Disconnect the water valve switch wire leads.
4. Remove the switch mounting screws. Then, remove the switch.
5. Replace in reverse order, making sure the switch insulator is in place.
6. Check the water fill cycle. Adjust as required.

To replace the holding switch, proceed as follows:

1. Remove the front cover.
2. Remove the three mounting-plate-to-support-housing attaching screws.
3. Disconnect the holding switch wire leads.
4. Remove the switch mounting screws. Then, remove the switch.
5. Replace in reverse order.

To replace the signal arm shut-off switch, proceed as follows:

1. Remove the front cover.
2. Remove the three mounting-plate-to-support-housing attaching screws.
3. Disconnect the switch wire leads.
4. Raise the signal arm.
5. Remove the switch mounting screws.
6. Replace in reverse order.

To replace the icemaker thermostat, proceed as follows:

1. Remove the front cover.
2. Remove the three mounting-plate-to-support-housing attaching screws.
3. Loosen the retaining clip mounting screw on the thermostat.
4. Disconnect the thermostat wire leads. Then, remove the thermostat.
5. Apply Alumilastic® to the sensing surface of the replacement thermostat to insure positive bond to the mold.
6. Replace in reverse order.

To replace the mold heater, proceed as follows:

1. Remove the front cover.
2. Remove the icemaker.
3. Remove the three mounting-plate-to-support-housing attaching screws.
4. Remove the four mold-to-support-housing screws.
5. Detach the thermostat from the mold.
6. Detach the mold heater from the wire leads.
7. Separate the mold from the support housing.

NOTE: Use care not to destroy the thermostat gasket located between the mold and the support housing.

8. Use a flat-bladed screwdriver to pry the inoperative heater from the mold groove.
9. Clean the remaining Alumilastic® from the mold groove.
10. Apply a layer of Alumilastic® in the mold groove.
11. Install the replacement mold heater. Thread the four screws supplied with the replacement heater into the holes provided in the mold. These secure the heater in place.
12. Replace the parts in reverse order of removal. Be sure the thermostat gasket is in place. Bond the thermostat to the mold with Alumilastic®.

TROUBLESHOOTING

Determining the causes of problems in icemakers is not difficult. Figure 12-31 shows the schematic for one type of icemaker. Refer to it while reading about icemaker problems.

Complaint: Icemaker Fails to Start

Cause

1. The signal arm is locked in the raised position.

WATER FILL ADJUSTING SCREW
1 FULL TURN = 18 cm³ CHANGE

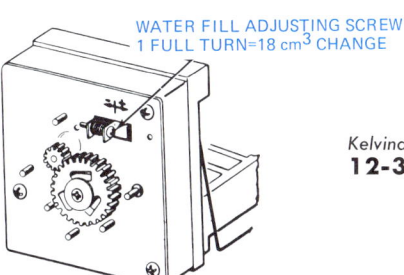

Kelvinator
12-30. Adjusting the water fill level.

311

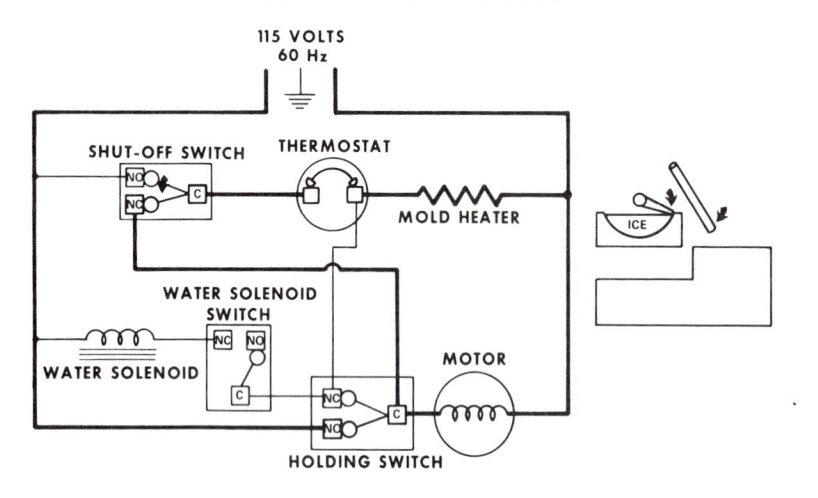

REFRIGERATION AND AIR CONDITIONING TECHNOLOGY

When the ejector blades reach the ice in the mold, the motor will stall. It will remain in this position until the ice has thawed loose. During this time the mold heater remains energized.

Kelvinator

12-31. The schematic for one type of icemaker. Here, the icemaker is shown midway through its cycle.

2. An open circuit in the wiring or components. Check the terminals from the cabinet wiring to the icemaker.

3. Check the operation of the icemaker with a test service cord.

4. Check the mold temperature at mounting screws. If it is above 15° F. [−9.4°C], the freezer air temperature is not cold enough to switch the icemaker thermostat to the closed position. If the mold is below 9° F. [−12.8°C], manually start the icemaker by rotating the timing gear. If the motor fails to start, check the motor for continuity. If the motor starts the thermostat, the shut-off switch, or holding switch is inoperative. Check the hold switch first. With the ejector blades in the starting position, check terminals C and NC for continuity. Replace the switch if it is open. Check the shut-off switch linkage. Then check the signal arm in the lowest position. Replace the switch if it is open. If the holding switch and the shut-off switch are operative, replace the thermostat.

Complaint: Icemaker Fails to Complete the Cycle

Cause

1. With the ejector blades at the ten-o'clock position, and the holding switch plunger depressed, check terminals C and NO for continuity. Replace the switch if it is open.

2. With the ejector blades at the twelve-o'clock position, check the shut-off switch terminals C and NC for continuity. Replace the switch if it is open.

3. With the blades at the four-o'clock position, check the mold heater and icemaker thermostat for continuity. Replace the heater if it is open. If the heater shows continuity, replace the thermostat.

4. Check the motor operation with a test cord. Replace the motor if it fails to start.

Complaint: Icemaker Fails to Stop at the End of the Cycle

Cause

1. With the ejector blades in the starting position, check the holding switch terminals C and NO for continuity. Replace the switch if it is closed.

Complaint: Icemaker Continues to Eject When the Container is Full

Cause

1. Loose linkage to the signal arm shut-off switch. The switch should open when the arm is in a raised position. Adjust if required.

2. Check the shut-off switch terminals C and NO for continuity with the signal arm raised. Replace the switch if it is closed.

Complaint: Icemaker Produces Undersized Ice Pieces

Cause

1. Icemaker mold not level. Check.

2. Partial restriction in the supply line or the water valve.

3. Insufficient water pressure to the water valve.

4. Water valve switch is not adjusted for the proper water fill of 145 cubic centimetres. Adjust.

5. Insufficient thermal bond between the thermostat and the mold. Rebond with Alumilastic®.

6. Check the thermostat calibration by replacing with a new thermostat.

Complaint: Water Overflows the Mold

Cause

1. Icemaker is not level. Check.

2. Evidence of water siphoning off the top edge of the mold. Wipe a thin layer of silicone

CHAPTER 12—REFRIGERATORS

grease on the top edge of the mold to prevent capillary action.

3. Fill tube is not properly positioned in the fill trough. Check.

4. Water leaks into the mold after the cycle is completed. Replace the water valve and check the water pressure to the valve.

5. Water fill exceeds volume capacity of the mold. Adjust and test-cycle the icemaker for proper water fill.

6. With the ejector blades in the starting position, check terminals C and NC. Replace the switch if it is closed.

Complaint: Water Fails to Enter the Mold

Cause

1. Insufficient water pressure to the water valve.

2. Restriction in the supply line or the water valve.

3. Water supply line valve is closed.

4. Water valve solenoid coil is open or shorted.

5. Inoperative water valve switch. With the plunger out, check terminals C and NC for continuity. Replace the switch if the terminals check out open.

DEFROSTING SYSTEM ANALYSIS

Moisture forms frost. There are two sources for the moisture in a refrigerator—air and food. Moisture from the air is the primary source of frost. Whenever the door is opened, heavy, cool, dry air comes out. This air is replaced by lighter, warm, moist air. Obviously, the more times the door is opened, the greater the volume of moist air entering the refrigerator cabinet. Unusual frost forms if the door is opened too frequently or is left ajar.

Removing heat causes air to contract. This creates a partial vacuum. The pressure inside is lower than room pressure. This results in heat leakage. Heat is drawn in through those openings that are not properly sealed.

Excessive moisture-laden air can enter around the gasket and through the trim holes. Fig. 12-32. Careful attention is given to the sealing of all openings in the cabinet and doors of each refrigerator and freezer. All seams in the outer case are sealed with sealer. Plug buttons are fitted into openings in the outer case and outer door panels. Sealing gaskets and fasteners are used for trim and nameplates. These prevent moisture-laden air from entering the cabinet.

Generally, small amounts of moisture-laden air seeping into the cabinet will not adversely affect the ability of the refrigerator or freezer to function properly. In automatic defrost models, small leaks may result in only light frosting on the freezer shelves or frozen food packages and/or sweating inside the fresh food compartment.

In many instances, the major source of air leaking into the cabinet can be found at the door gasket seal. Use of a dollar bill to check for a satisfactory door gasket seal does not apply to later models (since 1957) that do not employ a positive-action mechanical door latch. Formerly, a dollar bill could be used to test this door gasket seal. The door was opened. A dollar bill was placed flat against the doorframe of the refrigerator. The door was then closed. If the bill could not

12-32. Proper sealing of all seams and around the door improves the efficiency of the refrigerator.

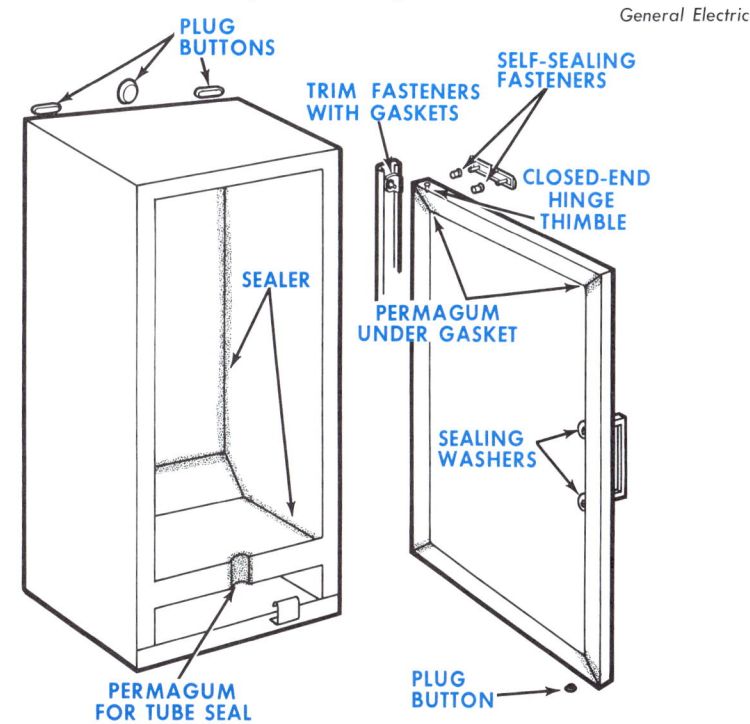

General Electric

313

be pulled out, the seal was effective.

The only method recommended for checking the door gasket seal on modern refrigerators and freezers is to use a 150-W floodlamp from within the cabinet. The lamp cord will be sealed by the gasket when the door is closed. With the door closed, and with subdued light surrounding the cabinet, direct the light toward one length of gasket at a time and inspect along the entire length of the gasket for light showing between the gasket and the cabinet face. A satisfactory seal is present only when there is neither light nor translucency.

Moisture from food is a secondary source of frost. Almost all food has a high moisture (water) content. Ice, for example, is 100% water. As heat is removed from an ice tray filled with water, some of the water vaporizes. This process continues even after the water is frozen. Sublimination, or "ice cube shrinkage," is especially noticeable in automatic defrost models where cold air is circulated by an evaporator fan.

This same phenomenon occurs in food that is not adequately covered. As air is circulated over the food, especially uncovered beverages, moisture is removed by vaporization. Moisture from foods inadequately covered or sealed causes dehydration. Texture and taste deteriorate. The dehydration of foods, especially meats, is often called "freezer burn." It is caused by the frozen food not being tightly wrapped and sealed. Use only moisture/vapor-proof wrapping materials. These exclude virtually all air from the food package. For example, ice crystals sometimes form inside the plastic wrapper of bread stored in the freezer for even a few days. Such crystals are moisture that has been extracted from the bread (along with the heat) and trapped by the airtight wrapper.

Frost Accumulation

Warm moisture-laden air that enters the cabinet is circulated across the evaporator. When the moisture-laden air contacts the evaporator, heat is removed. Moisture condenses, forming frost. Under light usage conditions, frost accumulation will be heavier at the inlet and lighter at the outlet. Under moderate to heavy usage, frost accumulation will be more evenly distributed throughout the evaporator.

When the evaporator has collected a heavy coating of frost, its efficiency is reduced. Therefore, the evaporator surfaces must be defrosted periodically to maintain good heat transfer.

Frost Removal

After six or eight hours (depending on the model) of accumulated compressor run time, the defrost control opens the circuit contacts to the compressor and evaporator fan. It closes the contacts to the defrost heater circuit. The condenser fan, on most recent models, remains on during the defrost cycle. When the defrost heater circuit contacts are closed, the heater (or heaters) will normally remain energized for approximately 50% of the time alloted by the defrost control.

Frost removal from the evaporator is accomplished by all three means of heat transfer: radiation, conduction, and convection. The heater (or heaters), together with the reflector shields, begins to melt the frost by means of radiation from the heating element. As some of the frost nearest the heater is removed, heat is absorbed by the aluminum evaporator tubing. It is conducted away toward unfrosted areas of the evaporator. Heat always flows from hot to cold. Then, as the surrounding air is heated, convection aids in frost removal. Convection plays a more important role in side-by-side models than top-mounts, due to the evaporator configuration. The upper portion of the evaporator in side-by-side models is defrosted, mainly by convection.

Heater operation is terminated when the predetermined temperature is sensed by the defrost thermostat. With approximately 50% of the alloted defrost cycle time remaining, the melted frost is permitted to drain away from the evaporator. The defrost cycle ends when the defrost control opens the contacts to the heater circuit and closes the contacts to the compressor circuit.

Excessive frost accumulation and/or insufficient heater operation time can result in incomplete defrosting of the evaporator. This condition can usually be attributed to extremely heavy usage or a door left ajar, along with high humidity. In some instances, it is possible for more frost to accumulate during the compressor run cycle than can be removed during the defrost cycle. When this occurs, the partially melted frost refreezes. This results in residual ice in the evaporator.

As frost is melted, the defrost water is drained away from the evaporator through a tube that carries the water to the evapora-

CHAPTER 12—REFRIGERATORS

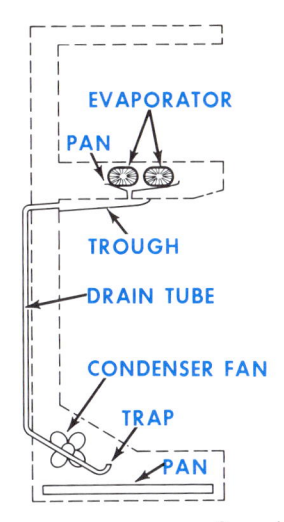

12-33. Note that the drain tube is hooked at the bottom to catch some water and prevent warm air from traveling up the tube and into the interior of the refrigerator.

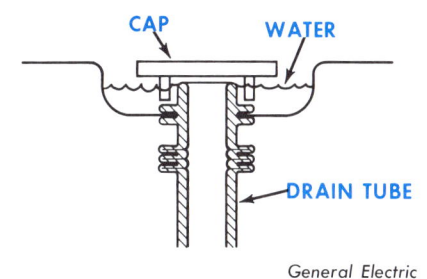

12-34. Another arrangement for preventing hot air from entering the interior of the cabinet by way of the drain tube.

Fig. 12-34. The defrost water contains traces of bacteria. These will grow inside the drain tube and eventually stop the flow of water. When this occurs, the defrost water spills over from the drain tube and becomes a nuisance. When a service call results, the customer should be made aware that uncovered foods in the fresh food and freezer compartments are the source of the food bacteria that clogged the drain.

A handy tool for clearing a clogged drain tube is a baster. Fig. 12-35. It can be purchased in the housewares department of most discount stores. To clear the drain, insert the tip of the baster into the drain tube opening. Force air through the drain by pumping the rubber bulb. If the drain is ice-clogged, hot tap water can be drawn into the baster. The water can then be squirted over the ice until the opening has cleared enough for the tip of the baster to be inserted. Then, hot tap water can be drawn into the baster and forced through the drain tube.

Parts of the Defrost System

The defrost system consists of only three functional components. These are the defrost thermostat, the defrost heater, and the defrost control.

If there is trouble with the defrost system, the first thing to check is the defrost thermostat. It is a simple device. A bimetallic disc, pushing against a transfer pin, holds the contacts open. As the evaporator temperature is

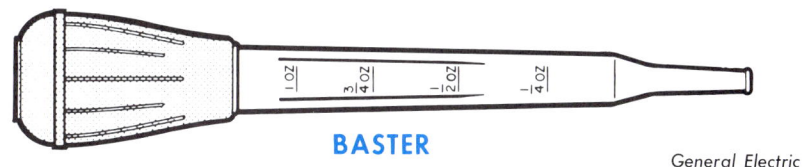

12-35. A baster can be used to help unclog a drain pipe.

12-36. The cutaway view of a defrost thermostat.

tion pan in the machine compartment. On most models, a trap is formed into the lower portion of the drain tube. It holds enough water to act as a seal in preventing warm moisture-laden air from entering the cabinet through the drain tube. Fig. 12-33. On other models, not having a trap in the drain tube, a cap is fitted over the inlet to the drain tube and into a recess where residual water seals the drain tube.

315

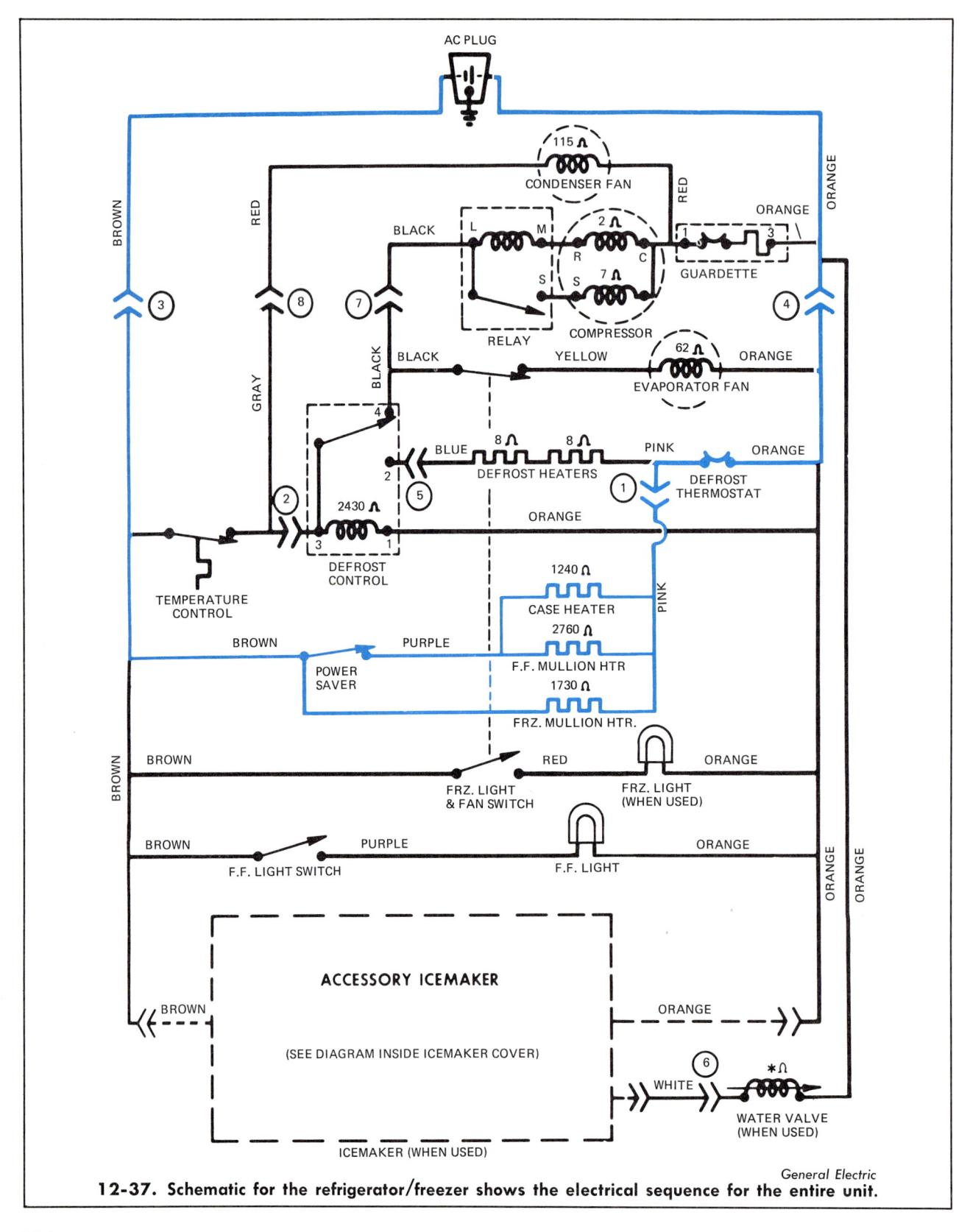

REFRIGERATION AND AIR CONDITIONING TECHNOLOGY

12-37. Schematic for the refrigerator/freezer shows the electrical sequence for the entire unit.

General Electric

lowered, the bimetal disc warps. The spring-loaded contact arm pushes the transfer pin out of the way, allowing contacts to snap closed. Fig. 12-36, page 315.

The thermostat design, incorporating the transfer pin, provides a fail-safe feature. In other words, when a defrost thermostat fails to function properly, it normally fails with the contacts open. It rarely fails with the contacts closed.

Defrost thermostats are precision built and tested to within ±6° F. [3.3°C] of the specified limit. A unique characteristic of a bimetal disc is that its calibration is fixed and does not change. This provides for reliability in excess of the life expectancy of the refrigerator or freezer. Continual life tests at the factory of 100,000 cycles, corresponding to 100 years, reveal that the calibration of a defrost thermostat does not drift out of tolerance (±6° F. [3.3°C]). A slight "creep" of about 2° F. [1.1°C] occurs at about 28,000 cycles during the life of all defrost thermostats as the components "wear in." A defrost thermostat can, however, be incorrectly calibrated from the beginning. However, defrost thermostats never get weak during the life of the refrigerator or freezer. Therefore, if the defrost system has functioned properly for several months before failing, disregard the possibility of an incorrectly calibrated defrost thermostat. Do not suspect that the defrost thermostat has the wrong calibration unless residual ice is found in the evaporator.

The only practical method for checking the defrost thermostat in the field is to test it for continuity. The contacts should be closed at all times—except during the later part of the defrost cycle and for the first ten minutes thereafter, when the compressor resumes operation. You can determine that the defrost thermostat contacts are closed on side-by-side models by feeling the mullion for heat. On side-by-side models, the mullion heater is in series with the defrost thermostat. Thus, if the mullion heater is warm, the thermostat contacts must be closed. Fig. 12-37.

12-38. Defrost heater encased in glass.
General Electric

Never replace a defrost thermostat unless it is known to be inoperative. Never substitute a defrost thermostat unless it is a recommended field correction or a factory authorized supersedure. Under no circumstances should a defrost thermostat be bypassed, other than momentarily for testing purposes.

Check the defrost heater or heaters. Radiant (glass sheath)-type defrost heaters are used on virtually all GE or Hotpoint models of recent years. If the heaters are connected in series and one becomes open, the whole heater arrangement will be inoperative. Figure 12-38 shows the defrost heat design. Checking the heater circuit with an ohmmeter does not conclusively establish that the heaters are operative or inoperative. For example, the glass sheath may be broken, the glass darkened, the element coils bunched together, or the jumper lead between the heaters shorted to the ground. In this case, the ohmmeter may indicate the correct resistance value for the circuit. An infinity reading on the ohmmeter when it is applied to the heater circuit may be due to a detached wire at one heater terminal or loose terminals in the R-E-D connector. When in doubt, these items should be visually inspected. Fig. 12-39.

Where replacement defrost heaters are furnished in sets of two or three, always use the complete set—even though only one heater is actually inoperative. Never substitute defrost heaters unless it is a factory-authorized supersedure. When replacing defrost heaters, avoid handling

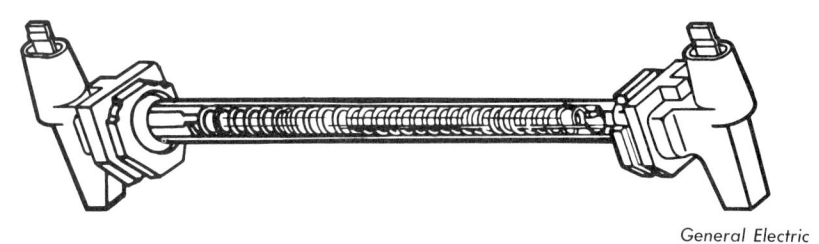

12-39. Location of wires inside the walls of a refrigerator.
General Electric

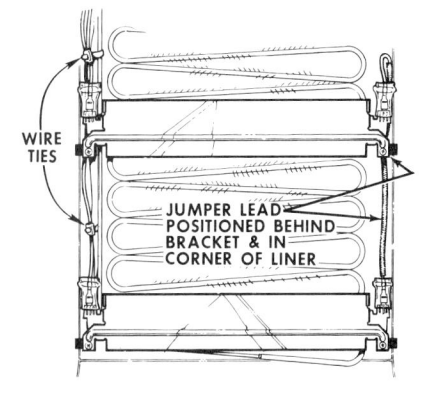

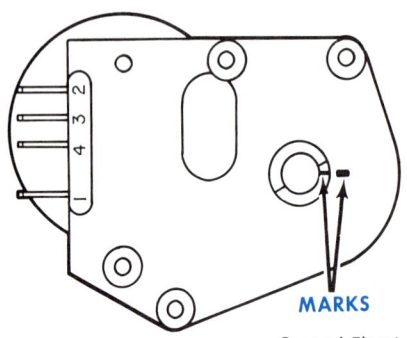

REFRIGERATION AND AIR CONDITIONING TECHNOLOGY

12-40. Defrost control in a refrigerator. 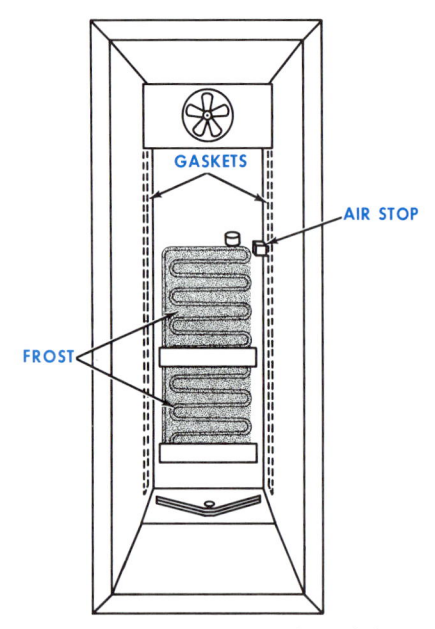
General Electric

12-41. Location of the evaporator coils inside a refrigerator.
General Electric

chanical characteristics can be checked only by an operational test. To do this, manually advance the defrost control to the start of the defrost cycle (where the marks align). Then, with the temperature control contacts closed, wait approximately five minutes for the control to advance itself automatically.

When replacing a defrost control, check that the wiring is connected correctly and that the device will function properly.

Remove all frost from the evaporator. Use a heat gun, if necessary. Partially melted frost remaining on the evaporator will refreeze. It will form a more solid mass of ice that may never be cleared by subsequent automatic defrosting. If in doubt, visually inspect the evaporator. Fig. 12-41.

Check the location and position of the coil cover gaskets and air stops. These items are necessary to direct the air flow across

12-42. Newer devices are available to check the refrigerator operation without moving or opening the door.
General Electric

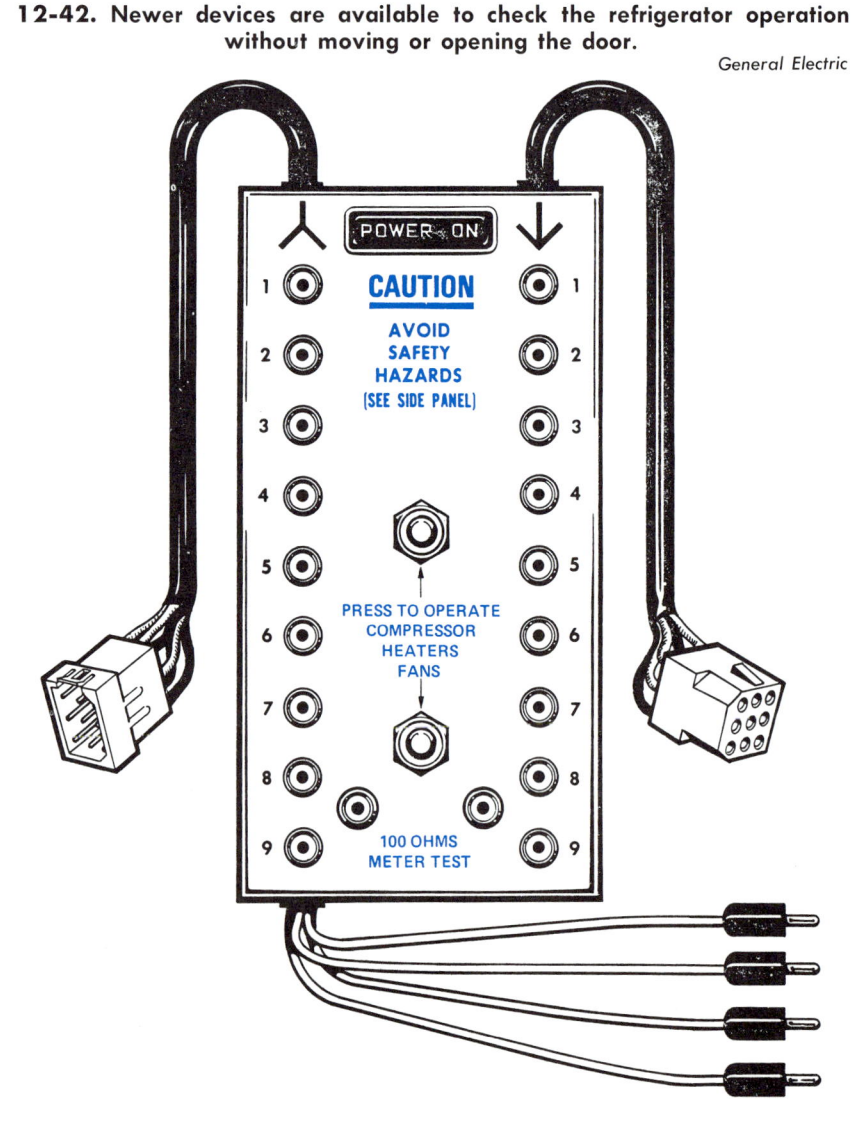

the glass sheath. Handling leaves salt deposits, resulting in glass embrittlement. Rinse any fingerprints from the glass with water and dry with a clean paper towel.

Check the defrost control next. Fig. 12-40. The defrost control is an electromechanical device that is subject to electrical and mechanical failures. An ohmmeter can be used to check the electrical characteristics. The me-

318

CHAPTER 12—REFRIGERATORS

the evaporator. If coil cover gaskets and air stops are missing or mispositioned, moisture-laden air will bypass the evaporator. Frost accumulation will appear in unusual locations. This will stall the evaporator fan, frost the control console, and cause defrosting problems.

Check for excessive heat leakage. Check that a door is not ajar due to a binding condition. Check that food packages have not been improperly placed so as to interfere with the door closing fully. Look for a poor door gasket seal caused by a racked cabinet, improper latch adjustment, or torn door gaskets. If any of these conditions exist, more frost may be accumulated than can be removed during the defrost cycle.

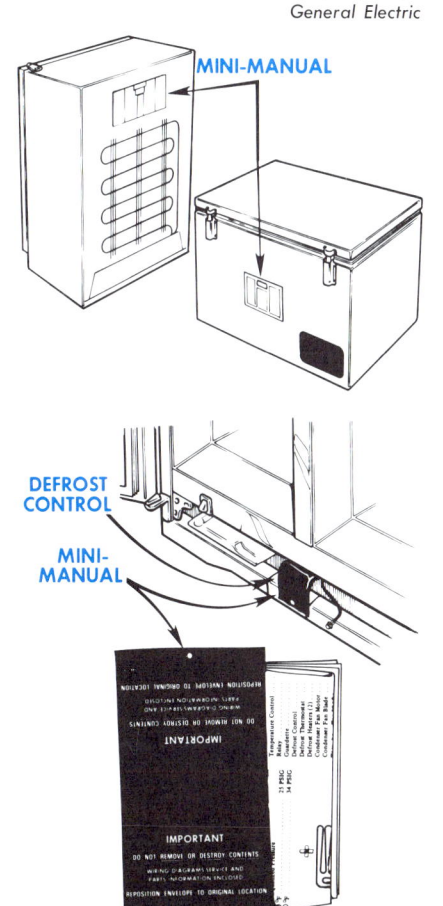

Consider the following when checking for the frost problem solution:

Room Ambient Temperature

Room temperature and, especially, humidity contribute to a defrosting problem. High ambient temperatures will indirectly impose a greater load on the refrigerator or freezer. This results in longer run time and greater frost load. High humidity, even with a low room temperature, can likewise result in an abnormal frost load.

Usage Conditions

Obviously, the heavier the usage, the longer the run time, and the greater the frost load. Under such conditions, more frost may be accumulated than can be removed during the defrost cycle. This results in residual ice.

Cabinet Sealing

Missing drain caps and leaks from torn door gaskets or at the tube seal can contribute to a defrost problem. If leaks in the outer case bottom seams are suspected, RTV® sealer can be applied to the perimeter of the bottom, from underneath the cabinet.

Length of Heater Operation Time

Determine the length of time the defrost heater circuit is allowed to remain energized. As a general rule, with a heavily loaded evaporator, the heater should be energized for at least one-half of the defrost cycle time alloted.

Refrigerant Charge

A short refrigerant charge will allow greater-than-normal frost accumulation at the inlet of the evaporator and no frost at the outlet. This imbalance can result in incomplete defrosting of the evaporator. A lack of frost (cold mass) at the defrost thermostat will permit the thermostat to open sooner than normal.

RAPID ELECTRICAL DIAGNOSIS

GE and Hotpoint have a quick and accurate device for diagnosing electrical faults in a refrigerator. The hand-held device is a result of down-sizing a computer used in the factory to check production models of refrigerators. The device is called "Big Red" (Rapid Electrical Diagnosis unit). It allows the technician to check almost every electrical component in the refrigerator within six minutes. It is not necessary to unplug the refrigerator, unload any food, move the refrigerator from the wall, or, usually, open the door.

Multiconnectors are designed into the refrigerator wiring harness. They are located behind the front grille. Fig. 12-42. A mini-manual contains information that makes the device useful for a particular refrigerator model. All the latest models with the RED feature have a packet or envelope containing the necessary information. Fig. 12-43. The packet contains the pictoral and schematic wiring diagrams, RED

12-43. All late model GE and Hotpoint refrigerators and freezers have a mini-manual that gives the service technician the information needed to service the unit properly.
General Electric

319

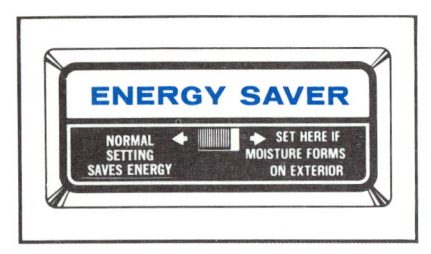

REFRIGERATION AND AIR CONDITIONING TECHNOLOGY

12-44. The energy saver cuts down on electric bills by eliminating one of the defrost heaters. *General Electric*

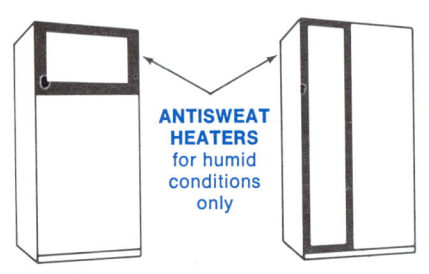

12-45. Antisweat heaters are located around the freezer section doors. *General Electric*

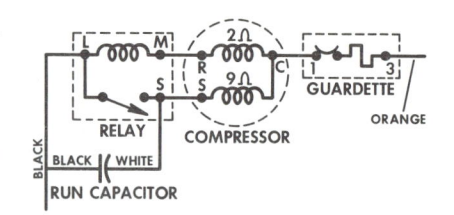

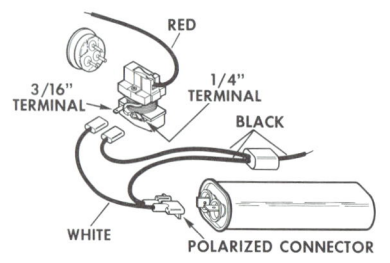

12-46. The electrical diagram of the capacitor run type of refrigerator compressor. The capacitor has a polarized plug so it cannot be connected incorrectly. *General Electric*

component circuits, machine wiring diagram, and components energized through the RED system.

Use of the RED may reduce the technician's service time by 20%. It reduces the additional service callbacks for the same problem by 50%. All No-Frost GE models 16 cubic feet and larger have this feature. Hotpoint uses the same system.

ENERGY-SAVER SWITCH

An energy-saver switch is the latest development in refrigerator design. Fig. 12-44. The energy-saver switch can reduce operating costs. It can take some of the heaters out of the circuit. Controls for the case and mullion antisweat heaters are incorporated into the energy-saver switch. The heaters prevent moisture formation on the outside of the case in humid conditions. The normal setting may be used about 80% of the time due to low humidity in the winter and air conditioning in the summer. The switch can be set to an alternate setting only if moisture forms on the outside of the refrigerator. Fig. 12-45. The power saver switch is shown in the schematic diagram in Fig. 12-37. Notice how it is in series with the 1240-ohm and the 2760-ohm heaters. Only the freezer mullion heater is in the circuit and not affected by the power saver switch.

CONDENSER FAN MOTOR

A low power/high efficiency condenser fan motor is standard equipment on the low-energy 16- and 18-cubic feet models and all side-by-side models since 1977 in the GE and Hotpoint line. The new motor can be used as a standard replacement part for virtually all models for the past several years. The motor has a cast iron housing. It looks identical to the previous motor but will run at a slightly higher speed (1500 r/min) compared to the previous motor's speed of 1350 r/min. Power consumption is 21 W, compared to 26 W for the previous model.

RUN CAPACITOR

The new models have the condenser fan modified to produce the same or better performance at lower energy consumption. They also have a 30-minute defrost control and high-density fiberglass cabinet and door insulation. A 920 Btu compressor that uses a run capacitor and a special relay have also been incorporated into the energy-saver designs. The run capacitor, located in the rear of the machine compartment above the defrost water evaporation pan, is mounted to the outer case back with a clamp.

The run capacitor, rated at 15 microfarads, 220 volts, is oil filled. It has a built-in pressure disconnect device that opens the capacitor connections if the internal pressure exceeds a safe limit. The pressure disconnect device is a nonresetting safety feature that prevents the capacitor case from rupturing due to a short or overheating. The capacitor has polarized connections that mate with a polarized wiring harness connector. This ensures that the inner foil plate of the capacitor is connected to the "hot" (black wire) side of the power cord. Fig. 12-46.

This is not an ordinary run capacitor. Thus, a label is affixed to the back of the cabinet just above the location of the capacitor that contains the following message:

> **Attention Service Technician**
> Special run capacitor required for compressor motor circuit of this refrigerator. Use only genuine replacement part for substitute-testing or replacement.
> Refrigerator Part Nos.
> WR62X29—capacitor
> WR7X137—relay

The run capacitor is connected across the relay contacts and in series with the start winding. See Fig. 12-46. When the relay "picks up" during the starting, the capacitor is shorted by the closed relay contacts. As the compressor motor reaches approximately 75% of full speed, the relay "drops out." The contacts open and the capacitor is then in series with the start winding. Thus, both windings are used while the compressor is running. The capacitor, therefore, is used to improve the running efficiency of the compressor. The compressor will start and run without the capacitor in the circuit. However, it will not run as efficiently.

The relay has two spade terminals on the left side, near the bottom. The terminal nearest the compressor case is $3/16''$ wide. It is for the white wire from the capacitor. The other terminal is $1/4''$ wide. It is for the black wire from the defrost control. This insures against these two leads being accidentally reversed. This could quickly burn out the start winding.

Testing the Run Capacitor

Measure the line voltage at the wall receptacle with the compressor running. Then measure the line voltage across the capacitor terminals with the leads connected.

With a good capacitor, the voltage measured across the capacitor will be at least 10% greater than the measured line voltage, if the line voltage is at least 110 V.

With a shorted capacitor, no voltage or voltage less than line voltage will be measured across the capacitor.

With an open capacitor, the voltage measured across the capacitor will be greater than the line voltage (but less than 10% greater), if line voltage is at least 110 V.

The next method available for testing is the ohmmeter method. Disconnect the power cord. Remove the leads from the capacitor and discharge the capacitor by shorting the terminals with an insulated screwdriver. Turn to the R $\times$ 10K (or higher) scale on the ohmmeter. Place the probes on the capacitor terminals and observe the ohmmeter reaction. Reverse the probes (reverse polarity) and again observe the ohmmeter reaction.

With a good capacitor, the ohmmeter should register zero ohms and deflect slowly back toward infinity.

With a shorted capacitor, the ohmmeter should register zero ohms and *not* deflect back toward infinity.

With an open capacitor, the ohmmeter should not register any resistance, but should remain at infinity.

NOTE: If the capacitor passes the ohmmeter test, check from each terminal to the case. This will make sure that the capacitor is not shorted to ground. If the capacitor is shorted to ground, it must be replaced.

Another test for the capacitor is to substitute another capacitor. Substitute a known good replacement capacitor of the same size and rating for the original. If as much as 0.6 A *less* current draw is observed with the replacement capacitor, the original capacitor should be replaced.

The replacement high-side will include the guardette, relay, and run capacitor.

TYPES OF COMPRESSORS

Four sizes of compressors are used on GE and Hotpoint refrigerators and freezers.

- $1/8$ hp 420 Btu—dial defrost units.
- $1/4$ hp 740 Btu—automatic defrost units.
- $1/3$ hp 920 Btu—top mount no frost.
- $1/3$ hp 1100 Btu—side-by-side.

The primary differences are the size of the motor shaft and the length of the stroke. The motors are two-pole types run at 3600 r/min. They are hermetically sealed.

The scotch yoke compressor is an exclusive GE type. Fig. 12-47. Note the unique oil cooler loop that circulates oil outside the compressor for cooler operation. Fig. 12-48. The motor is spring mounted top and bottom inside the sealed case. This absorbs vibration. The motor is matched to the refrigerator size. The oil pump has positive displacement. It circulates the oil for a constantly cooling action. The rotating parts are self-aligning. The compressor unit is sealed in steel with a lubricant that never needs changing.

REFRIGERATION AND AIR CONDITIONING TECHNOLOGY

12-47. The scotch yoke keeps the compressor oil cooler.

OIL COOLER LOOP

General Electric
12-48. Operation of the scotch yoke on a compressor.

USE AND CARE OF A REFRIGERATOR/FREEZER

Installation

Allow ⅝" clearance at both sides and at the top for ease of installation. When a new home is planned, place a water supply near the refrigerator location. It will simplify connection of an optional automatic icemaker.

The refrigerator should not be installed where the temperature will go below 60° F. [15.6°C]. The refrigerator will not run frequently enough under 60° F. [15.6°C] ambient temperature to maintain proper temperatures inside.

The floor should be strong enough to support the refrigerator/freezer *plus its load.*

Adjusting screws are located behind the grille. They can be adjusted to level the unit. Fig. 12-49. Set these screws so the refrigerator is firmly positioned on the floor and the front is raised just enough that doors close easily when opened about halfway. Specially designed door hinges lift the door slightly when opened. The force of gravity then pulls the door closed automatically.

Figure 12-50 shows how to remove the grille. Turn the adjusting screws clockwise to raise the refrigerator, counterclockwise to lower it. To replace the grille, set the tabs on hooks at both ends and push toward the refrigerator until the grille snaps into place. Fig. 12-51.

Electrical Connection

For safety, this appliance must be properly grounded. The power cord of this refrigerator/freezer is usually equipped with a three-pronged plug that mates with a standard three-prong (grounding) wall receptacle. Fig. 12-52. The plug should be checked to make sure it is in a properly grounded receptacle. Where a two-prong wall receptacle is found (some older refrigerators are also equipped with a two-pronged plug), it is the re-

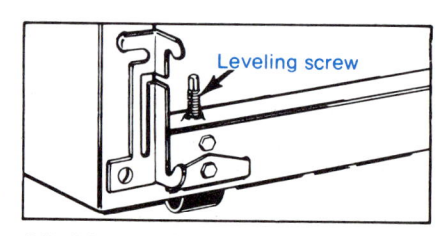

12-49. Leveling screw on a refrigerator.

12-50. Remove the grille to gain access to the leveling screw and remove obstructions to the free flow of air around the condenser.

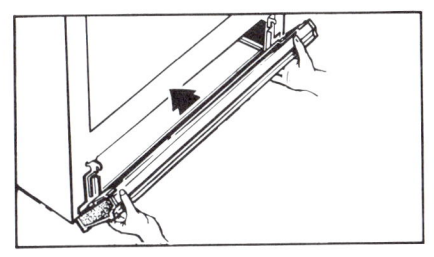

CHAPTER 12—REFRIGERATORS

12-51. Replacing the grille on a refrigerator.

12-52. Preferred method of connecting an AC plug to the wall receptacle. *General Electric*

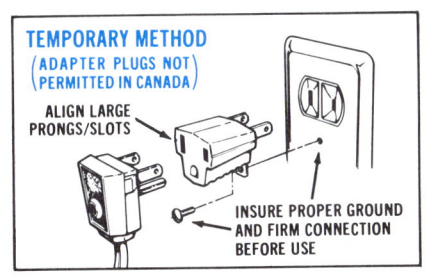

12-53. Temporary method of connecting the refrigerator's three-prong plug to a two-prong wall socket. *General Electric*

12-54. Power saver switch. *General Electric*

sponsibility and obligation of the owner to have it replaced with a properly grounded three-prong wall receptacle. The third prong should NEVER be cut or removed from a plug.

Use of an adapter plug such as shown in Fig. 12-53 is not recommended. In fact, they are not allowed in Canada. However, if the customer chooses to use an adapter, where local codes permit, a temporary connection may be made to a properly grounded two-prong wall receptacle by use of a UL-approved adapter. The larger slot in the wall outlet is the ground connection (white wire in the electrical system). The black wire in the home electrical system is "hot." It should be connected to the smaller of the two slots in the wall receptacle. One thing to remember is that the wall receptacle cover plate screw should be attached to a metal wall box. An uninsulated box will not provide a ground.

When the plug will be frequently removed from the wall socket, make sure an adapter is NOT used. Frequent disconnection of the power cord places undue strain on the adapter and leads to eventual failure of the adapter ground terminal. The refrigerator should always be plugged into its own individual electrical outlet. In most instances, refrigerators require 115-V, 60-Hz, single-phase alternating current (AC).

Temperature Controls

Manufacturers of refrigerators and freezers design their controls so they will be easily read by the average person. Before checking the various components and circuits of the refrigerator, check that the temperature controls are properly set. They may be the cause of the problem.

Power-Saver Switch

Some refrigerators are equipped with a power-saver switch located on the left side of the refrigerator near the top of the fresh food compartment. Fig. 12-54. This switch allows the owner to turn heaters on to eliminate moisture on the exterior of the cabinet. In some environments, moisture will not form. Thus, the extra heaters are not needed. Power is saved.

Other Maintenance

To replace the burned-out light bulb, unplug the refrigerator. In some models, you may then need to remove the top shelf temporarily. Replace the bulb with one that is made for appliance use. Make sure the bulb is the same size as the one it is replacing.

For the most efficient operation, remove the grille and either sweep away or vacuum the dust in and around the condenser coils, compressor, and fan. This area should be cleaned once a year.

The refrigerator and freezer sections should be cleaned at least once a year. It is recommended that the refrigerator be unplugged before cleaning. Clean door gaskets and the inside of the refrigerator with a sponge that has been wetted with a solution of baking soda and water. About one tablespoon of baking soda in one quart of water is sufficient.

TOUCH-UP AND REFINISHING PROCEDURE

Vinyl gaskets are used on all models. Lacquer repairs can be made on all areas of the cabinet, except the pilaster that is in contact with the vinyl gasket. Since

323

prolonged contact of the vinyl gasket with lacquer will soften the lacquer, a special touch-up enamel must be used on the pilaster.

Lacquer Refinishing

The following steps are necessary to touch up or refinish any part of a white acrylic cabinet, *except the pilaster area.* When refinishing a cabinet other than white, follow the same outline, but substitute the color of lacquer required.

1. Sand out the spot to be repaired with 360 or 400 wet-or-dry sandpaper. Finish sanding to a feather edge with 600 wet-or-dry sandpaper. Water or naphtha may be used to facilitate the sanding.

Wipe area dry. Then, hand rub with a fine rubbing compound (du Pont VZ-1090 or equivalent) an area extending at least 6″ beyond the edges of the proposed lacquer repair spot. Wipe compound off and wash area with naphtha. Dry with a clean cloth.

2. Clean the bare metal only with Sol-Kleen® cleaner and rust remover. The Sol-Kleen® should not be allowed to contact the painted surface. This cleaner is reduced with two parts of water and applied by wiping with a clean rag wet with the reduced cleaner. Deep-seated rust may be more readily removed by the use of steel wool wet with the cleaner. It is important to note that the Sol-Kleen® cleaner should be used to prepare the surface of the metal whether there is rust present or not.

3. Before cleaner is allowed to dry, wipe the surface dry with a clean rag.

4. Wipe over surface thoroughly a second time with a solution of 50-50 alcohol and water. Wipe again with clean dry rags, preferably new cheesecloth.

5. Allow to dry for at least 10 to 15 minutes.

6. Apply primer surfacer, reduced by approximately equal parts of lacquer thinner, to build bare metal area up to surrounding surface. Any imperfections that the primer surfacer has not filled should be knifed out with lacquer-type putty glaze.

7. If no putty glaze is used, allow to dry about 30 minutes. If necessary, sand out lightly with 360 wet-or-dry sandpaper. If putty glaze is used, it will be necessary to allow about 3 hours before sanding with abrasive and water. Remove sanding residue thoroughly by wiping with a clean rag wet with naphtha. Wiping with a tack rag will insure a surface free from dust and lint.

8. Finish the repair with two or more coats of lacquer, reduced approximately one part of lacquer to one and one-half parts of thinner. Finally, apply a mist coat of lacquer thinner to flow out the surface.

9. The patched area should be allowed to dry three or more hours before being rubbed with compound or polish. The above procedure also applies to the complete refinishing of the cabinet, except the pilaster area. All damaged areas should be repaired as outlined above in steps (1) through (6). The overall surface of the cabinet should be sanded thoroughly and cleaned as outlined in step (7). The cabinet should be given two or more coats of touch-up lacquer and polished as outlined in steps (8) and (9).

Touch-Up of Pilaster Area

Lacquer repairs cannot be made on cabinet pilaster areas, because the vinyl door gasket will soften the lacquer on prolonged contact.

Use the special DUX® touch-up enamels. These are available through your local authorized distributor.

Proceed as follows:

1. The damaged area should be wiped with naphtha to remove waxes and other soils. Sand the area to a feather edge with 400 sandpaper. Finish sanding with 600 sandpaper. Water or naphtha may be used to facilitate the sanding.

CAUTION: Sanding through to bare metal should be avoided, where possible, as priming is not permitted on pilaster repairs. Wipe sanded area with tack rag to remove dust. Then, hand rub with a fine rubbing compound (du Pont VZ-1090 or equivalent) an area around the proposed repair spot. Wipe off compound. Wash area with naphtha and dry with a clean cloth.

2. Reduce the touch-up enamel with reducer in the ratio of approximately one to one.

3. Use a narrow spray pattern

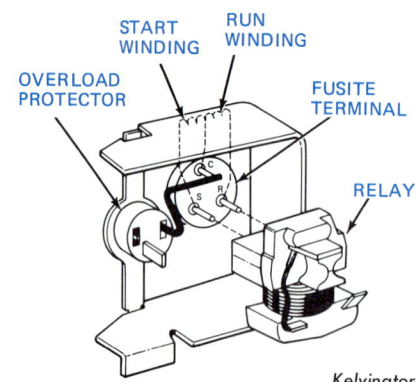

12-55. Fusite thermal relay and overload protector for A-line compressor.

and 10–15 pounds air pressure. Spray on several dusting coats, to build up sanded area. Allow one minute for drying between coats, to prevent sagging. Spray on a wet coat extending beyond the damaged area. Then, spray a mist coat of reducer to flow out the enameled area.

4. Bake out the repair for *1 minute* with a 375-W, infrared heat lamp. Surface temperature should not exceed 120° F. [48.9°C]. If the fill is insufficient, the area should be sanded again with 600 sandpaper, and the process repeated.

5. If necessary, the repair may be rubbed with a fine compound or polish.

Epoxy Repair of the "ABS" Inner Wall

The "ABS" inner wall can be fractured through abuse, misuse, or carelessness by the user. Also, the cold ban trim may separate along the back edge from the inner walls of the PC (provisions compartment) or FC (freezer compartment). Where a fracture or separation occurs, repairs can be made with an epoxy repair package.

TROUBLESHOOTING THE ELECTRICAL COMPONENTS OF A REFRIGERATOR

Electrical components may vary on the different models of refrigerators. All refrigerators, however, have essentially the same components. Similar components have basically the same method of operation. Thus, for example, the compressors in different brands of refrigerators operate in much the same way.

This discussion of troubleshooting procedures is focused on one particular model of refrigerator. Nonetheless, due to the similarity of refrigerator components and their methods of operation, this information can be very helpful. It will provide you with an information base for evaluating problems in other refrigerator models.

Relay and Overload Protector

The starting relay on the A-line compressor is of the push-on type. It mounts on the start (S) and run (R) terminals of the compressor as shown in Fig. 12-55.

The starting relay is a magnetic switch with starting contacts. Its magnetic coil is in series with the run winding of the motor. The relay coil carries the main-winding current. The relay armature holds the start winding contacts open, except during the start period.

At the moment of starting, when the thermostat closes the electrical circuit, a surge of electric current passes through the main motor winding and the relay coil. This energizes the relay coil and pulls up the relay armature, allowing the start winding contacts to close.

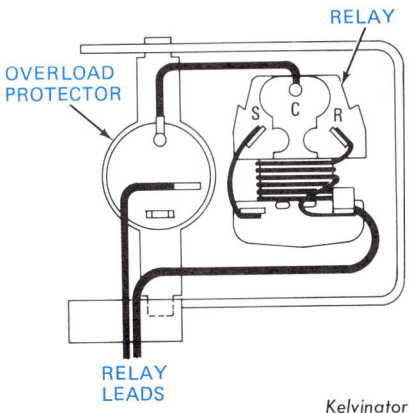

12-56. Starting relay and overload protector. *Kelvinator*

The current through the start winding introduces a second, out-of-phase, magnetic field in the stator and starts to motor. As the motor speed increases, the main winding current is reduced. At a predetermined condition, the main winding current, which is also the current through the relay coil, drops to a value below that needed to hold up the relay armature. The armature drops, opens the start winding contacts, and takes the start winding out of the circuit.

In series with the motor windings is a separate bimetallic overload protector. It is held in place on the compressor by a spring clip. The short wire lead on the overload protector connects to the common (C) terminal on the compressor. Fig. 12-56.

Should the current in the motor windings increase to a dangerous point, the heat developed by passage of current through the bimetallic disc will cause it to deflect and open the contacts. This breaks the circuit to the motor windings. The motor stops before any damage can occur.

The dome-mounted overload protector provides added protection for the compressor motor. In addition to protecting against excessive current, it also protects against excessive temperature rise.

After an overload or a temperature rise has caused the overload protector to break the circuit, the bimetallic disc cools and returns the contacts to the closed position. The time required for the overload switch to reset varies with room temperature and compressor dome temperature.

The overload protector is specially designed with the proper electrical characteristics for the compressor motor and its appli-

REFRIGERATION AND AIR CONDITIONING TECHNOLOGY

cation. Any replacement must be made with an exact replacement with the same part number. Never substitute an overload protector with another unauthorized part number. Use of the wrong protector can result in a burned-out motor. If the relay is inoperative, change both relay and protector. If the protector is inoperative, change only the protector.

When the thermostat cuts off after a normal cycle, or when the service cord is pulled from the electrical outlet during a running cycle, about 8 minutes is required for *unloading*. (A longer period is required if it occurs during a pull-down.) Unloading is the reduction of the pressure differential between the highside and the lowside of the system. During this unloading period, the overload will trip if the service cord is plugged into the electrical outlet.

To check for an *open* overload protector, short across its terminals. See Fig. 12-56. If the compressor starts, replace the overload. If the compressor does not start, look for other causes. (A few such causes might be voltage of less than 100 V at the compressor terminals during the starting interval, an inoperative relay, or an inoperative compressor.)

The relay is current operated and designed for a specific compressor and motor current value. Thus, the correct size relay, represented by the part number, is a necessity. Never substitute a starting relay with another unauthorized relay.

The relay cannot be adjusted or properly repaired in the field.

The compressor may repeatedly start and run for a few seconds, and then cycle on. In such a case, the start relay contacts may be stuck closed. The excessive current then trips the overload protector.

Figure 12-57 shows the relay-overload protector and the terminal block. Note the colors of the wires. Figure 12-58 shows how the overload protector makes contact.

Compressor Motor Electrical Check

When checking for electrical trouble, always be sure there is a *live* electrical circuit to the cabinet and that the temperature selector dial is not in the off position.

When the sealed unit will not start, and the cabinet temperature is warm the trouble may be in the relay, overload, thermostat, wiring, or in the compressor itself.

If the compressor will not run, make a test lamp check across the power terminals on the relay and overload protector. Check the wiring diagram for connections.

The test lamp should light to show a live circuit if the thermostat knob is in the normal operation position and not in the off position. If this check does not show a live circuit, the thermostat and wiring should be checked for an open circuit. Pay

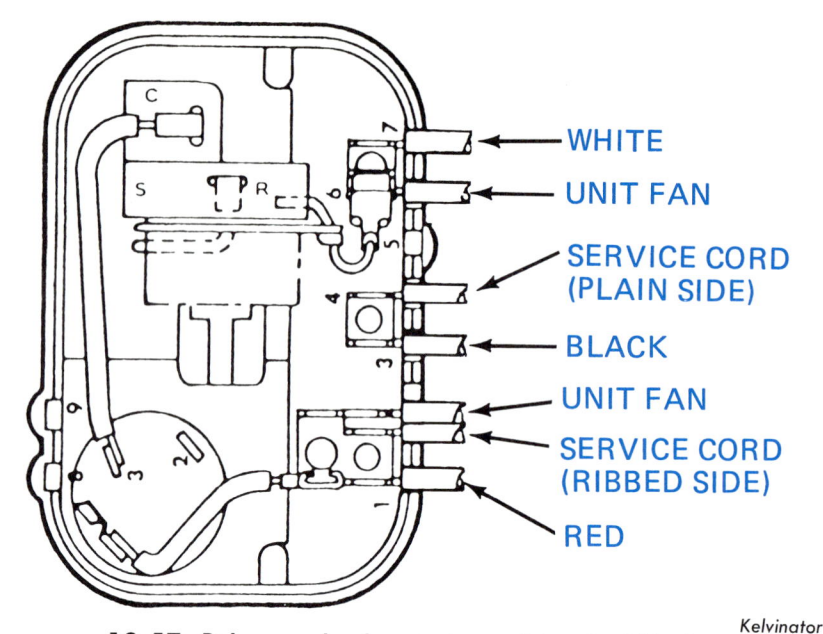

12-57. Relay overload protector and terminal block. *Kelvinator*

- WHITE
- UNIT FAN
- SERVICE CORD (PLAIN SIDE)
- BLACK
- UNIT FAN
- SERVICE CORD (RIBBED SIDE)
- RED

12-58. Overload protector, sectional view. *Kelvinator*

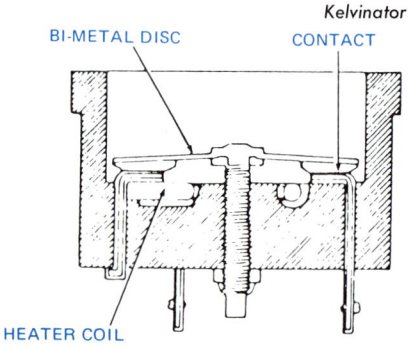

- BI-METAL DISC
- CONTACT
- HEATER COIL

326

CHAPTER 12—REFRIGERATORS

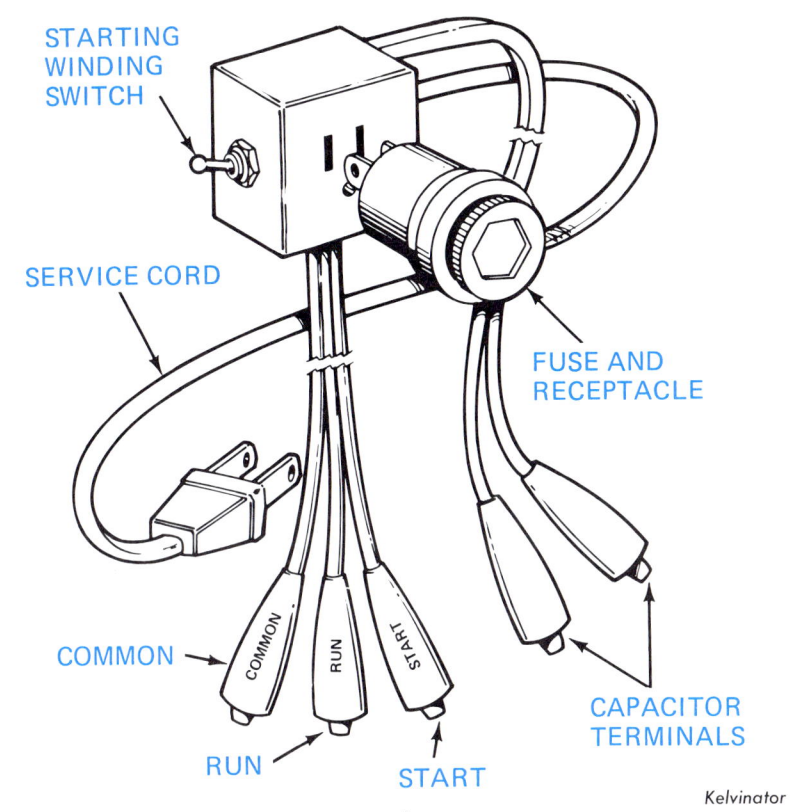

12-59. Manual test set.
Kelvinator

particular attention to all terminal connections.

A thermostat check can be made by using a piece of wire as a temporary bridge across the two thermostat terminals. If the compressor does not start and runs with the bridge, the thermostat is at fault and should be changed.

If the test lamp check shows power supply at the relay terminals, check the compressor with a manual test set. See Fig. 12-59.

The compressor motor may not start and run with either the test set or the regular electrical accessories. In this case, check the line voltage to see that there is not more than 10% variation from the normal 115 V. If the voltage is correct and the compressor will not start and run, change the compressor.

The manual test set can be obtained from a distributor parts department. It incorporates a hand-operated switch for the energizing of the motor start winding.

OPERATING THE MANUAL START TESTER

Connect the test leads to their corresponding motor terminals. Thus, C goes to common, S to start, and R to run. Plug the tester service cord into a live wall receptacle of proper voltage supply.

Hold the manual starting switch lever in the starting position with the thumb and insert the fused receptacle. Release the starting switch within three or four seconds.

If the compressor is all right it should start and run in this length of time.

CAUTION: Do not keep the starting switch closed for more than a few seconds. Otherwise, you may burn out the motor start winding.

If the compressor starts and runs with the test set, replace the relay.

If the compressor does not start and run, disconnect the test set immediately. This will prevent the burning out of the main motor winding, since it is always in the circuit when the test set is connected.

IDENTIFICATION OF REFRIGERATOR PARTS IN TROUBLESHOOTING

Thermostats

The thermostat is mounted in the *control center* on the right-hand side of the refrigerator in the manual defrost models. Fig. 12-60. Moving the knob from one setting to another changes both the cut-in and the cut-out

12-60. Cutler-Hammer thermostat.
Kelvinator

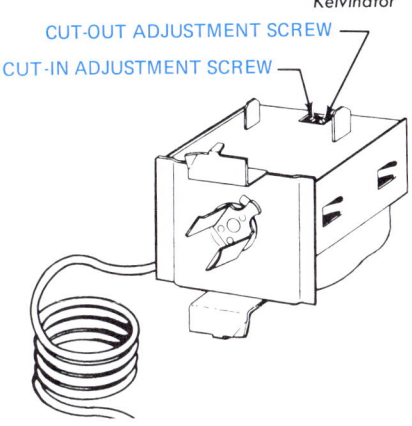

327

REFRIGERATION AND AIR CONDITIONING TECHNOLOGY

Table 12-A.
Altitude Adjustment

Altitude above Sea Level (in Feet)	Variable Cut-In Turns Counterclockwise	
	Cut-In Screw	Cut-Out Screw
2,000	1/8	1/8
4,000	5/16	5/16
6,000	1/2	1/2
8,000	3/4	3/4
10,000	7/8	7/8

Table 12-B.
Altitude Adjustment

Altitude above Sea Level (in Feet)	Range Screw Adjustment (Number of Turns Clockwise)
2,000	1/8
4,000	1/4
6,000	3/8
8,000	1/2
10,000	5/8

temperatures. The temperature changes in degrees are determined by the knob setting. Compressor operation is controlled by the thermostat. This one has nine points of temperature selection between the coldest and the off position. The thermostat thermal element is secured to the evaporator with a clamp and screw.

Altitude adjustments are made in accordance with Table 12-A. To determine the thermostat cut-in and cut-out temperatures, use a refrigeration tester. Position the test bulb firmly on the evaporator surface opposite the thermal element clamp. The refrigerator should be operated through a normal cycle and the cut-in and cut-out temperatures recorded.

The thermostat on the top-freezer, cycle-defrost model looks like the one on the manual defrost models. The thermostat is a variable cut-in type, however. When the thermostat knob is changed from one setting to another, both cut-in and cut-out temperatures change. The amount of temperature change is determined by the knob setting. Turn the screws to the left to raise the cut-out and cut-in temperatures. Turn the screws to the right to lower cut-out and cut-in temperatures. Both the cut-in and cut-out screws must be adjusted counterclockwise to compensate for altitudes above 1000'. See the adjustment in Table 12-A.

The GE thermostat is shown in Fig. 12-61. Note the changes for altitude that are shown in Table 12-B. This type of thermostat is used on some top-freezer models with no-frost capability. The thermostat shown in Fig. 12-60 is made by Cutler-Hammer and will be located on other models of this type of refrigerator. The thermal element of the thermostat is inserted into a thermal well attached to the provisions cabinet evaporator plate. Fig. 12-62. The thermal well protects the sending unit or sensing part of the thermostat.

THERMOSTAT REPLACEMENT

To replace the thermostat, disconnect the service cord from the electrical outlet. Remove the thermostat knob and one Phillips-head screw on the bottom side of the control center. Slide the control center housing off. Remove two thermostat mounting screws and conductors. Unwrap the thermal element from the bracket and remove. Figure 12-63 is an exploded view of a Kelvinator thermostat and its mounting bracket.

Figure 12-64 shows the location of the thermostat on top-freezer models with no-frost control and fiberglass insulation. The previous models had foam insulation.

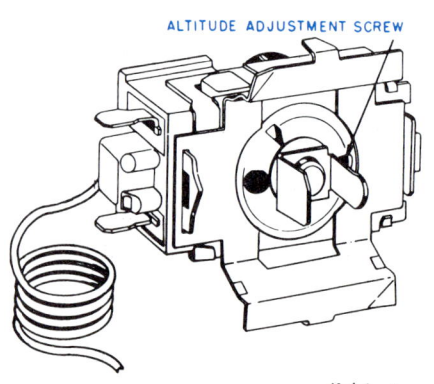

Kelvinator
12-61. General Electric thermostat.

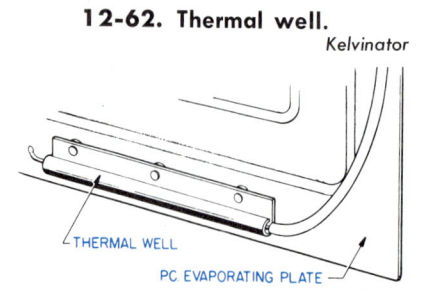

12-62. Thermal well.
Kelvinator

12-63. Thermostat and its mounting bracket.
Kelvinator

CHAPTER 12—REFRIGERATORS

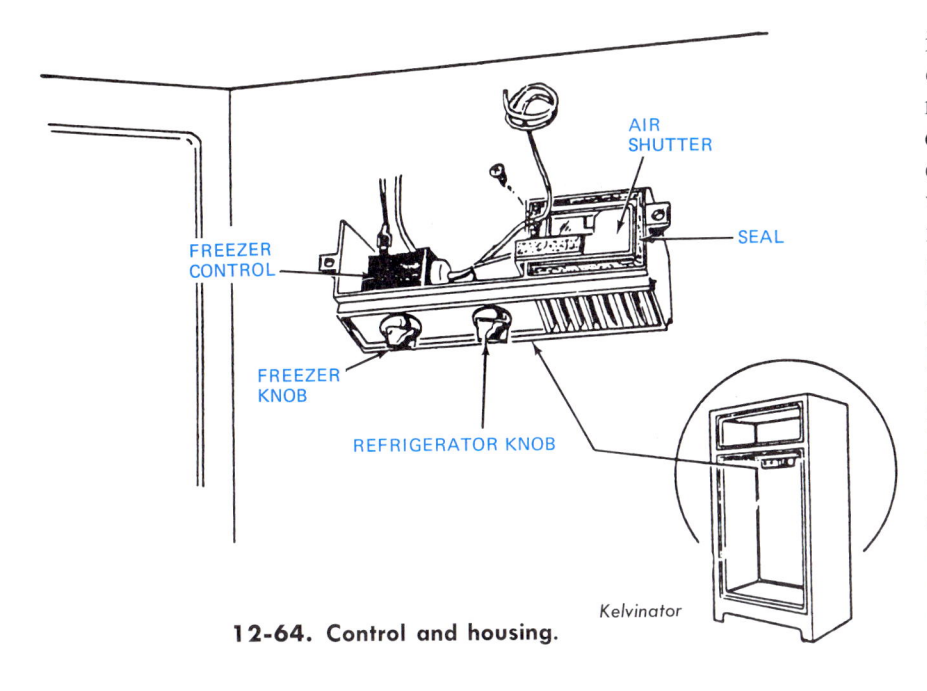

12-64. Control and housing. *Kelvinator*

With this model, note that a temperature adjustment to one compartment may require adjustment to the other. For example, if the provision compartment knob is turned to a colder setting, the freezer compartment knob may have to be adjusted to a warmer setting.

The thermostat on the top-freezer, no-frost model is located in the provision compartment. It is a variable cut-in, cut-out type. Changing the setting on the thermostat knob changes both cut-in and cut-out temperatures. The temperature setting is determined by the knob setting. The thermostat thermal element senses air temperature in the air duct that cools the provision compartment. The thermostat thermal element extends upward from the control housing with a 1" diameter coil at the end. Here again, a temperature adjustment to one compartment may require an adjustment to the other compartment.

In the bottom-freezer, no-frost models the thermostat is located in the provisions compartment. It is a variable cut-in, cut-out type. Changing the knob setting changes the cut-in and cut-out temperatures. The knob setting determines the temperature. The thermostat thermal element extends upward from the control housing with a 1" diameter coil at the end. The thermostat cycles the compressor on demand by sensing the freezer air.

The side-by-side models have the same thermostat as shown in Fig. 12-60. The adjustments for altitude are located in the same place. Such thermostats are made by either Cutler-Hammer or General Electric. Information has already been given for these no-frost, foam-insulated models in the information relative to Fig. 12-60.

To replace this thermostat, disconnect the service cord from the wall outlet and remove the light shield. The thermostat is located in the control center. The freezer compartment is equipped with a manual temperature control. The dial control meters the amount of cold air allowed to enter the provisions compartment. The lower numbers on the dial are warmer temperatures and the higher numbers are colder temperatures. When the dial is set at a low number, the orifice is completely open, allowing the maximum volume of air to enter the provision compartment. This satisfies the thermostat in the provision compartment and cycles the fan and compressor off. Because of the short run cycle, the freezer compartment operates at a warmer temperature. If the control dial is set at a higher number, the orifice size is reduced. This increases the length of the run time and reduces the freezer temperatures.

The side-by-side, no-frost, foam-insulated models have a thermostat that is slightly different in appearance. See Fig. 12-65. The thermostat is located in the freezer compartment. It is the cut-in, cut-out type. Changing the knob changes cut-in and cut-out temperatures. The thermostat thermal element is coiled and positioned inside the thermostat cover housing as shown in Fig. 12-65. The range screw adjustment is located on top of the thermostat body.

The power-saver switch is a single-pole, single-throw switch that is wired in series with the vertical column drier and the crossrail drier. The switch has two positions—on and off. When the switch is *on*, the drier coil circuit is *open*. When the switch is *off*, the drier coil is *in* the circuit. This feature allows the consumer to reduce operating costs by turn-

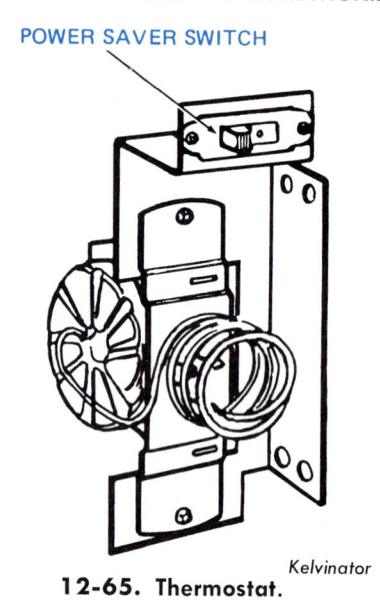

REFRIGERATION AND AIR CONDITIONING TECHNOLOGY

12-65. Thermostat. *Kelvinator*

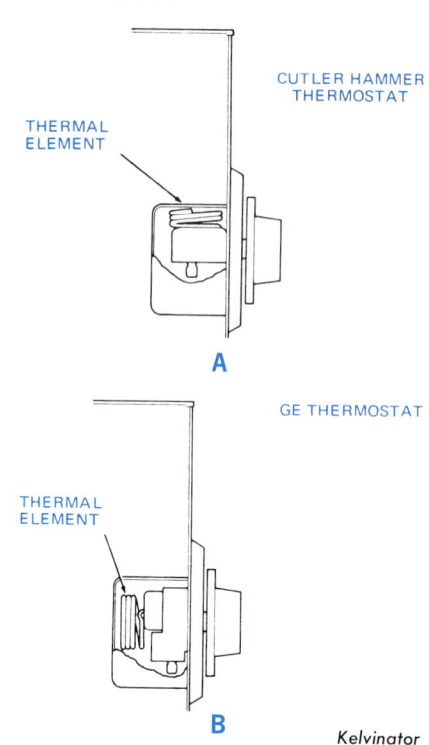

12-66. Thermostat locations. (A) Location of the Cutler-Hammer thermostat. (B) Location of the GE thermostat. *Kelvinator*

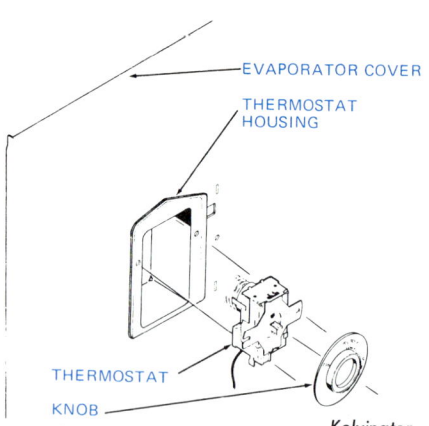

12-67. Thermostat and housing assembly. *Kelvinator*

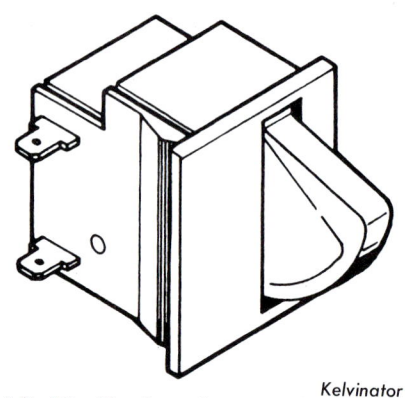

12-68. Single-pole, single-throw provision compartment door switch. *Kelvinator*

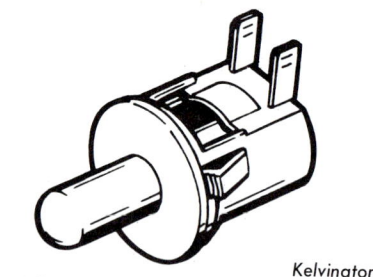

12-69. Freezer compartment and provision compartment door switch. *Kelvinator*

ing on the power saver in dry weather.

The side-by-side, no-frost, foam-insulated model has a thermostat that resembles that shown in Fig. 12-66. It is located in the freezer compartment and is a cut-in, cut-out type. Adjusting the thermostat knob changes both cut-in and cut-out temperatures.

The thermostat thermal element is coiled inside the thermostat housing. Figure 12-66 shows both thermostat models. Cut-in and cut-out adjustments for altitude are made in accordance with Table 12-A.

Failure of the freezer fan motor or ice accumulation on the freezer evaporator will cause the freezer temperature to rise. It will also cause the provision compartment temperature to rise. Reduced air flow across the freezer evaporator (fan not working) reduces refrigerant pressure and temperature inside the freezer evaporator. This means the refrigerant pressure and temperature inside the freezer evaporator will drop considerably. Refrigerant vaporization also is reduced. This causes the evaporator to retain more of the system refrigerant charge.

To replace the thermostat, make sure the power cord is disconnected from the wall outlet. Remove the thermostat knob and the screw in the thermostat cover plate. Disengage the thermostat cover plate from the evaporator cover. Fig. 12-67. Remove the thermostat mounting screw and pull the thermostat forward out of the thermostat housing. Disconnect the wire harness from the thermostat. Reassemble in the reverse order when replacing.

LIGHT SWITCHES

There are several types of light switches the repairperson will need to become familiar with. Figure 12-68 shows a provision compartment door switch. Figure 12-69 shows a freezer compartment and provision compart-

330

CHAPTER 12—REFRIGERATORS

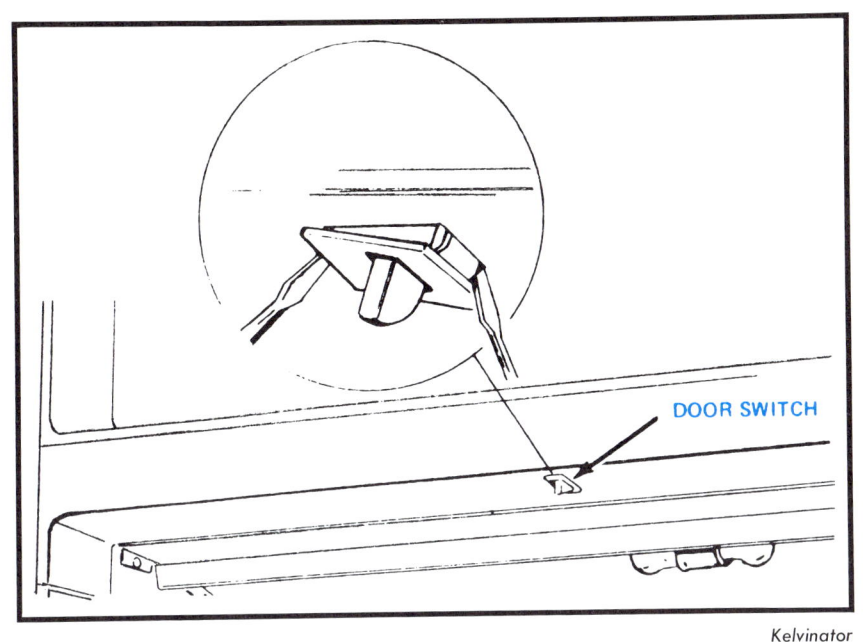

12-70. Freezer and provision compartments switch.

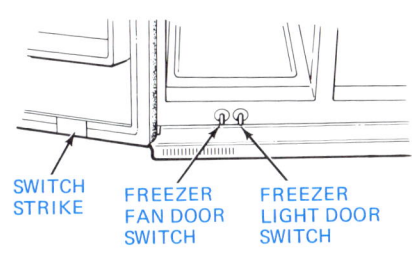

12-71. Light and fan switch location.

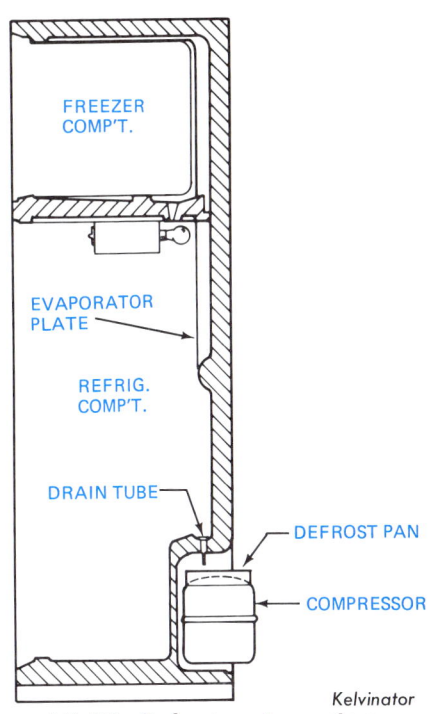

12-72. Defrost water system.

ment door switch. These are single-pole, single-throw types. When the door is closed, the light circuit is off. Figure 12-70 shows the placement of the door switch shown in Fig. 12-68. Push-on terminal connectors secure the wiring harness leads to the switch terminals. To replace the switch shown in Fig. 12-70, insert a small screwdriver under the switch flange. Pry the switch from the cold ban trim. Disconnect the wires from the switch. Install a replacement switch.

Some refrigerators have as many as five door switches. There are three light switches and two fan switches. The switches are all single-pole, single-throw plunger types. The light switches are *normally closed* (NC). The fan switches are *normally open* (NO). Thus, they cannot be interchanged. Figure 12-71 shows switch locations on some models.

TUBE HEATERS

The tube heater between the inlet and outlet lines of the evaporator plate is on during the off-cycle of the thermostat. This assures defrosting of the evaporator plate during each off-cycle.

To check the tube heater, disconnect the service cord from the receptacle. Disconnect the tube heater leads from the thermostat in the control center. With an ohmmeter, check the heater at the leads for continuity. If there is no continuity, replace the heater. Figure 12-72 shows the location of the evaporator plate and the defrosting pan. The tube heater is located only on the top-freezer, cycle-defrost models. Note the defrost water drain system. Heat from the compressor warms the pan and evaporates the water.

The thermostat cut-in temperatures are always above 32° F. [0°C]. Regardless of the thermostat setting, it will not go below 32° F. [0°C]. The freezer evaporator must be defrosted manually. Twice-a-year complete defrosting is recommended. Frost should not be allowed to build up more than $\frac{1}{4}''$ in thickness. Some excess frost can be removed without defrosting by scraping with a plastic spatula or scraper. To completely defrost the freezer evaporator, turn the thermostat to the off position. Open both doors and allow the frost to melt.

331

REFRIGERATION AND AIR CONDITIONING TECHNOLOGY

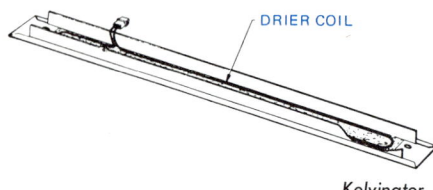

12-73. Crossrail and drier coil. *Kelvinator*

PERIMETER TUBE AND DRIER COIL

To reduce the possibility of condensation forming on the cabinet exterior in areas with high humidity, a perimeter tube extends across the top and down both sides of the cabinet. This tube is located in the cabinet front U-channel. When the compressor operates, warm refrigerant flowing through the perimeter tube warms the cabinet front exterior. See Fig. 12-81.

The perimeter tube is not replaceable. It must be repaired. This repair procedure is discussed on pages 352–355.

In addition to the perimeter tube, these models have a low-wattage electric drier coil on aluminum foil adhered on the back side of the removable crossrail. Fig. 12-73. Removal of the crossrail will be shown later in this chapter.

The top-freezer, no-frost, fiberglass-insulated models have the perimeter heater extending around the front cabinet flange of the freezer compartment. This 19-W heater is across the line and is on whenever the cabinet is plugged in. To check the heater, remove the bottom freezer cold ban trim. Locate the divider terminal block and disconnect the black and red wires with No. 1 stamped on them. Using an ohmmeter, check for continuity. If there is no reading, replace the heater coil.

The bottom freezer, no-frost, refrigerator models have a 17-W heater coil. The terminals are marked the same as on the previously described models.

The vertical column and crossrail drier coils are wired in series in the side-by-side models. The two heaters total 554 ohms and draw 24 W. The vertical column drier coil on aluminum foil adhered on the back side is removable. Figure 12-74 shows how to remove the vertical column so the drier coil can be reached. Starting at the bottom end, progressively pry the vertical column from the spring clips and

12-74. Removing vertical column. *Kelvinator*

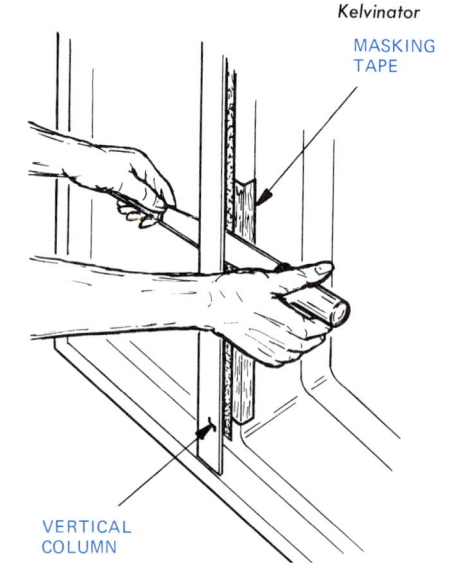

12-75. Cold ban trim removal. *Kelvinator*

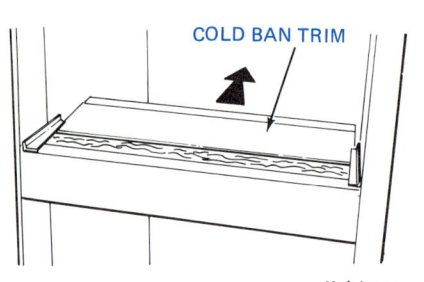

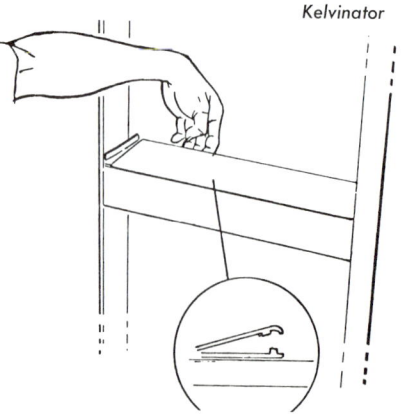

12-76. Removing the cold ban trim. *Kelvinator*

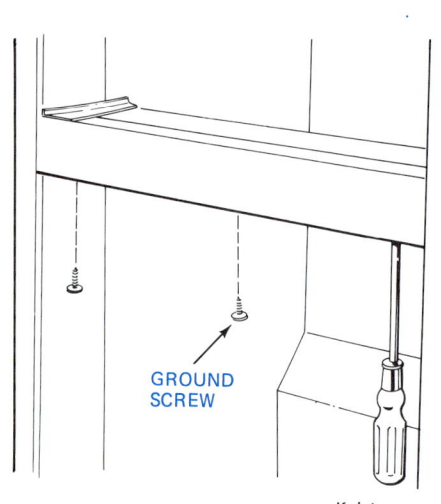

12-77. Removing the screws. *Kelvinator*

12-78. Prying the vertical column with a screwdriver. *Kelvinator*

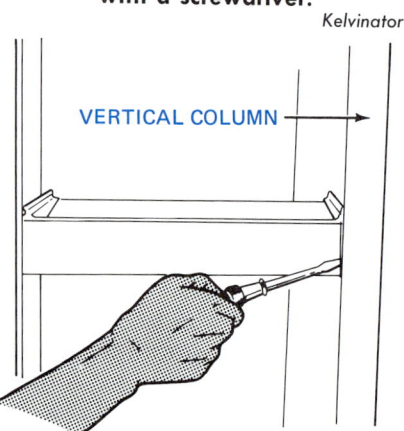

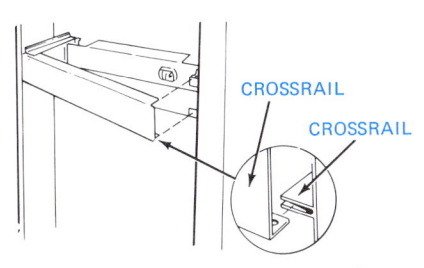

CHAPTER 12—REFRIGERATORS

12-79. Crossrail removal. *Kelvinator*

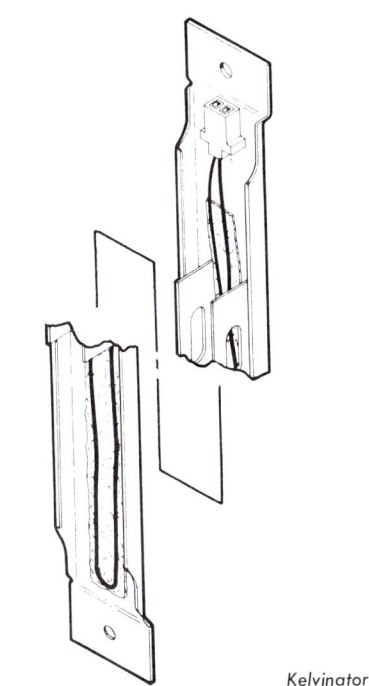

12-80. Vertical column and drier coil. *Kelvinator*

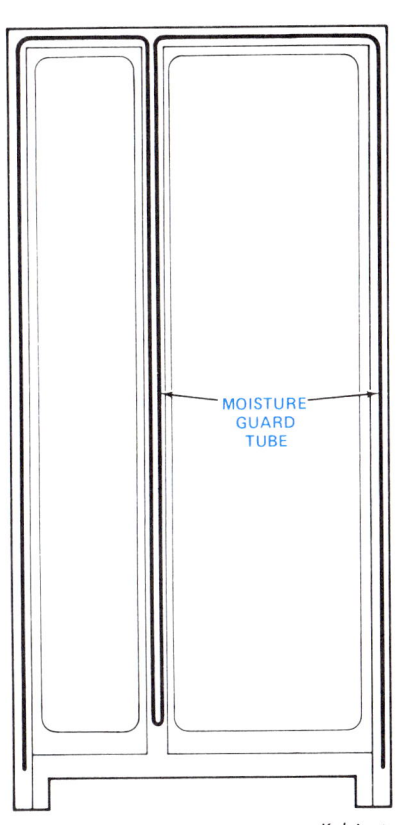

12-81. Perimeter tube. *Kelvinator*

U-channel. Keep in mind that the top end of the vertical column is offset. It extends behind the upper cabinet flange. Therefore, removal of the vertical column must always begin by disengaging the bottom end from the U-channel.

Disconnect the drier coil from the harness receptacle located behind the vertical column in the upper end of the U-channel. Remove the vertical column.

FREEZER DIVIDER CROSSRAIL

The freezer compartment crossrail contains a crossrail drier coil and a light socket. To remove these and other associated parts, remove the freezer shelf located directly behind the crossrail. Lift up on the rear edge of the cold ban trim to disengage the locking tabs of the cold ban. Fig. 12-75. Holding up the rear edge of the cold ban trim, slide the trim rearward to remove as shown in Fig. 12-76. Remove three screws on the underside of the crossrail. See Fig. 12-77. Pry the vertical column far enough to remove the crossrail. Fig. 12-78. Remove the crossrail as shown in Fig. 12-79.

The side-by-side model has the drier coil located on the vertical column. Disconnect the inoperative drier coil from the disconnect receptacle and remove the inoperative drier coil from the vertical column. Fig. 12-80. Remove the protective paper from the aluminum foil. Press the replacement drier coil firmly into place in the vertical column. Connect the drier coil to the disconnect receptacle. Be sure to connect the ground wire before assembling the vertical column to the cabinet.

Location of the perimeter tube is shown for a side-by-side refrigerator in Fig. 12-81. This tube is not replaceable. It must be repaired. The repair procedure is explained on pages 352–355.

DEFROST TIMERS

Automatic defrosting uses resistance heat to defrost the evaporator. A radiant heater is attached to the bottom of the evaporator with two aluminum straps.

A defrost timer and defrost termination thermostat is wired in series with the defrost heater. The defrost timer initiates and terminates a 21-minute defrost cycle every 6 hours. A termination thermostat cycles the defrost heater off at a predetermined temperature, prior to termination of the 21-minute defrost cycle. The aluminum drain positioned below the evaporator is defrosted by the defrost heater.

In addition to the thermostat, the refrigerator is equipped with an automatic defrost timer. This is located in the control center in the provisions compartment.

The automatic defrost timer is driven by a self-starting electric motor. This is geared to turn the shaft and cam one revolution

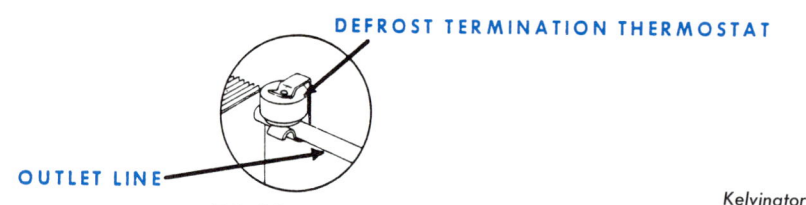

REFRIGERATION AND AIR CONDITIONING TECHNOLOGY

12-82. Termination thermostat.

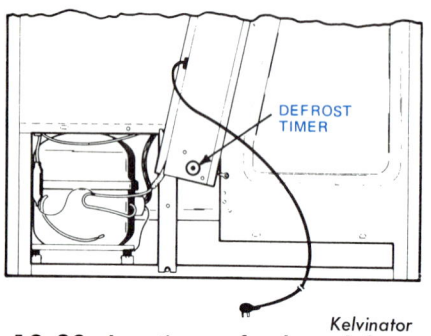

Kelvinator

12-83. Location of the defrost timer.

Kelvinator

every 6 hours, initiating an automatic 21-minute defrost cycle.

The defrost termination thermostat is a temperature-sensing device attached to the outlet line of the evaporator. Fig. 12-82. It is wired in series with the defrost timer, and evaporator defrost heater. It senses the rise in evaporator temperature during a defrost cycle. It cycles the defrost heater off after all frost is melted.

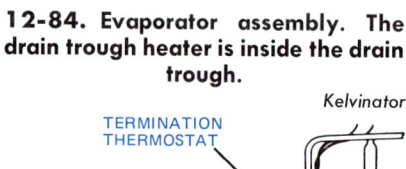

It is calibrated to permit a defrost cycle only when the temperature is below a prescribed temperature.

To remove the termination thermostat, remove the evaporator cover. Disconnect the conductors and snap the thermostat off the outlet line.

Some refrigerator models have a twenty-five-minute defrost cycle. Figure 12-83 shows the defrost timer location on the side-by-side models. Figure 12-84 shows the location of the termination thermostat, defrost heater, and drain trough on the evaporator assembly of a side-by-side refrigerator. On the top-freezer, no-frost, foam-insulated model, the defrost heater and termina-

12-84. Evaporator assembly. The drain trough heater is inside the drain trough.

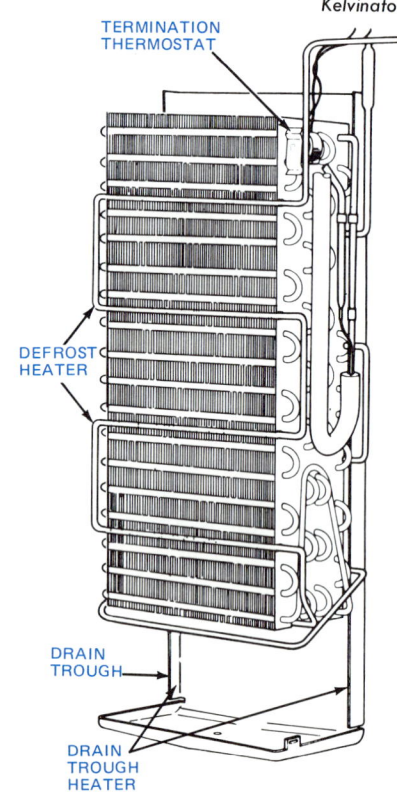

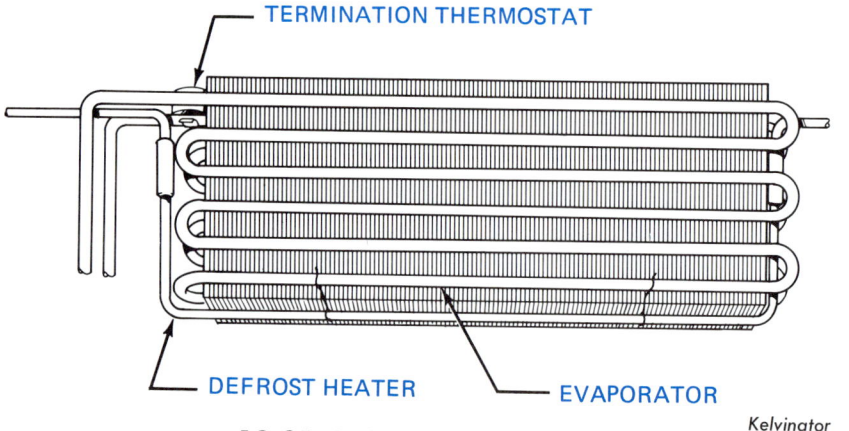

Kelvinator

12-85. Radiant defrost heater.

Kelvinator

12-86. Defrost timer.

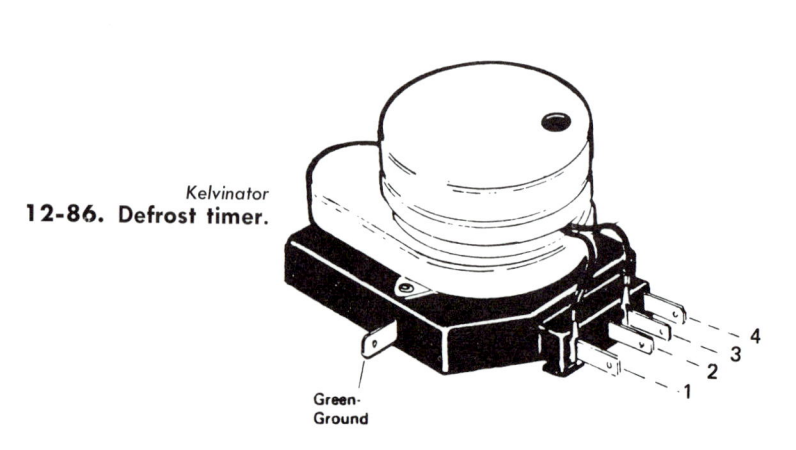

Kelvinator

334

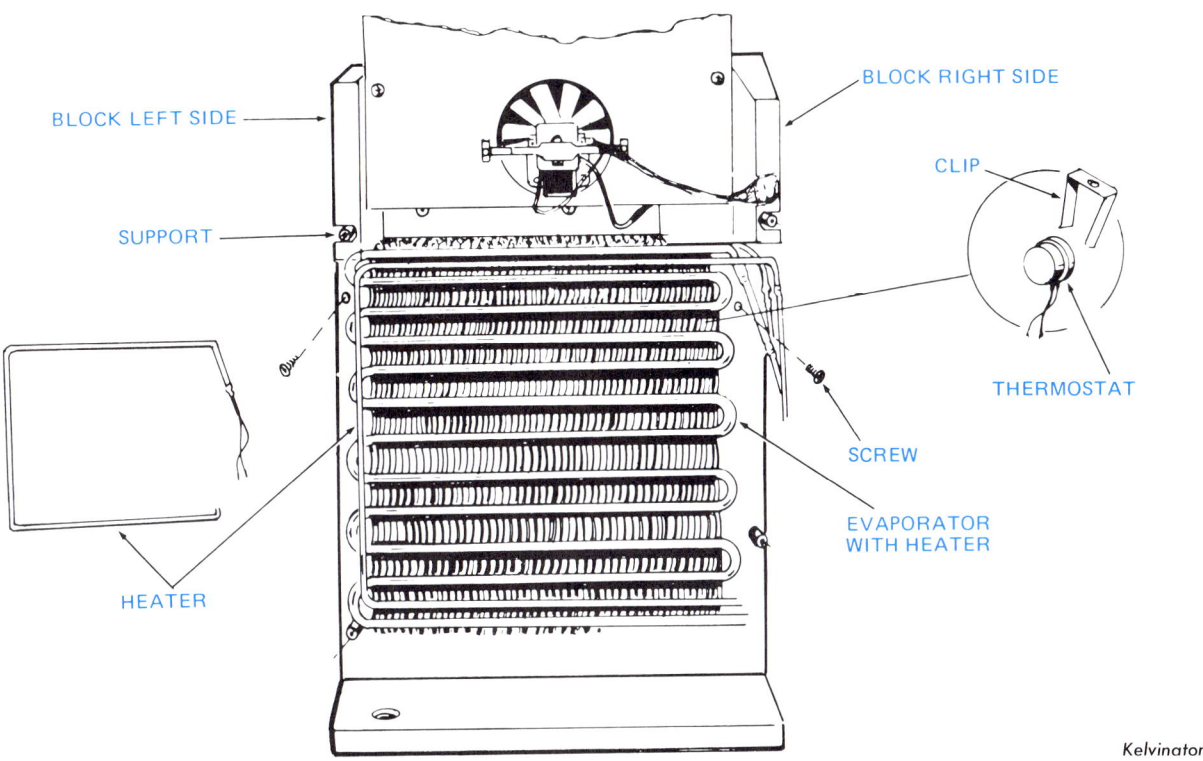

CHAPTER 12—REFRIGERATORS

12-87. Evaporator, defrost heater, termination thermostat, and fan motor assembly.

tion thermostat are located as shown in Fig. 12-85. Figure 12-86 shows the defrost timer on top-freezer, no-frost, fiberglass-insulated models.

DEFROST HEATERS

The defrost heater is wired in series with the termination thermostat and contacts two (2) to three (3) of the defrost timer. The defrost heater is energized during that period of the defrost cycle when the termination thermostat contacts are closed. Contacts open at 60 ± 5° F. [15.6 ± 2.8°C]. The length of time the heater is energized will depend on the amount of frost that has accumulated on the evaporator.

The defrost heater is embedded in the front and the rear surface of the aluminum fin and copper tube evaporator shown in Fig. 12-84. This is a side-by-side model. Figure 12-87 shows the location of the heater and termination thermostat in the evaporator. Defrost water collects in an aluminum trough below the evaporator. Then it drains through a plastic tube into a water evaporating pan. Heat dissipated by the condenser evaporates the defrost water. The termination thermostat and radiant defrost heater are shown on the single-pass, aluminum-finned, tube-type evaporator. Fig. 12-88.

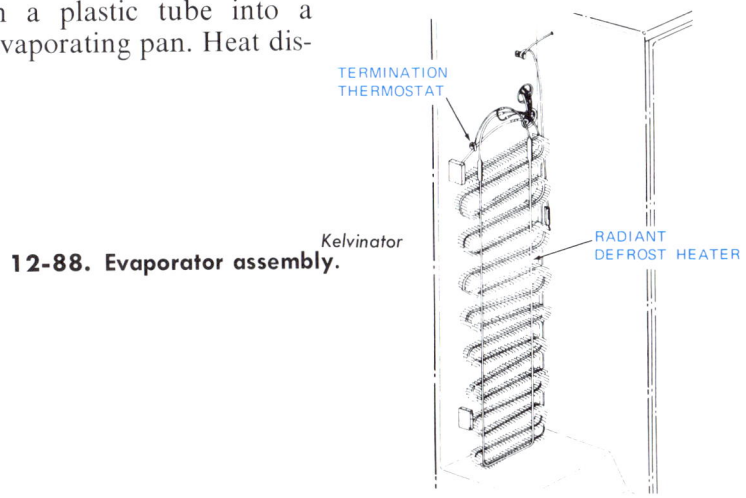

12-88. Evaporator assembly.

335

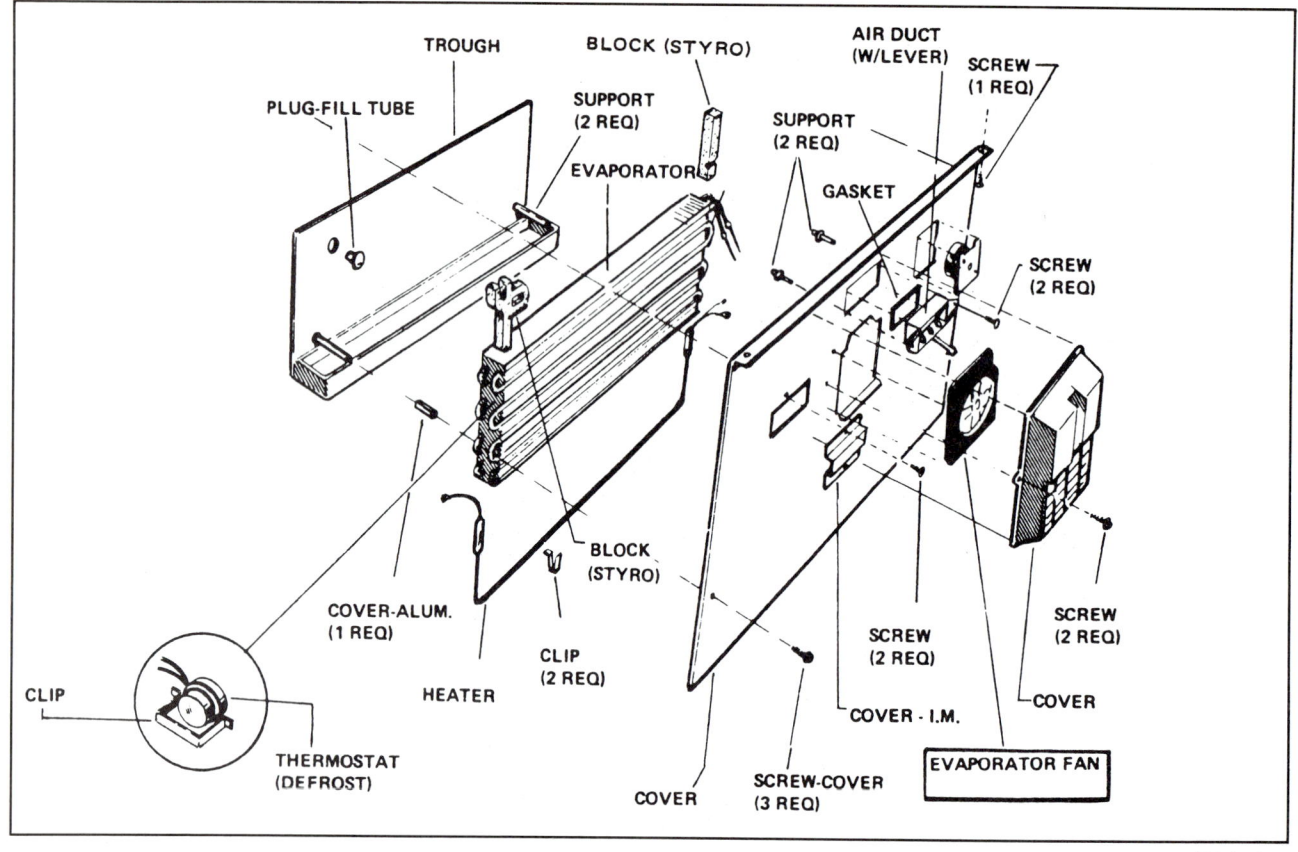

REFRIGERATION AND AIR CONDITIONING TECHNOLOGY

12-89. Evaporator and defrost heater. — *Kelvinator*

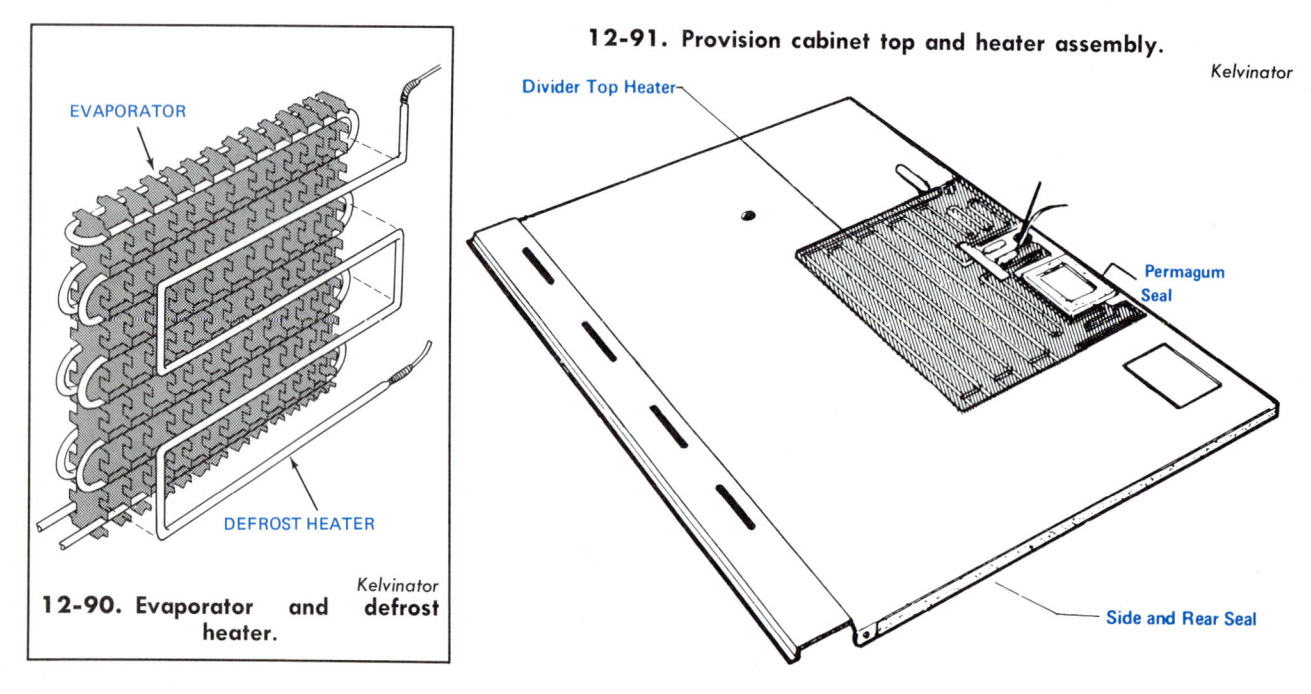

12-90. Evaporator and defrost heater. — *Kelvinator*

12-91. Provision cabinet top and heater assembly. — *Kelvinator*

336

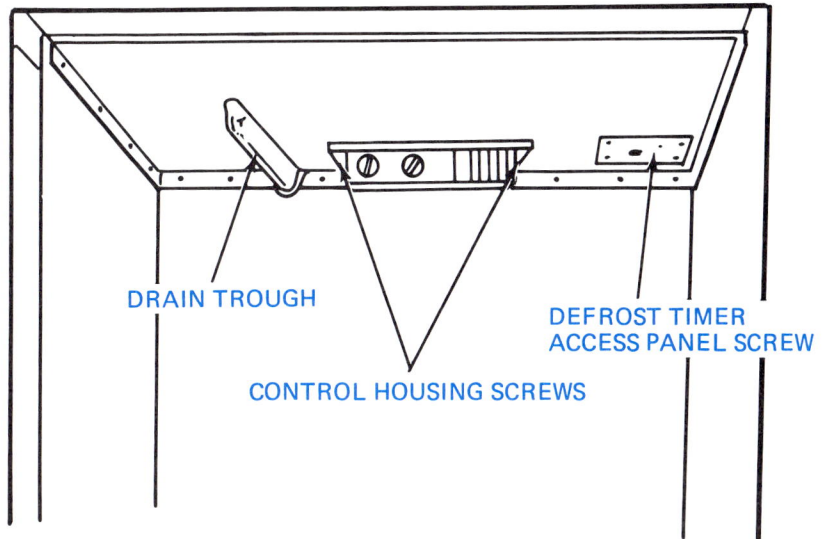

CHAPTER 12—REFRIGERATORS

12-92. Defrost timer and drain trough location. *Kelvinator*

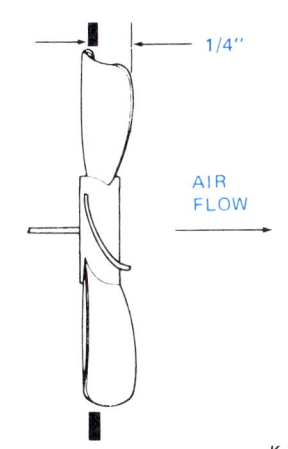

12-94. Position of the fan blade. *Kelvinator*

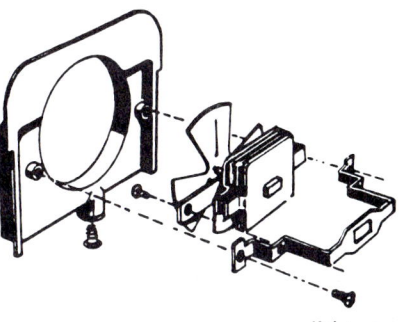

12-93. Fan, motor, and bracket assembly. *Kelvinator*

12-95. Fan and fan motor bracket assembly. *Kelvinator*

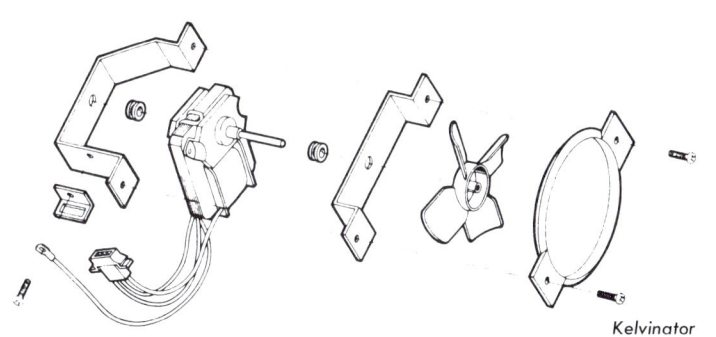

Figure 12-89 shows the location of the defrost heater on the bottom-freezer, no-frost models. The placement of the heater on the top-freezer, no-frost model is slightly different. Figs. 12-90 and 12-91. Figure 12-92 shows the defrost timer and drain-trough location.

FAN AND MOTOR ASSEMBLIES

The fan and motor assembly are located behind the provisions compartment air duct directly above the evaporator in the freezer compartment. The suction-type fan pulls air through the evaporator and blows it through the provisions compartment air duct and freezer compartment fan grille. One type of fan assembly is shown in Fig. 12-93. It is a shaded-pole motor with a molded plastic fan blade. For maximum air circulation, the location of the fan blade on the motor shaft is most important. Mounting the fan blade too far back or too far forward on the motor shaft, in relation to the evaporator cover, will result in improper air circulation. The freezer compartment fan must be positioned with the lead edge of the fan 1/4" in front of the evaporator cover. Fig. 12-94.

The fan assembly shown in Fig. 12-95 is used on the top-freezer, no-frost, fiberglass-insulated models. The freezer fan and motor assembly are located in the divider partition directly under the freezer air duct.

To remove the fan and motor assembly, loosen the screw at the top of the air duct. Remove the screw on the right side of the air

337

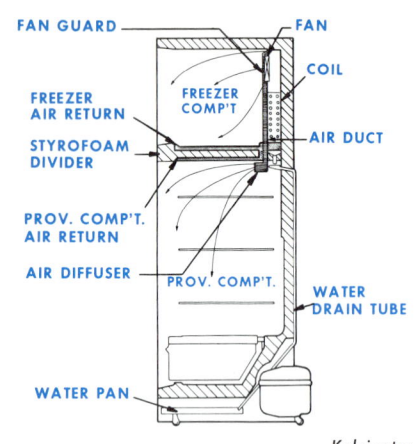

REFRIGERATION AND AIR CONDITIONING TECHNOLOGY

12-96. Air circulation pattern in a top-freezer, no-frost model refrigerator.

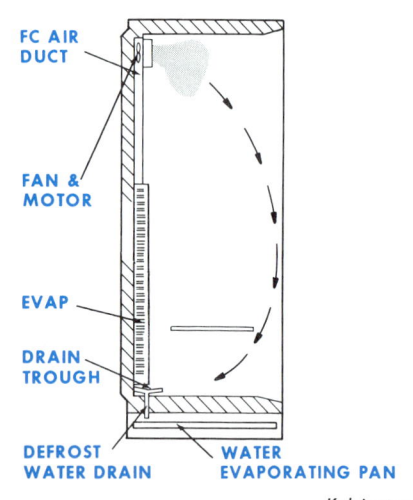

12-97. Air circulation pattern in a side-by-side model.

duct. Lift the air duct up to disengage from the fan shroud. Remove the fan shroud screw which is located next to the control assembly in the food compartment. Lift the motor and fan assembly up and out of the divider. The fan blade has a friction fit that slides on the motor shaft. If the fan blade is removed, prior to reinstallation of the blade hold the fan motor so you are viewing the end of the motor shaft. Install the fan blade with the word *front* facing you. This word is stamped on one of the blades. Position the fan blade hub at the end of the motor shaft.

The fan is wired in series with the thermostat and contacts of the defrost timer. The freezer compartment fan motor operates when the thermostat contacts are closed, except during the defrost cycle.

AIR CIRCULATION

No-frost refrigerators operate on the principle that moisture or frost transfers or migrates to the coldest surfaces in the freezer compartment. For example, a small amount of water spilled from an ice cube tray in the freezer compartment will freeze immediately. However, in time this ice will evaporate and transfer to the colder surfaces of the freezer evaporator coil.

Top-Freezer Models

Top-freezer, no-frost models with a single evaporator in the freezer compartment have forced-air cooling in the freezer and provision compartments. The fin-and-tube-type evaporator is located on the back wall of the freezer compartment. A suction-type circulating fan pulls the air from the freezer and provision compartments across the evaporator surfaces. The cold air is forced into a fan cover and discharged into the provision compartment as shown in Fig. 12-96.

Air from the freezer and provision compartments flows to the evaporator through air return ducts into the divider between the freezer compartment and the provision compartment. The air circulating fan in the freezer compartment operates only when the compressor is running. During the defrost period, however, the compressor and circulating fan do not operate. The automatic defrost timer opens the electrical circuit to the fan motor and compressor.

Side-by-Side Models

All side-by-side models have forced-air cooling in the freezer and provision compartments. Fig. 12-97. The fin-and-tube-type evaporator is located on the back wall of the freezer compartment. A suction-type circulating fan pulls the air from the freezer and provision compartments across the evaporator surfaces. The cold air is forced up through a duct in the back wall and discharged into the freezer and provision compartments. Fig. 12-97. Air from the provision compartment then flows to the evaporator through an air duct in the separating wall.

The air circulating fan operates only when the compressor is running. During the defrost period the compressor and circulating fan do not operate. The automatic defrost timer opens the electrical circuit to the fan motor and the compressor.

The solid metal shelf in the bottom freezer compartment must be in place to maintain proper air flow and temperatures in the freezer compartment.

AIR SHUTTERS

The provision compartment temperatures are controlled by an air shutter. The air shutter meters the amount of cold air allowed to enter the provision compartment. Therefore, temperatures can be changed by opening the shutter or closing it.

CHAPTER 12—REFRIGERATORS

Different refrigerators use different types of air shutters.

DRAIN-TROUGH HEATER

The drain-trough heater is shown in Fig. 12-84. It is wired in parallel with the defrost heater and in series with the defrost termination thermostat and contacts two (2) to three (3) of the defrost timer. It is energized during that period of the defrost cycle when the termination thermostat contacts are closed. Contacts open at $60 \pm 5°$ F. [$15.6 \pm 2.8°C$].

To check the drain-trough heater, remove the harness channel on the rear of the cabinet. Disconnect the drain heater from the harness. Use an ohmmeter to make a continuity check.

To replace the drain-trough heater, remove the shelves, shelf supports, evaporator cover, and evaporator mounting screws. Swing the bottom of the evaporator forward and away from the drain trough. Remove the screw from the right-hand front cover

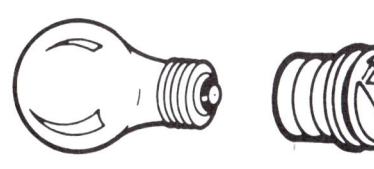

12-98. Light socket for screw-in bulb.

12-99. Light socket that is mounted in the liner.
Kelvinator

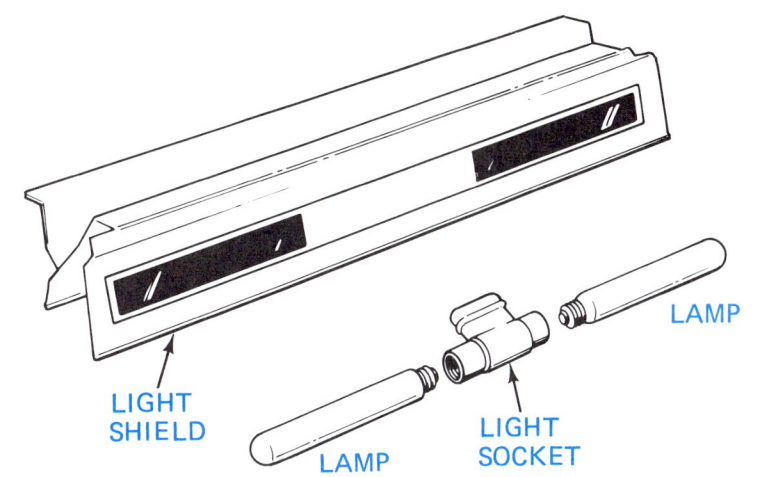

12-100. Light socket and shield.
Kelvinator

of the drain trough. Remove the harness channel on the rear of the cabinet. Remove the mastic sealer from the tubing-harness entrance hole. Disconnect the drain-trough heater leads from the harness. Swing the bottom of the drain trough forward. Pull the drain trough out from behind the evaporator. Install a new heater on the drain trough and assemble in reverse order. Make sure you press mastic sealer firmly into the tubing-harness entrance hole to form a positive air seal.

LIGHT BULB SOCKETS

Light bulb sockets vary in shape and connections according to the model in which they are installed. Figure 12-98 shows the light socket as a molded fixture that snaps into the back of the control center bracket. Figure 12-99 shows another type of light socket. This one is mounted in the liner. The sockets have a standard base for a 40-W appliance bulb. To remove the molded light socket, pull downward and twist clockwise. The light shield attaches to the bottom of the control housing. To remove, press on the sides and pull down.

The provision compartment has a double-socket, molded light bulb receptacle that uses standard base appliance bulbs. The socket is mounted in the provision compartment liner. The light shield is a lift-off, white plastic cover. The top door light in the freezer compartment holds a single bulb and has a small base, $3/4'' \times 5''$. The socket is molded vinyl and snaps in the liner. The lower door light in the freezer compartment is the same as the top light and mounts in the crossrail.

Figure 12-100 shows how the side-by-side models use the double light socket with the lamps mounted under a long light shield. The light socket and the shield assemblies are secured to the provision compartment and the freezer compartment inner wall with screws. Removal of the screws releases the shield and the socket for individual replacement.

339

REVIEW QUESTIONS

1. How is the refrigerator door adjusted?
2. Why are two evaporators needed in a cycle defrost refrigerator-freezer?
3. What is the purpose of a thermostat?
4. What is the temperature range of an icemaker in a refrigerator?
5. What is the purpose of the mold heater?
6. What is an ice stripper?
7. What happens to the water formed when frost is melted in the defrost cycle?
8. What is the tolerance of the defrost thermostat?
9. How do you test the run capacitor?
10. How long does it take for a refrigerator to unload when the thermostat cuts off?
11. How do you check the compressor motor's electrical circuits?
12. How does one adjust the thermostat in a refrigerator-freezer?
13. What is a perimeter tube?
14. How is the automatic defrost timer driven?
15. No-frost units operate on the principle that moisture or frost transfers or migrates to the coldest surfaces in the freezer compartment. What does this mean when a small amount of water is spilled inside the freezer compartment when ice cube trays are being reloaded?

13

Servicing

SAFETY

Safe practices are important in servicing refrigeration units. Such practices are common sense, but must be reinforced to make one aware of the problems that can result when a job is done incorrectly.

Handling Cylinders

Refrigeration and air conditioning servicepersons must be able to handle compressed gases. Accidents occur when compressed gases are not handled properly.

Oxygen or acetylene must never be used to pressurize a refrigeration system. Oxygen will explode when it comes into contact with oil. Acetylene will explode under pressure, except when properly dissolved in acetone as used in commercial acetylene cylinders.

Dry nitrogen or dry carbon dioxide are suitable gases for pressurizing refrigeration or air conditioning systems for leak tests or system cleaning. However, the following specific restrictions must be observed:

- Nitrogen (N_2). Commercial cylinders contain pressures in excess of 2,000 pounds per square inch at normal room temperature. Carbon dioxide (CO_2). Commercial cylinders contain pressures in excess of 800 pounds per square inch at normal room temperature. Cylinders should be handled carefully. Do not drop them or bump them.
- Keep cylinders in a vertical position and securely fastened to prevent them from tipping over.
- Do not heat the cylinder with a torch or other open flame. If heat is necessary to withdraw gas from the cylinder, apply heat by immersing the lower portion of the cylinder in warm water. Never heat a cylinder to a temperature over 110° F. [43°C].

Pressurizing

Pressure testing or cleaning refrigeration and air conditioning systems can be dangerous! Extreme caution must be used in the selection and use of pressurizing equipment. Follow these procedures:

- Never attempt to pressurize a system without first installing an appropriate pressure-regulating valve on the nitrogen or carbon dioxide cylinder discharge. This regulating valve should be equipped with two functioning pressure gages. One gage indicates cylinder pressure. The other gage indicates discharge or downstream pressure.
- Always install a pressure relief valve or frangible-disc type pressure relief device in the pressure supply line. This device should have a discharge port of at least $1/2''$ NPT size. This valve or frangible-disc device should be set to release at 175 psig.
- A system can be pressurized up to a *maximum* of 150 psig for leak testing or purging. Fig. 13-1.

Tecumseh hermetic-type com-

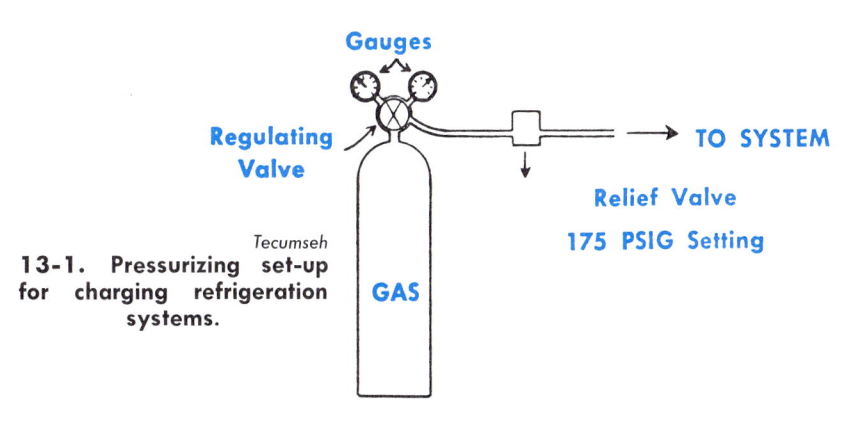

Tecumseh
13-1. Pressurizing set-up for charging refrigeration systems.

341

pressors are low-pressure housing compressors. The compressor housings (cans or domes) are not normally subjected to discharge pressures. They operate instead at relatively low suction pressures. These Tecumseh compressors are generally installed on equipment where it is impractical to disconnect or isolate the compressor from the system during pressure testing. Therefore, do not exceed 150 psig when pressurizing such a complete system.

> *When flushing or purging a contaminated system, care must be taken to protect the eyes and skin from contact with acid-saturated refrigerant or oil mists. The eyes should be protected with goggles. All parts of the body should be protected by clothing to prevent injury by refrigerant. If contact with either skin or eyes occurs, flush the exposed area with cold water. Apply an ice pack if the burn is severe, and see a physician at once.*

Working with Refrigerants R-12 and R-22

R-12 and R-22 are considered to be nontoxic and noninflammable. However, any gas under pressure can be hazardous. The latent energy in the pressure alone can cause damage. In working with R-12 and R-22, observe the precautions that apply when working with other pressurized gases.

Never completely fill any refrigerant gas cylinder with liquid. Never fill more than 80% with liquid. This will allow for expansion under normal conditions.

Make sure an area is properly ventilated before purging or evacuating a system that uses R-12 or R-22. In certain concentrations and in the presence of an open flame, such as a gas range or a gas water heater, R-12 and R-22 may break down and form a small amount of harmful phosgene gas. This gas was used in World War I for poison gas warfare.

Lifting

Lifting heavy objects can cause serious problems. Strains and sprains are often caused by improper lifting methods. Figure 13-2 indicates the right and the wrong way to lift heavy objects. In this case, a compressor is shown.

To avoid injury, learn to lift the safe way. Bend your knees, keep your back erect, and lift gradually with your leg muscles.

The material you are lifting may slip from your hands and injure your feet. To prevent foot injuries, wear the proper shoes.

Electrical Safety

Many Tecumseh single-phase compressors are installed in systems requiring off-cycle crankcase heating. This is designed to prevent refrigerant accumulation in the compressor housing. The power is on at all times. Even if the compressor is not running, power is applied to the compressor housing where the heating element is located.

Another popular system uses a run capacitor that is always connected to the compressor motor windings, even when the compressor is not running. Other devices are energized when the compressor is not running. That means there is electrical power applied to the unit even when the compressor is not running. This calls for an awareness of the situation and the proper safety procedures.

13-2. Safety first. Lift with the legs not the back.

Tecumseh

CHAPTER 13—SERVICING

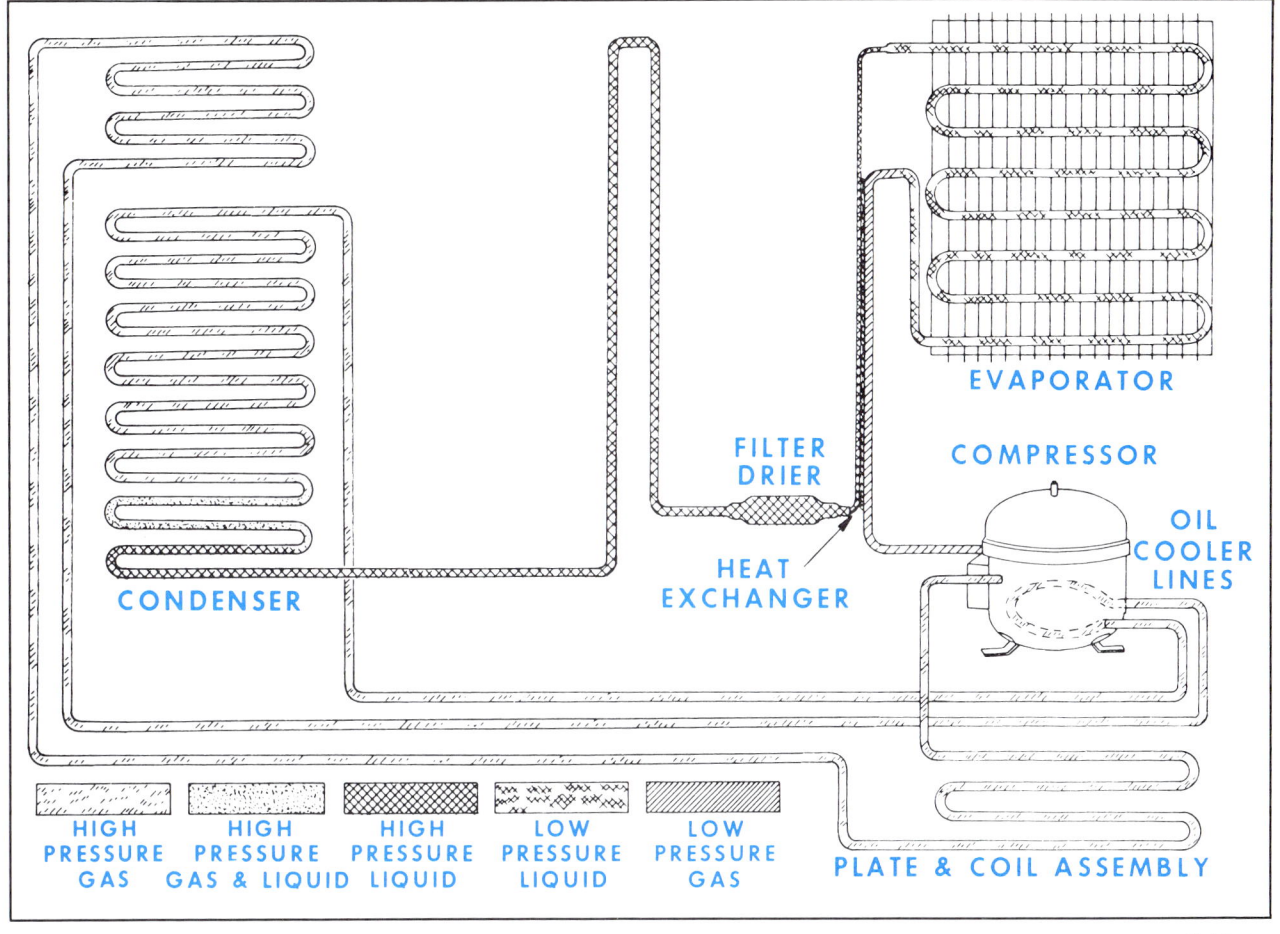

13-3. Refrigerating system with various pressures located.

Be safe. Before you attempt to service any refrigeration system, make sure that the main circuit breaker is open and all power is off.

SERVICING THE REFRIGERATOR SECTION

The refrigerant cycle is a continuous cycle, which occurs whenever the compressor is operating. Liquid refrigerant is evaporated in the evaporator by the heat that enters the cabinet through the insulated walls and by product load and door openings. The refrigerant vapor passes from the evaporator, through the suction line, to the compressor dome, which is at suction pressure. From the top interior of the dome, the vapor passes down through a tube into the pump cylinder. The pressure and temperature of the vapor are raised in the cylinder by compression. The vapor is then forced through the discharge valve into the discharge line and the condenser. Air passing over the condenser surface removes heat from the high pressure vapor, which then condenses to a liquid. The liquid refrigerant flows from the condenser to the evaporator through the small diameter liquid line (capillary tube). Before it enters the evaporator, it is subcooled in the heat exchanger by the low temperature suction vapor in the suction line. Fig. 13-3.

Sealed Compressor and Motor

All models are equipped with a compressor with internal spring suspension. Some compressors have a plug-in magnetic starting relay, with a separate motor overload protector. Others have

343

REFRIGERATION AND AIR CONDITIONING TECHNOLOGY

a built-in metallic motor overload protector. When ordering a replacement compressor, you should always give the refrigerator model number and serial number and the compressor part number. Every manufacturer has a listing available to servicepersons.

Condenser

Side-by-side and top-freezer models with a vertical, natural draft, wire-tube type condenser have a water evaporating coil connected in series with the condenser. The high-temperature, high-pressure, compressed refrigerant vapor passes first through the water evaporating coil. There, part of the latent heat of evaporation and sensible heat of compression are released. See Fig. 13-3. The refrigerant then flows back through the oil cooling coil in the compressor shell. There, additional heat is picked up from the oil. The refrigerant then flows back to the main condenser, where sufficient heat is released to the atmosphere. This results in the condensation of refrigerant from a high-pressure vapor to high-pressure liquid.

Filter Drier

A filter drier is located in the liquid line at the outlet of the condenser. Its purpose is to filter or trap minute particles of foreign materials and absorb any moisture in the system. Fine mesh screens filter out foreign particles. The desiccant absorbs the moisture.

Capillary Tube

The capillary tube is a small diameter liquid line connecting the condenser to the evaporator. Its resistance, or pressure drop, due to the length of the tube and its small diameter, meters the refrigerant flow into the evaporator.

The capillary tube allows the high-side pressure to unload, or balance out, with the low-side pressure during the *off-cycle*. This permits the compressor to start under a no-load condition.

The design of the refrigerating system for capillary feed must be carefully engineered. The capillary feed must be matched to the compressor for the conditions under which the system is most likely to operate. Both the high side (condenser) and low side (evaporator) must be specifically designed for use with a capillary tube.

Heat Exchanger

The heat exchanger is formed by soldering a portion of the capillary tube to the suction line. The purpose of the heat exchanger is to increase the over-all capacity and efficiency of the system. It does this by using the cold suction gas leaving the evaporator to cool the warm liquid refrigerant passing through the capillary tube to the evaporator. If the hot liquid refrigerant from the condenser were permitted to flow uncooled into the evaporator, part of the refrigerating effect of the refrigerant in the evaporator would have to be used to cool the incoming hot liquid down to evaporator temperature.

Freezer Compartment and Provision Compartment Assembly

Liquid refrigerant flows through the capillary and enters the freezer evaporator. Expansion and evaporation starts at this point. See Fig. 13-3.

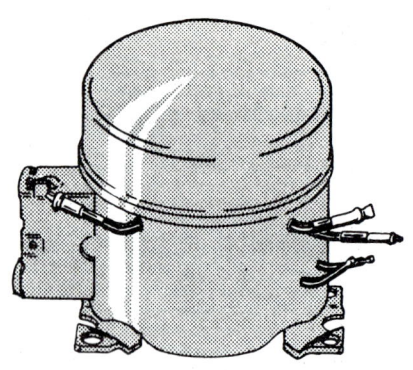

Kelvinator
13-4. N-line replacement compressor for top-freezer models.

COMPRESSOR REPLACEMENT

Replacement compressor packages are listed by the manufacturer. Check with the refrigerator manufacturer to be sure you have the proper replacement. Refer to the compressor number in the refrigerator under repair. Compare that number to the suggested replacement number.

Replacement compressors are charged with oil and a holding charge of nitrogen. A replacement filter drier is packaged with each replacement compressor. It must be installed with the compressor. Figure 13-4 shows the N-line replacement compressor designed for top-freezer Kelvinator models. The A-line replacement compressor is shown in Fig. 13-5. It is used on Kelvinator chest or upright freezers.

The new relay-overload protector assembly supplied with the N-line replacement compressor should always be used. The new motor overload protector supplied with the A-line replacement compressor should always be used. Transfer the relay, the relay cover, and the cover clamp from the original compressor to

CHAPTER 13—SERVICING

13-5. A-line replacement compressor.
Kelvinator

the replacement compressor. If a relatively small quantity of refrigerant is used, a major portion of it will be absorbed by the oil in the compressor when the refrigerator has been inoperative for a considerable length of time. When opening the system, use care to prevent oil from blowing out with the refrigerant.

TROUBLESHOOTING COMPRESSORS

There are several common compressor problems. Table 13-A lists these problems and their solutions.

TROUBLESHOOTING REFRIGERATOR COMPONENTS

Compressor Will Not Run
Cause
1. Inoperative thermostat. Replace.
2. Service cord is pulled from the wall receptacle. Replace.
3. Service is pulled from the harness. Disconnect.
4. No voltage at the wall receptacle. House fuse blown.
5. Faulty cabinet wiring. Repair or replace.
6. Relay leads. Disconnect.
7. Relay loose or inoperative. Tighten or replace.
8. Compressor windings open. Replace compressor.
9. Compressor stuck. Replace.
10. Low voltage, causing compressor to cycle on overload. Voltage fluctuation should not exceed 10%, plus or minus, from the nominal rating of 115 volts.

Compressor Runs, but There Is No Refrigeration
Cause
1. System is out of refrigerant. Check for leaks.
2. Compressor is not pumping. Replace.
3. Restricted filter drier. Replace.
4. Restricted capillary tube. Replace.
5. Moisture in the system. Pump down and recharge.

Compressor Short Cycles
Cause
1. Erratic thermostat operation. Replace.
2. Faulty relay. Replace.
3. Restricted air flow over the condenser. Remove restrictions.
4. Low voltage. Fluctuation exceeds 10%.
5. Inoperative condenser fan. Repair or replace.
6. Compressor draws excessive wattage. Replace.

Compressor Runs Too Much or 100%
Cause
1. Erratic thermostat or thermostat is set too cold. Replace or reset to normal position.
2. Refrigerator exposed to unusual heat. Relocate.
3. Abnormally high room temperature. If outside temperature is cooler, open windows to lower temperature. Turn on fans to move the air.
4. Low pumping capacity compressor. Replace.
5. Door gaskets not sealing. Check with 100-W lamp.
6. System is undercharged or overcharged. Correct the charge.
7. Interior light stays on. Check door switch.
8. Non-condensables are in the system. Evacuate and recharge.
9. Capillary tube kinked or partially restricted.
10. Filter drier or strainer partially restricted. Replace.
11. Excessive service load. Remove part of the load.
12. Restricted air flow over the condenser. Remove restriction.

Noise
Cause
1. Tubing vibrates. Adjust tubing.
2. Internal compressor noise. Replace.
3. Compressor vibrating on the cabinet frame. Adjust.
4. Loose water-evaporating pan. Tighten.
5. Rear machine compartment cover missing. Replace.
6. Compressor is operating at a high head pressure due to restricted air flow over the condenser. Reduce or remove restrictions.
7. Inoperative condenser fan motor. Check fuses, circuit breaker, and condition of fan motor. Replace, if necessary.

To Replace the Compressor
Following is the method for replacing the compressor on the Kelvinator refrigerator. The material on pages 346–348 applies to the servicing of the Kelvinator refrigerator.

Table 13-A.
Compressor Troubleshooting and Service

Complaint	Possible Cause	Repair
Compressor will not start. There is no hum.	1. Line disconnect switch open. 2. Fuse removed or blown. 3. Overload protector tripped. 4. Control stuck in open position. 5. Control off due to cold location. 6. Wiring improper or loose.	1. Close start or disconnect switch. 2. Replace fuse. 3. Refer to electrical section. 4. Repair or replace control. 5. Relocate control. 6. Check wiring against diagram.
Compressor will not start. It hums, but trips on overload protector.	1. Improperly wired. 2. Low voltage to unit. 3. Starting capacitor defective. 4. Relay failing to close. 5. Compressor motor has a winding open or shorted. 6. Internal mechanical trouble in compressor. 7. Liquid refrigerant in compressor.	1. Check wiring against diagram. 2. Determine reason and correct. 3. Determine reason and replace. 4. Determine reason and correct, replace if necessary. 5. Replace compressor. 6. Replace compressor. 7. Add crankcase heater and/or accumulator.
Compressor starts, but does not switch off of start winding.	1. Improperly wired. 2. Low voltage to unit. 3. Relay failing to open. 4. Run capacitor defective. 5. Excessively high discharge pressure. 6. Compressor motor has a winding open or shorted. 7. Internal mechanical trouble in compressor (tight).	1. Check wiring against diagram. 2. Determine reason and correct. 3. Determine reason and replace if necessary. 4. Determine reason and replace. 5. Check discharge shut-off valve, possible overcharge, or insufficient cooling of condenser. 6. Replace compressor. 7. Replace compressor.
Compressor starts and runs, but short cycles on overload protector.	1. Additional current passing through the overload protector. 2. Low voltage to unit (or unbalanced if three-phase). 3. Overload protector defective.	1. Check wiring against diagram. Check added fan motors, pumps, etc., connected to wrong side of protector. 2. Determine reason and correct. 3. Check current, replace protector.

CHAPTER 13—SERVICING

Table 13-A. (Continued)

Complaint	Possible Cause	Repair
	4. Run capacitor defective.	4. Determine reason and replace.
	5. Excessive discharge pressure.	5. Check ventilation, restrictions in cooling medium, restrictions in refrigeration system.
	6. Suction pressure too high.	6. Check for possibility of misapplication. Use stronger unit.
	7. Compressor too hot—return gas hot.	7. Check refrigerant charge. (Repair leak.) Add refrigerant if necessary.
	8. Compressor motor has a winding shorted.	8. Replace compressor.
Unit runs, but short cycles on.	1. Overload protector.	1. Check current. Replace protector.
	2. Thermostat.	2. Differential set too close. Widen.
	3. High pressure cut-out due to insufficient air or water supply, overcharge, or air in system.	3. Check air or water supply to condenser. Reduce refrigerant charge, or purge.
	4. Low pressure cut-out due to: a. Liquid line solenoid leaking. b. Compressor valve leak. c. Undercharge. d. Restriction in expansion device.	4. a. Replace. b. Replace. c. Repair leak and add refrigerant. d. Replace expansion device.
Unit operates long or continuously.	1. Shortage of refrigerant. 2. Control contacts stuck or frozen closed. 3. Refrigerated or air conditioned space has excessive load or poor insulation. 4. System inadequate to handle load. 5. Evaporator coil iced. 6. Restriction in refrigeration system. 7. Dirty condenser. 8. Filter dirty.	1. Repair leak. Add charge. 2. Clean contacts or replace control. 3. Determine fault and correct. 4. Replace with larger system. 5. Defrost. 6. Determine location and remove. 7. Clean condenser. 8. Clean or replace.
Start capacitor open, shorted, or blown.	1. Relay contacts not operating properly. 2. Prolonged operation on start cycle due to:	1. Clean contacts or replace relay if necessary. 2.

(Continued on next page)

347

Table 13-A. (Continued)
Compressor Troubleshooting and Service

Complaint	Possible Cause	Repair
	a. Low voltage to unit. b. Improper relay. c. Starting load too high. 3. Excessive short cycling. 4. Improper capacitor.	a. Determine reason and correct. b. Replace. c. Correct by using pump-down arrangement if necessary. 3. Determine reason for short cycling as mentioned in previous complaint. 4. Determine correct size and replace.
Run capacitor open, shorted, or blown.	1. Improper capacitor. 2. Excessively high line voltage (110% of rated maximum).	1. Determine correct size and replace. 2. Determine reason and correct.
Relay defective or burned out.	1. Incorrect relay. 2. Incorrect mounting angle. 3. Line voltage too high or too low. 4. Excessive short cycling. 5. Relay being influenced by loose vibrating mounting. 6. Incorrect run capacitor.	1. Check and replace. 2. Remount relay in correct position. 3. Determine reason and correct. 4. Determine reason and correct. 5. Remount rigidly. 6. Replace with proper capacitor.
Space temperature too high.	1. Control setting too high. 2. Expansion valve too small. 3. Cooling coils too small. 4. Inadequate air circulation.	1. Reset control. 2. Use larger valve. 3. Add surface or replace. 4. Improve air movement.
Suction line frosted or sweating.	1. Expansion valve oversized or passing excess refrigerant. 2. Expansion valve stuck open. 3. Evaporator fan not running. 4. Overcharge of refrigerant.	1. Readjust valve or replace with smaller valve. 2. Clean valve of foreign particles. Replace if necessary. 3. Determine reason and correct. 4. Correct charge.
Liquid line frosted or sweating.	1. Restriction in dehydrator or strainer. 2. Liquid shutoff (king valve) partially closed.	1. Replace part. 2. Open valve fully.
Unit noisy.	1. Loose parts or mountings. 2. Tubing rattle. 3. Bent fan blade causing vibration. 4. Fan motor bearings worn.	1. Tighten. 2. Reform to be free of contact. 3. Replace blade. 4. Replace motor.

CHAPTER 13—SERVICING

1. Bleed the refrigerant slowly by cutting the process tube on the compressor with diagonal cutters. If the refrigerator has not been in operation for some time, oil may be discharged with the refrigerant. Use care when bleeding the refrigerant. Place a cloth over the process tube to prevent oil and refrigerant from splattering the room. Preferably, run the compressor, if operative, until the dome becomes warm. This will separate the refrigerant from the oil.

CAUTION: Ventilate the room while purging, especially when open-flame cooking or baking is being done in the kitchen.

2. Use diagonal pliers. Cut the discharge, suction, and oil cooler tubes. Crimp the tubes that remain on the compressor dome to prevent oil leakage during shipment.

3. Remove wire leads from the relay. Remove mounting cap screws. Then, remove compressor from machine compartment.

CAUTION: If original compressor showed signs of burnout, follow instructions for "Cleaning System after Burnout" (page 350).

4. Remove the filter drier by cutting the $\frac{1}{4}''$ inlet tube 1'' from the brazed connection. Use a file to score the capillary tube uniformly about 1'' from the brazed joint at the filter drier. Break off the capillary tube.

5. Transfer the rubber mounts from the inoperative compressor to the replacement compressor. Set the replacement compressor in place and install.

6. Remove the line caps and bleed off the holding charge of nitrogen. Use a suitable tool. Cut the suction, discharge, and oil cooler extension tubes to the required lengths. Swage the tubes as required. Join to tubes on cabinet for brazing.

7. Install the replacement filter drier (packaged with replacement compressor).

8. Braze the refrigerant tubes to the filter drier and compressor. Use silver solder (Easy Flo-45).

CAUTION: Do not remove the ends of the filter drier until all tubes have been processed for installation of the filter drier.

9. Install the hand valve and the charging hose to the compressor copper process tube.

NOTE: On "N"-Line compressors, silver solder a 4'' piece of $\frac{1}{4}''$ O.D. copper tubing into the steel process tube. Pressurize the system to 75 pounds per square inch with R-12 refrigerant. Leak test all low-side joints. Operate the compressor for a few minutes. Then leak test all high-side joints. Discharge, and evacuate system with a vacuum pump.

10. Close the hand valve. Remove the vacuum pump. Connect the charging cylinder to the hand valve. Purge the charging hose between the charging cylinder and the hand valve. Open the valve on the charging cylinder and allow liquid refrigerant to fill the charging hose up to the hand valve.

NOTE: Do not open the hand valve until the charging hose is full of liquid refrigerant and the amount of refrigerant in the charging cylinder has been recorded. Failure to follow this procedure results in an undercharged system.

11. Open the hand valve and charge the system. Then, close the hand valve. Refer to manufacturer's recommendations for the proper refrigerant charge.

12. Pinch off the copper process tube after the charge has been established. With the pinch-off tool on the process tube, remove the charging hand valve and the charging hose. Flatten the end of the tube and seal it with phos-copper. Then, remove pinch-off tool.

COMPRESSOR MOTOR BURNOUT

There are four major causes of motor burnout: low line voltage, loss of refrigerant, high head pressure, and moisture.

Low line voltage. When the motor winding in a motor gets too hot, the insulation melts and the winding short circuits. A blackened burned-out run winding is the result. Low line voltage causes the winding to get very hot because it is forced to carry more current at the same compressor load. When this current gets too high, or is carried for too many hours, the motor run winding fails. A burnout caused by low voltage is generally a slow burnout. This contaminates the system.

Loss of refrigerant. A second cause of motor burnout is loss of refrigerant. In a hermetically sealed motor compressor, the refrigerant vapor passes down around the motor windings. The cool refrigerant vapor keeps the motor operating at proper temperature. If there is a refrigerant leak, and there is little or no refrigerant to cool the motor, the windings become too hot. A burnout results. The overload protector may not always protect against this type of burnout since it requires the transfer of high heat from the motor through the refrigerant vapor to the compressor dome.

349

REFRIGERATION AND AIR CONDITIONING TECHNOLOGY

Kelvinator

13-6. Replacement filter drier.

High head pressure. High head pressure is a third cause of motor burnout. With high head pressure, the motor load is increased. The increased current causes the winding to overheat and eventually fail. Poor circulation of air over the high-side condenser can cause motor failure for this reason.

Moisture. The fourth major cause of motor burnout is moisture. It takes very little moisture to cause problems. In the compressor dome, refrigerant is mixed with lubricating oil, and heat from the motor windings and compressor operation. If there is any air present, the oxygen can combine chemically with hydrogen in the refrigerant and oil to form water. Just one drop of water can cause problems. When water contacts the refrigerant and oil in the presence of heat, hydrochloric or hydrofluoric acid is formed. These acids destroy the insulation on the motor winding. When the winding short circuits, a momentary temperature of over 3000° F. [1648°C] is created. Acids combine chemically with the insulation and oil in the compressor dome to create sludge. This quickly contaminates the refrigerating system. Sludge collects in various places throughout the system. It is very hard to dislodge. Purging the refrigerant charge or blowing refrigerant vapor through the system will not clean the system.

CLEANING SYSTEM AFTER BURNOUT

Remove the inoperative compressor and filter drier.

Flush the high side and low side of the system with R-12 liquid refrigerant. (Invert the refrigerant drum.)

Connect the high side and low side of the system. Also, connect the oil cooler tubes. Then, evacuate the system using a vacuum pump. Never use the new replacement compressor for this purpose. It will quickly become contaminated. Break the vacuum with refrigerant. Repeat the process. Then, and this is extremely important, repeat the process a third time. Thus, there are three purges and three evacuations.

Remove the vacuum pump and install a new replacement compressor and filter drier. Follow the procedures previously outlined. Use silver solder (Easy Flo-45) or phos-copper to make brazed connections.

REPLACING THE FILTER DRIER

If the compressor is not to be changed, follow these procedures to replace the filter drier.

1. To replace the filter drier, move the refrigerator to a location where the rear of the machine compartment is accessible.
2. Remove the machine compartment sound deadener baffle.
3. Cut the copper process tube on the compressor and bleed the refrigerant. Retain as much length as possible. Remove the filter drier by cutting $\frac{1}{4}''$ inlet tube 1″ from the brazed connection. Use a file to score the capillary tube uniformly approximately 1″ from the brazed joint at the filter drier. Then, break off the capillary tube.
4. Install the replacement filter drier with its inlet at the top. (Arrow indicates direction of flow.) Fig. 13-6. Braze the refrigerant tubes to the filter drier. Use silver solder (Easy Flo-45).

CAUTION: Do not remove caps from the replacement filter drier until all the refrigerant tubes have been processed for installation of the filter drier.

5. Install a process tube adapter kit. Fig. 13-7. If the adapter is not available, slip a $\frac{1}{4}''$ flare nut over the copper tube and flare the tube.
6. Install a hand valve and a charging hose. Connect the hose to the vacuum pump and evacuate the system. Fig. 13-8.
7. Shut off the vacuum pump and close the hand valve at the process tube adapter. Remove the vacuum pump and charging hose at the hand valve.
8. Connect a drum of R-12 refrigerant, or a charging cylinder, of the hand valve at the process tube adapter. Fig. 13-9. Purge the charging hose between the drum or charging cylinder and the hand valve. Open the valve on the drum, or charging cylinder. Allow the liquid refrigerant to fill the charging hose up to the hand valve at the process tube.

NOTE: Do not open the hand valve until the charging hose is full of liquid and the amount of refrigerant in the charging cylinder has been recorded. Failure to

350

CHAPTER 13—SERVICING

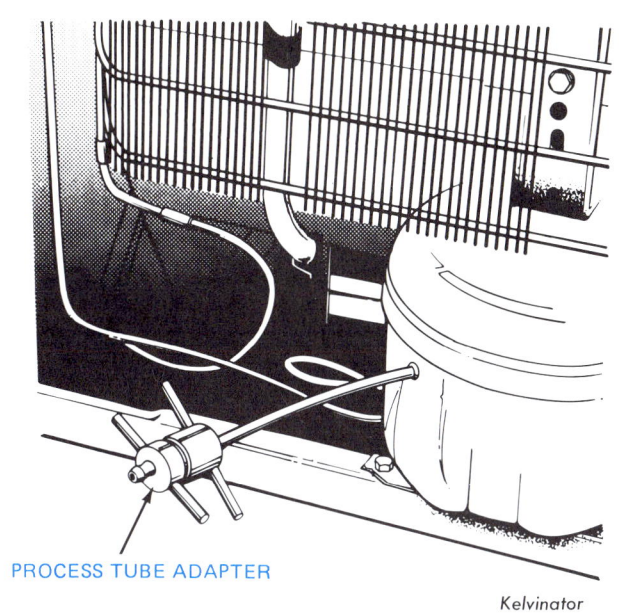

13-7. Process tube adapter.

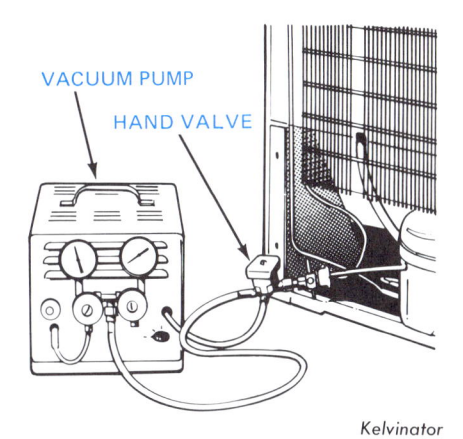

13-8. Vacuum pump and hand valve.

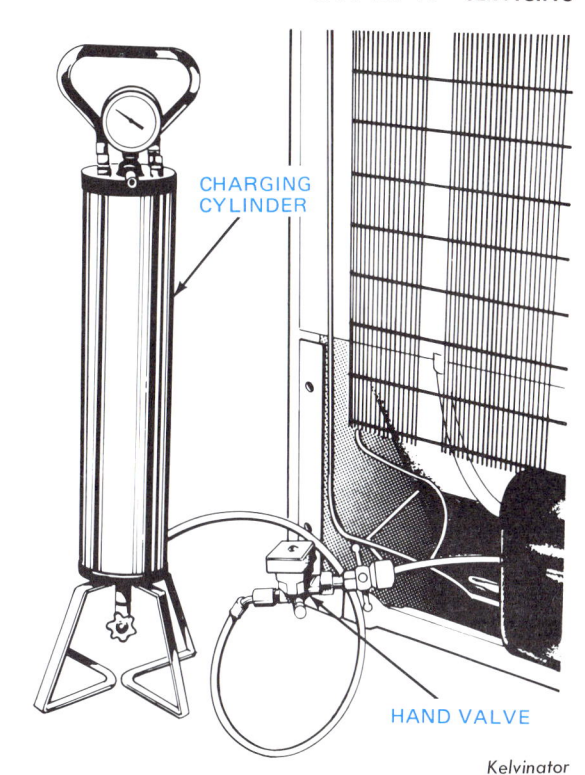

13-9. Charging cylinder.

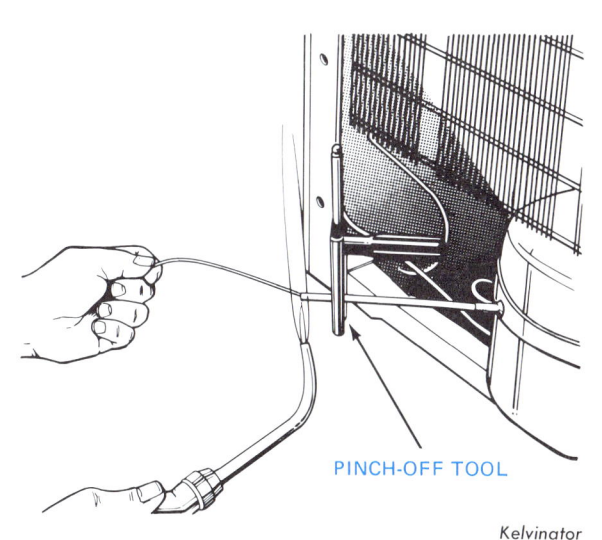

13-10. Brazing process tube with pinch-off tool in place.

follow this procedure results in an undercharge of refrigerant.

9. Open the hand valve and charge the system. Then, close the hand valve. If the system is charged from a refrigerant drum, operate the system until it has cycled to determine if the charge is proper. Refer to manufacturer's tables for operating pressures.

10. After the charge is established, pinch off the copper process tube with a pinch-off tool. With a pinch-off tool on the process tube, remove the charging adapter and hand valve. Seal the end of the tube with phos-copper. Then remove the pinch-off tool. Fig. 13-10.

351

REFRIGERATION AND AIR CONDITIONING TECHNOLOGY

TOP FREEZER MODEL REFRIGERATORS

SIDE BY SIDE REFRIGERATORS

Kelvinator

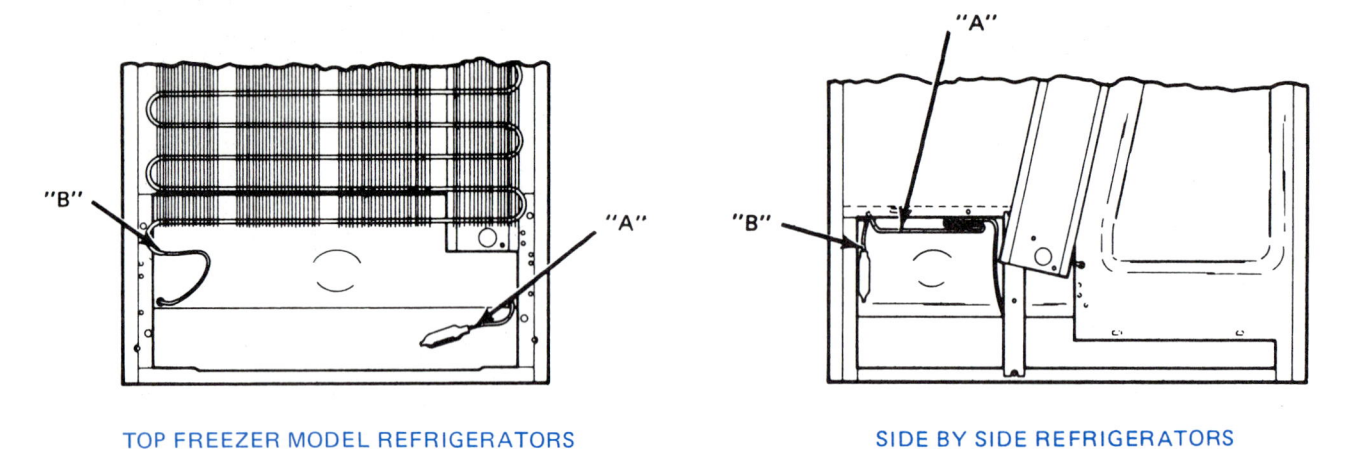

13-11. Repairing the perimeter tube. Cut and deburr the 3/16" O.D. perimeter tube at "A" and "B."

Kelvinator

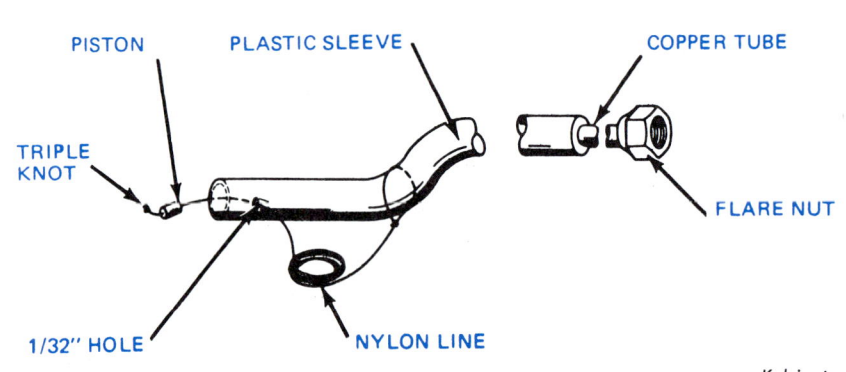

13-12. Repair tool made for repair of the perimeter tube.

REPLACING THE CONDENSER

To replace the condenser, bleed off the refrigerant as outlined under "Replacing the Filter Drier" (page 350). Cut the condenser and the filter drier inlet and outlet lines. Remove the condenser and the filter drier. Install a new condenser. Braze the refrigerant lines. Use silver solder (Easy Flo-45). Then, follow procedure for "Replacing the Filter Drier" (page 350).

REPLACING THE HEAT EXCHANGER

To replace the heat exchanger, follow the procedure given in "Replacing the Evaporator-Heat Exchanger Assembly" (page 356).

REPAIRING THE PERIMETER TUBE (FIBERGLASS INSULATED)

Top-Freezer and Side-by-Side Models

A perimeter tube, which is part of the refrigerating system, extends across the top and down both sides of the cabinet. Should a refrigerant leak develop in this tube, repairs are made as follows:

1. Use a tubing cutter. Cut and deburr the 3/16" O.D. perimeter tube at *A* and *B*. Fig. 13-11.

2. Use an ice pick to pierce a 1/32" diameter hole through the wall of the plastic sleeve 1 1/2" from the end. Thread one end of the nylon line through the pierced hole and piston. Tie a triple knot in the line. Loop the opposite end of the line around the plastic sleeve and tie. Fig. 13-12.

3. Insert the piston into the perimeter tube *A*. Slide the plastic sleeve onto the perimeter tube *A*. Insert the 3/16" O.D. copper tube. Flare the tube, and connect it to the refrigerant drum. Fig. 13-13.

4. Open the valve on the refrigerant drum. Blow the piston and the nylon line assembly through the perimeter tube.

5. Cut the nylon line that is looped around the plastic sleeve. Disconnect the refrigerant drum. Slide the plastic sleeve off the perimeter tube. Tie the nylon line to the electric drier coil (heater). Grasp the piston end of

352

CHAPTER 13—SERVICING

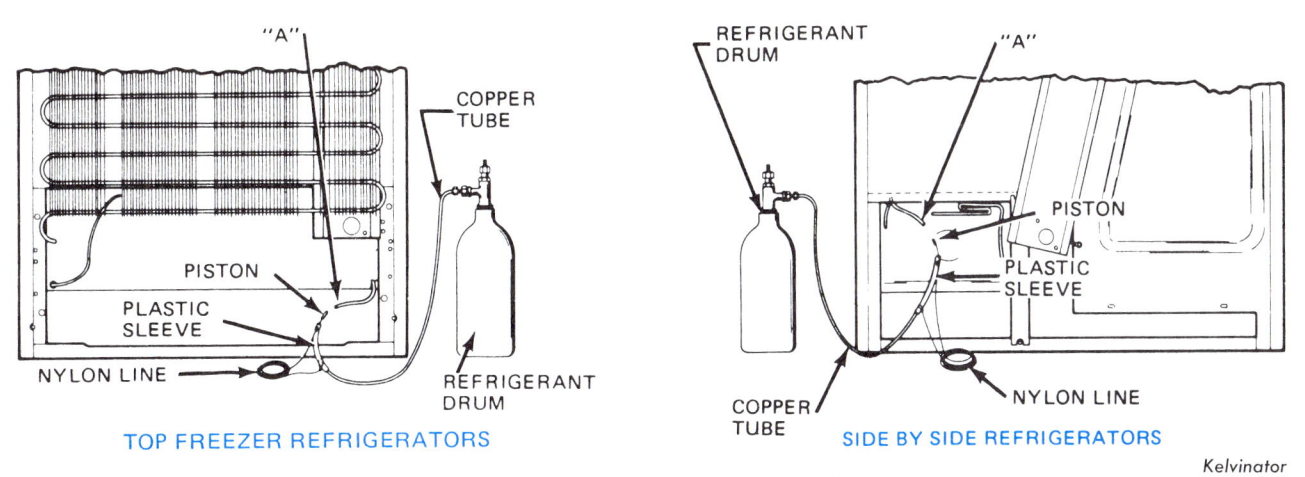

13-13. Repairing the perimeter tube.

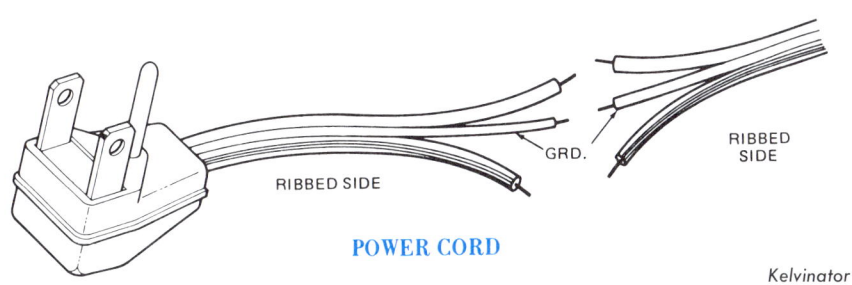

TOP FREEZER REFRIGERATORS

SIDE BY SIDE REFRIGERATORS

Kelvinator

POWER CORD

13-14. Splicing a new power cord plug onto the existing power cord line.

Kelvinator

SIDE BY SIDE REFRIGERATORS

Kelvinator

13-15. Power cord location and splices, side-by-side models.

13-16. Power cord splicing on top-freezer models.

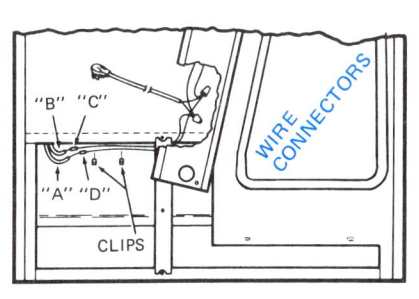

Kelvinator

the nylon line and pull the heater into the perimeter tube until the heater-lead connector on the opposite end rests against the end of the perimeter tube *A*. Remove the nylon line from the heater.

6. Swing the condenser aside (top freezer models). Remove the cabinet harness channel. Cut the power cord inside the channel about 3″ from the harness restraining grommet. On freezers and models where the power cord is not in the harness channel, cut the power cord 3″ inside the restraining strap.

Strip $5/8$″ of insulation off the power cord, as illustrated in Fig. 13-14. Splice one end of the white wire and both ends of the ribbed power cord together with wire connector. Splice one end of the black wire and both ends of the line side of the power cord together with a wire connector.

353

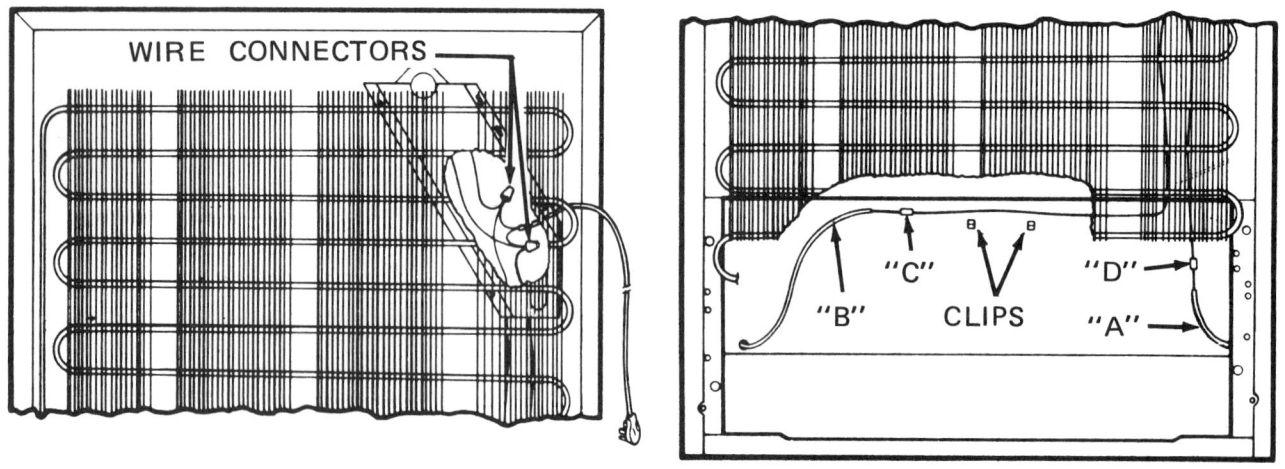

REFRIGERATION AND AIR CONDITIONING TECHNOLOGY

13-17. Another top-freezer refrigerator power cord splice.

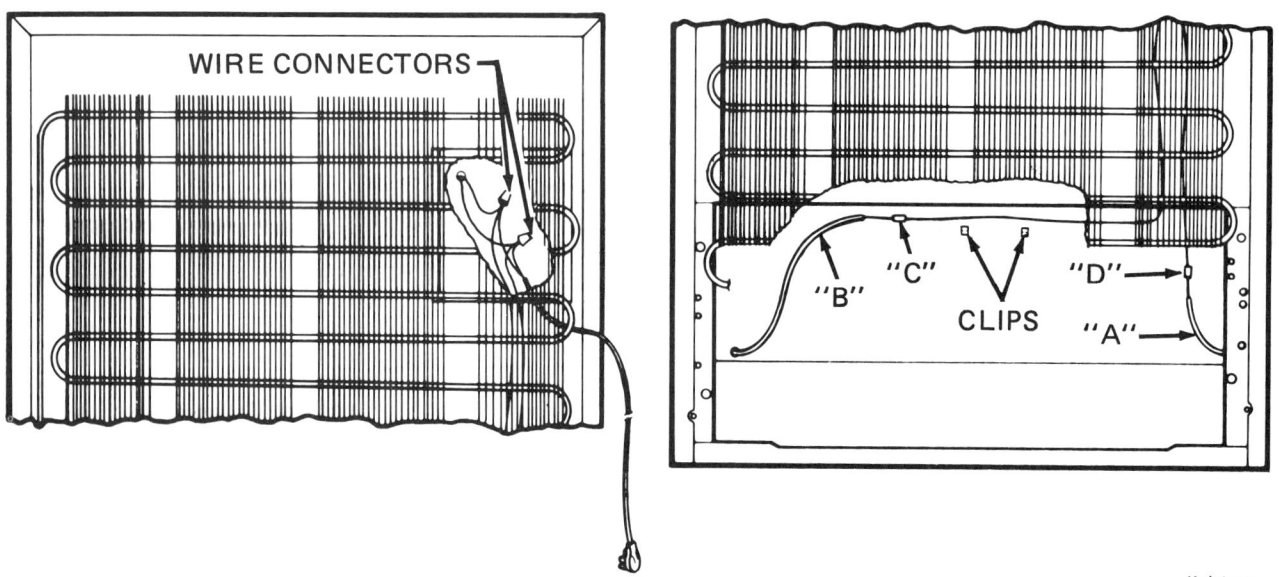

13-18. Top-freezer refrigerator cord splice location.

Figs. 13-15, 13-16, 13-17, and 13-18.

Connect the ground (green) wires together with a wire connector. Wrap the wire connectors with a piece of electrical tape. CAUTION: Ribbed ends of power cord must be spliced together to maintain polarity. Shape the perimeter tubes A and B as required for routing the heater leads along the flange of the cabinet to the harness channel.

Cut off the excess heater approximately 2" from the end of perimeter tube B.

NOTE: Exercise extreme care when stripping insulation to prevent damaging the heater resistance wire.

Cut the black and white wires to their proper length. Splice the white wire to the heater lead C and the black wire to heater lead D with wire connectors. Secure the leads to the cabinet flange with clips. Reinstall the harness channel.

7. Score the capillary tube and remove the filter drier. Use a

CHAPTER 13—SERVICING

TOP FREEZER MODEL REFRIGERATORS

SIDE BY SIDE REFRIGERATORS

Kelvinator

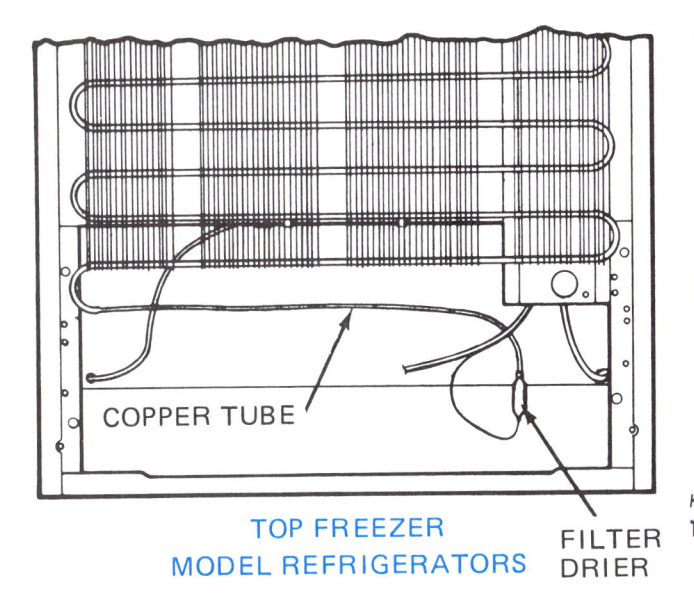

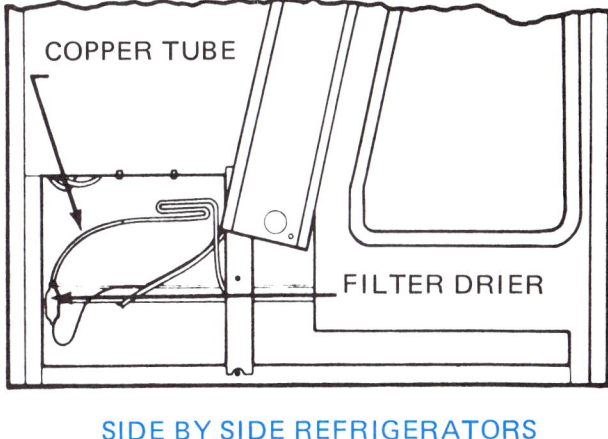

13-19. Side-by-side and top-freezer model refrigerator filter drier location.

copper tube and the filter drier to connect the condenser to the capillary tube. Fig. 13-19.

Foam-Insulated 12 and 14 Cubic-Foot Top-Freezer Models

1. Disconnect the service cord from the power supply.
2. Use a tubing cutter to cut and deburr the $\frac{3}{16}''$ O.D. hot tube at *A* and *B*. See Fig. 13-11.
3. Mount the auxiliary condenser on the bottom of the existing condenser, using bolts, nuts, and spacers.
4. Use silver solder to connect the refrigerant lines, that were removed from the hot tube, to the auxiliary condenser lines. NOTE: Refrigerant must enter the top of the auxiliary condenser.
5. Install a replacement drier. Evacuate and recharge the system using the nameplace charge.
6. Use an ice pick to pierce a $\frac{1}{32}''$ diameter hole through the wall of the plastic sleeve $1\frac{1}{2}''$ from the end. Thread one end of the nylon line through the pierced hole and piston. Tie a triple knot in the line. Loop the opposite end of the line around the plastic sleeve and tie. See Fig. 13-12.
7. Insert the piston into the hot tube *A*. Slide the plastic sleeve onto the hot tube *A*. Insert the $\frac{3}{16}''$ O.D. copper tube into the opposite end of the plastic sleeve. Install a flare nut on the copper tube and flare the tube. Connect to the refrigerant drum. Fig. 13-20.
8. Open the valve on the refrigerant drum. Blow the piston and nylon line assembly through the hot tube. (Use vapor pressure.)
9. Cut the nylon line that is looped around the plastic sleeve. Disconnect the refrigerant drum. Slide the plastic sleeve off the hot tube. Tie the nylon line to the electric heater. Grasp the piston end of the nylon line and pull the heater into the hot tube until the heater lead connector on the opposite end rests against the end of hot tube *A*. Remove the nylon line from the heater.
10. Remove the wiring cover from the back of the cabinet. Disconnect the power cord from the cabinet harness. Connect the adapter harness to the cabinet harness and power cord. See Fig. 13-19.
11. Shape the hot tubes *A* and *B* as required for routing the heater leads along the flange of the cabinet to the suction line. Cut the excess heater approxi-

13-20. Using vapor pressure from refrigerant drum to blow the piston and nylon line assembly through the hot tube.

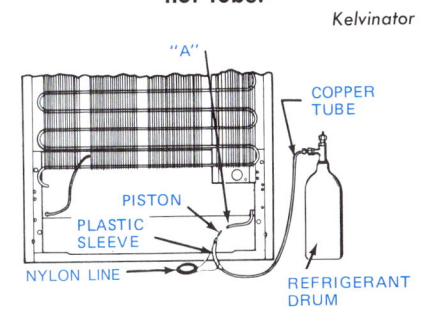

Kelvinator

355

mately 2″ from the end of the hot tube *B* and strip the insulation back ½″.

NOTE: Exercise extreme care when stripping insulation to prevent damaging the heater resistance wire.

12. Cut the adapter harness and heater lead to the proper length. Connect the adapter harness to the heater with connectors. Secure the leads to the suction line. Reinstall the wiring cover. See Fig. 13-19.

Foam-Insulated 19 Cubic-Foot Side-by-Side Models

1. Disconnect the service cord from the power supply.
2. Use a tubing cutter to cut and deburr the 3/16″ O.D. hot tube *A* and *B*. See Fig. 13-11.
3. Mount the auxiliary condenser on the bottom of the existing condenser, using bolts, nuts, and spacers.
4. Use silver solder to connect the refrigerant lines removed from the hot tube to the auxiliary condenser lines.

NOTE: Refrigerant must enter the top of the auxiliary condenser.

5. Install the replacement drier. Evacuate and recharge the system using the nameplate charge.
6. Remove the mullion to gain access to the mullion heater.
7. Cut 2″ from the bottom of the mullion hot tube at the U-bend. Deburr the remaining tubes *C* and *D*. See Fig. 13-21.
8. Use an ice pick to pierce a 1/32″ diameter hole through the wall of the plastic sleeve 1½″ from the end. Thread one end of the nylon line through the pierced hole and piston. Tie a triple knot in the line. Loop the opposite end of the line around the plastic sleeve and tie. See Fig. 13-12.

9. Insert the piston into hot tube *A*. Slide the plastic sleeve onto hot tube *A*. Insert the 3/16″ O.D. copper tube into the opposite end of the plastic sleeve. Install a flare nut on the copper tube and flare tube. Connect to the refrigerant drum. See Fig. 13-20.

10. Open the valve on the refrigerant drum. Blow the piston and the nylon line assembly through the hot tube to open the mullion hot tube *C*. Use vapor pressure.

11. Cut the nylon line looped around the plastic sleeve. Disconnect the refrigerant drum. Slide the plastic sleeve off the hot tube. Tie the nylon line to the electric heater. Grasp the piston end of the nylon line. Pull the heater into the hot tube until the heater lead connector on the opposite end rests against the end of hot tube *A*. Remove the nylon line from the piston.

12. Start at the remaining open mullion hot tube *D*. Repeat steps 8 through 11 to install the heater in the remaining part of the hot tube. Leave the free loop of wire at the end of the mullion heater.

13-21. Where to cut the hot gas tube for repair.
Kelvinator

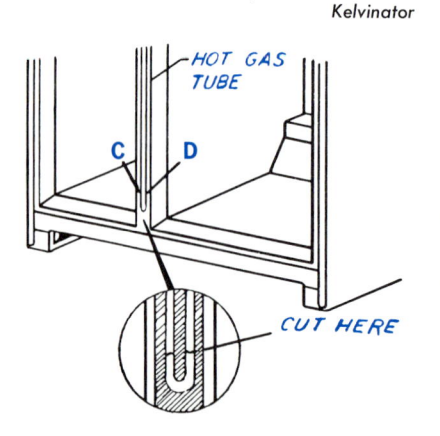

13. Reinstall the mullion.
14. Remove the wiring cover from the back of the cabinet harness. Disconnect the power cord from the cabinet. Connect the adapter harness to the cabinet harness and the power cord. See Fig. 13-22.

15. Shape hot tubes *A* and *B* as required for routing the heater leads along the flange of the cabinet to the suction line. Cut the excess heater approximately 2″ from the end of the hot tube *B* and strip the insulation back ½″.

NOTE: Exercise extreme care when stripping insulation to prevent damaging the heater resistance wire.

16. Cut the adapter harness and the heater lead to their proper lengths. Connect the adapter harness to the heater with connectors. Secure the leads to the suction line. Reinstall the wiring cover. See Fig. 13-22.

REPLACING THE EVAPORATOR-HEAT EXCHANGER ASSEMBLY

Top-Freezer, No-Frost Models

When an evaporator-heat exchanger assembly develops a refrigerant leak and the compressor has operated after the refrigerant escaped, air and moisture have entered the system. To protect the system, it must be flushed with liquid refrigerant and evacuated. A replacement filter drier must be installed in conjunction with the evaporator-heat exchanger assembly.

To replace the evaporator-heat exchanger assembly, move the refrigerator to a location where the front and rear are accessible. Remove the machine compartment cover on models so equipped. Then, follow steps (1) and (4) in the discussion of "To Replace the Compressor" (page 349). Use a tubing cutter to cut suction tube at the compressor. Swing the condenser aside. Disconnect the defrost heater and defrost termination thermostat leads from the cabinet harness. Bend the heat-exchanger upward. Remove the sealer from the tubing harness entrance hole. Remove the evaporator cover mounting screws and lay the cover on the bottom of the freezer compartment. Remove the evaporator mounting screws and RH and LH air barriers. Pull the evaporator-heat exchanger assembly forward out of the freezer compartment. Transfer the defrost heater and defrost termination thermostat to the replacement evaporator. Install the replacement evaporator-heat exchanger assembly. Install the evaporator RH and LH air barriers and evaporator cover. Press the sealer firmly into the tubing-harness entrance hole. Connect the defrost heater and termination thermostat leads to the harness. Swing the condenser into place and secure. Braze the suction tube to the compressor. Use silver solder (Easy Flo-45). Then, follow steps (4) through (10) in "Replacing the Filter Drier" (page 350). Install the machine compartment cover on models so equipped.

Side-by-Side Models

When an evaporator-heat exchanger assembly develops a refrigerant leak and the compressor has operated after the refrigerant escaped, air and moisture have entered the system. To protect the system, it must be flushed with liquid refrigerant and evacuated. A replacement filter drier must be installed in conjunction with the evaporator-heat exchanger assembly.

To replace the evaporator-heat exchanger assembly, move the refrigerator to a location where the front and rear are accessible. Remove the machine compartment cover. Then, follow steps (1) and (4) in "To Replace the Compressor." Use a tubing cutter to cut the suction tube at the compressor. Disconnect the defrost termination thermostat leads from the cabinet harness. Bend the heat-exchanger upward. Remove the sealer from the tubing-harness entrance hole. Remove the shelves, shelf supports, evaporator cover, and evaporator mounting screws. Pull the evaporator-heat exchanger assembly forward from the freezer compartment. Transfer the defrost heater and defrost termination thermostat to the replacement evaporator. Install the replacement evaporator-heat exchanger assembly. Connect the defrost heater and defrost termination thermostat leads to the cabinet harness. Install the sealer firmly into the tubing-harness entrance hole from the front and rear. Install the evaporator cover, shelf supports, and shelves. Braze the suction tube to the compressor. Use silver solder (Easy Flo-45). Then, follow steps (4) through (10) in "Replacing the Filter Drier." Install the machine compartment cover.

ADDING REFRIGERANT

CAUTION: Always introduce refrigerant into the system in a vapor state.

When the operation of a system indicates that it is short of refrigerant, it must be assumed that there is a leak in the system. Proceed to test the system with a leak detector. When the leak is located, it should be repaired. First, however, you must determine if the leak is repairable.

Unless the system has lost most of its refrigerant, the leak test can be made without the addition of extra refrigerant. If the system is completely out, sufficient refrigerant must be added to make a leak test. A new filter drier must be installed.

LOW-SIDE LEAK OR SLIGHT UNDERCHARGE

If a slight undercharge of refrigerant is indicated, without a leak being found, the charge can be corrected without changing the compressor.

In the case of a low-side refrigerant leak, resulting in complete loss of refrigerant, the compressor will run. However, there will be no refrigeration. Suction pressure will drop below atmospheric pressure. Air and moisture are

13-22. Location of the leads and power cord.
Kelvinator

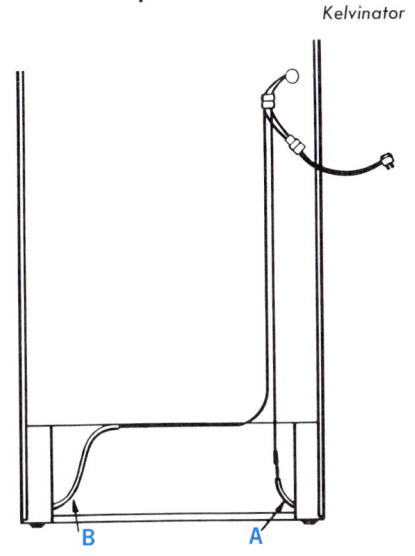

REFRIGERATION AND AIR CONDITIONING TECHNOLOGY

drawn into the system, saturating the filter drier.

It is not necessary to replace the compressor. The leak should be repaired. The system should be flushed with liquid R-12 refrigerant. A replacement filter drier should be installed. The system should be excavated and recharged.

The system may have operated for a considerable length of time with no refrigerant and the leak may have occurred in the evaporator. In this case, excessive amounts of moisture may have entered the system. In such cases the compressor may need to be replaced to prevent repetitive service.

HIGH-SIDE LEAK OR SLIGHT UNDERCHARGE

It is not necessary to change a compressor when a leak is found in the system. If a slight undercharge of refrigerant is indicated, without a leak being found, the charge can be corrected without changing the compressor.

It is recommended that the system be flushed with liquid refrigerant and evacuated. A replacement filter drier should be installed to protect the system against moisture.

OVERCHARGE OF REFRIGERANT

When the cabinet is pulled down to temperature, an indication of an overcharge is that the suction line will be colder than normal. The normal temperature of the suction line will be a few degrees cooler than room temperature. If its temperature is much lower than room temperature, the unit will run longer because the liquid is pulled into the heat exchanger. When the overcharge is excessive, the suction line will sweat or frost.

TESTING FOR REFRIGERANT LEAKS

If the system is diagnosed as short of refrigerant and has not been recently opened, there is probably a leak in the system. Adding refrigerant without first locating and repairing the leak will not permanently correct the difficulty. *The leak must be found.* Sufficient refrigerant may have escaped to make it impossible to leak test effectively. In such cases, add a $\frac{1}{4}''$ line piercing valve to the compressor process tube. Add sufficient refrigerant to increase the pressure to 75 pounds per square inch. Through this procedure, minute leaks are more easily detected before discharging the system and contaminating the surrounding air. NOTE: The line-piercing valve (clamp-on type) should be used only for adding refrigerant and for test purposes. It must be removed from the system after it has served its purpose. Braze-on type line-piercing valves may be left on the process tube to evacuate the system and recharge after repairs are completed.

Various types of leak detectors are available. Liquid detectors (bubbles), halide torches, halogen-sensing electronic detectors, and electronic transistor pressure-sensing detectors are used.

You can sometimes spot a leak by the presence of oil around it. To be conclusive, however, use a leak detector.

Liquid detectors (bubbles) can be used to detect small leaks in the following manner. Brush liquid detector over the suspected area and watch for the formation of bubbles as the gas escapes. If the leak is slight, you may have to wait several minutes for a bubble to appear. CAUTION: *Use the bubble method only when you are sure that the system has positive pressure. Using it where a vacuum is present could pull liquid detector into the system.*

When testing with the halide torch, be sure the room is free from refrigerant vapors. Watch the flame for the slightest change in color. A very faint green indicates a small leak. The flame will be unmistakably green to purple when large leaks are encountered. To simplify leak detection, keep the system pressurized to a minimum of 75 pounds per square inch.

For more sensitive testing, use an electronic leak detector. Halogen-sensing electronic detectors can detect minute refrigerant leaks, even though the surrounding air may contain small amounts of refrigerant.

An electronic, transistorized, pressure-sensing detector does not require that the system be pressurized with R-12 refrigerant. Dry air or nitrogen may be used to pressurize the system. The escaping pressure through a minute opening is detected.

In "Urethane" froth foam insulated models, R-11 refrigerant (halogenated hydrocarbons), is used as a blowing agent in the foaming process. Molecules of R-11 refrigerant are encased within the cellular formation of the insulation and between the inner and outer wall of the cabinet. Therefore, when checking for refrigerant leaks in these models with a halide torch or halogen electronic leak detector, a false indication of a refrigeration leak will be experienced where no leak actually exists.

This is particularly true where the cellular formation is disturbed or broken by moving the refrigerant lines or probing into the insulation. The electronic, transistorized, pressure-sensing detector is not affected by the presence of refrigerants in the air. Where refrigerant tubes are encased within the foam insulation, there is one continuous tube. All brazed joints are accessible for checking leaks.

A joint suspected of leaking can be enclosed in an envelope of cellophane film. Tightly tape both ends and any openings to make it gas-tight. Fig. 13-23. After about an hour, you can pierce one end of the film for your probe and pierce the other end for air to enter. If you get a response, the joint should be rebrazed. The component should be replaced if a leak is found at the aluminum-to-copper butt-weld joint.

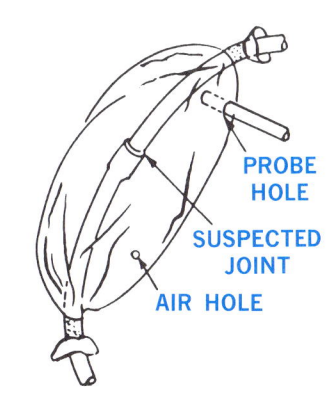

13-23. Leak detection envelope.

SERVICE DIAGNOSIS

To service refrigeration equipment properly, the serviceperson must possess the following:

- A thorough understanding of the theory of refrigeration.
- A good working knowledge of the purpose, design, and operation of the various mechanical parts of the refrigerator.
- The ability to diagnose and correct any trouble that may develop.

On the initial contact, always allow the customer to explain the problem. Many times the trouble can be diagnosed more quickly through the customer's explanation. Most of all, do not jump to conclusions until you have evaluated the information obtained from the customer. Then, proceed with your diagnosis.

Before starting a test procedure, connect the refrigerator service cord to the power source, through a wattmeter, combined with a voltmeter. Then, make a visual inspection and operational check of the refrigerator to determine the following.

- Is the refrigerator properly leveled?
- If the refrigerator is a static condenser model, is it located for proper dissipation of heat from the condenser? Check recommended spacing from rear wall and clearance above cabinet.
- Feel the condenser. With the compressor in operation, the condenser should be hot, with a gradual reduction in temperature from the top to the bottom of the condenser.
- Are door gaskets sealing on pilaster area?
- Does the door, PC or FC, actuate the light switch? (PC is the abbreviation for provisions compartment; FC is the abbreviation for freezer compartment.)
- Is FC fan guard in place?
- Is FC fan properly located on motor shaft?
- Is the thermostat thermal element properly positioned? The thermal element must not contact the evaporator.
- Observe the frost pattern on the evaporator.
- Check the thermostat knob setting.
- Check the air-damper control knob setting.
- Inscribe bracket opposite slotted shaft of defrost timer to determine if timer advances.
- Is condenser fan motor operating? NOTE: Condenser fan motor operates only when compressor operates.
- Are air ducts free of obstructions?

The service technician should inquire regarding the number of people in the family. This will help determine the service load and daily door openings. In addition, he or she should know the room temperature.

After this phase of diagnosis is completed, a thorough operational check should be made of the refrigeration system. Any components not previously checked should be checked in the following order.

Thermostat Cut-Out and Cut-In Temperatures

To check the cut-out and cut-in temperatures of the thermostat, use a refrigeration tester or a recording meter. Attach a test bulb to the thermal element on the top-freezer and side-by-side "no-frost" models. Replace the evaporator cover before starting the test.

When using a refrigeration tester equipped with several bulbs, place a #2 bulb in FC air (center) and a #3 bulb in PC air (center).

Allow the system to operate through a complete cut-out and cut-in cycle, with the thermostat set on middle position.

For accurate reading or recording, the temperature of the thermostat thermal element must not be reduced more than 1° F. [0.5°C] per minute through the final 5° F. [2.7°C] prior to cut-out.

Erratic operation of the thermostat will affect both the FC and PC air temperature.

Freezer and Provision Compartment Air Temperatures

Freezer and provision compartment temperatures are affected by the following:
- Improper door seal.
- Frost accumulation on the PC and the FC evaporators.
- Service load.
- Ambient temperature.
- Percentage of relative humidity.
- Thermostat calibration (cut-in and cut-out).
- Location of the FC fan blade on motor shaft.
- Compressor efficiency.

From this, it is evident that temperatures are not always the same in every refrigerator, even under identical conditions. However, an average FC air temperature of 0 ± 6° F. [−17.8 ± 3.3°C] with a PC air temperature of 36° F. [2.2°C] to 42° F. [5.6°C], should be obtained.

Freezer Compartment Too Warm

Cause
1. Inoperative fan motor.
2. Improperly position fan.
3. Evaporator iced up.
4. Defrost heater burned out.
5. Inoperative defrost timer.
6. Inoperative defrost temperature termination thermostat.
7. Wire loose at defrost timer.
8. Fan guard missing.
9. Excessive service load.
10. Abnormally low room temperatures.
11. FC or PC door left open.
12. Thermostat out of calibration.
13. PC or FC door gasket not sealing. Check with 100-W lamp.
14. Thermostat thermal element touching the evaporator.
15. Inoperative condenser fan motor.
16. Shortage of refrigerant (side-by-side models only).
17. Restricted filter drier or capillary tube.

Provision Compartment Too Warm

Cause
1. Inoperative fan motor.
2. Improperly positioned fan.
3. Fan guard missing.
4. PC air inlet air duct restricted.
5. PC to FC return air duct restricted.
6. Air-flow control on *warmer* position.
7. Thermostat out of calibration.
8. Thermostat knob set at warm setting.
9. Thermostat thermal element touching the evaporator.
10. Evaporator iced up.
11. Inoperative defrost timer.
12. Inoperative defrost heater.
13. Inoperative defrost temperature termination thermostat.
14. Loose wire at defrost timer.
15. Excessive service load, resulting from too much food in compartment.
16. Inoperative condenser in the fan motor.
17. PC or FC door left open.
18. Inoperative or erratically opening FC and/or FC door switch.
19. Shelves covered with foil wrap or paper, retarding air circulation.
20. Restricted capillary tube or filter drier.

Evaporator Blocked with Ice

Cause
1. Inoperative defrost timer.
2. Defrost timer terminates too early.
3. Defrost timer incorrectly wired. Check wiring.
4. Inoperative fan motor.
5. Inoperative termination thermostat.
6. Inoperative defrost heater.
7. PC or FC door left open.
8. FC drain plugged. Clean.
9. FC drain sump or drain trough heater burned out. Replace.

Line Voltage

It is essential to know the line voltage at the appliance. A voltage reading should be taken the instant the compressor starts and while the compressor is running. Line voltage fluctuation should not exceed 10%, plus or minus, from nominal rating. Low voltage will cause overheating of the compressor motor windings. This will result in compressor cycling on thermal overload, or the compressor may fail to start.

Inadequate line wire size and overloaded lines are the most common reasons for low voltage at the appliance.

Wattage

Wattage is a true measure of power. It is the measure of the rate at which electrical energy is consumed. Therefore, wattage readings are useful in determining compressor efficiency, proper refrigerant charge, and the presence of a restriction. They also help detect the malfunction of an electrical component.

Amperes, measured with an Amprobe®, multiplied by the voltage is not a true measurement of power in an alternating current (AC) circuit. It gives only "volt-amperes" or "apparent power." This value must be multiplied by the power factor (phase angle), to obtain the true or actual (AC) power. The actual power is indicated by a wattmeter.

$$\text{Watts} = \text{Volts} \times \text{Amperes} \times \text{Power Factor}$$

or

$$\text{Power Factor} = \frac{\text{Watts (W)}}{\text{Volts} \times \text{Amperes (VA)}} = \frac{\text{Actual Power}}{\text{Apparent Power}}$$

Thus, the power factor may be expressed as the ratio of the actual watts to the apparent watts. The apparent watts is the product of the amperes and volts as indicated by an ammeter and a voltmeter. The power factor varies from 0 to 1.00 (or 100%). On resistance heaters such as the drier coil and drain heater, the actual watts are equal to the amperes multiplied by the volts. Thus, the power factor is 100% or 1.00. On electric motors, because of their magnetic reaction, the actual watts are not equal to amperes multiplied by volts. This means the power factor is less than 1.00. For this reason, a wattmeter should be used.

Compressor Efficiency

A low-capacity pumping compressor causes excessive or continuous compressor operation, depending on the ambient temperature and service load. Recovery of cabinet temperature will be slow. If cycling does occur, wattage will generally be below normal. Condenser temperature will be near normal.

Refrigerant Shortage

A loss of refrigerant results in the following:
- Excessive or continuous compressor operation.
- Above-normal PC temperature.
- A partially frosted evaporator (depending on amount of refrigerant lost).
- Below-normal FC temperature.
- Low suction pressure (vacuum).
- Low wattage. The condenser will be "warm to cool" again, depending on the amount of refrigerant lost.

When refrigerant is added, the frost pattern will improve. The suction and discharge pressures will rise. The condenser will become hot. The wattage will increase.

The refrigerator should then be turned off and thoroughly leak tested.

It is not always necessary to change a compressor when a leak is found in the system. If a slight undercharge of refrigerant is indicated, without a leak being found, the charge can be corrected without changing the compressor.

It is recommended, however, that the filter drier be replaced to protect the system against moisture. This is essential if all the refrigerant has leaked out or if moisture may have entered the system. Refer to "Replacing the Filter Drier" (page 350).

Restrictions

Restrictions are classified as follows: total or partial, as a result of foreign matter, oil, or moisture in the capillary tube or drier.

A permanent or total restriction completely stops the flow of refrigerant through the system. The result is continuous compressor operation, low wattage, low suction pressure (vacuum), and a cool condenser. A cool condenser indicates liquid refrigerant is trapped in the condenser.

A partial restriction results in the following:
- A partially frosted evaporator.
- Excessive or continuous compressor operation.
- Above-normal PC temperature.
- Below-normal FC temperature.

REFRIGERATION AND AIR CONDITIONING TECHNOLOGY

- Low wattage.
- Low suction pressure (vacuum), depending on the amount of restriction.
- The lower (or outlet) one-half or one-third of the condenser will be cool. Such coolness indicates liquid refrigerant trapped in the condenser.

To make sure that the trouble is a partial restriction and not a shortage of refrigerant, cover the condenser. Allow the compressor to operate to increase the discharge pressure and temperature. (Note the increase in wattage.) The increase in discharge pressure will force refrigerant through the restricted area. Frosting of the evaporator will occur. The frost pattern will not improve with a shortage of refrigerant.

A total or partial "moisture" restriction always occurs at the outlet of the capillary tube. If moisture is suspected, turn the refrigerator "off." Allow all system temperatures to rise above 32° F. [0°C] or manually initiate a defrost cycle. If moisture is present, the restriction will be released. The system should be discharged. A replacement drier should be installed and the system evacuated and recharged.

Defrost Timer Termination

Manually initiate a defrost cycle. Do this by turning the slotted shaft of the timer clockwise. NOTE: Rotate the timer shaft slowly into the defrost cycle or part of the defrost timer will be missed. If the length of the defrost cycle is not in accord with specifications, change the defrost timer.

Computing Percent Run Time

That period of operation between the cut-in and cut-out points is called the *pulldown* cycle, on-cycle, or running cycle.

The period of time between the cut-out and cut-in point is called the *warm up* cycle, or off-cycle.

A complete cycle of operation is equal to the on-cycle plus the off-cycle. Such a cycle usually is timed in minutes.

The percent running time is computed by the following formula:

$$\frac{\text{On-Cycle Time}}{\text{On-Cycle Time} + \text{Off-Cycle Time}} \times 100 = \text{Percent Running Time}$$

START AND RUN CAPACITORS

Capacitor Ratings

Never use a capacitor with a lower rating than specified on the original equipment. The voltage rating and the microfarad rating are important. A higher voltage rating than that specified is always usable. However, a voltage rating lower than that specified can cause damage. Make sure the capacitance marked on the capacitor in MFD, or microfarads, is as specified. Replace with a capacitor of the *same* size in μF, MF, UF, or MFD. All these abbreviations are used to indicate microfarads.

Start Capacitor and Bleeder Resistors

The development of high power factor, low-current, single-phase compressor motors that require start and run capacitors used with potential type relays created electrical peculiarities. These did not exist in previous designs.

In some situations, relay contacts may weld together, causing compressor motor failure.

This phenomenon occurs due to the high voltage in the start capacitor discharging (arcing) across the potential relay contacts. To eliminate this, start capacitors are equipped with bleeder resistors across the capacitor terminals. Fig. 13-24.

Bleeder resistor equipped capacitors may not be available. Then, a 2-W 15,000-ohm resistor can be soldered across the capacitor terminals.

If the relay solenoid opens, the start capacitor is left in the circuit too long. Normally, it is in the circuit less than ten seconds. If it stays longer, it is subject to excessive heat buildup. It will spray its contents on the equipment nearest it. It is a good idea to mount the capacitor where it will cause little damage if it does malfunction.

Run Capacitors

The marked terminal of run

13-24. Bleeder resistor across the capacitor terminals.
Tecumseh

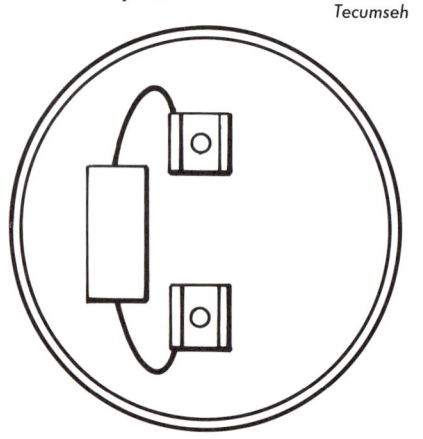

CHAPTER 13—SERVICING

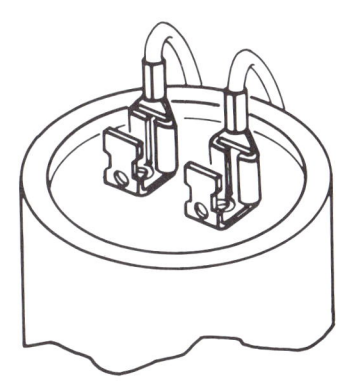

QUICK—CONNECT

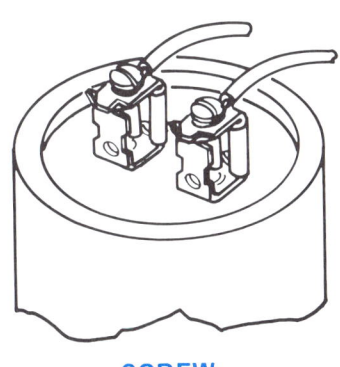

SCREW

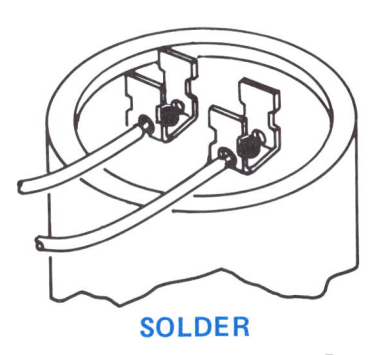

SOLDER

Tecumseh

13-25. There are three ways to attach terminals to an electrolytic capacitor.

capacitors should be connected to the "R" terminal of the compressor and thus to L_2. Check the wiring diagram for the correct terminal.

The run capacitor is in the circuit whenever the compressor is running. It is an oil-filled electrolytic capacitor that can take continuous use. The start capacitor is a dry type. It has in it a substance that can react quickly if too long in the circuit. The oil-filled type is a wet electrolytic. It will take longer circuit use.

There are at least three ways of attaching leads to the terminals of electrolytic capacitors. Fig. 13-25.

PERMANENT SPLIT-CAPACITOR (PSC) COMPRESSOR MOTORS

The permanent split-capacitor (PSC) motor eliminates the need for potentially troublesome and costly extra electrical components. Start capacitors and potential motor starting are needed for capacitor start motors.

Conditions that affect the PSC motor starting include the following:

• *Low voltage reduces motor starting and running torque.* Torque varies as the square of the voltage. Low voltage can prevent starting, cause slow starting, light flicker, and TV screen flip-flop.

The minimum voltage required to start a 230-volt or 230/208-volt PSC compressor is 200 volts locked rotor (LRV) measured at the compressor terminals. This cannot be measured accurately after the compressor starts. It can be measured only when on locked rotor.

• *Circuit breaker or fuse trips.* Branch circuit fuses or circuit breakers sized too small will cause nuisance tripping, incorrectly diagnosed as compressor "no start."

• *Unequalized system pressure.* The maximum equalized pressure against which a PSC compressor is designed to start is 170 psig. System pressure may not be equalized within the three-minute design limitation due to improper refrigerant metering device, excessive refrigerant charge, and rapid cycling of room thermostat.

Because of these, the compressor will not start.

• *Starting load too great.* A number of conditions can cause too great a starting load on a PSC compressor motor. System refrigerant charge may be excessive. Liquid refrigerant may have migrated to the compressor and formed a high liquid level in the crankcase.

Figure 13-26 shows how the run capacitor is wired into the compressor electrical system. Note how the run capacitor is in series with the start winding and in parallel with the main winding. The snap-acting disc opens the circuit when too much current is drawn through the windings. The heater provides the heat needed to cause quick action of the snap-acting disc.

The charts shown in Fig. 13-27 give wire size, locked rotor volts, and circuit breaker or fuse size requirements for compressors.

Table 13-B lists PSC compressor motor troubles and corrections.

FIELD TESTING HERMETIC COMPRESSORS

Before a hermetic compressor is returned to the source, it must be tested. A replacement service charge is made on all in-warranty and out-of-warranty compressors and units that are returned when no defect is found. To avoid this unnecessary expense and loss of time, all compressors must be tested before they are returned for repair.

REFRIGERATION AND AIR CONDITIONING TECHNOLOGY

Most authorized wholesalers have a compressor test stand. The following equipment is needed:

- Variable high-voltage transformer (high potential) for checking for grounds. GE Model 9892115G1 or equal is suggested.
- Capacitor check and analyzer for checking start and run capacitors for shorts, opens, intermittents, and capacitance. A suggested model (Model M-1, Mike-o-Meter) is manufactured by Sprague Products Company, North Adams, Massachussets.
- Voltmeter capable of measuring the three voltages shown below.

    a. 115-volt, 60-hertz, single-phase.

    b. 230-volt, 60-hertz, single-phase.

    c. 220-volt, 60-hertz, three-phase.

- Two sets of service test cords.

    a. One 12-gage, two-conductor, stranded copper, insulated test cord of suitable length with alligator test clips on one end for testing a single-phase compressor.

    b. One 12-gage, three-conductor, stranded copper insulated test cord of suitable length with alligator clips on one end for testing a three-phase compressor.

- One ammeter having a range of 40 amperes adjustable from 0–10, 0–20, and 0–40 amperes for 220 volts, single-phase usage.
- One by-pass line, including connector tube clamps for discharge and suction line, 200 psi pressure gage and control valves as shown in Fig. 13-28.
- Hose adapters.
- Four start capacitors having the following ratings:

    a. 100 μF @ 125-volts AC

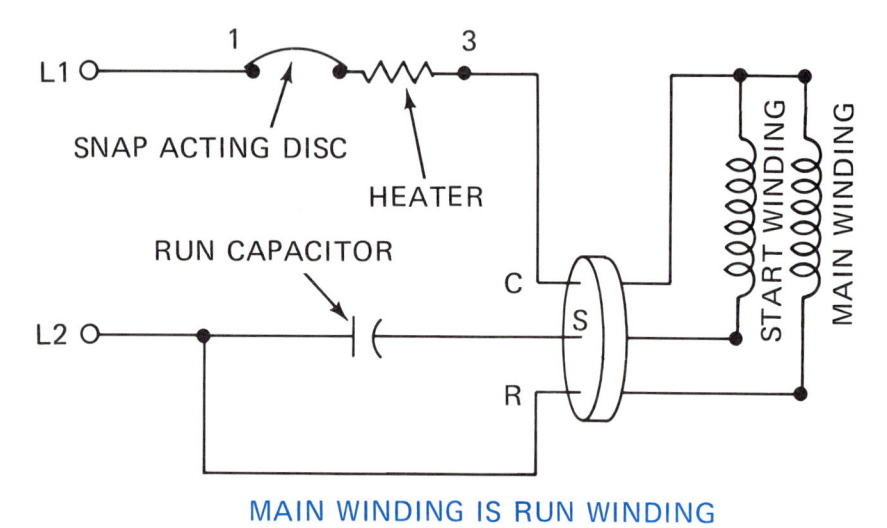

13-26. The PSC motor eliminates the need for potentially troublesome and costly extra electrical components such as start capacitors and potential motor starting relays.

Tecumseh

13-27. Wire size, locked rotor volts, and circuit breaker or fuse size requirements for compressors.

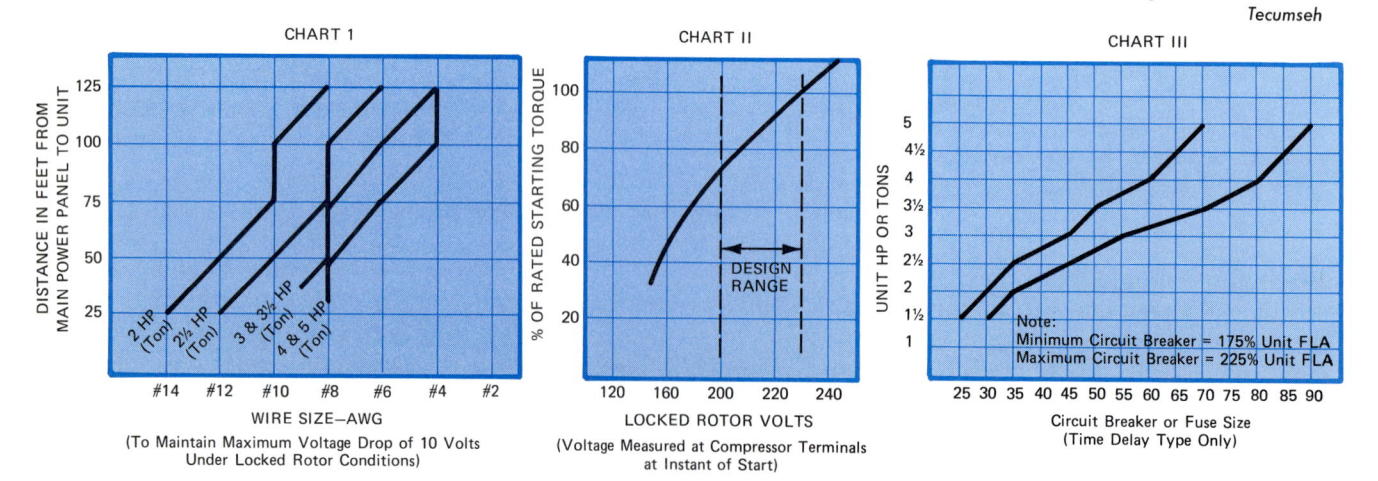

Tecumseh

CHART 1 — Wire Size–AWG (To Maintain Maximum Voltage Drop of 10 Volts Under Locked Rotor Conditions)

CHART II — Locked Rotor Volts (Voltage Measured at Compressor Terminals at Instant of Start)

CHART III — Circuit Breaker or Fuse Size (Time Delay Type Only)

Note:
Minimum Circuit Breaker = 175% Unit FLA
Maximum Circuit Breaker = 225% Unit FLA

Table 13-B.
PSC Compressor Motor Troubles and Corrections

Causes	Corrections
Low Voltage 1. Inadequate wire size. 2. Watt-hour meter too small. 3. Power transformer too small or feeding too many homes. 4. Input voltage too low. (Note: Starting torque varies as the square of the input voltage.)	1. Increase wire size. 2. Call utility company. 3. Call utility company. 4. Call utility company.
Branch Circuit Fuse or Circuit Breaker Tripping 1. Rating too low.	1. Increase size to a minimum of 175% of unit FLA (Full Load Amperes) to a maximum of 225% of FLA.
System Pressure High or Not Equalized 1. Pressures not equalizing within three minutes. 2. System pressure too high. 3. Excessive liquid in crankcase (split-system applications).	1. a. Check metering device (capillary tube or expansion valve). b. Check room thermostat for cycling rate. Off cycle should be at least five minutes. Also check for "chattering." c. Has some refrigerant dryer or some other possible restriction been added? 2. Make sure refrigerant charge is correct. 3. Add crankcase heater and suction line accumulator.
Miscellaneous 1. Run capacitor open or shorted. 2. Internal overload open.	1. Replace with new, properly-sized capacitor. 2. Allow two hours to reset before changing compressor.

rating (for $\frac{1}{2}$, $\frac{1}{3}$, and $\frac{3}{4}$ hp, 115-volt rated compressors).

b. 50 µF @ 250-volts AC rating (for $\frac{1}{3}$, $\frac{1}{2}$, and $\frac{3}{4}$ hp, 230-volt rated compressors).

c. 100 µF @ 250-volts AC rating (for 1, 1$\frac{1}{2}$, 1$\frac{3}{4}$, and 2 hp, 230-volt rated compressors).

d. 200 µF @ 230-volts AC rating (for 3, 4, and 5 hp, 230-volt rated compressors).

• Continuity test cord and lamp as shown in Fig. 13-29.

• Electrical instrument requirements can be covered by the Model #85 tester manufactured by Airserco Manufacturing Company, 3875 Bigelow Boulevard, Pittsburgh, PA 15213.

Warranty Test Procedure

Several checks must be made to make sure the compressor is operating correctly. The high potential test is one of the tests to be performed. Following are the voltage requirements.

• Use 950 test volts for any compressor having rating up to $\frac{1}{2}$ hp.

• Use 1450 test volts for any compressor having a rating of $\frac{1}{2}$ hp and greater.

METHOD OF TESTING

With the transformer adjusted to the correct specified voltage, attach one lead of the tester as a ground by holding it against a nonpainted portion of the compressor housing. Touch the other lead for one second to any one of the compressor terminal posts. The high potential ground tester will then indicate if a ground is present.

If a ground is indicated, do not check further. Remove the compressor and attach a tag to it. Note on the tag "grounded compressor." Return the compressor to the manufacturer for replacement.

A run test is made if the compressor is not showing a ground. The procedure is as follows:

1. Check the compressor assembly for correctness of wiring and for loose or broken terminals or joints. Change or repair where possible.

2. Install a by-pass line from the suction to the discharge line as shown in Fig. 13-30. The control valve in the line is put in *full open* position.

3. Remove all electrical components.

4. Place an ammeter in L_1 leg of the power supply.

5. Hook up the compressor motor to the power supply as follows:

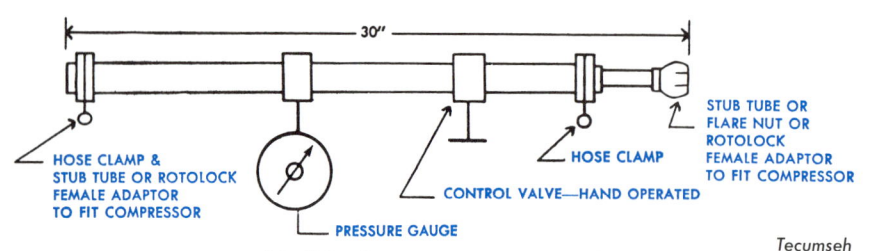

REFRIGERATION AND AIR CONDITIONING TECHNOLOGY

13-28. Compressor test stand.

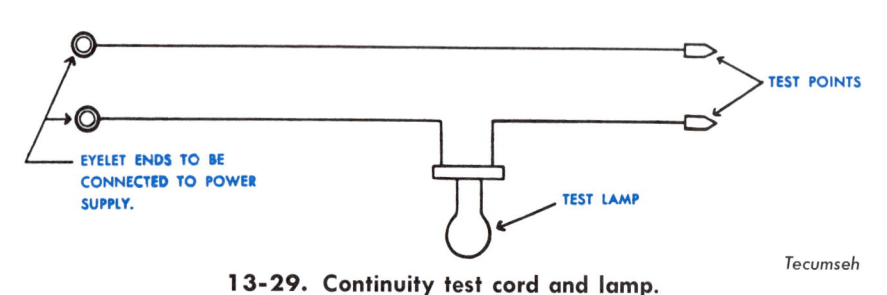

13-29. Continuity test cord and lamp.

Single Phase
a. L₁ to C (common) and L₂ to R (run) terminals.
b. Using proper start capacitor from the manufacturer's chart, connect one capacitor terminal to R. Let the lead from the other capacitor terminal dangle loose. Fig. 13-31.
c. Energize the compressor and momentarily touch the loose start capacitor lead to S. (CAUTION: Do not touch the start capacitor lead to S for more than a few seconds or the start winding will burn.)

Three Phase
a. Connect L₁, L₂, and L₃ to the three compressor terminals.
b. Energize the compressor.
6. Observe the ammeter.
  a. If the compressor does not run and draws *locked rotor amperes* (LRA—check the serial plate for this information), attach a tag noting that the compressor is "stuck." Return the compressor for replacement.
  b. If the compressor runs, but is abnormally noisy, attach a tag noting that the compressor is "noisy." Return the compressor for replacement.
  c. If the compressor does not run and draws no amperes, attach a tag noting that the compressor has "open winding." Return the compressor for replacement.
7. If the compressor runs normally, close the control valve in the bypass line until the gage reads approximately 175 psig.

Read the amperes. If the current is more than the rated FLA, attach a tag noting that "high amperes" are the problem. Return the compressor for replacement.

8. If the current and sound are normal, close the control valve. Stop the compressor and clock the rate of pressure fall on the bypass line gage.

The following pressures are for Tecumseh compressors. For other compressors, the manufacturer's recommendation must be checked.
• 25 psig per minute initial rate of discharge pressure drop should not be exceeded on models AE, T, AT, AK, AJ, AR, AU, and ISM.
• 40 psig per minute initial rate of discharge pressure drop should not be exceeded on models AB, AH, B, C, P, and AP.
• 80 psig per minute initial rate discharge pressure drop should not be exceeded on models F and PF.

If pressure changes occur in excess of those indicated, tag the compressor and mark it "internal leak." Return the compressor to the manufacturer for replacement.

9. If the compressor tests normal, open the control valve in the bypass line. Remove the bypass, and immediately seal the discharge and suction tubes.

10. Check the resistance of the motor windings against the val-

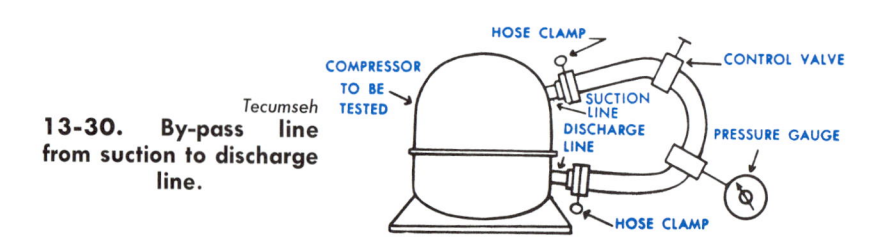

13-30. By-pass line from suction to discharge line.

13-31. Location of start capacitor in a circuit.

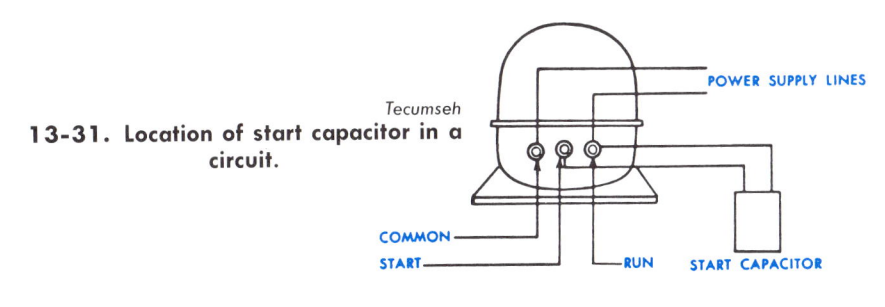

ues furnished by the manufacturer.

11. If the compressor checks out normal, return it to the customer.

Resistance Checks

The run (main) windings of a single-phase hermetic compressor motor consist of large-diameter wire having very low resistance. The accurate measurement of run windings resistance requires a digital ohmmeter. Less accurate meters are not sensitive enough to give correct readings.

Motors are sometimes diagnosed as grounded when the problem actually lies with the ohmmeter. It may not be sensitive to the normal 1- or 2-ohm run winding resistance.

There is a test procedure that can circumvent this measurement problem. This technique also enables the serviceperson to determine the operating position of an internal line break overload if one is installed in the motor. The procedure is as follows:

1. Remove all electrical connections from the compressor terminals.
2. Measure the resistance across the *run* and *start* terminals of the motor. This measures the combined resistance of the run and start windings. If the measured resistance approaches the value given in the manufacturer's specifications, the windings can be considered normal. If no resistance is read, the windings are open. The motor should be rejected.
3. If Step 2 indicates no problems, measure the resistance of the start winding only by checking across the terminal S (start) and C (common). The ohms should approach the range given by the manufacturer. If no resistance is read, the internal common lead to the motor is open.

If the motor has an external overload, then an open common lead means the motor is defective.

If the motor has an internal line break overload, then the overload may be open because the motor is overly hot.

It is not uncommon for the overload to remain open for more than an hour if tripped by a hot motor. Motors cool slowly. The internal overload will not close until the compressor dome (and thus the motor iron) is cool to the touch—below 130° F. [54°C].

4. If Steps 2 and 3 indicate no problems, check each terminal in turn for a ground to the compressor housing. First, file a shiny spot to ensure a good electrical connection.

5. If Steps 2, 3, and 4 indicate no problems, the problem may not be with the single-phase motor. Three-phase motors are more difficult to diagnose, since all three windings are run windings with low resistance. The internal overload, if present, is across all three windings. Here, the suggested procedure is as follows:

   a. Check for continuity between each of the three terminal pairs.

   b. If the circuit is open and an internal overload is present, be sure the motor is cool before rejecting the motor.

   c. If *Step a* indicates no problems, check for ground between each terminal and the housing.

Testing Electrical Components

The individual electrical components in a compressor may be tested as follows:

1. Check the rating and part number of each component to ensure that the component is correct for the compressor model. If the wrong component is being used, replace with the correct one.
2. Using an ohmmeter, check the external overload (at room temperature) for continuity across the terminals. If defective, replace with the proper component.
3. The start and run capacitor should be checked on the recommended capacitor tester according to the manufacturer's instructions. An alternate, but less precise method, uses a good ohmmeter. With the meter on the R × 1 range, continuity indicates a shorted capacitor. Replace. With the meter on the R × 100,000

scale, no needle deflection indicates an open capacitor. Replace.

4. It is difficult to check the starting relay without special equipment. Use the ohmmeter to make continuity checks.

While being checked, current relays must be held upright in their normal operating position.

Contacts should be open between terminals 1 (or L) and S. Therefore, there should be no continuity.

Terminals 2 (or L) and M should indicate continuity through the operating coil. (If there is no terminal 2, use terminal 1.)

Test the potential relay as follows:

Contacts should be closed between terminals 1 and 2. Therefore, there should be continuity.

There should be continuity through the operating coil between terminals 2 and 5.

5. If all the above tests prove satisfactory, change the relay. The new relay will eliminate any faulty electrical characteristics, such as improper pickup and dropout. These cannot be determined in the previous tests.

6. As a final check, connect the new relay to the compressor and the capacitors previously checked. If the compressor fails to start at serial plate voltage, the compressor should be considered inoperative because of internal defects. It must be replaced. If a capacitor checker other than that recommended was used, try new capacitors before rejecting the compressor.

REVIEW QUESTIONS

1. Why is oxygen not used to pressurize a refrigeration system?
2. Why can pressure testing of a refrigeration system be dangerous?
3. Why should you wear goggles when purging a contaminated system?
4. Why do you not fill a gas cylinder completely full?
5. What is meant by off-cycle crankcase heating?
6. Where is the filter drier located in the liquid line?
7. What is a heat exchanger?
8. What are the four major causes of motor burnout on a compressor?
9. How do you repair the perimeter tube if it develops a break?
10. What is the mullion on a refrigerator cabinet?
11. Why do you not use the bubble method for finding a leak when a vacuum is present?
12. What color of flame indicates a leak with a halide torch?
13. Why can you get a false reading of a refrigerant leak in a refrigerator when the insulation is urethane foam?
14. How do you test the potential relay?

14

Freezers

TYPES OF FREEZERS

There are two types of domestic freezers—the upright freezer and the chest-type freezer. The essential parts of the upright freezer are shown in Fig. 14-1. Notice that the evaporator coils are built into the shelves as part of that unit. This means the shelves are not adjustable. Notice that the condenser coils (17) in Fig. 14-1 are welded to the outside of the cabinet. This prevents sweating and aids the dissipation of heat over a large surface. The primary convenience of this type of freezer is that the frozen food is visible, easily arranged, and easily removed.

The chest-type freezer provides a different storage arrangement. Fig. 14-2. In some instances, the condenser coils are mounted on the back of the chest-type freezer.

The electrical diagram for a chest-type freezer is shown in Fig. 14-3. Note that the thermostat controls the on-off operation of the compressor. The light switch completes the circuit from one side of the power cord to the other through the light. Some models have a mercury switch that operates when the lid is up. This type of freezer does not have automatic defrost. Defrosting must be done manually.

INSTALLING A FREEZER

It may be necessary to remove the freezer door for passage through narrow doors. With each freezer there is an instruction sheet explaining the step-by-step procedure for door removal.

Screw-type levelers are used to adjust the level of the freezer. Upright freezer models use a screw-type leveler that can be moved up and down by turning to the left or right. Fig. 14-4.

14-1. The parts of an upright home freezer. (1) polyurethane foam-insulated cabinets, (2) wraparound steel cabinet, (3) baked-on enamel finish, (4) magnetic door seal, (5) key ejecting lock, (6) "bookshelf" door storage, (7) slide-out basket, (8) juice can shelf, (9) steel shelves, (10) fast two-way freezing level, (11) temperature control knob, (12) door stops, (13) interior light, (14) "power-on" light, (15) defrost water drain, (16) adjustable leveling legs, (17) coils welded to outer walls, (18) sealed compressor, and (19) RED.

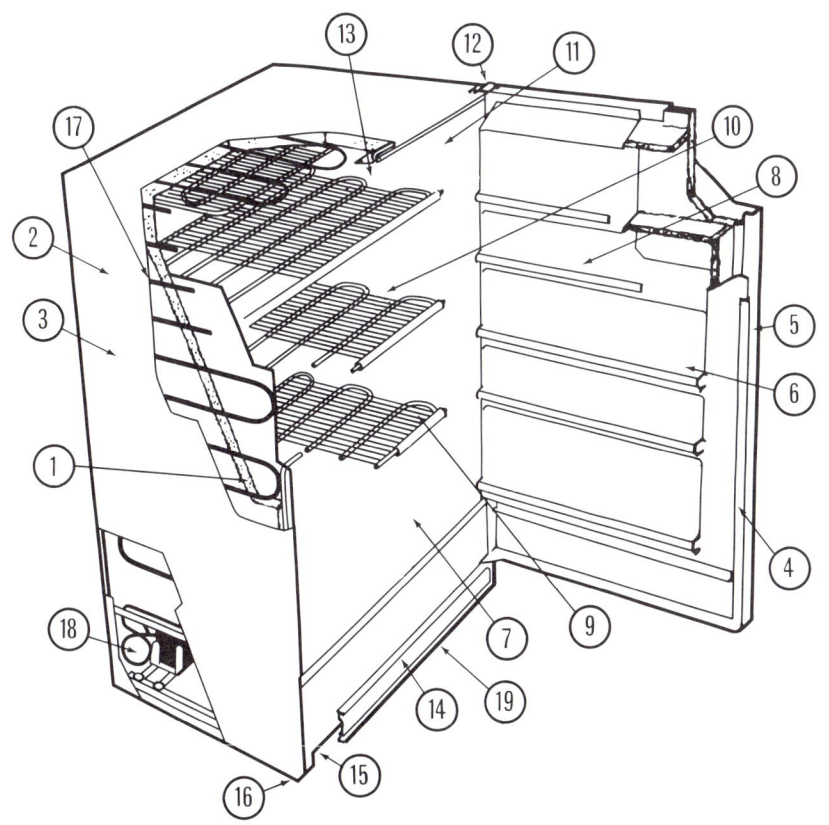

369

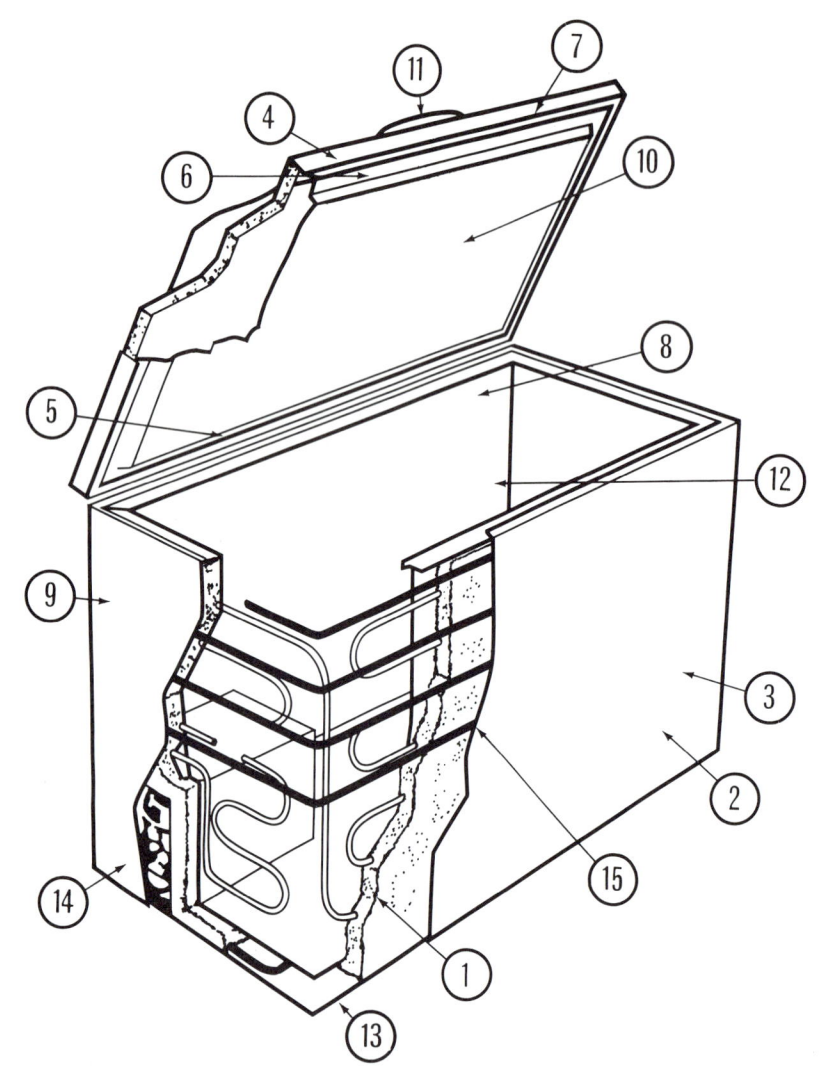

REFRIGERATION AND AIR CONDITIONING TECHNOLOGY

14-2. The parts of a chest-type freezer. (1) polyurethane foam insulation, (2) wraparound steel cabinet, (3) baked-on enamel finish, (4) self-adjusting lid, (5) spring-loaded hinges, (6) vinyl lid gasket, (7) safety lock and self-ejecting key, (8) lift-out wire baskets, (9) temperature control knob, (10) automatic interior light, (11) "power-on" light, (12) vertical cabinet divider, (13) defrost water drain, (14) sealed compressor, and (15) wraparound condenser.

The cabinet must be level side-to-side with a very slight tilt towards the rear. This will aid in obtaining a tight door gasket seal. If the cabinet is tilted towards the front, the weight of the door, plus the door food load, will result in poor gasket seal. NOTE: Caution the user that he or she is not to slam the door. If the door is slammed, the air pressure may be sufficient to open the door slightly. Chest freezers are leveled by putting metal or wood shims between the floor and the freezer as required.

Do not locate the freezer adjacent to a stove or other heat source. Avoid an area that is exposed to direct sunlight for long periods.

FREEZER COMPONENTS

Wrapped Condenser

The wrapped condenser incorporates a precooler condenser in series (through an oil cooler) with the main condenser. The condenser, made of $\frac{1}{4}''$ steel tubing, is clamped to the cabinet wrapper. Thermal-mastic is applied to each pass for maximum heat dissipation.

A wrapped condenser depends on natural convection of room air for dissipation of heat. Restricted air circulation around the cabinet will cause high operating temperatures and reduced capacity. The wrapped condenser reduces the possibility of moisture condensing on the cabinet shell during extremely humid weather. It also eliminates the need for periodic cleaning of the condenser. Figure 14-5 shows the condenser layout on an upright freezer.

Cold Ban Trim

Upright models have four-piece sectional cold ban trim strips that extend around the periphery of the freezer storage compartment. These trim strips are replaceable.

Starting at the lower corners, force the side trims toward the opposite side of the freezer as shown in Fig. 14-6.

Use a small flat screwdriver to release the cold ban trim from the cabinet U-channel. Then, pull the trims down and out from the overlapping top trim strip.

Remove the top and bottom cold trim strips by grasping one

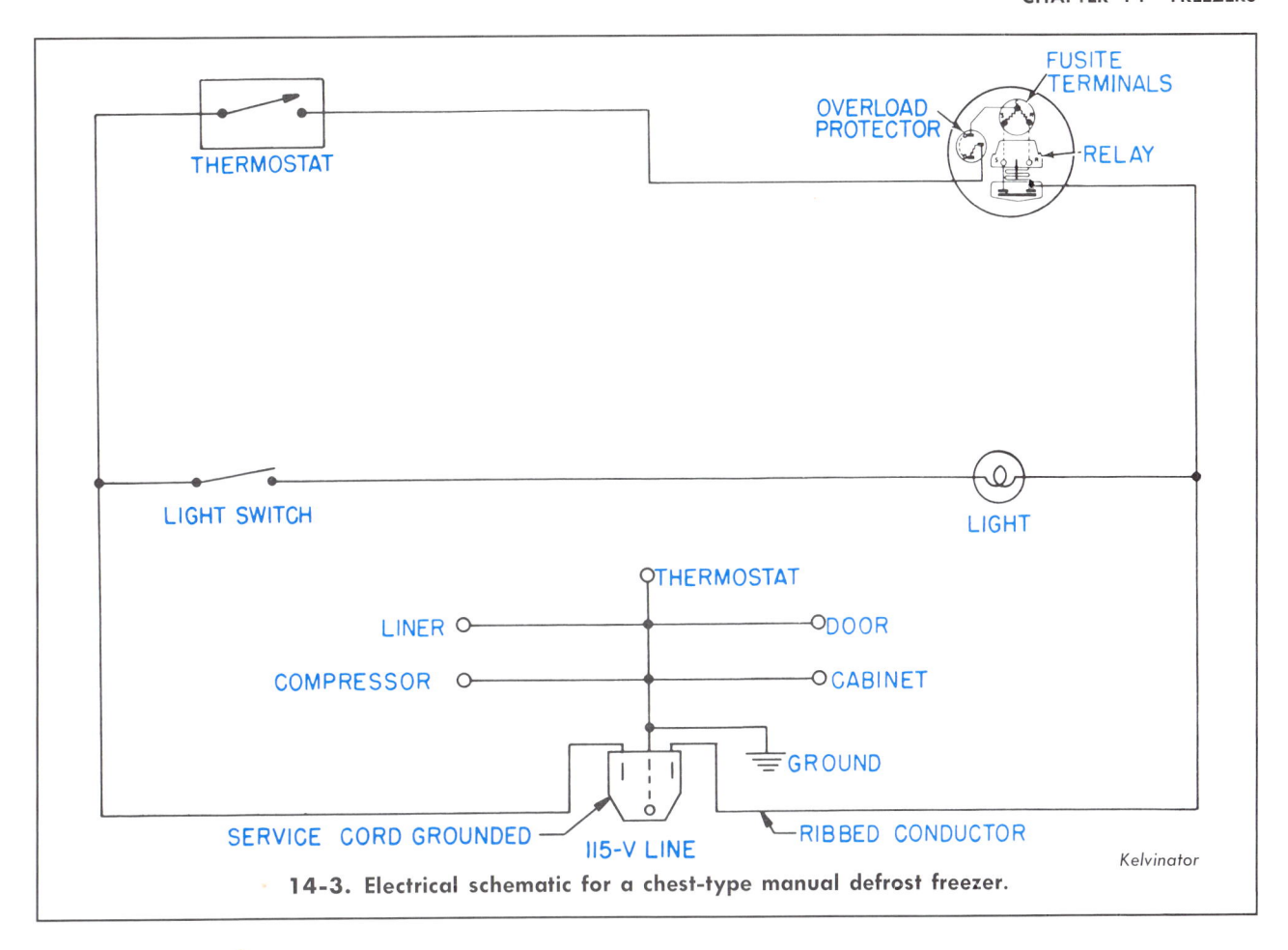

CHAPTER 14—FREEZERS

14-3. Electrical schematic for a chest-type manual defrost freezer.

end and pulling the trim out of the cabinet U-channel.

Before installing replacement cold ban trim, be sure the fiberglass filler insulation sections are in place.

Install the bottom cold ban trim. Squeeze one end of the trim and press the front flanges of the trim into the U-channel as shown in Fig. 14-7. Then, use the palm of the hand to press on the rear edge of trim, forcing the lock tabs on the trim over the flange of the freezer liner.

Install the right-hand and the left-hand side cold ban trims. Then install the top cold ban trim.

Shelf Fronts

The lower portions of the door shelves are formed as a unit with the inner panel. The door shelf fronts are removable. To remove, push down on the end caps and tilt out slightly at the top. See Fig. 14-8. Slide the end caps up. Pull outward on the bottom to free the caps from the slots in the door inner panel.

Vacuum Release

Some models have a vacuum release in the bottom edge of the

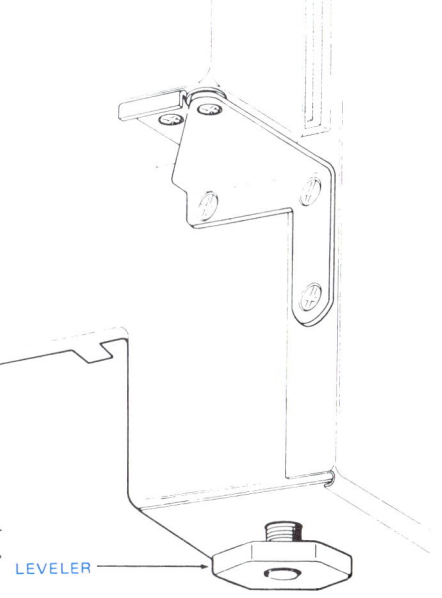

14-4. Adjustment screw for the upright model freezer.

371

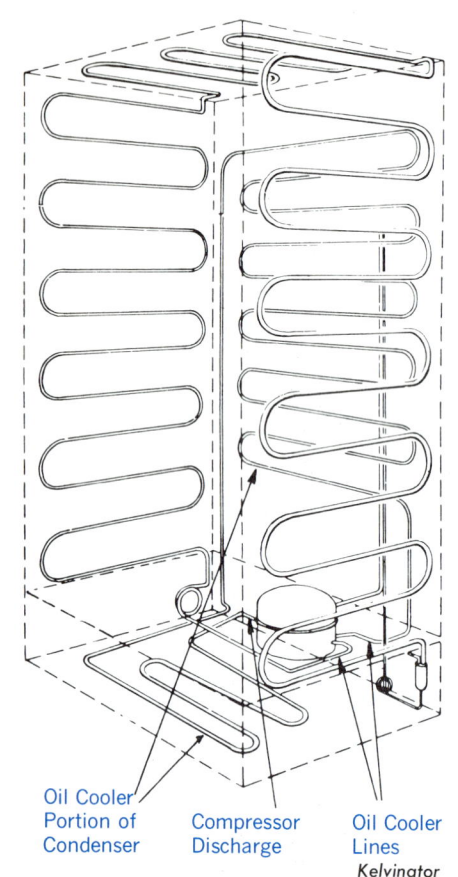

REFRIGERATION AND AIR CONDITIONING TECHNOLOGY

14-5. Wrapper condenser for the upright model freezer.

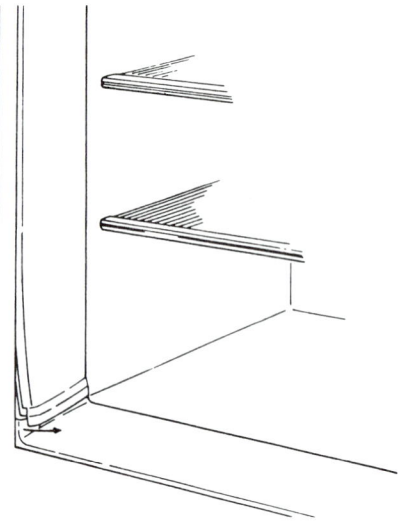

14-6. Removing cold ban trim.

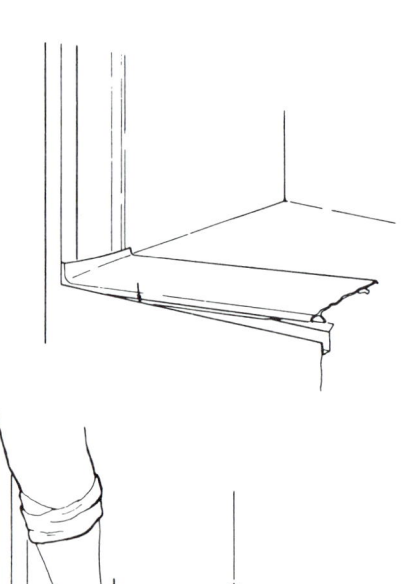

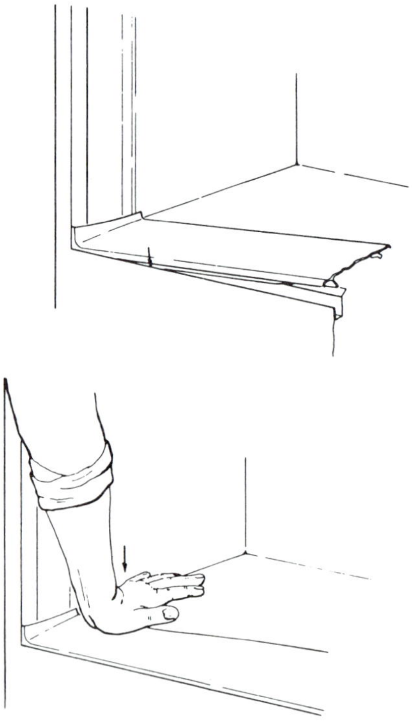

14-7. Installing cold ban trim.

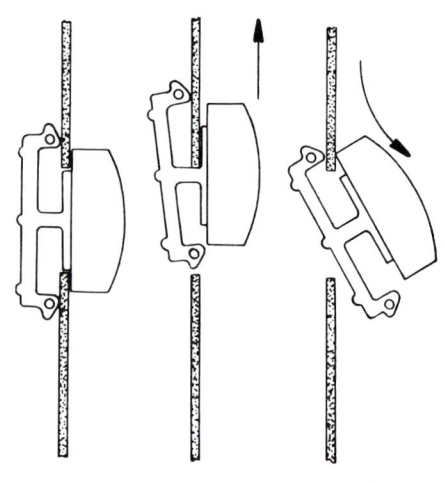

14-8. Removing shelf fronts in a chest-type freezer.

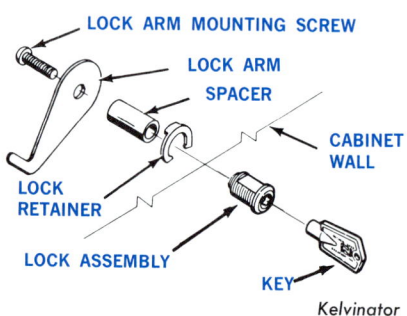

14-9. Door lock assembly.

door. This speeds up equalization of the air pressures, permitting successive door openings.

Some models do not incorporate a vacuum release device. If they have a good airtight gasket seal and the freezer is in operation, the door cannot be opened the second time when two door openings are required in quick succession. This is due to a difference in air pressure between the freezer interior and the room atmosphere. Opening the door the first time results in spillage of cold air from the freezer. This cold air is replaced by warm air. When the door is closed, this warm air is cooled, reducing its specific volume, thus creating a vacuum. Leaving the door closed for about one and one-half minutes will allow the air pressure to equalize.

Lock Assembly

Some models have a lock. The lock assembly is mounted in the door outer panel. It is held in place by the lock retainer. Figure 14-9 shows an exploded view of a typical lock assembly.

To replace the lock, remove the door inner panel. Then remove the lock retainer and lock assembly. Replace in the reverse order of removal. Remember that the lock key is self-ejecting.

Hinges

Chrome-plated steel hinges are used on all models. Figs. 14-10 and 14-11. The hinge pins ride in nylon thimbles placed in

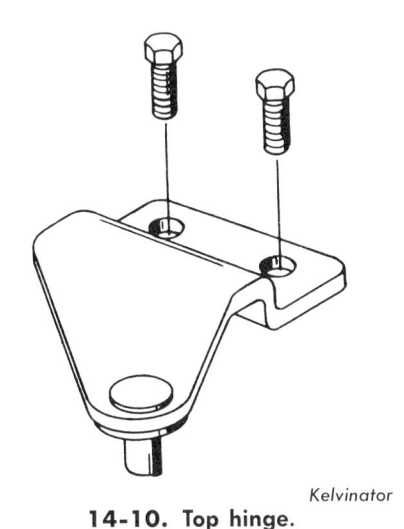

CHAPTER 14—FREEZERS

14-10. Top hinge. *Kelvinator*

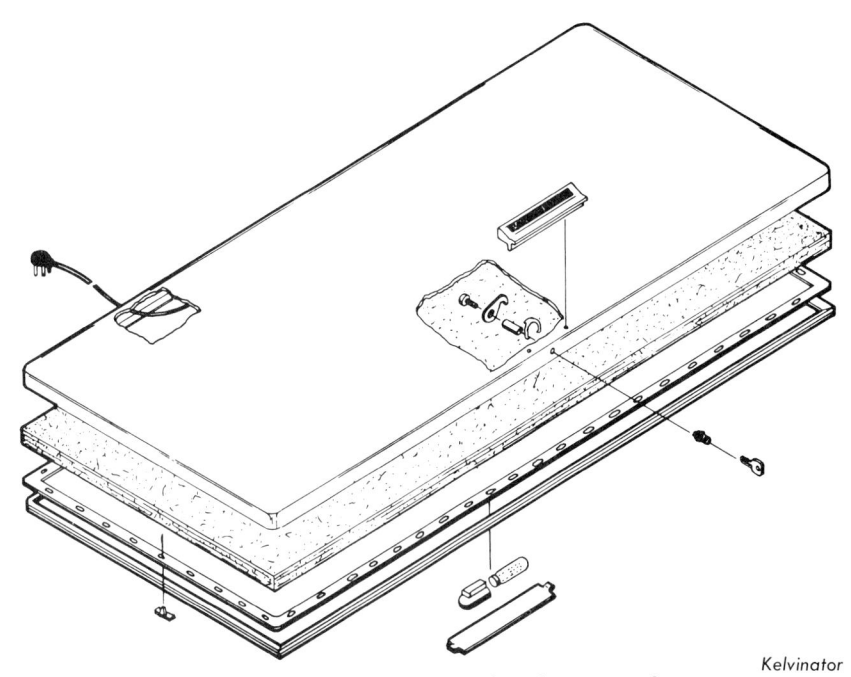

14-12. Freezer lid construction for chest-type freezer. *Kelvinator*

the door panel piercings. Nylon spacers are placed over the hinge pins to form the weight-bearing surfaces.

To replace hinges, remove the top hinge screws and hinge. Then remove the door from the freezer.

HINGE ADJUSTMENT

Some hinges have enlarged mounting holes to permit hinge adjustment. Shims may be added or removed from behind the bottom hinge to eliminate hinge bind or to improve the door gasket seal at the hinge side.

14-11. Bottom hinge assembly. *Kelvinator*

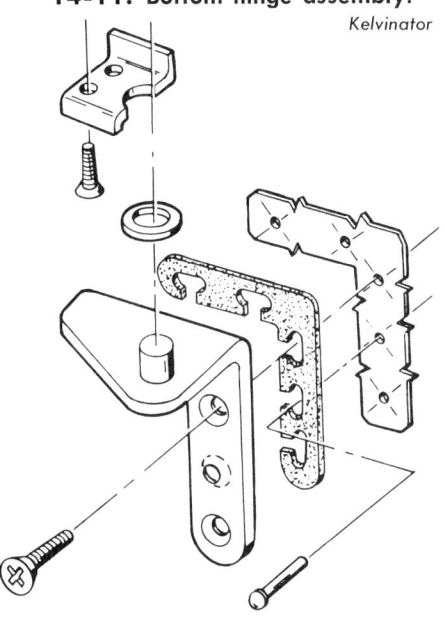

Chest models have spring-loaded hinges. These hinges incorporate a strong coil spring. The force of this spring counterbalances the weight of the lid and lifts the lid to the open position.

The hinge butt is fastened to the cabinet with four screws. These screws engage a fixed tapping plate inside the wrapper wall. The hinge leaf is fastened to the lid's outer panel with four screws that engage a fixed tapping plate. Pressed fiber shims are used under the hinge butt.

The hinge butt holes are slotted vertically to allow adjustment of the lid and to secure proper gasket fit. The hinge leaf holes are slotted horizontally to allow adjustment of the lid either to the left or to the right.

Lid

Most chest-type freezers have flexible lids. Even when the lid is lifted at one corner, it will seal properly on closing under its own weight. The lid's outer panel is drawn from one piece of steel. The edge is turned back to form a flat flange. This gives strength and furnishes a plane surface for support of the gasket and the lid's inner panel. Tapping plates for the hinges are welded in place.

The inner lid panel, gasket, handle, lock assembly, and insulation may be seen in Fig. 14-12.

Thermostats

Freezers and refrigerators have the same theory of operation. The start relay for the compressor operates the same as the start relay for a refrigerator. However,

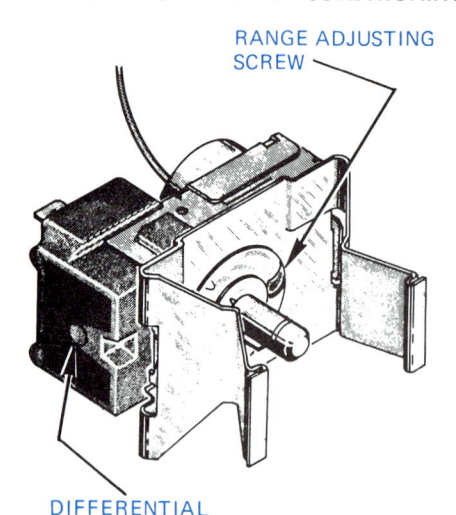

REFRIGERATION AND AIR CONDITIONING TECHNOLOGY

14-13. GE thermostat. *Kelvinator*
(RANGE ADJUSTING SCREW, DIFFERENTIAL ADJUSTING SCREW)

14-15. Cutler-Hammer thermostat. *Kelvinator*
(CUT-OUT ADJUSTMENT SCREW, CUT-IN ADJUSTMENT SCREW)

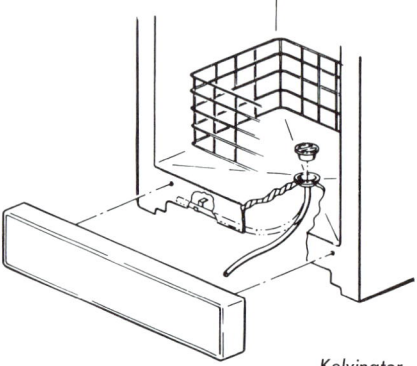

14-14. Ranco thermostat. *Kelvinator*
(DIFFERENTIAL ADJUSTMENT SCREW, RANGE ADJUSTMENT SCREW)

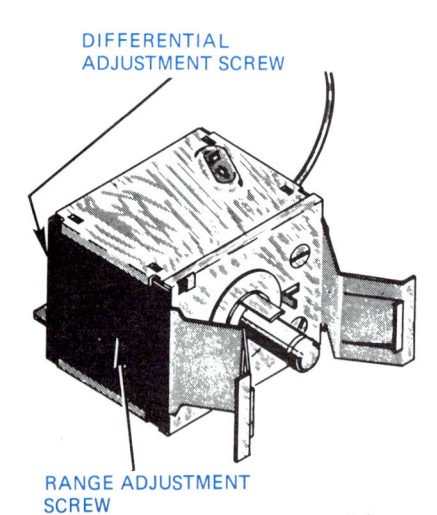

14-16. Drain system for a manual-defrost model upright freezer. *Kelvinator*

the thermostats are somewhat different.

On upright freezers the thermostat is mounted in the upper right-hand corner of the storage compartment in all manual defrost models. The thermostat knob in the manual defrost models is numbered 1 through 6 or *coldest* and *off*.

On all other models, remove the right-hand side cold ban trim and filler insulation. Loosen and remove the thermostat thermal element clamp from beneath the refrigerated shelf. Straighten the thermal element and attach a three-foot length of cord to the end of the thermal element. Use tape. Remove the light shield and thermostat knob. Disconnect the wire leads and remove the thermostat from the mounting bracket. Pull the thermal element out of the insulation. NOTE: Cord taped to the end of the thermal element feeds into the insulation cavity as the thermal element is pulled out. Remove the cord from the inoperative thermostat and tape it to the replacement thermostat thermal element. Pulling on the opposite end of the cord, thread the thermal element through the insulation and hole in the liner. Attach the thermal element to the refrigerated shelf. Mount the thermostat. Replace the light shield and knob.

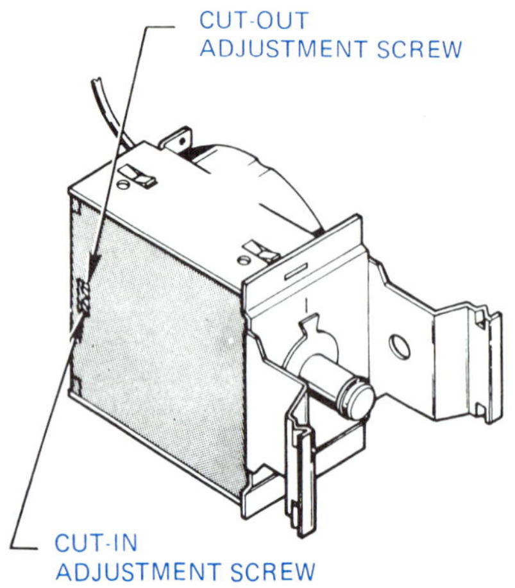

On the chest models the thermostat is located on the left end of the cabinet near the top of the unit compartment. The dial is marked *off, normal,* and *cold.* To stop the compressor during a normal running cycle, pull the service cord from the electrical outlet or turn the thermostat to the *off* position.

To replace the thermostat, first disconnect the power cord from the electrical outlet, then remove the knob. Remove the thermostat mounting screws. Pull the thermostat into view in the machine compartment opening. Disconnect the wire leads from the thermostat terminals.

Remove the mastic sealer from around the thermal element where it enters the thermal well trough in the cabinet outer wrapper in the machine compartment.

Before removing the thermal element from the thermal well, wrap a small piece of tape around the thermal element next to the opening of the thermal well. Remove the thermostat from the machine compartment. Wrap a piece of tape on the new thermostat thermal element at the same location as the tape on the inoperative thermostat. Push the thermal element into the thermal well. To insure the correct length of thermal element in the well for positive contact, the

CHAPTER 14—FREEZERS

tape on the thermostat thermal element should be at the entrance of the well. Replace the mastic sealer. Connect the wire leads. Install the thermostat mounting screws and knob.

Figures 14-13, 14-14, and 14-15 show three types of thermostats. The thermostats are set at the factory in accordance with the manufacturer's specifications for cut-in and cut-out. No adjustment should be made unless it is absolutely proven the thermostats are not in accordance with specifications.

If a higher or lower range than is obtainable by the selector knob is desired, adjust the range (altitude) adjustment screw.

On GE thermostats, the range adjustment screw is reached through the small hole in the face of the thermostat. See Fig. 14-13. Turn the screw to the left to lower the cut-out and cut-in. Turn the screw to the right to raise the cut-out and cut-in temperatures. Altitude adjustments are made by turning the range screw to the right.

The range adjustment screw on the Ranco thermostats is located behind a removable cover. See Fig. 14-14. Turn the screw to the left to lower the cut-out and cut-in temperatures, and to the right to raise cut-out and cut-in temperatures. Altitude adjustments are made by turning the range screw to the right.

Cutler-Hammer thermostats have cut-in and cut-out temperature adjustment screws. See Fig. 14-15. Turn the screws to the left to raise the cut-out and cut-in temperatures and to the right to lower the cut-in and cut-out temperatures.

Both cut-out and cut-in screws must be adjusted counterclockwise to compensate for altitudes above 1000 feet.

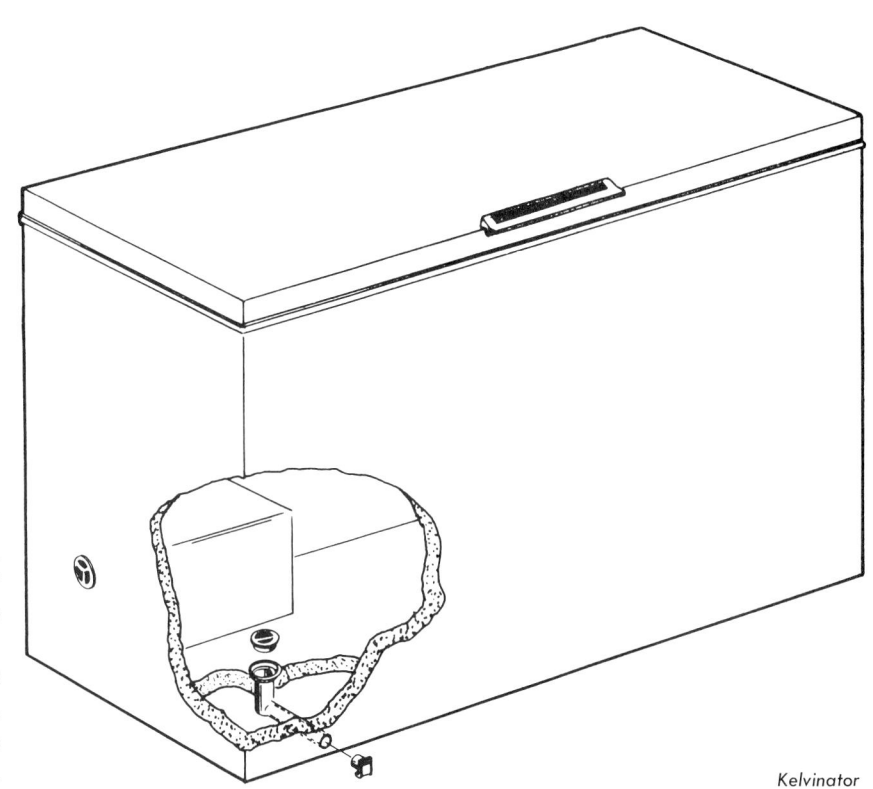

Kelvinator

14-17. Drain system for a chest model freezer.

Drain System

Manual defrost models have a defrost water drain and tube assembly for draining defrost water into a shallow pan. The drain tube is located behind the removable front grille. Fig. 14-16.

Chest models must be defrosted manually. A drain and tube assembly is located in the bottom left-hand corner of the storage compartment. Fig. 14-17.

Remove the drain plug from the inside bottom of the compartment. Place a shallow pan under the drain tube in front of the freezer and remove the cap. An alternate method is to insert a $\frac{1}{2}''$ male garden hose adapter into the drain tube and attach a garden hose. Remove the hose and adapter when defrosting is completed. Replace the drain plug and cap.

Wrapper Condenser

All compressors have internal spring suspension with four-point external mounting. The compressors have plug-in magnetic starting relays. These mount directly over the compressor Fusite® terminal assembly and a separate motor overload protector.

The wrapper condenser incorporates a precooler condenser in series with the main condenser. A wrapper condenser depends on the natural convection of the room air for dissipation of heat.

The high-temperature, high-pressure discharge refrigerant vapor is pumped into the precooler condenser. This is located

375

REFRIGERATION AND AIR CONDITIONING TECHNOLOGY

on the back wall of the freezer where it releases part of its latent heat of vaporization and sensible heat of compression. From the precooler condenser, the refrigerant passes back to the machine compartment and through the cooler coil in the compressor dome (where additional heat is picked up from the oil). It then passes back to the main condenser, where additional heat is released to the atmosphere. This results in condensation of the refrigerant from the high-pressure vapor to the high-pressure liquid.

Ample condenser area is provided to keep the surface temperature of the cabinet only 10 to 15° F. [5.5 to 8.3°C] above room temperatures. Heat released by the condenser helps reduce the possibility of moisture condensation on the cabinet surface in humid areas. The wrapper-type condenser eliminates service calls caused by plugged or dirty condensers. For the life of the freezer, the wrapper condenser remains efficient. A filter drier is located in the liquid line at the outlet of the condenser.

Evaporator Coil

Liquid refrigerant flows through the capillary tube and into the evaporator coil where expansion and evaporation of the refrigerant takes place. The evaporator coil (lowside) is a pattern of zig-zag passes of tubing. The evaporator coil is designed to produce adequate refrigeration and maintain uniform storage temperatures throughout the cabinet. Figs. 14-18 and 14-19.

Replacing the Compressor

With a relatively small amount of refrigerant used in the freezer, a major portion of it will be absorbed by the oil in the compressor when the freezer has been inoperative for a considerable length of time. When opening the system, use care to prevent the oil from blowing out with the refrigerant.

When replacing a compressor on a freezer that is in operation, disconnect the service cord from the electrical outlet. Allow the lowside to warm up to room temperature before removing the compressor. Placing an electric lamp inside the cabinet will help raise the temperature.

The procedure for replacement of the freezer compressor is the same as that for the refrigerator compressor. Check the earlier part of the chapter for a step-by-step procedure.

The procedures for replacing the filter drier and heat exchanger, cleaning the capillary tube, replacing the evaporator,

14-18. Refrigerant systems for a chest model freezer. *Kelvinator*

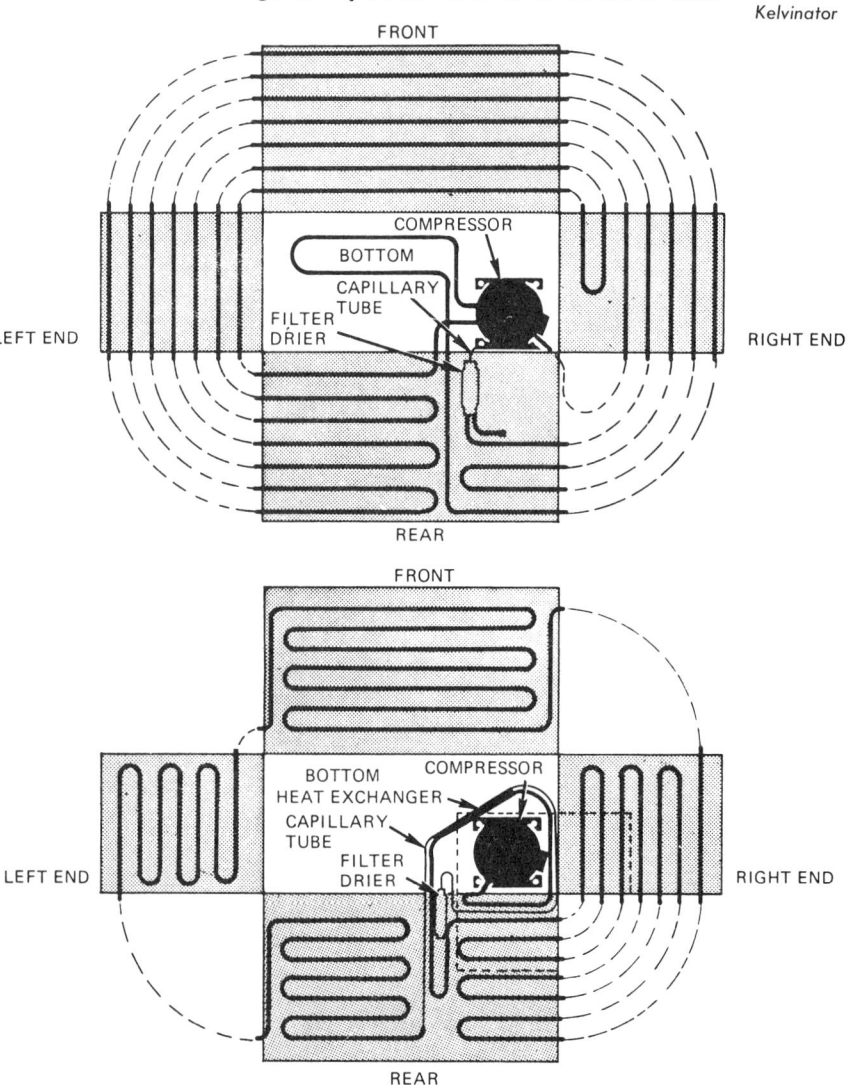

CHAPTER 14—FREEZERS

grid (or shelf), as well as the condenser are covered in Chapter 13.

Repairing the Condenser

When a refrigerant leak is found in any portion of the internal wrapper-type condenser (including the precooler) on models with removable liners, the condenser must be repaired. Fig. 14-20. On the foam-insulated models, an external natural-draft condenser is installed on the rear of the freezer.

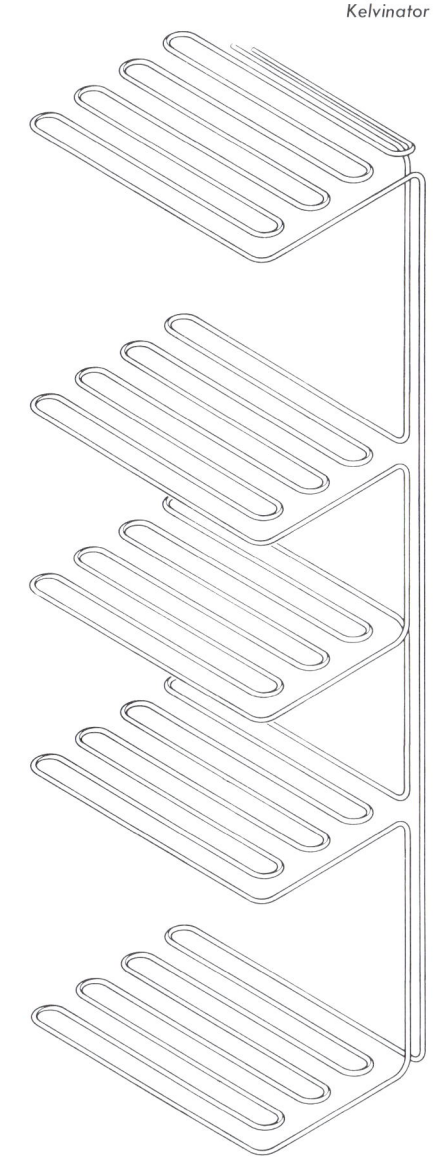

14-19. Evaporator for a manual-defrost upright model.
Kelvinator

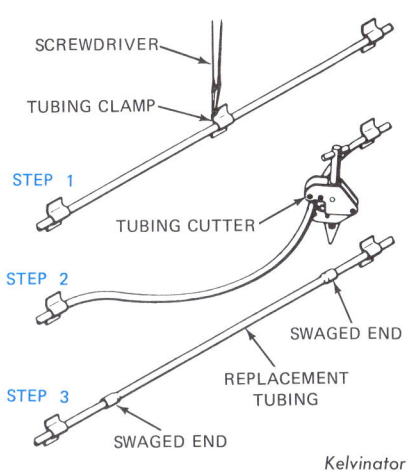

14-20. Repairing a condenser on the upright model. Step 1: Loosen the tubing within problem area. Step 2: Use tubing cutter to remove problem area. Cut in two places. Step 3: Swage the new pipe ends so they fit over the existing tubing.

Installing the Drier Coil

Install the drier coil in the U-channel as shown in Fig. 14-21. Secure the drier coil to the outer shell with short strips of mastic sealer spaced approximately 6" apart.

Thread the drier coil wire leads down through the wiring harness grommet in the lower right-hand front corner of the freezer U-channel. Route the wire lead to the rear of the machine compartment and splice into the service cord.

Install a length of mastic sealer, supplied with the drier coil, on the lower front flange and up 2" on each side of the freezer liner. Install the bottom cold ban trim and press the rear edge firmly into the mastic sealer. Then, install the remaining cold ban trim section.

All chest models have the wrapper-type condenser attached to the inner surface of the

14-21. External condenser on an upright model freezer.
Kelvinator

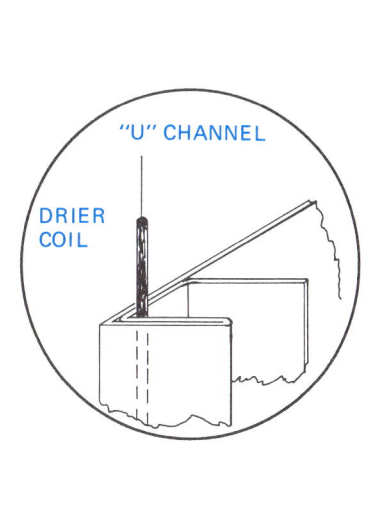

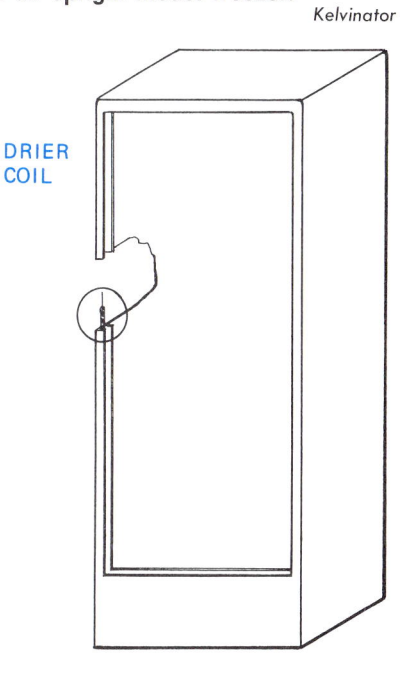

377

REFRIGERATION AND AIR CONDITIONING TECHNOLOGY

cabinet outer shell. The condenser, encased with foam insulation, is not accessible for repair.

When there is evidence of an internal refrigerant leak, the evaporator (lowside) and condenser (highside) should be disconnected from the system and individually pressurized and leak tested.

CAUTION: Do not disturb or puncture the cellular formation of foam insulation when testing for leaks. R-11 refrigerant entrapped in the cellular formation will be released, indicating a refrigerant leak. (R-11 is used to make the foam insulation. It is trapped inside the insulation and will be released when disturbed.)

When a refrigerant leak is found in any part of the internal wrapper-type condenser (including the precooler), the freezer is repaired by installing an external natural draft type condenser on the rear of the freezer.

Remember that the condenser outlet line must be positioned so it has a gradual slope downward from the point it leaves the freezer to where it enters the drier. The filter drier must be vertical or at a 30° to 45° angle, with the outlet end down so a liquid refrigerant seal is maintained.

COMPLETE RECHARGE OF REFRIGERANT

In the case of a major refrigerant leak that is repairable, such as a broken or cracked refrigerant line in the machine compartment, the unit will run. However, there may be no refrigeration or only partial refrigeration. Suction pressure will drop below the atmospheric pressure. Thus, with a leak on the lowside, air and moisture are drawn into the system, saturating the filter drier.

If there is reason to believe the system contains an appreciable amount of moisture, the compressor and drier filter should be replaced. The system should be blown out with liquid R-12 refrigerant and evacuated, after the leak has been repaired.

OVERCHARGE OF REFRIGERANT

When the cabinet is pulled down to temperature, an indication of an overcharge is that the suction line will be cooler than normal. The normal temperature of the suction line will be a few degrees cooler than room temperature. If its temperature is much lower than room temperature, the unit will run longer because the liquid is pulled beyond the accumulator into the heat exchanger. When the overcharge is excessive, the suction line will sweat or frost.

Restricted Capillary Tube

The capillary tube is restricted when the flow of liquid refrigerant through the tube is completely or partially interrupted. Symptoms are similar to those of a system that has lost its refrigerant. However, the major part of the refrigerant charge will be pumped into the highside (condenser), the same as with a moisture restriction. The suction pressure will range slightly below normal to very low (2″ to 20″ vacuum), depending on the amount of restriction.

TESTING FOR REFRIGERANT LEAKS

If the system is diagnosed as short of refrigerant and has not been recently opened, there is probably a leak in the system. Adding refrigerant without first locating and repairing the leak or replacing the faulty component would not permanently correct the difficulty. *The leak must be found.*

Sufficient refrigerant may have escaped to make it impossible to leak test effectively. In such cases, add a $\frac{1}{4}''$ line piercing valve to the compressor process tube. Add sufficient refrigerant to increase the pressure to 75 pounds per square inch. Through this procedure, minute leaks are more easily detected before the refrigerant is discharged from the system and contaminates the surrounding air. The line piercing valve should be used for adding refrigerant and for test purposes only. It must be removed from the system after it has served its purpose. Braze a 4″ piece of $\frac{1}{4}''$ O.D. copper tube into the compressor process tube. Evacuate the system and recharge after repairs are completed.

Troubleshooting Freezers

Table 14-A lists troubleshooting procedures for upright freezer models. Table 14-B on pages 382 to 384 lists troubleshooting procedures for chest-type freezer models.

PORTABLE FREEZERS

Many uses can be found for portable freezers. For example, much of the ice cream sold during the summer months is sold from refrigerated trucks. It is essential that these trucks have reliable portable freezing units. Figs. 14-22 and 14-23 on pages 384 and 385.

The compressor is run by 110-120 V, 60 Hz AC. It runs dur-

CHAPTER 14—FREEZERS

Table 14-A.
Troubleshooting Freezers: Upright Models

Trouble	Probable Cause	Remedy or Repair
Product too cold.	Temperature selector knob set too cold.	Set warmer.
	Thermostat bulb contact bad.	If the bulb contact is bad, the bulb temperature will lag behind the cooling coil temperature. The unit will run longer and make the freezer too cold. See that the bulb makes good contact with the bulb well.
	Thermostat is out of adjustment.	Readjust or change the thermostat.
Product too warm.	Thermostat selector knob set too warm.	Set cooler.
	Thermostat contact points dirty or burned.	Replace thermostat.
	Thermostat out of adjustment.	Readjust or change the thermostat.
	Loose electrical connection.	This may break the circuit periodically and cause the freezer to become warm because of irregular or erratic operation. Check the circuit and repair or replace parts.
	Excessive service load or abnormally high room temperature.	Unload part of the contents. Move unit to a room with lower temperature or exhaust excess room heat.
	Restricted air circulation over wrapped condenser.	Allow 6″ clearance above the top and $3\frac{1}{2}$″ clearance at the sides and between the back of the cabinet and the wall.
	Excessive frost accumulation on the refrigerated shelves (manual defrost models).	Remove the frost.
	Compressor cycling on overload protector.	Check the protector and line voltage at the compressor.
Unit will not operate.	Service cord out of wall receptacle.	Plug in the service cord.
	Blown fuse in the feed circuit.	Check the wall receptacle with a test lamp for a live circuit. If the receptacle is dead but the building has current, replace the fuse. Determine the cause of the overload or short circuit.
	Bad service cord plug, loose connection or broken wire.	If the wall receptacle is live, check the circuit and make necessary repairs.

(Continued on next page)

Table 14-A. (Continued)
Troubleshooting Freezers: Upright Models

Trouble	Probable Cause	Remedy or Repair
	Inoperative thermostat.	Power element may have lost charge or points may be dirty. Check the points. Short out the thermostat. Repair or replace the thermostat.
	Inoperative relay.	Replace.
	Stuck or burned-out compressor.	Replace the compressor.
	Low voltage. Cycling on overload.	Call utility company, asking them to increase voltage to the house. Or, move unit to a separate household circuit.
	Inoperative overload protector.	Replace.
Unit runs all the time.	Thermostat out of adjustment.	Readjust or change the thermostat.
	Short refrigerant charge (up to 4 oz.). Cabinet temperatures abnormally low in lower section.	Not enough refrigerant to flood the evaporator coil at the outlet to cause the thermostat to cut-out. Recharge and test for leaks.
	Restricted air flow over the wrapper condenser.	Provide proper clearances around the cabinet.
	Inefficient compressor.	Replace.
Unit short cycles.	Thermostat erratic or out of adjustment.	Readjust or change the thermostat.
	Cycling on the relay.	This may be caused by low or high line voltage that varies more than 10% from the 115 volts. It may also be caused by high discharge pressures caused by air or noncondensable gases in the system. Correct either condition.
Unit runs too much.	Abnormal use of the cabinet.	Heavy usage requires more operation. Check the usage and correct or explain.
	Shortage of refrigerant.	Unit will run longer to remove the necessary amount of heat and it will operate at a lower than normal suction pressure. Put in the normal charge and check for leaks.
	Overcharge of refrigerant.	Excessively cold or frosted suction line results in lost refrigeration effort. Unit must run longer to compensate for the loss. Purge off excessive charge.

CHAPTER 14—FREEZERS

Table 14-A. (Continued)
Troubleshooting Freezers: Upright Models

Trouble	Probable Cause	Remedy or Repair
Too much frost on refrigerated surfaces—lowside.	Restricted air flow over the condenser.	This can result if the cabinet is enclosed. This will obstruct the air flow around the cabinet shell. Restricted air flow can also be caused by air or noncondensable gases in the system. This results in a higher head pressure. The higher head pressure produces more re-expansion during the suction stroke of the compressor. Consequently, less suction vapor is taken. Increased running time must compensate for loss of efficiency. Correct the condition.
	High room or ambient temperature.	Any increase in temperature around the cabinet will increase the refrigeration load. This will result in longer running time to maintain cabinet temperature.
	Abnormally heavy usage in humid weather.	Do not leave the freezer door open any longer than necessary to load or remove products.
	Poor door gasket seal.	This permits the entrance of moisture by migration, which freezes out of the air as frost on the refrigerated surfaces.

ing the night when the truck is out of service. A small fan circulates the cold air. The fan runs on the truck battery during the day. At night, it is plugged into line current (120 volts, AC).

This unit uses a hermetically-sealed compressor designed for use with R-22. About 10 or 12 ounces of R-22 are used for a full charge.

Troubleshooting

If the machine runs short of refrigerant, it should be allowed to warm to room temperature and checked for leaks with a halide torch. For finding small leaks, at least 90 pounds of internal pressure are needed. It may be necessary to add refrigerant to obtain this pressure. If so, connect the suction line service opening to a drum of R-22, making sure the drum remains upright so only gas will enter the unit.

Never connect in this manner a drum that is warmer than any part of the system. The gas will condense in the system, resulting in overcharge and waste of refrigerant.

If the unit is charged due to leaks, or any major repairs are made on the system, it is recommended that a new drier be installed. When replacing the original drier, be certain that the replacement drier has a good filter and strainer incorporated with the drying agent.

If a gas leak has allowed air to enter the system, the system must be evacuated or thoroughly purged with R-22. A new drier must be installed before charging. Air remaining in the system cannot be purged off. It permeates the complete system and is not trapped in the high side as in other systems using a liquid receiver.

Allow pressure to build to approximately 100 pounds in the unit and shut off the charging valve immediately. After the unit

Table 14-B.
Troubleshooting Freezers: Chest Models

Trouble	Probable Cause	Remedy or Repair
Product too cold.	Temperature selector knob set too cold.	Set warmer.
	Thermostat bulb contact is bad.	If the bulb contact is bad, the bulb temperature will lag behind the cooling coil temperature. The unit will then run longer and make the freezer too cold. See that the bulb makes good contact with the bulb well.
	Thermostat is out of adjustment.	Readjust or change the thermostat.
Product too warm.	Temperature selector knob set too warm.	Set colder.
	Thermostat contact points dirty or burned.	Clean or replace the thermostat.
	Thermostat out of adjustment.	Readjust or change the thermostat.
	Loose electrical connection.	This may break the circuit periodically and cause the freezer to become warm because of irregular or erratic operation. Check the circuit and repair or replace the parts.
	Excessive service load or abnormally high room temperature.	Remove excess load. Reduce room temperature or move unit to a cooler room.
	Excessive frost accumulation on the liner walls.	Remove frost from walls.
	System short of refrigerant.	Recharge and test for leak.
	Restricted air circulation over the wrapper condenser.	Allow 2″ clearance at the ends and 3″ clearance at the back of the cabinet and wall for chest freezers.
Unit will not operate.	Service cord pulled from the wall receptacle.	Plug in the service cord.
	Blown fuse in the feed circuit.	Check the wall receptacle with a test lamp for a live circuit. If the receptacle is dead but building has current, replace the fuse. Try to determine the cause of the overload or short circuit.
	Bad service cord plug, loose connection, or broken wire.	If the wall receptacle is live, check the circuit and make necessary repairs.
	Inoperative thermostat.	Power element may have lost charge or points may be dirty. Check the points. Short out the thermostat. Repair or replace the thermostat.

Table 14-B. (Continued)
Troubleshooting Freezers: Chest Models

Trouble	Probable Cause	Remedy or Repair
Unit runs all the time.	Inoperative relay.	Replace relay.
	Stuck or burned-out compressor.	Replace compressor.
	Low voltage. Cycling on overload.	Call utility company, asking them to increase voltage. Or, change unit to a different circuit in house.
	Inoperative overload protector.	Replace.
	Thermostat out of adjustment.	Readjust or change the thermostat.
	Short refrigerant charge (up to 4 oz.).	Cabinet temperatures abnormally low in the lower section. Not enough refrigerant to flood the evaporator coil at the outlet to cause the thermostat to cut-out. Recharge and test for leaks.
	Restricted air flow over the wrapper condenser.	Provide proper clearances around the cabinet.
	Inefficient compressor.	Replace compressor.
Unit short cycles.	Thermostat erratic or out of adjustment.	Readjust or change.
	Cycling on the relay.	This may be caused by low or high line voltage that varies more than 10% from 115 volts. It may also be caused by high discharge pressures caused by air or noncondensable gases in the system. Correct either condition.
Unit runs too much.	Abnormal use of the cabinet.	Heavy usage requires more operation. Check the usage and correct or explain.
	Shortage of refrigerant.	Unit will run longer to remove the necessary amount of heat and it will operate at a lower than normal suction pressure. Put in the normal charge and check for leaks.
	Overcharge of refrigerant.	Excessively cold or frosted suction line results in lost refrigerant effort. Unit must run longer to compensate for the loss. Purge off the excess charge.
	Restricted air flow over the condenser.	This can be the result of enclosing the cabinet. This will cause obstruction to the air flow around the cabinet shell. Inefficient compression can be caused by air or noncondensable gases in the system. This results in a higher head pressure.

(Continued on next page)

REFRIGERATION AND AIR CONDITIONING TECHNOLOGY

Table 14-B. (Continued)
Troubleshooting Freezers: Chest Models

Trouble	Probable Cause	Remedy or Repair
Too much frost on refrigerated surfaces—lowside.		The higher head pressure produces more re-expansion during the suction stroke of the compressor and, consequently, less suction vapor is taken. Increased running time must compensate for the loss of efficiency. Correct the condition.
	High room or ambient temperature.	Any increase in temperature around the cabinet location will increase the refrigeration load. This will result in longer running time to maintain the cabinet temperature.
	Abnormally heavy usage in humid weather.	Do not leave the freezer lid open any longer than necessary to load and unload.
	Poor gasket seal.	This permits the entrance of moisture by migration, which freezes out the air as frost on the refrigerated surfaces.

is started, add R-22 slowly until back pressure is between 10 and 16 pounds, depending on the ambient temperature. (A high ambient temperature will produce a higher head and back pressure.) The back pressure will then remain about the same until the eutetic (contents of the freezer) is completely frozen.

The charge should be checked again when the cabinet is approximately −15° F. [−26°C] or colder. Then, with the condensing unit running, the suction line should frost out of the cabinet about 6″ to 8″. The desired frost line can be obtained by adding or purging off refrigerant a little

Kari-Kold
14-22. Portable freezer used in a truck to transport ice cream and milk.

CHAPTER 14—FREEZERS

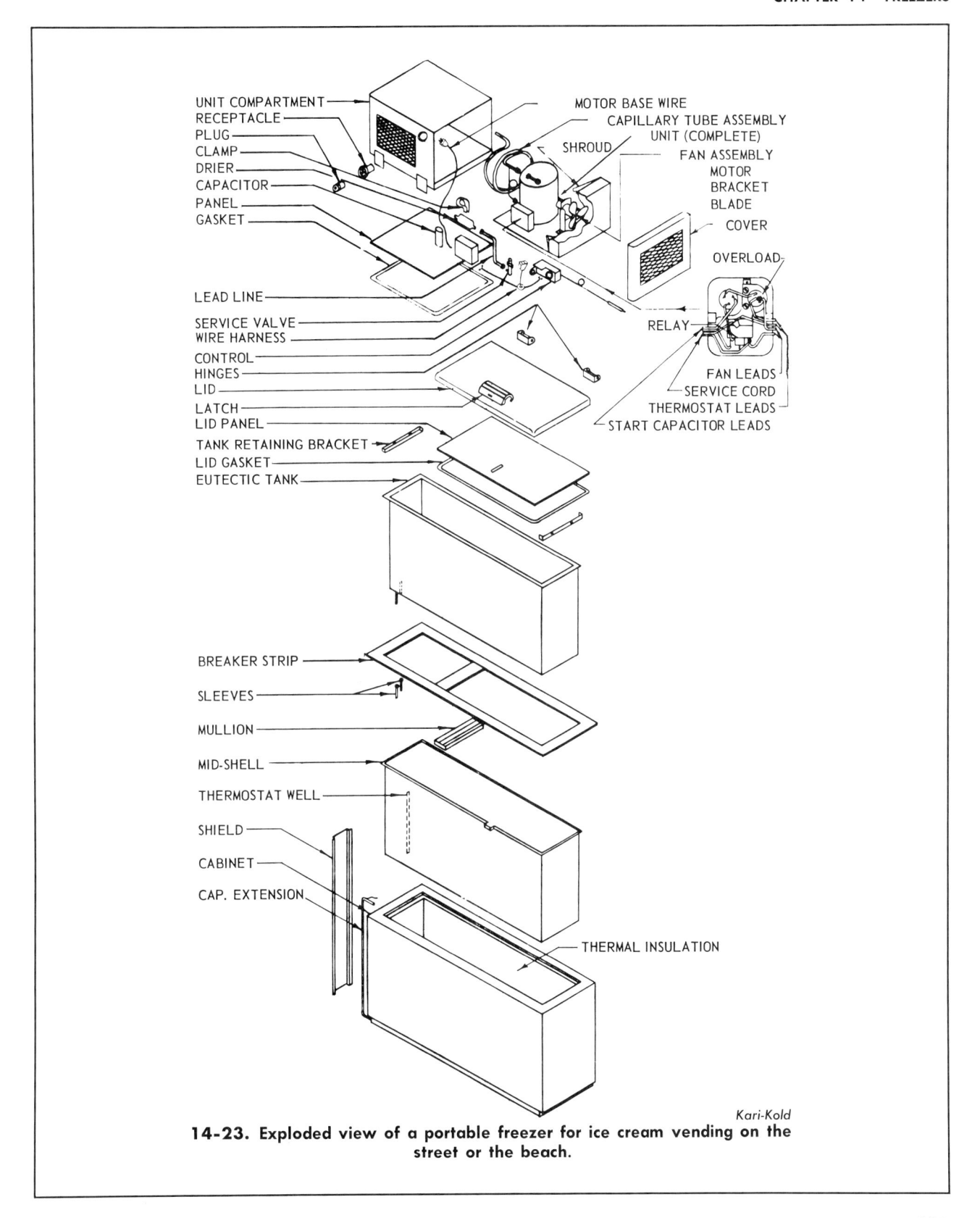

14-23. Exploded view of a portable freezer for ice cream vending on the street or the beach.

Kari-Kold

REFRIGERATION AND AIR CONDITIONING TECHNOLOGY

14-24. Ice cream dispenser capable of being mounted on a truck and used to transport frozen goods. *Kari-Kold*

at a time, allowing time for the system to equalize.

If the compressor will not start but the condenser fan is running, check the head and back pressures. If the pressures are not equal, a capillary tube may be clogged with moisture or foreign material. Heating the end of the capillary tube where it enters the cabinet will usually begin to equalize pressures if the restriction is due to moisture freezing. Evacuate the system, install a new drier, and recharge the system as described already. If the capillary tube is clogged with material other than frozen moisture, it should be replaced.

When the compressor does not start and the head and back pressures are approximately equal, check for trouble as follows:

- Check the line voltage by holding the voltmeter leads on contacts of the motor base plug. Take a reading when the overload protector clicks in and the compressor is trying to start. This reading should be 100 volts or more. If less, the trouble is probably in the supply line.
- Replace the capacitor, if the unit has one.
- Replace the relay and/or overload.

If, after these checks, the compressor will not start, the unit should be returned to the manufacturer.

Figure 14-24 shows an ice cream vending unit. Most of the mechanical parts are located on top of the unit to prevent damage when the unit is handled frequently.

This type of freezer, in various sizes, can be mounted in a variety of vehicles. The cabinet provides for economical operation that can pay for itself in dry ice savings alone. The unit is plugged in at night. The smaller units rely upon insulation to hold the cold air. Other units plug into the vehicle's battery.

Small trucks can be fitted with portable freezers. These are useful to dairies servicing school cafeterias and other large food dispensing operations where milk must be kept cool and ready for servicing at a specific time. Some units can handle between 400 and 700 bottles or cartons of milk. They can be rolled into a cafeteria line.

REVIEW QUESTIONS

1. What are the two types of domestic freezers?
2. What is a wrapped condenser?
3. How do you remove the hinges on domestic freezers?
4. In what way do freezer and refrigerator thermostats differ?
5. How is the heat dissipated with a wrapped condenser?
6. What type of condenser is used when repairing a foam-insulated freezer?
7. Why must you be careful not to puncture the cellular formation of foam insulation when testing for leaks in a freezer?
8. When does the suction pressure drop below the atmospheric pressure in a defective freezer?
9. What indicates an overcharge of refrigerant in a freezer?
10. What symptoms are observed when the capillary tube is restricted and partially interrupts the flow of refrigerant?
11. How much internal pressure is needed in a freezer system to aid in locating small leaks?
12. When the compressor does not start and the head and back pressures are approximately equal, what should be checked first?

15

Commercial Refrigeration Systems

Every commercial air conditioning and refrigeration installation must be carefully planned. The selection of a system depends upon the job to be done. The store size, type of business, and departments to be emphasized relate to the selection of a refrigeration system.

REFRIGERATION CYCLE

The refrigeration cycle is the basis of all cooling in compact units. The refrigeration cycle is shown in Fig. 15-1.

The liquid refrigerant begins its cycle from the storage container called the *receiver*. Here, the temperature of this liquid refrigerant is usually slightly above room temperature. The pressure generated by the liquid in the receiver is sufficient to cause the liquid to travel through the *liquid line* to the *metering device*.

This metering device may be a capillary tube or an expansion valve. It reduces the pressure of the refrigerant as the liquid enters the lower pressure *evaporator coil*.

Just downstream from the metering device, the refrigerant still exists largely in the liquid state. However, as a result of the pressure reduction, it is now at a considerably lower temperature than it was upstream of the metering device. Heat from the medium being cooled passes through the walls of the evaporator. This heat is absorbed by the refrigerant. More and more of the low temperature liquid boils off and changes into a gas. By the time the refrigerant reaches the evaporator outlet, it has been vaporized.

The evaporator need not be a serpentine coil as shown. In household refrigerators and milk coolers, for example, the evaporator is in plate form. The refrigerant flows through circuits and passages blown or shaped into the plate.

From the evaporator outlet, the refrigerant vapor is drawn through the *suction line*. From here, it is drawn into the compressor by the action of the much lower pressure created in the low side of the compressor. In the compressor, sometimes referred to as the pump, the low-pressure, low-temperature refrigerant vapor is compressed into high-pressure, high-temperature vapor.

This high-temperature vapor passes into the *condenser*. Here it is cooled through the coil walls, by water or air. This cooling action causes the high pressure gas to condense or liquefy. The liquid refrigerant flows into the receiver. Here it is ready once more to repeat the cycle just described.

In the condenser, the heat that the refrigerant picked up from the medium being cooled is rejected to the atmosphere. This is true in the case of an air-cooled condenser. In the case of a water-cooled condenser, it is picked up by the water. The water may then be wasted to the sewer or it may be reclaimed for use in a cooling tower.

Basically, then, the refrigeration cycle consists of alternately vaporizing and condensing the refrigerant. In so doing, it is possible to absorb heat in the refrigerant at the evaporator and reject it from the refrigerant at the condenser. The refrigerant undergoes no chemical change. Unless a leak develops in the system, the refrigerant is reuseable indefinitely.

The condenser, receiver, and liquid line of a system are jointly referred to as the *high side* of the

CHAPTER 15—COMMERCIAL REFRIGERATION SYSTEMS

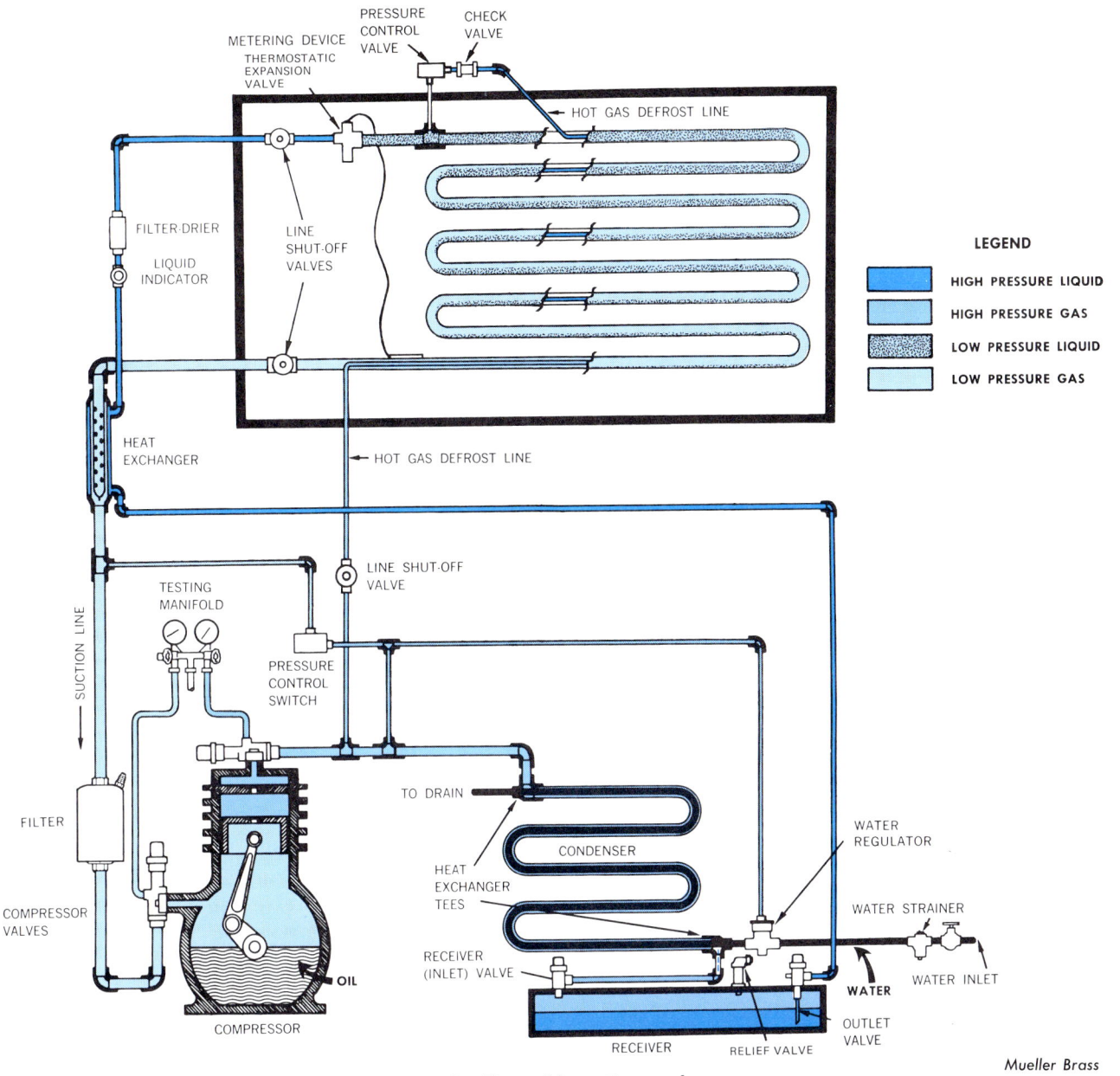

15-1. The refrigeration cycle.

unit. The evaporator and suction line are called the *low side* because here the pressures are much lower than in the high side.

The physical assembly that contains the compressor, condenser, and receiver is called the *condensing unit*.

AIR-COOLED REFRIGERATION EQUIPMENT

One basic reason for choosing the air-cooled condensing process is the fact that air is free. Also, in many areas of the country, water is of poor quality, expensive, or not available. Other features, such as reduced mainte-

389

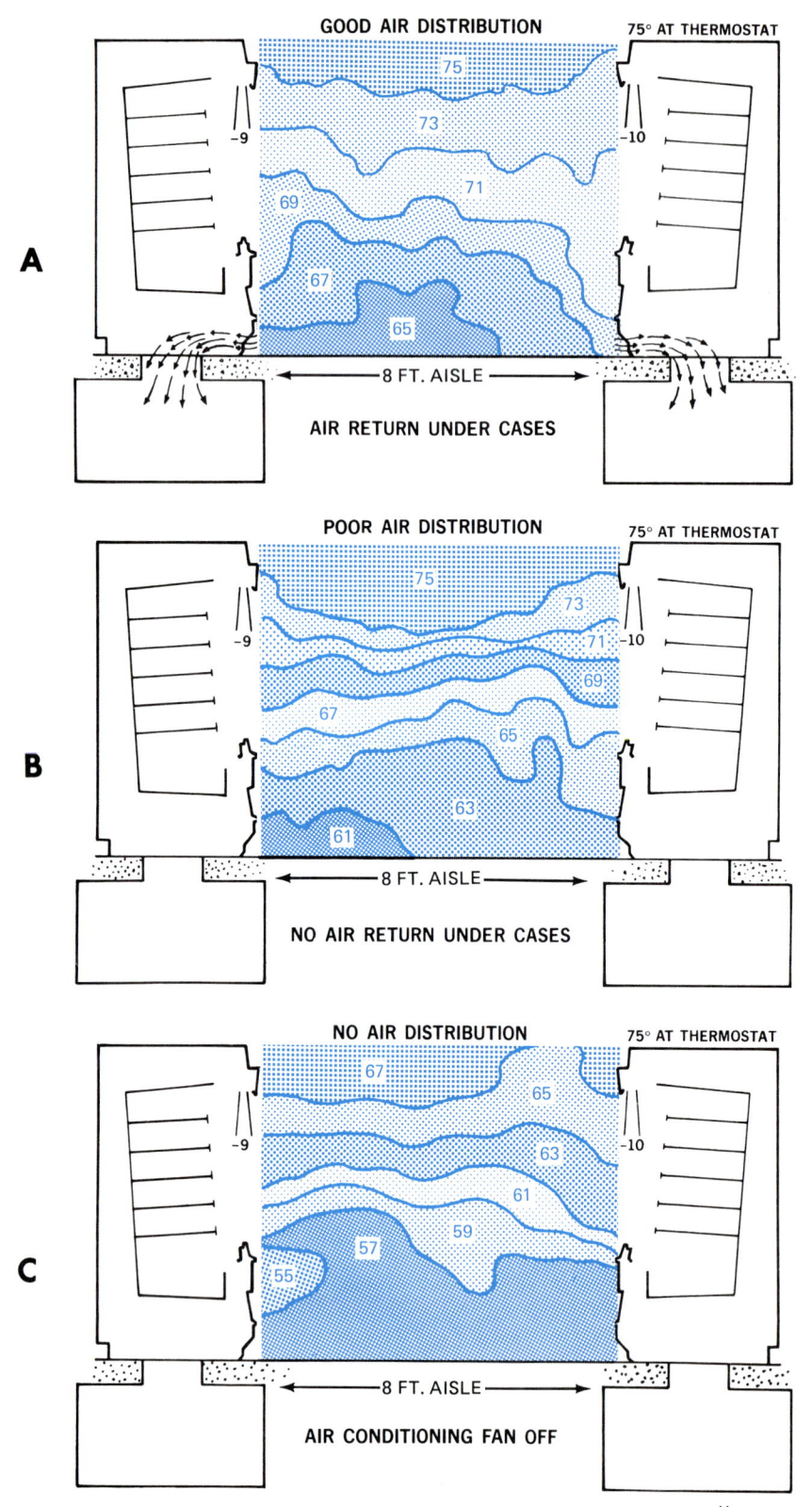

REFRIGERATION AND AIR CONDITIONING TECHNOLOGY

15-2. Temperatures in frozen food aisles with good, poor, and no air return.

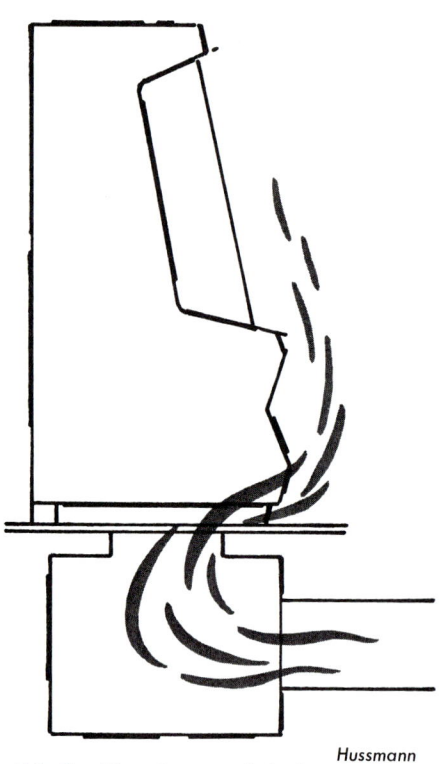

15-3. Air return path in front of refrigeration cases.

nance costs, often make air-cooled units more attractive.

Air-cooled equipment for indoor and outdoor commercial uses is available. Sizes range from the small 1/3 hp condensing units to large systems of over 250 hp.

Indoor Installation

Probably the simplest, most well-known food store refrigeration method is the standard "back room" installation. This installation is simple and direct. Performance is good. The condensers can be racked with the compressors in the equipment room. Or, they can be located outdoors. The advantages of each method vary with the installation. It should be noted that saving space is important in planning. It costs too much to

CHAPTER 15—COMMERCIAL REFRIGERATION SYSTEMS

15-4. Outside view of a mechanical center.

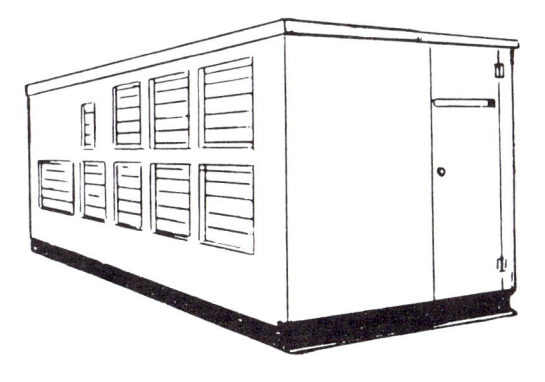

Supermarket Refrigeration

One of the most frequently encountered methods of refrigeration in the supermarket is the ice cream freezer. The open multideck ice cream merchandiser is typical of those used today.

One of the problems with the supermarket is its cold and hot zones. This means the area around the freezers or frozen foods tends to be cold. This sometimes causes people to avoid the area. To prevent a detour around such areas, several steps have been taken to improve the situation. Figure 15-2 shows how the temperatures vary between the freezers. In-aisle temperatures can be controlled with the proper installation. A cold air return system under the frozen-food cases will help prevent low temperatures in the aisle. Fig. 15-3.

Recirculating this cold air in the air conditioning system within the store reduces cooling costs. The refrigeration units are designed for a store temperature of 75° F. [23.9°C]. Installation costs can be reduced if the refrigeration lines are run in the air returns. This also ensures ease of service. Fig. 15-3. Such a procedure depends upon local codes.

REFRIGERATION UNITS

Self-Contained Compressor and Condenser Units

In the operation of freezer and cooler cases, it is sometimes appropriate to use a self-contained compressor and condensing unit. Details on the operation of this type of unit will be found in Chapter 7.

Mechanical Centers

In large stores, several coolers and freezers are used. Then, it is necessary to stack the compressors and condensers so that less room is used to house the refrigeration units. Figure 15-4 shows one of these self-contained units. Some of these units are mounted on the roof to make use of all available space. Figure 15-5

15-5. Inside view of a mechanical center.

391

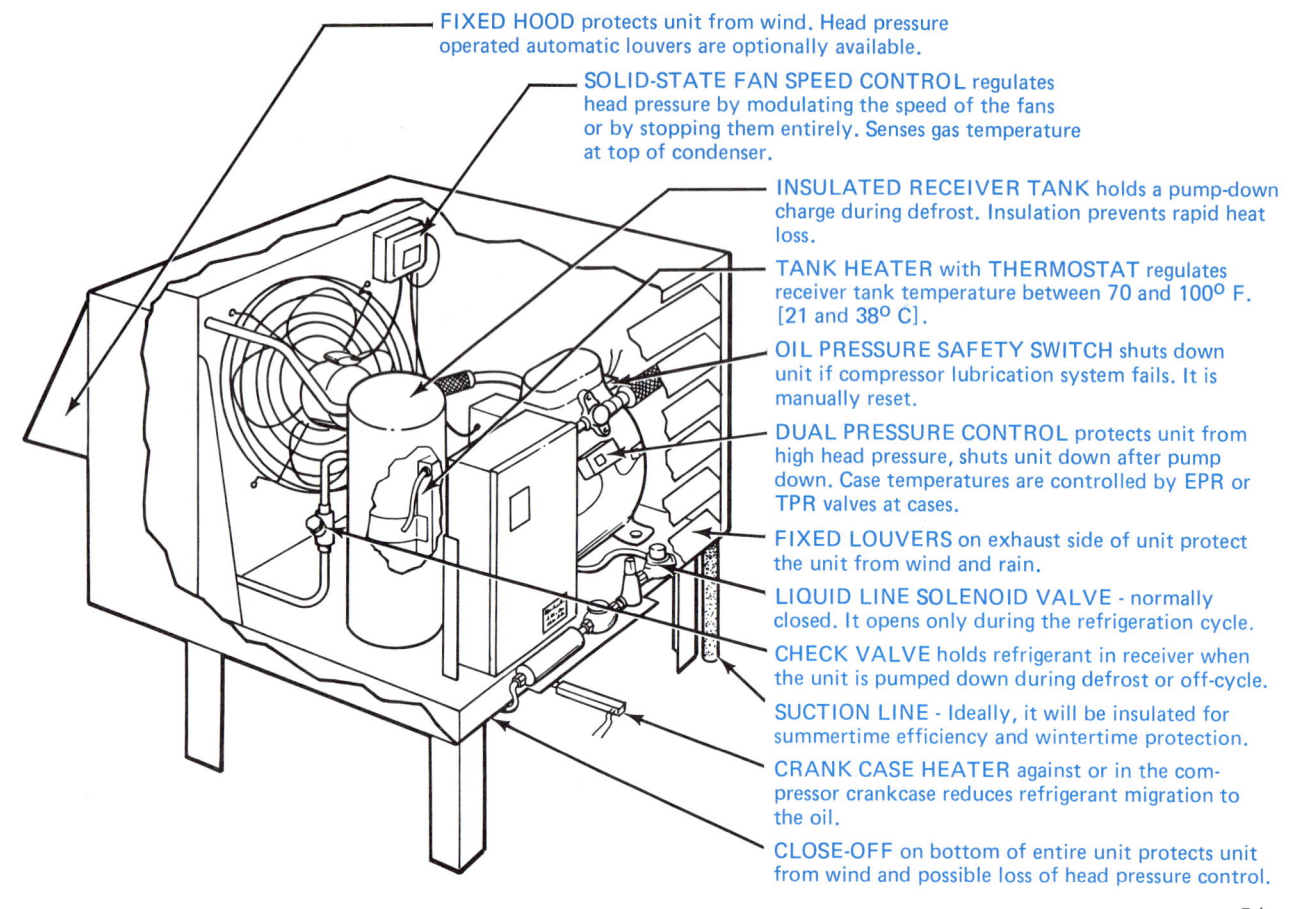

REFRIGERATION AND AIR CONDITIONING TECHNOLOGY

FIXED HOOD protects unit from wind. Head pressure operated automatic louvers are optionally available.

SOLID-STATE FAN SPEED CONTROL regulates head pressure by modulating the speed of the fans or by stopping them entirely. Senses gas temperature at top of condenser.

INSULATED RECEIVER TANK holds a pump-down charge during defrost. Insulation prevents rapid heat loss.

TANK HEATER with THERMOSTAT regulates receiver tank temperature between 70 and 100° F. [21 and 38° C].

OIL PRESSURE SAFETY SWITCH shuts down unit if compressor lubrication system fails. It is manually reset.

DUAL PRESSURE CONTROL protects unit from high head pressure, shuts unit down after pump down. Case temperatures are controlled by EPR or TPR valves at cases.

FIXED LOUVERS on exhaust side of unit protect the unit from wind and rain.

LIQUID LINE SOLENOID VALVE - normally closed. It opens only during the refrigeration cycle.

CHECK VALVE holds refrigerant in receiver when the unit is pumped down during defrost or off-cycle.

SUCTION LINE - Ideally, it will be insulated for summertime efficiency and wintertime protection.

CRANK CASE HEATER against or in the compressor crankcase reduces refrigerant migration to the oil.

CLOSE-OFF on bottom of entire unit protects unit from wind and possible loss of head pressure control.

15-6. Outdoor condensing unit with weather hood.

Tyler

shows the interior of a mechanical center. These units are usually ½ to 20 hp. They may be remote, air-cooled, or water-cooled. They are available in configurations other than that shown in Fig. 15-5.

Outdoor Condensing Units with Weather Hoods

In some applications, a number of units are not needed and a mechanical center is not called for. Then, an outdoor condensing unit is used. Fig. 15-6. These units come as a factory package. They are easily hooked up with the refrigeration unit inside the store. They may be located on the roof or on a slab outside the store. Possible loss of heat pressure control caused by wind is prevented by enclosing the underside of the unit.

It has been generally believed that it was necessary artificially to maintain summerlike conditions on the condenser of refrigeration systems. This was because a large pressure difference had to exist between the high and low sides. It has been found, however, that if adequate measures are taken to assure a solid column of liquid refrigerant to the expansion valves of cases, no problems will be experienced. The advantage is in lowering the condenser temperature. This makes the compressor run more efficiently, saving energy costs.

Liquid lines exposed to temperatures above 60° F. [15.6°C] must be insulated. This would apply to any lines that run overhead and to some lines close to the floor.

SELF-CONTAINED WALK-IN COOLERS

Many installations call for a low- or medium-temperature walk-in cooler. Such coolers are lightweight and easy to install.

392

CHAPTER 15—COMMERCIAL REFRIGERATION SYSTEMS

The air-cooled unit has condensers mounted high, requiring minimum cleaning. The front grille is easily removed for access to the refrigeration unit. A semi-hermetic unit is used in the low-temperature cooler. A full hermetic unit is used in the medium-temperature unit. Electric heaters give fast defrost. There is no drain required. The defrost water is then automatically evaporated.

Coolers installed outdoors need a roof to protect them from the sun, rain, and snow. Fig. 15-8. A prefab aluminum roof is available. A cooler can be set so that the unit projects through an outside wall of a building. An optional weather package consists of louvers, crankcase heater, fan control, and temperature control. Fig. 15-9.

SELF-CONTAINED UNIT CORNER SECTION

The complete high and low sides of the refrigeration system are contained in the $2' \times 2'$ triangular corner section. Fig. 15-10. Air-cooled units are standard. However, water-cooled units are available. Units are made to fit to the right or left of the door section. Viewed from the inside,

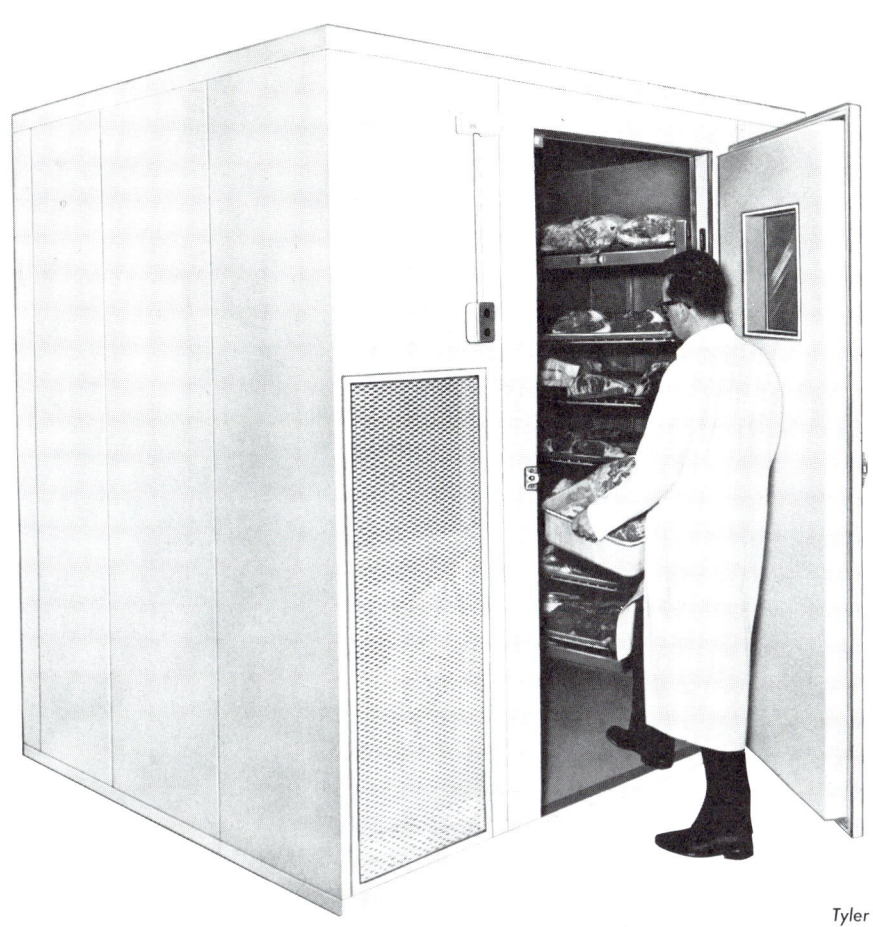

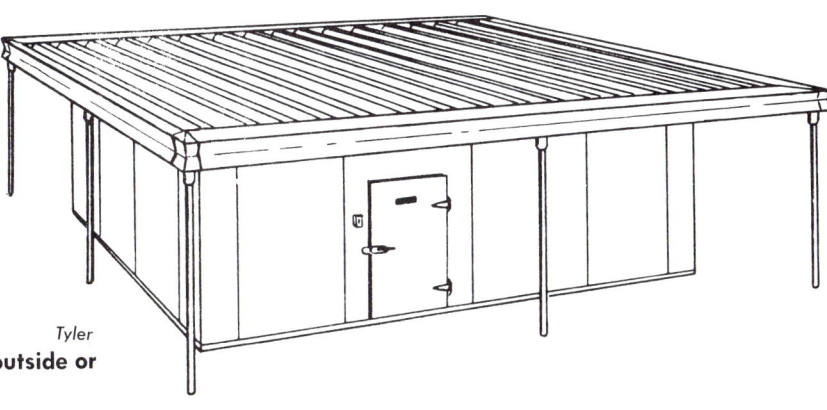

Tyler
15-7. Self-contained walk-in cooler.

The walls, inside and out, are made of 22-gage, zinc-coated steel with $2\frac{1}{2}''$ of polyurethane-foamed cores sandwiched together under pressure. This helps with structural strength and also serves as insulation. Maximum insulations can be obtained with a minimum of space. Fig. 15-7.

Refrigeration System

The refrigeration system comes completely charged, ready to operate. It fits into a wall corner section. Air-cooled or water-cooled condensing is available.

Tyler
15-8. Roof for walk-in cooler located outside or on a roof.

393

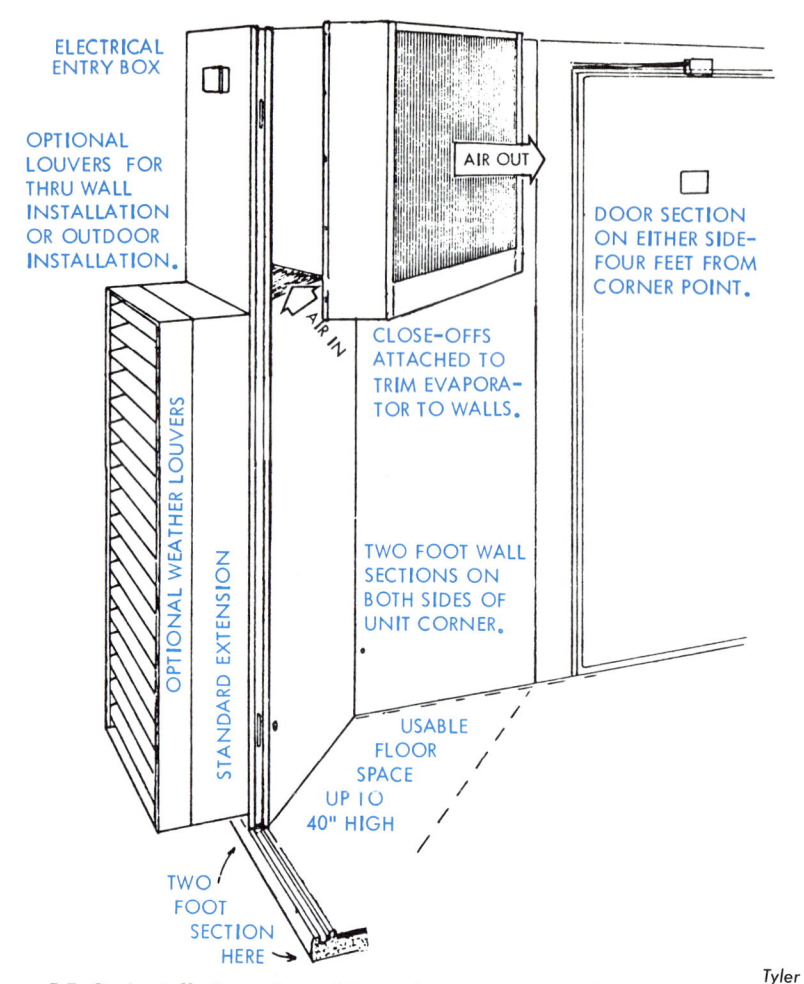

REFRIGERATION AND AIR CONDITIONING TECHNOLOGY

15-9. Installation of a refrigeration unit in a walk-in cooler. *Tyler*

right is standard and left is optional. The electrical box is on the front of the corner, as is the grille. Low-temperature units have a vacuum release device to allow air pressure to equalize quickly after the door is closed. Without this device, contracting warm air creates a partial vacuum. This prevents immediate door reopening.

Clearance

An air-cooled unit can be placed against a wall (side without the grille). However, this makes it hard to service the electrically heated drain pan. Therefore, it is good, but not necessary, to have free access on the side. The pan can be serviced by removing the condensing fan. The area in front of the grille must be clear for adequate air circulation. The water-cooled unit must have clearance on the side for the drain and water supply. Low-temperature units and medium-temperature units with the optional 1-hp compressor have projected grilles to accomodate the larger compressor bodies.

Setting Controls

When the defrost is dependent on the Dual Pressure Control (off cycle defrost), the cut-in setting must be at 36 or 37 pounds with R-12. This is most accurately determined with a pressure gage because the indicator on the control is not that precise. Gage setting of the control is also the best way of setting this control when a timed defrost control is used.

Coolers cut in at 36 to 37 pounds with R-12. Freezers have three defrosts per day at 50 minutes each.

The following factors affect defrost.

1. Store humidity and temperature, which varies through the year. Therefore, the amount of frost to be removed varies through the year.

2. Voltage to the defrost heaters may vary, causing the amount of available heat to vary.

3. Door usage (coolers and reach-ins) varies.

4. Operating temperature of the case (particularly with multishelf cases, low and normal temperature) may be lower than necessary. This will cause a heavier frost load on the coil.

5. Drafts across the case permit too much warm, moist air to enter the case, thereby loading the coil.

The defrost period must be long enough to clear the coil completely. Defrosting must occur frequently enough to keep refrigerated air circulating freely through the case.

A recording temperature gage is the most efficient method of determining when a case coil is becoming choked with frost.

CHAPTER 15—COMMERCIAL REFRIGERATION SYSTEMS

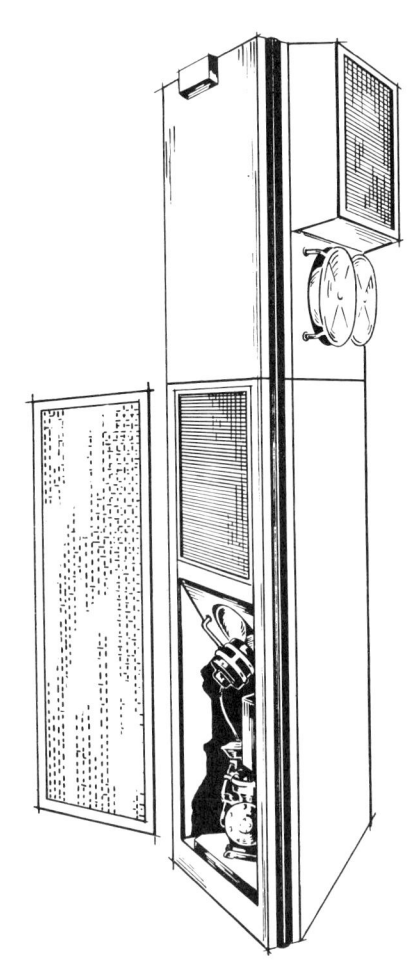

Tyler
15-10. Refrigeration unit for walk-in cooler.

Penn
15-11. Automatic water valve for water-cooled units only.

sides of the main spring and lifting upwards.

Valve Inspection of Water-Cooled Units

The valve seat and rubber disc, after long periods of operation, may become worn or pitted. To inspect the valve seat and rubber disc, relieve tension on the main spring by turning the adjusting screw clockwise. Remove the four round head screws extending through the main spring housing from the end of the valve opposite the bellows. Screws on the side must be removed on valves of 1″ or larger. The complete housing unit is removed as a unit. Next, remove the center assembly screw that allows access to all internal parts. Once the valve has been inspected, it is best to refer to the manufacturer's recommendations for repair or replacement.

Electrical

The units use single-phase or three-phase 230 V. Three-phase must be specially ordered. The door heaters and lights use 115 V. The fans must have single-phase or three-phase, according to their method of winding. Check the nameplates for this information. The cooler condensing fans use 220 V and 50 W. The air flows over the motor. In the freezer, the motor uses 220 V and is rated at $\frac{1}{4}$ hp. The air flows away from the motor in the freezer. The evaporator fans use 220 V at 9 W. They blow the air over the motor. The condensing fan cycles with the compressor in the cooler. The two evaporator fans inside run continuously.

In the freezer, the condensing fan cycles with the compressor. The inside fans are controlled by three switches:

1. The *door switch* turns off the fans when the door is opened.
2. The *defrost switch* turns off the fans during the defrost cycle.
3. The *fan delay switch* keeps the fans off after defrost until the coil is below freezing. The contacts open at 40° F. [4.4°C] and close at 20° F. [−6.7°C]. This prevents droplets of water from being blown off the coils and freezing on the ceiling and stored products.

Maintenance

The normal temperature condensing fan must be oiled once a year. An SAE 20 nondetergent oil is recommended. It should be used sparingly to avoid the col-

When the temperature "sags" before a defrost period, another defrost per day is indicated. Make sure, however, that the factors causing the temperature sag (case or ambient temperature, humidity, usage, or drafts) have been controlled as much as possible.

Water-cooled units have automatic water valves. They open or close automatically, according to pressure. They open on increased pressure and close on pressure decrease. Fig. 15-11.

Manual flushing may be accomplished by inserting a screwdriver or similar tool under two

395

REFRIGERATION AND AIR CONDITIONING TECHNOLOGY

lection of dust. The cooler condensing fan motor has one hole for oil. The inside fans do not need oiling. The fan blades should be cleaned occasionally. The low-temperature unit has a ball bearing motor that is permanently sealed. No oil is ever needed.

Periodic cleaning of the condenser and condensing compartment will eliminate 75% of potential difficulties. Accumulations of dust and dirt hamper free airflow. This causes poor heat exchange and may affect the proper functions of the controls. A thorough cleaning with a powerful vacuum cleaner every month (more often in extremely dusty conditions) is ideal preventive maintenance. Use a whisk broom on the condenser fins, stroking downward against the fins to loosen the accumulated dust. If the surroundings permit, compressed air may be used to remove the dirt.

One of the first things to check in any maintenance schedule or program is the proper electrical power. The medium-temperature cooler uses 220-V, 1Ø, 60-Hz power for operation with R-12 refrigerant to keep the cooler at 35 or 40° F. [1.7 or 4.4°C]. The low-temperature freezer can be used on 220-V single-phase or three-phase. It uses R-502 refrigerant to produce temperatures of 0° F. [−17.8°C], −5° F. [−20.4°C], or −10° F. [−23°C]. The drain eliminator uses 350 W of power to evaporate the water. Ecologically and economically, it is better to provide a drain to remove the water. The cost of installing a drain can be balanced against the cost of electrical power over one to three years' time.

SECTIONAL WALK-IN STORAGE COOLERS

Walk-in coolers and freezers are used for many applications. They may have several doors. The number of doors will depend upon the projected use of the cooler.

Sweating

Sweating is the formation of condensation on a cooler wall. This is a problem with coolers that are set at different temperatures and located adjacent to one another. The temperature gradient is shown in Fig. 15-12. This is a typical wall of two coolers with different temperatures. Note the dew point temperature is 37° F. [2.7°C] when the medium-temperature cooler has 40° F. [4.4°C] at 90% relative humidity. The low-temperature cooler has a 0° F. [−17.8°C] temperature. An insulating air film is provided on

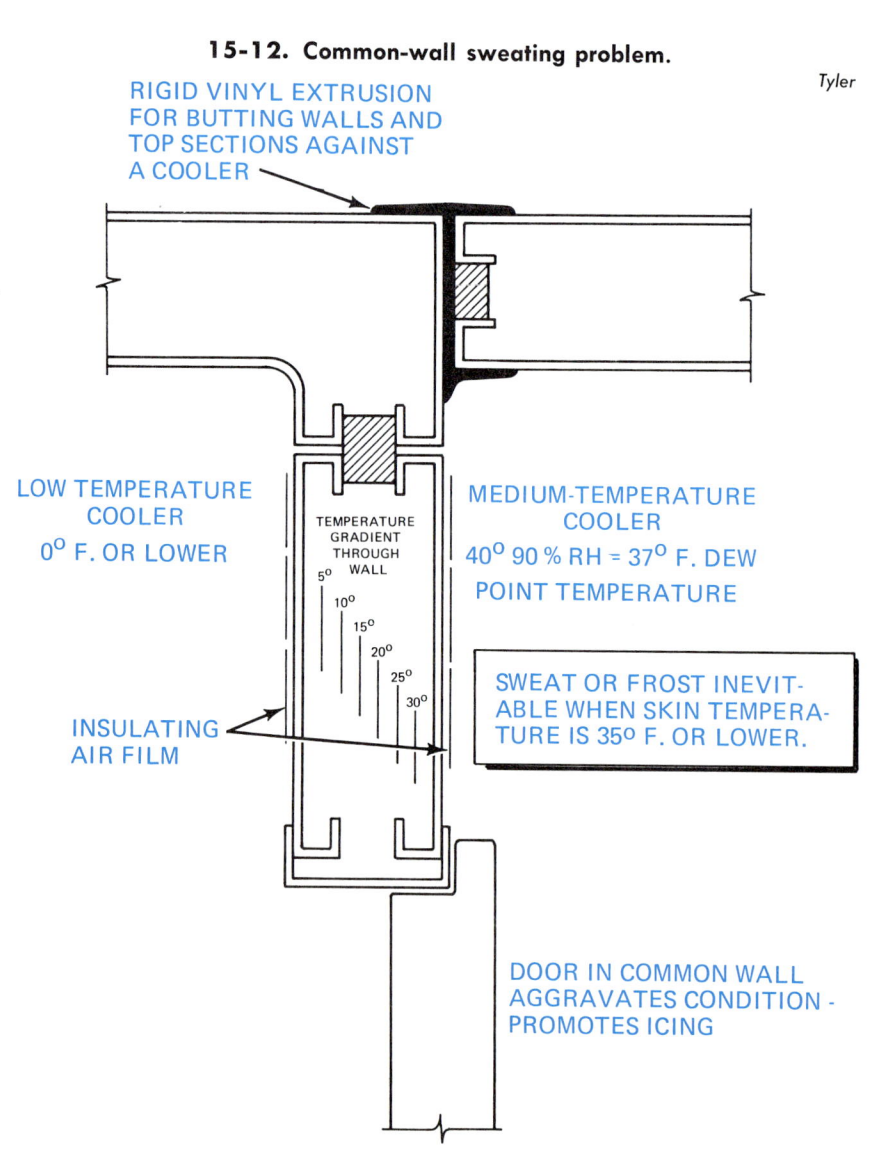

15-12. Common-wall sweating problem.
Tyler

396

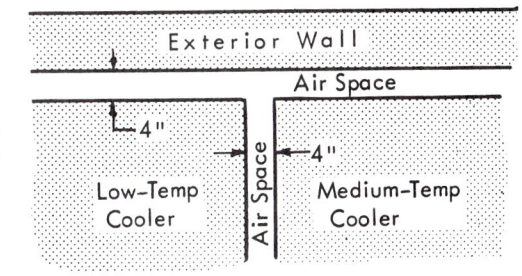

15-13. Cooler against outside walls.

PROVIDE FOR ADEQUATE AIR CIRCULATION IN AIR SPACE TO PREVENT CONDENSATION

both sides of the wall separating the coolers. This makes the low-temperature side warmer than 0°F. [−17.8°C] and the medium-temperature side cooler than 40°F. [4.4°C]. This makes condensation on the warmer side virtually inevitable.

Sweating is difficult to remedy. Electric heating wires could be arranged on or under the surface. However, this would not be practical for a large area. Another method would be to sweep the wall with high-velocity air from the coil. However, shelving or stacked boxes would interfere. Frost would appear wherever the air did not reach. There is no advantage in increasing the thickness or efficiency of the insulation. The temperature gradient would still exist.

It is much better to prevent sweating than to try to remedy it. There is no sure cure for common wall sweating. The best method is to provide separate coolers with a 4″ space between them. This space should be arranged so that natural air circulation will take place. Fig. 15-13. This is to prevent condensation on the outside of a low-temperature cooler. If adequate circulation is not available, use a continuously running fan and suitable ducts or baffles.

Coolers placed against outside walls also require a 4″ space to provide for air circulation. Fig. 15-13. Condensation may occur on either side of the cooler wall. Fans with suitable ducts or baffles may be needed, even with the spacing.

Vacuum Release Valve

Foamed walk-in coolers are very nearly airtight. When the door is opened on a low-temperature cooler, warm air enters. When the door is closed, warm air is also admitted. Closing the door quickly reduces the volume of warm air. This reduction causes a partial vacuum. The vacuum prevents the immediate reopening of the door. To eliminate this nuisance, a vacuum-release air valve is installed on the door section of the low-temperature coolers. The valve allows outside air to enter the cooler as long as a negative pressure exists inside the cooler.

Floors

Special consideration must be given to floors in walk-in coolers. They are assembled easily and must be placed on a surface made to take the load and the temperature. Figure 15-14 shows a cross-section of a floor in a

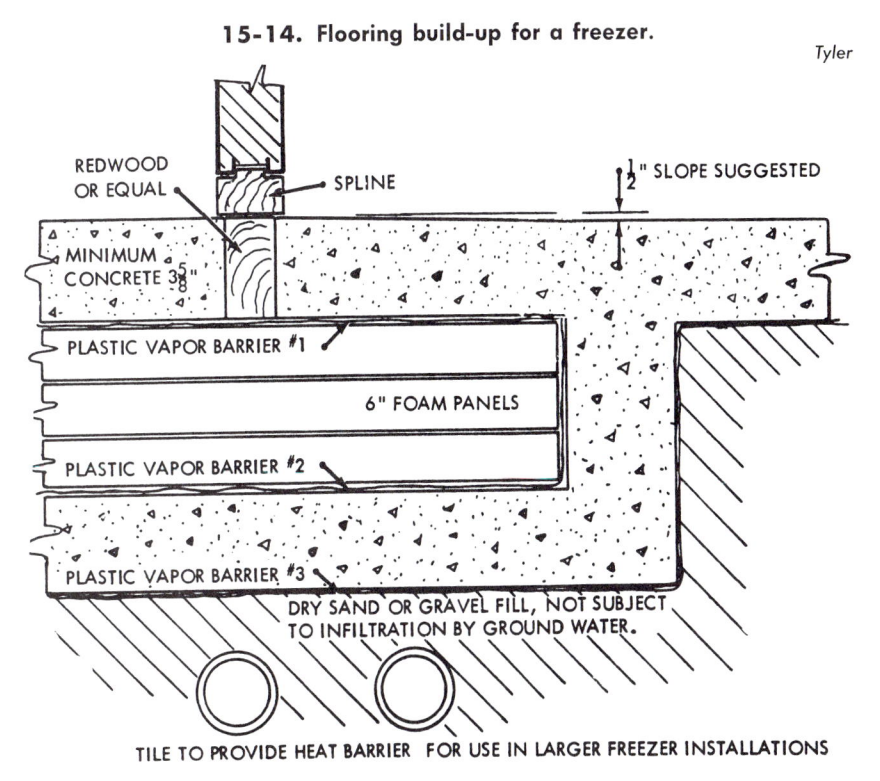

15-14. Flooring build-up for a freezer.

TILE TO PROVIDE HEAT BARRIER FOR USE IN LARGER FREEZER INSTALLATIONS low-temperature floorless cooler. A medium-temperature cooler can be placed on existing solid concrete floors. However, with a temperature of 0° F. [−17.8°C] or less, special preparations are needed. These must be made to ensure that the subsoil does not freeze and cause the floor to heave. Also, there would be a continual extra operating expense because of the heat loss through a noninsulated floor.

Electrical

Electrical considerations take some planning in the walk-in cooler. Figure 15-15 shows the typical wiring diagram for a cooler door section. Note the threshold heater and door frame heaters. A vaporproof light is also shown.

Coolers of this type use remote compressors. The evaporator unit comes in many designs. Electric defrost will provide a frost-free evaporator in a range of temperatures from 32 to −20° F. [0 to −29°C]. During the defrost cycle, a constant supply of electrical power to the evaporator and drain pan heating elements assures adequate defrosting, regardless of the amount of frost.

During the cooling cycle, an electric defrost system operates like any standard refrigeration system. At the beginning of the defrost cycle, the timer stops the evaporator fan or fans and compressor. It also energizes the electric heaters. After defrosting the evaporator, the defrost termination thermostat reverses the switches in the timer to end the defrost cycle.

The fan delay feature of the thermostat keeps the fans from starting until the evaporator has cooled down enough to prevent heat or water droplets from being blown into the cold room.

The timer also has a defrost cycle termination fail-safe built into the time control dial. If the system is not returned to the refrigeration cycle by the increase in temperature, the fail-safe will reverse the switches.

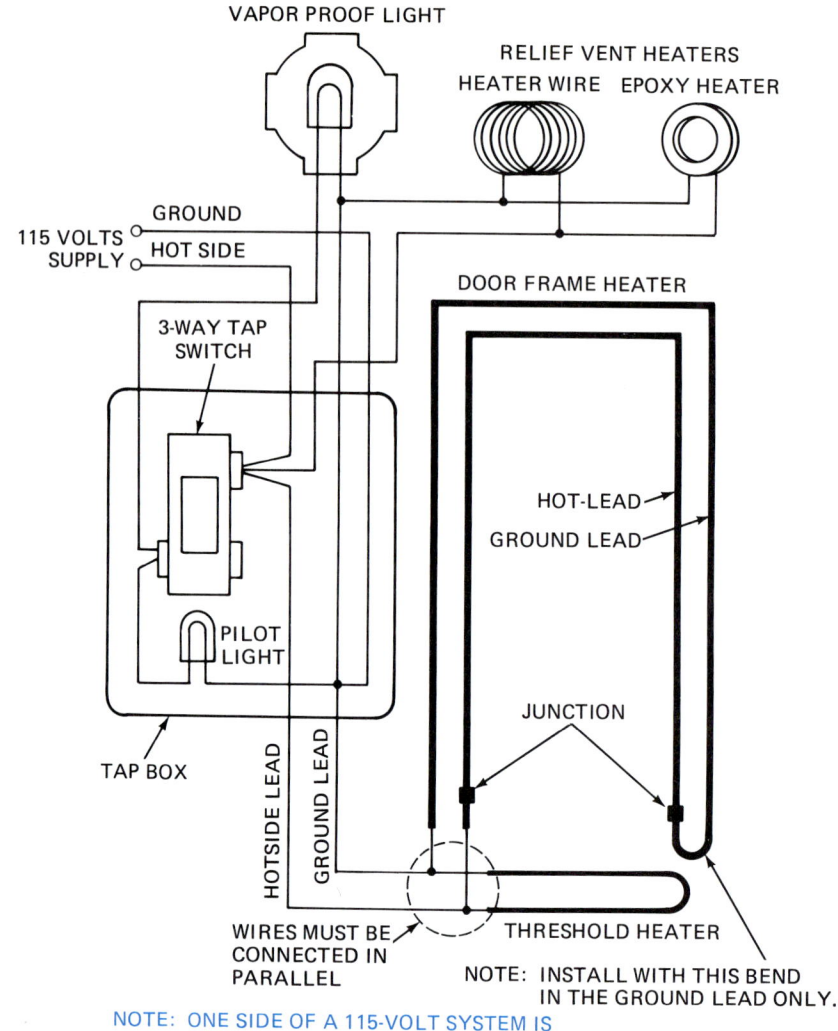

Maintenance

If a heater or heaters must be replaced, the following procedure is suggested:

1. Shut off all power to the unit. Disconnect from the terminal block the leads of the heater(s) to be removed.

2. To remove the drain pan heater, the drain pan must first be lowered. Then pull the heater out of the retaining clip, starting at the drain end. To install the new heater, reverse the procedure. Then kink down the heater

15-15. Typical wiring diagram for cooler door sections.

CHAPTER 15—COMMERCIAL REFRIGERATION SYSTEMS

clip in several places to keep the heater in place.

3. To remove the heater(s) from the inlet face of the coil, first remove both end panels of the unit. Next, remove the spring wire clips holding the heaters in place. Then pull the heater straight out the front. Locate the new heater in the unit so that it will not interfere with replacing the end panels. Then carefully drive as deeply as possible into the slots where the heater was removed. Install the heater spring clips.

4. Always be sure that the heater leads or seals are routed or bent such that they do not contact the hot part of the heater. Also be sure to connect lead wires to the same terminals as used by the heater which was removed.

Other maintenance amounts to little more than oiling the motor. All the motors are provided with oil holes. Reoiling is recommended every six months with SAE 20 nondetergent oil.

The evaporator units have motors that need lubrication once every year. The motors are easily removed through the fan openings. The only tool required is a crescent wrench.

Filters are used over the fans. They are of polyester plastic with an electrostatic characteristic. They can be cleaned by removing light dirt with a vacuum cleaner or by removing the filter and running warm water over it.

To remove heavy or oily deposits from the filter, wash it in warm water containing a good detergent. Then rinse thoroughly, since a detergent film impairs the filter's effectiveness.

On multiple units, the thermostats are wired in series. Thus, the defrost is not terminated until all units are free of frost.

Figure 15-16 shows the two different types of timers that may be found on walk-in coolers. Both models have a manual defrost spot on them. Instructions come with the units.

Hot Gas Defrost

So far, we have concentrated on the electric defrost for the walk-in coolers. It is possible to use hot gas to defrost. Figure 15-17 shows the installation of the expansion valve and the necessary hookup for accomplishing hot gas defrost. Figure 15-18 shows the assembled by-pass kit for the unit shown in Fig. 15-17.

When a defrost is initiated, the hot gas valve reverses the refrigerant flow, sending hot gas to the coil through the suction line. This hot gas heats the thermostat bulb. It causes the SPDT thermostat to turn off the fans and turn on the drain pan heater. Refrigeration resumes when the defrost is terminated. The thermostat soon switches the fans on and the drain pan heater off. Thermostat operation is at ap-

15-16. Two types of timers for defrosting.

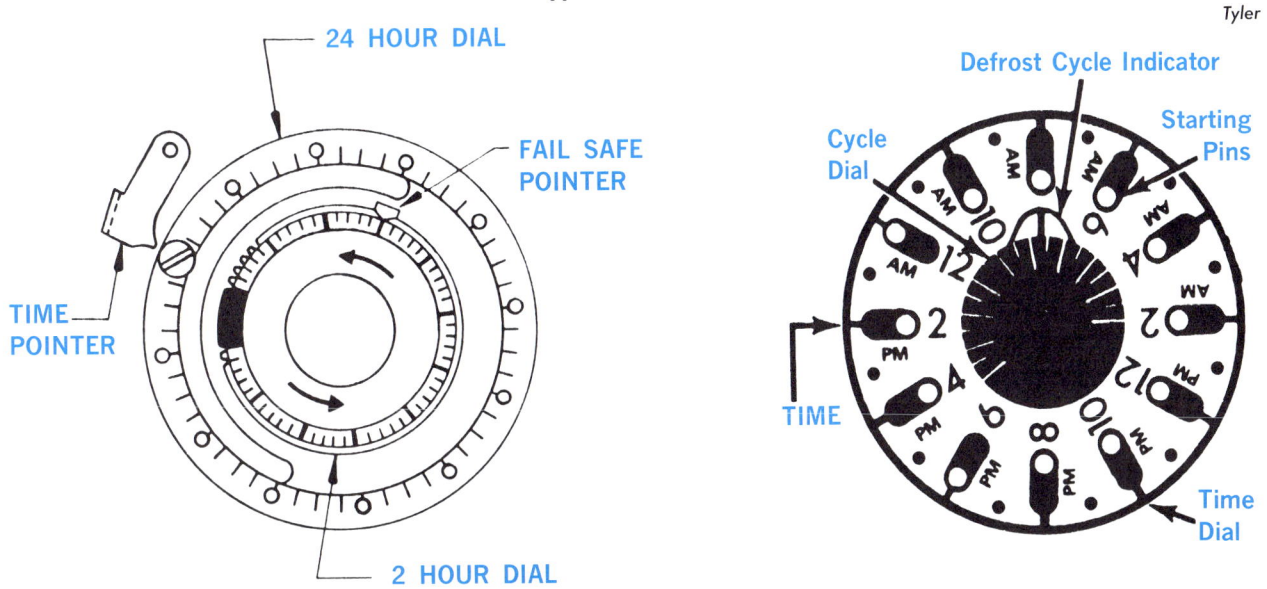

Tyler

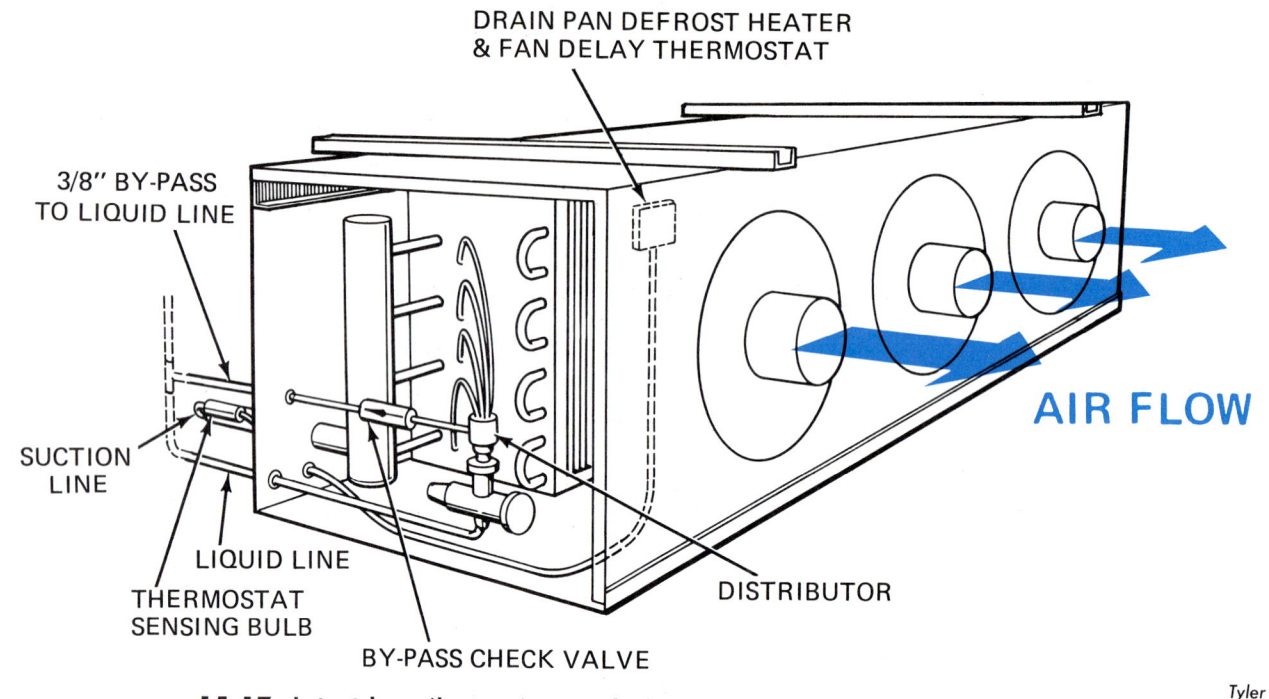

REFRIGERATION AND AIR CONDITIONING TECHNOLOGY

15-17. Latent heat (hot gas) as applied to low-temperature unit cooler coils.

Tyler

proximately 50 to 30° F. [10 to −1.1°C]. The defrost frequency is four to six times per day with each duration being 15 to 20 minutes. Of course, this may vary with conditions.

COMMERCIAL REFRIGERATION UNITS

Commercial refrigeration units include the following:
- Produce cases.
- Meat cases.
- Dairy and deli cases.
- Frozen food cases.

Each of these units has a compressor, condensing unit, and/or mechanical center to keep it cool or cold. The refrigeration unit is usually located outside the store. In some cases the equipment room arrangement is used to house the mechanical units. This leaves the inside of the store free of the noise generated by the compressor and the fans in the condenser.

Heat recovery systems are also popular because they conserve energy. The heat generated by the unit is recirculated by the heating/air conditioning system to keep the store at a comfortable level. This recirculated heat can save on energy costs. Exhausting it to the outside air is wasteful. As energy becomes more expensive, it will be more economical to ensure that all units recover part of the wasted energy.

Setup of Cases

Many problems can be caused by improper installation of cases. Thus, it is a good idea to check the precautions noted by the manufacturer. Refrigerant lines, for instance, may be run from case to case in the area designed for them. Notches are provided in the ends for this purpose. The tubing *should not* be run through these cases to another machine if the cases are controlled by a pressure control.

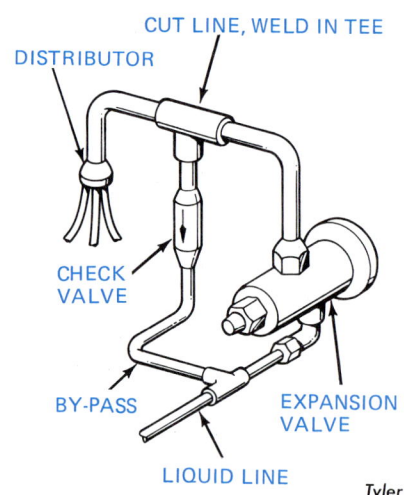

15-18. Assembled by-pass kit for unit shown Fig. 15-17.

Tyler

400

CHAPTER 15—COMMERCIAL REFRIGERATION SYSTEMS

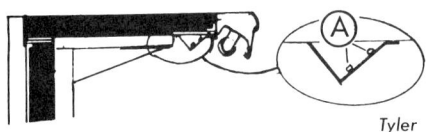

15-19. Antisweat wire location on case.
Tyler

Since the pressure in a system responds to the coldest location, tubing from one system running through a second system may be chilled to the point that its pressure control will respond to pressures in the second system. This is called cross-controlling. *Avoid it.*

When using a torch to solder the tubing, make sure the torch and hot solder do not touch the fiberglass drain pan. This can be prevented by using a wet cloth to cover the pan. At the same time, make sure the expansion valve feeler bulb is protected from heat. Remove it if necessary. Replace the bulb tightly, though.

Seal the refrigerant lines where they enter the case. Permagum sealer is furnished. This sealing is necessary to prevent condensation, air leaks, and other problems. Be sure the sealing has been done before leaving the job.

Tighten the flare nuts on expansion valves in cases before charging the system. They might have been loosened from the vibration that occurs in shipment.

Tighten all the electrical screws in the control boxes. This will eliminate possible burning of these terminals a short time after installation.

Rubbing alcohol (isopropyl) can be used to remove tape and other stains on the case.

Shelves are placed in the proper order when shipped to help with air circulation. If they are removed, make sure they are replaced in their proper place to assure the proper flow of cold air.

ANTISWEAT HEATERS

Continuous electrical wireways have been incorporated into the case. There are knockouts for supply conduits from under the case. When using 1⅛″ conduit, use the supplied reinforcing washer to add strength to the raceway connection. The base cladding kick-plate covers the wireways and can be removed to speed running the wires.

Antisweat wire prevents sweat from appearing in areas where cool surfaces would condense moisture from the air or where the customer may contact a cool surface. Fig. 15-19. The antisweat heater may be 25 ohms per foot on the 8′ case and 11 ohms per foot on the 12′ case.

Fans, antisweat heaters, and lighting are the only electrical circuits to be considered in the case. The defrost circuit and other electrical controls are enclosed in a control box. This box is usually located elsewhere than on the case. Figure 15-20 shows the entire electrical wiring diagram of one case model.

LOCATION

Selecting a location for a case can make a difference in carefree operation. Three conditions must be considered before a case is placed into position.

1. The case should be located away from direct sunlight and drafts.
2. The case should be located where the customer will be most likely to stop and make a selection.
3. There should be adequate aisle space around the case.

ELECTRICAL CONNECTIONS

Figure 15-21 shows a 4′ × 6′ self-contained dairy, deli, and produce display case. This case requires a separate 30-A, 115-V circuit. This will supply the case fan, antisweat wire, case canopy, and shelf lights. It also supplies the condensing unit. The voltage range for the welded steel compressor is 103–126 V. Voltage lower than this will cause problems in starting and in operation. There are, of course, units that operate on 208–230 V, single-phase, 60 Hz.

This unit is self-contained. It has its own pull-out compressor. Temperature adjustment is obtained by turning the knob on top of the thermostat case. The knob should be turned to the right for cooler and to the left for warmer temperatures.

The factory settings of the thermostat are: Cut-in, 28° F. [−2.2°C] and Differential, 15° F. [−9.4°C].

Defrost control is during four 30-minute periods per day. These settings will give an average temperature of 38° F. [3.3°C]. A high-pressure control is in series with the thermostat. This is a safety device. If the timer is properly set when the unit is started, the defrost periods will occur at 6:00 A.M., 12:00 noon, 6:00 P.M., and 12:00 midnight. The frost removed from the refrigeration coil melts during defrost and drains into a condensate pan. The pan has a thermostatically controlled electric heater. This heated pan vaporizes the water to the air, eliminating the need for external waste outlets. This applies only to self-contained units. With this feature, the unit can be placed in almost any location after a store

401

REFRIGERATION AND AIR CONDITIONING TECHNOLOGY

is built. It becomes a flexible unit capable of being used anywhere there is 30-A, 120-V service. Smaller units use a hermetic compressor that has a ¾ hp rating for the 4' model. It has a capacitor-start induction-run motor. The larger 6' case uses a 1.5 hp unit. The motor is a permanent split-capacitor type.

Maintenance of Self-Contained Equipment

This case is equipped with an air-cooled condensing unit. The condenser (like a car radiator) has two condensing fans that force air through the condenser to cool the refrigerant. The fins on the condenser will accumulate dust. This impedes air flow and, thus, proper cooling. A regular cleaning program should be set up to keep the condenser clean. Once a month is the average time between recommended cleanings. However, in extremely dusty locations, such cleaning should be more frequent. A whisk broom and a tank-type vacuum cleaner make an effective combination in removing dust from the fins.

Detecting Improper Refrigeration Charge

The refrigerant charge in a capillary tube system is *critical*. This means that the amount of refrigerant in the system must be correct within an ounce or so. An undercharge or overcharge will result in poor performance, if not in a total loss of refrigeration.

An *overcharge* of refrigerant will cause the suction line to frost over more than usual. In short, the system will show the same symptoms as an expansion valve with the valve too far open.

To correct, simply purge the refrigerant from the high-side service valve. Do this with care so as not to let out too much refrigerant. CAUTION: An iced-up coil will show the same symptoms as an overcharge. Make sure the coil is clean before you purge.

An *undercharge* is most obviously indicated by poor refrigeration in the display area. The heat exchanger may also be ab-

15-20. Condensing unit wiring diagram for four and six foot self-contained merchandiser.

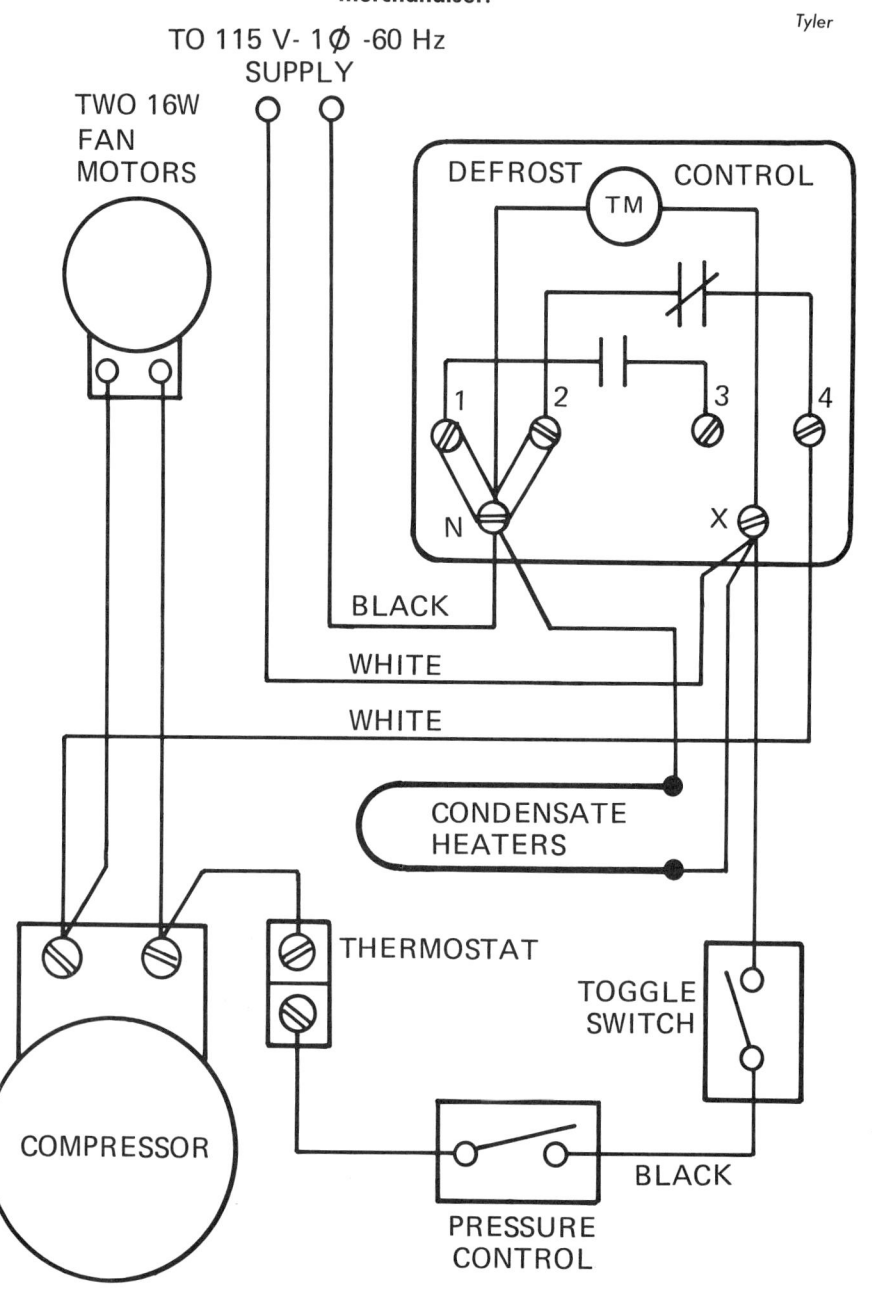

Tyler

402

normally warm. An undercharge may cause the coil to partially ice up. This is because part of the coil will be just at, or just below, the freezing point. If such is the case, any accumulation of moisture on this portion of the coil will result in "glazed" ice, which will not be removed during the defrost period.

To generalize, it might be said that adjusting the change of a capillary tube system can be done in much the same manner as adjusting the expansion valve of a system equipped with such a valve.

Measuring Head Pressure and Back Pressure

HEAD PRESSURE

Gages can be used to measure the head pressure to check the R-502 refrigerant charge.

Room Temperature	Average psi
70° F. [21.1°C]	200 pounds
80° F. [26.7°C]	230 pounds
90° F. [32.2°C]	260 pounds

BACK PRESSURE

Gages can be used to check average conditions for R-502 refrigerant using the thermostat settings shown above. Note that the following readings show at cut-in and cut-out.

Cut-in at approximately 40 pounds
Cut-out at approximately 70 pounds

Low Voltage Problems

Low voltage at the case is indicated by the following symptoms:

• The compressor will have difficulty starting and may cycle on overload several times before starting. This may happen if the unit comes on too soon after stopping.
• The compressor may not start at all.
• Fluorescent lights will flicker.

The above symptoms are especially true of 115-V circuits and to a lesser extent of 208- to 230-V circuits.

Checking the voltage with an accurate voltmeter *while the unit is starting* will give a true reading of the actual voltage. Reasons for low voltage may be:

• The electrical lines from the store fuse cabinet may be too small.
• The power company may have overloaded the lines and may not be able to deliver full voltage.
• The connections may be loose.

Low voltage conditions can be corrected by the following:

• If the lines are undersized, rerun the lines from the store fuse cabinet using the properly sized wire according to good wiring practices.
• If the low voltage is with the supply, contact the power company to correct the deficiency. If this is not possible, the only practical solution is to install a booster transformer on the case. Contact the manufacturer of the unit for the proper size and specifications.
• If the connections are loose, tighten them.

Table 15-A lists guidelines for troubleshooting refrigerated cases.

OPEN PRODUCE CASES

Fresh fruits and vegetables are still living, even after harvesting. They continue the processes of respiration and transpiration after harvesting. Respiration is the process of self-feeding to provide energy for maintaining life. For example, asparagus and sweet corn generate considerable heat after picking. Transpiration is the loss of water in vapor form from the plant tissues. Post-harvest life can be maintained by slowing the rate of water loss. Refrigeration lowers the rates of respiration and transpiration. Most types of produce should be kept at storage temperatures close to freezing prior to display. There are, of course, exceptions. For example, cucumbers can keep relatively cool by themselves. They can be injured by temperatures below 40° F. [4.4°C].

Refrigerated produce cases provide a refrigerated place to display produce attractively. The refrigeration coil is below the display. Fans are used to circulate air through the display. This moving air will pick up moisture

Tyler
15-21. A versatile minicase, multi-shelf merchandiser for dairy, deli, and produce.

403

Table 15-A.
Troubleshooting Refrigerated Cases

Problem	Possible Cause	Remedy
Unit fails to run. (Condenser fans do not run.)	Blown fuse or tripped breaker. System on defrost. Defective time clock. Defective thermostat.	Check fuses, settings, time clock, and thermostat.
Unit fails to start (tries but cycles on the overload).	Insufficient time allowed for pressure to equalize in unit. If equalizing time is more than 4 minutes, check for dirty condenser, low line voltage, bad condenser fan(s), system overcharged, or bad relay or capacitor.	Voltage must not be below 10% of standard. Check for balanced pressures with gage. Thermostat setting must allow over 4 minutes before recycling. Capillary tube may be restricted or plugged.
Unit runs but gives poor or no refrigeration.	Coil iced up. Dirty condenser. High head/non-condensibles Case fan(s) not working. Moisture in system. Unit undercharged. Bad condensing unit. Capillary tube plugged. Drafts around case. Merchandise poorly stacked.	Displayed merchandise must not extend beyond the shelves. Case must not be exposed to sun. Eliminate drafts.
Coil ices up repeatedly.	Defrost timer set for less than 30 minutes and 4 times per day defrost. Drafts around case. Store poorly heated. Unit undercharged. Time clock faulty.	In some situations, increase defrost length.

from unwrapped produce and carry it to the coil. Fig. 15-22.

It is necessary to replace this moisture by using a water spray several times during the day and to cover the produce with a wet cloth cover during the night. The alternative to sprinkling is to wrap the produce.

Maintaining Case Air Flow

To provide proper refrigeration in a produce case, the front of the case must *not* be blocked with packages. One way to improve air circulation is to use air deflectors below elevating screens in the case. These deflectors will direct the air flow into the display. Fig. 15-22. This prevents cool air from "short circuiting" the display. They are always recommended with hump screens.

Defrosting

The case environment temperature is 40°F. [4.4°C]. A straight-time defrost control time clock is recommended for the defrost control. The low-pressure control should be set for an off-cycle defrost (36 pounds high event setting). The high event is the high-pressure limit for the low-pressure control. This procedure will assure the best humidity performance. If desired, the time clock can be omitted. The time clock control is a safety measure. It helps to keep the coil free of ice. It does this even with

15-22. Air circulation in a produce case.

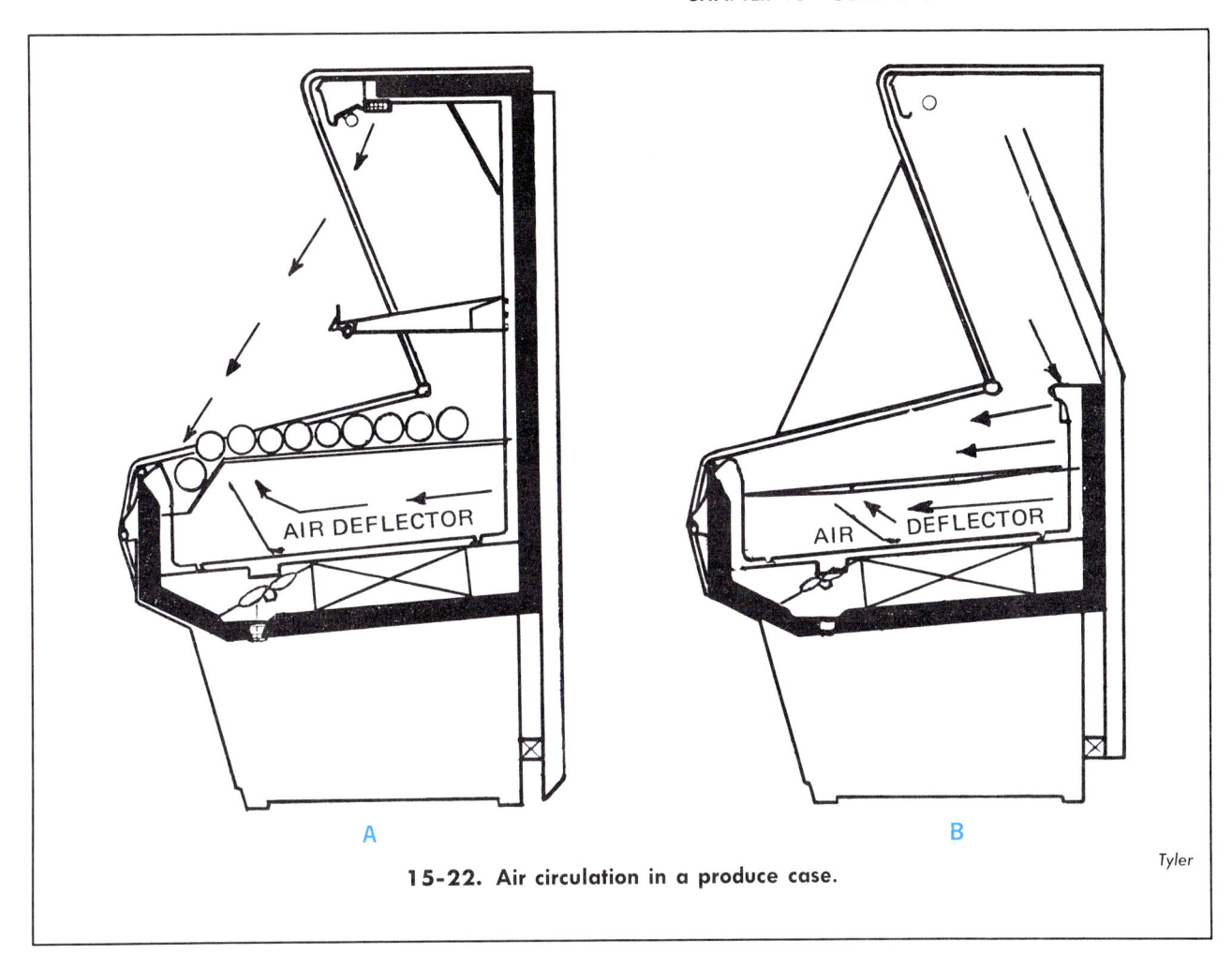

improper pressure control settings.

A solenoid-reset, defrost-control time clock is recommended for the defrost control on some models. The control is basically an on-off timer switch with an added solenoid trip mechanism. The solenoid is wired in series with the "line" and the terminating thermostat(s) on the coil(s). Thus, the thermostat senses the end of the defrost and terminates the cycle. The *timed termination* function acts as a "fail safe."

Typically, during the defrost cycle, the coil fans are off, the compressor is off, and the defrost heaters are on. The reverse occurs during the refrigeration cycle.

To determine the correct size for the condensing unit it is necessary to know the Btu per hour rating of the condenser. The condensing unit selection is based on 90° F. [32.2°C] air entering the condenser. The aim is to produce an environment in the cooler of 40° F. [4.4°C] at a suction temperature of 25° F. [−3.9°C] when the refrigerant is R-12. Manufacturers will furnish the Btu rating of each compressor-condensing unit and the horsepower requirements to maintain the proper temperature inside the cooler. It may be self-contained or be a remote refrigeration unit.

Meat-Display Cases

OPEN MEAT-DISPLAY CASES

Meat-display cases are enclosed or open. Fig. 15-23. They are designed to keep meat on display at 30° F. [−1.1°C]. Note the airflow pattern and the temperatures in the display case. Fig. 15-24.

Generally, the temperature of displayed meat in refrigerated sales cases will run higher than the circulated air temperature in the cases. The surface temperature of the meat will be higher than the center temperature due to the radiant heat. Fig. 15-25.

Two major sources of radiant heat are display lights and ceiling surfaces. Two other important heat sources are bad display practices. These include overloading the case with meat and allowing voids to develop in the display. Fig. 15-26. Such practices cut the efficiency of the refrigeration unit. This inefficiency raises the surface temperature of the meat. Bacteria and mold grow rapidly when surface temperatures rise above 45° F. [7.2°C]. This discolors the meat prematurely and causes unnecessary meat department losses. Following are two examples of how heating factors add to raise the surface temperature of meat.

EXAMPLE 1

Case air temperature over and around meat product	30.0° F. [−1.1°C]
Add for 150 footcandles from heat-rejecting overhead flood lamps	+ 5.0° F. [2.8°C]
Add for 80° F. [26.7°C] ceiling temperature	+ 3.5° F. [1.9°C]
Add for high load conditions	+ 2.0° F. [1.1°C]
Add for voids in display	+ 2.0° F. [1.1°C]
Net surface temperature of meat (typical internal temperature 38° F. [3.3°C])	42.5° F. [5.8°C]

EXAMPLE 2

Case air temperature over and around meat product	34.0° F. [1.1°C]
Add for 150 footcandles from standard overhead R-40 flood lamps	+8.0° F. [4.4°C]
Add for 100° F. [37.8°C] ceiling temperature	+5.0° F. [2.8°C]
Add for high load conditions	+2.0° F. [1.1°C]
Add for voids in display	+2.0° F. [1.1°C]
Net surface temperature of meat (typical internal temperature 45° F. [7.2°C])	51.0° F. [10.5°C]

Tyler
15-23. An open meat case with vertical mirrors or a glass sliding door canopy. The counter can be filled from the workroom.

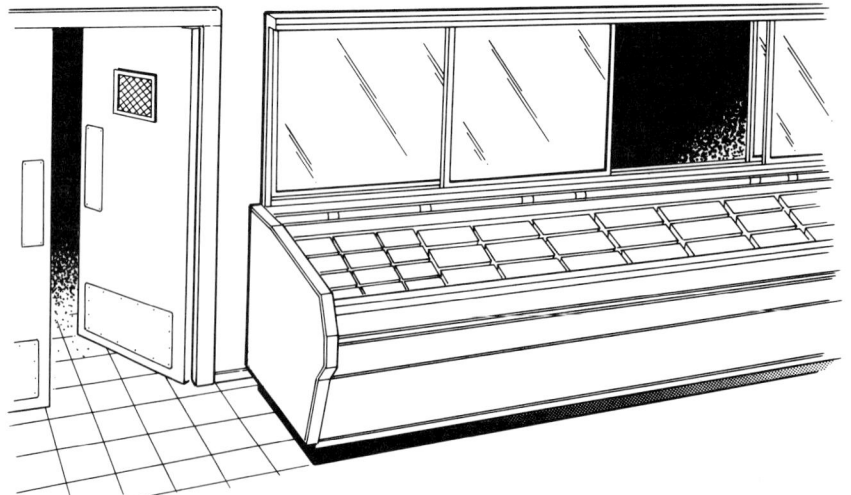

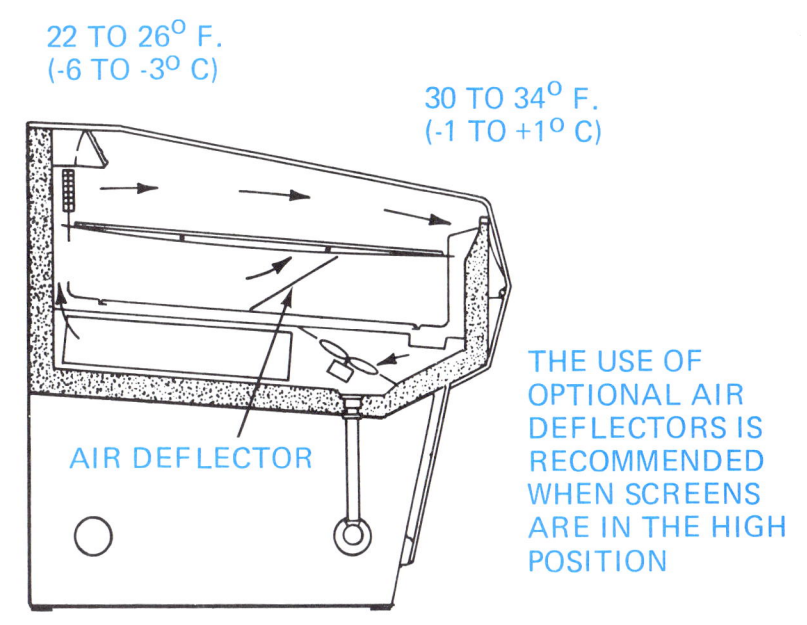

CHAPTER 15—COMMERCIAL REFRIGERATION SYSTEMS

22 TO 26° F.
(-6 TO -3° C)

30 TO 34° F.
(-1 TO +1° C)

15-24. Air flow and temperatures in a top display case.

AIR DEFLECTOR

THE USE OF OPTIONAL AIR DEFLECTORS IS RECOMMENDED WHEN SCREENS ARE IN THE HIGH POSITION

To measure radiant heat, place two accurate dial thermometers side by side in a case. Cover one of the thermometer stems with black friction tape. Expose the other thermometer stem. The temperature difference is the amount of radiant heat. Reduce the amount of radiant heat by changing the display lighting. Determine the source of the high ceiling temperature. Eliminate the ceiling high temperature source. Encourage the butcher to maintain all meat below the case load lines. Instruct the butcher to fill all product voids. In some instances, case screens can be covered to keep the refrigerated air over the display.

Figure 15-27 gives information

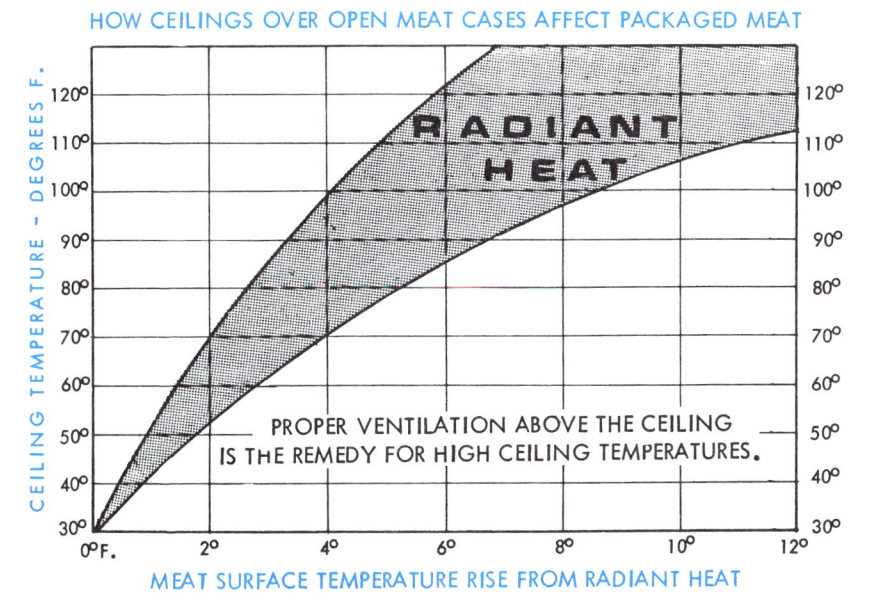

HOW CEILINGS OVER OPEN MEAT CASES AFFECT PACKAGED MEAT

RADIANT HEAT

PROPER VENTILATION ABOVE THE CEILING IS THE REMEDY FOR HIGH CEILING TEMPERATURES.

MEAT SURFACE TEMPERATURE RISE FROM RADIANT HEAT

15-25. The effects of ceiling temperature on packaged meat.

407

REFRIGERATION AND AIR CONDITIONING TECHNOLOGY

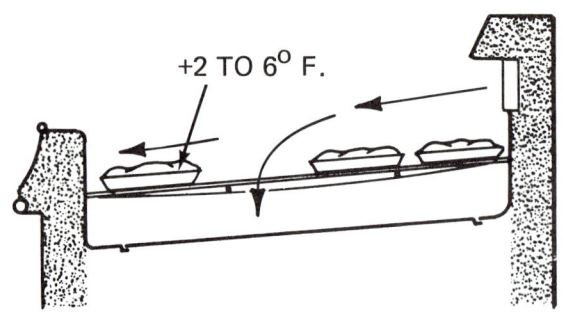

VOIDS IN DISPLAY RAISE SURFACE TEMPERATURE OF PACKAGE IN FRONT OF VOID 2 TO 6° F.

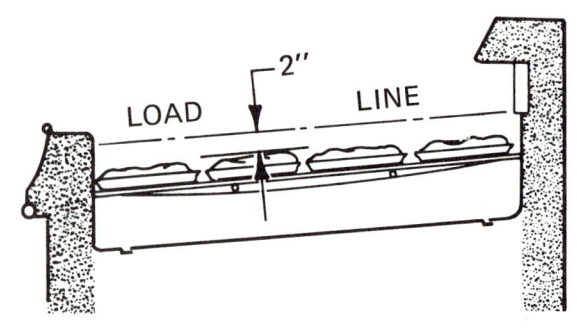

ALL MAKES & STYLES OF DISPLAY CASES - PRODUCT 2" BELOW LOAD LINE IMPROVES SURFACE TEMPERATURE 2 TO 6° F.

15-26. Voids and display load line in a meat case.

on some of the problems associated with display heat sources.

DEFROST

Defrosting can be nonelectric with a straight time defrost control time clock. Defrosting can also be done with a solenoid reset defrost control time clock. It is an on-off timer with an added solenoid trip mechanism. A thermostat turns off the coil fans during defrost.

CHECKLIST FOR OPEN MEAT CASES

Open meat cases should be checked for correct operation. Meat is a large investment for the store and its temperature is very important. Table 15-B lists general guidelines for checking open meat cases.

Enclosed Meat Cases

Some meat cases are enclosed. They require the butcher to serve

HEATING EFFECT OF VARIOUS TYPES OF LIGHTING AND INTENSITIES ON PACKAGED MEAT IN OPEN DISPLAY CASES

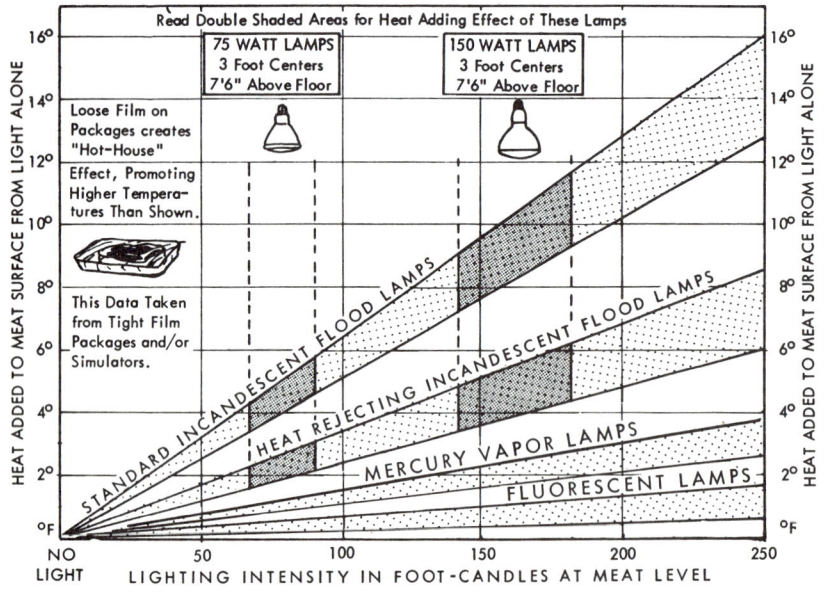

Read Double Shaded Areas for Heat Adding Effect of These Lamps

75 WATT LAMPS — 3 Foot Centers — 7'6" Above Floor

150 WATT LAMPS — 3 Foot Centers — 7'6" Above Floor

Loose Film on Packages creates "Hot-House" Effect, Promoting Higher Temperatures Than Shown.

This Data Taken from Tight Film Packages and/or Simulators.

STANDARD INCANDESCENT FLOOD LAMPS
HEAT REJECTING INCANDESCENT FLOOD LAMPS
MERCURY VAPOR LAMPS
FLUORESCENT LAMPS

HEAT ADDED TO MEAT SURFACE FROM LIGHT ALONE (°F)
LIGHTING INTENSITY IN FOOT-CANDLES AT MEAT LEVEL

15-27. The effects of lighting on packaged meat.

NOTES ON CHART: A wide temperature range is shown for each type of lighting. This data is conservative - in many situations, the package warm-up may be more than indicated. Brand names of heat rejecting lamps: GE - Cool Beam; Sylvania - Cool Lux; Westinghouse - Low Temperature Flood. Standard flood lamps are clear PAR 38 and R-40 types. Blue, pink or other hues in these types reduce the amount of light drastically, but not the heat which is about the same as shown.

CHAPTER 15—COMMERCIAL REFRIGERATION SYSTEMS

15-28. An enclosed meat case.

the customer. Fig. 15-28. Sliding doors on the back-side of the case give easy access to the display. Some refrigeration requirements for this type of case differ from those for the open type. The main noticeable difference is the air circulation pattern over the display. Fig. 15-29. The correct compressor size is important. Oversizing is as detrimental as undersizing. Select the compressor size from the manufacturer's specification guide sheets on each case.

15-29. Air circulation in a top display case. A double-duty case has slightly lower temperatures in the storage section.

34 to 38°F.

PRESSURE CONTROL SETTINGS

A cut-in of 36 psig using R-12 refrigerant assures complete clearance of frost on the off cycle and a cut-out of 16 to 19 psig. Satisfactory temperatures can be obtained under these cut-out and cut-in settings. These are the pressures at which defrosting is turned on and off. Running this type of equipment at lower than 35° F. [1.7°C] results in wide temperature fluctuations and a large humidity spread. It is better to operate at a slightly higher and steadier temperature and thus avoid the drying effect of these fluctuations. These cases differ markedly from open meat cases, which cycle every few minutes and operate at colder temperatures.

It should be noted that 36 to 40° F. [2.2 to 4.4°C] temperatures will not necessarily result in any detrimental effects to the meat. This is because the radiant heat sources are within a closed case and can be minimized. Use of fluorescent lighting can help keep the product temperature close to the air temperature. Therefore, the meat may be maintained at a temperature that is the same, or slightly colder, than that in an open case.

DEFROSTING

The standard piping and coil arrangement for these cases is designed for pressure control or thermostatic control in which the control shuts off the unit each time the pressure or temperature control is satisfied. The ½-ton expansion valve feeds the lower

REFRIGERATION AND AIR CONDITIONING TECHNOLOGY

Table 15-B
Checklist for Open Meat Cases
CHECK AND RECORD ALL OPERATING CONDITIONS *BEFORE ANY CHANGES OR ADJUSTMENTS ARE MADE*

(A) Check and record pressure control *cut-in* _____ and *cut-out* _____.
(B) Check and record defrosts per 24 hours _____, time or fail-safe _____. Minimum reset _____ psig.
(C) Check and record head pressure _____ psig and refrigerant charge.
(D) Check and record case air temperatures.* (entering, center of display, return air) Use chart like the one shown in (L) below.
(E) Check and record product temperature _____ °F. Stem of thermometer placed in center of product that has been in the case at least four hours without a defrost.
(F) Check and record footcandles of light at product level. _____ FC. Lamp type _____, Watts _____, Brand _____.
(G) Check coil for ice or blockage _____. Drain pan for ice or water _____.
(H) Check for correct fan blades and motors. Check each blade's position in the fan panel to be assured of maximum airflow performance. See this drawing.
(I) Check coil to see that it is being fed properly. _____.
(J) Check electric defrost heaters on cases so equipped to see that heaters are not drawing any current. _____.
(K) Check for drafts in area of case, using a smoke gun. _____.
(L) Check case air velocity using a velometer.** Check entering air and air at center of the display just above the product level. Record readings as shown below.

	LH End	2'	4'	6'	8'	10'	RH End	LH End	2'	4'	6'	RH End
Entering Air												
Center of Display												

*HOW TO TAKE CASE AIR TEMPERATURES: Use a stainless-steel stem thermometer. Hold it horizontally in front of the *entering air* grids, reading the temperature at 2-foot intervals the full length of the case. Do not insert the stem into the rear duct. *Center of Display*—place the thermometer stem at product level with the stem in the air flow. (Do not let the stem touch the surface of the product.) *Return Air*—place the thermometer at the return duct in the air flow. Do not insert the stem inside the front duct. All the above temperatures are to be taken every 2 feet, the full length of the case, left to right, for easy reporting and averaging of recorded temperatures.

coil first. Fig. 15-30. The refrigerant passes from the lower coil through the check valve, then goes to the top coil. The check valve permits rapid warmup of the top coil independently of the bottom coil. The bottom coil, by design, does not get above freezing. Therefore, weekly defrosting (accomplished by the normal weekly case cleaning) or daily automatic defrosts are recommended. This design gives maximum product shelf life with the best temperatures and humidity.

If the coiling were built without the check valve, it would take 30 minutes or more to clear the bottom coil of frost. This is because of its location at this coldest part of the case. During this long off-cycle defrost, the product would warm excessively and the humidity would rise to 100%. When the machine did run again, falling temperatures and decreasing humidity would dry

410

CHAPTER 15—COMMERCIAL REFRIGERATION SYSTEMS

Table 15-B. (Continued)
Checklist for Open Meat Cases
AFTER RECORDING CASE OPERATION, EVALUATE COLLECTED DATA AND MAKE NECESSARY CHANGES AND ADJUSTMENTS.

(A)	*Cut-in* 20 to 22 #, *cut-out* 9 to 12 # for Tyler Cases.
(B)	Two defrosts per 24 hours, 40 # pressure reset, 60-minute fail-safe. Electric defrost—52 # pressure reset. 36-minute fail-safe.
(C)	Head pressure 90 psig or more with a clear refrigerant sight glass.
(D)	Entering air 20 to 23° F., center of display 26 to 29° F., return air 30 to 35° F.
(E)	Product center of display 36 to 44° F. (after product has been in the case 4 hours.)
(F)	100 to 150 footcandles at product level. Less light is better for displayed meat.
(G)	Evenly frosted coil with no blockage. Drain pan clear of all ice and water.
(H)	Fan blades and motors, check Spec Guide Parts List. Fan blade centers should be flush with the plane of the fan panel *when the blade is pulled to its upper position.*
(I)	The last pass of the coil should be partly flooded or there should be a slight frost on the suction line at the TXV bulb.
(J)	Check defrost heaters with an Amprobe to be sure they are not on or getting low voltage from feedback or cross wiring.
(K)	Draft in area of meat case display not to exceed 15 feet per minute (fpm).
(L)	Velocities double-duty case entering air: 270 to 330 fpm Center of case at display load level: 70 to 90 fpm Velocities top display case entering air: 260 to 280 fpm Center of case at display load level: 70 to 90 fpm (These values for 1974 Tyler Cases. See manufacturers' specifications for other cases.)

**An Alnor Velometer, Jr. was adapted for taking case air velocities by adding a collecting scoop. Plans available from Tyler Tech Staff. Take velocities in this manner: hold scoop against lower part of plastic air grid starting at the left end and recording readings for every 2 feet of the case. Take a center of display case readings at product level every 2 feet of the case.

the meat, shortening its display life.

Simply adding a liquid line solenoid valve and thermostat to a case and operating it on a pump-down cycle will not give desirable results.

When the thermostat is satisfied, it shuts off the liquid line solenoid. With the compressor still running, the liquid remaining in the top coil is boiled away first. This area at the top is the warmest. The bottom coil refrigerant boils away last in this coldest part of the case, pulling the temperature down further. Freezing may result in the front of the case. A longer than desirable cycle results. Setting thermostats for 32° F. [0°C] also accentuates the problem, since closed meat cases are not designed to operate at freezing. Operating these cases at 36 to 40° F. [2.2 to 4.4°C] will give the best overall results in obtaining maximum meat shelf life.

When cases are to be operated from parallel or other large, constantly running systems, use a suction line solenoid valve and a thermostat. Do *not* use a liquid

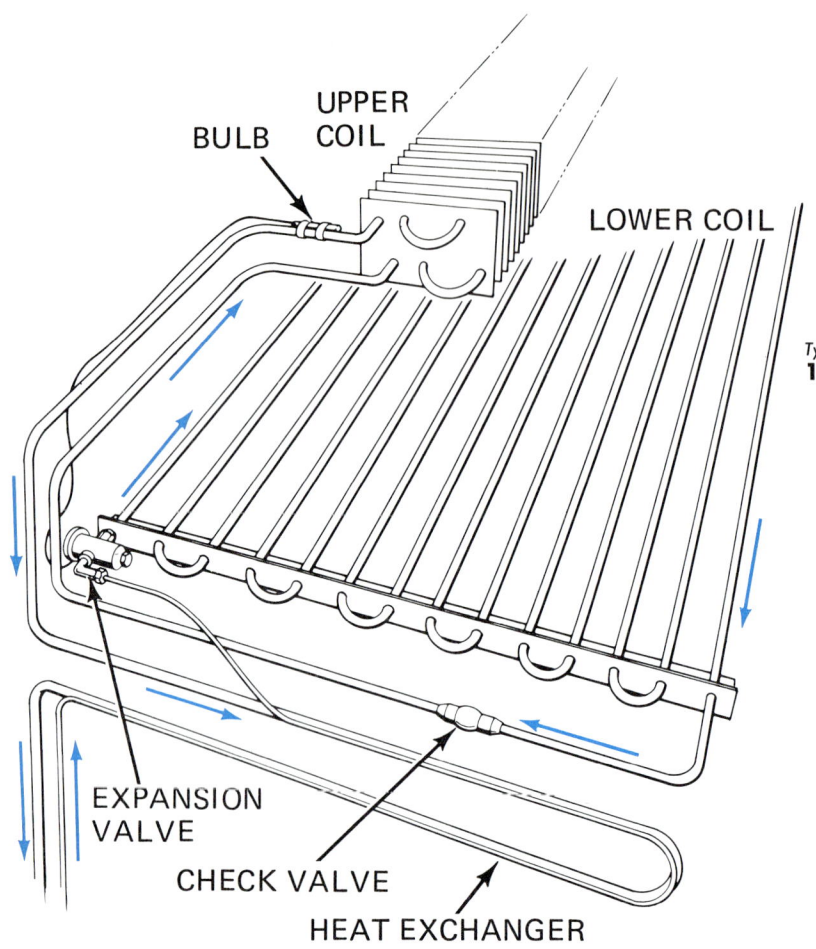

REFRIGERATION AND AIR CONDITIONING TECHNOLOGY

15-30. U-case coil design and off-cycle defrosting.

line solenoid valve. There will be problems if a liquid line solenoid is used. *Shutting off the suction line will allow the case to off-cycle defrost as intended.*

ICE CREAM DISPLAY

The parts of an open low-temperature refrigerator are shown in Fig. 15-31. This unit is designed for merchandising ice cream. These refrigerators have been designed for use only in air conditioned stores where the temperature and humidity are maintained at or below 75° F. [23.9°C] dry bulb temperature and 50% relative humidity.

The vertical open design is sensitive to air disturbances. Air currents passing around this refrigerator will seriously impair its operation. It is important to make sure no air conditioning units, electric fans, open doors, or windows create air currents around the case. The upper shelves are adjustable. The lower shelf is stationary. The adjustable shelves can be staggered, but not removed. If they are removed, unsatisfactory performance may result. These upper shelves are tilted back at an angle. This presents the most favorable view of the product to the customer. Fig. 15-32.

Merchandise should not be placed in this refrigerator for at least 6 hours after the unit is placed in operation. At no time should stock extend beyond the load limits indicated at the ends of the refrigerator. See Fig. 15-33. The air discharge return flues must be unobstructed at all times. Otherwise, operation will be seriously affected.

Some frost formation may be noticed on packages in the rear of the case. This frosting is very light. It does not detract from the merchandising unless the product stays in the rear too long. Packages that are in the case too long will tend to stick together.

Clearance-Top Access Panels

Two removable panels are provided in the top of the case. These allow access to the air curtain fans. No canopy or permanent display should be erected that interferes with the removal of the panels and servicing of the fans.

CHAPTER 15—COMMERCIAL REFRIGERATION SYSTEMS

Clearance from Walls

A minimum of 1" must be allowed between the back exterior surface of the case and walls or other cases at its back. Sweating may occur, even with the required clearance. Here, humidity is high. For these conditions, some method of forced ventilation, such as a built-in fan ventilation kit, must be used.

The ventilation kit improves the store aisle temperature, especially when many multideck refrigerators are concentrated in any portion of a store.

Refrigerant

This case uses Refrigerant 502. The liquid line is $\frac{3}{8}$" O.D. and the suction line is $1\frac{1}{8}$" O.D. Refrigerant piping may be brought out of the refrigerator through the refrigerant line outlet. This is located at the front and rear of

15-31. Cutaway view of supermarket refrigeration case. (1) Wiring harness-lights, (2) wiring harness assembly for drip pipe heater, (3) wiring harness, 120 volt, fan and heaters, (4) defrost terminating thermostat (Klixon), (5) bottom coil assembly, (6) heat exchanger assembly, (7) motor mount bracket, (8) wiring harness main fans, (9) light channel, front, (10) package guard, (11) front shelf support, (12) bottom rubrail, (13) top rubrail, (14) front panel, (15) color band assembly, (16) trim, (17) trim top front, (18) lamp shield, (19) coil assembly, rear, (20) rear shelf support, (21) rear discharge panel, (22) shelf assembly (16"), (23) shelf discharge grille, (24) product stop (16" shelf), (25) shelf assembly (14"), (26) product stop for rear of 14" shelves, (27) top discharge panel, (28) wiring harness, air curtain, (29) canopy and exterior top, (30) canopy trim, (31) display pan assembly, (32) splash guard assembly, (33) splash guard assembly (short section), (34) refrigeration thermostat.

Hussmann

REFRIGERATION AND AIR CONDITIONING TECHNOLOGY

Schaefer

15-32. Vertical ice cream freezer with doors.

They should be insulated for a minimum of 30' from the refrigerator. Additional insulation for the balance of liquid and suction lines is required wherever condensation and drippage would be objectionable.

Expansion Valve Adjustment

Expansion valves must be adjusted to feed the evaporator fully. Before attempting to adjust valves, make sure the evaporator is either clear or only lightly covered with frost. The fixture should be within 10° F. [5.5°C] of its expected operating temperature before adjusting.

Adjust the valves as follows. See Fig. 15-34. Attach two sensing probes (either thermocouple or thermistor) to the evaporator. Attach one under the clamp holding the expansion valve bulb. Attach the other by taping it securely to one of the return bends approximately two-thirds through the evaporator circuit. Some "hunting" of the expansion valve is normal. The valve should be adjusted so that during

15-33. Side view of case showing the discharge air and return air vents and load limits.

Hussmann

the right hand end of the case base. Lines can be routed to the rear outside of the case through the space provided by the recessed base.

Pressure Drop

Pressure drop can rob the system of capacity. To keep the pressure drop to a minimum, keep the refrigerant line run as short as possible, using the minimum number of elbows. Where elbows are needed, use long radius elbows only.

Insulating Refrigerant Line

For refrigerators with electric defrost, suction and liquid lines should be clamped together.

the hunting the greatest difference between the two probes is 3 to 5° F. [1.6 to 2.7°C]. With this adjustment, during a portion of the hunting the temperature difference between the probes will be less than 3° F. [1.6°C]. At times it will be as low as 0° F. [−17.8°C]. Adjust the valve stem no more than one-half turn at a time. Wait for at least 15 minutes before making another adjustment.

Refrigerator Temperature Control

The refrigerator temperature should be controlled by a thermostat. One refrigerator thermostat is required per condensing unit. For mixed multiplexing, one is required for each refrigerator system.

Defrost Controls

The standard type of defrost is electric. It is time initiated and temperature terminated.

The thermostat for defrost control for the case shown in Fig. 15-33 is adjusted as follows:

1. Adjust expansion valves to feed the evaporator fully.
2. Measure discharge air temperature at the moiré grille offset in the interior top panel. Set thermostat to close the liquid line solenoid at the discharge air temperature shown. The thermostat must have a differential of 3 to 6° F. [1.6 to 3.3°C] and have its bulb mounted just above the rear evaporator.
3. For electric defrost, the defrost should be time initiated and temperature terminated.
4. Defrost termination is controlled by defrost termination thermostats (one in each refrigerator). They are located behind the raceway at the left end of the case. Initially set the thermostat dials at 60° F. [15.6°C]. After the refrigerator has been in operation for at least four hours and has a normal frost load on the evaporators, trip the timer for a defrost. Check discharge air temperature in each refrigerator at the moiré grille offset in the interior top panel. Adjust the thermostat in each refrigerator to close its contacts when discharge air in that refrigerator reaches the temperature shown. Defrost will terminate when all thermostats have closed contacts.
5. No other refrigeration systems connected to the same compressor(s) should be on defrost at the same time as this unit just described.

Thermostats

The bulbs of both the refrigeration thermostat and the defrost thermostat are located in the rear air flue above the back coil. Fig. 15-35. Wire defrost termination

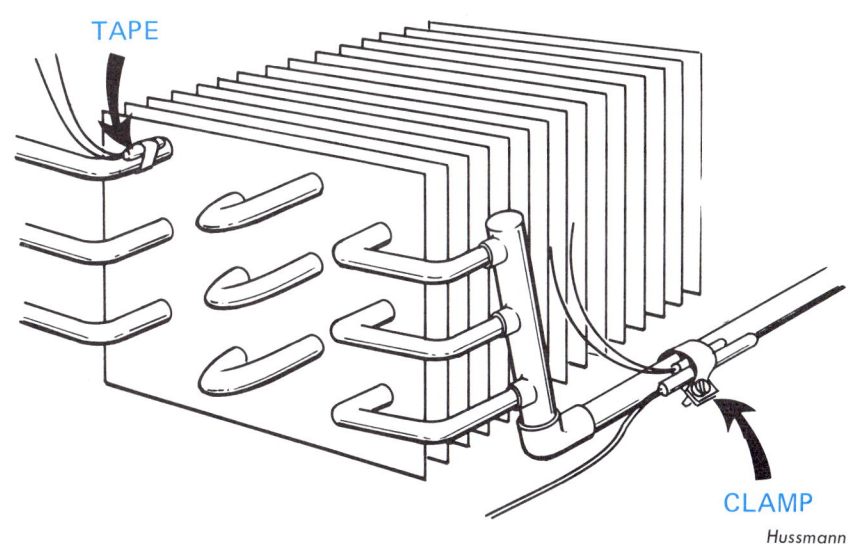

15-34. Expansion valve adjustment.

15-35. Location of thermostats.

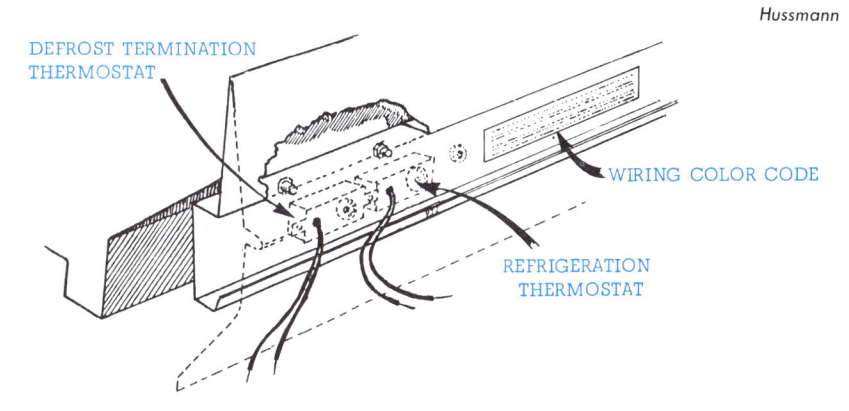

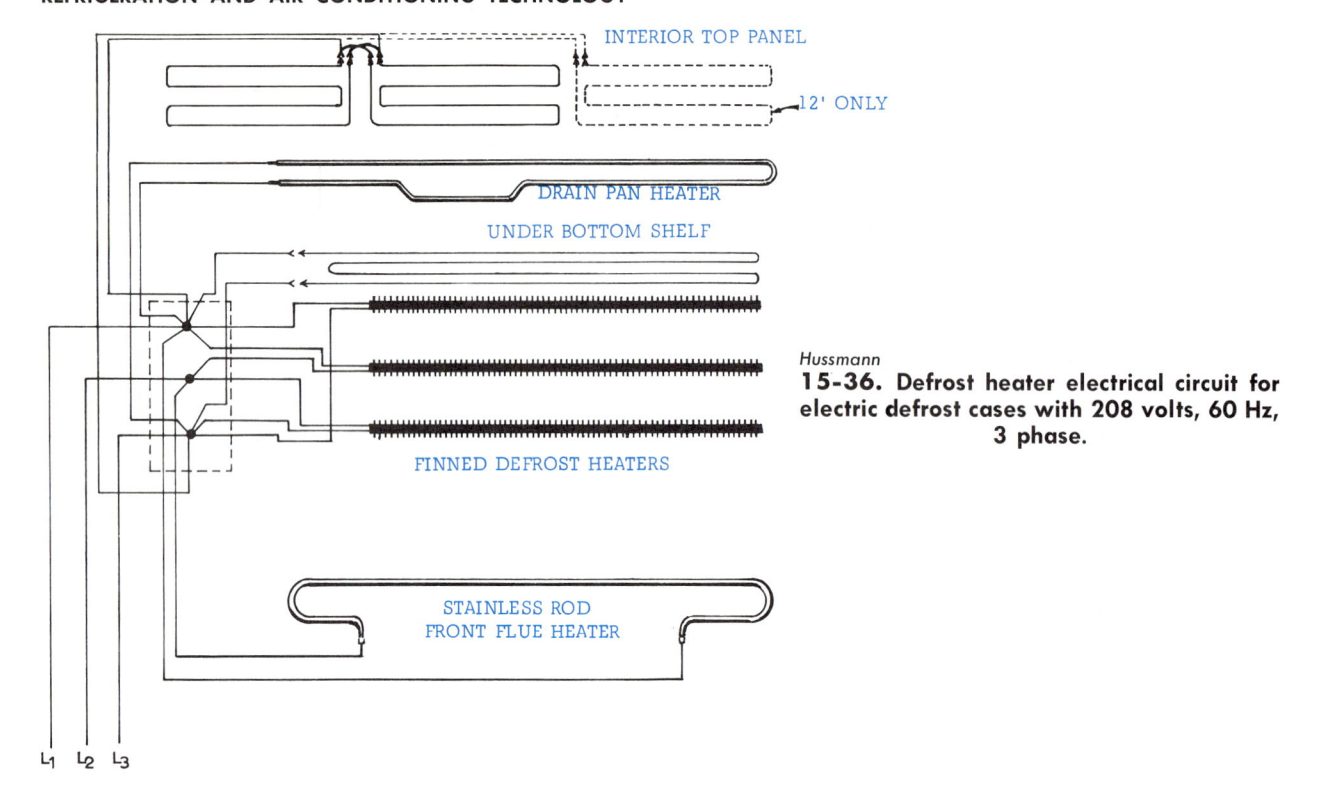

REFRIGERATION AND AIR CONDITIONING TECHNOLOGY

Hussmann
15-36. Defrost heater electrical circuit for electric defrost cases with 208 volts, 60 Hz, 3 phase.

thermostats in series for all refrigerators are connected to the same condensing unit or refrigeration system. Only one refrigeration thermostat is required per condensing unit or system.

Figure 15-36 shows the electrical wiring circuit for electric defrost cases. Supplemental electric heat is necessary when cases are equipped for Koolgas® defrost. A stainless rod heater is located in the return air flue and is controlled through a factory-installed thermostat. The leads are located in the electrical wireway. They are to be connected in parallel with fans and anti-sweat circuit.

Electrical Controls

When more than one case is connected in line it is called multiplexing. This indoor unit has the three thermostats for defrost wired in series. Fig. 15-37. This means each thermostat switch must be closed before the defrost circuit is energized. The heavy lines are to be installed by an electrician. The defrost circuit requires 28.7 A on the 220-V line for the 12′ case. Thus, the size of the wire is quite large. Figure 15-37 shows three-phase, 60-Hz (cycle) power supplied to the refrigerator.

Installing Drip Pipes

A 1″ MPT pipe extends through the front of the refrigerator base at the center of the refrigerator. An external water-seal pipe section is supplied with each case. It must be used as part of the drip pipes to prevent air leakage or insect entrance.

Poorly installed drip pipes can seriously interfere with the operation of refrigeration equipment. They can cause problems and product losses. The following recommendations are made for installing drip pipes:

1. Never use pipe smaller than 1″ diameter for drip pipes.
2. Always provide as much downhill slope (fall) as possible. The minimum is $\frac{1}{8}$″ per foot.
3. Never use two water seals in any one drip line. Double water seals in series will cause a lock and prevent draining.
4. Prevent drip pipes from freezing. Where drip pipes are located in a dead air space, between refrigerators and a cold wall, provide means to prevent freezing. Never install drip pipes in contact with uninsulated suction lines. Suction lines should be insulated with a nonabsorbent insulation such as Armstrong's Armaflex®.
5. Avoid long runs of drip pipes. Long runs make it impossible to provide the fall necessary for good drainage.

CHAPTER 15—COMMERCIAL REFRIGERATION SYSTEMS

6. Provide a suitable air break between the flood rim of the floor drain and the outlet of drip pipes.

OUTDOOR INSTALLATION

By mounting equipment outdoors you can save space. If the refrigeration unit's compressor and associated equipment are located on the roof or along the store, the equipment room can be eliminated. Outdoor condensing units also offer the ideal answer for the remodeled store with a crowded equipment room. The units are designed for complete protection from the elements. They can operate during summer or winter.

An upright merchandiser with a close coupled remote refrigeration system is shown in Fig. 15-38.

Installation costs are low. No piping is required. Only one electrical connection is required at the bottom of the cabinet to supply power. This unit can be mounted on the roof or outside on the ground or concrete slab. It has 15' of flexible liquid suction line for direct connection to the outside condensing unit. Quick-connect lines make it easy to install.

By placing the condensing unit outside, you eliminate the additional heat from the compressor. This reduces the air conditioning load. It also reduces the horsepower requirements needed to operate the air conditioning system. The compressor unit is compact and has a baked acrylic finish for weather resistance. The inside cabinet is sufficient to display merchandise that needs a low temperature. The refrigerant is R-502. It operates on 208 or 230 V on single-phase or three-phase power.

RECLAIMING HEAT AND AIR CONDITIONING

To provide the most efficient environmental system, designers

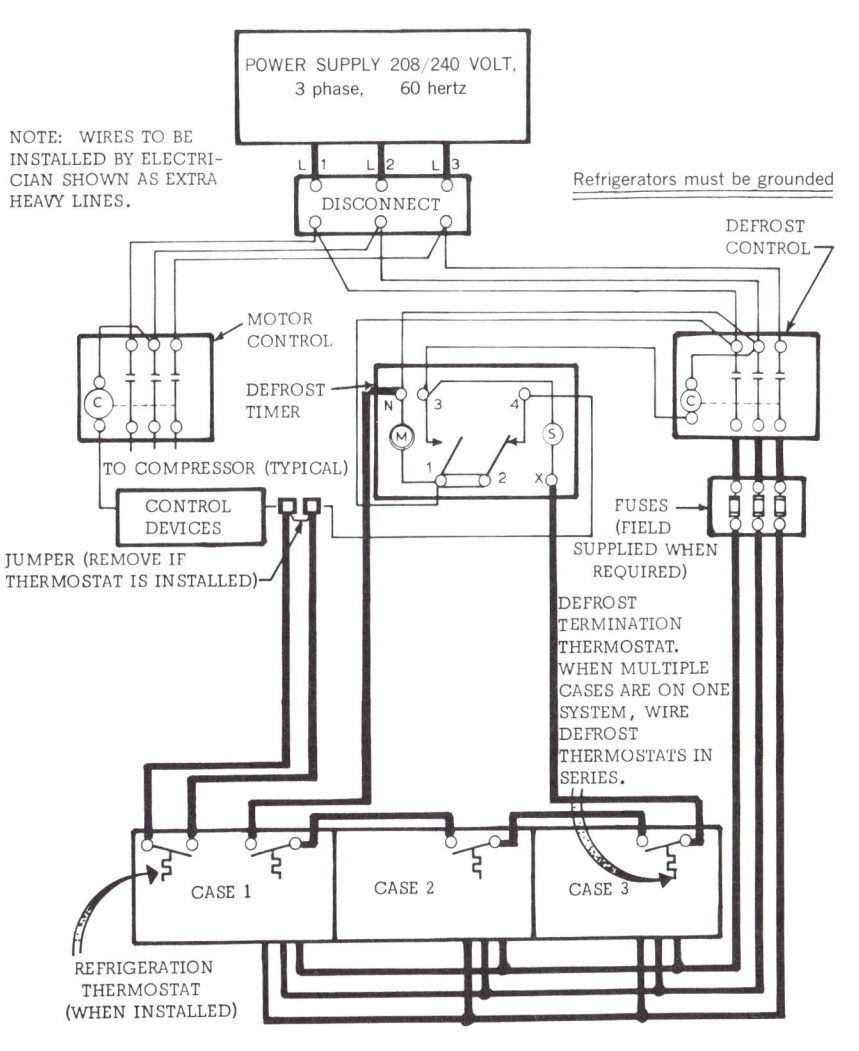

Hussmann

15-37. Conventional multiplexing—indoor unit. This shows a typical wiring diagram of condensing unit and a control panel.

REFRIGERATION AND AIR CONDITIONING TECHNOLOGY

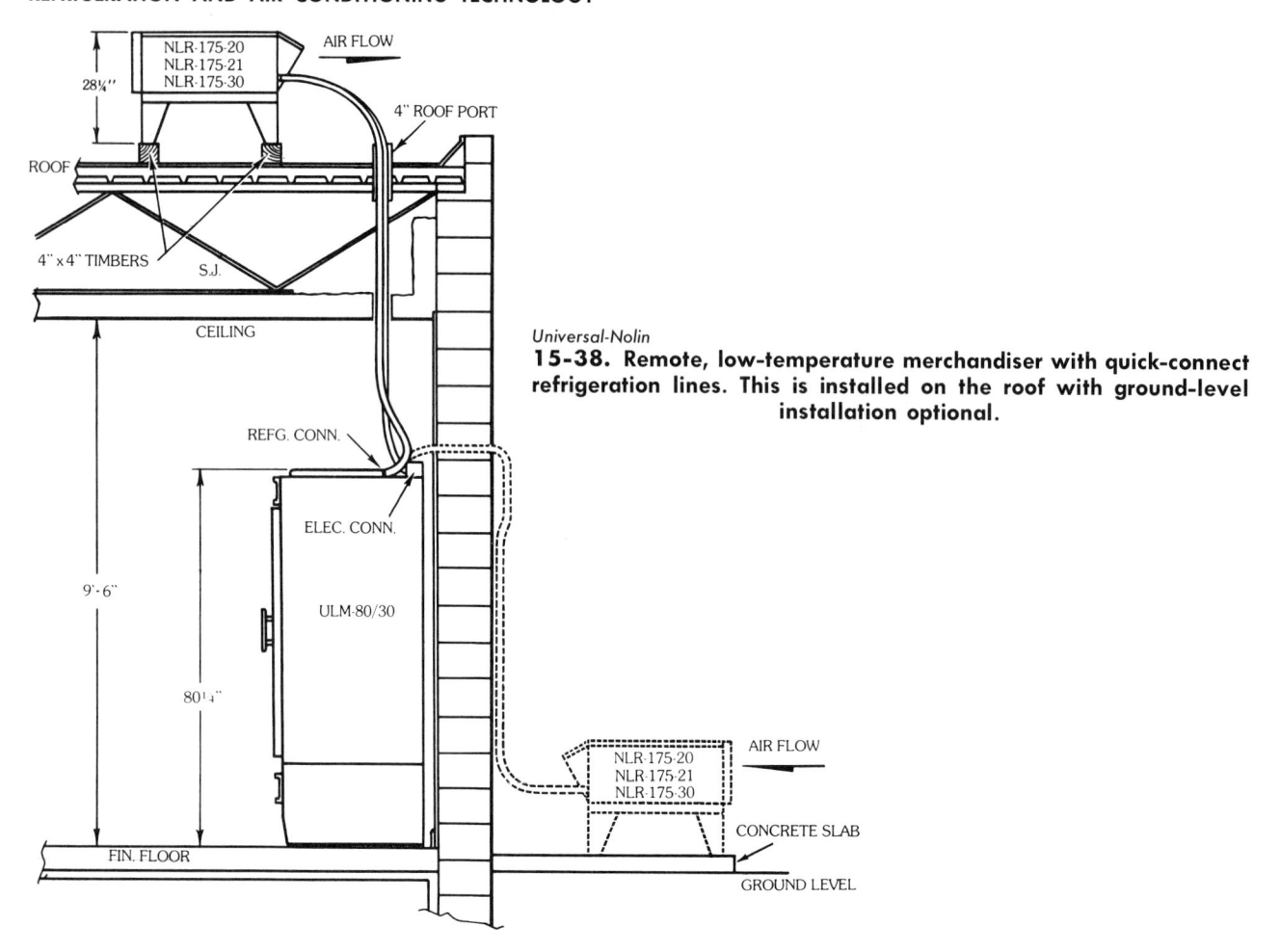

Universal-Nolin

15-38. Remote, low-temperature merchandiser with quick-connect refrigeration lines. This is installed on the roof with ground-level installation optional.

have made use of heat discharged through condensers to heat stores. To cool the stores, they have used air conditioning provided by the open freezers and refrigerators. Dehumidification has been utilized, too. If the air conditioning unit is used to dehumidify and it becomes too cold in the store, heat from refrigeration units will warm the store.

Heating

Each year millions of Btu of energy are lost. They are lost to the air through the refrigeration condensers. This heat is now being reclaimed by the use of a three-way thermostatically controlled valve. It routes the discharge gas to a heating coil in the air handler. Here it is used to heat the store. Fig. 15-39. The heat is free and will provide all or part of the store's heating requirement. In most areas a small amount of supplementary heat is provided for additional heat on the coldest days.

Cooling

The cold air from the refrigeration units is collected through vents located underneath each case. Fig. 15-40. This air is routed to the air handler and distributed to warmer areas of the store. (In the winter the air is heated with "free heat" before being distributed to the store.) The amount of "free air conditioning" available is contingent upon the store's equipment and ambient conditions. In almost all cases, air conditioning sizing is significantly less than with conventional air conditioning systems. This helps control the air spill from open multideck cases.

Dehumidification

Heating-cooling dehumidification systems provide year-round maximum controlled relative humidity. With humidity over 50%, moisture collects in the case. Ice

418

CHAPTER 15—COMMERCIAL REFRIGERATION SYSTEMS

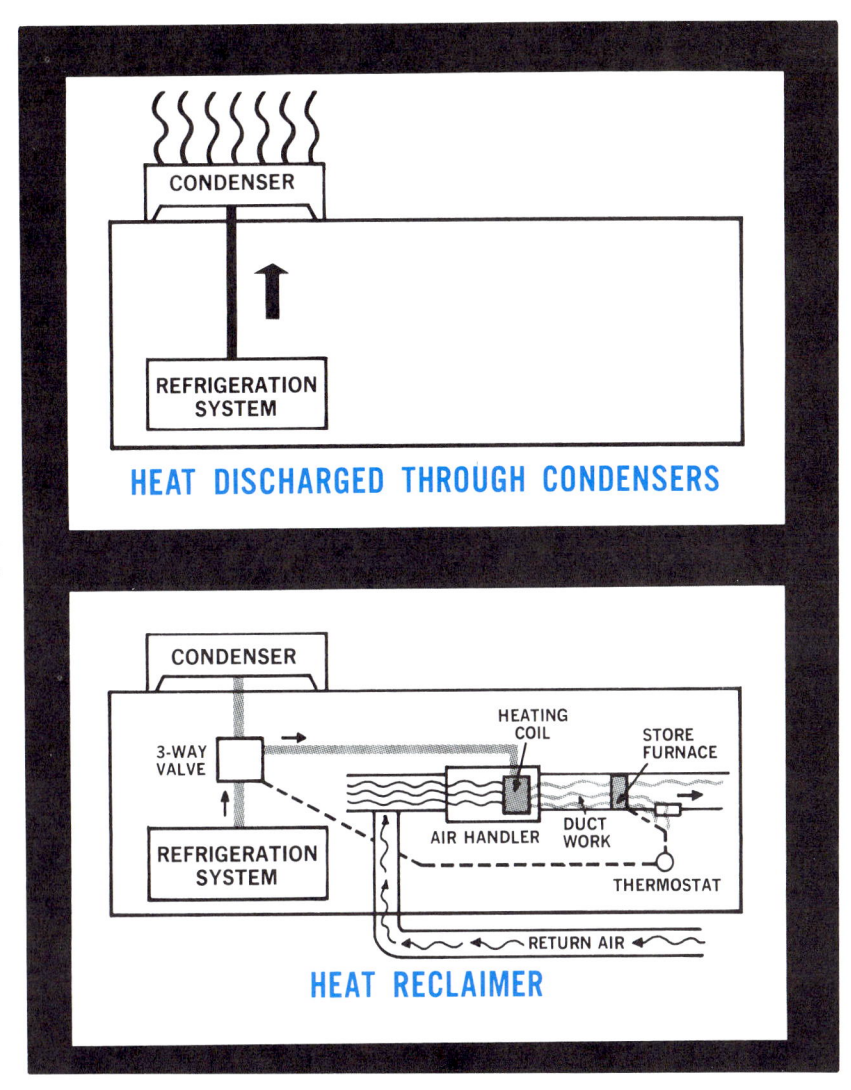

15-39. Heat recirculation. *Hussmann*

and frost build up on the product, the cases, and the coil. Case efficiency goes down and power demands increase. Whenever the store humidity rises above 50%, the air conditioner kicks on to dehumidify the air. If this causes temperatures that are too low, free heat will restore the balance. Cold air return from the cases is already at or below the 50% relative humidity level. This covers most of the dehumidification load.

Humidity comes from the outdoor weather. It enters the store through doors and leaks in the construction. It is also drawn in by exhaust fans. The custom design of heating-cooling dehumidification systems insures that the proper amount of outside air and humidity is brought into the store controlled by the system, rather than uncontrolled through doors, leaks, and other sources. The humidity is removed before it reaches the cases. The system compensates for exhaust fans and minimizes the uncontrolled leakage of outside air. Fig. 15-41.

HEAT RECLAIMER

A heat reclaimer recirculates heat throughout the store. It lowers heating bills in the winter. Many types of installations now have heat reclaimers. Those that do not can be modified to handle the extra heat generated by refrigeration.

Two important keys to achiev-

419

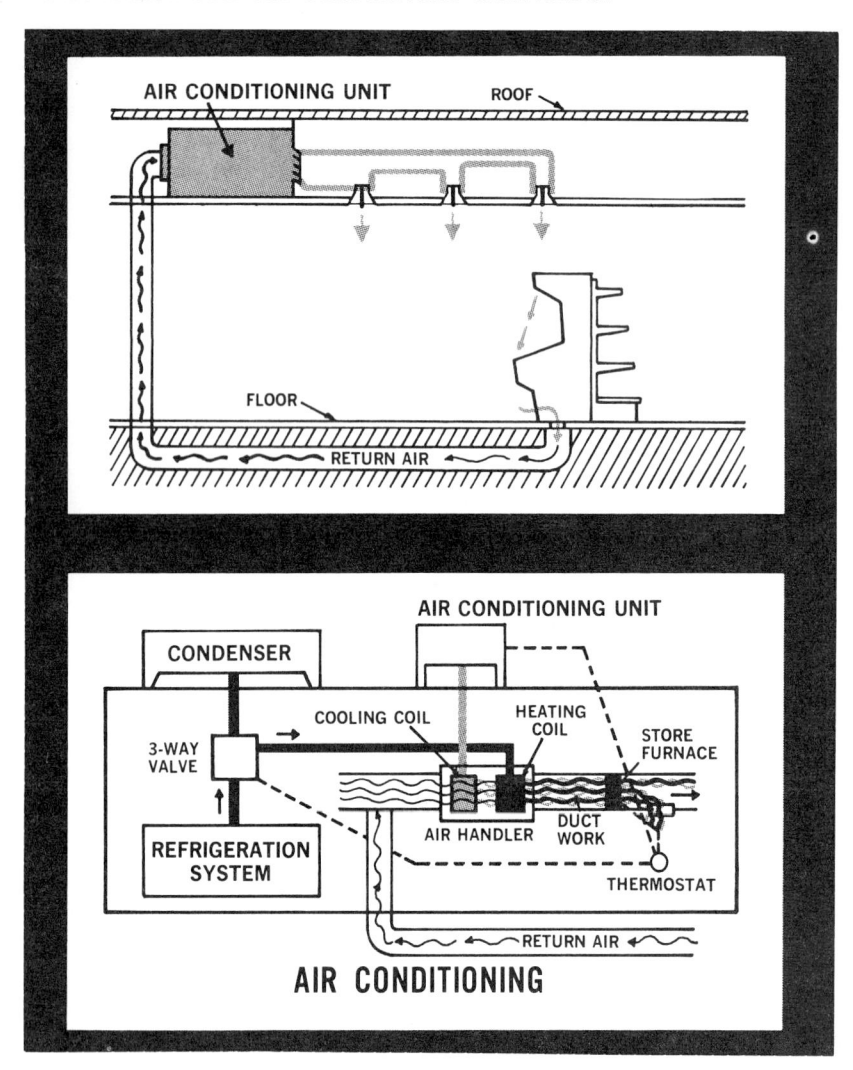

REFRIGERATION AND AIR CONDITIONING TECHNOLOGY

15-40. Use of case cold air to augment air conditioning. *Hussmann*

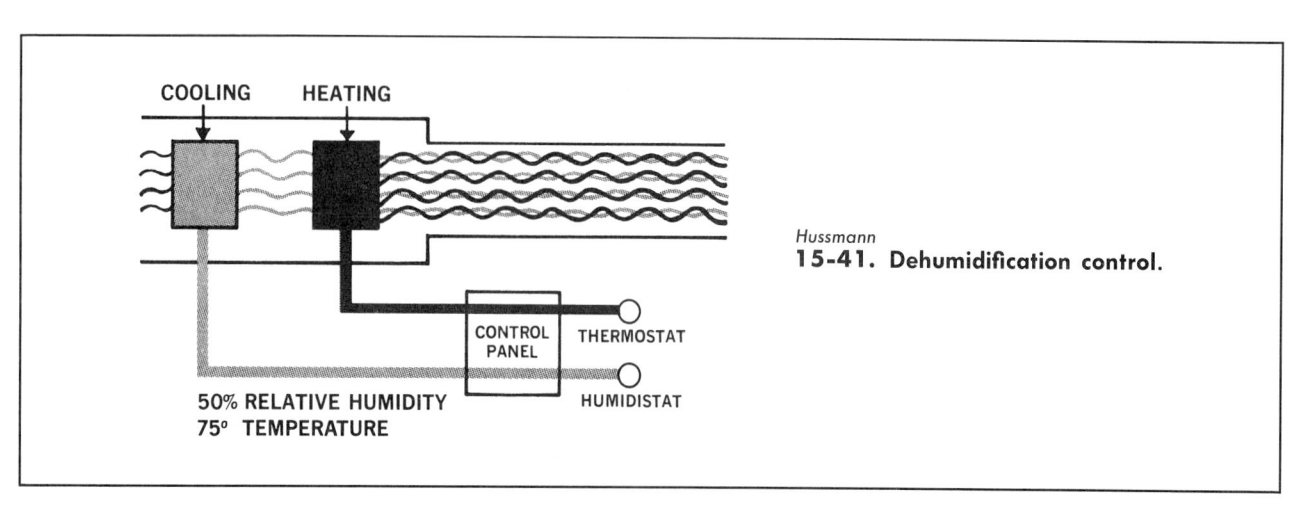

15-41. Dehumidification control. *Hussmann*

15-42. Heat recovery system.

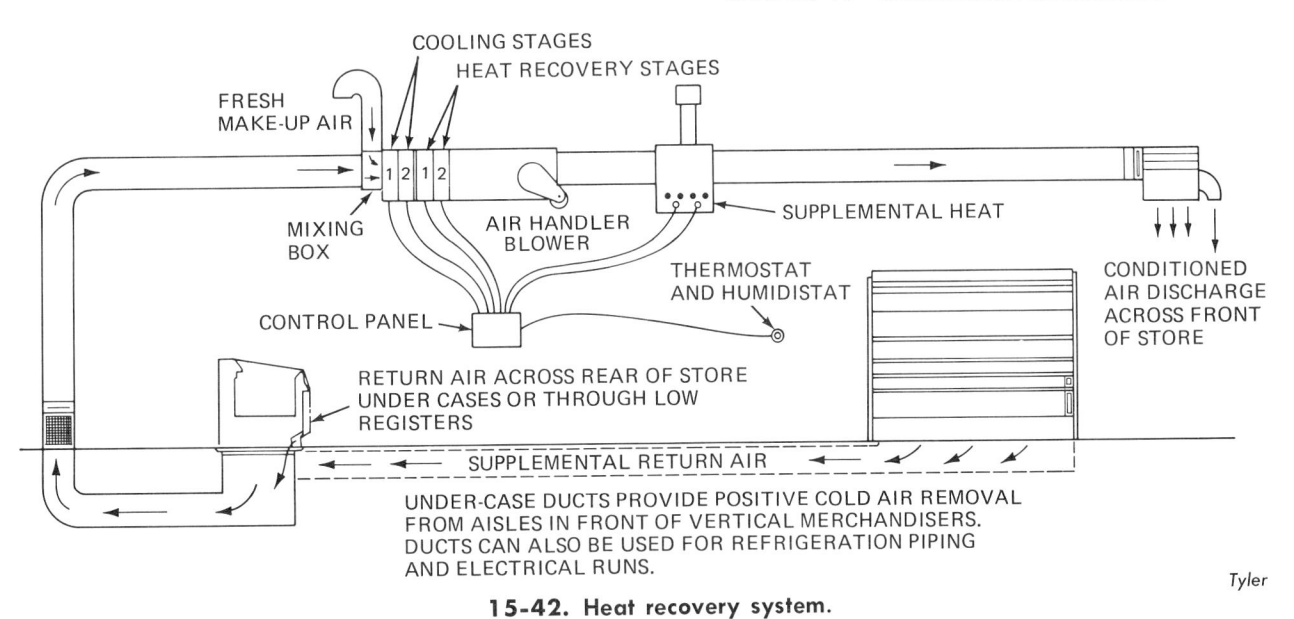

ing a really comfortable supermarket are system control and air distribution.

Some comfort control systems feature a solid state thermostat. It operates up to eight stages of heating and cooling. Such a control simplifies controlling of heating and cooling components. A humidistat included with the thermostat makes possible automatic dehumidification.

Air Handlers with Heat Recovery and Cooling Coils

The heat recovery portion of the coil is custom-made for the store. That is because each heat circuit is sized for the condensing unit it serves.

Heat recovery stages must be *upstream* from furnaces or heaters. Figure 15-42 shows one of the typical applications of an environmental system. This one has heat recovery and cooling stages in the store air handler. Note that the cooling stages come first. This is so that more moisture can be wrung from the air when cooling and heating operate together to reduce humidity.

Equipment for make-up heat is always *downstream* of the heat recovery coil since cool air is needed to condense refrigerant gases. Most of the system discharge air should occur at the front of the store. Low air returns are also essential for good air distribution.

Heat Recovery Methods

The heat removed from refrigerated cases and coolers in many stores is of sufficient quantity to reduce the heating bill. There are several methods by which it can be economically recovered and returned to the sales area. The more common methods are the direct air method, the condenser water recovery method, and the use of a secondary condenser coil. The last method is the most satisfactory.

The use of direct air requires complex duct work, louvers, damper controls, and powerful blower and filter systems. The direct system can transmit noise and dirt.

The efficiency of water recovery is limited, primarily because of the low temperature difference between inlet and outlet water. Water with a temperature above 90° F. [32.2°C] allows unnecessarily high condensing temperatures.

The most effective and economical heat recovery method uses the secondary condenser. This is nothing more than a remote condenser located in the store air-handling system. This is in addition to whatever the primary means of condensing may be. The primary means may be either air-cooled, remote air-cooled, or water-cooled.

Comfort Control Panel

Figure 15-43 indicates how the control panel works. With the thermostat set at 75° F. [23.9°C],

REFRIGERATION AND AIR CONDITIONING TECHNOLOGY

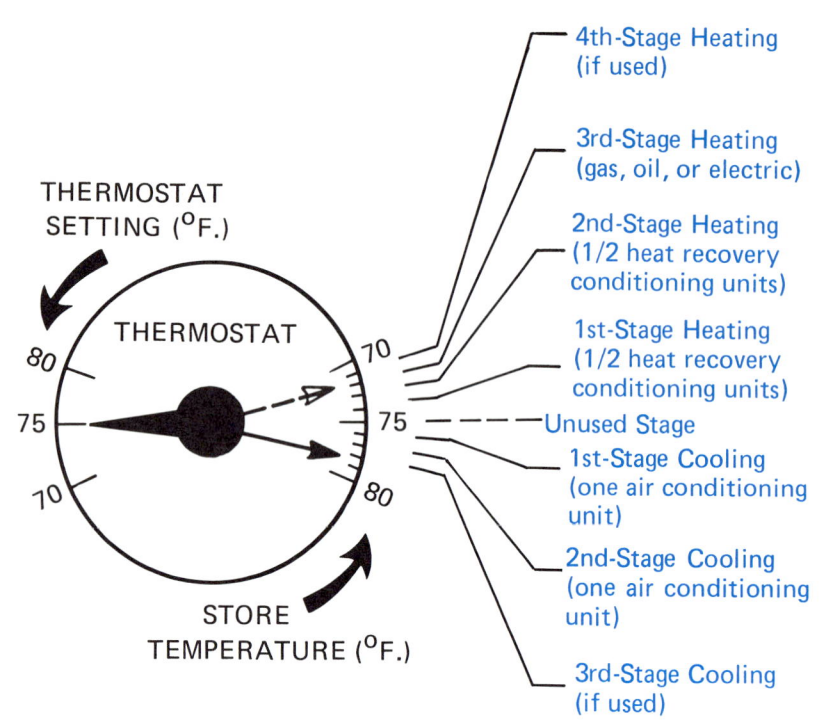

Tyler
15-43. Comfort control panel's thermostat.

the actual temperature at 78° F. [25.6°C], and total spread at 8° F. [4.4°C], all three cooling stages would be activated. As the temperature drops, the cooling stages cease at the indicated temperatures. At an actual temperature of 75° F. [23.9°C], all three of the cooling stages would be inactive.

Flushing Clock Operation

Solenoid and/or diverting valves do not always close tightly. Thus, the refrigerant may migrate to either the heat recovery coil or the remote condenser if either is inactive for long periods of time. This could affect system operation. To counteract this, two clocks can be installed. One clock breaks the heat recovery circuit. The other clock activates the circuit for very short time periods. This flushes the coils. This is a simpler device than a solenoid or capillary bleed circuit, which would tie into the refrigeration piping.

Humidity Control

The humidistat is not a solid state device. The sensing unit—which is the heart of the device—is very primitive. It uses human hair to sense the humidity. One hundred and fifty strands of specially selected human hair are attached to levers that operate an electric switch. The switching action takes place in a relative humidity change of 4 to 5%.

With this control panel, the humidity control is connected to an overriding relay. Thus, the first stage of heating is activated by an increase in humidity. The thermostat will respond to the added heat in the store by calling for one or more stages of cooling. The excess moisture in the air is "squeezed" out as it passes over the air conditioning coil. With heat recovery, the only extra cost is for running the air conditioner. That is less expensive and more effective than dehumidifying with display case coils. Furthermore, the overall effect is far better from the standpoint of the shopper.

With a practical means of controlling humidity, the next step is to select a range that is desirable. At 75° F. [23.9°C], the comfort range is from 30 to 50% rh. Humidity below 30% causes wood in furniture to shrink. Humidity above 55% increases refrigeration costs because more frost accumulates on case coils. This added frost requires more running time for compressors. As the frost accumulates, longer defrost times are needed to melt the frost. Humidity of 55% also causes produce packaging and contents to deteriorate faster. This increases store housekeeping costs. A setting of 45 to 50% rh will keep excessive humidity from affecting personnel, equipment, and stock. Yet, it will avoid the extra cost of running the air conditioning unit. Late spring and early fall months are the time when excessive store humidity may be more of a problem.

REVIEW QUESTIONS

1. What is a receiver?
2. Name two types of metering devices for refrigerant.
3. What are the condenser, receiver, and liquid line of a system called?
4. What physical assembly contains the compressor, condenser, and receiver?
5. What are three factors that affect defrost cycles?
6. What causes sweating?
7. How does a vacuum release air valve operate?
8. Evaporator units have motors that need lubrication. How often should the motors be oiled?
9. What happens when hot gas defrost is used and the hot gas is energized?
10. What is a psychrometric chart?
11. How is a psychrometric chart used in designing cooling systems?
12. What is transpiration?
13. Why do produce departments spray their vegetables or cover them with a wet cloth at night?
14. What are two major sources of radiant heat in a store?
15. What is multiplexing?
16. What is the minimum downhill slope needed for a drain pipe?
17. Why must heat recovery stages be upstream from furnaces or heaters?
18. What is a humidistat?

16

Psychrometrics and Air Movement

TEMPERATURE

Temperature is defined as the thermal state of matter. Matter receives or gives up heat as it is contacted by another object. If no heat flows upon contact, there is no difference in temperature. Figure 16-1 shows the different types of dry-bulb thermometers. The centigrade scale is now referred to as degrees *Celsius* (°C). (*Centi* is metric for "100.") The Celsius scale is divided into 100 degrees, from the freezing point of water to the boiling point.

Degrees Fahrenheit

American industry and commerce still use the Fahrenheit scale for temperature measurement. However, the metric scale (degrees Celsius, °C) is becoming rapidly accepted. The Fahrenheit scale divides into 180 parts the temperature range from the freezing point of water to its boiling point. The Fahrenheit temperature scale measures water at its freezing point of 32° and its boiling point of 212°. The pressure reference is sea level, or 14.7 pounds per square inch.

Degrees Celsius

In laboratory work and in the metric system, the temperature is measured in degrees Celsius. It ranges from the freezing point of water (0°C) to its boiling point (100°C). Again the pressure reference is sea level.

Absolute Temperature

Absolute temperatures are measured from *absolute zero.* This is the point at which there is no heat. On the Fahrenheit scale, absolute zero is −460°. Temperatures on the absolute Fahrenheit scale (Rankine) can be found by adding 460° to the thermometer reading. On the Celsius scale, absolute zero is −273°. Any temperatures on the absolute Celsius scale (Kelvin) can be found by adding 273° to the thermometer reading. Fig. 16-1.

Absolute zero temperature is the base point for calculations of heat. For example, if air or steam is kept in a closed vessel, the air or steam pressure will change roughly in direct proportion to its absolute temperature. Thus, if 0° F. air (460° absolute) is heated to 77° F. (537° absolute), without increasing the volume, the pressure will increase to roughly 537/460 times the original pressure. A more formal statement of the important physical law involved states:

> *At a constant temperature, as the absolute temperature of a perfect gas varies, its absolute pressure will vary directly. Or, at a constant pressure, as the absolute temperature of a perfect gas varies, the volume of the gas will vary directly.*

This statement is known as the *Perfect Gas Law.* It can be expressed mathematically by the following equation.

$$PV = TR$$

P = absolute pressure
V = volume
T = absolute temperature
R = a constant, depending on the units selected for P, V, and T.

CONVERTING TEMPERATURES

It is sometimes necessary to convert from one temperature scale to another. In converting from the Fahrenheit to the Celsius scale, one degree Fahrenheit is equal to 5/9 of one degree Celsius. Or, one degree Celsius is equal to 9/5 of one degree Fahrenheit. Equations facilitate converting from one scale to the other.

EXAMPLE. Convert 77° F. to °C. Use the following formula:
°C = $\frac{5}{9}$ (° F. − 32)

$$\frac{5}{9} (77 - 32)$$
$$\frac{5}{9} (45)$$
$$\frac{5 (45)}{9}$$
$$\frac{225}{9} = 25$$
$$77° F. = 25°C$$

EXAMPLE. Convert 25°C to ° F. Use the following formula:
° F. = $\frac{9}{5}$ (°C) + 32

$$\frac{9}{5} (25) + 32$$
$$\frac{9 (25)}{5} + 32$$
$$\frac{225}{5} = 45 + 32 = 77$$
$$25°C = 77° F.$$

Temperature conversion tables are available. Using them, it is easy to convert temperatures from one temperature scale to another.

A calculator can be used for the above temperature conversions. If a calculator is used, the number .55555555 can be substituted for $\frac{5}{9}$. The number 1.8 can be substituted for $\frac{9}{5}$.

PRESSURES

All devices that measure pressure must be exposed to *two* pressures. The measurement is always the *difference* between two pressures, such as gage pressure and atmospheric pressure.

Gage Pressure

On an ordinary pressure gage, one side of the measuring element is exposed to the medium under pressure. The other side of the measuring element is exposed to the atmosphere. Atmospheric pressure varies with altitude and climatic conditions. Thus, it is obvious that gage pressure readings will not represent a precise, definite value unless the atmospheric pressure is known. Gage pressures are usually designated as *psig* (pounds per square inch gage). Pressure values that include atmospheric pressure are designated *psia* (pounds per square inch absolute).

Atmospheric Pressure

A *barometer* is used to measure atmospheric pressure. A simple mercury barometer may be made with a glass tube slightly more than 30″ in length. It should be sealed at one end and filled with mercury. The open end should be inverted in a container of mercury. Fig. 16-2. The mercury will drop in the tube until the weight of the atmosphere on the surface of the mercury in the pan just supports the weight of the mercury column in the tube. At sea level and under certain average climatic conditions, the height of mercury in the tube will be 29.92″. The space above the mercury in the tube will be an almost perfect vacuum, except for a slight amount of mercury vapor.

A mercury *manometer* is an accurate instrument for measuring pressure. The mercury is

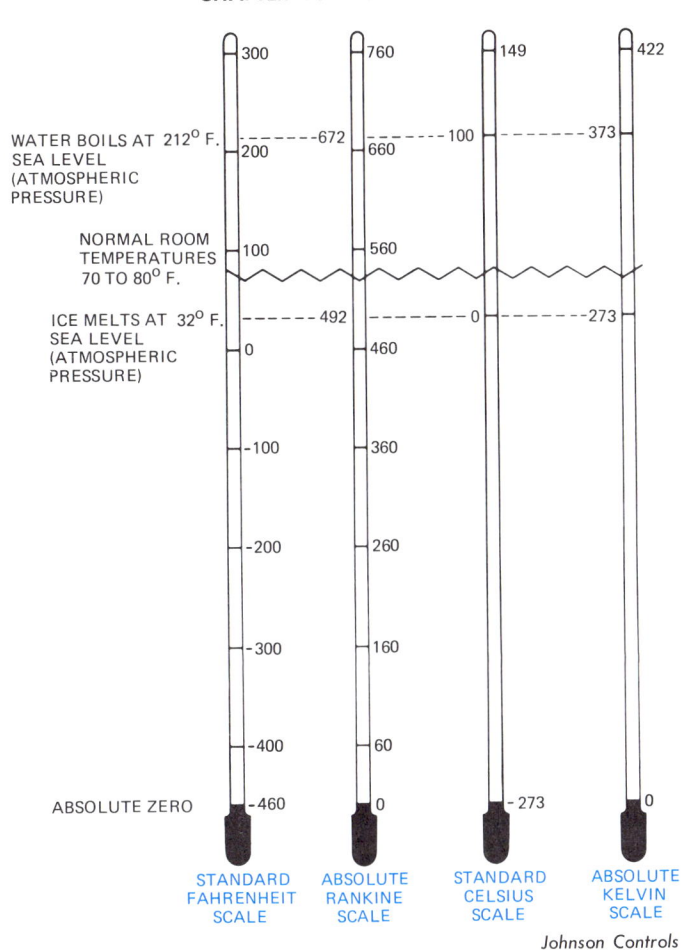

16-1. Standard dry-bulb thermometer scales.

Johnson Controls

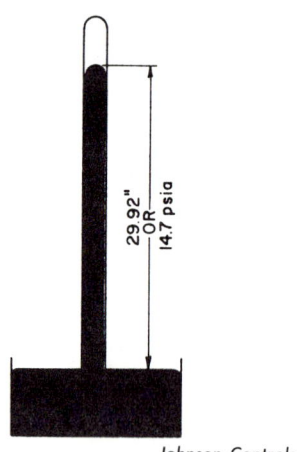

REFRIGERATION AND AIR CONDITIONING TECHNOLOGY

16-2. Simple mercury barometer.

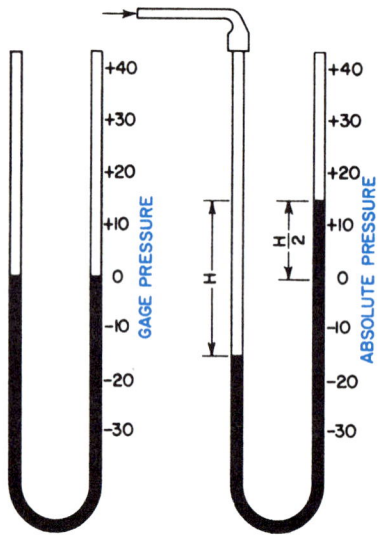

16-3. Mercury Manometer at rest (left) and with Pressure applied (right). H is the height from the top of one tube to the top of the other; $H/2$ is one half of H.

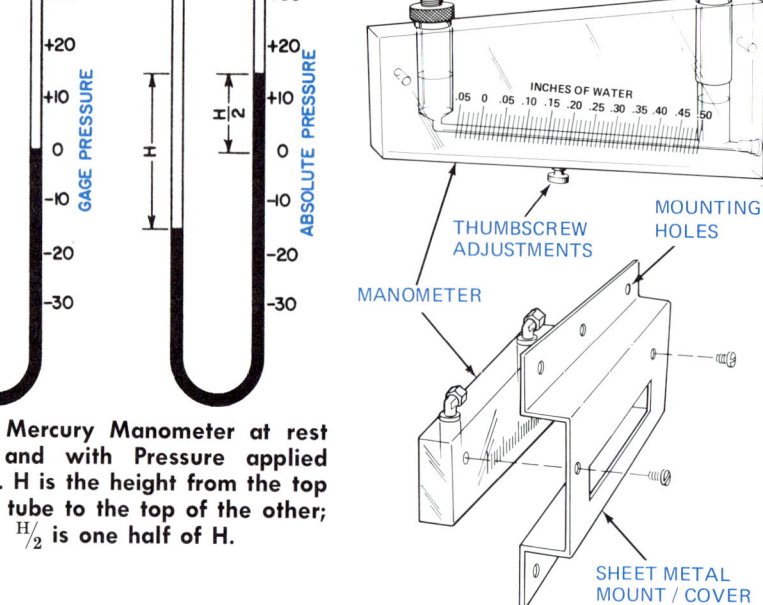

16-4. Well-type manometer.

16-5. Incline manometer.

placed in a glass U-tube. With both ends open to the atmosphere, the mercury will stand at the same level in both sides of the tube. Fig. 16-3. A scale is usually mounted on one of the tubes with its zero point at the mercury level.

Pressure Measuring Devices

Low pressures, such as in an air distribution duct, are measured with a manometer using water instead of mercury in the U-tube. Fig. 16-3. The unit of measurement, *one inch of water*, is often abbreviated as 1″ H$_2$O or 1″ w.g. (water gage). Water is much lighter than mercury. Thus, a water column is a more sensitive gage of pressure than a mercury column. Figure 16-4 shows a well-type manometer. It indicates 0″ w.g. Figure 16-5 shows an incline manometer. It spreads a small range over a longer scale for accurate measurement of low pressures. Most manometers of this type use a red oily liquid in place of water to provide a more practical and useful instrument. Figure 16-6 shows a magnahelic mechanical manometer. It is designed to eliminate the liquid in the water gage. It is calibrated in hundredths of an inch water gage for very sensitive measurement.

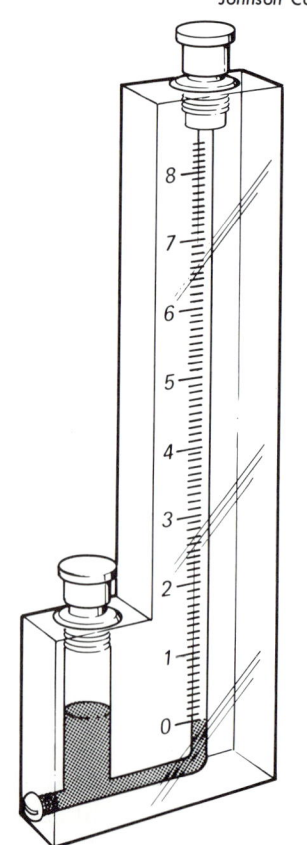

Figure 16-7 shows a standard air pressure gage used for adjusting control instruments. It is calibrated in inches of water and psig. It is apparent from the scale that one psig is equal to 27.6″ w.g.

Bourdon (spring) tube gages and metal diaphragm gages are also used for measuring pressure. These gages are satisfactory for most commercial uses. They are not as accurate as the manometer or barometer because of the mechanical methods involved. The Bourdon tube gage is discussed in Chapter 2.

Another unit of pressure measurement is the "atmosphere." Zero gage pressure is one atmosphere (14.7 pounds per square

CHAPTER 16—PSYCHROMETRIC AND AIR MOVEMENT

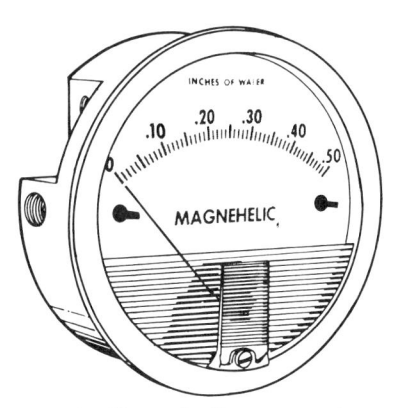

16-6. Magnahelic manometer.

16-7. Inches of water gage. *Johnson Controls*

inch) at sea level. For rough calculations, one atmosphere can be considered 15 *psig*. A gage pressure of 15 *psia* is approximately two atmospheres. The volume of a perfect gas varies inversely with its pressure as long as the temperature remains constant. Thus, measuring pressures in atmospheres is convenient in some cases, as may be observed from the following example:

A 30-gallon tank, open to the atmosphere, contains 30 gallons of free air at a pressure of one atmosphere. If the tank is closed and air pumped in until the pressure equals two atmospheres, the tank will contain 60 gallons of free air. The original 30 gallons now occupies only one-half of the volume it originally occupied.

Conversion charts for pressures (psi to in. Hg) may be found in various publications. The reference section of an engineering data book is a good source.

Properties of Air

Air is composed of nitrogen, oxygen, and small amounts of water vapor. Nitrogen makes up 77%, while oxygen accounts for 23%. Water vapor can account for 0 to 3% under certain conditions. Water vapor is measured in grains or, in some cases, pounds per pound of dry air. Seven thousand grains of water equal one pound.

Temperature determines the amount of water vapor that air can hold. Hotter temperatures mean that air has a greater capacity to hold water suspended. Water is condensed out of air as it is cooled. Outside, water condensation becomes rain. Inside, it becomes condensation on the window glass.

Thus, dry air acts somewhat like a sponge. It absorbs moisture. There are four properties of air that account for its behavior under varying conditions. These properties are dry bulb temperature, wet bulb temperature, dewpoint temperature, and relative humidity.

Dry Bulb Temperature

There are certain amounts of water vapor per pound of dry air. They can be plotted on a psychrometric chart. *Psychro* is a Greek term meaning "cold." A psychrometer is an instrument for measuring the aquous vapor in the atmosphere. A difference between a wet bulb thermometer and a dry bulb thermometer is an indication of the dryness of the air. A psychrometer, then, is a hygrometer, which is a device for measuring water content in air. A psychrometric chart indicates the different values of temperature and water moisture in air.

Wet Bulb Temperature

A wet bulb thermometer is the same as a dry bulb thermometer, except that it has a wet cloth around the bulb. Fig. 16-8. The thermometer is swung around in the air. The temperature is read after this operation. The wet bulb temperature is lower than the dry bulb temperature. It is the lowest temperature that a water-wetted body will attain when exposed to an air current. The measurement is an indication of the moisture content of the air.

Dewpoint Temperature

The dewpoint of air is reached when the air contains all the moisture it can hold. The dry bulb and wet bulb temperatures are the same at this point. The air is said to be at 100% relative humidity when both thermometers read the same. Dewpoint is important when designing a humidifying system for human comfort. If the humidity is too high in a room, the moisture will condense and form on the windows.

Relative Humidity

Relative humidity (rh) is based on the percentage of humidity contained in the air, based on the saturation condition of the air. A reading of 70% means that the air contains 70% of the moisture it can hold. Relative humidity lines

427

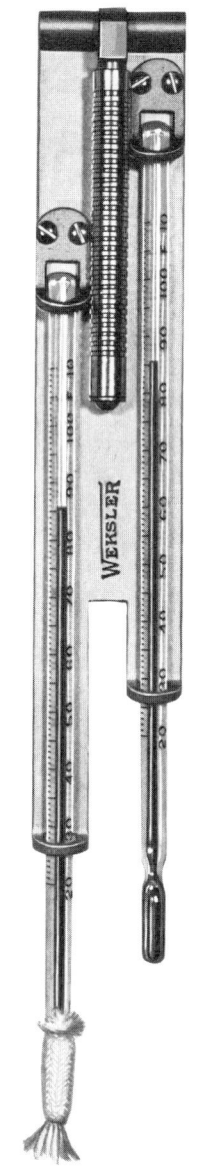

16-8. Wet bulb and dry bulb thermometers mounted together. Note the knurled and ringed rod between the two at the top of the scales. It is used to hold the unit and twirl it in the air. *Weksler*

heat will in time remove most of the moisture in the living space. The addition of moisture is accomplished in a number of ways. Humidifiers are used to spray water into the air or large areas of water are made available to evaporate. Showers and running water also add moisture to a living space.

In summer, however, the amount of moisture per pound on the outside is greater than on the inside, especially when the room is air conditioned. This means the vapor pressure is greater on the outside than the inside. Under these conditions, moisture will enter the air-conditioned space by any available route. It will enter through cracks, around doors and windows, and through walls. In winter, the moisture moves the other way—from the inside to the outside.

The percentage of relative humidity is never more than 100%. When the air is not saturated, the dry bulb temperature will always be higher than the wet bulb temperature. The dewpoint temperature will always be the lowest reading. Also, the greater the difference between the dewpoint temperature and the dry bulb temperature, the lower will be the percentage of relative humidity. The wet bulb reading can never be higher than the dry bulb reading. Nor can the dewpoint reading be higher than the dry bulb reading. Consider the following example.

Saturated air (100% humidity)
Temperature is: Dry bulb 90° F.
Wet bulb 90° F.
Relative humidity 100%
Dewpoint 90° F.

Unsaturated air (less than 100% humidity)
Temperature is: Dry bulb 80° F.
Wet bulb 75° F.
Relative humidity 80%
Dewpoint 73° F.

Temperature is: Dry bulb 90° F.
Wet bulb 75° F.
Relative humidity 50%
Dewpoint 69° F.

Manufacturers of humidifiers furnish a dial similar to the thermostat for controlling the humidity. A chart on the control tells what the humidity setting should be when the temperature outside is at a given point. Table 16-A gives an example of what the settings should be.

The relationship between humidity, wet bulb temperature, and dry bulb temperature has much to do with the designing of air conditioning systems. There are three methods of controlling the saturation of air.

1. Keep the dry bulb temperature constant. Raise the wet bulb temperature and the dewpoint temperature to the dry bulb temperature.

This can be done by adding moisture to the air. This will raise the dewpoint temperature to the dry bulb temperature. This automatically raises the wet bulb temperature to the dry bulb temperature.

2. Keep the wet bulb temperature constant. Lower the dry bulb temperature. Raise the dewpoint temperature to the wet bulb temperature.

This is done by cooling the dry bulb temperature to the level of the wet bulb temperature. The idea here is do it without adding or removing any moisture. The

on the psychrometric chart are sweeping curves, as shown in Fig. 16-13.

To keep the home comfortable in winter it is sometimes necessary to add humidity. Hot air

CHAPTER 16—PSYCHROMETRIC AND AIR MOVEMENT

Table 16-A.
Permissible Relative Humidity (in Winter)

Outside Temperature		Brick Wall 12 inches thick Plastered inside	Single Glass	Double Glass
°F.	°C	Percentage		
−20	−29.0	45	7	35
−10	−23.0	50	10	40
0	−17.8	60	18	45
10	−12.2	64	25	50
20	−6.7	67	30	55
30	−1.1	74	38	60
40	4.4	80	45	65
50	10.0	85	50	70
60	15.6	90	55	75

Table 16-C.
Saturated Vapor per Pound of Dry Air
(Barometer Reading at 29.92 inches per square inch)

Temperature		Weight in Grains
°F.	°C	
25	−3.9	19.1
30	−1.1	24.1
35	1.7	29.9
40	4.4	36.4
45	7.2	44.2
50	10.0	53.5
55	12.8	64.4
60	15.6	77.3
65	18.3	92.6
70	21.1	110.5
75	23.9	131.4
80	26.7	155.8
85	29.4	184.4
90	32.2	217.6
95	35.0	256.3
100	37.8	301.3
105	40.6	354.0
110	43.3	415.0

Table 16-B.
Activity-Heat Relationships

Activity	Total Heat in Btu per hour
Person at rest	385
Person standing	430
Tailor	480
Clerk	600
Dancing	760
Waiter	1,000
Walking	1,400
Bowling	1,500
Fast walking	2,300
Severe exercise	2,600

ing or occupied space give off moisture as they work. They also give off heat. Such moisture and heat must be considered in determining air conditioning requirements. Table 16-B indicates some of the heat given off by the human body when working.

PSYCHROMETRIC CHART

The psychrometric chart holds much information. Fig. 16-9. However, it is hard to read. It must be studied for some time. The dry bulb temperature is located in one place and the wet bulb in another. If the two are known, it is easy to find the relative humidity and other factors relating to air being checked. Both customary and metric psychrometric charts are available.

Air contains different amounts of moisture at different temperatures. Table 16-C shows the amounts of moisture that air can hold at various temperatures.

An explanation of the various quantities shown on a psychrometric chart will enable you to understand the chart. The different quantities on the chart are shown separately on the following charts. These charts will help you see how the psychrometric chart is constructed. Fig. 16-10.

Across the bottom, the vertical lines are labeled from 25 to 110° F. in increments of 5° F. These temperatures indicate the dry bulb temperature. Fig. 16-11.

The horizontal lines are labeled from 0 to 180° F. This span of numbers represents the grains of moisture per pound of dry air (when saturated). Fig. 16-11.

The wet bulb, dewpoint, or saturation temperature is indicated by the outside curving line on the left side of the graph. Fig. 16-12, page 432.

dewpoint is automatically raised to the wet bulb temperature.

3. Keep the dewpoint temperature constant and the wet bulb temperature at the dewpoint temperature.

This can be done by cooling the dry bulb and wet bulb temperatures to the dewpoint temperature.

People and Moisture

People working inside a build-

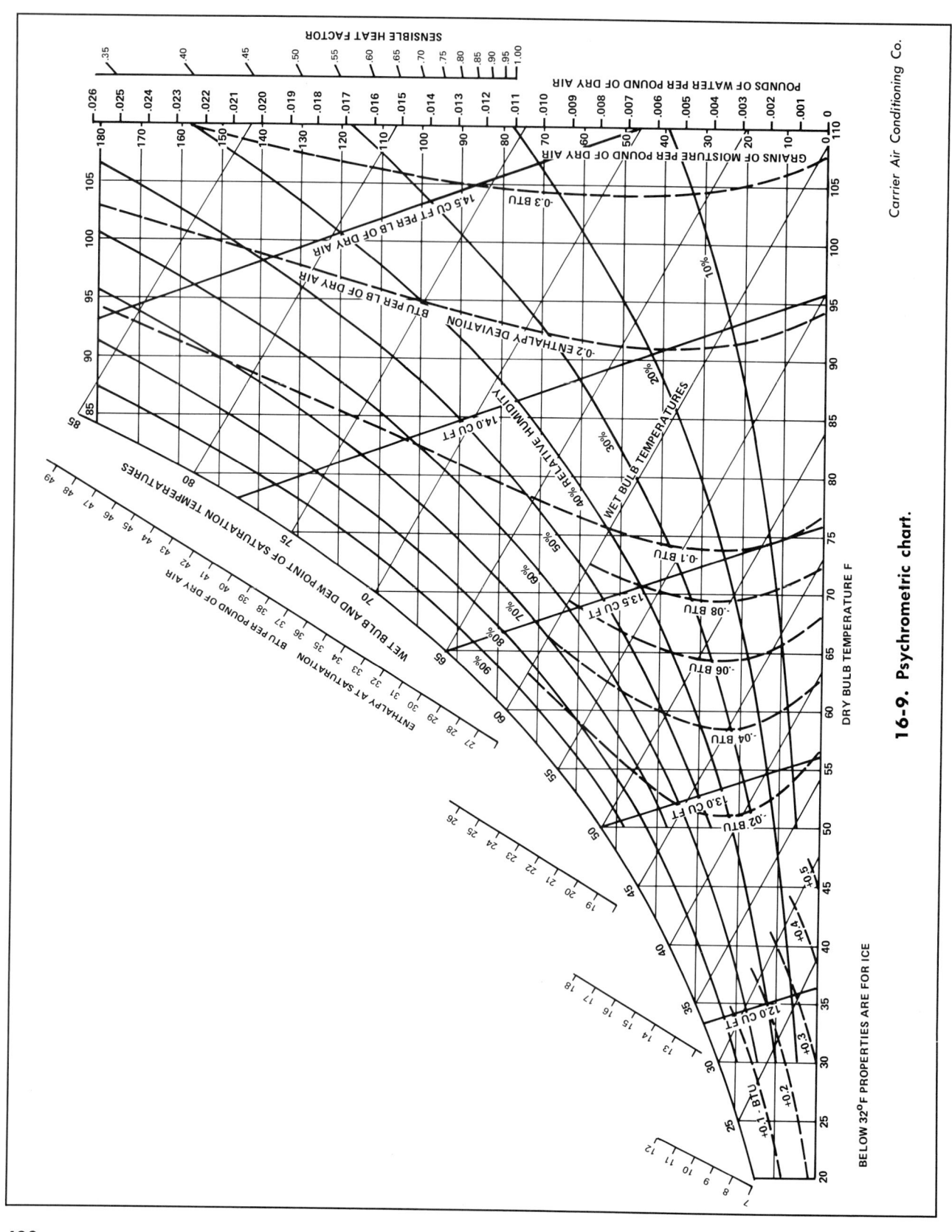

16-9. Psychrometric chart.

CHAPTER 16—PSYCHROMETRIC AND AIR MOVEMENT

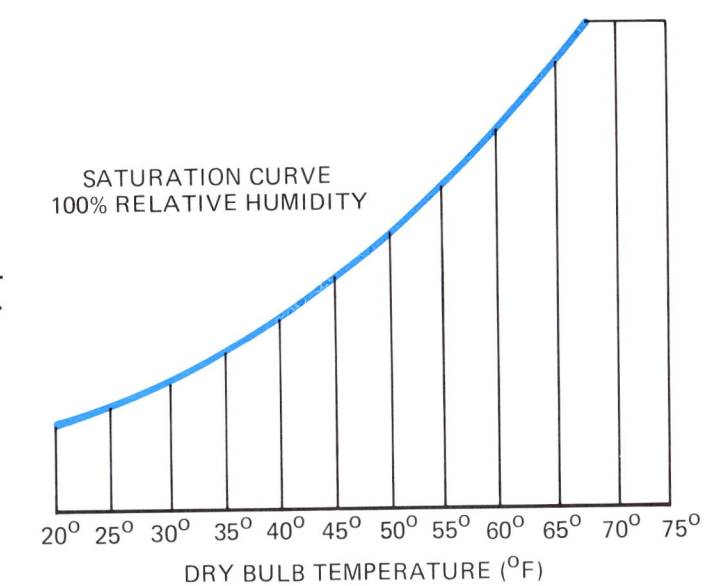

16-10. Temperature lines (dry bulb) on psychrometric chart. Only a portion of the chart is shown.

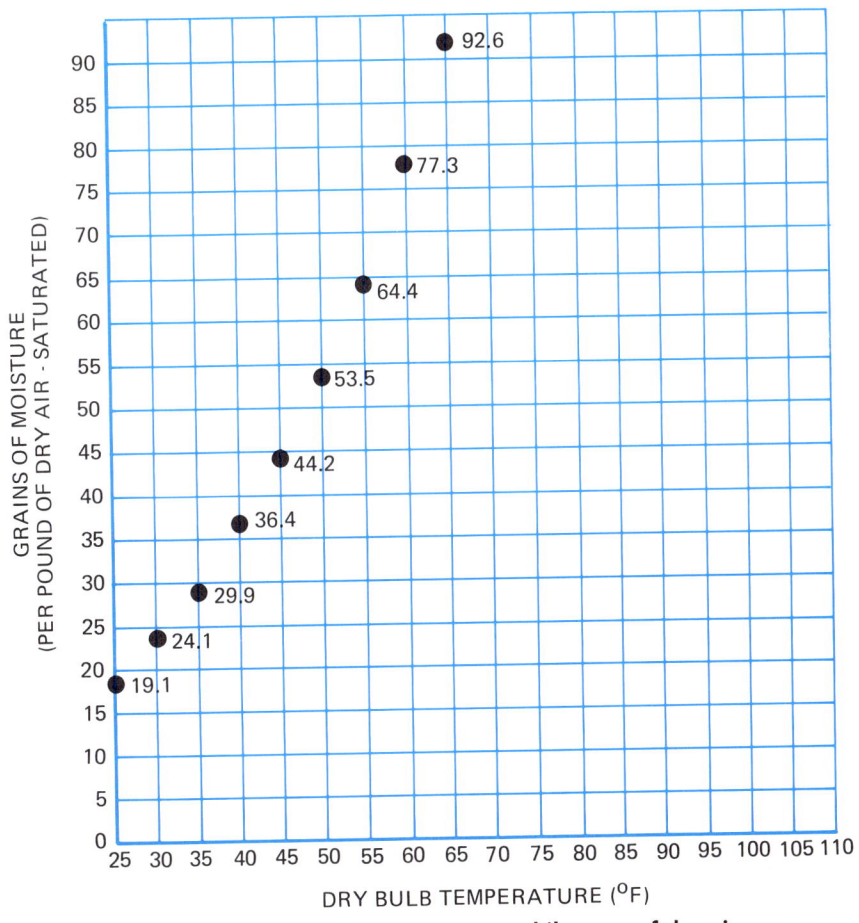

SATURATION CURVE
100% RELATIVE HUMIDITY

DRY BULB TEMPERATURE (°F)

16-11. The moisture content in kilograms per kilogram of dry air are measured on the vertical column. Here, only a portion of the chart is shown.

GRAINS OF MOISTURE (PER POUND OF DRY AIR - SATURATED)

DRY BULB TEMPERATURE (°F)

At 100% saturation the wet bulb temperatures are the same as the dry bulb and the dewpoint temperatures. This means the wet bulb lines start from the 100% saturation curve. Diagonal lines represent the wet bulb temperatures. The point where the diagonal line of the wet bulb crosses the dry bulb's vertical line is the dewpoint. The temperature of the dewpoint will be found by running the horizontal line to the left and reading the temperature on the curve since the wet bulb and dewpoint temperatures are on the curve.

The percentage of relative humidity is indicated by the curving lines within the graph. These lines are labeled 10%, 20%, etc. Fig. 16-13.

The pounds of water per pound of dry air are shown in the middle column of numbers on the right. Fig. 16-14.

The grains of moisture per pound of dry air are shown in the left-hand column of the three

431

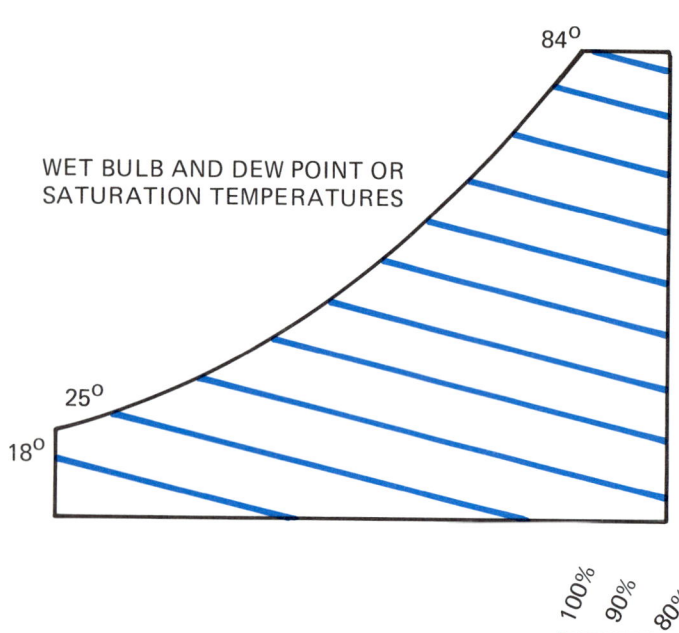

REFRIGERATION AND AIR CONDITIONING TECHNOLOGY

16-12. Wet bulb temperature lines on psychrometric chart.

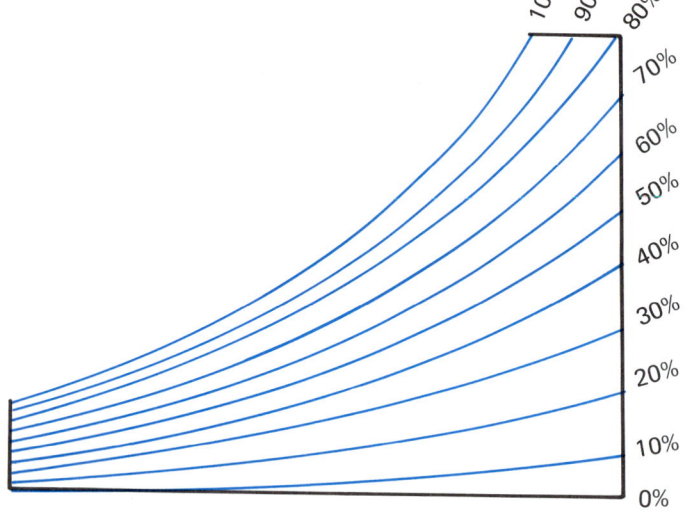

16-13. Relative humidity lines on a psychrometric chart.

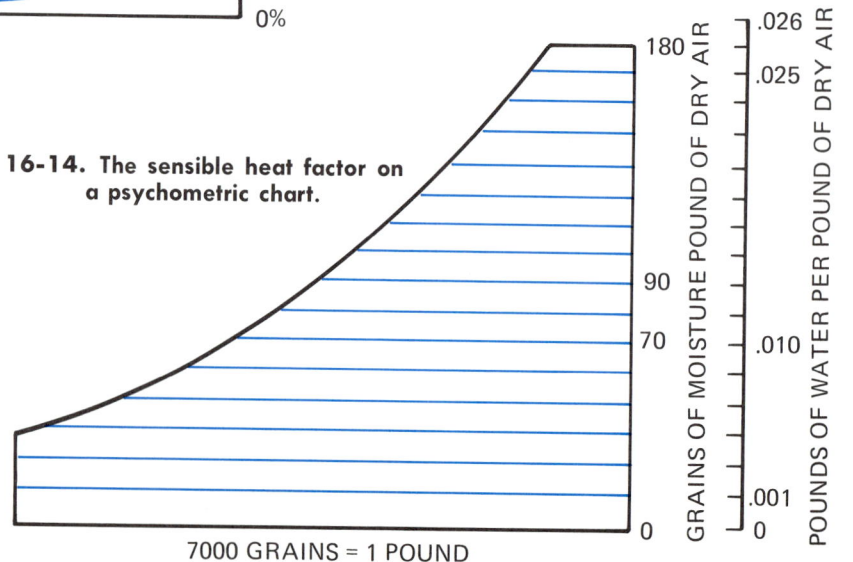

16-14. The sensible heat factor on a psychometric chart.

columns of numbers on the right. Fig. 16-14.

Table 16-C shows that one pound of dry air will hold 19.1 grains of water at 25° F. [−3.9°C]. One pound of dry air will hold 415 grains of water at 110° F. It can be seen that the higher the temperature, the more moisture the air can hold. This is one point that should be remembered. To find the weight per grain, divide 1 by 7000 to get

432

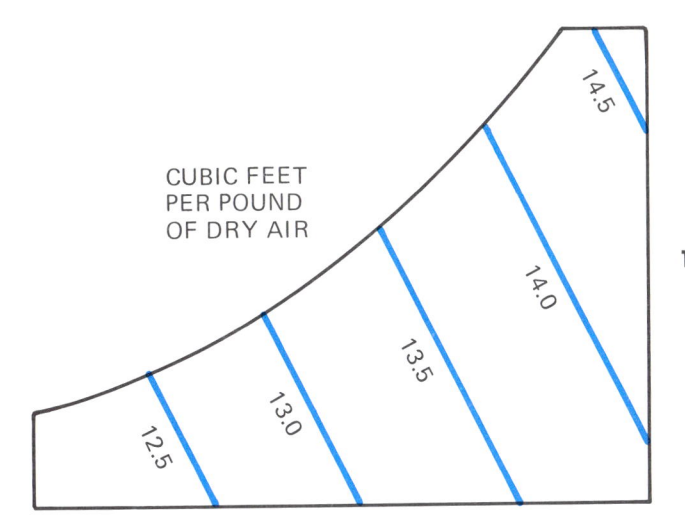

CHAPTER 16—PSYCHROMETRIC AND AIR MOVEMENT

16-15. Air volume lines on a psychrometric chart.

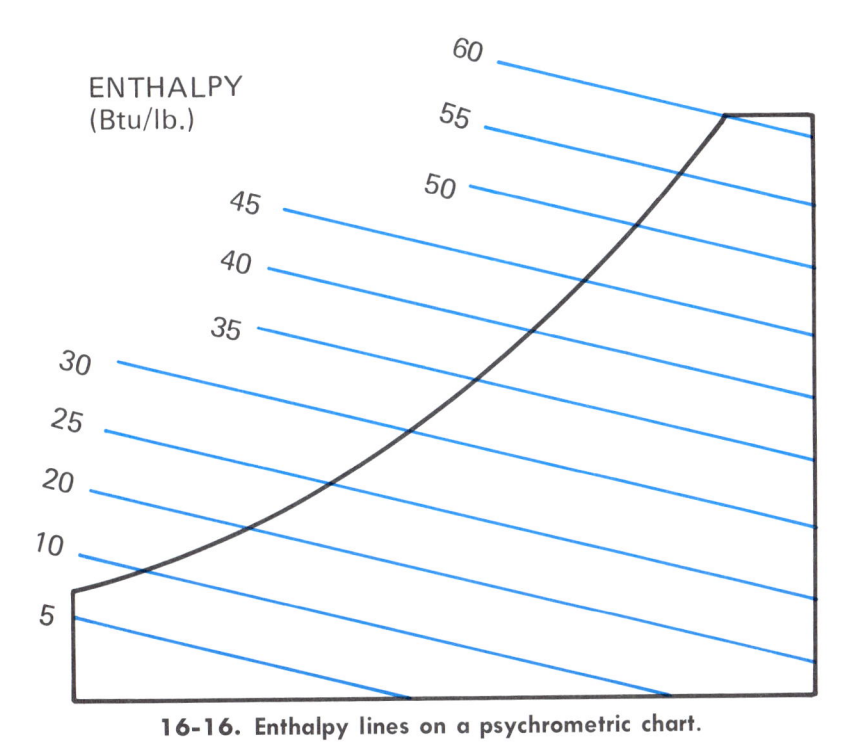

16-16. Enthalpy lines on a psychrometric chart.

intervals of one-half cubic foot per pound.

Enthalpy is the total amount of heat contained in the air above 0° F. [−17.8°C]. Fig. 16-16. The lines on the chart that represent enthalpy are extensions of the wet bulb lines. They are extended and labeled in Btu per pound. This value can be used to help determine the load on an air conditioning unit.

AIR MOVEMENT

Convection, Conduction, and Radiation

Heat always passes from a warmer to a colder object or space. The action of refrigeration depends upon this natural law. The three methods by which heat can be transferred are convection, conduction, and radiation.

Convection is heat transfer that takes place in liquids and gases. In convection, the molecules carry the heat from one point to another.

Conduction is heat transfer that takes place chiefly in solids. In conduction, the heat passes from one molecule to another

.00014 pounds per grain. Therefore, on the chart .01 pounds corresponds with about 70 grains.

The volume of dry air (cubic feet per pound) is represented by diagonal lines. Fig. 16-15. The values are marked along the lines. They represent the cubic feet of the mixture of vapor and air per pound of dry air. The chart indicates that volume is affected by temperature relationships of the wet and dry bulb readings. The lines are usually at

433

without any noticeable movement of the molecules.

Radiation is heat transfer in wave form, such as light or radio waves. It takes place through a transparent medium such as air, without affecting that medium's temperature, volume, and pressure. Radiant heat is not apparent until it strikes an opaque surface, where it is absorbed. The presence of radiant heat is felt when it is absorbed by a substance or by your body.

Convection can be used to remove heat from an area. Thus, it can be used to cool. Air or water can be cooled in one place and circulated through pipes or radiators in another location. In this way the cool water or air is used to remove heat.

COMFORT CONDITIONS

The surface temperature of the average adult's skin is 80° F. [26.7°C]. The body can either gain or lose heat according to the surrounding air. If the surrounding air is hotter than the skin temperature, the body gains heat and the person may become uncomfortable. If the surrounding air is cooler than the skin temperature, then the body loses heat. Again, the person may become uncomfortable. If the temperature is much higher than the skin temperature or much cooler than the body temperature, then the person becomes uncomfortable. If the air is about 70° F. [21.1°C] then the body feels comfortable. Skin temperature fluctuates with the temperature of the surface air. The total range of skin temperature is between 40 and 105° F. [4.4 and 40.6°C]. However, if the temperature rises 10° F. [5.5°C], the skin temperature rises only 3° F. [1.7°C]. Most of the time the normal temperature of the body ranges from 75 to 100° F. [23.9 to 37.8°C]. Both humidity and temperature affect the comfort of the human body. However, they are not the only factors that cause a person to be comfortable or uncomfortable. In heating or cooling a room, the air velocity, noise level, and temperature variation caused by the treated air must also be considered.

Velocity

When checking for room comfort, it is best to measure the velocity of the air at the distance of 4″ to 72″ from floor level. Velocity is measured with a velometer. See Fig. 16-19. Following is a range of air velocities and their characteristics.

• Slower than 15 feet per minute (fpm): stagnant air.
• 20 fpm to 50 fpm: acceptable air velocities.
• 25 fpm to 35 fpm: the best range for human comfort.
• 35 fpm to 50 fpm: comfortable for cooling purposes.

Velocities of 50 fpm or higher call for a very high speed for the air entering the room. A velocity of about 750 fpm or greater is needed to create a velocity of 50 fpm or more inside the room. When velocities greater than 750 fpm are introduced noise will also be present.

Sitting and standing levels must be considered when designing a cooling system for a room. People will tolerate cooler temperatures at the ankle level than at the sitting level, which is about 30″ from the floor. Variations of 4° F. [2.2°C] are acceptable between levels. This is also an acceptable level for temperature variations between rooms.

To make sure that the air is properly distributed for comfort, it is necessary to look at the methods used to accomplish the job.

Terminology

The following terms apply to the movement of air. They are frequently used in referring to air conditioning systems.

Aspiration is the induction of room air into the primary airstream. Aspiration helps eliminate stratification of air within the room. When outlets are properly located along exposed walls, aspiration also aids in absorbing undesirable currents from these walls and windows. Fig. 16-17.

Cubic feet per minute (cfm) is the measure of a volume of air. Air flow in cubic feet per minute of a register or grille is computed by multiplying the face velocity times the free area in square feet. EXAMPLE: A register with 144 square inches (1 square foot) of free area and a measured face velocity of 500 feet per minute would be delivering 500 cubic feet per minute (cfm).

Decibels (db) are units of measure of sound level. It is important to keep this noise at a minimum. In most catalogs for outlets, there is a line dividing the noise level of the registers or diffusers. Lower total pressure loss provides a quieter system.

Drop is generally associated with cooling where air is discharged horizontally from high sidewall outlets. Since cool air has a natural tendency to drop, it will fall progressively as the velocity decreases. Measured at the point of terminal velocity, drop is the distance in feet that the air has fallen below the level of the outlet. Fig. 16-18.

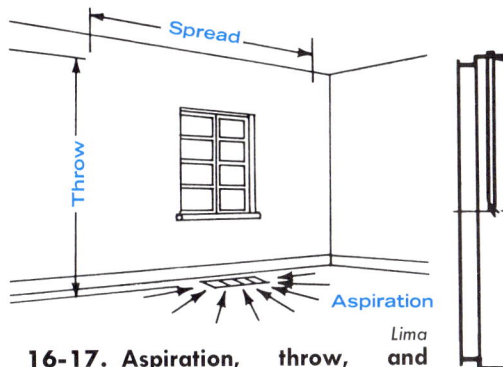

16-17. Aspiration, throw, and spread. *Lima*

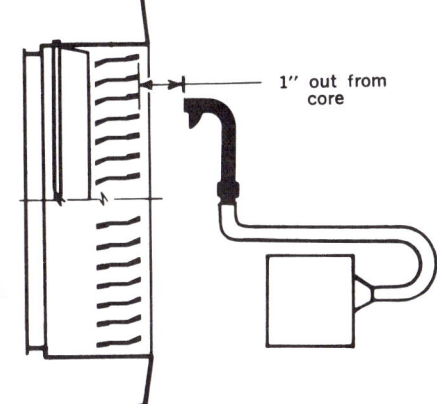

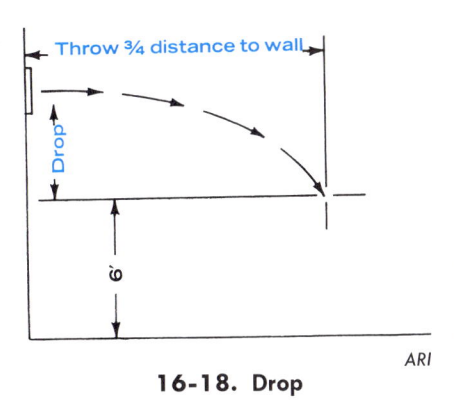

16-18. Drop *ARI*

16-19. Air measurement at the grille. *Lima*

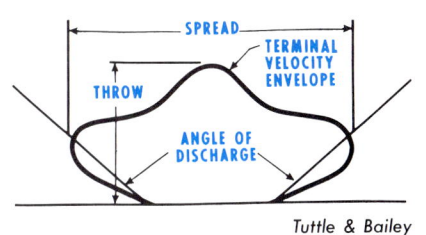

V_K METER recommended is Velometer with 2220A or 6070 jet probe

$$CFM = A_k \times V_k$$

16-20. Typical airstream pattern. *Tuttle & Bailey*

Diffusers are outlets that have a widespread, fan-shaped pattern of air.

Effective area is the smallest net area of an outlet utilized by the airstream in passing through the outlet passages. It determines the maximum, or jet, velocity of the air in the outlet. In many outlets, the effective area occurs at the velocity measuring point and is equal to the outlet area. Fig. 16-19.

Face velocity is the average velocity of air passing through the face of an outlet or a return.

Feet per minute (fpm) is the measure of the velocity of an airstream. This velocity can be measured with a velocity meter that is calibrated in feet per minute.

Free area is the total area of the openings in the outlet or inlet through which air can pass. With gravity systems, free area is of prime importance. With forced-air systems, free area is secondary to total pressure loss, except in sizing return air grilles.

Noise criteria (nc) is an outlet sound rating in pressure level at a given condition of operation, based on established criteria and a specific room acoustic absorption value.

Occupied zone is that interior area of a conditioned space that extends to within 6″ of all room walls and to a height of 6′ above the floor.

Outlet area is the area of an outlet utilized by the airstream at the point of the outlet velocity as measured with an appropriate meter. The point of measurement and type of meter must be defined to determine cfm accurately.

Outlet velocity (V_k) is the measured velocity at the started point with a specific meter.

Perimeter systems are heating and cooling installations in which the diffusers are installed to blanket the outside walls. Returns are usually located at one or more centrally located places. High sidewall or ceiling returns are preferred, especially for cooling. Low returns are acceptable for heating. High sidewall or ceiling returns are highly recommended for combination heating and cooling installations.

Registers are outlets that deliver air in a concentrated stream into the occupied zone.

Residual velocity (V_R) is the average sustained velocity within the confines of the occupied zone, generally ranging from 20 to 70 fpm.

Sound power level (L_w) is the total sound created by an outlet under a specified condition of operation.

Spread is the measurement (in feet) of the maximum width of the air pattern at the point of terminal velocity. Fig. 16-20.

Static pressure (sp) is the outward force of air within a duct. This pressure is measured in inches of water. The static pressure within a duct is comparable to the air pressure within an automobile tire. Static pressure is measured by a manometer. See Figs. 16-3 through 16-6.

Temperature differential (ΔT) is the difference between primary

REFRIGERATION AND AIR CONDITIONING TECHNOLOGY

supply and room air temperatures.

Terminal velocity is the point at which the discharged air from an outlet decreases to a given speed, generally accepted as 50 feet per minute.

Throw is the distance (measured in feet) that the airstream travels from the outlet to the point of terminal velocity. Throw is measured vertically from perimeter diffusers and horizontally from registers and ceiling diffusers. Fig. 16-17.

Total pressure (tp) is the sum of the static pressure and the velocity pressure. Total pressure is also known as impact pressure. This pressure is expressed in inches of water. The total pressure is directly associated with the sound level of an outlet. Therefore, any factor that increases the total pressure will also increase the sound level. The undersizing of outlets or increasing the speed of the blower will increase total pressure and the sound level.

Velocity pressure (vp) is the forward moving force of air within a duct. This pressure is measured in inches of water. The velocity pressure is comparable to the rush of air from a punctured tire. A velometer is used to measure air velocity. See Fig. 16-19.

DESIGNING A PERIMETER SYSTEM

After the heat loss or heat gain has been calculated, the sum of these heat losses or heat gains will determine the size of the duct systems and the heating and cooling unit.

The three factors that insure proper delivery and distribution of air within a room are location of outlet, type of outlet, and size of outlet.

Supply outlets, if possible, should always be located to blanket every window and every outside wall. Fig. 16-21. Thus, a register is recommended under each window.

The outlet selected should be a diffuser whose air pattern is fan-shaped to blanket the exposed walls and windows.

The *American Society of Heating, Refrigeration and Air-Conditioning Engineers* (ASHRAE) furnishes a chart with the locations and load factors needed for the climate of each major city in the United States. The chart should be followed carefully. The type of house, the construction materials, house location,

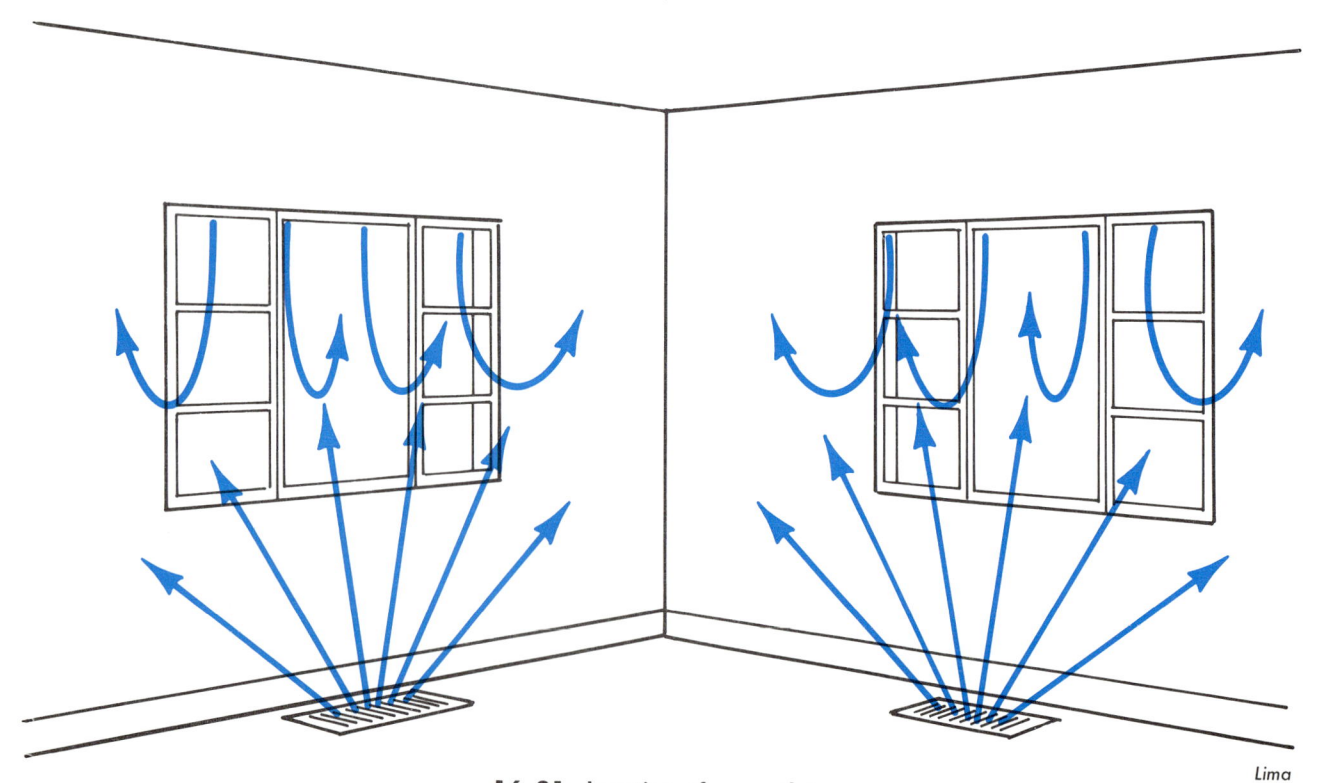

16-21. Location of an outlet.

room sizes, and exposure to sun and wind are important factors. With such information, you can determine how much heat will be dissipated. You can also determine how much cooling will be lost in a building.

The *ASHRAE Handbook of Fundamentals* lists the information needed to compute the load factors.

Calculate the heat loss or heat gain of the room, divide this figure by the number of outlets to be installed. From this you can determine the Btu/h required of each outlet. Refer to the performance data furnished by the manufacturer to determine the size the outlet should be. For residential application, the size selected should be large enough so that the Btu/h capacity on the chart falls to the side where the quiet zone is indicated and there is still a minimum vertical throw of 6' where cooling is involved.

Locating and Sizing Returns

Properly locating and sizing return air grilles is important. It is generally recommended that the returns be installed in high sidewall or the ceiling. They should be in one or more centrally located places, depending upon the size and floor plan of the structure. Although such a design is preferred, low returns are acceptable for heating.

To minimize noise, care must be taken to size correctly the return air grille. The blower in the equipment to be used is rated in cfm by the manufacturer. This rating can usually be found in the specification sheets. Select the grille or grilles necessary to handle this cfm.

The grille or grilles selected should deliver the necessary cfm

Tuttle & Bailey
16-22. This flow path diagram shows the pronounced one-sided flow effect from an outlet opening before corrective effect of air-turning devices.

Tuttle & Bailey
16-23. A control grid is added to equalize flow in the takeoff collar.

for the air to be conditioned. Thus, the proper size must be selected. The throw should reach approximately three-quarters of the distance from the outlet to the opposite wall. Fig. 16-18. The face velocity should not exceed the recommended velocity for the application. See Table 16-D. The drop should be such that the airstream will not drop into the occupied zone. The occupied zone is generally thought of as 6' above floor level.

The sound caused by an air outlet in operation varies in direct proportion to the velocity of the air passing through it. Air velocity depends partially on outlet size. Table 16-E lists recommendations for outlet velocities within safe sound limits for most applications.

Air Flow Distribution

Bottom or side outlet openings in horizontal or vertical supply ducts should be equipped with adjustable flow equalizing devices. Figure 16-22 indicates the pronounced one-sided flow effect from an outlet opening. This is before the corrective effect of air-turning devices. A control grid is added in Fig. 16-23 to equalize flow in the takeoff collar. A Vectrol® is added in Fig. 16-24 to turn air into the branch duct and provide volume control. Air-turning devices are recom-

REFRIGERATION AND AIR CONDITIONING TECHNOLOGY

Table 16-D.
Register or Grille Size Related to Air Capacities (in cfm)

Register or Grille Size	Area in Sq. Feet	250 fpm	300 fpm	400 fpm	500 fpm	600 fpm	700 fpm	750 fpm	800 fpm	900 fpm	1000 fpm	1250 fpm
8 × 4	.163	41	49	65	82	98	114	122	130	147	163	204
10 × 4	.206	52	62	82	103	124	144	155	165	185	206	258
10 × 6	.317	79	95	127	158	190	222	238	254	285	317	396
12 × 4	.249	62	75	100	125	149	174	187	199	224	249	311
12 × 5	.320	80	96	128	160	192	224	240	256	288	320	400
12 × 6	.383	96	115	153	192	230	268	287	306	345	383	479
14 × 4	.292	73	88	117	146	175	204	219	234	263	292	365
14 × 5	.375	94	113	150	188	225	263	281	300	338	375	469
14 × 6	.449	112	135	179	225	269	314	337	359	404	449	561
16 × 5	.431	108	129	172	216	259	302	323	345	388	431	539
16 × 6	.515	129	155	206	258	309	361	386	412	464	515	644
20 × 5	.541	135	162	216	271	325	379	406	433	487	541	676
20 × 6	.647	162	194	259	324	388	453	485	518	582	647	809
20 × 8	.874	219	262	350	437	524	612	656	699	787	874	1093
24 × 5	.652	162	195	261	326	391	456	489	522	587	652	815
24 × 6	.779	195	234	312	390	467	545	584	623	701	779	974
24 × 8	1.053	263	316	421	527	632	737	790	842	948	1053	1316
24 × 10	1.326	332	398	530	663	796	928	995	1061	1193	1326	1658
24 × 12	1.595	399	479	638	798	951	1117	1196	1276	1436	1595	1993
30 × 6	.978	245	293	391	489	587	685	734	782	880	978	1223
30 × 8	1.321	330	396	528	661	793	925	991	1057	1189	1371	1651
30 × 10	1.664	416	499	666	832	998	1165	1248	1331	1498	1664	2080
30 × 12	2.007	502	602	803	1004	1204	1405	1505	1606	1806	2007	2509
36 × 8	1.589	397	477	636	795	953	1112	1192	1271	1430	1589	1986
36 × 10	2.005	501	602	802	1003	1203	1404	1504	1604	1805	2005	2506
36 × 12	2.414	604	724	966	1207	1448	1690	1811	1931	2173	2414	3018

*Based on LIMA registers of the 100 Series.

mended for installation at all outlet collars and branch duct connections.

Square unvaned elbows are also a source of poor duct distribution and high pressure loss.

Non-uniform flow in a main duct, occurring after an unvaned ell, severely limits the distribu-

CHAPTER 16—PSYCHROMETRIC AND AIR MOVEMENT

Table 16-E.
Outlet Velocity Ratings

Area	Rating (in feet per minute)
Broadcast studios	500 fpm
Residences	500 to 750 fpm
Apartments	500 to 750 fpm
Churches	500 to 750 fpm
Hotel bedrooms	500 to 750 fpm
Legitimate theatres	500 to 1000 fpm
Private offices, acoustically treated	500 to 1000 fpm
Motion picture theatres	1000 to 1250 fpm
Private offices, not acoustically treated	1000 to 1250 fpm
General offices	1250 to 1500 fpm
Stores	1500 fpm
Industrial buildings	1500 to 2000 fpm

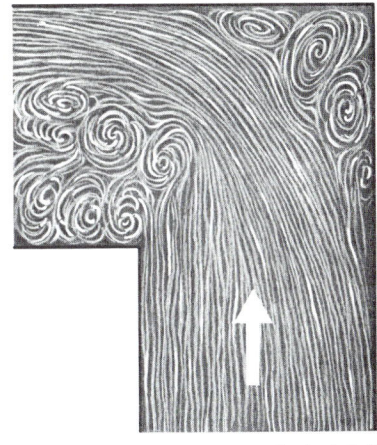

16-25. Note the turbulence and piling up of air flow in an ell.

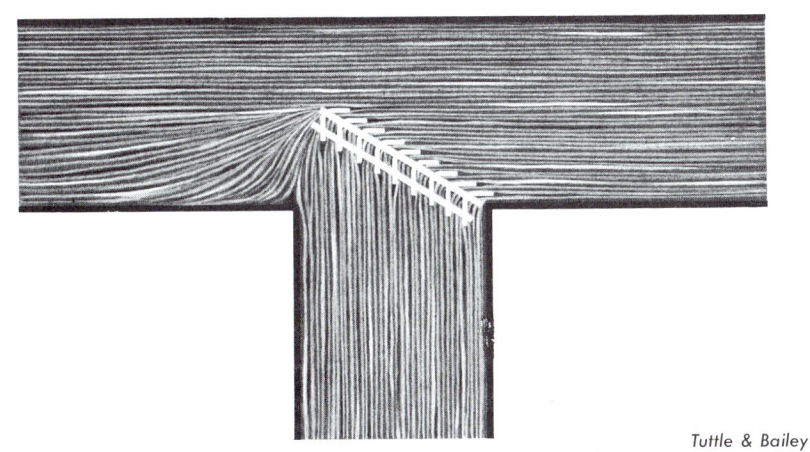

16-24. A Vectrol® is added to turn air into the branch duct and provide volume control.

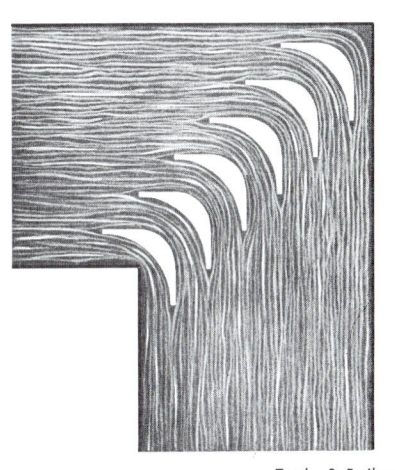

16-26. A ducturn reduces the pressure loss in square elbows by as much as 80%.

tion of air into branch ducts in the vicinity of the ell. One side of the duct may be void, thus starving a branch duct. Conversely, all flow may be stacked up on one side. This requires dampers to be excessively closed, resulting in higher sound levels.

Flow diagrams show the pronounced turbulence and piling up of air flow in an ell. Fig. 16-25. *Ducturns* reduce the pressure loss in square elbows as much as 80%. Their corrective effect is shown in Fig. 16-26.

SELECTION OF DIFFUSERS AND GRILLES

The selection of a linear diffuser or grille involves job condition requirements, selection judgement, and performance data analysis.

Diffusers and grilles should be selected and sized according to the following characteristics:
- Type and style.
- Function.
- Air volume requirement.
- Throw requirement.
- Pressure requirement.
- Sound requirement.

Air Volume Requirement

The air volume per diffuser or grille is that which is necessary for the cooling, heating, or ventilation requirements of the area served by the unit. The air volume required, when related to throw, sound, or pressure design limitations, determines the proper diffuser or grille size.

439

Generally, air volumes for internal zones of building spaces vary from 1 to 3 cfm per square foot of floor area. Exterior zones will require higher air volumes of 2.5 to 4 cfm per square foot. In some cases, only the heating or cooling load of the exterior wall panel or glass surface is to be carried by the distribution center. Then, the air volume per linear foot of diffuser or grille will vary from 20 to 200 cfm, depending on heat transfer coefficient, wall height, and infiltration rate.

Throw Requirement

Throw and occupied area air motion are closely related. Both should be considered in the analysis of specific area requirements. The minimum-maximum throw for a given condition of operation is based upon a terminal velocity at that distance from the diffuser. The residual room velocity is a function of throw and terminal velocity. Throw values are based on terminal velocities ranging from 75 to 150 fpm with corresponding residual room velocities of 35 to 50 fpm.

The diffuser or grille location, together with the air pattern selected, should generally direct the air path above the occupied zone. The air path then induces room air along its throw as it expands in cross-section. This equalizes temperature and velocity within the stream. With the throw terminating at a partition or wall surface, the mixed air path further dissipates energy.

Ceiling mounted grilles and diffusers are recommended for vertical down pattern. Some locations in the room may need to be cooler than others. Also, some room locations may be harder to condition because of air flow problems. They are used in areas adjacent to perimeter wall locations that require localized spot conditioning. Ceiling heights of 12' or greater are needed. The throw for vertical projection is greatly affected by supply air temperature and proximity of wall surfaces.

Sidewall mounted diffusers and grilles have horizontal values based on a ceiling height of 8' to 10'. The diffuser or grille is mounted approximately 1' below the ceiling. For a given listed throw, the room air motion will increase or decrease inversely with the ceiling height. For a given air pattern setting and room air motion, the listed minimum-maximum throw value can be decreased by 1' for each 1' increase in ceiling height above 10'. Throw values are furnished by the manufacturer.

When sidewall grilles are installed remote from the ceiling (more than 3' away), reduce rated throw values by 20%.

Sill mounted diffusers or grilles have throw values based on an 8' to 10' ceiling height. This is with the outlet installed in the top of a 30" high sill. For a given listed throw, the room air motion will change with the ceiling height. For a given air pattern setting and room air motion, the listed minimum-maximum throw value can be decreased by 2' for each 1' increase in ceiling height above 10'. Decrease 1' for each 1' decrease in sill height.

The minimum throw results in a room air motion higher than that obtained when utilizing the maximum throw. Thus, 50 fpm, rather than 35 fpm, is the air motion. The listed minimum throw indicates the minimum distance recommended. The minimum distance is from the diffuser to a wall or major obstruction, such as a structural beam. The listed maximum throw is the recommended maximum distance to a wall or major obstruction. Throw values for sidewall grilles and ceiling diffusers and the occupied area velocity are based on flush ceiling construction providing an unobstructed airstream path. The listed maximum throw times 1.3 is the complete throw of the airstream. This is where the terminal velocity equals the room air velocity. Rated occupied area velocities range from 25 to 35 fpm for maximum listed throws and 35 to 50 fpm for minimum listed throw values.

Cooled-air drop or heated-air rise are of practical significance when supplying heated or cooled air from a sidewall grille. If the throw is such that the airstream prematurely enters the occupied zone, considerable draft may be experienced. This is due to incomplete mixing. The total air drop must be considered when the wall grille is located a distance from the ceiling. Cooled air drop is controlled by spacing the wall grille from the ceiling and adjusting the grilles upward 15°. Heated-air rise contributes significantly to temperature stratification in the upper part of the room.

The minimum separation between grille and ceiling must be 2' or more. The minimum mounting separation must be 2' or more. The minimum mounting height should be 7'.

Pressure Requirement

The diffuser or grille minimum pressure for a given air volume reflects itself in ultimate sys-

Table 16-F.
Recommended NC Criteria

NC Curve	Communication Environment	Typical Occupancy
Below NC 25	Extremely quiet environment, suppressed speech is quite audible, suitable for acute pickup of all sounds.	Broadcasting studios, concert halls, music rooms.
NC 30	Very quiet office, suitable for large conferences; telephone use satisfactory.	Residences, theatres, libraries, executive offices, directors' rooms.
NC 35	Quiet office; satisfactory for conference at a 15 ft. table; normal voice 10–30 ft.; telephone use satisfactory.	Private offices, schools, hotel rooms, courtrooms, churches, hospital rooms.
NC 40	Satisfactory for conferences at a 6–8 ft. table; normal voice 6–12 ft.; telephone use satisfactory.	General offices, labs, dining rooms.
NC 45	Satisfactory for conferences at a 4–5 ft. table; normal voice 3–6 ft.; raised voice 6–12 ft.; telephone use occasionally difficult.	Retail stores, cafeterias, lobby areas, large drafting & engineering offices, reception areas.
Above NC 50	Unsatisfactory for conferences of more than two or three persons; normal voice 1–2 ft.; raised voice 3–6 ft.; telephone use slightly difficult.	Photocopy rooms, stenographic pools, print machine rooms, process areas.

tem fan horsepower requirements. A diffuser or grille with a lower pressure rating requires less total energy than a unit with a higher pressure rating for a given air volume and effective area. Diffusers and grilles of a given size having lower pressure ratings usually have a lower sound level rating at a specified air volume.

Sound Requirement

Diffusers and grilles should be selected for the recommended noise criteria rating for a specific application. The data for each specific diffuser or grille type contains an NC rating.

Table 16-F lists recommended noise criteria (NC) and area of application.

Air Noise

Air noise is typically generated by high velocities in the duct or diffuser. The flow turbulence in the duct and the excessive pressure reductions in the duct and diffuser system also generate noise. Such noise is most apparent directly under the diffuser. Room background levels of nc 35 and less provide little masking effect. Any noise source stands out above the background level and is easily detected.

Typically, air noise can be minimized by the following procedures.

- Limiting branch duct velocities to 1200 fpm.
- Limiting static pressure in branch ducts adjacent to outlets to 0.15 inches H_2O.
- Sizing diffusers to operate at outlet jet velocities up to 1200 fpm, (neck velocities limited to 500 to 900 fpm), and total pressures of 0.10 inches H_2O.
- Using several small diffusers (and return grilles) instead of one or two large outlets or inlets that have a higher sound power.
- Providing low-noise dampers in the branch duct where pressure drops of more than 0.20 inches of water must be taken.
- Internally lining branch ducts near the fan to quiet this noise source.
- Designing background sound levels in the room to be a minimum of nc 35 or nc 40.

CASING RADIATED NOISE

Casing noise differs from air noise in the way it is generated. Casing noise is generated by volume controllers and pressure reducing dampers. Inside terminal boxes are sound baffles, absorbing blankets, and orifice restrictions to eliminate line of sight through the box. All these work to reduce the generated noise before the air and air noise discharge from the box into the outlet duct. During this process, the box casing is vibrated by the internal noise. This causes the casing to radiate noise through the suspended ceiling into the room. Fig. 16-27.

LOCATING TERMINAL BOXES

In the past, terminal boxes and ductwork were separated from the room by dense ceilings. These ceilings prevented the system noise from radiating into the room. Plaster and taped sheetrock ceilings are examples of dense ceilings. Current architectural practice is to utilize lightweight (and low-cost) decorative suspended ceilings. These ceilings are not dense. They have only one-half the resistance to noise transmission that plaster and sheetrock ceilings have. Exposed tee-bar grid ceilings with 2 × 4 glass fiber pads, and perforated metal pan ceilings are examples. The end result is readily apparent. Casing radiated noise in lightweight, modern building is a problem.

Controlling Casing Noise

Terminal boxes can sometimes be located over noisy areas (corridors, toilet areas, machine equipment rooms), rather than over quiet areas. In quiet areas casing noise can penetrate the suspended ceiling and become objectionable. Enclosures built around the terminal box (such as sheetrock or sheet lead over a glass fiber blanket wrapped around the box) can reduce the radiated noise to an acceptable level. However, this method is cumbersome and limits access to the motor and volume controllers in the box. It depends upon field conditions for satisfactory performance, and is expensive. Limiting static pressure in the branch ducts minimizes casing noise. This technique, however, limits the flexibility of terminal box systems. It hardly classifies as a control.

Vortex Shedding

Product research in controlling casing noise has developed a new method of reducing radiated noise. The technique is known as vortex shedding. When applied to terminal boxes, casing radiated noise is dramatically lowered. Casing radiation attenuation vortex shedders (CRA) can be installed in all single- or dual-duct boxes up to 7000 cfm, both constant volume or variable volume, with or without reheat coils. CRA devices provide unique features and the following benefits.

- No change in terminal box size. Box is easier to install in tight ceiling plenums to insure minimum casing noise under all conditions.
- Factory-fabricated box and casing noise eliminator, a one-piece assembly, reduces cost of installation. Only one box is hung. Only one duct connection is made.
- Quick-opening access door is provided in box. This assures easy and convenient access to all operating parts without having to cut and patch field-fabricated enclosures.
- Equipment is laboratory tested and performance rated. Engineering measurements are made in accordance with industry standards. Thus, on-job performance is insured. Quiet rooms result and owner satisfaction is assured.

RETURN GRILLES

Performance

Return air grilles are usually selected for the required air volume at a given sound level or pressure value. The intake air velocity at the face of the grille depends mainly on the grille size and the air volume.

The grille style and damper setting have a small effect on this intake velocity. The grille style, however, has a very great effect on the pressure drop. This, in turn, directly influences the sound level.

The intake velocity is evident only in the immediate vicinity of the return grille. It cannot influ-

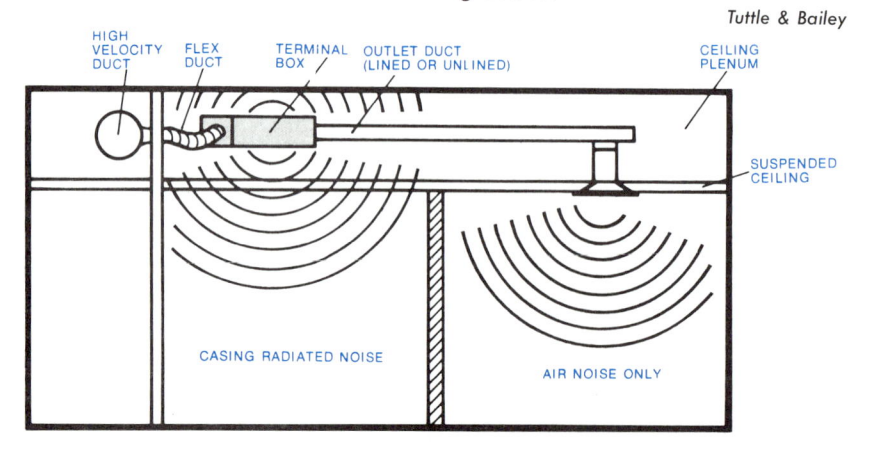

16-27. Casing noises.

Tuttle & Bailey

ence room air distribution. Recent ASHRAE research projects have developed a scientific computerized method of relating intake grille velocities, measured 1″ out from the grille face, to air volume. Grille measuring factors for straight, deflected bar, open, and partially closed dampers are in the engineering data furnished with the grille.

It still remains the function of the supply outlets to establish proper coverage, air motion, and thermal equilibrium. Because of this, the location of return grilles is not critical and their placement can be largely a matter of convenience. Specific locations in the ceiling may be desirable for local heat loads or smoke exhaust or a location in the perimeter sill or floor may be desirable for an exterior zone intake under a window wall section. It is not advisable to locate large centralized return grilles in an occupied area. This is because the large mass of air moving toward the grille can cause objectionable air motion for nearby occupants.

Return Grille Sound Requirement

Return air grilles should be selected for static pressures. These pressures will provide the required nc rating and conform to the return system performance characteristics. Fan sound power is transmitted through the return air system as well as the supply system. Fan silencing may be necessary or desirable in the return side. This is particularly so if silencing is being considered on the supply side.

Transfer grilles venting into the ceiling plenum should be located remote from plenum noise source. The use of a lined sheet metal elbow can reduce transmitted sound. Lined elbows on vent grilles and lined common ducts on ducted return grilles can minimize "cross talk" between private offices.

TYPES OF REGISTERS AND GRILLES

The spread of an unrestricted airstream is determined by the grille bar deflection. Grilles with vertical face bars at 0° deflection will have a maximum throw value. As the deflection setting of vertical bars is increased, the airstream covers a wider area and the throw decreases.

Registers are available with adjustable valves. An air leakage problem is eliminated if the register has a rubber gasket mounted around the grille. When it pulls up tightly against the wall, an airtight seal is made. This helps to eliminate noise. The damper has to be cam-operated so that it will stay open and not blow shut when the air comes through.

On some registers, a simple tool can be used to change the direction of the deflection bars. This means a number of deflection patterns can be had by adjusting the bars in the register.

FIRE AND SMOKE DAMPERS

Ventilating, air conditioning, and heating ducts provide a path for fire and smoke, which can travel throughout a building. The ordinary types of dampers that are often installed in these ducts depend on gravity-close action or spring and level mechanisms. When their releases are activated, they are freed to drop inside the duct.

A fusible link attachment to individual registers also helps control fire and smoke. Figure 16-28 shows a fusible-link type register. The link is available with melting points of 160° F. [71.1°C] or 212° F. [100°C]. When the link melts, it releases a spring that forces the damper to a fully closed position. The attachment does not interfere with damper operation.

Smoke Dampers for High-Rise Buildings

Fire and smoke safety concepts in high-rise buildings are increasingly focusing on providing safety havens for personnel on each floor. This provision is to optimize air flow to or away from the fire floor or adjacent floors. Such systems require computer-actuated smoke dampers. Dampers are placed in supply and return ducts that are reliable. They must be tight closing, and offer minimum flow resistance when fully open.

CEILING SUPPLY GRILLES AND REGISTERS

Some ceiling grilles and registers have individually adjustable vanes. They are arranged to provide a one-way ceiling air pattern. They are recommended for applications in ceiling and sidewall locations for heating and cooling systems. They work best

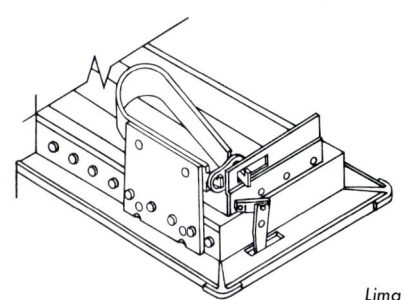

16-28. Register with fusible link for fire control.

REFRIGERATION AND AIR CONDITIONING TECHNOLOGY

Tuttle & Bailey
16-29. Ceiling grille.

Tuttle & Bailey
16-32. Round diffusers with flush face, fixed pattern for ceiling installation.

Tuttle & Bailey
16-33. Control grid with multiblade devices to control airflow in a diffuser collar.

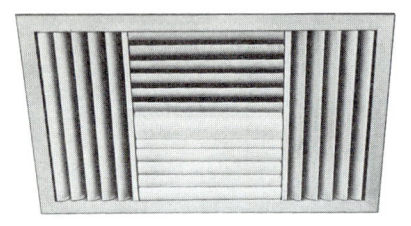

Tuttle & Bailey
16-30. Vertical face vanes in a four-way ceiling supply grille.

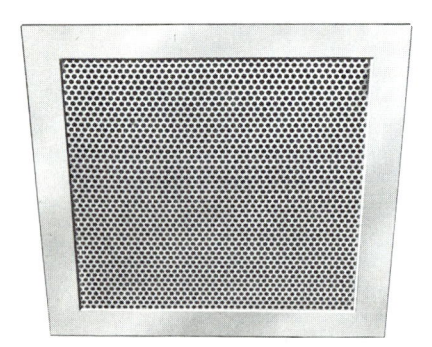

Tuttle & Bailey
16-31. Perforated face adjustable diffuser for full flow and a deflector for ceiling installation.

where the system has 0.75 to 1.75 cfm per square foot of room area. Fig. 16-29.

Some supply ceiling grilles and registers have individually adjustable curved vanes. They are arranged to provide a three-way ceiling air pattern. The vertical face vanes are a three-way diversion for air. A horizontal pattern with the face vanes also produces a three-way dispersion of air. These grilles and registers are recommended for applications in ceiling locations for heating and cooling systems handling 1.0 to 2.0 cfm per square foot of room area.

Figure 16-30 shows a grille with four-way vertical face vanes. Horizontal face vanes are also available. They, too, are adjustable individually for focusing an airstream in any direction. Both the three-way and four-way pattern grilles can be adjusted to a full or partial downblow position. The curved streamlined vanes are adjusted to a uniform partially closed position. This deflects the air path while retaining an effective area capacity of 35% of the neck area. In the full downblow position, grille effective area is increased by 75%.

Perforated adjustable diffusers for ceiling installation are recommended for heating and cooling. Fig. 16-31. They are also recommended for jobs requiring on-the-job adjustment of air diffusion patterns.

Full-flow square or round necks have expanded metal air pattern deflectors. They are adjustable for 4-, 3-, 2-, or 1-way horizontal diffusion patterns. This can be done without change in the air volume, pressure, or sound levels. This deflector and diffuser has high diffusion rates. The result is rapid temperature and velocity equalization of the mixed air mass well above the zone of occupancy.

They diffuse efficiently with six to eighteen air changes per hour.

CEILING DIFFUSERS

There are other designs in ceiling diffusers. The type shown in Fig. 16-32 is often used in a supermarket or other large store. Here, it is difficult to mount other means of air distribution. These round diffusers with a flush face and fixed pattern are for ceiling installation. They are used for heating, ventilating, and cooling. They are compact and simple flush diffusers. High induction rates result in rapid temperature and velocity equalization of the mixed air mass. Mixing is done above the zone of occupancy.

Grids are used and sold as an

444

CHAPTER 16—PSYCHROMETRIC AND AIR MOVEMENT

Tuttle & Bailey
16-34. Antismudge ring.

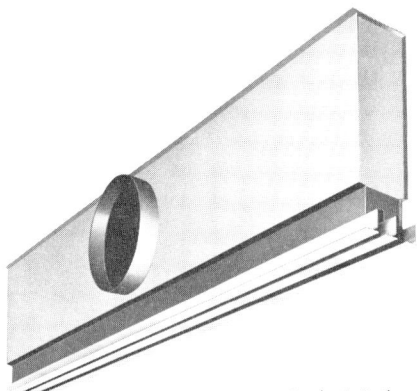

Tuttle & Bailey
16-35. High-capacity air channel diffuser with fixed pattern for suspended grid ceilings.

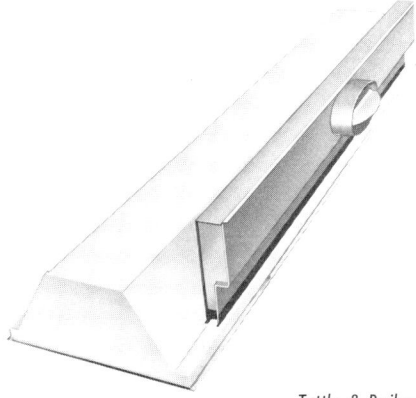

Tuttle & Bailey
16-36. Single-side diffuser with side inlet.

Tuttle & Bailey
16-37. Dual-side diffuser with side inlet.

accessory to these diffusers. The grid, Fig. 16-33, is a multiblade device designed to insure uniform air flow in a diffuser collar. It is individually adjustable. The blades can be moved to control the airstream precisely.

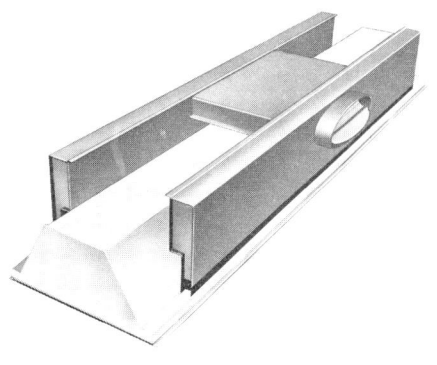

For maximum effect, the control grid should be installed with the blades perpendicular to the direction of approaching air flow. Where short collars are encountered, a double bank of control grids is recommended. The upper grid is placed perpendicular to the branch duct flow. The lower grid is placed parallel to the branch duct flow. The control grid is attached to the duct collar by means of mounting straps. It is commonly used with volume dampers.

Antismudge Rings

The antismudge ring is designed to cause the diffuser discharge air path to contact the ceiling in a thin layered pattern. This minimizes local turbulence, the cause of distinct smudging. Fig. 16-34.

For the best effect, the antismudge ring must fit evenly against the ceiling surface. It is held in position against the ceiling by the diffuser margin. This eliminates any exposed screws.

Air Channel Diffusers

Air channel supply diffusers are designed for use with integrated air handling ceiling systems. They are adaptable to fit between open parallel tee bars. They fit within perforated or slotted ceiling runners. The appearance of the integrated ceiling remains unchanged regardless of the size of the unit. They are painted-out to be invisible when viewing the ceiling. These high capacity diffusers provide a greater air handling capability. Fig. 16-35.

Luminaire Diffusers

The luminaire is a complete lighting unit. The luminaire diffuser fits close to the fluorescent lamp fixtures in the ceiling. The single-side diffuser with side inlet is designed to provide single-side concealed air distribution. Fig. 16-36. These diffusers are designed with oval-shaped side inlets and inlet dampers. They provide effective single-point dampering.

Dual-side diffusers with side inlet are designed to provide concealed air distribution. Note the crossover from the oval side inlet to the other side of the diffuser. This type of unit handles more air and spreads it more evenly when used in large areas. Fig. 16-37. This type of diffuser is also available with an insulation jacket when needed.

Room Air Motion

Figure 16-38 illustrates the air flow from ceiling diffusers. The top view illustrates the motion from the diffuser. The side view shows how the temperature differential is very low. Note that

REFRIGERATION AND AIR CONDITIONING TECHNOLOGY

OPERATING CONDITIONS
SUPPLY CFM = 280 (140 EA. DIFF.)
RETURN CFM = 210 (105 EA. DIFF.)
CFM/FT² FLOOR = 2 (APPROX.)
AIR CHANGE/HR. = 13
SUPPLY AIR TEMP. = 52°F.
AVG. ROOM TEMP. = 72°F.
ONE WAY PATTERN.

16-38. Room air motion.

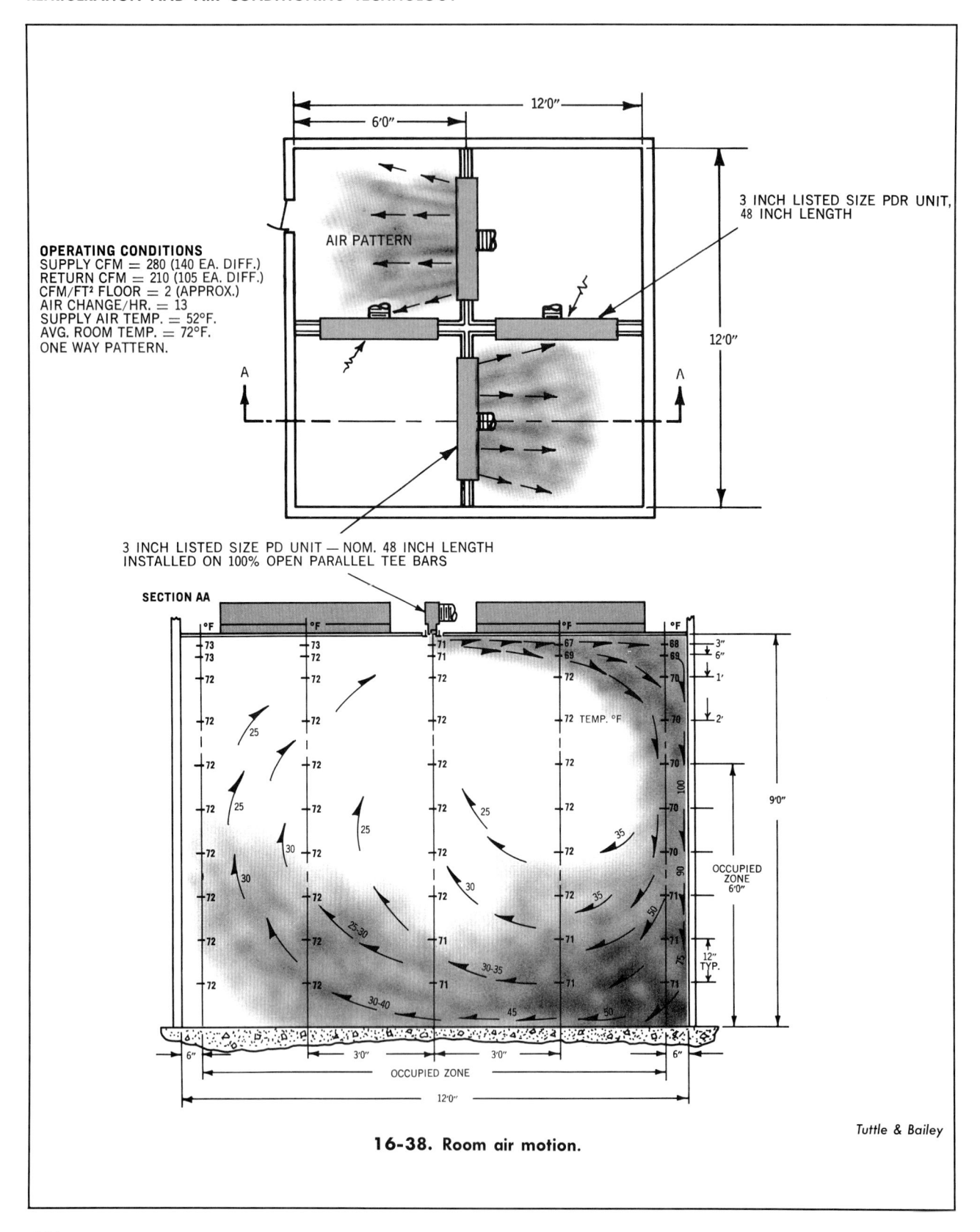

Tuttle & Bailey

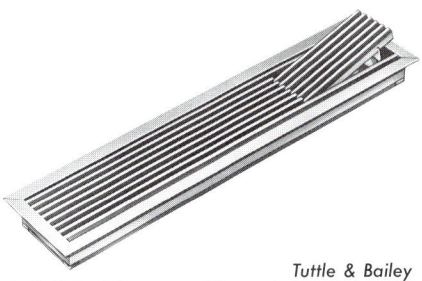

16-39. Linear grille with a hinged access door — Tuttle & Bailey

16-40. Debris screen for linear grilles. — Tuttle & Bailey

the temperature is 68° F. near the ceiling and sidewall and 73° F. on the opposite wall near the ceiling.

LINEAR GRILLES

Linear grilles are designed for installation in the sidewall, sill, floor, and ceiling. They are recommended for supplying heated, ventilated, or cooled air and for returning or exhausting room air. Fig. 16-39.

When installed in the sidewall near the ceiling, the linear grilles provide a horizontal pattern above the occupied zone. Core deflections of 15° and 30° direct the air path upward to overcome the drop effect resulting from cool primary air.

When installed in the top of a sill or enclosure, the linear grilles provide a vertical up-pattern. This is effective in overcoming uncomfortable cold downdrafts. It also offsets the radiant effect of glass surfaces. Core deflections of 0° and 15° directed toward the glass surface provide upward airflow to the ceiling, and along the ceiling toward the interior zone.

When installed in the ceiling, linear grilles provide a vertical downward air pattern. This pattern is effective in projection heating and in cooling the building perimeter from ceiling heights above 13' to 15'. Application of downflow primary air should be limited to insure against excessive drafts at the end of the throw. Core deflections of 0°, 15°, and 30° direct the air path angularly downward as required.

Debris screens can be integrally attached. Fig. 16-40.

REVIEW QUESTIONS

1. Define temperature.
2. What is the difference between absolute zero and zero on the Fahrenheit scale?
3. What is the formula for converting degrees Fahrenheit to degrees Celsius?
4. What is a manometer?
5. What does psig stand for?
6. What is the meaning of psia?
7. What is the difference between a wet bulb thermometer and a dry bulb thermometer?
8. Define dewpoint.
9. What is relative humidity?
10. What is a psychrometric chart?
11. How is a psychrometric chart used in designing cooling systems?
12. What is enthalpy?
13. What is the purpose of ducturns?
14. Name at least three factors to be considered when selecting and sizing diffusers and grilles.
15. Name two ways to minimize air noise.
16. What is vortex shedding?
17. How do you minimize or eliminate cross talk between private offices due to the air distribution system?
18. What is a fusible link?

17
Comfort Air Conditioning

WINDOW UNITS

Room air conditioners are appliances that cool, dehumidify, filter, and circulate air. Fig. 17-1. There are several manufacturers of window-mounted air conditioning units. No attempt is being made here to represent all of them. Some general information is needed to help the repairperson make recommendations to those who request information on various units.

Today, energy conservation is important. Thus, most manufacturers specify the energy savings their unit will produce. The best way to determine the amount of energy consumed is to check the electrical requirements of the unit. Check this figure with the Btu generated by that power. Room air conditioners are usually rated according to the number of Btu they will produce in cooling. The smallest unit is about 5000 Btu. Thus, it is possible to determine which unit can produce a greater amount of cooling in terms of Btu per kilowatt hour.

Figure 17-2 shows a room air conditioner with the cover removed. The compressor, fan, evaporator, and condenser are visible. The three main parts of an air conditioner are the fan, filter, and cooling element. The fan pulls warm moist air into the unit. The air is moved through the filter, where dust particles are removed. Next, the air is passed over the cooling element, where it is cooled and dehumidified. The fan then returns the conditioned air to the room. This conditioned air has been cooled, dehumidfied, and filtered.

Mounting

Window units are mounted in several different ways. Some are mounted in windows and some are mounted in a hole in the wall. The hole-in-the-wall mount is usually designed into the building at the time it is constructed. Different window types require that adapters be fitted to the air conditioning unit. Figure 17-3 shows a model with expanding side panels. A universal mounting kit is shown with flush mounting in Fig. 17-4. It indicates how the units will look inside and outside. Figures 17-5 and 17-6 show two mounting methods. Figure 17-7 illustrates a kit for mounting a unit in a sliding or casement window from 21 $\frac{1}{4}''$ to 36" high. A Mylar® slide-up weathertight wing is pulled up to fill the space around the air conditioner.

17-1. Window-mounted air conditioning unit.
Admiral

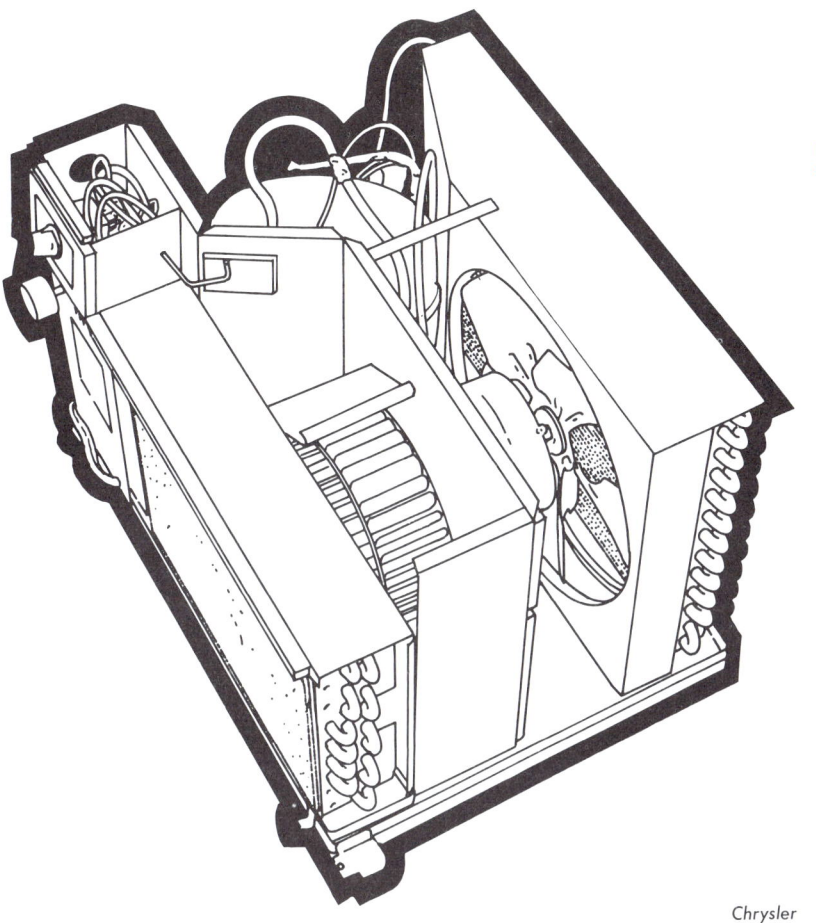

CHAPTER 17—COMFORT AIR CONDITIONING

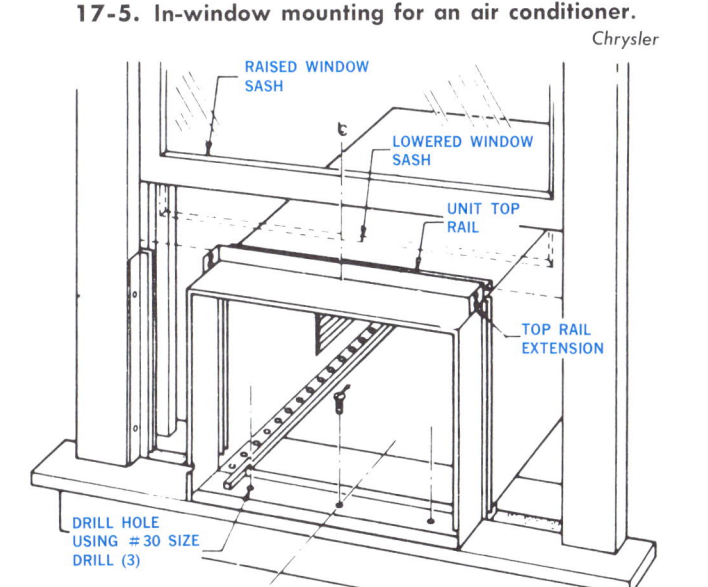

17-3. Window air conditioner with expanding side panels for window mounting.

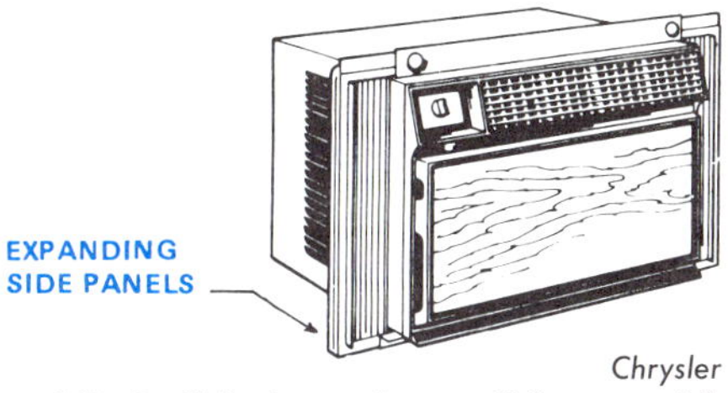

17-4. Universal mounting kit for flush mounting of window air conditioners.

17-2. Window air conditioner with cover removed.

17-5. In-window mounting for an air conditioner.

The wall mounting of a small unit is shown in Fig. 17-8. A telescopic wall sleeve is available or an in-wall fixed sleeve is constructed when the house or office is built or remodeled.

The main advantage of the window unit is ease of installation. It is ready for use when plugged into the wall plug.

17-6. Another arrangement for in-window mounting of an air conditioner.

449

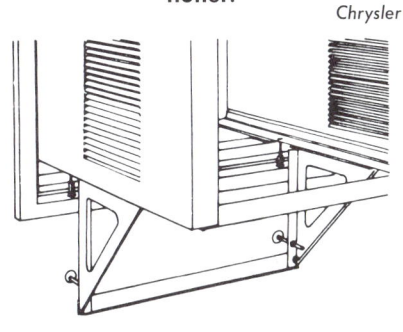

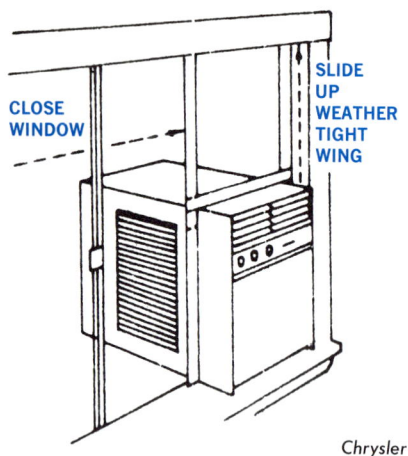

REFRIGERATION AND AIR CONDITIONING TECHNOLOGY

17-7. Mounting an air conditioner in a sliding or casement window. *Chrysler*

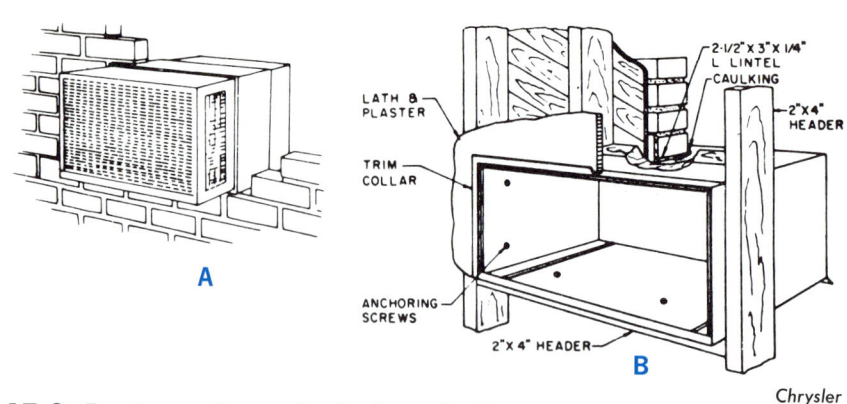

17-8. Two types of mounting in the wall. (A) Using a telescopic wall sleeve and (B) using a fixed in-wall sleever. *Chrysler*

17-9. Types of electrical plugs found on air conditioners for home use. *Chrysler*

17-10. Wiring an air conditioner with a fixed electrical connection. *Chrysler*

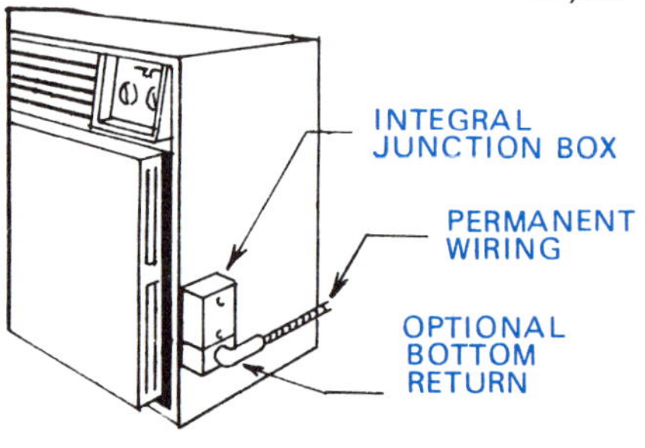

Electrical Plugs

Some units demand more current than can be safely furnished by a 120-V, 15-A circuit. Thus, a 230–240 V plug must be installed. Different plugs are used for different outlet sizes. Figure 17-9 shows several different plugs. Note the difference as in the arrangement of the slots. It is obvious that the plug must be fitted to the correct socket. The correct socket will have wiring of the proper size to handle the load. Plugs vary in design because the power demands of the air conditioners will vary with the Btu rating. For example, one of the plugs shown in Fig. 17-9 requires 50 A. In most cases, this would require a special circuit for the air conditioner. In most cases with high current demands, the air conditioner is also capable of heating with an electric heating coil installed for cold weather use.

Larger air conditioners do not come with plugs. They must be wired directly to the service. Wire of the correct size must be run to the junction box mounted on the side of the conditioner unit. Fig. 17-10.

Maintenance

The units are designed for ease of maintenance. They usually require a filter change or cleaning at least once a year. Where dust is a problem, such maintenance should be more frequent. At this time, the condenser coil should be brushed with a soft brush and flushed with water. The filters should be vacuumed and then washed to remove dust. The outside of the case should be wiped clean with a soapy cloth. Needless to say, the cleaner the filter, the more efficient the unit.

Low-Voltage Operation

Electrical apparatus is designed to produce at full capacity at the voltage indicated on the rating plate. Motors operated at lower than rated voltage cannot provide full horsepower without shortening their service life. Low voltage can result in energy that is insufficient to energize relays and coils.

The ARI (Air Conditioning and Refrigeration Institute) certifies cooling units after testing

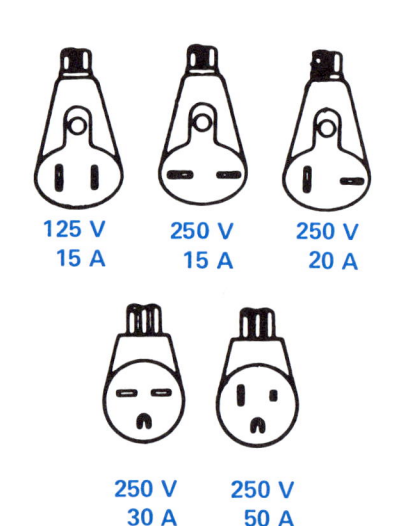

450

CHAPTER 17—COMFORT AIR CONDITIONING

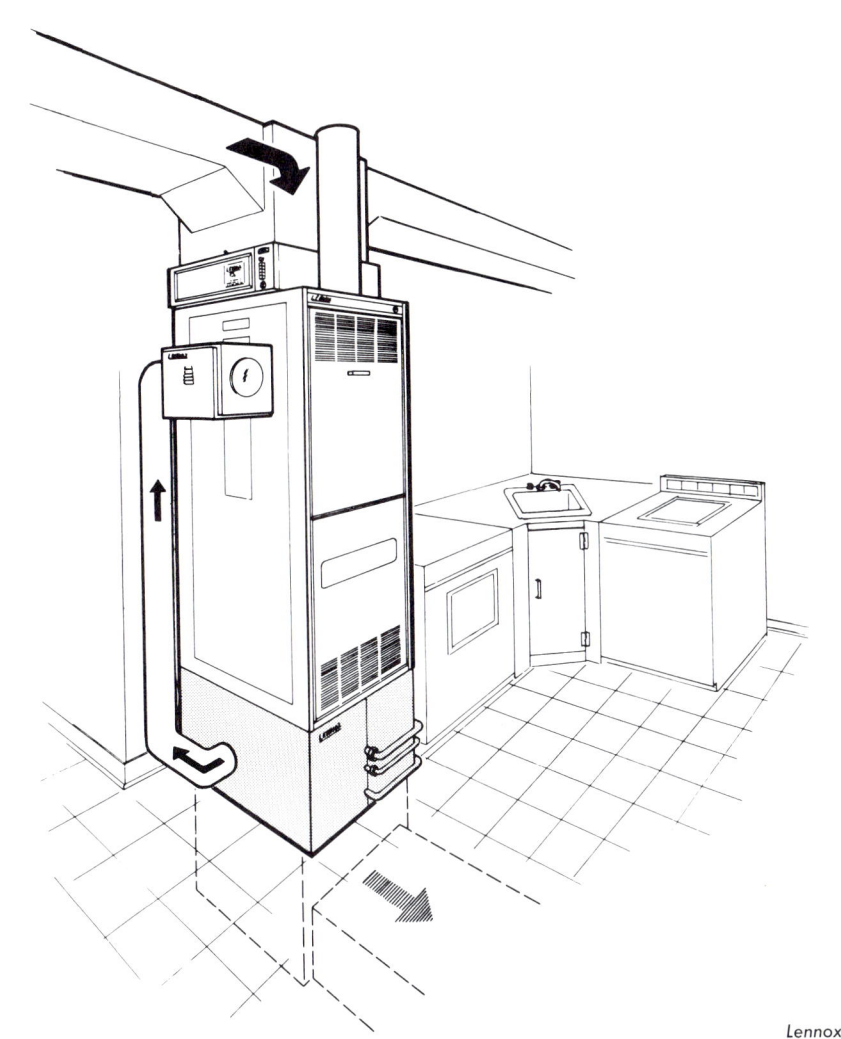

17-11. Utility room installation of a down-flow evaporator on an existing furnace.
Lennox

them. The units are tested to make sure they will operate with 10% above or 10% below rated voltage. This does not mean that the units will operate continuously without damage to the motor. A large proportion of air conditioning compressor burnouts can be traced to low voltage. Because a hermetic compressor motor is entirely enclosed within the refrigerant cycle, it is very important that it not be abused either by overloading or low voltage. Both of these conditions can occur during peak load conditions. A national survey has shown that the most common cause of compressor low voltage is the use of undersized conductors between the utility lines and the condensing unit.

Low voltage becomes extremely important when it is necessary to plug into an existing circuit. The existing load on the circuit may be sufficient to load the circuit. In this case, the air conditioning unit will result in too much load, blowing the fuse or tripping the circuit breaker. However, in some cases, the fuse does not blow and the circuit breaker does not trip. This reduces the line voltage, since the wire for the circuit is too small to handle the current needed to operate all devices plugged into it. Check the circuit load before adding the air conditioner to the line. This will prevent damage to the unit.

Troubleshooting

To troubleshoot this type of air conditioning equipment, a troubleshooting table (Table 17-E) has been provided at the end of this chapter. The general troubleshooting procedures listed there are used for hermetically sealed compressors.

EVAPORATORS FOR ADD-ON RESIDENTIAL USE

One of the more efficient ways of adding whole-house air conditioning is by adding an evaporator coil in the furnace. The evaporator coil becomes an important part of the whole system. It can be added to the existing furnace to make a total air conditioning and heating package. There are two types of evaporators—down-flow and up-flow.

The down-flow evaporator is installed beneath a down-flow furnace. Fig. 17-11. Lennox makes down-flow models in 3-, 4-, and 5-ton sizes with an inverted "A" coil. Fig. 17-12. Condensate runs down the slanted side to the drain pan. This unit can be installed in a closet or utility room wherever the furnace is located. This type of unit is shipped factory assembled and tested.

451

REFRIGERATION AND AIR CONDITIONING TECHNOLOGY

17-12. Evaporator units for add-on cooling for up-flow and down-flow furnaces.

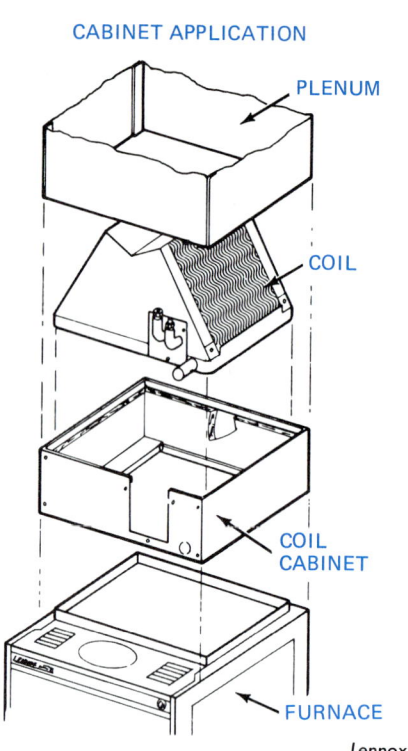

17-13. Installation of up-flow evaporators. (A) Basement installation with an oil furnace, return air cabinet, and automatic humidifier. (B) Closet installation of an evaporator coil with electric furnace and electronic air cleaner.

17-14. Installation of an evaporator coil on top of an existing furnace installation.

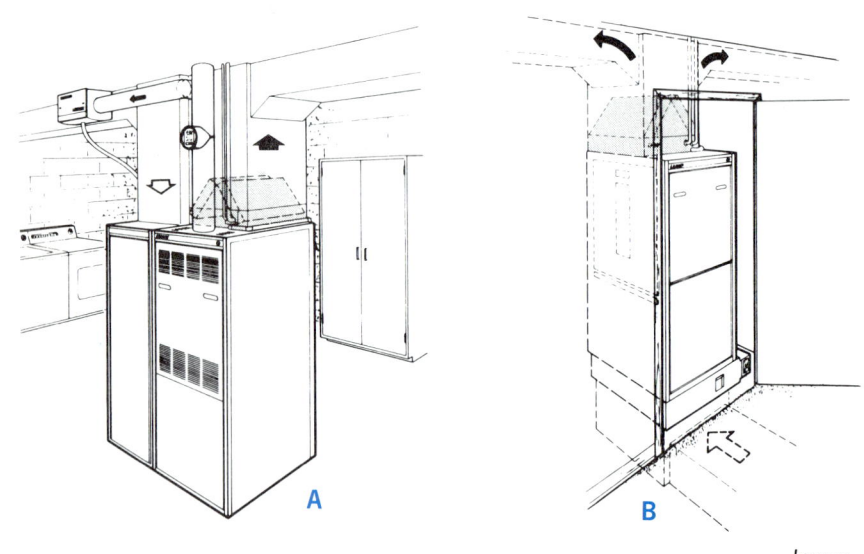

Up-flow evaporators are installed on top of the furnace. They are used in basement installations and in closet installations. Fig. 17-13. The adapter base and coil in Fig. 17-14 are shown as they would fit onto the top of an up-flow existing furnace. The plenum must be removed and replaced once the coil has been placed on top of the furnace.

In most cases, use of an add-on top evaporator means the fan motor must be changed to a higher horsepower rating. The evaporator in the plenum makes it more difficult to force air through the heating system. In some cases, the pulley size on the blower and the motor must be changed to increase the cfm moving past the evaporator.

Some motors have sealed bearings. Some blower assemblies, such as that shown in Fig. 17-15, have sealed bearings. However, some have sleeve bearings. In such cases, the owner should know that the motor and blower must be oiled periodically to operate efficiently.

Figure 17-16 shows how the evaporator coil sits on top of the furnace, making the up-flow type of air conditioning operate properly. The blower motor is located below the heater and plenum.

The evaporator is not useful unless it is connected to a compressor and condenser. These are usually located outside the

452

CHAPTER 17—COMFORT AIR CONDITIONING

17-15. Motor-driven blower unit.
Lennox

Lennox
17-16. Cutaway view of the furnace, blower, and evaporator coil on an air conditioning and heating unit.

house. Figure 17-17 shows the usual outdoor compressor and condenser unit. This unit is capable of furnishing 2.5 to 5 tons of air conditioning, ranging in capacities from 27,000 to 58,000 Btu. Note that this particular unit has a U-shaped condenser coil that forms three vertical sides. The extra surface area is designed to make the unit more efficient in heat transfer. The fan, which is thermostatically operated, has two-speeds. It changes to low speed when the outside temperature is below 75° F. [23.9°C].

Like most compressors designed for residential use, this compressor is hermetically sealed. The following safety devices are built in: a suction-cooled overload protector, a pressure relief valve, and a crankcase heater. Controls include high-

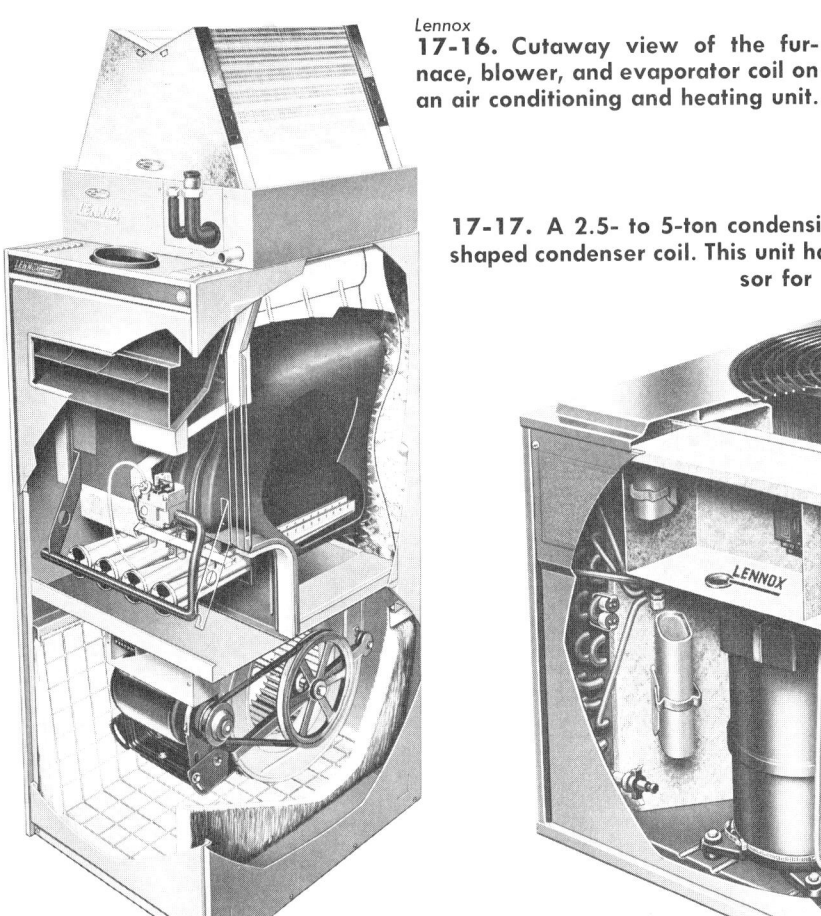

17-17. A 2.5- to 5-ton condensing unit with a one-piece, wraparound, U-shaped condenser coil. This unit has a two-speed fan and sealed-in compressor for quiet operation.
Lennox

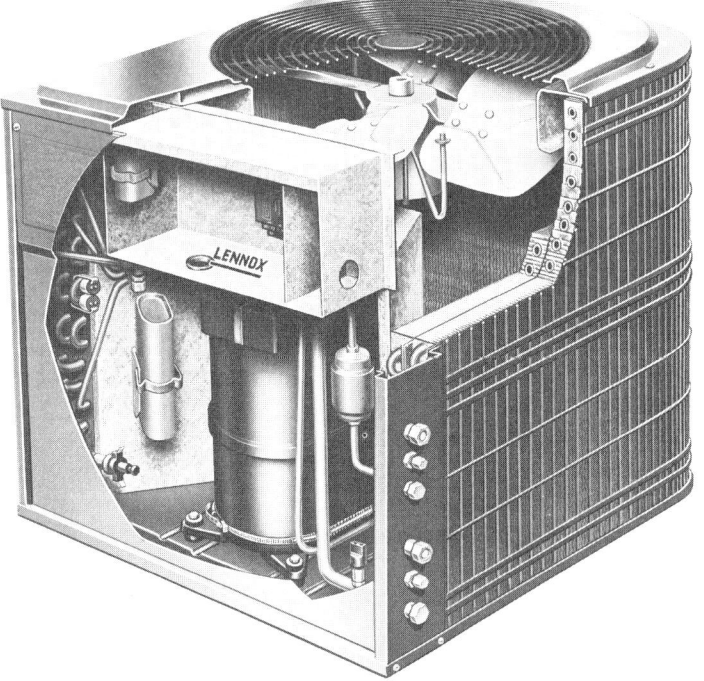

453

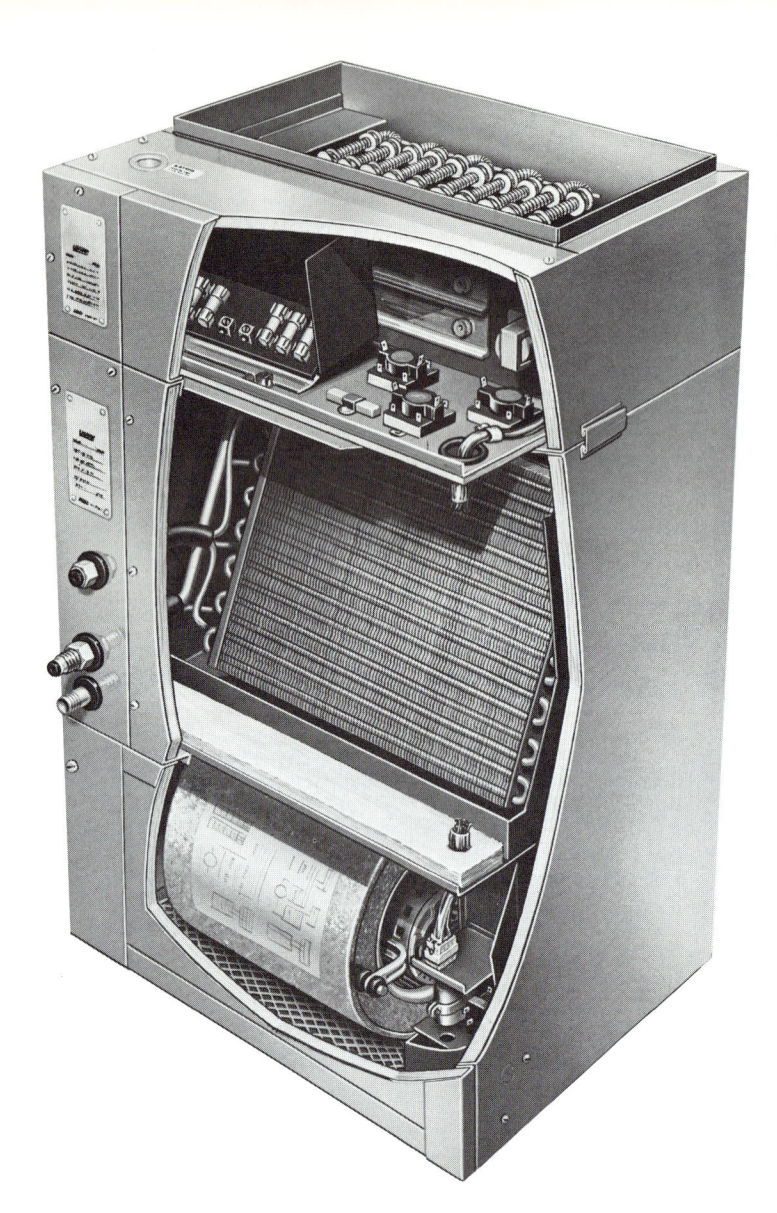

Lennox

17-18. This unit fits in small closets or corners. It has the possibility of producing hot water or electric heat as well as cooling. Note the heating coils on top of the unit. It comes in three sizes, 1.5, 2, and 2.5 tons.

and low-pressure switches. They automatically shut off the unit if discharge pressure becomes excessive or suction pressure falls too low.

In apartments where space is at a premium, a different type of unit is used. It differs only in size. Fig. 17-18. These compact units have a blower, filter, and evaporator coil contained in a small package. They have electric heat coils on top. In some cases, hot water is used to heat in the winter.

Figure 17-19 shows the various ways in which these units may be mounted. The capacity is usually 18,000 to 28,000 Btu. Note location of the control box. This is important since most of the maintenance problems are caused by electrical, rather than refrigerant, malfunctions. This type of

17-19. Typical installations of the blower, coil, and filter units. (A) A closet installation with electric heat section. (B) A utility room installation. (C) A wall-mounted closet installation with hot water section.

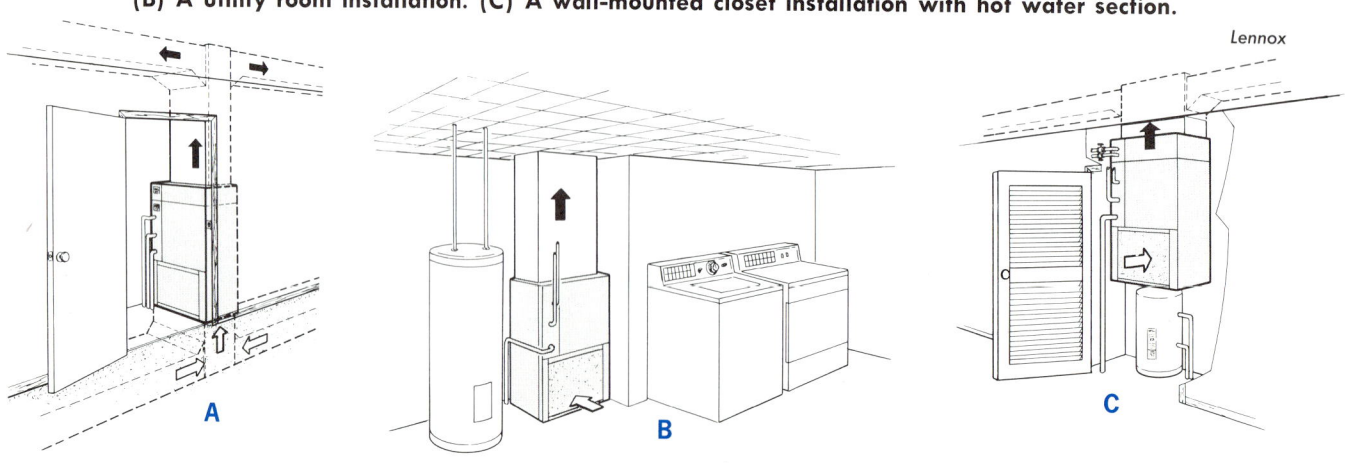

Lennox

A B C

CHAPTER 17—COMFORT AIR CONDITIONING

unit allows each apartment tenant to have his or her own controls.

Troubleshooting

To troubleshoot this type of air conditioning equipment, a troubleshooting table (Table 17-E) has been provided at the end of this chapter. The general troubleshooting procedures listed there are used for hermetically sealed compressors.

REMOTE SYSTEMS

A remote system designed for home or commercial installation can be obtained with a complete package. It has the condensing unit, correct operating refrigerant charge, refrigerant lines and evaporator unit.

The charge in the line makes it important to have the correct size of line for the installation. Metering control of the refrigerant in the system is accomplished by the exact sizing (bore and length) of the liquid line. The line must be used as delivered. It cannot be shortened.

Lennox has the RFC (Refrigerant Flow Control) system. It is a very accurate means of metering refrigerant in the system. It must never be tampered with during installation. The whole principle of the RFC system involves matching the evaporator coil to the proper length and bore of the liquid line. This is believed by the manufacturer to be superior to the capillary tube system. The RFC equalizes pressures almost instantly after the compressor stops. Therefore, it starts unloaded, eliminating the need for any extra controls. In addition, a precise amount of refrigerant charge is added to the system at the factory, resulting in trouble-free operation.

The condensing unit is shipped with a complete refrigerant charge. The condensing unit and evaporator are equipped with flared liquid and suction lines for quick connection. The compressor is hermetically sealed.

The unit may be built in and weatherproofed as a rigid part of the building structure. Fig. 17-20. The condensing unit can be free-standing on rooftops or slabs at grade level.

Figure 17-21 shows the condensing unit designed for the apartment developer and volume builder. It comes in 1-, 1.5-, and 2-ton sizes. Cooling capacities range from 17,000 through 28,000 Btu. An aluminum grille protects the condenser while offering low resistance to air discharge. The fan is mounted for less noise. It also reduces the possibility of air recirculation back through the condenser when it is closely banked for multiple installations. When mounted at grade level, this also keeps hot air discharge from damaging grass and shrubs.

SINGLE-PACKAGE ROOFTOP UNITS

Single-rooftop units can be used for both heating and cooling for industrial and commercial installations. Figure 17-22 shows such a unit. It can provide up to 1.5 million Btu if water heat is used. It can also include optional equipment to supply heat up to 546,000 Btu using electricity. They can use oil, gas, or propane for heating fuels. These units require large amounts of energy to operate. It is possible to conserve energy by using more sensitive controls. Highly sensitive controls monitor

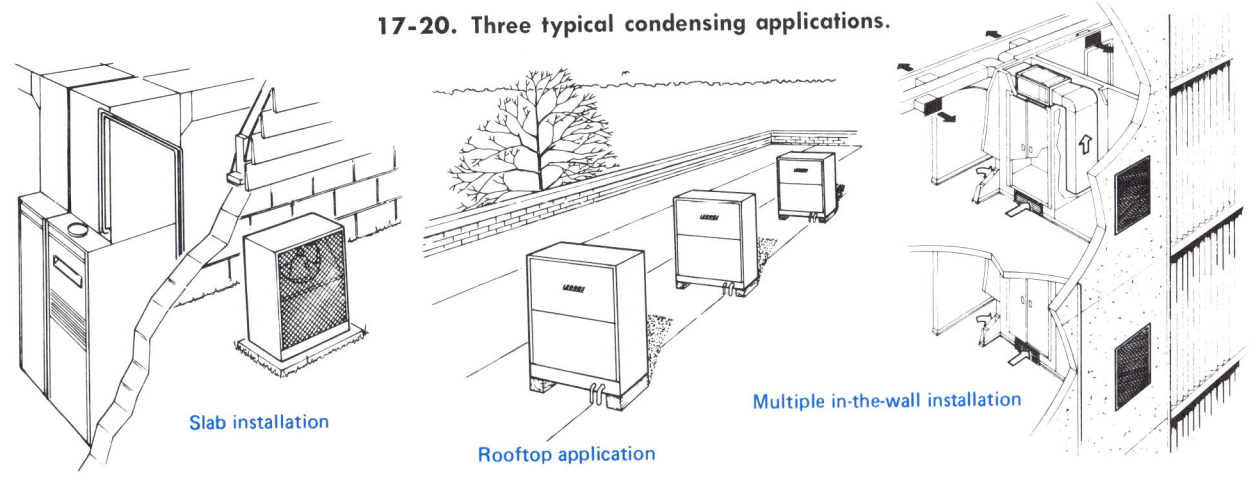

17-20. Three typical condensing applications.

Slab installation

Rooftop application

Multiple in-the-wall installation

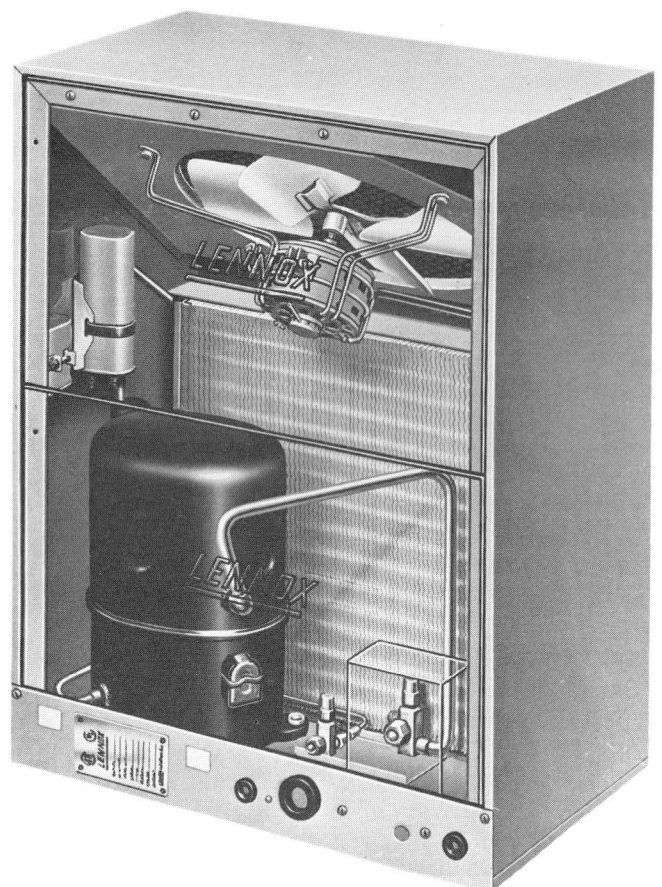

17-21. Apartment house or residential condensing unit. It can be installed through-the-wall, free-standing at grade level, or on the roof.

17-22. Single-zone rooftop system. This unit is used for industrial and commercial market applications. Cooling capacity ranges from 8 through 60 tons.

the room temperature and the supply air. They send signals to the control module. It, in turn, cycles the mechanical equipment to match the output to the load condition.

An optional device for conserving energy is available. It has a "no load" band thermostat that has a built-in differential of 6° F. [3.3°C]. This gives the system the ability to "coast" between the normal control points without consuming any primary energy within the recommended comfort setting range.

Another feature that may become more prevalent in the future is a refrigerant heat-reclaim coil. It can reduce supermarket heating costs significantly. A reheat coil can be factory installed downstream from the evaporator coil. It will use the condenser heat to control humidity and prevent overcooling.

A unit of this size is designed for a large store or supermarket. Figure 17-23 shows how the rooftop model is mounted for efficient distribution of the cold air. Since cold air is heavy, it will settle quickly to floor level. Hot air rises and stays near the ceiling in a room. Thus, it is possible for this warmer air to increase the temperature of the cold air from the conditioner before it comes into contact with the room's occupants.

Smoke Detectors

Photocell detectors detect smoke within the system. They actuate the blower motor controls and other devices to perform the following:

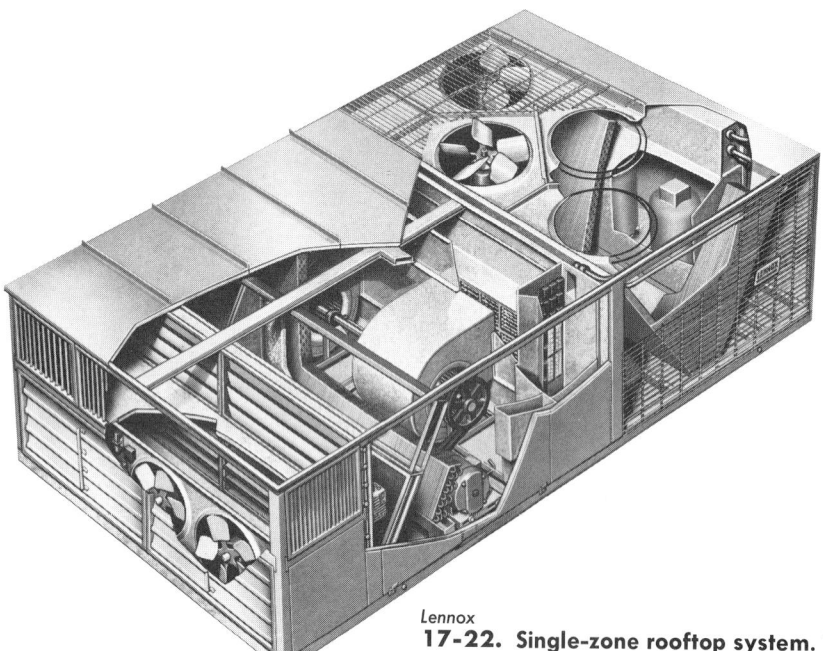

CHAPTER 17—COMFORT AIR CONDITIONING

1. Shut off the entire system.
2. Shut down the supply blower, close outside air and return air dampers and runs.
3. Run supply air blower, open outside air dampers, close return air dampers, and stop return air blower or exhaust fans.
4. Run supply air blower, open outside air dampers, close return air dampers, and run return air blower or exhaust fan. Actuation occurs when smoke within the unit exceeds a density that is sufficient to obscure light by a factor of 2% or 4% per foot. A key switch is used for testing. Two detectors are used. One is located in the return air section. The other is located in the blower section downstream from the air filters.

Firestats

Firestats are furnished as standard equipment. Manual reset types are mounted in the return air and supply airstream. They will shut off the unit completely when either firestat detects excessive air temperatures.

On this type of unit, the blowers are turned by one motor with a shaft connecting the two fans. There are three small fan motors and blades mounted to exhaust air from the unit. There are four condenser fans. The evaporator coil is slanted. There are two condenser coils mounted at angles. There are two compressors. The path for the return air is through the filters and evaporator coil back to the supply air ductwork.

Return Air Systems

Return air systems are generally one of two types: the ducted return air system or the open-plenum return air system (sandwich space). Fig. 17-24. The ducted return air system duct is lined with insulation, which greatly reduces noise.

The open-plenum system eliminates the cost of return air returns or ducts and is extremely flexible. In a building with relocatable interior walls, it is much easier to change the location of a ceiling grille than reroute a ducted return system.

Acoustical Treatment

Insulating the supply duct reduces duct loss or gain and prevents condensation. Use 1½ pounds density on ducts that

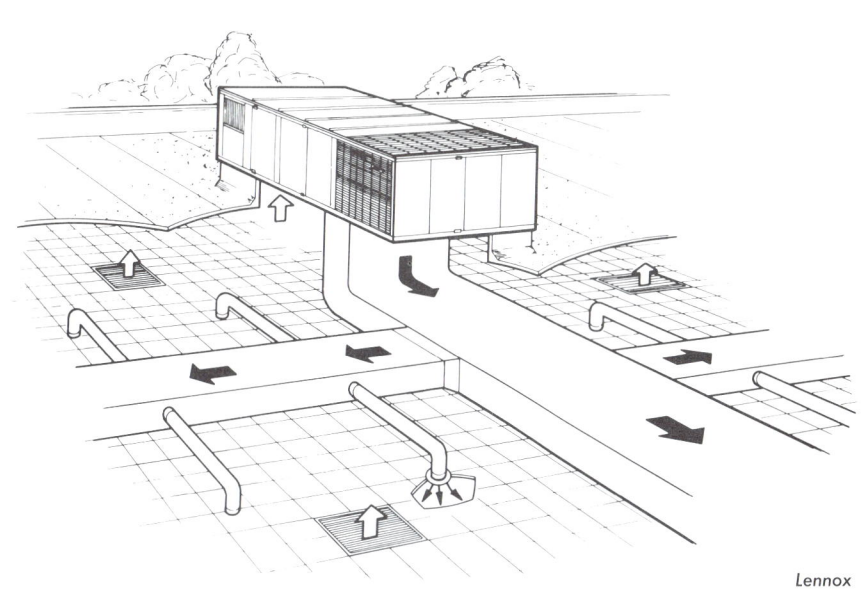

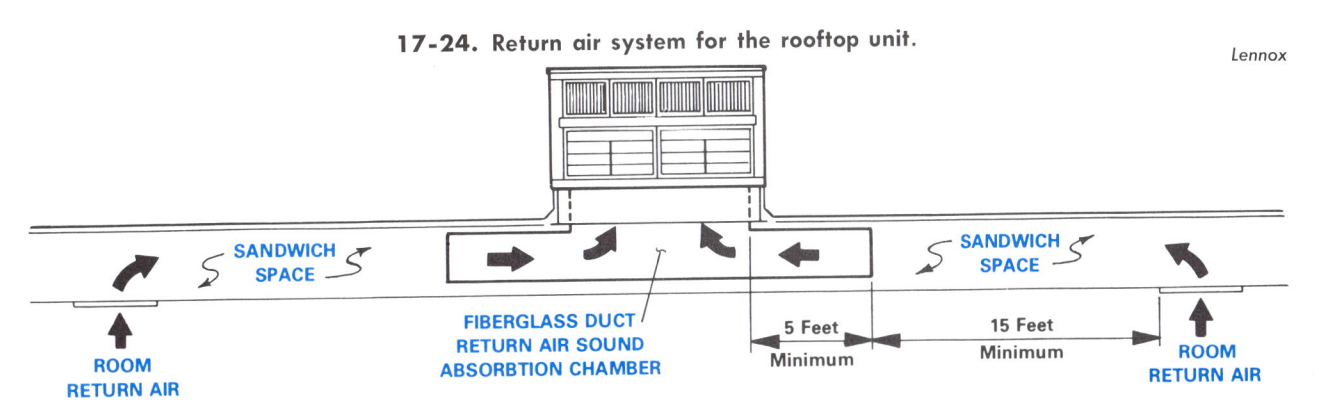

17-23. Typical rooftop installation of the single-zone system.

17-24. Return air system for the rooftop unit.

457

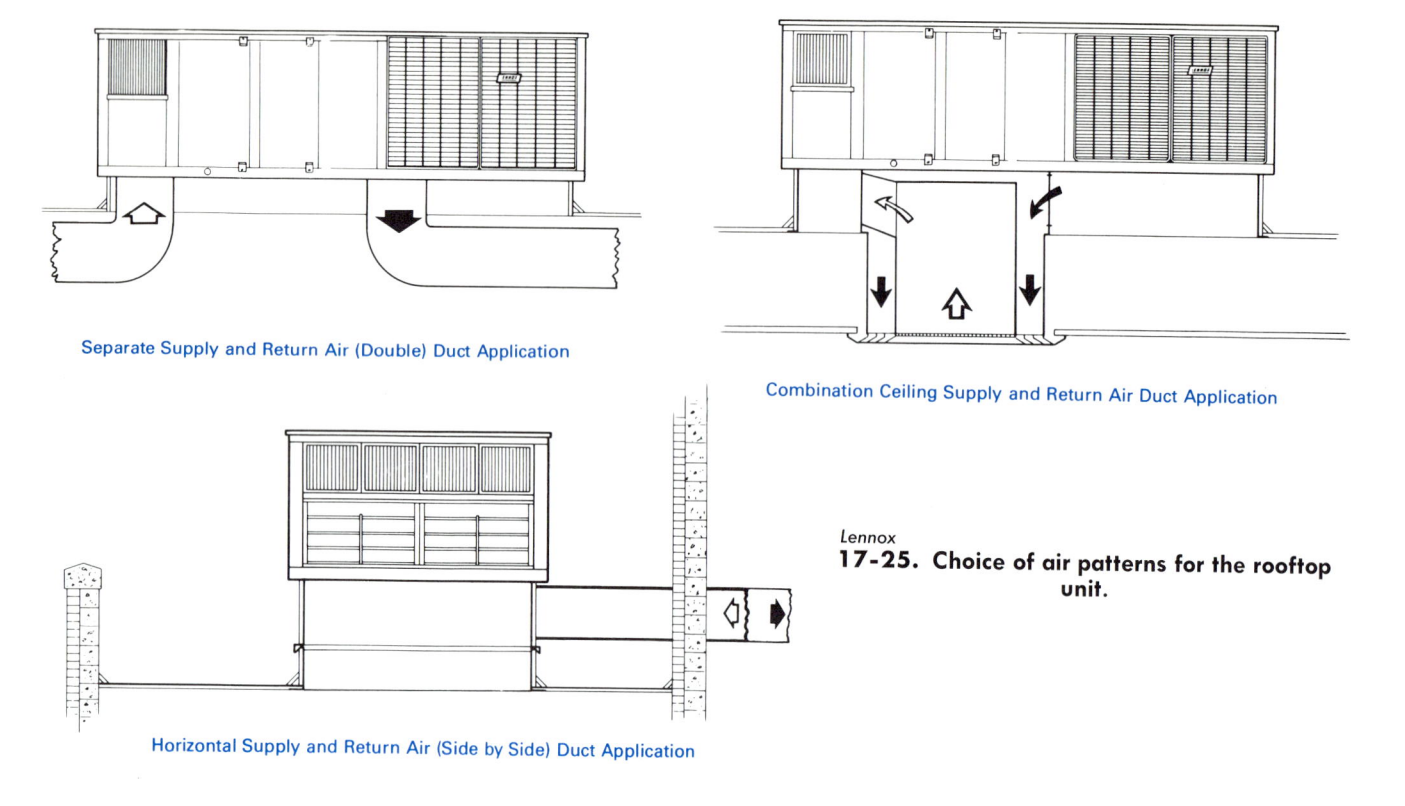

REFRIGERATION AND AIR CONDITIONING TECHNOLOGY

Separate Supply and Return Air (Double) Duct Application

Combination Ceiling Supply and Return Air Duct Application

Horizontal Supply and Return Air (Side by Side) Duct Application

Lennox
17-25. Choice of air patterns for the rooftop unit.

deliver air velocities up to 1500 fpm.

Three-pound density or neoprene-coated insulation is recommended for ducts that handle air at velocities greater than 1500 fpm. Insulation can be ½" or 1" thick on the inside of the duct.

Where rooftop equipment uses the sandwich space for the return air system, a return air chamber should be connected to the air inlet opening. Such an air chamber is shown in Fig. 17-24. This reduces air handling sound transmission through the thin ceiling panels. It should be sized not to exceed 1500 fpm return air velocity. The duct can be fiberglass or a fiberglass-lined metal duct. A ceiling return air grille should not be installed within 15' of the duct inlet.

Volume Dampers

Volume dampers are important to good system design. Lengths of supply runs vary and usually have the same cubic-foot measurements. Therefore, balancing dampers should be used in each supply branch run. The installer must furnish and install the balancing dampers. Dampers should be installed between the supply air duct and the diffuser outlet.

There are several ways in which rooftop conditioners can be installed. Figure 17-25 shows three installation methods.

Refrigerant Piping

Figure 17-26 shows how the unit just discussed is hooked up for refrigerant flow. Note how the two compressors are hooked into the operation of the unit. Note also the location of the reheat condenser coil, if it is installed in this type of unit.

Troubleshooting

To troubleshoot this type of air conditioning equipment, a troubleshooting table (Table 17-E) has been provided at the end of this chapter. The general troubleshooting procedures listed there are used for hermetically sealed compressors.

REFRIGERANT PIPE SIZES

In some installations on rooftop or slab, the unit does not come self-contained. This means the condensing unit may be located on a slab and the evaporator coil in some other location. Fig. 17-27. In this instance, it is necessary to make sure the refrigerant piping design is correct.

458

CHAPTER 17—COMFORT AIR CONDITIONING

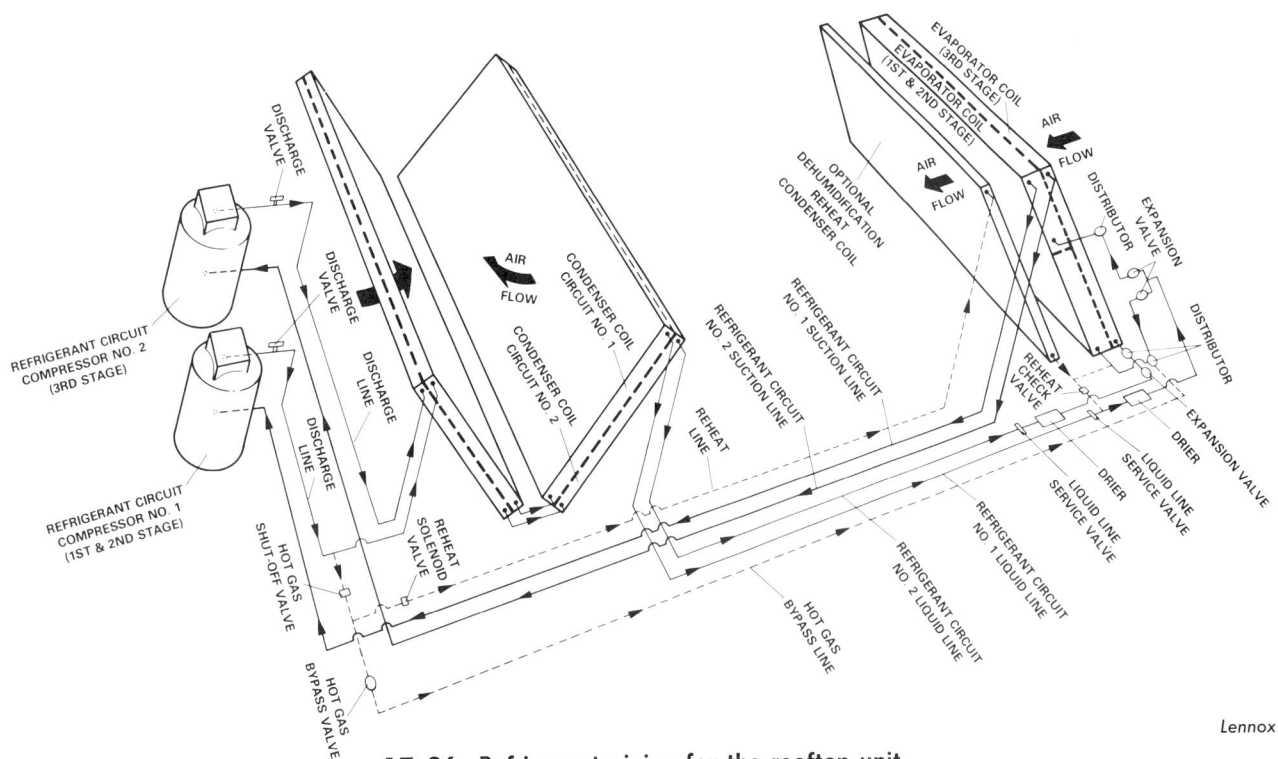

17-26. Refrigerant piping for the rooftop unit.

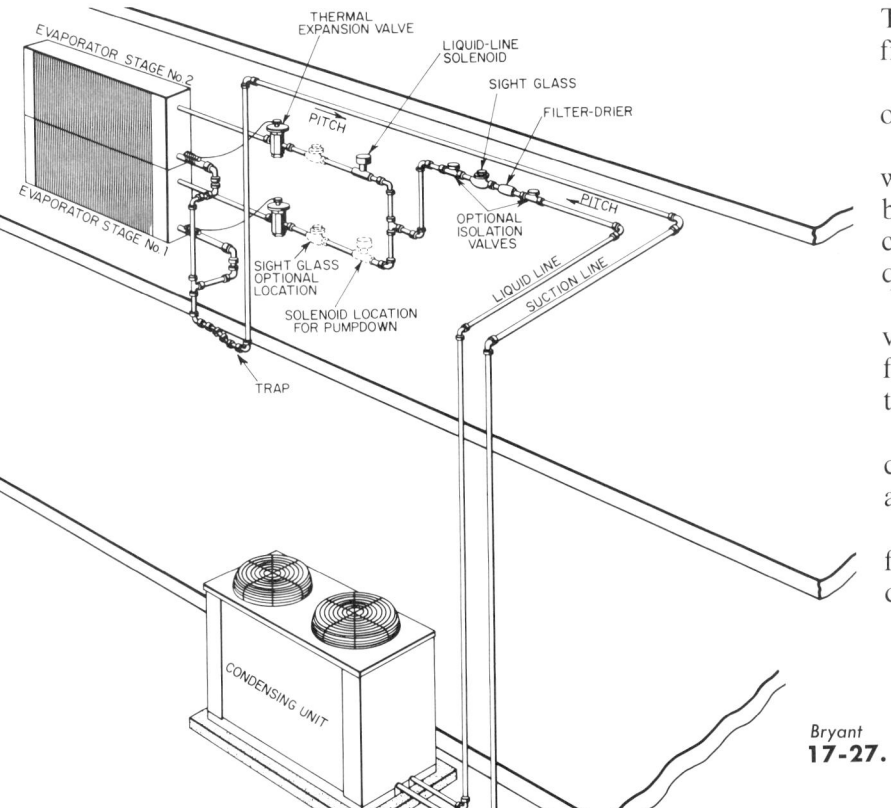

Lennox

The principal objectives of refrigerant piping are to:

- Insure proper feed to evaporators.
- Provide practical line sizes without excessive pressure drop because pressure losses decrease capacity and increase power requirements.
- Protect compressors by preventing excessive lubricating oil from being trapped in the system.
- Minimize the loss of lubricating oil from the compressor at all times.
- Prevent liquid refrigerant from entering the compressor during operation and shutdown.

Bryant
17-27. Refrigerant pipe sizes.

459

REFRIGERATION AND AIR CONDITIONING TECHNOLOGY

In general, larger pipe sizes have lower pressure drops, lower power requirements, and high capacity. The more economical smaller pipe sizes provide sufficient velocities to carry oil at all loads.

Liquid-Line Sizing

Liquid-line sizing presents less of a problem than suction-line sizing for the following reasons:

1. The smaller liquid-line piping is much cheaper than suction-line piping.

2. Compressor lubricating oil and fluorinated hydrocarbon refrigerants, such as R-22 in the liquid state, mix well enough that, in normal comfort air conditioning uses, positive oil return is not a problem.

3. Vertical risers, traps, and low velocities do not interfere with oil return in liquid lines.

Although liquid-line sizing offers more latitude than suction-line sizing, high-pressure drops should be avoided to prevent flash gas formation in the liquid line. Flash gas interferes with expansion valve operation. It also causes liquid distribution problems where more than one evaporator coil is being used. Where applications requirements are such that flash gas is unavoidable, there are methods of making allowance for it. Liquid refrigerant pumps and separation tanks can be used. (See *ASHRAE Guide and Data Book* for details.)

The acceptable pressure drop depends upon the amount of subcooling the condenser unit offers and the inherent losses resulting from liquid lift, if present. It is advisable to have the liquid slightly subcooled when it reaches the expansion valve. This helps avoid flash gas formation and provides stable operation of the expansion valve.

Suction-Line Sizing

The importance of suction-line design and sizing cannot be overemphasized. Lubricating oil does not mix well with the cold refrigerant vapor leaving the evaporator(s). It must be returned to the compressor either by entrainment with the refrigerant vapor or by gravity.

Traps and areas where oil may pool must be kept to a minimum. This is because large quantities of oil may become "lost" in the system. Piping should be level or with a slight pitch in the direction of the compressor.

Suction-line evaporator takeoffs should be designed so that oil cannot drain into idle coils. The common suction for multi-evaporator coils should be lower than the lowest evaporator outlet. Where an application requires that a common suction be above one or more of the coils on a multi-evaporator coil application, a suction riser with top loop connection is recommended.

Systems requiring a suction riser are more difficult to design. Sizing the pipe for minimum gas velocity at minimum system capacity (minimum displacement and suction temperature) may result in excessive pressure losses at full load. Excessive pressure losses in a suction riser may be compensated for by increasing pipe sizing in horizontal or down runs to reduce total system pressure losses. It can also be compensated for by using double-suction risers. In comfort air conditioning, the use of double-suction risers is the *exception*, rather than the rule. However, where necessary, it proves a valuable tool. (See *ASHRAE Guide Book* for details.)

Sizing Procedure

Use the following procedure for selecting the proper refrigerant pipe size:

1. Determine the "measured" length of the straight pipe. Do this separately for the liquid line and the suction line.

2. The fittings cause additional friction above that created by the measured length of straight pipe. To account for this extra friction, the equivalent method is used. This technique consists of adding to the measured length an additional length of straight pipe that has the same pressure loss as the fittings.

The pressure loss of the fittings depends upon the number, type, and diameter of the fittings used. The dependence of the pressure loss in the fitting on the diameter of the fittings presents a real problem. To determine the pipe diameter, it is necessary to know the total equivalent length of pipe, the sum of the measured straight pipe length, and the fitting losses in equivalent pipe length. On the other hand, since the fitting pressure losses do depend upon diameter, it is necessary to know the diameter of the pipe to determine the fitting losses in equivalent pipe length.

The result is a situation in which it seems that the solution is needed to solve the problem. Fortunately, if the pipe diameter can be estimated with reasonable accuracy, the fitting losses can be determined. In turn, the correct pipe diameter can be determined.

A pipe diameter selection based upon 1.5 times the measured straight pipe length can be

Table 17-A.
Copper Tubing Suction Line Sizes (in inches) for Pressure Drop Corresponding to 2° F.

Equivalent Pipe Length (ft.)	\multicolumn{15}{c}{R-22 Refrigerant Systems (tons)}
	2	3	3.5	4	5	7.5	10	15	20	25	30	40	50	60	80	100
500	1 1/8	1 3/8	1 3/8	1 3/8	1 5/8	2 1/8	2 1/8	2 5/8	2 5/8	2 5/8	3 1/8	3 1/8	3 5/8	3 5/8	4 1/8	5 1/8
400	1 1/8	1 3/8	1 3/8	1 3/8	1 3/8	1 5/8	2 1/8	2 1/8	2 5/8	2 5/8	3 1/8	3 1/8	3 5/8	3 5/8	4 1/8	5 1/8
300	1 1/8	1 1/8	1 3/8	1 3/8	1 3/8	1 5/8	2 1/8	2 1/8	2 5/8	2 5/8	2 5/8	3 1/8	3 1/8	3 5/8	3 5/8	4 1/8
200	7/8	1 1/8	1 1/8	1 1/8	1 3/8	1 5/8	1 5/8	2 1/8	2 1/8	2 5/8	2 5/8	2 5/8	3 1/8	3 1/8	3 5/8	3 5/8
150	7/8	1 1/8	1 1/8	1 1/8	1 3/8	1 3/8	1 5/8	1 5/8	2 1/8	2 1/8	2 5/8	2 5/8	2 5/8	3 1/8	3 1/8	3 5/8
100	7/8	7/8	1 1/8	1 1/8	1 1/8	1 3/8	1 3/8	1 5/8	2 1/8	2 1/8	2 1/8	2 5/8	2 5/8	2 5/8	3 1/8	3 5/8
80	7/8	7/8	1 1/8	1 1/8	1 1/8	1 3/8	1 3/8	1 5/8	2 1/8	2 1/8	2 1/8	2 1/8	2 5/8	2 5/8	3 1/8	3 1/8
60	7/8	7/8	7/8	7/8	1 1/8	1 1/8	1 3/8	1 5/8	1 5/8	2 1/8	2 1/8	2 1/8	2 5/8	2 5/8	2 5/8	3 1/8
50	7/8	7/8	7/8	7/8	1 1/8	1 1/8	1 3/8	1 3/8	1 5/8	2 1/8	2 1/8	2 1/8	2 1/8	2 5/8	2 5/8	3 1/8
40	7/8	7/8	7/8	7/8	7/8	1 1/8	1 1/8	1 3/8	1 5/8	1 5/8	2 1/8	2 1/8	2 1/8	2 5/8	2 5/8	2 5/8
30	5/8	7/8	7/8	7/8	7/8	1 1/8	1 1/8	1 3/8	1 3/8	1 5/8	1 5/8	2 1/8	2 1/8	2 1/8	2 5/8	2 5/8
20	5/8	7/8	7/8	7/8	7/8	1 1/8	1 1/8	1 3/8	1 3/8	1 5/8	1 5/8	2 1/8	2 1/8	2 1/8	2 1/8	2 5/8

Bryant

Table 17-B.
Valve Losses in Equivalent Feet of Pipe (Screwed, Welded, Flanged, and Flared Connections)

Nominal Pipe Size (inches)	Globe and Lift Check	60°—Y	45°—Y	Angle	Gate	Swing Check	Y—Type Strainer Flanged End	Y—Type Strainer Screwed End
3/8	17	8	6	6	0.6	5	—	—
1/2	18	9	7	7	0.7	6	—	3
3/4	22	11	9	9	0.9	8	—	4
1	29	15	12	12	1.0	10	—	5
1 1/4	38	20	15	15	1.5	14	—	9
1 1/2	43	24	18	18	1.8	16	—	10
2	55	30	24	24	2.3	20	27	14
2 1/2	69	35	29	29	2.8	25	28	20
3	84	43	35	35	3.2	30	42	40
3 1/2	100	50	41	41	4.0	35	48	—
4	120	58	47	47	4.5	40	60	—
5	140	71	58	58	6.0	50	80	—

Bryant

Table 17-C.
Fitting Losses in Equivalent Feet of Pipe
(Screwed, Welded, Flared, and Brazed Connections)

Nominal Pipe Size (inches)	Smooth Bend Elbows 90° Standard*	90° Long Radius	90° Street*	45° Standard*	45° Street*	180° Standard*	Flow-thru Branch	Smooth Bend Tees No Reduction	Straight-thru Flow Reduced	Reduced	Sudden Enlargement*** d/D 1/4	1/2	3/4	Sudden Contraction** d/D 1/4	1/2	3/4
3/8	1.4	0.9	2.3	0.7	1.1	2.3	2.7	0.9	1.2	1.4	1.4	0.8	0.3	0.7	0.5	0.3
1/2	1.6	1.0	2.5	0.8	1.3	2.5	3.0	1.0	1.4	1.6	1.8	1.1	0.4	0.9	0.7	0.4
3/4	2.0	1.4	3.2	0.9	1.6	3.2	4.0	1.4	1.9	2.0	2.5	1.5	0.5	1.2	1.0	0.5
1	2.6	1.7	4.1	1.3	2.1	4.1	5.0	1.7	2.3	2.6	3.2	2.0	0.7	1.6	1.2	0.7
1 1/4	3.3	2.3	5.6	1.7	3.0	5.6	7.0	2.3	3.1	3.3	4.7	3.0	1.0	2.3	1.8	1.0
1 1/2	4.0	2.6	6.3	2.1	3.4	6.3	8.0	2.6	3.7	4.0	5.8	3.6	1.2	2.9	2.2	1.2
2	5.0	3.3	8.2	2.6	4.5	8.2	10.0	3.3	4.7	5.0	8.0	4.8	1.6	4.0	3.0	1.6
2 1/2	6.0	4.1	10.0	3.2	5.2	10.0	12.0	4.1	5.6	6.0	10.0	6.1	2.0	5.0	3.8	2.0
3	7.5	5.0	12.0	4.0	6.4	12.0	15.0	5.0	7.0	7.5	13.0	8.0	2.6	6.5	4.9	2.6
3 1/2	9.0	5.9	15.0	4.7	7.3	15.0	18.0	5.9	8.0	9.0	15.0	9.2	3.0	7.7	6.0	3.0
4	10.0	6.7	17.0	5.2	8.5	17.0	21.0	6.7	9.0	10.0	17.0	11.0	3.8	9.0	6.8	3.8
5	13.0	8.2	21.0	6.5	11.0	21.0	25.0	8.2	12.0	13.0	24.0	15.0	5.0	12.0	9.0	5.0

* R/D approximately equal to 1. ** R/D approximately equal to 1.5. *** Enter table for losses at smallest diameter "d."

Bryant

used to obtain an estimate of the pipe diameter.

3. In the majority of applications, the design conditions are a pressure loss equal to 2° F. [1.1°C], a suction temperature of 40° F. [4.4°C], and a condensing temperature of 105° F. [40.6°C].

4. Once the estimated pipe diameter is determined, the actual equivalent length (corrected if necessary) can be determined. Then the final pipe diameter can be determined.

For example, the following demonstrates the procedure that was just outlined. A Bryant unit (566B360RCU) is used for the example. This 360 size unit has two separate 180,000 Btu/hr (15-ton) condensing sections. Each section is piped individually. The liquid and suction lines for each condensing section should be sized for 15 tons. This example covers only suction-line sizing. However, the same procedure is used for the liquid lines.

In the majority of comfort air conditioning applications, the design conditions closely approximate 105°/40°/2°. The following design conditions are used to demonstrate the use of the correction factors:

Condensing temperature 110° F.
Suction temperature 35° F.
Maximum friction drop 2.5° F.

The measured straight pipe length in suction or liquid line equals 100′.

Number and type of fitting in each line:

    10 standard 45° elbows.
    4 gate valves.
    4 standard 90° elbows.

The measured straight pipe length of the suction line is equal to 100′. Therefore, use 150′ as a

CHAPTER 17—COMFORT AIR CONDITIONING

Table 17-D.
Friction Drop Correction Factors— Liquid and Suction Lines

Friction Drop (° F.)	0.5	1.0	1.5	2.0	2.5	3.0	4.0	5.0	6.0
Multiplier	4.0	2.0	1.3	1.0	0.8	0.7	0.5	0.4	0.3

Bryant

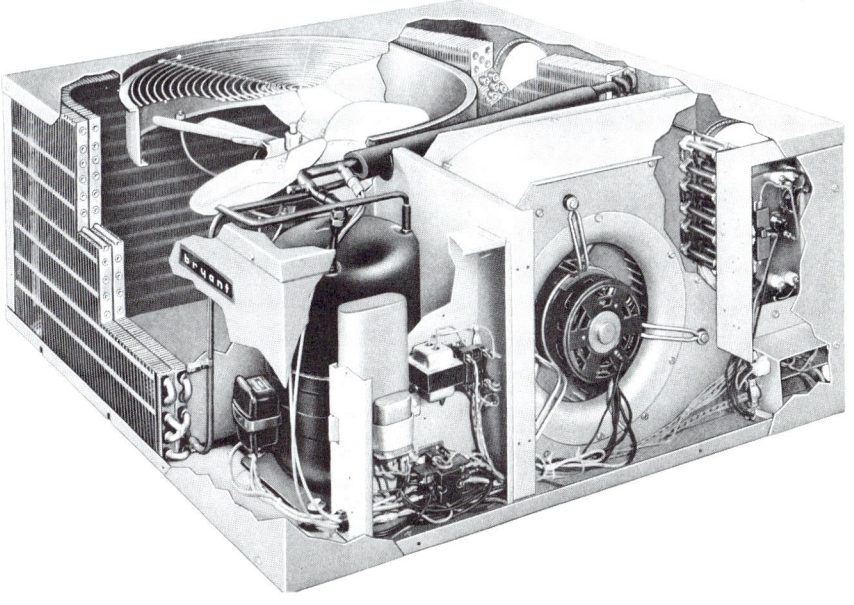

first approximation of the total equivalent length. For a combination of 150' and 15 tons, Table 17-A, on page 461, gives an estimated pipe diameter of $1\frac{5}{8}''$.

Once the estimated pipe diameter is obtained, obtain the following fitting losses in equivalent lengths from Tables 17-B, on page 461, and 17-C. Use the 2" pipe size for $1\frac{5}{8}''$ pipe.

10 standard 45° elbows
$\qquad 10 \times 2.6 = 26'$
4 gate valves
$\qquad 4 \times 2.3 = 9.2'$
4 standard 90° elbows
$\qquad 4 \times 5.0 = \underline{20'}$
$\qquad\qquad\qquad\quad 55.2'$

Actual total equivalent pipe length:

100' measured pipe length
+ 55.2' fitting losses = 155.2'

Correct the nominal tonnage for the 35° F. suction and 110° F. condensing temperatures. The factor is 1.13. The factor is given in the equipment manufacturer's specifications for the unit. Instead of using 15.0 tons, use 17.0 tons (1.13 × 15 = 16.95) for the final pipe size selection.

In addition, the acceptable friction loss is 2.5° F., instead of 2.0° F. The correction factor from Table 7-D is 0.8. Thus, use 124' (0.8 × 155.2) for the final pipe size selection.

A combination of 124' and 17 tons gives a final pipe size selection of $2\frac{1}{8}''$ diameter. See Table 17-A. The $2\frac{1}{8}''$ is used since it is given for 20 tons. Seventeen tons are more than 15 tons, so you move to the next highest value in the Table. It is better to have a larger pipe than a smaller one. You also use the 150' equivalent pipe length, since a 100' length would be too small and not allow for errors in the original estimates.

Troubleshooting

To troubleshoot this type of air conditioning equipment, a troubleshooting table (Table 17-E) has been provided at the end of this chapter. The general procedures listed there are used for hermetically sealed compressors.

MOBILE HOMES

Some units are now available for mobile home installations. Fig. 17-28. Such a unit will furnish from 2 to 4 tons of cooling. The unit is 3' by 3' and will occupy a very small area outside the mobile home. Electric heat can be added to provide a comfortable year-round condition. The noise problem is minimized by ducting the condenser exhaust fan upward.

The cooling coil, blower, compressor, and all other refrigera-

17-28. Mobile home self-contained air conditioning unit.
Bryant

REFRIGERATION AND AIR CONDITIONING TECHNOLOGY

tion components are contained in a low-silhouette weatherproof cabinet.

If a mobile home owner decides to move, the heating and cooling unit can be disconnected from the mobile home, transported to a new homesite, and easily reconnected.

A flexible insulated duct with round flanges simplifies hookups to mobile home ductwork. Conventional metal ductwork can be attached if desired.

This type of unit has a relatively large cooling capacity. Thus, it can be used on smaller homes, vacation cottages, and other small buildings. It is delivered as a complete package. All that is needed is the electrical power source, a thermostat connection, and a hookup to the ductwork of the building or mobile home.

Troubleshooting

To troubleshoot this type of air conditioning equipment, a troubleshooting table (Table 17-E) has been provided at the end of this chapter. The general troubleshooting procedures listed there are used for hermetically sealed compressors.

Each manufacturer publishes manuals for use with the equipment the manufacturer makes. As you get more involved in the troubleshooting of specific types of equipment you will build your own library of troubleshooting manuals. Many of these contain wiring and piping schematics.

Other chapters in this book detail the proper operation of this type of equipment. Knowing the details of equipment operation will help you use the manuals more effectively. Familiarity with trade magazines will lead you to articles on problems with specific equipment.

Table 17-E.
Troubleshooting Hermetic Compressor Type Air Conditioning Equipment*

Trouble	Probable Cause	Remedy or Repair
Compressor will not start. (No hum.)	Open line circuit.	Check the wiring, fuses, and receptacle.
	Protector open.	Wait for reset. Check current drawn from line.
	Contacts open on control relay.	Check control, and check pressure readings.
	Open circuit in the motor stator.	Replace the stator or the whole compressor.
Compressor will not start. However, it hums intermittently. Cycles with the protector.	Not wired correctly.	Check wiring diagram and actual wiring.
	Line voltage low.	Check line voltage. Find where line voltage is dropped. Correct.
	Start capacitor open.	Replace start capacitor.
	Relay contacts do not close.	Check by manually operating. Replace if defective.
	Start winding open.	Check stator leads. Replace compressor if the leads are OK.
	Stator winding grounded. (Usually blows fuse.)	Check stator leads. Replace compressor if leads are OK.
	Discharge pressure too high.	Remove cause of excessive pressure. Discharge shutoff and receiver valves should be open.

Table 17-E. (Continued)

Trouble	Probable Cause	Remedy or Repair
Compressor starts. Motor will not speed up enough to have start winding drop out of circuit.	Compressor too tight.	Check oil level. Correct the binding cause. If this cannot be done, replace compressor.
	Start capacitor weak.	Replace the capacitor.
	Line voltage low.	Increase the voltage.
	Wired incorrectly.	Rewire according to wiring diagram.
	Relay defective.	Check operation. If defective, replace.
	Run capacitor shorted.	Disconnect the run capacitor and check for short.
	Start and run windings shorted.	Check winding resistances. If incorrect, replace the compressor.
	Start capacitor weak.	Check capacitors. Replace those defective.
	Discharge pressure high.	Check discharge shutoff valves. Check pressure.
Compressor starts and runs. However, it cycles on the protector.	Tight compressor.	Check oil level. Check binding. Replace if necessary.
	Low line voltage.	Increase the voltage.
	Additional current being drawn through the protector.	Check to see if fans or pumps are wired to the wrong connector.
	Suction pressure is too high.	Check compressor. See if it is the right size for the job.
	Discharge pressure is too high.	Check ventilation. Check for overcharge. Also check for obstructions to air flow or refrigerant flow.
	Protector is weak.	Check current. Replace protector if it is not clicking out at right point.
	Run capacitor defective.	Check capacitance. Replace if found defective.
	Stator partially shorted or grounded.	Check resistance for a short to the frame. Replace if found shorted to ground (frame).
	Inadequate motor cooling.	Correct air flow.
	Compressor tight.	Check oil level. Check cause of binding.
	3Ø line unbalanced	Check each leg or phase. Correct if the voltages are not the same between legs.
	Discharge valve leaks or is damaged.	Replace the valve plate.

(Continued on next page)

REFRIGERATION AND AIR CONDITIONING TECHNOLOGY

Table 17-E. (Continued)
Troubleshooting Hermetic Compressor Type Air Conditioning Equipment*

Trouble	Probable Cause	Remedy or Repair
Start capacitors burn out.	Short cycling.	Reduce the number of starts. They should not exceed 20 per hour.
	Prolonged operation with start winding in circuit.	Reduce the starting load. Install a crankcase pressure limit valve. Increase low voltage if this is found to be the condition. Replace the relay if it is found to be defective.
	Relay contacts sticking.	Clean the relay contacts. Or, replace the relay.
	Wrong relay or wrong relay setting.	Replace the relay.
	Wrong capacitor.	Check specifications for correct size capacitor. Be sure the MFD and WVDC are correct for this compressor.
	Working voltage of capacitor too low.	Replace with capacitor of correct WVDC.
	Water shorts out terminals of the capacitor.	Place capacitor so the terminals will not get wet.
Run capacitors burn out. They spew their contents over the surfaces of anything nearby. This problem can usually be identified with a visual check.	Excessive line voltage.	Reduce line voltage. It should not be over 10% of the motor rating.
	Light load with a high line voltage.	Reduce voltage if not within 10% overage limit.
	Voltage rating of capacitor too low.	Replace with capacitors of the correct WVDC.
	Capacitor terminals shorted by water.	Place capacitor so the terminals will not get wet.
Relays burn out.	Low line voltage.	Increase voltage to within 10% of limit.
	High line voltage.	Reduce voltage to within 10% of the motor rating.
	Wrong size capacitor.	Use correct size capacitor. The proper MFD rating should be installed.
	Short cycling.	Decrease the number of starts per hour.
	Relay vibrating.	Make sure you mount the relay rigidly.
	Wrong relay.	Use the recommended relay for the compressor motor.

* These are general problems that can be identified with any hermetic compressor. Problems with the electrical switches, valves, and tubing are located by using the knowledge you have acquired previously in the theory and applications sections of this book.

466

CHAPTER 17—COMFORT AIR CONDITIONING

REVIEW QUESTIONS

1. How often should a window unit's air filter be cleaned?
2. What causes a large percentage of motor burnouts in air conditioners?
3. Name two types of evaporators for residential use.
4. What is a plenum?
5. Where are single-rooftop units used?
6. Why are smoke detectors needed to work in conjunction with an air conditioner?
7. What is a firestat?
8. What are two types of return air systems?
9. What is a volume damper?
10. What are two reasons why liquid-line sizing presents less of a problem than suction-line sizing?

18

Commercial Air-Conditioning Systems

There are several types of commercial air conditioning systems. This chapter discusses the following systems: expansion-valve air conditioning systems, packaged cooling units, rooftop heating and cooling units, direct multizone systems, evaporative cooling systems, absorption-type air conditioning systems, chilled water air conditioning, chillers, and console-type air conditioning systems.

EXPANSION-VALVE AIR CONDITIONING SYSTEM

This type of condensing unit can be installed in singles or multiples. Such units are used in residential, apartment, motel, and commercial applications. These units are applicable only to expansion valve systems. The low height and upward discharge of air make it easy to conceal the unit among shrubs on a slab at ground level or out of sight on a roof.

Compressor

The compressor is hermetically sealed. Built-in protection devices protect from excessive current and temperature. It is suction-cooled. Overload-protection is by an internal pressure relief. A crankcase heater is standard equipment on all of these units. It ensures proper compressor lubrication at all times. The crankcase heater is thermostatically controlled and temperature actuated to operate only when required. Rubber mounts help to reduce the noise associated with compressors.

Condenser

The condenser coil is U-shaped to provide a large surface area for heat exchange. The joints in the compressor are silver soldered. The compressor is tested for leaks at 450 to 500 psi. Refrigerant lines come precharged. This unit comes with a drier. Solid-state controls prevent compressor short-cycling and also allow time for suction and discharge pressure to equalize. This permits the compressor to start in an unloaded condition. An automatic reset control will shut off the compressor for 5 minutes.

Expansion Valve Kit

An expansion valve kit has been developed to match the evaporator unit. The expansion valve is equipped with a bleed port. This permits pressures to equalize after the compressor stops. This means the compressor can restart in an unloaded condition. Flare fittings permit connections on the valve in a simple field installation.

Since single-phase models require large amounts of current, they can cause lights to blink when they start up. A positive temperature coefficient (PTC) kit can eliminate some of the start-up problems. It consists of a solid-state circuit with a ceramic thermistor. The thermistor provides extra starting torque to solve most compressor hard-starting problems such as low voltage or light dimming. It switches itself out of the circuit after start-up.

A start kit consisting of a start capacitor and potential relay must be installed in some models when used with certain evaporator units and expansion valve kits. The added capacitance is taken from the circuit when the coil energizes almost instantly. However, it does help with the starting load. Since the coil is energized whenever the air conditioner system is operating, it is possible for the coil to become open. This causes the start capacitor to "blow-up" and spread its

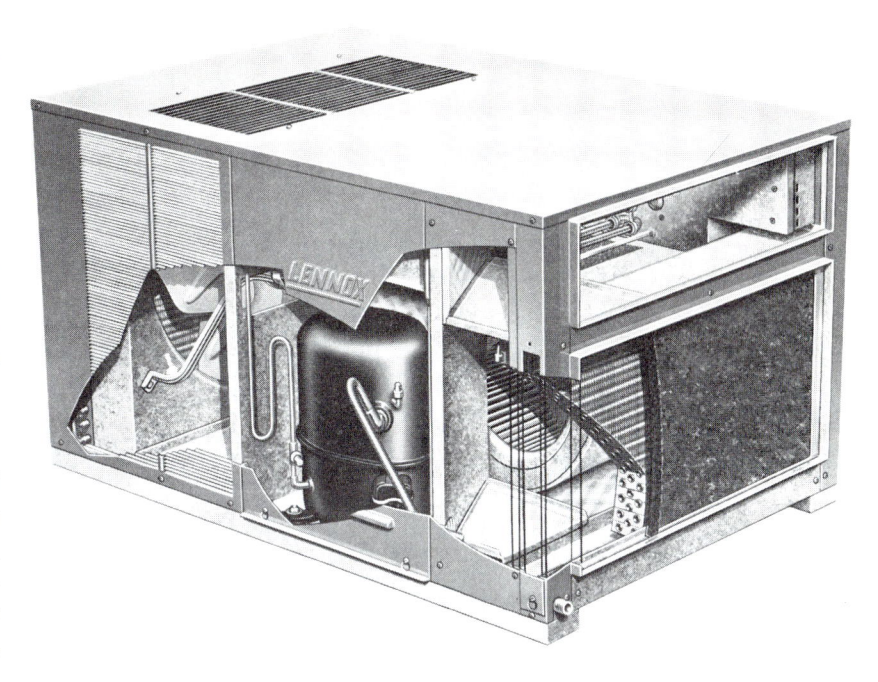

CHAPTER 18—COMMERCIAL AIR-CONDITIONING SYSTEMS

18-1. Horizontal, single-package air conditioner.
Lennox

contents inside the control housing. Replacement of the capacitor and the coil of the relay is necessary to repair the equipment for proper operation. Usually the entire relay must be replaced, since the coil is not always available separately.

TROUBLESHOOTING

To troubleshoot this type of air conditioning equipment, refer to Table 17-E (pages 464–466). The general troubleshooting procedures listed there are used for hermetically sealed compressors.

PACKAGED COOLING UNITS

A 2- to 5-ton packaged unit is available from several manufacturers. Lennox makes a self-contained unit that can be mounted on a slab or on the rooftop. Fig. 18-1. This one is designed for the residential replacement market. The compressor, control box, filter, condenser coil, and evaporator coil are in one package. The blower is also located in the package. The only component inside the building is the ductwork. Return air enters in the lower opening through the evaporator coil and is discharged out the top opening.

One of the advantages of a unit of this type is its completeness. It comes ready to connect to the ductwork and the electrical outlets. The blower is located outside the house. The inside noise is that of the air moving through the ductwork. The air filter is also located outside. It is a vacuum-cleanable type with polyurethane coating. It is coated with oil to increase efficiency. If the filter is washed, it should be re-oiled.

Up to 58,000 Btu can be added with an optional field-installed heating unit. If electric heat is desired, it is possible for this type of unit to heat and cool, using the same ductwork. Figure 18-2 shows typical unit installations.

18-2. Typical installations of the horizontal single-package air conditioner. (A) A rooftop installation. (B) A unit on a slab at grade level.
Lennox

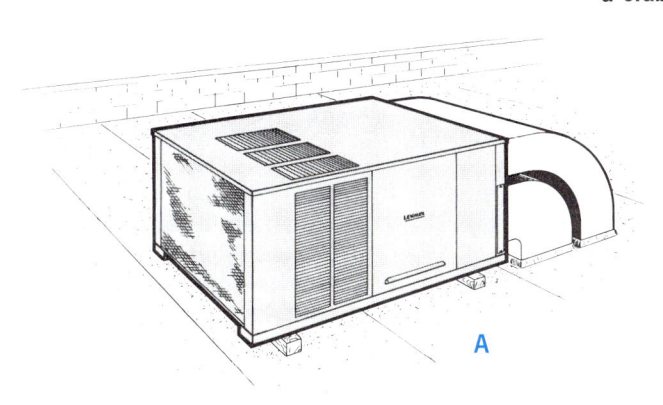

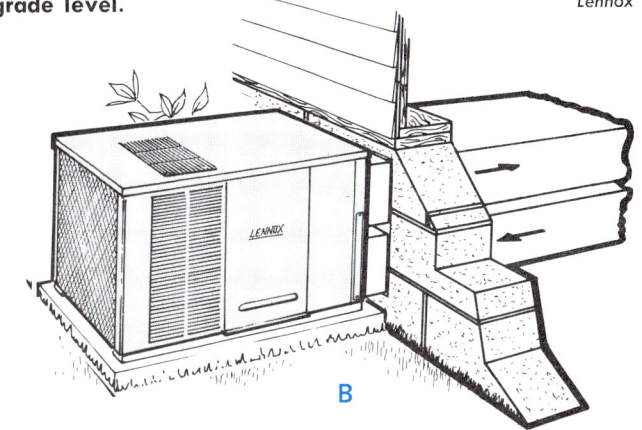

469

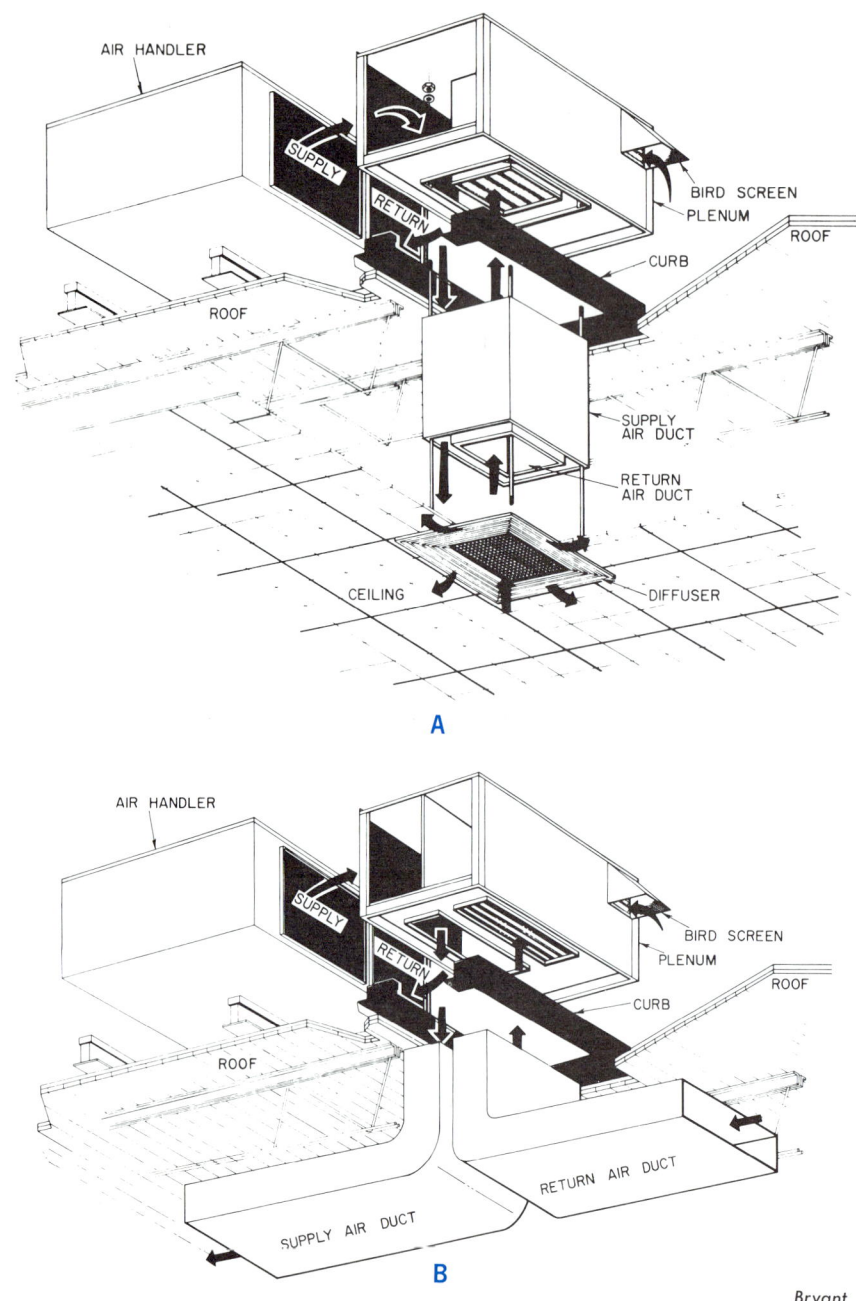

REFRIGERATION AND AIR CONDITIONING TECHNOLOGY

18-3. Typical system installations on the rooftop. (A) A concentric duct arrangement. (B) A side-by-side duct arrangement.
Bryant

Rooftop Heating and Cooling Units

Rooftop heating and cooling units are made by many manufacturers. Some units are delivered with a full refrigerant charge. This means there are no refrigerant lines to connect. This cuts labor costs and installation time. Since the unit is on the rooftop, no inside room has to be allocated for the heating and cooling equipment. This unit uses gas for heating up to 112,500 Btu. The cooling can reach 60,000 Btu or 5 tons.

Typical installation systems are shown in Fig. 18-3. One of the advantages is that heat and cooling can be added rather quickly in the construction phase. The installers can work in comfort and thus improve their efficiency during the construction phase of the building. The unit can be set in place on a slab at ground level. The duct, gas, and electrical connections can be made to it at that location.

Electrical

Since this unit has the ability to deliver 5 tons of air conditioning, it needs some type of electrical control to ensure that proper operation is obtained. Figure 18-4 shows the 230-V, 60-Hz, single-phase unit's electrical schematic. The same unit is manufactured with 208/230-V, 60-Hz and 3-phase wiring. The same unit can be operated on 460 V, 60 Hz, 3 phase. The motors and some controls must be changed to take the higher voltages. This does not mean that a unit that operates on 460 V will operate on 220 V when the supply is changed. Certain parts must be changed. In some cases, relay coils and the compressor motor must be rated at the voltage present for operation. These units are factory wired for both high voltage and low voltage. Connections to the thermostat, located inside the living space, are connected to terminals marked W, J, G, X, and Y. Note the part of the circuit responsible for the gas heating. This part would be eliminated if the unit was used only for cooling. For instance, if cooling is called for, the contacts on 7K would be making contact be-

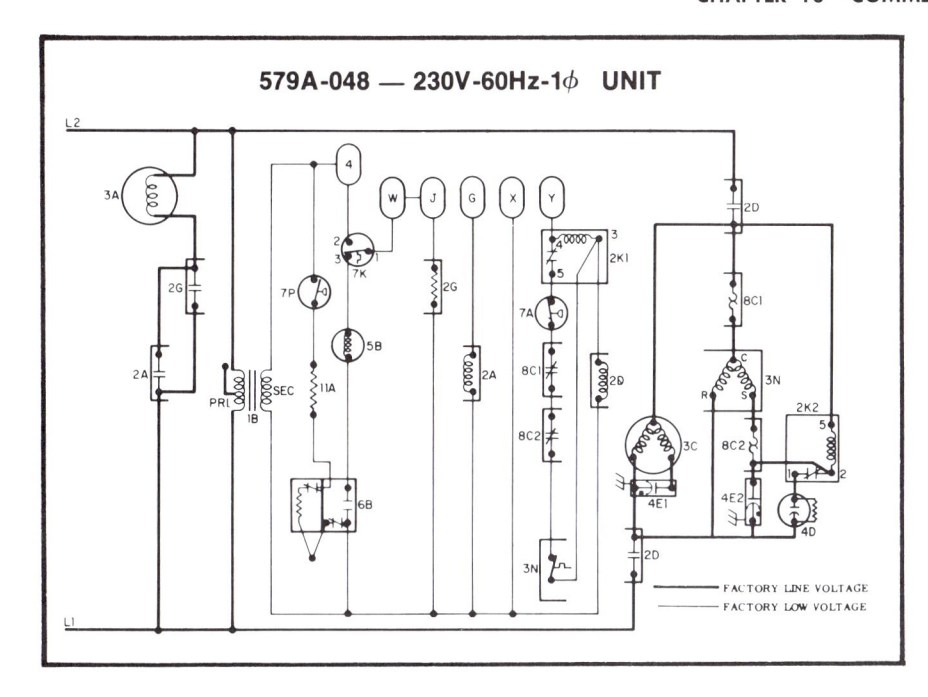

CHAPTER 18—COMMERCIAL AIR-CONDITIONING SYSTEMS

579A-048 — 230V-60Hz-1φ UNIT

LEGEND
1B-Transformer
2A-Blower Relay (Cooling)
   S.P.S.T. N.O.
2D-Contactor D.P.S.T. N.O.
2G-Blower Relay (Heating)
   S.P.S.T. N.O.
2K1-Impedance Relay
   S.P.S.T. N.C.
2K2-Start Relay S.P.S.T. N.C.
3A-Evaporator Blower Motor
3C-Condenser Fan Motor
3N-Compressor Motor
4D-Start Capacitor
4E1-Run Capacitor (Fan)
4E2-Run Capacitor (Compressor)
5B-Magnetic Gas Valve
6B-Pilot (Reignition)
7A-High-Pressure Switch
   S.P.S.T. N.C.
7K-Limit Switch S.P.D.T.
7P-Pressure Switch S.P.S.T. N.O.
8C1-2-Compressor Overload
11A-Resistor

Bryant

18-4. Wiring diagram for 230-volts, 60-hertz, single-phase unit with its parts labeled.

tween points 1 and 2, instead of point 3, as shown. Fig. 18-4.

Sequence of Operation

Only the cooling operations are shown in Fig. 18-5. This rooftop conditioner operates in the following manner. Most others operate in the same way. This sequence permits operation on the 208–230 V units. With thermostat system switch and fan switch in *auto* position, the operation sequence is as follows:

When there is a demand for cooling by the thermostat, terminal R "makes" to terminals Y1 and G through the thermostat. This thermostat switching action electrically connects blower motor contactor (2M1) and cooling relay (2A) across the 24-V secondary of the control transformer (1B). This causes the blower motor contactor (2M1) and cooling relay (2A) to become energized.

The contacts of the energized blower motor contactor close to energize blower motor (3E). This starts the blower motor instantly.

The contacts of cooling relay (2A) close to energize compressor contactor (2M2). As the compressor contactor pulls in, the compressor (3L) starts. At the same time the compressor contactor (2M2) is energized, the condenser fan motor (3C2) is placed in operation. The condenser fan motor (3C1) does not start until the high-side pressure reaches 280 psig, at which point the low-ambient pressure switch (7P) closes to complete the line voltage circuit to the fan motor (3C1) and starts this motor. When the compressor discharge pressure drops to 178 psig, the low-ambient pressure switch (7P) will reopen and the condenser fan motor (3C1) will stop. This provides high-side pressure for low-ambient operation down to 32° F. [0°C]. When the pressure builds back up to 280 psig, the low-ambient pressure switch (7P) will close again. This restarts the condenser fan motor (3C1).

During this time of operation, only one-half of the evaporator coil is being used. Should the indoor temperature continue to rise, the thermostat will make between R and Y2, at which time the liquid line solenoid valve (5B) is energized and opens. This permits the refrigerant to flow to the other half of the evaporator coil.

Keep in mind that two-stage cooling is not available on all models. Those with single-stage cooling will not have half of the evaporator coil operating.

As the conditioned space temperature drops, the second stage contacts R to Y2 will open within the thermostat and close the liquid line solenoid valve (5B). The unit will continue to operate at

471

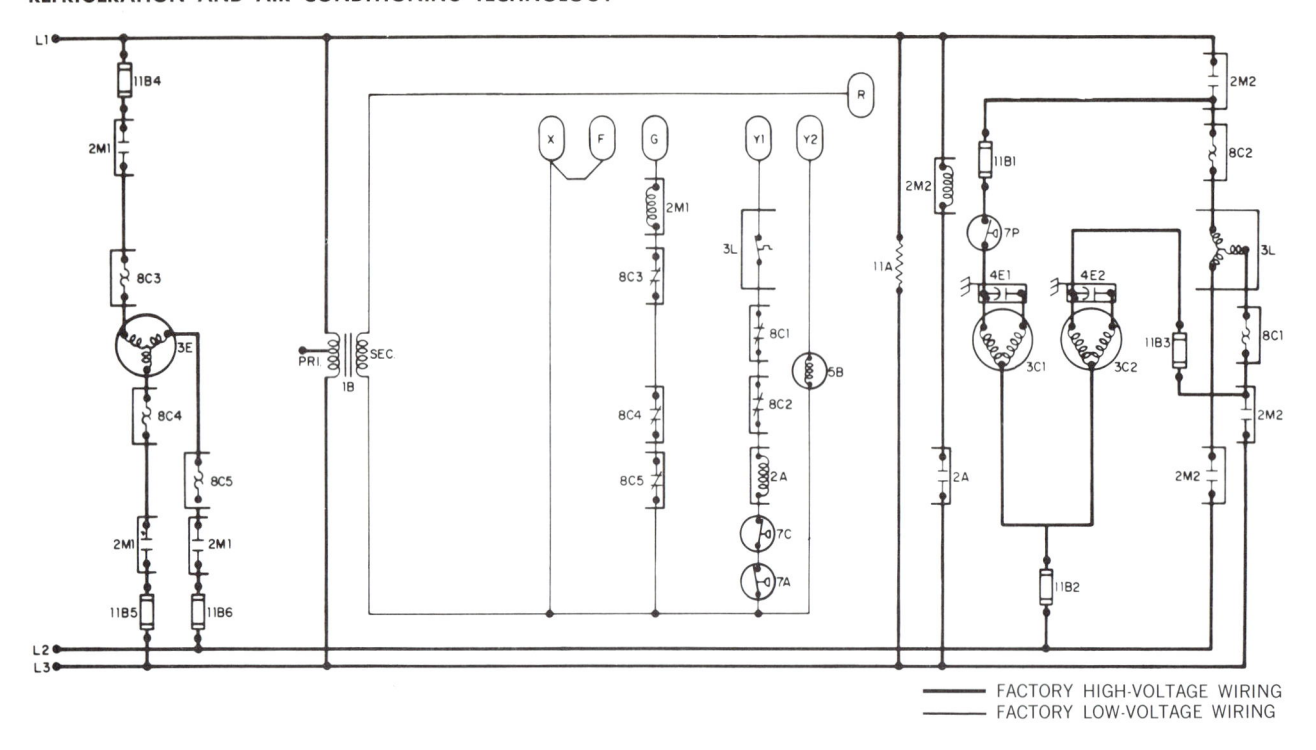

REFRIGERATION AND AIR CONDITIONING TECHNOLOGY

LEGEND

1B-Transformer (Tapped Primary)
2A-Cooling Relay N.O.
2M1-Blower Motor Contactor N.O.
2M2-Compressor Contactor N.O.
3C1 & 2-Fan Motors
3E-Blower Motor
3L-Compressor with Internal Overload
4E1 & 2-Run Capacitor
5B-Liquid Line Solenoid Valve
7A-High-Pressure Switch N.C.
7C-Low-Pressure Switch N.C.
7P-Low-Ambient Pressure Switch N.O.
8C1 & 2-Compressor Overloads
8C3, 4, & 5-Blower Motor Overload
11A-Crankcase Heater
11B1, 2, & 3-Fan Motor Fuses
11B4, 5, & 6-Blower Motor Fuses

Bryant

18-5. Wiring diagram for a 208-230 volt operation.

two-thirds capacity. As the temperature within the conditioned space continues to drop and reaches the thermostat setting, contacts R to Y1 and G will open. At this time, compressor (3L), condenser fan motors (3C1 and 3C2) and the blower motor (3E) will stop. After all motors have stopped, the unit remains in standby position ready for the next call for cooling by the thermostat.

Compressor Safety Devices

Several safety devices protect the compressor in abnormal situations. For instance, the high-pressure switch interrupts the compressor control circuit when the refrigerant high-side pressure becomes excessive. A low-pressure switch interrupts the compressor control circuit when the refrigerant low-side pressure becomes too low. The compressor is protected from overloads by current-operated circuit breakers. Thermal devices embedded in the windings of the compressor motor open the circuit when too much heat is generated by the windings. Some manufacturers place a 5-minute delay device in series with the compressor motor. Thus, the motor cannot be restarted for 5 minutes after shutdown.

When any of the above safety devices are actuated, current in the Y1 leg is interrupted and shuts off the compressor and condenser fan motors. Fig. 18-5.

Maintenance

Before performing any maintenance on the unit, make sure the main line switch is open or in the "off" position. Label the switch so someone will not turn it on while you are working.

472

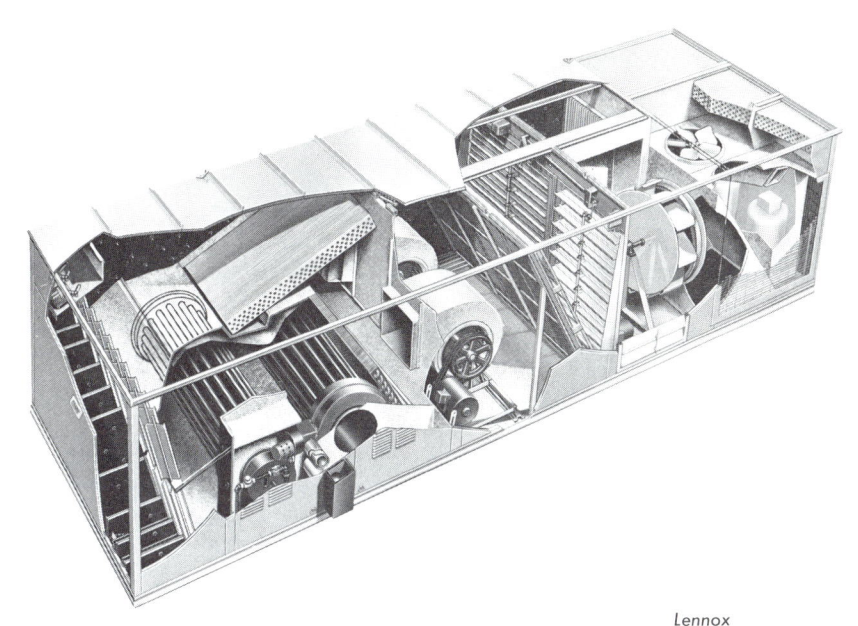

CHAPTER 18—COMMERCIAL AIR-CONDITIONING SYSTEMS

18-6. Direct multizone system for rooftop mounting. *Lennox*

The components should be checked and serviced as follows:

- *Blower motor.* Oil according to the manufacturer's recommendations. The rating plate will usually give lubrication instructions.
- *Electrical connections.* The electrical connections should be checked periodically and retightened.
- *Pulley alignment and belt tension.* Check the blower and motor pulley for alignment. Also check the belt for proper tension. It should have approximately 1″ of sag under normal finger pressure.
- *Blower bearings.* Blowers are equipped with prelubricated bearings and need no lubrication. If, however, there are a blower unit and blower motor without sealed bearings, a few drops of oil must be added occasionally.
- *Condenser and evaporator coils.* Coils should be inspected occasionally and cleaned as necessary. Be careful not to bend the soft aluminum fins. WARNING: *Make sure the main line disconnect switch is in the Off position before cleaning the coils.*
- *Filters.* System air filters should be inspected every 2 months for clogging because of dirt. When necessary, replace disposable-type filters.

Special Instructions

1. Do not rapid-cycle the unit. Allow at least 5 minutes before turning on the unit after it has shut off.
2. If a general power failure occurs, the electrical power supply should be turned off at the unit disconnect switch until the electrical power supply has been restored.
3. Air filters should be cleaned or replaced at regular intervals to ensure against restricted airflow across the cooling coil.
4. During the off season, the main power supply may be left On or turned Off. Leaving the power turned on will keep the compressor crankcase heaters energized.
5. If power has been off during the winter, it must be turned on for at least 12 hours before spring startup of the unit. This allows the crankcase heaters to vaporize any liquid refrigerant that may be condensed in the compressor.

Troubleshooting

To troubleshoot this type of air conditioning equipment, refer to Table 17-E (pages 464–466). The general troubleshooting procedures listed there are used for hermetically sealed compressors.

DIRECT MULTIZONE SYSTEM

The direct multizone system unit is roof mounted and can be used for cooling and heating. Fig. 18-6. It can use chilled water for cooling up to 550,000 Btu per hour.

Air distribution is 12- or 16-zone multizone control at the unit or double duct with independent mixing dampers at each zone. Figure 18-7 shows the typical applications of such a unit with a zone distribution system using mixing dampers located at the unit. A double duct distribution system with zone damper boxes can be used. Mixing dampers are remote from the unit. The net weight of the unit is 2525 pounds.

Figure 18-8 shows the location of the parts inside the unit. Figure 18-9 shows how the refrigerant piping is laid out for the unit when two compressors are used for cooling purposes. Note that this unit uses an accumulator. There are certain conditions under which the capacity of such

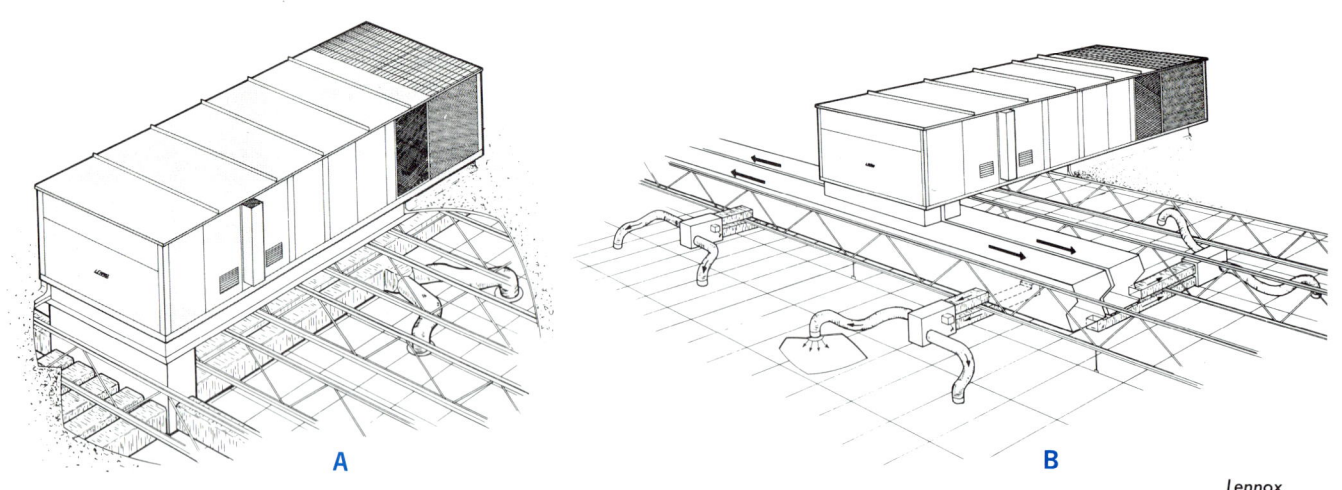

REFRIGERATION AND AIR CONDITIONING TECHNOLOGY

18-7. Typical installations of the multizone system unit. (A) A zone distribution system with mixing dampers located at the unit. (B) A double-duct distribution system with zone damper boxes and mixing dampers remote from the unit.

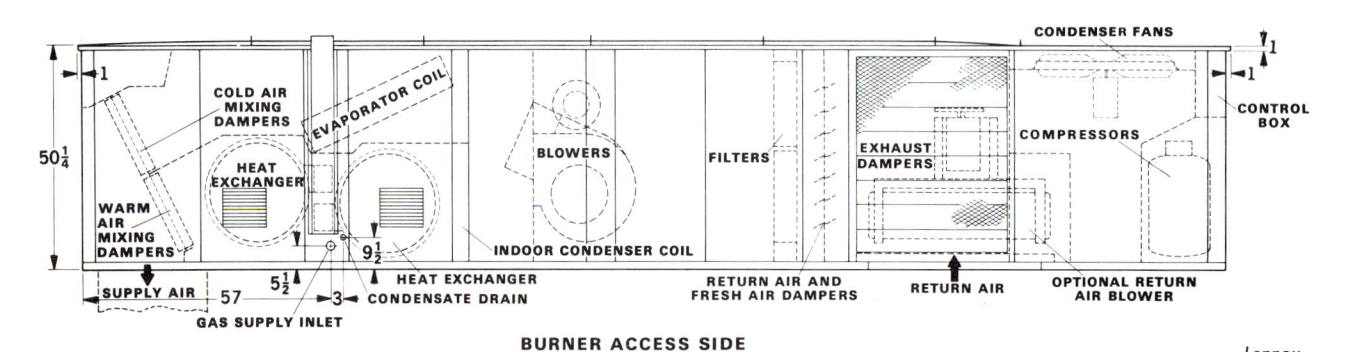

Lennox

18-8. Location of the component parts to the multizone unit.

Lennox

18-9. Refrigerant piping for the multizone system.

Lennox

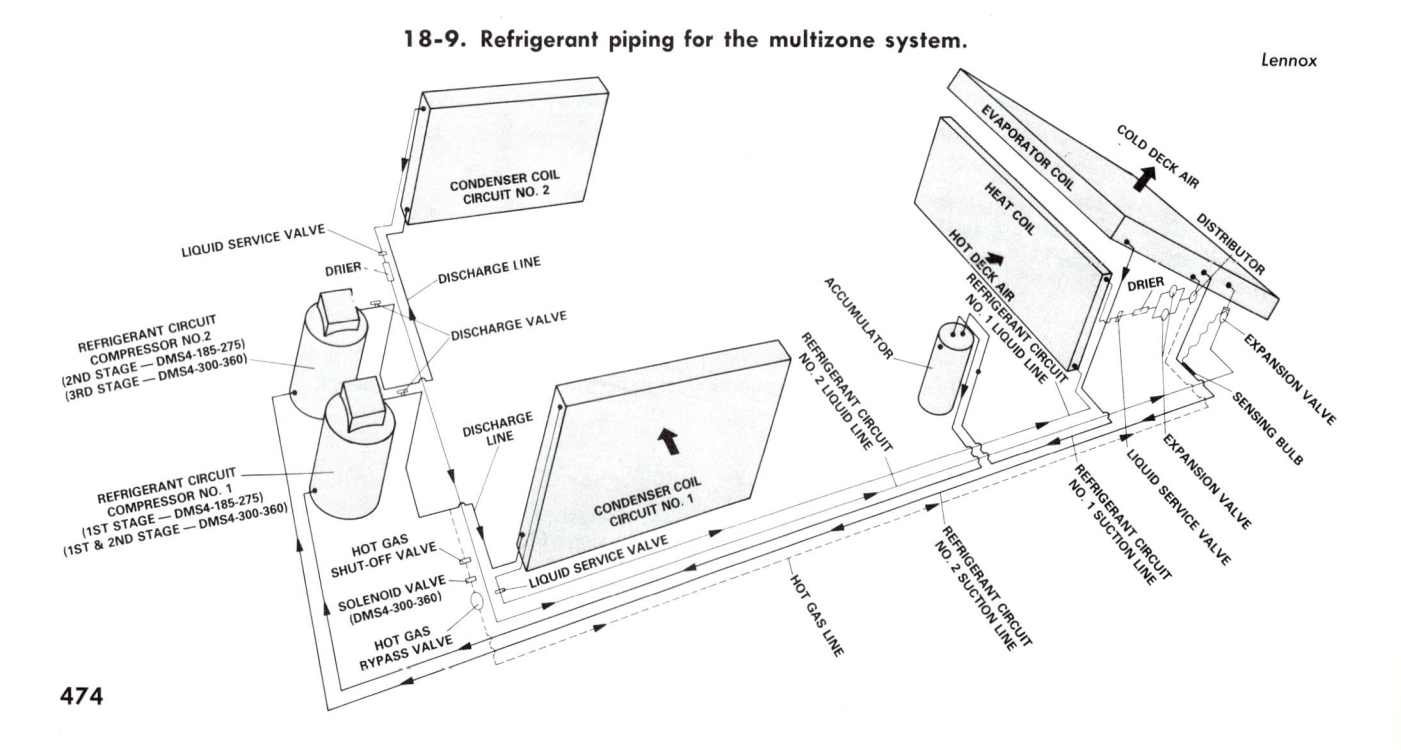

474

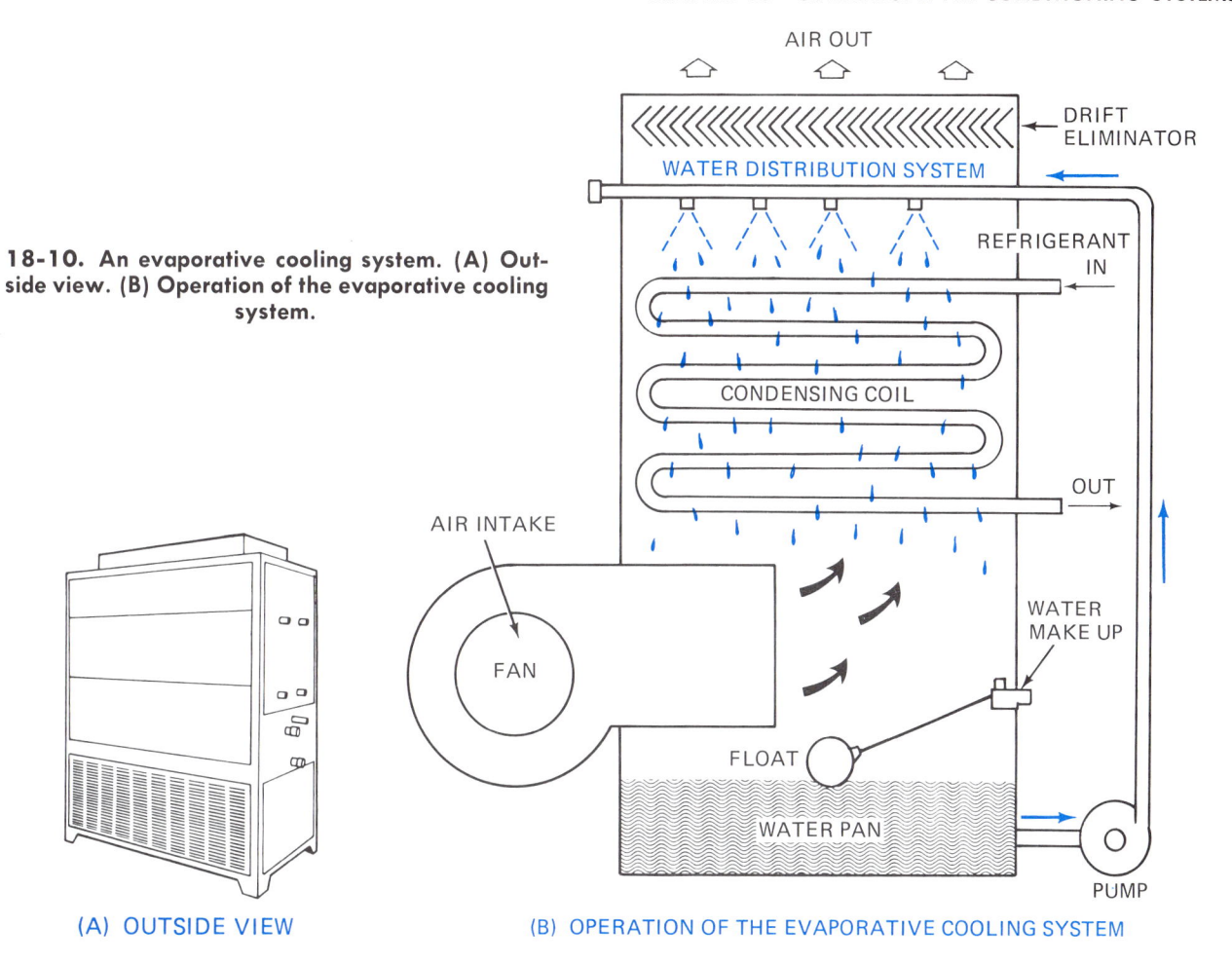

CHAPTER 18—COMMERCIAL AIR-CONDITIONING SYSTEMS

18-10. An evaporative cooling system. (A) Outside view. (B) Operation of the evaporative cooling system.

(A) OUTSIDE VIEW

(B) OPERATION OF THE EVAPORATIVE COOLING SYSTEM

a unit must be rated. These conditions are the temperature of the evaporator air, the condenser coil air temperature, the speed of the blower motor, and its volume of air delivered.

Troubleshooting

To troubleshoot this type of air conditioning equipment, refer to Table 17-E (pages 464–466). The general troubleshooting procedures listed there are used for hermetically sealed compressors.

EVAPORATIVE COOLING SYSTEM

In some locations, it is possible to use the cooling tower principles to condense the refrigerant. This method has the usual problems with water and tower fungi. Those problems have been discussed in Chapter 9.

The condensing coil is cooled by air drawn in from outside the tower and forced upward over the coil. Water is pumped continuously to a distribution system and sprayed so it drops in small droplets over the condensing coil. Fig. 18-10. The water is reused since it cools as it drops through the moving airstream. In some systems the water is pumped up and into a trough. The water drips down over the condenser coils and cools them.

In some cases the water moves through tubes that surround the refrigerant-carrying tubes. The airstream then removes the heat and discharges it into the surrounding air. This means the cooling tower should be mounted outside a building. In some instances it is possible to mount the tower inside. However, a duct is then needed to carry the discharged air outside. As shown in Fig. 18-10(B), the water is carried off and must be replaced as it, too, evaporates. The pan is filled to level when the float moves down and allows the water makeup valve to open. If the condenser temperature reaches

475

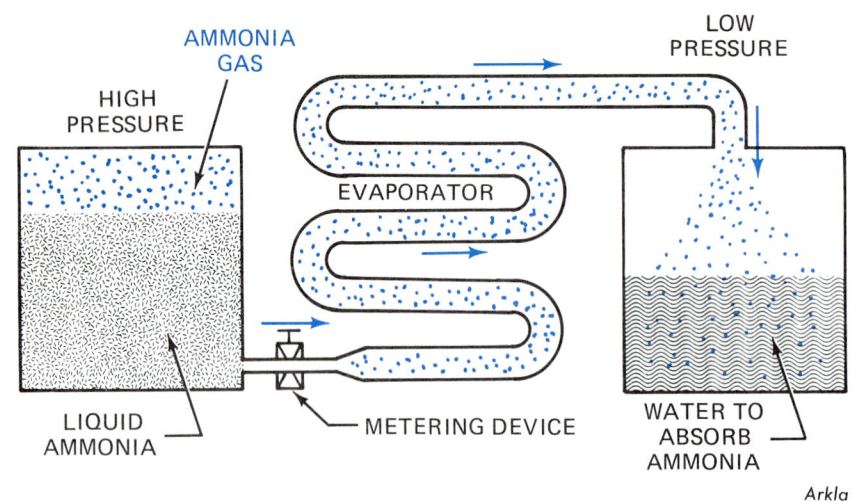

REFRIGERATION AND AIR CONDITIONING TECHNOLOGY

18-11. A simplified system of absorption of refrigeration using ammonia as the refrigerant. *Arkla*

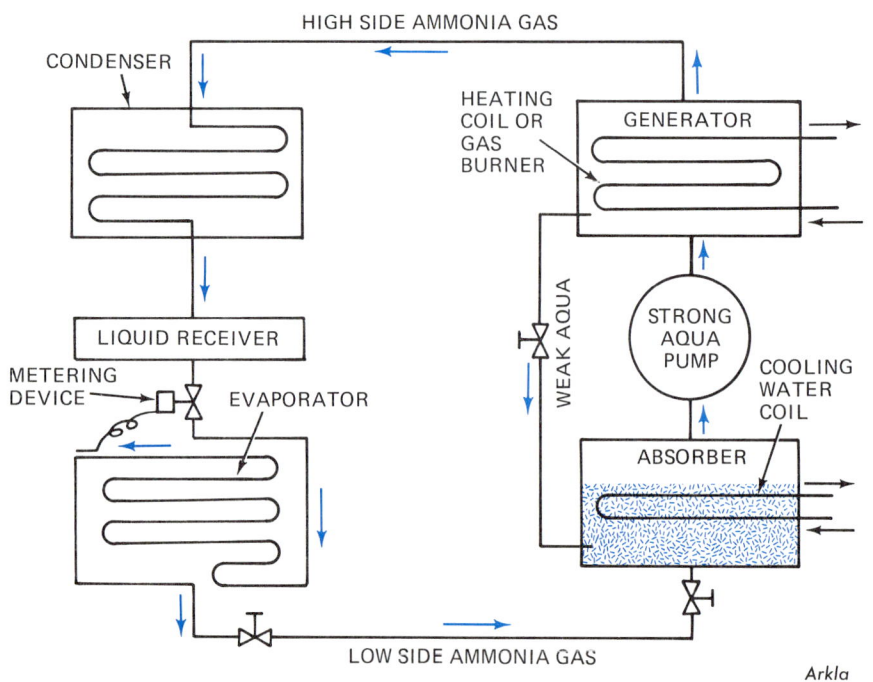

18-12. The ammonia absorption system of refrigeration used in large installations. *Arkla*

or exceeds 100° F. [37.8°C], the thermostat turns on the water and the fan.

Problems with this system center in the electrical control system and the water system. The controls, fan motor, and pump motor are electrically operated. Thus, troubleshooting involves the usual electrical circuit checks.

ABSORPTION-TYPE AIR CONDITIONING SYSTEMS

A boiling refrigerant in an evaporator *absorbs* heat. The evaporator pressure must be low for boiling to take place. To produce the low pressure, it is necessary to remove the refrigerant as soon as the boiling refrigerant vaporizes. Vapors can be absorbed quickly by another liquid. However, the other liquid must be able to absorb the vapor when it is cool. It will then release the absorbed heat when it is heated.

Ammonia is one of the refrigerants most commonly used in absorption-type air conditioning systems. Ammonia vapors are absorbed quickly by large amounts of cool water. In fact, it can absorb vapor as quickly as a compressor.

High-pressure ammonia can be fed as a pure liquid through a metering device directly into an evaporator. See Fig. 18-11. Refrigeration takes place until the high-side liquid ammonia is exhausted or the water in the absorber tank is saturated. Once saturated, it no longer absorbs ammonia. If the ammonia tank and the absorber are large enough, these components can be used as part of an air conditioning system.

A system can be devised to handle large installations. See Fig. 18-12. In this system, *some* of the ammonia is removed from the water. This leaves a weak water solution of ammonia. This solution flows by gravity to the absorber.

The water in the absorber absorbs the ammonia. Such absorption continues until ammonia represents 30% of the water-ammonia solution.

Such a strong (30%) solution of ammonia is called strong aqua. *Aqua* means 'water.' The strong aqua is pumped up to the generator. The absorber operates at low-side evaporator pressure.

476

CHAPTER 18—COMMERCIAL AIR-CONDITIONING SYSTEMS

That is why the pump is necessary. The generator has a high-side pressure.

Air is driven out of water by heat. Ammonia also can be driven out of water by applying heat. The high-temperature ammonia vapor rises and moves to the condenser. Weak condensed liquid flows back to the absorber through the force of gravity. In the condenser, the latent heat is removed from the ammonia vapor. Condensed ammonia liquid flows through the liquid receiver to the evaporator. In the evaporator, the ammonia boils at reduced pressure. Latent heat is absorbed. The liquid ammonia changes into a vapor. In changing to a vapor, the ammonia produces refrigeration.

Ammonia is only one refrigerant used for this type of system. Lithium bromide and water also can be made into a refrigerant.

Figure 18-13 shows a typical absorber system. Several manufacturers make packaged units for absorber systems.

CHILLED WATER AIR CONDITIONING

To produce air conditioning for large areas such as department stores and office buildings, it is necessary to use another means of cooling the air. Chilled water is used to produce the cooling needed to reduce the interior temperature of offices and stores. To understand the function of the chilled water, it is necessary to look at the total system. Fig. 18-14.

The refrigerating machine is the chiller. Water is supplied to the chiller. There, its temperature is reduced to about 48°F. [8.9°C]. The chilled water then flows to the coils in the fan coil unit. The fan coil unit is located in the space to be conditioned. In some cases, a central air handling system is used. Pumps are used to move the water between the chiller and the air handling equipment. The water is heated by the room air that is pulled over the chilled water coils. Thus, the water reaches a temperature of about 55°F. [12.8°C]. In some installations it reaches

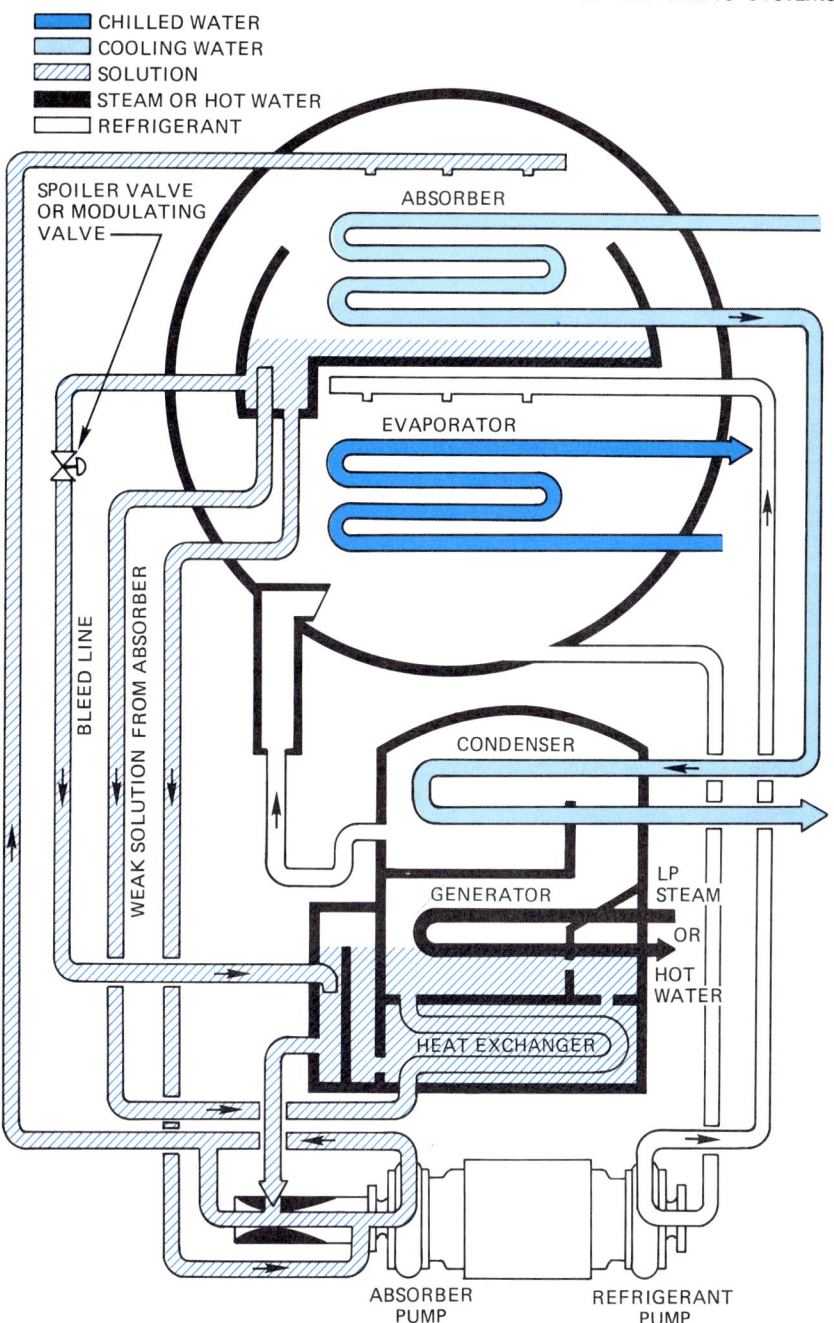

18-13. A typical absorption system used in commercial air-conditioning applications.

Worthington Compressors

477

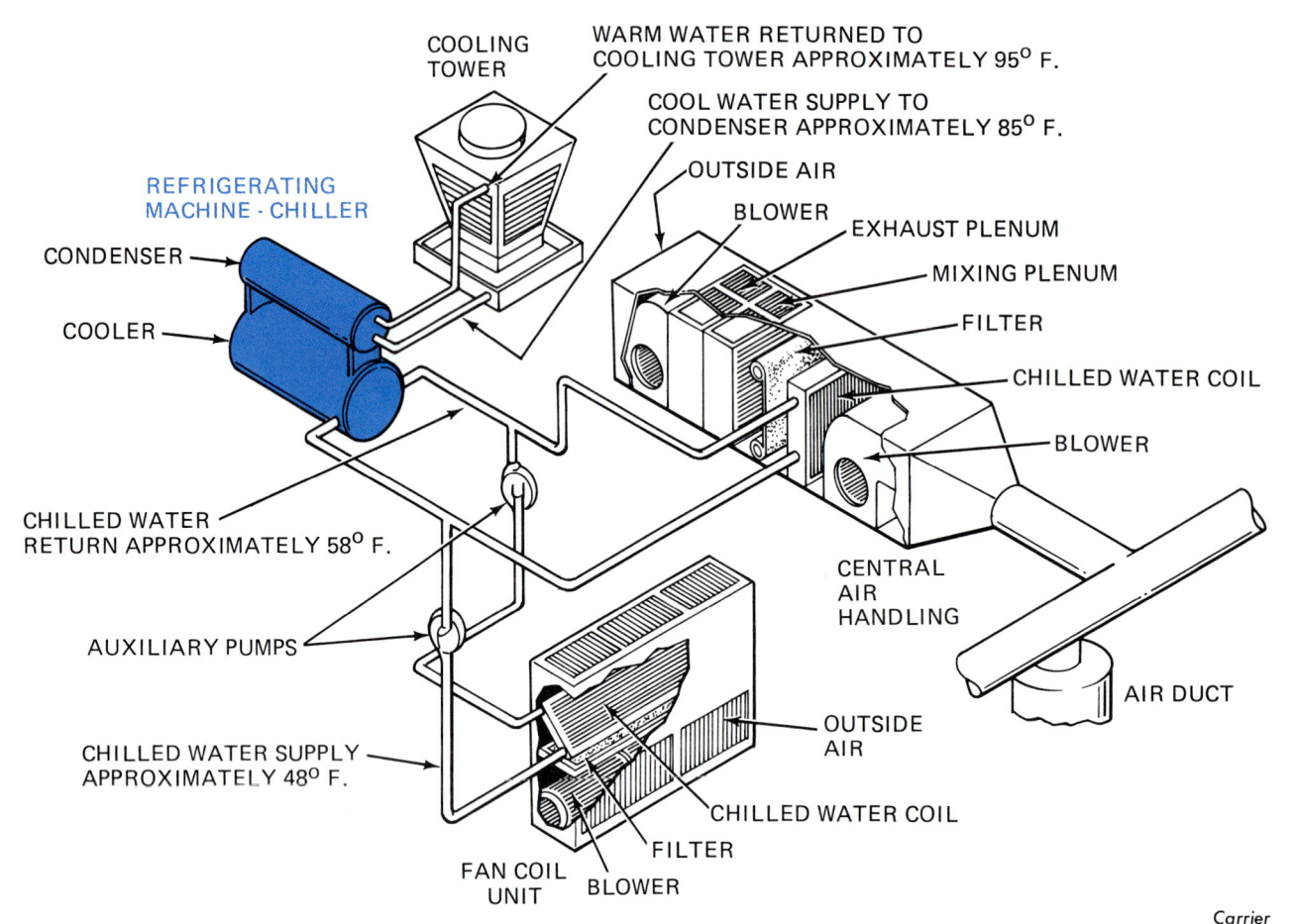

REFRIGERATION AND AIR CONDITIONING TECHNOLOGY

18.14. Complete air-conditioning system using chilled water.

Carrier

58° F. [14.4°C]. The water absorbs about 10° F. [5.5°C] of heat as it is exposed to the room air being drawn into the unit by blowers.

The heated water is then pumped back to the chiller. There, the water is chilled again by the machine removing the absorbed heat. Once chilled, the 48° F. [8.9°C] water is again ready to be pumped back to the fan coil unit or the central air handling system. This process of recirculation is repeated as needed to reduce the temperature of the space being conditioned.

A cooling tower is used to remove the heat to the outside of the building. Cooling towers were discussed in Chapter 9. The condenser water is cooled by the cooling tower. Fig. 18-14.

Figure 18-15 shows the refrigeration cycle of a chiller.

Refrigerant Cycle

When the compressor starts, the impellers draw large quantities of refrigerant vapor from the cooler at a rate determined by the size of the guide vane opening. This compressor suction reduces the pressure within the cooler. This causes the liquid refrigerant to boil vigorously at a fairly low temperature (typically 30 to 35° F. [−1.1 to 1.7°C]).

The liquid refrigerant obtains the energy necessary for the change to vapor by removing heat from the water in the cooler tubes. The cold water can then be used for process chilling or air conditioning, as desired.

After removing heat from the water, the warm refrigerant vapor is compressed by the first stage impeller. It then mixes with flash economizer gas and is further compressed by the second stage impeller.

Compression raises the tem-

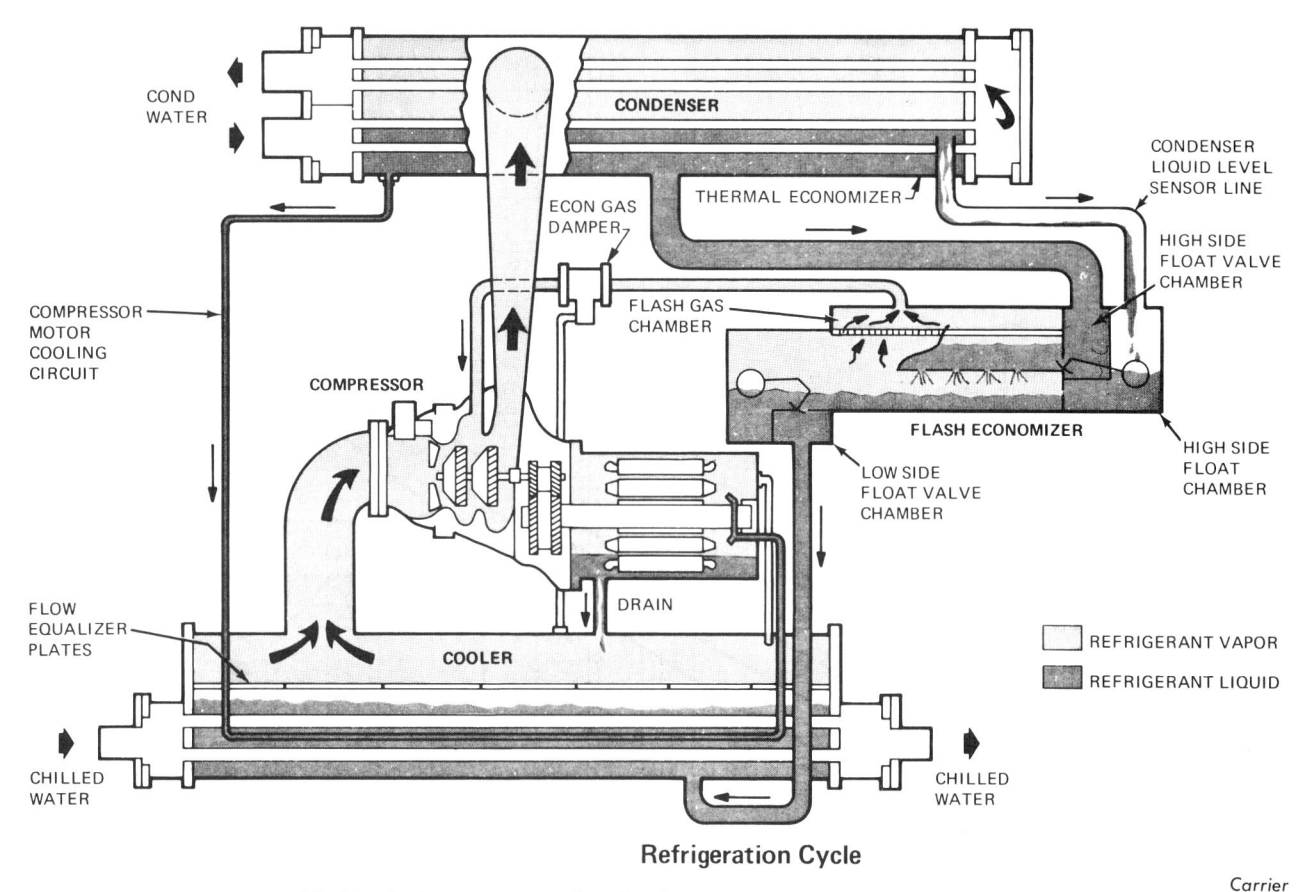

CHAPTER 18—COMMERCIAL AIR-CONDITIONING SYSTEMS

Refrigeration Cycle

18-15. Refrigeration cycle of a hermetic-type centrifugal chiller.

Carrier

perature of the refrigerant vapor above that of the water flowing through the condenser tubes. The compressed vapor is then discharged into the condenser at 95 to 105° F. [35 to 40.6°C]. Thus, the relatively cool condensing water (typically 75 to 85° F. [23.9 to 29.4°C]) removes some of the heat, condensing the vapor into a liquid.

The liquid refrigerant then drains into a valve chamber with a liquid seal. This prevents gas from passing into the economizer. When the refrigerant level in the valve chamber reaches a preset level, the valve opens. This allows liquid to flow through spray pipes in the economizer.

The pressure in the economizer is midway between those of the condenser and cooler pressures. At this low pressure, some of the liquid refrigerant flashes to gas, cooling the remaining liquid.

The flash gas is piped through the compressor motor for cooling purposes. It is then mixed with gas already compressed by the first stage impeller.

The cooled liquid refrigerant in the economizer is metered through a low-side valve chamber into the cooler. Pressure in the cooler is lower than economizer pressure. Thus, some of the liquid flashes, cooling the remainder to cooler (evaporator) temperature. The cycle is now complete.

Figure 18-16 shows a cutaway view of a chiller. This is how Fig. 18-15 looks in a packaged unit. These chillers are available from 425 to 2500 tons of refrigeration.

Such large systems must be operated by a person with specialized knowledge of the unit. The manufacturers publish training manuals that detail the operation, maintenance, and repair of the units. Some of the training manuals are as long as this textbook. Thus, there can be pre-

479

REFRIGERATION AND AIR CONDITIONING TECHNOLOGY

sented here only a brief discussion of the information you will need to operate and maintain such a cooling operation.

A typical chiller installation is shown in Fig. 18-17. Note the piping and wiring systems. As can be observed, the electrical system is rather complicated.

Control System

Chiller safety controls are electronic. Chiller capacity controls may be either solid state (transistorized) or pneumatic (air-pressure controlled).

Chiller operating capacity is determined by the position of the guide vanes at the entrance to the compressor suction. As cooling needs change, the vanes open and close automatically.

A thermistor probe in the chilled water line constantly monitors chilled water temperature. The probe signals any temperature change to a capacity control module in the machine

18-16. Cutaway view of a centrifugal hermetic-type chiller.

Carrier

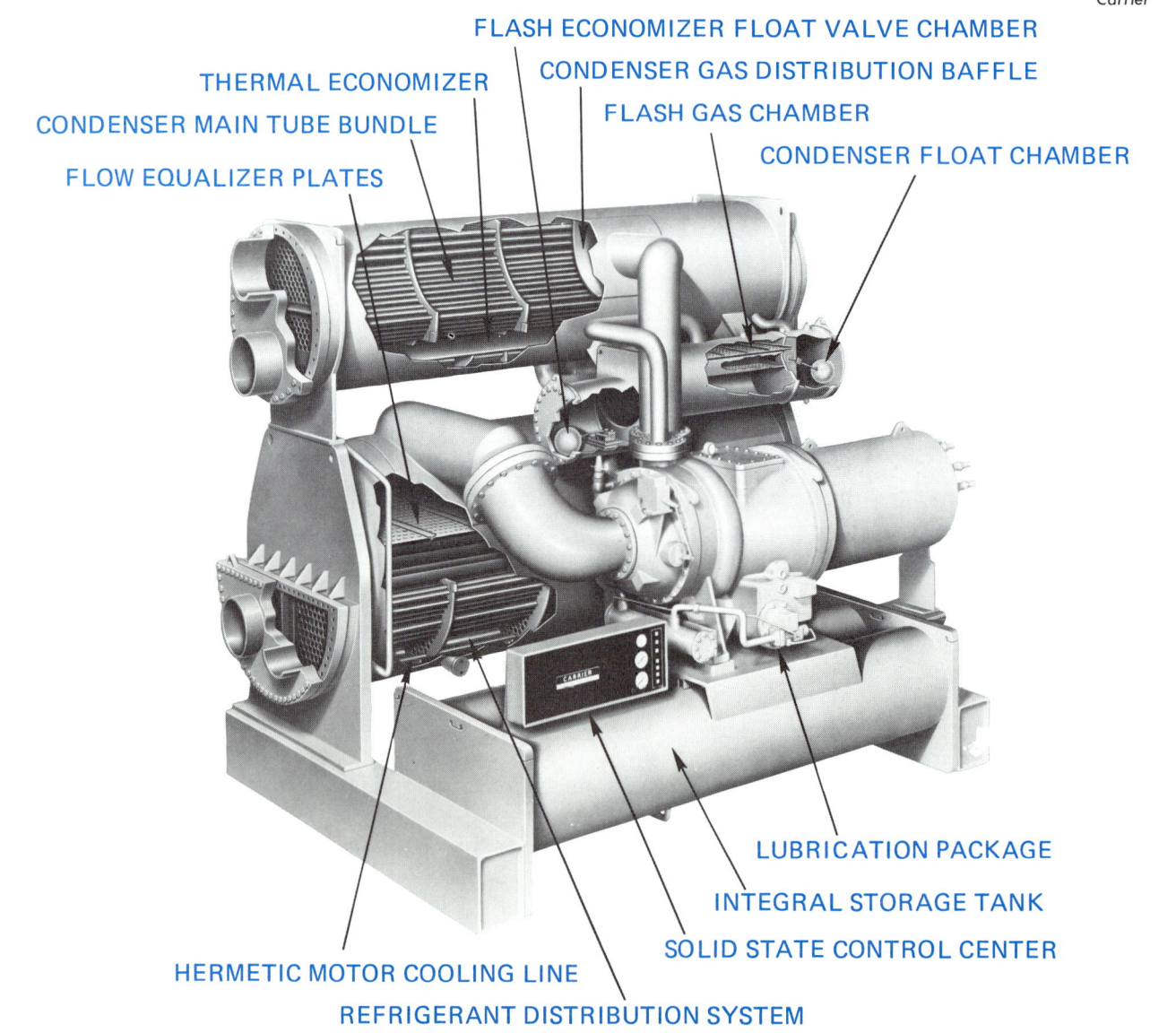

- FLASH ECONOMIZER FLOAT VALVE CHAMBER
- THERMAL ECONOMIZER
- CONDENSER GAS DISTRIBUTION BAFFLE
- CONDENSER MAIN TUBE BUNDLE
- FLASH GAS CHAMBER
- FLOW EQUALIZER PLATES
- CONDENSER FLOAT CHAMBER
- LUBRICATION PACKAGE
- INTEGRAL STORAGE TANK
- SOLID STATE CONTROL CENTER
- HERMETIC MOTOR COOLING LINE
- REFRIGERANT DISTRIBUTION SYSTEM

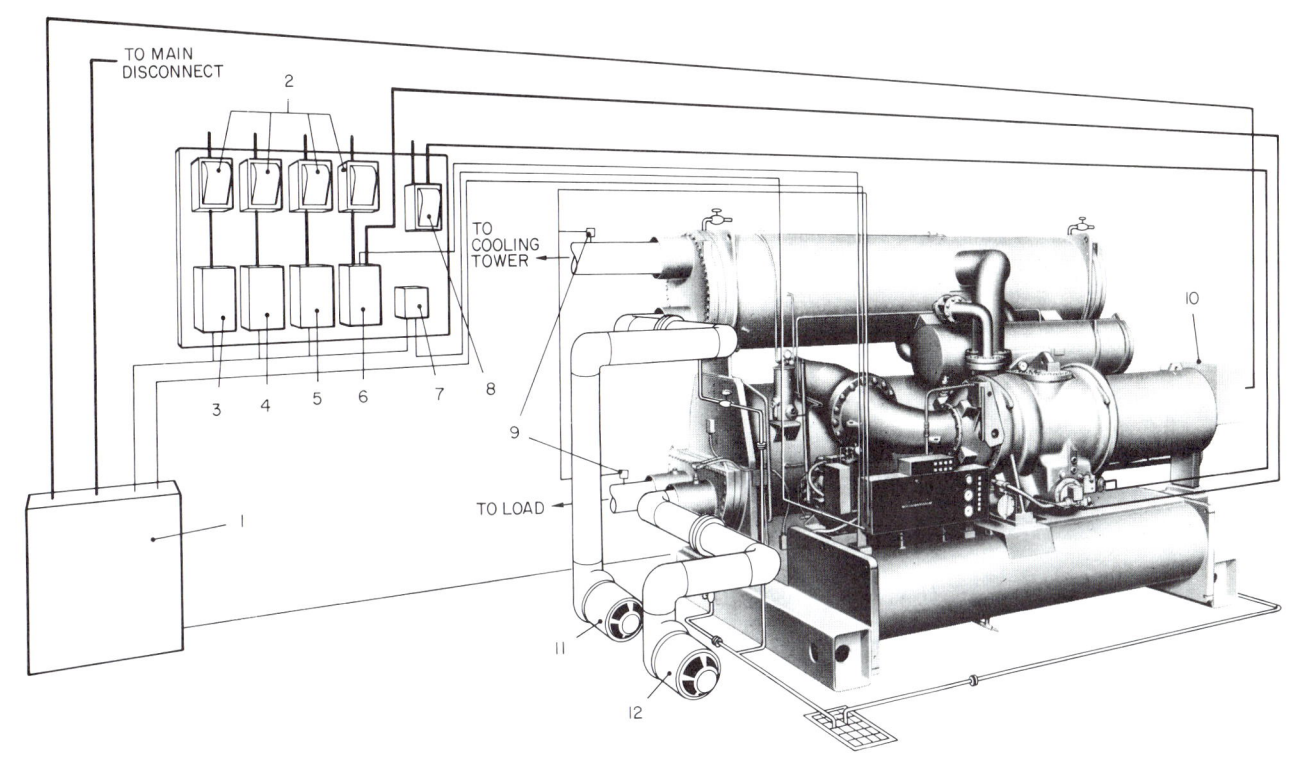

CHAPTER 18—COMMERCIAL AIR-CONDITIONING SYSTEMS

1 — Compressor Motor Starter
2 — Fused Disconnect
3 — Cooling Tower Fan Starter
4 — Condenser Water Pump Starter
5 — Cooler Water Pump Starter
6 — Lube Oil Pump Starter
7 — Pilot Relay
8 — Fused Disconnect for Oil Heater and Heater Controller (115 v)
9 — Water Flow Switches
10 — Compressor Motor Terminal Box
11 — Condenser Water Pump
12 — Cooler Water Pump

NOTES:
1. Separate 115-volt source for controls, unless transformer is furnished with compressor motor starter.
2. Wiring and piping shown do not include all details for a specific installation. Refer to certified electrical drawings.
3. Wiring must conform to applicable local and national codes.
4. Pipe per standard techniques.

→ 5. Oil cooler water source must be clean and noncorrosive. City water or chilled water may be used. Recommended flow condition is: 30 gpm water at 85 – 100 F. Vary gpm to obtain 100 F leaving water.
→ 6. Oil heater must be on separate circuit providing continuous service.

Carrier

18-17. Typical piping and wiring diagram for the chiller in Fig. 18-16.

control center. The module, in turn, initiates a response from the guide vane actuator.

When chilled water temperature drops, the vane actuator causes the guide vanes to move towards the closed position. The chiller capacity decreases. Conversely, a rise in chilled water temperature causes the guide vanes to open and increase chiller capacity. If the water temperature continues to rise, the vanes open further to compensate. Built-in safeguards prevent motor overloads. To minimize start-up demand, control interlocks keep the guide vanes closed (at minimum capacity position) until the compressor reaches run condition.

CHILLERS

Chillers are divided by type according to the compressors they use. Thus, there are reciprocating compressors and centrifugal compressors.

Reciprocating compressors may have single-acting or double-acting arrangements. They are made with from one to sixteen cylinders. These cylinders may be arranged in a V-, W-, or radial pattern. Each cylinder arrangement is designed for a spe-

481

REFRIGERATION AND AIR CONDITIONING TECHNOLOGY

cific requirement. Reciprocating compressors have already been discussed in Chapter 7.

There are two types of centrifugal compressors for chillers. There are the hermetic centrifugal compressor and the open-drive centrifugal compressor. The hermetic centrifugal compressor has been discussed on page 162.

The open-drive centrifugal compressor is another type of chiller. Centrifugal compressors are used in units that produce 200 tons or more of refrigeration. Centrifugal compressors used in industrial applications may range from 200 to 10,000 tons. Flexible under varying loads, they are efficient at loads of less than 40% of their design capacity.

Large volumes of refrigerant are used in centrifugal compressors. They operate at relatively low pressures. The refrigerants used are R-11, R-12, R-113, R-114, and R-500.

Centrifugal compressors operate most efficiently with a high-molecular weight, high-specific volume refrigerant.

Reciprocating Chillers

Reciprocating chillers are made in sizes up to 200 tons. They cannot handle large quantities of refrigerant. Thus, more than one compressor must be used. That is why the compressors are stacked on a large frame. Usually no more than two compressors are used in a refrigeration circuit. Thermostatic expansion valves, discussed earlier, are used as metering devices.

Components Used with Chillers

Some components of the

Table 18-A.
Troubleshooting a Console Model (Self-Contained) Air Conditioner

Symptom and Possible Cause	Possible Remedy
Unit Fails to Start	
1. Start switch off.	1. Place start switch in start position.
2. Reset button out.	2. Push reset button.
3. Power supply off.	3. Check voltage at connection terminals.
4. Loose connection in wiring.	4. Check external and internal wiring connections.
5. Valves closed.	5. See that all valves are opened.
Motor Hums, but Fails to Start	
1. Motor is a single-phase on a three-phase circuit.	1. Test for blown fuse and/or tripped circuit breaker.
2. Belts too tight.	2. See that the motor is floating freely on trunnion base. See that the belts are in the pulley groove and not binding.
3. No oil in bearings. Bearings tight from lack of lubrication.	3. Use proper oil for motor.
Unit Fails to Cool	
1. Thermostat set incorrectly.	1. Check thermostat setting.
2. Fan not running.	2. Check electrical circuit for fan motor. Determine if fan blade and motor shaft revolve freely.
3. Coil frosted.	3. Dirty filters restrict air flow through unit. Check for an obstruction at air grille. Fan not operating. Check fan operation. Attempting to operate unit at too low a coil temperature.
Unit Runs Continuously, but no Cooling	
1. Shortage of refrigerant.	1. Check liquid refrigerant level. Check for leaks. Repair and add refrigerant to proper level.
Unit Cycles too Often	
1. Thermostat differential too close.	1. Check differential setting of thermostat and adjust setting.
Unit Vibrates	
1. Unit is not level.	1. Level all sides.
2. Shipping bolts not removed.	2. Remove all shipping bolts and steel bandings.
3. Belts jerking.	3. Motor not floating freely.
4. Unit suspension springs not balanced.	4. Adjust unit suspension until unit ceases vibrating.

Table 18-A. (Continued)
Troubleshooting a Console Model (Self-Contained) Air Conditioner

Symptom and Possible Cause	Possible Remedy
Condensate Leaks 1. Drain lines not properly installed. 2. Slime formation in pan and drain lines. Slime sometimes present on evaporator fins.	1. Drain pipe sizes, proper fall in drain line, traps, and possible obstruction due to foreign matter should be checked. 2. This formation is largely biological and usually complex in nature. Different localities produce different types. It is largely a local problem. Check Chapter 9 ("Cooling Water Problems"). Periodic cleaning will tend to reduce the trouble, but will not eliminate it totally. Filtering air thoroughly will also help. However, at times some capacity must be sacrificed when doing this.
Noisy Compressor 1. Too much vibration in unit. 2. Slugging oil. 3. Bearing knock. 4. Oil level low in crankcase.	1. Check for vibration point. 2. Low suction pressure. 3. Liquid in crankcase. 4. Pump-down system. Add oil if too low.

The aspects of troubleshooting detailed in Table 18–A are as comprehensive as they can be within the limits of this text. Obviously, there is no substitute for experience. Working with air conditioning and refrigeration systems will give you this experience. Thus, you will sharpen your troubleshooting skills.

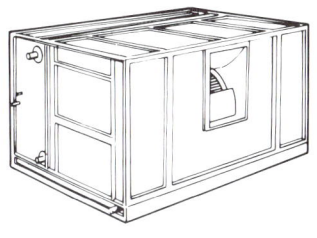

CENTRAL STATION

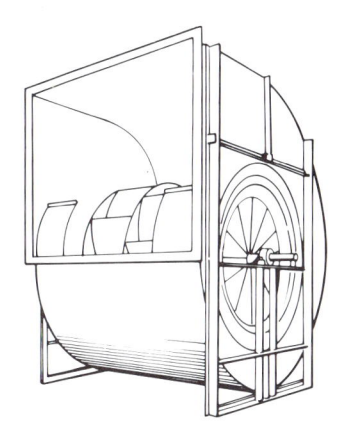

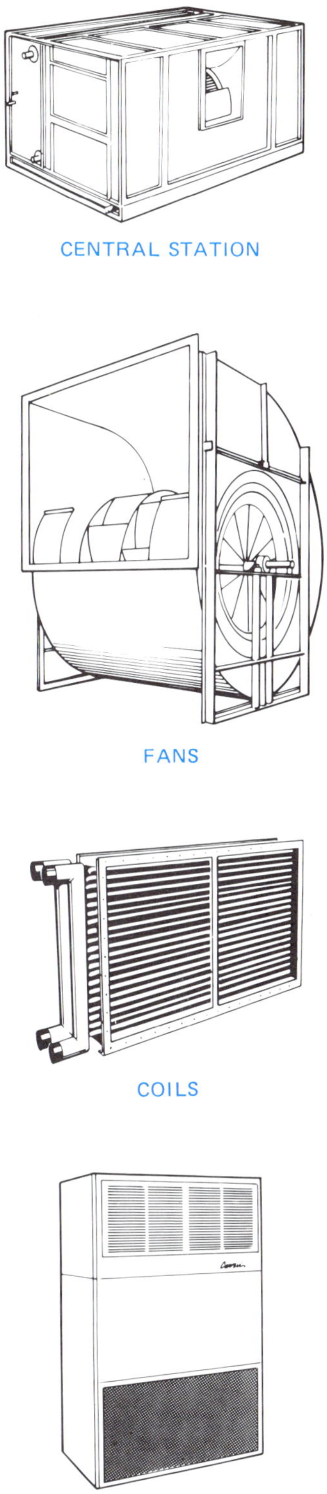

FANS

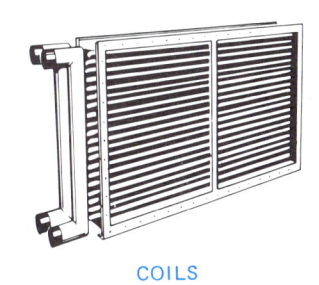

COILS

PACKAGED FAN COILS

Carrier

18-18. Air handling equipment used with a chiller system.

chiller system are huge. They are capable of handling large volumes of air. Such rugged air handling equipment is necessary for this type of installation. See Fig. 18-18.

Air terminals are used to distribute the air when it reaches its destination.

Figure 18-19 shows the whole system with the entire unit hooked up to furnish cool air to a room or building. Installation of this type of equipment requires a thorough knowledge of plumbing, electricity, and air conditioning refrigeration. Trained specialists are needed to handle problems that arise from the operation of these large systems.

CONSOLE-TYPE AIR CONDITIONING SYSTEMS

The console air conditioner is a self-contained unit. These units come in 2 to 10 hp sizes. They are used in small commercial buildings, restaurants, stores, and banks. They may be water-cooled or air-cooled.

Figure 18-19 shows an air-cooled console air conditioner.

483

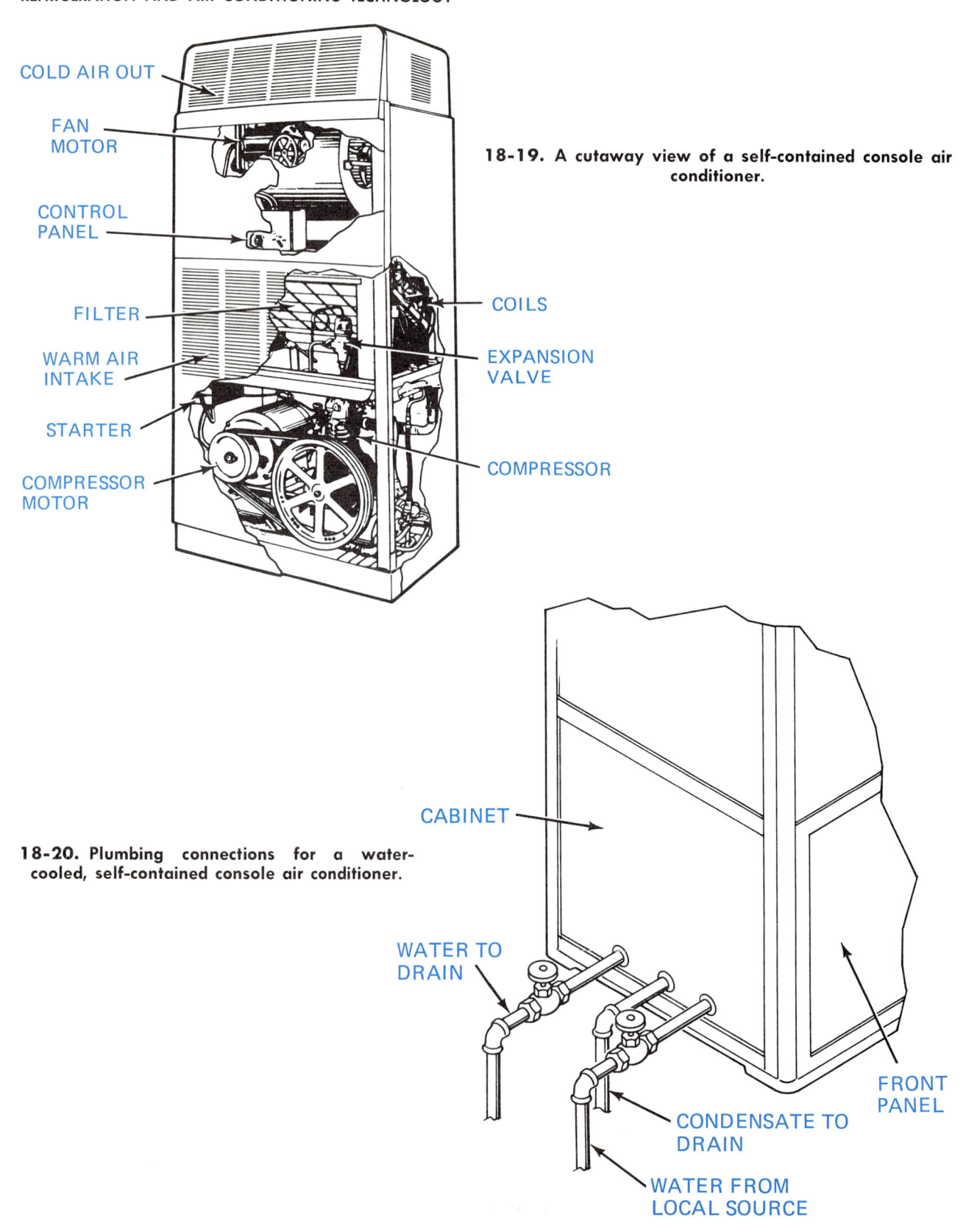

REFRIGERATION AND AIR CONDITIONING TECHNOLOGY

18-19. A cutaway view of a self-contained console air conditioner.

18-20. Plumbing connections for a water-cooled, self-contained console air conditioner.

CHAPTER 18—COMMERCIAL AIR-CONDITIONING SYSTEMS

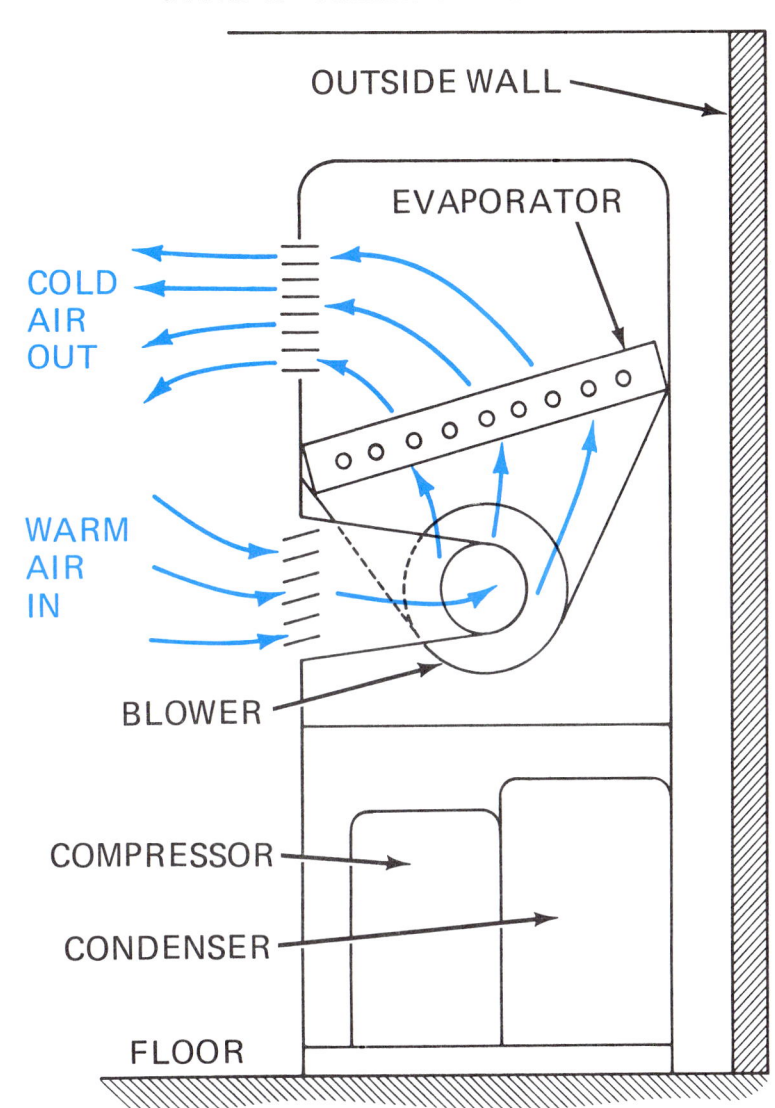

18-21. A self-contained console air conditioner, showing air flow over the evaporator.

You should be able to vent to the outside the hot air produced by the compressor and the condenser.

There are also water-cooled console air conditioners. They will require connections to the local water supply as well as a water drain and condensate drain. See Fig. 18-20. Note the location of the parts in Fig. 18-19. Water is used to cool the compressor. In both models, the evaporator coil is mounted in the top of the unit. See Fig. 18-21. Air blown through the evaporator is cooled and directed to the space to be conditioned. In some areas, a water-cooled model is not feasible.

Since the evaporator coil also traps moisture from the air, this condensate must be drained. This dehumidifying action accounts for large amounts of water on humid days. If outside air is brought in, the condensate will be more visible than if inside air is recirculated.

Installation

The console air conditioner is produced by the factory ready for installation. It must be moved to a suitable location and hooked to electrical and plumbing connections. Once located and connected, it must be checked for level. Electrical and plumbing

485

REFRIGERATION AND AIR CONDITIONING TECHNOLOGY

18-22. A service log used with an open-drive centrifugal refrigeration machine (chiller).

work must conform to local codes.

Check the unit for damage that may have occurred during shipping and installation. Note the type of compressor and the type of cooling used. The compressor is usually hermetic. Refrigerant is usually controlled by a thermostatic expansion valve. Once installed, check the operating conditions. Check and record the temperature in and the temperature out. The difference in temperature is important. Check the electrical circuits so no overload is produced by adding the unit to the line. Record your observations for future use in servicing.

Service

The unit is easily serviced since the component parts are located in one cabinet. Remove panels to gain access to the compressor, valves, blowers, filter, evaporator, and motors. A maintenance schedule should be set up and followed. Most maintenance consists of changing filters and checking pressures. The servicing of the refrigerating unit has already been described in detail on pages 482–483. The servicing of evaporators has also been described on pages 226–253.

Cleaning the filters, cleaning the inside of the cabinet with a vacuum, and cleaning the evaporator fins are the normal service procedures. Water connections and electrical control devices should be checked for integrity. Clean the fan motor. Oil the bearings on the blowers and motors whenever specified by the manufacturer. If there are problems with a water-cooled condenser, refer to pages 203–214.

Scheduled maintenance is very important for all types of air conditioning units. For some units, a log is kept to make sure the various components are checked periodically. Check the log first for any abnormal readings. Fig. 18-22.

Troubleshooting

Table 18-A, on pages 482 and 483, lists basic troubleshooting procedures for the console model (self-contained) air conditioner.

REVIEW QUESTIONS

1. List three types of commercial air-conditioning systems.
2. What is the major advantage of a packaged cooling unit?
3. How often should the system air filters be inspected for dirt and clogging?
4. How long must power be on before starting the unit in the spring?
5. Where is the direct multizone system mounted?
6. What refrigerant is most commonly used in an absorption-type air-conditioning system?
7. What is the meaning of the term *strong aqua?*
8. What is a chiller?
9. How are cooling towers used to remove heat from a building?
10. How do you determine the operating capacity of a chiller?
11. Where is the hot air produced by the console model's compressor and condenser vented?
12. What maintenance procedures are necessary for a console air-conditioning unit?

19

Heat Pumps, Gas-Fired Air Conditioners, and Solar Air Conditioners

GAS AIR CONDITIONING

There are three main types of gas air conditioning cycles used today: compression, absorption, and dehumidification.

In both compression and absorption cycles, air temperature and humidity are tailored to meet variations in surrounding air conditions and changes in room occupancy. Both of these cycles evaporate and condense a refrigerant. They require energy for operation. Mechanical energy is used in the compression type. Heat energy is used in the absorption type. The dehumidification cycle is used primarily in industrial and commercial applications. Dehumidification reduces the moisture content of the air.

Absorption Cooling Cycle

The absorption type of air conditioning equipment works on two basic principles: a salt solution absorbs water vapor and the evaporation of water causes cooling.

In this particular discussion, the absorption cooling cycle is appropriate since it is used in the gas-fired air conditioners. Most gas-fueled air conditioning equipment uses a solution of lithium bromide in water. Lithium bromide (LiBr) is a colorless, saltlike compound that dissolves in water, even to a greater extent than does common salt.

A solution of lithium bromide and water can absorb still more water. Note that, in Fig. 19-1, a tank of absorbing solution (tank B) is connected with a tank of water (tank A). The air in the system is almost completely evacuated. The partial vacuum aids the evaporation process. Water vapor is drawn from the evaporator to the absorber. Evaporation of the water in the evaporator causes the water remaining in it to cool about 10° F. [5.5°C]. The evaporator effect in the evaporator is greatly hastened if the water is sprayed through several shower-bath sprinkler heads. A coil of pipe through which a material such as water passes can be placed within the shower of evaporating water. The water entering the coil of pipe at 55° F. [12.8°C] will be cooled to about 45° F. [7.2°C].

Since the absorber (B) shown in Fig. 19-1 continually receives water, it would soon overflow if the excess water that comes to it as water vapor were not removed. To avoid overflow, the solution that has absorbed water is pumped to a generator (C).

In the generator, the solution is heated directly by a natural-gas flame. It may be heated indirectly by a steam coil. The steam is made in a gas-fueled boiler.

When the solution is heated, some of the water evaporates and passes into the condenser (D). The concentrated solution that remains is sprayed back into the absorber (B). Here, it again absorbs water vapor that comes from the evaporator.

Water vapor in the condenser (D) is cooled by a separate coil of pipe through which water passes. The condensed water is returned to the evaporator (A).

Careful engineering is needed to make the system work well and economically. Attention must be given to temperatures, pressures, and heat transfer in all parts of the system. Practical machines with very few moving parts have now been developed.

488

CHAPTER 19—HEAT PUMPS, GAS-FIRED AIR CONDITIONERS, AND SOLAR AIR CONDITIONERS

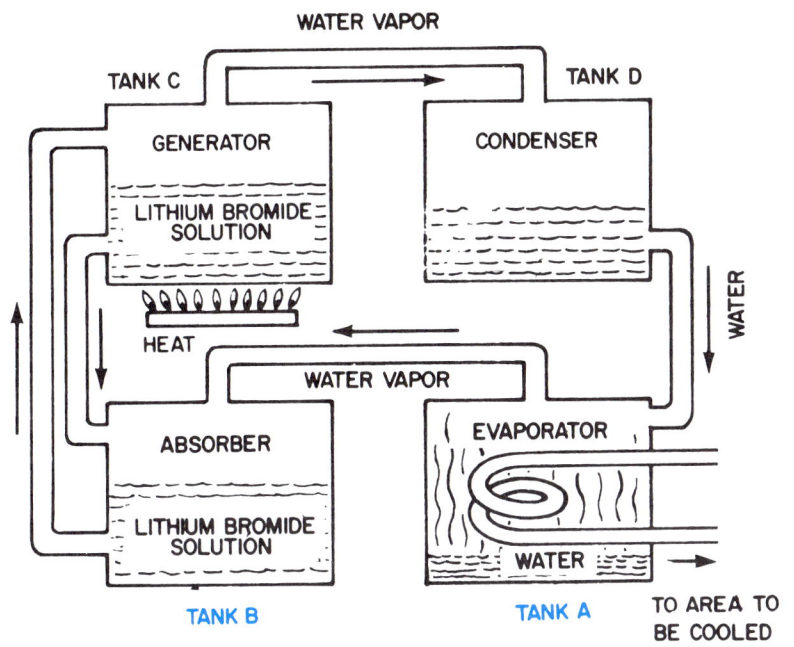

19-1. Lithium bromide system of refrigeration.

the ammonia. A pump is needed for circulating the chilled water.

GAS-FIRED CHILLERS

Chillers operate on natural gas or propane gas. Fig. 19-2. Gas is used for the major job of cooling. Electricity is used for the smaller energy requirements of fans, motors, and controls. This means electrical power requirements are only about 20% of those of a completely electrical unit.

Gas units are available in 3-, 4-, and 5-ton capacities. They use heat as a catalyst. They have no compressor. This means they have fewer moving parts than other types of systems. They are designed for outside installation. They cool by circulating the re-

Absorption units may also use ammonia as the refrigerant. In such systems, heat from a natural-gas flame is used to boil an ammonia-water solution. The operation of the lithium bromide cycle discussed above generally applies. In a system using ammonia, the temperature of the evaporator can go below the freezing point of water. Ammonia is referred to as R-717.

Ammonia Refrigerant in a Gas-Fired System

Ammonia is also used as a refrigerant in a gas-fired system. As the ammonia is moved through the system, it changes its state, becoming strong and weak, a vapor and a liquid. Chilled water is used as a circulation coolant. Very few electrical pumps are needed. Fans are still needed to remove the collected heat. A pump is needed for circulating

19-2. Gas-fired air conditioning unit.

Arkla

489

REFRIGERATION AND AIR CONDITIONING TECHNOLOGY

Table 19-A.
Antifreeze Needed to Prevent Damage to Water in a Chilled-Water Cooler

Lowest Expected Outdoor Temp. °F. (Freezing Point of Mixture)	Permanent Antifreeze Percentage by Volume
25	10%
15	20%
5	30%
0	33%
−5	35%
−10	40%
−20	45%

ARKLA

frigerant, which is plain tap water, through a matching coil. The coil is added to a new or existing furnace in the house. As the chilled water produced by the unit circulates through the coil, it absorbs heat from the conditioned space. The water, bearing the absorbed heat, is then returned to the unit outdoors, where the heat is dissipated to the outside air. Table 19-A shows the amount of permanent antifreeze required when the outside temperature is below freezing. A defoaming agent also must be added.

Table 19-B shows the specifications of Arkla 3-, 4-, and 5-ton units. Note that the refrigerant is R-717 (ammonia). Also, note the amount of gas needed to produce 3 tons of air conditioning—79,000 Btu.

Gas-fired units may be connected in 5-ton multiples to provide up to 30 tons of air conditioning. Figure 19-3 shows how they are doubled up to provide 10 tons. For some units, it takes 250,000 Btu of gas input per hour to provide 120,000 Btu per hour of cooling. That means 48% efficiency if the electrical energy needed is not accounted for in the figuring. For the unit referred to, the operating voltage is 230 V, with 60-Hz, single-phase operation. Wiring size is not too large, since there is a maximum of 8 A drawn for the condenser fan motors and 33 A for the solution pumps. Normal running current for the solution pump motors is only 5 A. Locked rotor current of 33 A occurs only if the motor is stuck or jammed so it cannot start. The start currents may also reach this 33-A level under some load conditions.

The chilled water system uses stainless steel to prevent problems with rust and other ferrous metal piping problems. There is only one electrical, one gas, and one chilled water supply and return connection for each unit.

CHILLER-HEATER

Some gas-fired units furnish cooling for the summer and heat for the winter. The user changes the functions simply by changing the settings of a room thermostat. The All-Year® units are designed for outdoor installation. They operate on either natural or propane gas.

Changeover Sequence for Chilled Water Operation

When the thermostat calls for cooling, the hot water pump is off. The chilled water pump moves water from the chiller tank and pumps it up a "candy cane" shaped loop and out to the air handler. Fig. 19-4. As the water returns to the chiller tank to be cooled again, it passes through the water reservoir. The water does not flow through the tubes of the hot

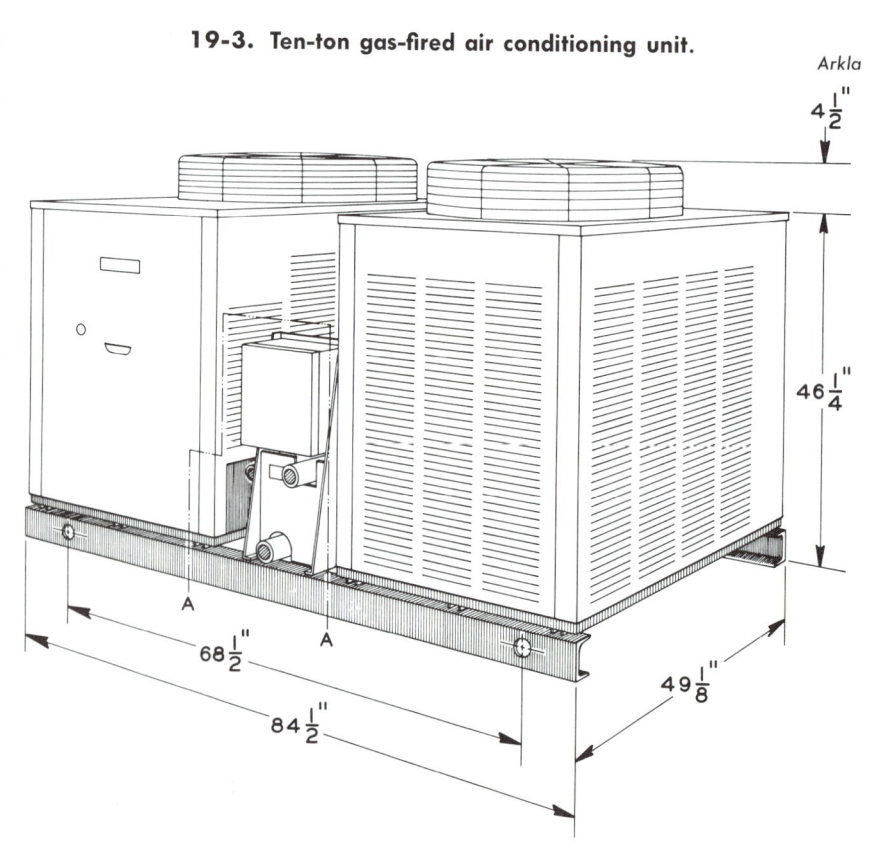

19-3. Ten-ton gas-fired air conditioning unit.

Arkla

490

CHAPTER 19—HEAT PUMPS, GAS-FIRED AIR CONDITIONERS, AND SOLAR AIR CONDITIONERS

Table 19-B.
Specifications for 3-, 4-, and 5-Ton Gas-Fuel Air Conditioning Units

SPECIFICATIONS Performance Ratings	MODEL 3 Ton ACB 36–00	4 Ton ACB 48–00	5 Ton ACB 60–00
Gas input, Btu/h	79,000	100,000	125,000
Delivered capacity, Btu/h*	36,000	48,000	60,000
Condenser entering air temperature	95° F.	95° F.	95° F.
Condenser air flow, cfm, approx.	4,000	6,000	6,000
Chilled water** entering temperature	55° F.	55° F.	55° F.
Chilled water** leaving temperature	45° F.	45° F.	45° F.
Chilled water** flow, gpm	7.2	9.6	12.0
maximum permissible flow, gpm	12.0	12.0	16.0
Allowable friction loss for piping and coil, feet of water	27	26	25
Maximum vertical distance, top of coil above unit base			
with rigid piping, feet	25	25	25
with flexible piping, feet	15	15	15
Do not use ferrous metal piping or tubing.			
Electrical Ratings			
Required voltage, 60 Hz, 1 phase	115	230	230†
Condenser fan motor horsepower	1/3	1/3	1/2
full load/locked rotor amperes, nominal	4.6/9.9	3.5/7.2	3.5/7.2
Pump drive motor horsepower	1/2	1/2	3/4
full load/locked rotor amperes, nominal	7.4/44.8	3.6/21.8	5.1/30.5
Operating wattage draw, typical	875	1,000	1,275
Branch circuit ampacity, minimum	13.85	8.00	9.88
Number and size of time delay fuses	1–20 amperes	2–15 amperes	2–15 amperes
Physical Data			
Refrigerant type	717	717	717
Unit chilled water** volume, gallons	3.0	3.0	5.0
Operating weight, pounds	550	750	775
Shipping weight, pounds	590	825	850
Inlet chilled water connection size, FPT	3/4″	1″	1″
Outlet chilled water connection size, FPT	3/4″	1″	1″
Gas inlet connection size, FPT	1/2″	1/2″	1/2″
Electric entrance knockouts, diameter	7/8″	7/8″	7/8″

*See tables below for additional capacity data.
**"Chilled water" is a solution of good quality tap water and 10% by volume of permanent antifreeze, with sufficient defoaming agent added to prevent foaming.
†ACB 60–00 equipped for 208-volt operation available on special order.

Once Through or Process Applications
Only water or glycol-water mixtures shall be circulated through the unit. In process application where cooling of other solutions is desired, they should be circulated through a secondary heat exchanger.

Refrigeration Capacity
The capacity of the air-cooled chillers varies with ambient air temperature and leaving chilled water temperature. Capacity characteristics of the units are shown in the tables below.

CAPACITY ACB36–00 IN BTU/H

Leaving Chilled Water Temperature (° F.)	Air temperature entering condenser (° F.)			
	90	95	100	105
48	36,720	36,180	34,920	32,652
46	36,650	36,036	34,200	31,320
45	36,576	36,000	33,840	30,600
44	36,500	35,820	33,720	29,592
42	36,360	35,532	32,220	27,432
40	36,000	35,028	30,600	23,400

CAPACITY ACB48–00 IN BTU/H

Leaving Chilled Water Temperature (° F.)	Air temperature entering condenser (° F.)			
	90	95	100	105
48	48,960	48,240	46,560	43,536
46	48,912	48,048	45,600	41,760
45	48,768	48,000	45,120	40,800
44	48,624	47,760	44,496	39,456
42	48,480	47,328	42,960	36,576
40	48,000	46,704	41,300	31,200

CAPACITY ACB60–00 IN BTU/H

Leaving Chilled Water Temperature (° F.)	Air temperature entering condenser (° F.)			
	90	95	100	105
48	61,200	60,300	58,200	54,420
46	61,140	60,060	57,000	52,200
45	60,960	60,000	56,400	51,000
44	60,780	59,700	55,620	49,320
42	60,600	59,160	53,700	45,720
40	60,000	58,380	51,000	39,000

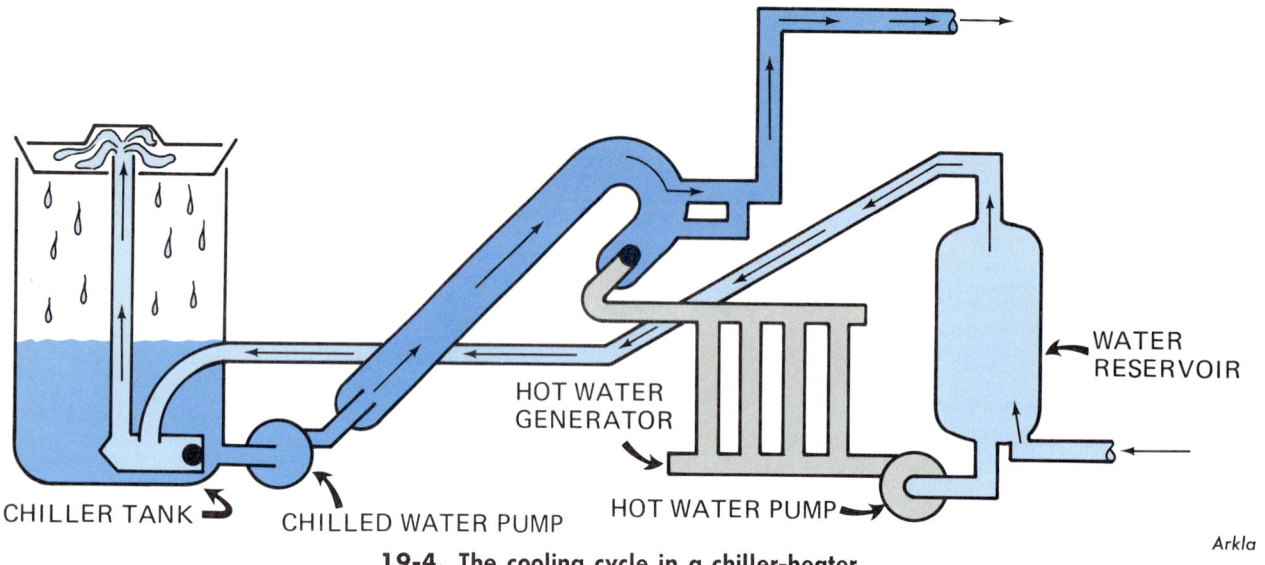

REFRIGERATION AND AIR CONDITIONING TECHNOLOGY

19-4. The cooling cycle in a chiller-heater.

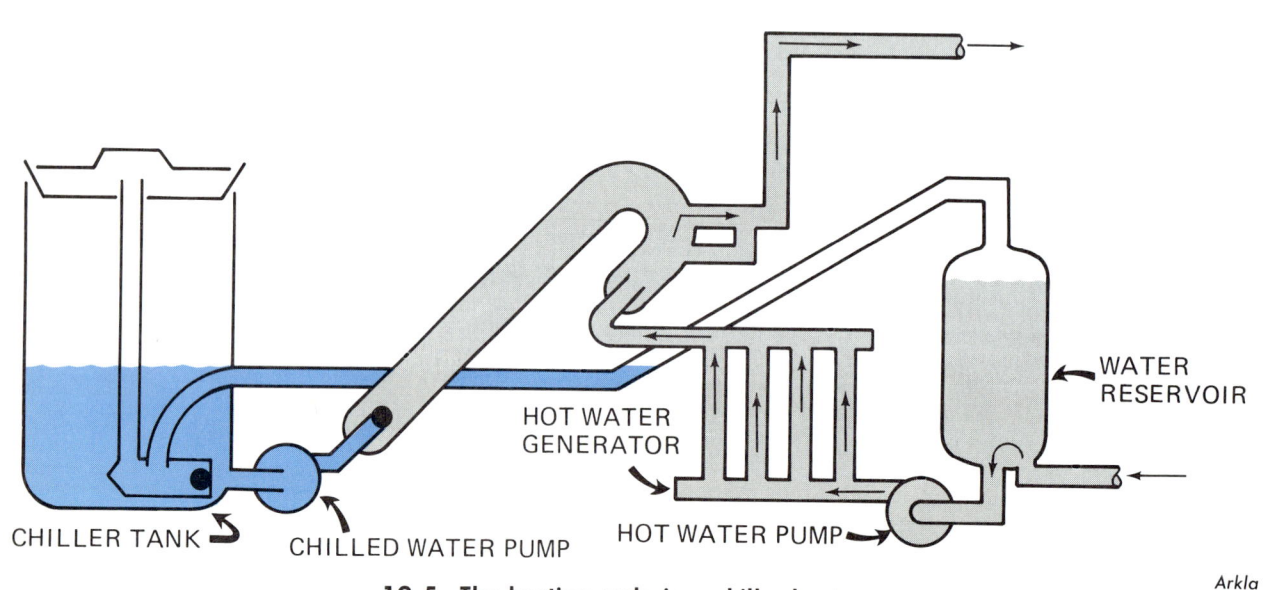

19-5. The heating cycle in a chiller-heater.

water generator as it returns to the chiller tank. The water in the generator is dormant because it is plugged by a check ball. This is held in place by the pressure from the discharge side of the chilled water pump.

Changeover Sequence for Hot Water Operation

When the thermostat is set for heating, the chilled water pump is off. The pressure from the hot water pump moves the check ball to seal off the water in the chiller tank. Fig. 19-5. Now the water in the chiller system is dormant. The hot water pump circulates the water from the hot water generator through the air handler and back to the generator through the water reservoir. Dur-

492

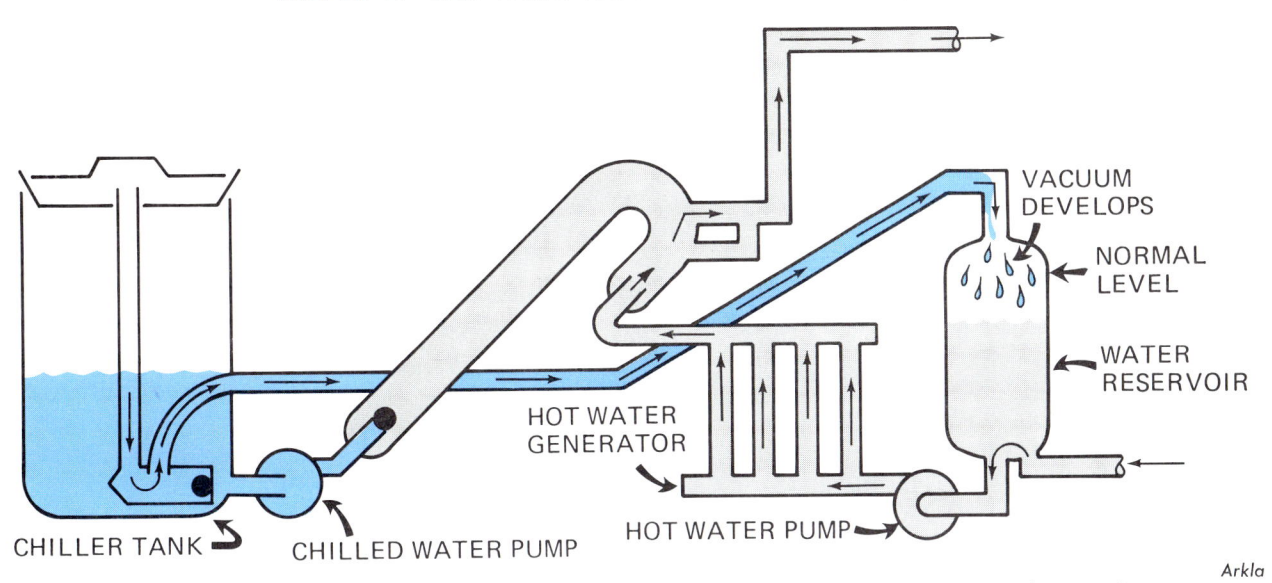

CHAPTER 19—HEAT PUMPS, GAS-FIRED AIR CONDITIONERS, AND SOLAR AIR CONDITIONERS

19-6. Self-leveling feature in a chiller-heater. Vacuum draws water up the air release tube.

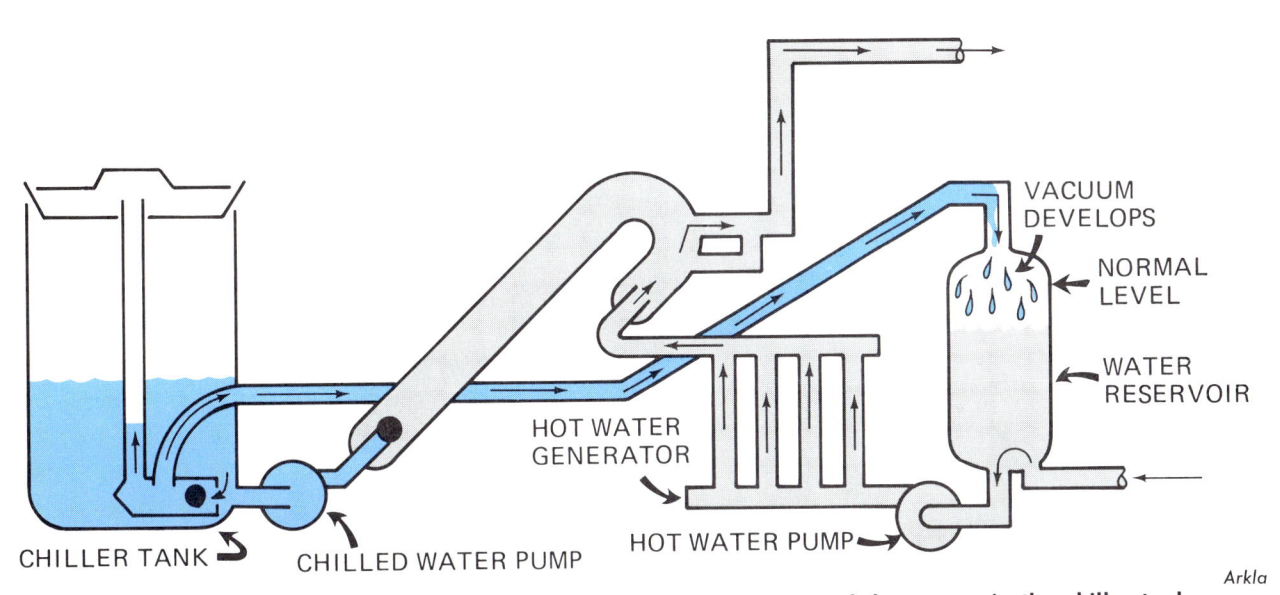

19-7. Self-leveling feature in a chiller-heater. Notice the direction of the arrows in the chiller tank.

ing the heating cycle the reservoir also serves as a place to relieve air from the system. The tube from the top of the reservoir passes through the chiller tank and runs up to the distribution pan, which is open to atmospheric pressure.

Self-Leveling Feature

Self-leveling of water between the chiller tank and the water reservoir during the heating cycle is another unique feature of this system.

Should the water level in the reservoir drop below normal, a vacuum is created in the top of the reservoir. The vacuum causes a negative pressure that acts as a suction to draw water up the air release tube, refilling the reservoir. Fig. 19-6. As the water level in the distribution tube falls below the level of the water in

493

REFRIGERATION AND AIR CONDITIONING TECHNOLOGY

the chiller tank, the second check ball is forced away from the seat to allow the water level to return to normal. Fig. 19-7. It should be remembered that antifreeze and a defoaming agent are necessary for this water system.

ABSORPTION REFRIGERATION MACHINE

The absorption refrigeration machine is used primarily in air conditioning applications. Chilled water is the output of the machine. The chilled water is then used to cool. This particular machine is available for capacities of 100 through 600 tons. These units are small, relatively lightweight, and vibration free. They can be located wherever a source of steam or very hot water is available. Lithium bromide, a salt solution, is used as the absorbent.

Absorption Operation Cycle

Figure 19-8 is a schematic diagram of an absorption cold generator. Note that the evaporator, absorber, concentrator, and condenser are enclosed in a single casing. The heat exchanger is located externally below the main shell.

• *Evaporator.* The evaporator pump circulates the refrigerant (water) from the refrigerant pump into the spray trees. To utilize the maximum surface for evaporation, the refrigerant is sprayed over the evaporator tubes. As the spray contacts the relatively warm surface of the tubes carrying the water to be chilled, a vapor is created. In this manner heat is extracted from the tube surface, chilling the fluid in the tubes. The vapor created in this process passes through eliminators to the absorber.

• *Absorber.* The lithium bromide solution (under proper conditions) keeps the pressure in the absorber section low enough to pull the refrigerant vapor from the high pressure evaporator. As the vapor flows into the absorber, it mixes with the absorbent solution being sprayed over the tube bundle.

• *Heat Exchanger.* The heat ex-

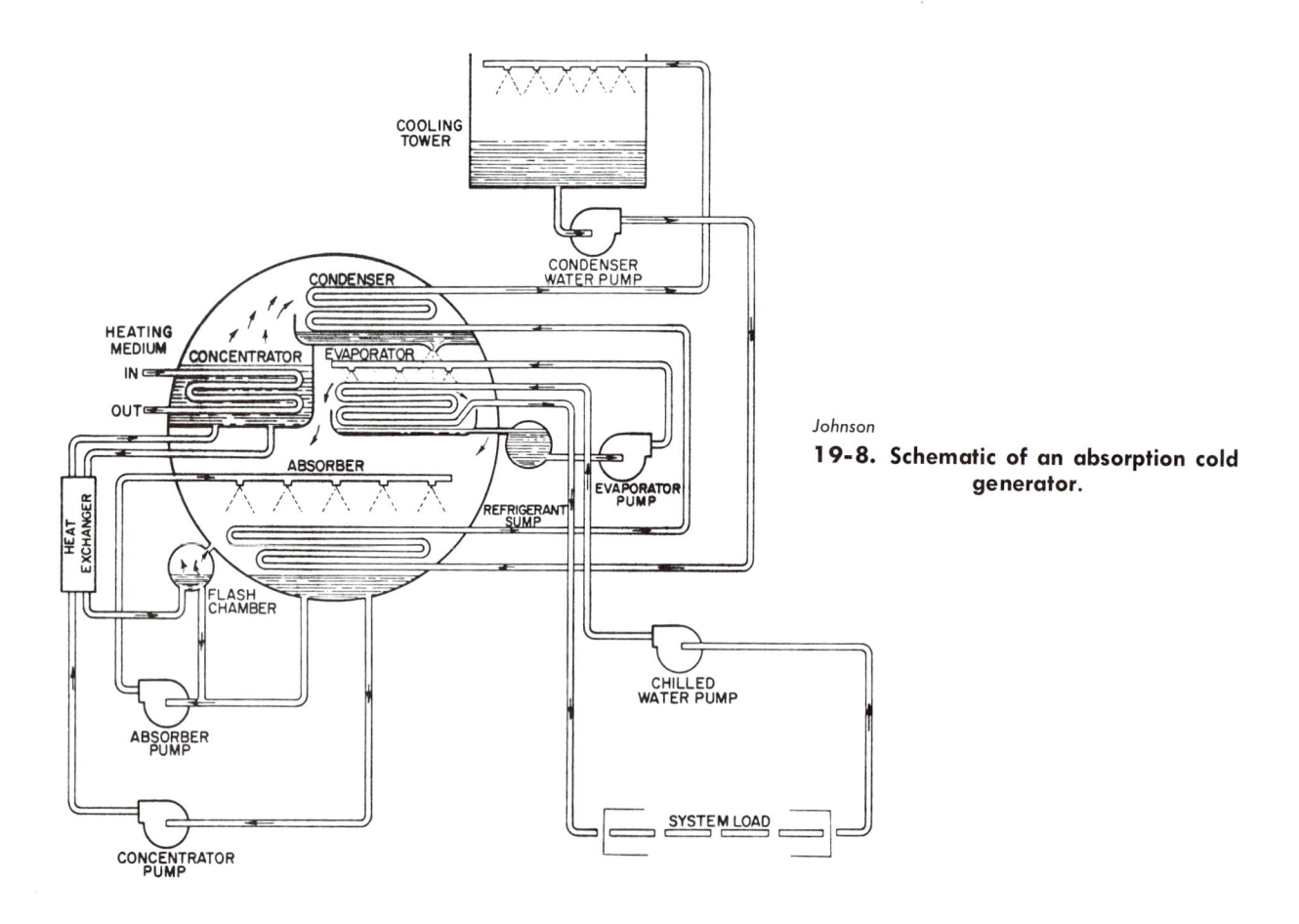

Johnson

19-8. Schematic of an absorption cold generator.

changer is used only as an economizer. The cool diluted solution from the concentrator pump is heated by the hotter concentrated solution moving from the concentrator to the absorber. Steam or hot water (heating medium) is conserved. The heat transfer in the heat exchanger brings the temperature of the diluted solution closer to the boiling point. It also brings the concentrated solution temperature closer to the absorber temperature.

• *Concentrator.* Steam or high temperature water entering the concentrator is controlled to boil off the same quantity of refrigerant picked up by the absorber. The refrigerant vapor is given up by boiling the solution in the concentrator. The vapor passes through eliminators to the tube surface of the condenser.

• *Condenser.* The refrigerant vapor from the concentrator is condensed on the tube surface of the condenser and falls into the pan below the tube bundle.

SOLAR AIR CONDITIONERS

Harnessing the sun's energy is nothing new. As far back as 1878 the sun was used to power a printing plant. In Egypt in 1913, solar energy was used to produce steam to operate an engine that drove an irrigation pump.

In the United States, solar energy powered the phone of a Georgia cotton grower in 1955.

Even then, the costs involved in harnessing solar energy were astronomical in comparison to the costs for the abundantly available fossil fuels—coal, oil, and natural gas. Thus, the research programs languished until the energy crisis of the early 1970s. Then, fossil fuel shortages, environmental concerns, and the rising costs of energy reawakened interest in solar energy. It was known that solar-heated water could provide the power for space heating and water heating. However, could it effectively cool a home? Could the costs of using solar energy be brought within the reach of the average person?

History of Solar Cooling

Prior to 1972, little, if any, research had been done regarding the use of solar energy for cooling. Therefore, no air conditioning equipment specially designed for use with solar energy was available. Currently, however, there are cooling systems that lend themselves to easy modification for use with solar energy. One is the absorption-type system manufactured by Arkla Industries. The other is the Rankine cycle.

Of the two, only the Arkla absorption cooling equipment can use solar-heated water directly to produce cooling.

The Rankine cycle needs an intermediate step. This involves replacing the electric motor in the conventional vapor compression refrigeration cycle with a turbine or using solar cells to produce electricity. In either case, making modifications for the Rankine cycle is more costly than making modifications of the absorption system.

Arkla Industries, aided by a grant from the National Science Foundation, is working on a cooling system specially designed for use with solar energy. At the present time the company is working on a new design in both the residential and medium-tonnage range.

Systems of Solar Cooling

There are two systems used in solar cooling: the *direct* and the *indirect.*

The direct system of application uses the absorption cooling system. It provides higher firing water temperatures directly from the storage tanks to the unit's generator. Fig. 19-9.

The indirect system is a closed system in which a heat exchanger transfers the heat from the solar-heated water storage tanks. This allows the use of an antifreeze fluid. Fig. 19-10.

Lithium-Bromide Water Absorption Cycle

The Arkla-Solaire® unit operates on the absorption principle. Fig. 19-11. It uses solar-heated water as the energy source. Lithium bromide and water are used as the absorbent/refrigerant solution. The refrigeration tonnage is delivered through a chilled water circuit that flows between the unit's evaporator and a standard fan-coil assembly located inside the conditioned space. The heat from the conditioned space is dissipated externally at the cooling tower.

The four main components of the Solaire® cooling unit are the generator, condenser, evaporator, and absorber.

When the solar-heated water enters the tubes inside the generator, the heat from the hot water vaporizes the refrigerant (water), separating it from the absorbent (lithium bromide).

The vaporized refrigerant vapor then flows to the absorber. There, it again liquifies and com-

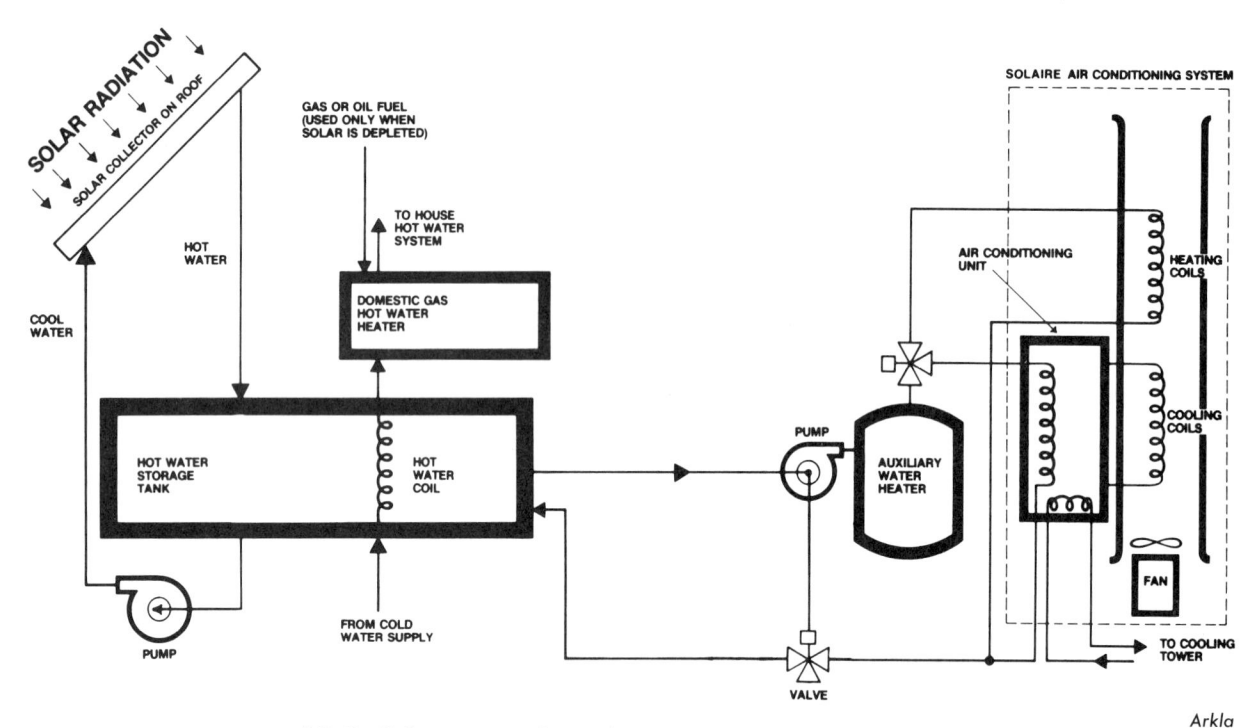

REFRIGERATION AND AIR CONDITIONING TECHNOLOGY

19-9. Solar-energy air conditioning unit—direct system

Arkla

19-10. Solar-energy air conditioning unit—indirect system.

Arkla

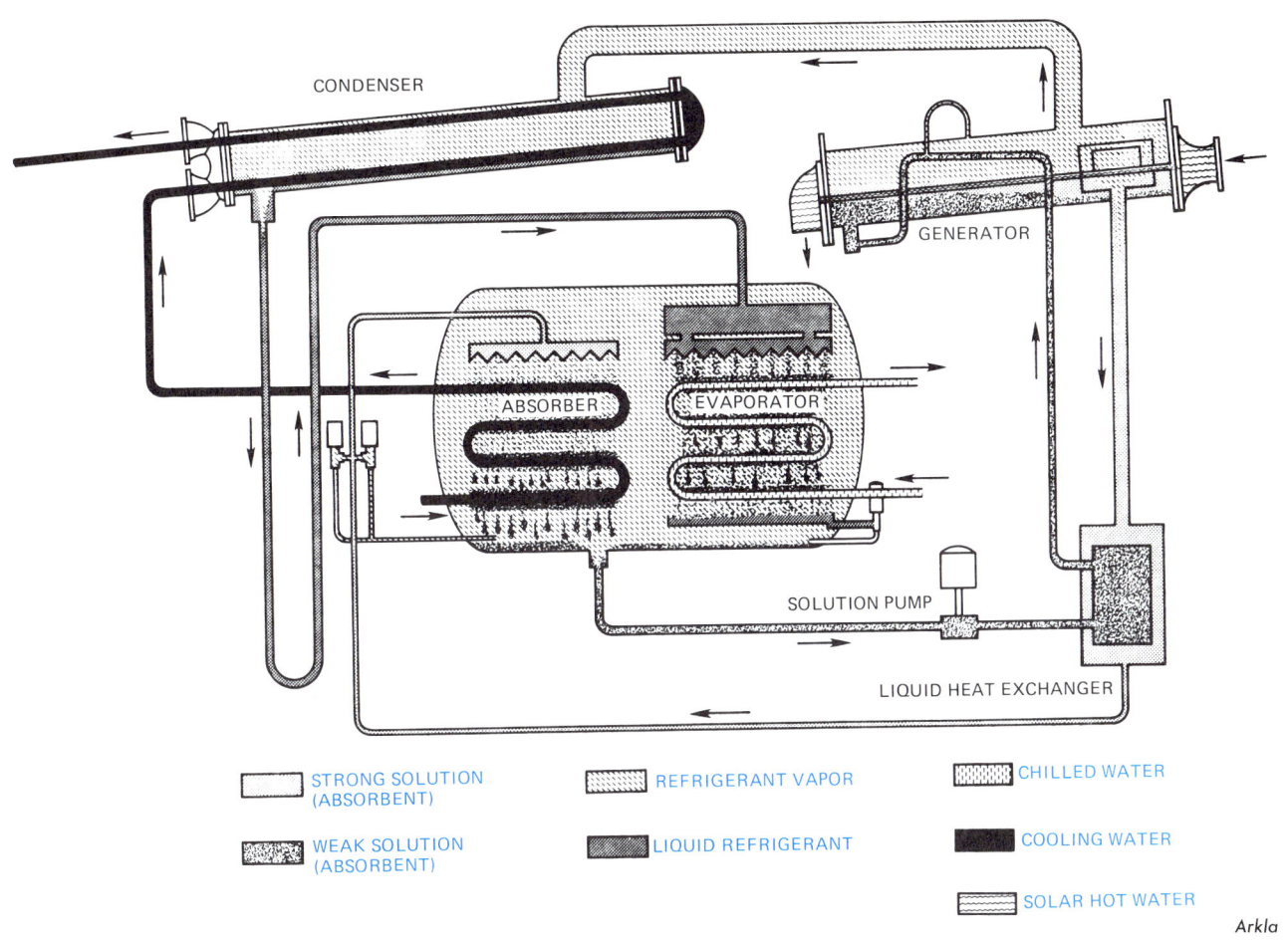

CHAPTER 19—HEAT PUMPS, GAS-FIRED AIR CONDITIONERS, AND SOLAR AIR CONDITIONERS

19-11. Solar air conditioning using lithium bromide and water absorption cycle.

bines with the circulating solution. The reunited lithium bromide and water solution then passes to the liquid heat exchanger. There, it is reheated before being returned to the generator.

Figure 19-12 shows a medium-tonnage air conditioning unit specially designed for use with solar energy. This is a nominally rated 25-ton unit that can operate with a firing water temperature as low as 190° F. [87.8°C].

There is also a three-ton absorption unit that operates with solar-heated water. It provides full capacity with a firing water temperature of 210° F. [98.9°C].

Solar Cooling Research Centers

Figure 19-13 pictures the Decade 80 Solar House, conceived by the Copper Development Association, Inc. Almost everything in this innovative four-bedroom house from the heating/cooling and sound systems to the door chimes and kitchen clock can be run on stored solar energy. Coupled with the solar energy collector system, the heating/cooling system is probably the most challenging technological innovation in the house.

The climate control system consists of two three-ton Solaire WF 501 units for cooling and two duct coils for heating. As designed, it is anticipated that the solar heated water will be able to operate the cooling cycle nearly 75% of the time. It should supply 100% of the heating requirements. A back-up water heater is installed near the 3000-gallon, hot water thermal energy stor-

497

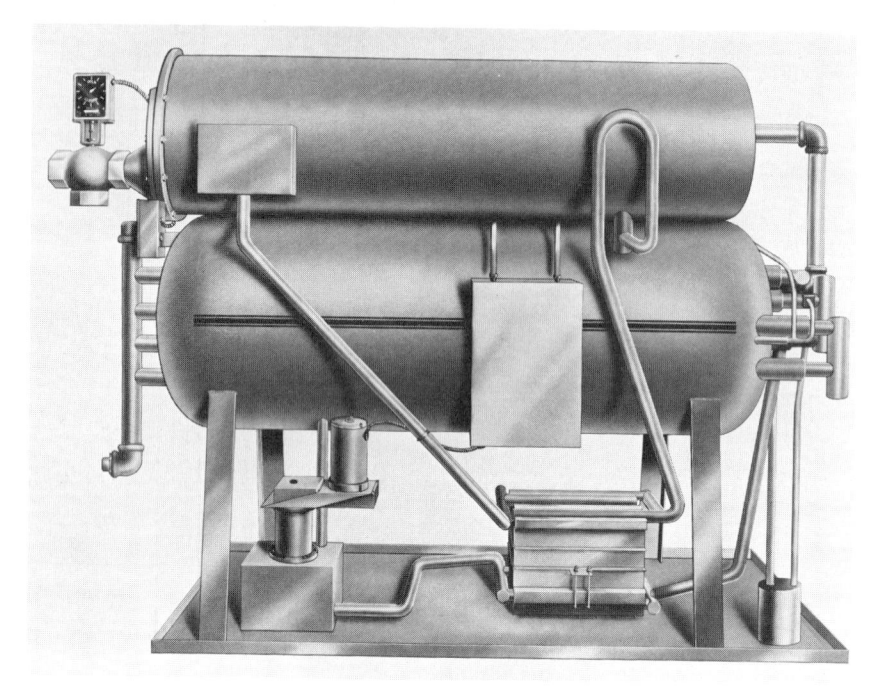

REFRIGERATION AND AIR CONDITIONING TECHNOLOGY

age. The backup unit will function automatically if the water temperature drops too low to operate the climate control system. The house is fully computerized for analysis of hard data on solar energy technology in a normal home environment.

Solar heating and cooling for homes are presently under development and engineering study at a number of locations. A Colorado State University solar research project consists of a residential-type structure. It is designed as a laboratory to test and evaluate the performance of solar equipment for heating and cooling. The solar cooling system is a three-ton lithium bromide absorption unit, modified to use hot water as the heat source.

The Marshall Space Flight Center Solar House is a simulated resi-

Arkla

19-12. A medium-tonnage air conditioning unit specially designed for use with solar energy. This is a 25-ton unit that operates with a firing temperature as low as 190° F.

19-13. Artist's rendering of a solar house in Tucson, Arizona. The four-bedroom home is cooled by two units fired by solar-heated water.

Arkla

19-14. This research solar energy house is cooled by a lithium bromide air conditioner. It uses solar heated water.

dence using three surplus office trailers with a free-standing roof. Fig. 19-14. It has the effective areas and the heating/cooling load equivalent of an average three-bedroom house. A 3500-gallon water tank is used as a heat storage reservoir. A lithium-bromide type of air conditioning system is used.

One of the first schools heated and cooled by solar energy is shown in Fig. 19-15. It uses a lithium bromide system for cooling. The control system's components include 10,000 square feet of collector area and 45,000 gallons of fluid for thermal storage. This experimental project is designed to provide solar heating and cooling and domestic hot water for the 32,000 square foot, one-story school.

A closer look at the solar energy collectors is shown in Fig. 19-16. This is an absorption air conditioner. It is used for both cooling and heating a portion of a training facility for Bell and Gossett ITT.

499

19-15. This school in Atlanta, Georgia, is cooled by solar energy.

19-16. An absorption air conditioner is used to heat this training facility. Solar water heats and cools the house located in Morton Grove, Illinois.

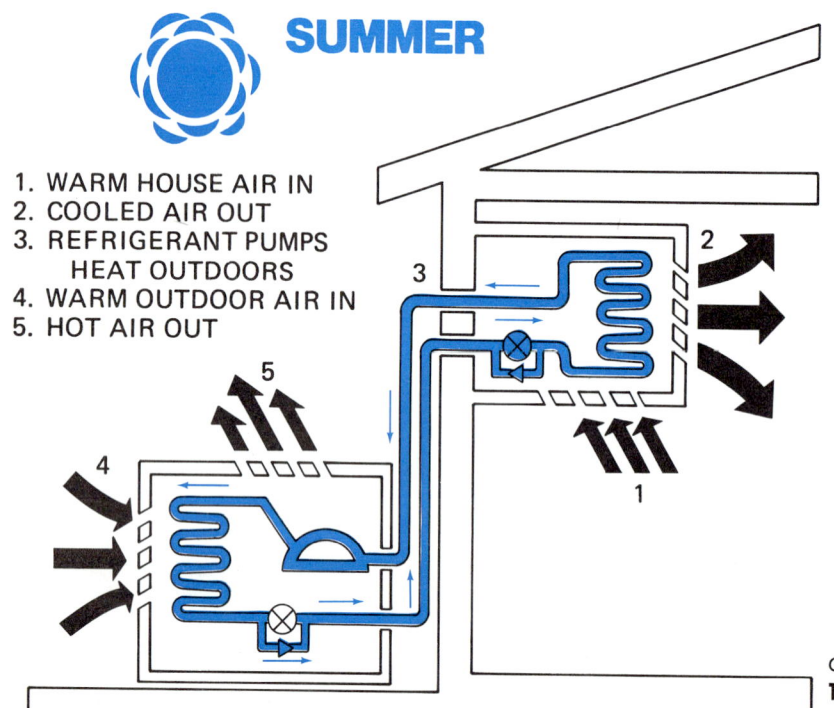

Arkla

SUMMER

1. WARM HOUSE AIR IN
2. COOLED AIR OUT
3. REFRIGERANT PUMPS HEAT OUTDOORS
4. WARM OUTDOOR AIR IN
5. HOT AIR OUT

HEAT PUMPS

The heat pump is primarily a central air conditioner. It can also act as a heating system.

During the cooling season the heat pump performs exactly like a central air conditioner. It removes heat from the indoor air and discharges it outside. Fig. 19-17.

During the heating season, the heat pump reverses its function. It changes from a cooling system to a heating system. It then removes the available heat from the outdoor air and discharges it inside the house. Fig. 19-18.

General Electric
19-17. Operation of a heat pump during the summer.

500

CHAPTER 19—HEAT PUMPS, GAS-FIRED AIR CONDITIONERS, AND SOLAR AIR CONDITIONERS

WINTER

1. COOL OUTDOOR AIR IN
2. COLD AIR OUT
3. REFRIGERANT PUMPS HEAT INDOORS
4. HOUSE AIR IN
5. WARMED HOUSE AIR TO ROOMS

General Electric

19-18. Operation of a heat pump during the winter.

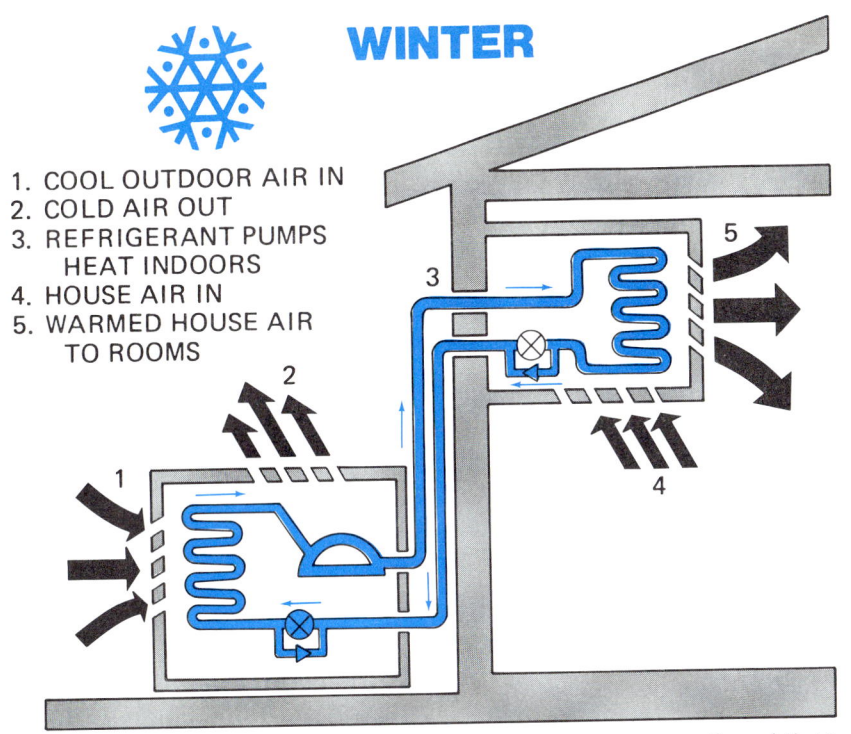

19-19. Outside unit for a heat pump made by Lennox.

Lennox

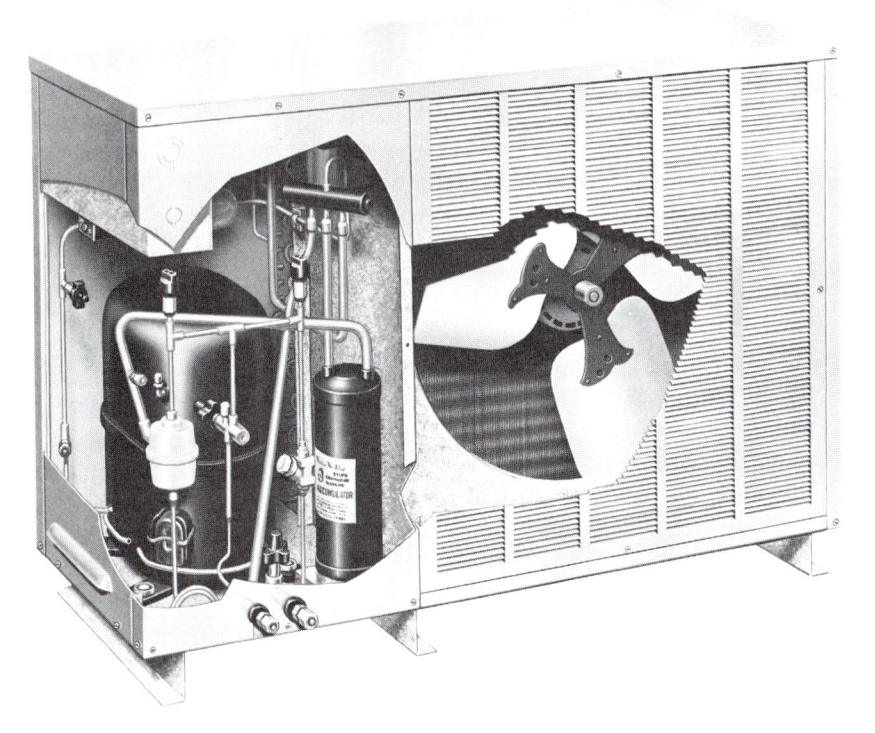

There is heat in outdoor air, even at 0° F. [-17.8°C]. In fact, heat is available in outdoor air down to -460° F. [-273°C].

Since the heat pump is a refrigeration machine, it needs only enough electrical power to run a compressor, an outdoor fan, and an indoor blower.

The result is a heating system with a seasonal efficiency of better than 150%. This means that for every kilowatt of electric power used, the heat pump will produce more than 1.5 kW of heat energy. Only the heat pump can give this level of efficiency.

Heat pumps are available in all sizes for apartments, homes, and commercial applications.

Heat pumps are not new. General Electric has been selling them since 1952. There are now various types of units on the market.

One unit, the Fuelmaster®, works with a heat pump. It can be used with gas, oil, and electric furnaces. Fig. 19-19. As can be seen from the illustration, the heat pump resembles a compressor-condenser unit. However, the control box is different. Fig. 19-20. The control box has relays and terminal strips factory-installed and wired. The heat pump delay and defrost limit control are included in the unit.

Operation

On mild temperature heating days, the heat pump handles all heating needs. When the outdoor temperature reaches the "balance point" of the home (heat loss equals heat pump heating capacity), the two-stage indoor thermostat activates the furnace (secondary heat source). When the furnace fires, a heat relay de-energizes the heat pump.

501

REFRIGERATION AND AIR CONDITIONING TECHNOLOGY

19-20. Control box for a heat pump add-on unit made by Lennox. *Lennox*

When the second stage (furnace) need is satisfied and plenum temperature has cooled to 90 to 100° F. [32.2 to 37.8°C], the heat pump delay turns the heat pump back on. It controls the conditioned space until the second stage (full heat) operation is required again.

When outdoor temperature drops below the setting of the low-temperature compressor monitor (field-installed option) the control shuts out the heat pump. The furnace handles all of the heating need. The low-temperature compressor monitor will be standard on models dated 1974 and after.

During the cooling season the heat pump operates in its normal cooling mode. It uses the furnace blower as the primary air mover. Fig. 19-21.

Defrost

During a defrost cycle, the heat pump switches from heating to cooling. To prevent cool air from being circulated when heating is needed, the control automatically turns on the furnace to compensate for the heat pump defrost cycle. (Most modern heat pump systems do the same thing with strip heating.) When supply air temperature climbs above 110 to 120° F. [43.3 to 48.9°C], the defrost limit control turns off the furnace and keeps indoor air from getting too warm.

After a defrost cycle, the air temperature downstream of the coil may go above the 115° F. [46.1°C] closing point of the heat pump delay. Then, the compressor will stop until the heat exchanger has cooled to 90° to 100° F. [32.2 to 37.8°C], as it does during normal cycling operation between furnace and heat pump.

Outdoor Thermostat

In a straight heat pump/supplementary electric heater application, at least one outdoor thermostat is required to cycle the heaters as the outdoor temperature drops. In the Fuelmaster® system, the indoor thermostat controls the supplemental heat source (furnace). The outdoor thermostat is not required.

Since the furnace is serving as the secondary heat source, the Fuelmaster® system does not require the home rewiring usually associated with supplemental electric strip heating.

Special Requirements of Heat Pump Systems

The installation, maintenance, and operating efficiency of the heat pump system are like those of no other comfort system. A heat pump system requires the same air quantity for heating and cooling. Because of this, the air-moving capability of an existing furnace is extremely important. It should be carefully checked before a heat pump is added. Heating and load calculations must be accurate. System design and installation must be precise.

The air distribution system and diffuser location are equally important. Supply ducts must be

Typical Fuelmaster Components

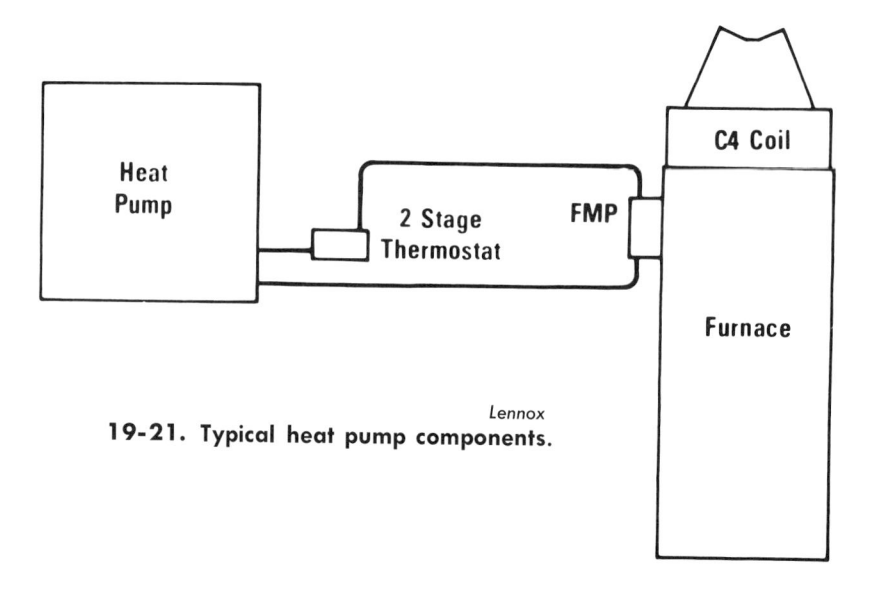

19-21. Typical heat pump components. *Lennox*

CHAPTER 19—HEAT PUMPS, GAS-FIRED AIR CONDITIONERS, AND SOLAR AIR CONDITIONERS

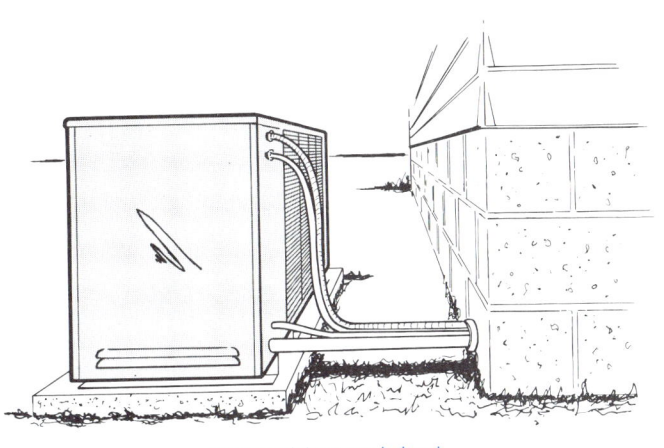

Unit on slab at grade level

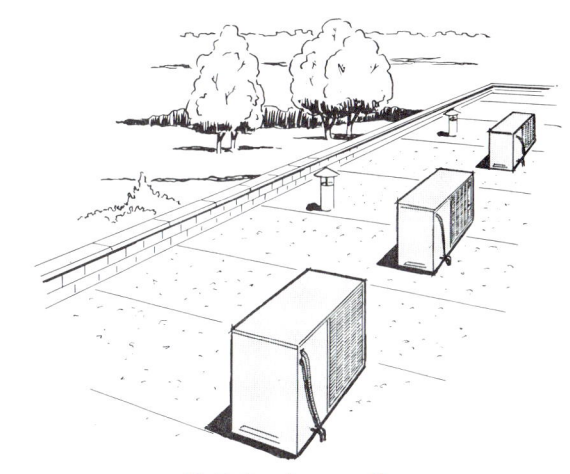

Multiple units on rooftop

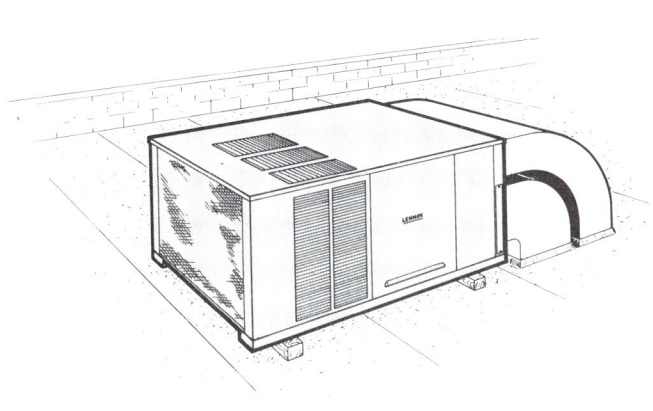

Rooftop installation

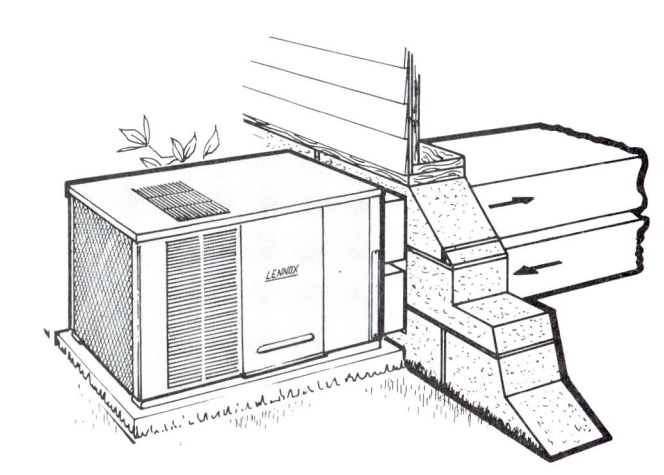

Unit on slab at grade level

Lennox

19-22. Typical installations of heat pumps.

properly sized and insulated. Adequate return air is also a must.

Heating supply air is cooler than with other systems. This is quite noticeable to homeowners accustomed to gas or oil heat. This makes diffuser location and system balancing critical. Typical installations of heat pumps are shown in Fig. 19-22.

Sizing Equipment

Home insulation, exposure, design and construction, climate, and living habits determine the efficiency of a heat pump system. Each heat pump installation is unique. Each job must be calculated carefully. Remember, there are no *rules of thumb* for heat pump sizing.

To determine the most economical operating cost, size the system for its cooling load. Use supplemental heating (second stage) to make up the difference between heat pump heating capacity and building heat loss.

Sizing the heat pump to handle the entire heat loss will result in oversized cooling capacity that will adversely affect dehumidification during cooling.

Defrost Cycle

During a heat pump heating cycle, the outdoor coil absorbs heat from outdoor air. To do this, the coil must be cooler than the air.

When air temperature falls below 40° F. [4.4°C], coil surface temperature is below freezing. Moisture in the air freezes on the coil. Frost or ice builds up, reducing the air passage through the coil and cutting heat pump output.

The heat pump defrost cycle removes this buildup. The system is reversed to a cooling cycle, which heats the coil and melts the ice. Supplemental heat is used to counteract the cooling effect of the cycle change.

Balance Point

The outdoor temperature at which heat loss of a building and heating output of a heat pump are equal is called the balance point. It is the lowest temperature at which the heat pump alone can handle the heating load.

REVIEW QUESTIONS

1. What are the three main types of gas air conditioning cycles used today?
2. What are the two basic principles of operation of the absorption-type air conditioner?
3. What type of refrigerant does an absorption-type unit use?
4. What type of refrigerant is used in a gas-fired system?
5. How can problems from rust and other ferrous metals be prevented in a chilled water system?
6. What is a concentrator?
7. What two types of systems are used in solar cooling?
8. What determines the efficiency of a heat pump system?

20

Load Estimation and Insulation

REFRIGERATION AND AIR CONDITIONING LOAD

The load for a refrigeration unit comes from many sources. This load comes from several heat sources. The more common sources of heat are:

- Heat from the outside leaks in through doors and windows or is conducted through the insulated walls.
- Transparent materials allow heat to penetrate them. This occurs when windows are used in a refrigerated space.
- Open doors and windows may allow heat to enter a refrigerated place. Cracks around the doors and windows also allow heat to enter the refrigerated space.
- The materials (products) stored in the refrigerated space give off heat. To lower the product temperature, it is necessary to lower the heat content of the materials.
- People who occupy a cooled space give off heat. This must be considered when figuring any load for a given refrigeration unit.
- Equipment inside the refrigerated space may give off heat. For example, motors, electric lights, electronic equipment, steam tables, urns, hair dryers, and similar items give off heat.

To obtain an accurate figure, it is necessary to consider all of the heat sources. This will, in turn, determine how long the equipment must run to maintain a given temperature in the space being cooled.

RUNNING TIME

The time necessary for the cooling equipment to maintain or come down to a certain temperature is called the *running time*. The time used for calculations is 24 hours. Equipment capacity is rated in Btu per hour. Therefore, 24 hours times the Btu per hour will produce the normal capacity of refrigeration equipment. A quick way to determine this is to use the following formula:

$$\frac{\text{Required Btu/hr}}{\text{Equipment Capacity}} = \frac{\text{Total cooling load in Btu for 24 hours}}{\text{Desired running time}}$$

Most equipment cannot run for 24 hours, since defrosting consumes some of the time. That is why the total cooling load for 24 hours is divided by the desired running time.

Moisture taken from the stored product causes frost to form on the evaporator coils or surface. This frost must be removed to maintain the efficiency of the unit. The defrost cycle must be determined by the amount of frost that will form on the coils. The refrigeration process stops while defrosting is being accomplished.

A system may be stopped long enough for the frost to melt. This is not the desired method in most cases. It may allow the product's temperature to rise, thus spoiling the product. To speed up the defrosting process, a heating element is usually introduced into the system. The heating element melts the frost rapidly. The water is drained to the outside of the unit. The *off-cycle* type of defrost is time consuming. It usually takes 8 hours. This means the loss of 8 hours of refrigeration or a total running time of 16 hours for the equipment. (It takes 8 hours for defrosting and about 8 hours to bring the temperature

505

back to the previous point.) Where heated defrost methods are used, it usually takes about 6 hours on the average. That produces an equipment running time of 18 hours. This is usually taken as average for calculations. Several heating methods are used. Electric heating elements may be mounted near the evaporator surfaces or hot gas may be recirculated to produce the same effect. This method is explored in detail in Chapter 10.

In air conditioning equipment, the temperature of the coil rarely gets below 40° F. [4.4°C]. Thus, no frost accumulates. This means, in most cases, that a defrost method is not necessary in air conditioning equipment. Air conditioning equipment is designed for continuous running. The running time is determined by the Btu actually needed to cool a room.

CALCULATING COOLING LOAD

The individual loads should be figured first. These are then totaled. This produces the load to be used for figuring running time and equipment design characteristics.

The four sources of load are wall gain load, air change load, product load, and miscellaneous load.

Wall Gain Load

Heat that leaks through the wall is the *wall gain load*. This heat comes from outside the refrigerated area. There is no perfect insulation. Thus, there is always some heat leakage through the walls. Heat always moves toward a less heated (cooled) area. So, if the inside of a space is cooler than the outside, there is always movement of heat from the warmer to the cooler area. Insulation is used to slow down this heat movement. Air conditioners and commercial refrigeration systems are always subject to wall gain or heat gain from outside the cooled area.

Air Change Load

The *air change load* originates when the door is opened to a refrigerated space. The warm moist air that enters the area must be cooled to the inside temperature. This cooling presents a load to the equipment.

In some cases, this is not a factor. In the case of chillers, there is no opening through which air can pass. Thus, this type of load does not exist. However, this is not the case with air conditioning equipment. The cracks around doors and windows also add to the load in an air conditioned space. In some cases air is introduced from outside to improve ventilation. This is especially true in air conditioning systems. The outside air must be cooled. Therefore, it presents a load to the air conditioner.

The introduction of air for ventilation is the ventilation load. The air that leaks around doors and windows is the infiltration load. Every air conditioning system must deal with this type of load.

Most commercial refrigerators have well-fitted door gaskets. Thus, they have very little infiltration load. Air changes here are the result of opening and closing the door or doors.

Product Load

Any material stored in a refrigerated space must be brought down to the temperature of the inside space if it is not already to that temperature. In some cases the temperature of the product is lower than the inside temperature. This means it can also add to the cooling process. However, in most cases the temperature of such a product is not taken into consideration. The refrigeration process it contributes to is gradual and, usually, slowly diminishing.

Once the product is cooled to the temperature of the refrigerated space it is not longer a part of the load.

Fruits and vegetables give off respiration heat the entire time they are in storage. They give off heat even though they reach the temperature of the storage area. There is no further decrease in their temperature, however.

In some instances a product will give off heat all the time it is stored. In this case it is best to place it in a chiller first, then transfer it to cold storage.

Air conditioning has no product load as such. However, there is often a *pull-down load*. This is thought of as a *product load*.

Miscellaneous Load

Heat from electrical equipment, electric lights, and people working in a refrigerated place is thought of as a *miscellaneous load*.

In an air conditioned space there is no miscellaneous load. It has been taken care of previously. The people in the air conditioned space make up the major part of the load. In fact, human occupancy is the primary load on most air conditioned spaces. There are exceptions in which electrical equipment or other types of equipment is the entire load. This would occur when the air conditioning system was designed solely for cooling some type of equipment for better operation.

CALCULATING HEAT LEAKAGE

It is difficult to estimate some loads with accuracy. Heat leakage can be estimated with some degree of accuracy. Heat leakage through walls, floors, and ceilings depends on the insulating material and its thickness. The formula for determining heat leakage is:

$$H = KA(t_1 - t_2)$$

H is heat leakage.

K is the heat transfer coefficient in Btu per square foot per degree Fahrenheit.

A is the area in square feet.

$t_1 - t_2$ is the temperature gradient through the wall. It is expressed in degrees Fahrenheit.

In this example, the wall is made of 6″ of concrete and 4″ of cork. What is the heat transfer coefficient of the wall? Assume that the concrete wall is an outside wall. It is exposed to air circulation from the environment. The transfer of heat through the air film next to a concrete and cork wall is about 4.2 Btu per square foot per hour for each degree difference in temperature between the inside and outside. This information can be found in handbooks for engineers. The resistance is the reciprocal of this, or

$$\frac{1}{4.2} = 0.24$$

(0.238095 rounded to 0.24)

The coefficient for concrete is about 8 Btu per square foot per hour per degree difference per inch of thickness, or $\frac{1}{8} = 0.125$.

The resistance is 0.125.

The resistance of a 6″ wall is six times greater (6 × 0.125), or 0.75. The coefficient for cork is about 0.31 Btu per square foot per hour per degree difference per inch of thickness, or $\frac{1}{0.31} = 3.225806$. That means it has a resistance of $\frac{1}{0.31}$, or 3.225806.

The resistance for a 4″ wall is 4 × 3.225806, or 12.903224. The inside wall contacts the still air.

Experience has shown that an average value of the coefficient for the film is about 1.4. The resistance then is $\frac{1}{1.4}$, or 0.714286.

The total resistance for this wall is 0.24 + 0.75 + 12.9 + 0.71 = 14.60. Therefore, the overall coefficient is $\frac{1}{14.60}$, or 0.068493. It is apparent that the principal resistance is offered by the cork. By using the same method, it is possible to obtain the overall coefficient for any type of wall.

Once you have found the heat transfer coefficient, the heat leakage can be found by the use of the formula. Suppose you have a room 20′ by 20′ and 8′ high. The walls of the room are of 6″ concrete and 4″ cork. The perimeter of the room is 4 sides × 20′, or 80′. The total wall surface is 8′ × 80′, or 640 square feet. If the outside temperature is 75° F. and the inside temperature is 30° F., the heat leakage through the walls is found by using the following formula:

$$H = 0.068493 \times 640 \times (75-30)$$
$$= 1972.5984 \text{ Btu per hour}$$

To find the total heat leakage, you must also figure the heat leakage from the floor and the ceiling. Once the ceiling and floor leakages have been added, then the refrigeration needed to cool the product must be added to obtain the total load on the refrigeration system.

CALCULATING PRODUCT COOLING LOAD

Product cooling load can be figured also. Heat emitted from the product to be cooled can be calculated. The amount of product per locker should be known. Most tables indicate that the average locker user will place an average of 2 to 2.5 pounds of product per day in the locker storage compartment. This means that, in a chill room having a 300-locker installation, the daily load would be 300 × 2.5, or 750 pounds. The initial temperature may be as high as 95° F. The final temperature can be assumed to be 36° F.

Various kinds of products will vary in terms of specific heat. However, an average specific heat of 0.7 is generally used as a value for making calculations. For a 300-locker unit, the daily product cooling load would be 0.7 × 750 (95–36), or 30,975 Btu per day.

Heat change loads are caused by the entrance of warm air when the doors are opened. This load is difficult to estimate accurately. It is affected by room usage, interior volume, whether or not the room is entered through an outside door, the size of the door, and how many times the door is opened.

EXAMPLE. When the temperature of meat reaches 35° F., the meat is moved to the cold storage room. During preparation for storage, the meat may warm up to 40 or 50° F. That means the meat must be cooled to 32° F. before it will begin to freeze. The average heat of fusion amounts to about 90 Btu per pound for meat when the latent heat of fusion is being removed. In this case, the heat of fusion is the heat that is removed at the freezing point before the meat is frozen. After freezing, the meat is subcooled to the quick-freeze temperature. It then has a specific heat on the average of 0.4.

It is possible to calculate the amount of heat removed from a pound of meat. Assume the final temperature of the quick freeze to be $-10°$ F. It is possible to obtain the amount of heat removed.

Capacity of the Machines Used in the System

The capacity of any refrigerating compressor depends on its running speed and the number and size of its cylinders. The efficiency of the compressor must be considered. The number of hours of operation per day and the suction and discharge pressure play an important role in the capacity of the machine. Capacity rating is usually based on conditions standardized by ASHRAE and ARI. (ASHRAE is the acronym for the American Society of Heating, Refrigeration, and Air Conditioning Engineers. ARI is the acronym for the Air Conditioning and Refrigeration Institute.) These standards call for compressor suction of 5° F. and

Usage factors are available from tables developed by ASHRAE. Table 20-A. These factors vary with the difference in temperature between the outside and inside of the cooler. An allowance is made for normal and heavy usage. There is no safety factor in this equation. To find the average daily load, divide the total loading factor by the desired operating time for the equipment. Equipment is selected from information supplied by manufacturers of the cooling units.

When determining the miscellaneous load, you will need a constant that has been found to be very reliable. This constant is that one watt of electrical energy produces 3.415 Btu. Thus, a 25-W bulb will generate 85.375 Btu (25 × 3.415).

When you have determined the load requirements, you will then need to check the manufacturer's recommendations for a particular unit to match the condensing unit to the load. For instance, Table 20-B shows that a 1-horsepower unit provides 8190 Btu per hour. A 2-horsepower unit provides 16,150 Btu per hour. For example, if a load calls for 16,000 Btu, then a 2-horsepower unit will suffice. Some allowance should be made for the load factor changing under maximum load conditions. It is, of course, wise to know exactly what the cooler will be used for before choosing the condensing unit.

The size of the condensing unit can also be found by dividing the Btu needed to cool the load by 12,000. This will give the horsepower rating of the condensing unit. This works since 12,000 Btu per horsepower is an industry standard. A ton of ice

---

Cooling the meat to the freezing point, remove 0.7
$(45 - 28)$ [0.7 = specific heat; 45 = average of 40
and 50; 28 = temperature just below freezing point] = 11.9 Btu

Freezing the meat (latent heat of fusion) = 90.0 Btu

Subcooling the meat to $-10°$ F. = 0.4 $(28 + 10)$ = 15.2 Btu
Total per pound 117.1 Btu

If you allow 2.5 pounds of meat per locker, the quick freezer in a 300-locker installation should have a capacity to freeze 750 pounds of meat per day (300 × 2.5). In this case, the product load would be 87,825 Btu per day (750 × 117.1).

Since miscellaneous loads cannot be accurately calculated, locker and freezer doors should be opened no more than is necessary. This will keep the load due to such openings to a minimum. Experience indicates that no more than 15% to 20% of the leakage load is caused by such openings.

19.6 pounds gage pressure and discharge pressure of 86° F. and 154.5 pounds gage pressure.

In the case of smaller coolers (1600 cubic feet), there is a shorter method for figuring the load. If these are used for general purpose cooling and storage, the product load is difficult to determine. It may vary. In these cases, an average is used. The wall gain load and the usage or service load are used to determine the total load.

Calculate the wall gain as previously shown. The usage load will equal the interior volume multiplied by a usage factor.

Table 20-A.
Usage Factors in Btu per 24 Hours per Cubic Foot Interior Capacity

Inside Volume	Type of Service	\multicolumn{9}{c}{Temperature Difference (Room temperature minus refrigerator temperature)}
		40° F.	45° F.	50° F.	55° F.	60° F.	65° F.	70° F.	75° F.	80° F.
15 cu. ft.	Normal	108	122	135	149	162	176	189	203	216
	Heavy	134	151	168	184	201	218	235	251	268
50 cu. ft.	Normal	97	109	121	133	145	157	169	182	194
	Heavy	124	140	155	171	186	202	217	233	248
100 cu. ft.	Normal	85	96	107	117	128	138	149	160	170
	Heavy	114	128	143	157	171	185	200	214	228
200 cu. ft.	Normal	74	83	93	102	111	120	130	139	148
	Heavy	104	117	130	143	156	169	182	195	208
300 cu. ft.	Normal	68	77	85	94	102	111	119	218	136
	Heavy	98	110	123	135	147	159	172	184	196
400 cu. ft.	Normal	65	73	81	89	97	105	113	122	130
	Heavy	95	107	119	130	142	154	166	178	190

Table 20-B.
Capacity of Typical Air-Cooled Condensing Units

Condensing Unit Horsepower	Capacity per Hour in Btu
1/3	2,460
1/2	4,010
3/4	5,820
1	8,190
1 1/2	12,050
2	16,150

melting in 1 hour will remove 12,000 Btu of heat from the area in which it is located.

AIR DOORS

One of the ways to minimize temperature losses and prevent entering warm moist air is by using an air door. An air door also provides protection against insects, dust, dirt, and fumes. It provides an invisible barrier affording people an unobstructed view of the work area. It ensures a constant interior temperature by preventing the entry of hot or cold air. Fig. 20-1.

An air door can be used to seal in cold air and save energy by preventing excessive operation of the refrigeration system.

Refrigeration coolers and freezers use the air door to maintain interior temperature. They stop the entry of warm moist air when doors are opened to the cooler. They lessen the frequency of expensive defrosting and prevent the refrigeration system from overloading. It is easier for personnel to see who is coming in or going out of the cooler. Thus, accident prevention is a good by-product of the air door installation.

The main component of an air door is a fan mounted in a unit. The fan controls the volume of air directed downward. This air seals off an area from any temperature change for short periods. Air doors are used in meat packing plants, food processing plants, supermarkets, commissaries, restaurants, hospitals, cold storage plants, and breweries.

INSULATION

Insulation is needed to prevent the penetration of heat through a wall or air hole into a cooled space. There are several insulation materials, such as wood, plastic, concrete and brick. Each has its application. However, more effective materials are being developed and made available.

Sheet Insulation

Vascocel® is an expanded, closed cell, sponge rubber that is made in a continuous sheet form 36" wide. It comes in a wide range of thicknesses (3/8", 1/2", and 3/4"). This material is designed primarily for insulating oversize pipes, large tanks and vessels, and other similar medium- and low-temperature areas. Because of its availability on continuous rolls, this material lends itself ideally to application on large air ducts and irregular shapes.

20-1. Using an air door to keep cold air inside a refrigerated area.
Virginia Chemicals

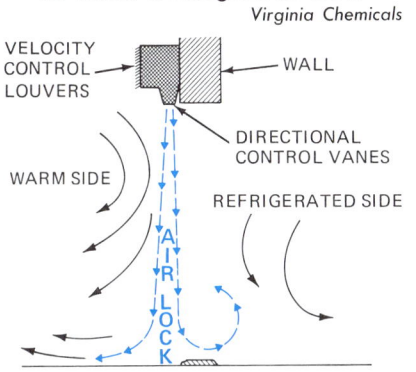

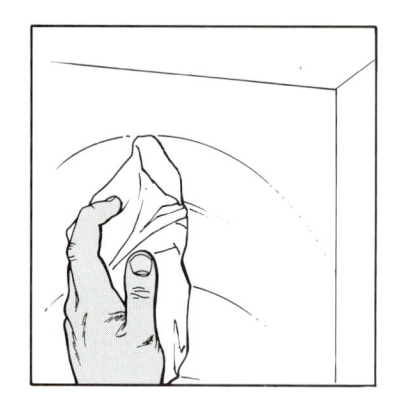

REFRIGERATION AND AIR CONDITIONING TECHNOLOGY

20-2. Installing sheet insulation. (A) Prepare the surfaces for application of the sheet insulation by wiping with a soft, dry cloth to remove any dust or foreign matter. Use a solvent to remove grease or oil. (B) Apply the adhesive in a thin, even coat to the surface to be insulated. (C) Position the sheet of insulation over the surface and then simply smooth it in place. The adhesive is a contact type. The sheet must be correctly positioned before it contacts the surface. (D) Check for adhesion of ends and edges. The surface can be painted.

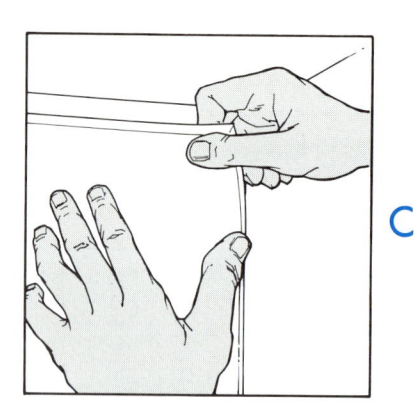

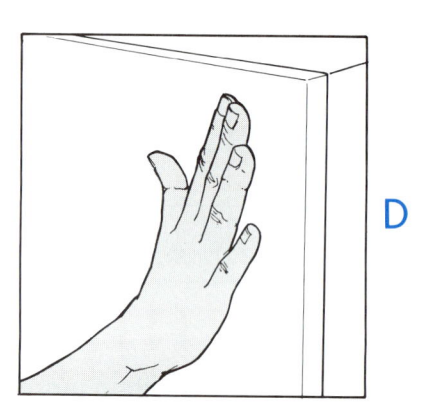

This material is similar to its companion product, Vascocel® tubing. It may be cut and worked with ordinary hand tools such as scissors or a knife. The sheet stock is easily applied to clean, dry surfaces with an adhesive. Fig. 20-2. The *k* factor (heat transfer coefficient) of this material is 0.23. It has some advantages over other materials. It is resistant to water penetration, water absorption, and physical abrasion.

Tubing Insulation

Insulation tape is a special synthetic rubber and cork compound designed to prevent condensation on pipes and tubing. It is usually soft and pliable. Thus, it can be molded to fit around fittings and connections. There are many uses for this type of insulation. It can be used on hot or cold pipes or tubing. It is used in residential buildings, air conditioning units, and commercial installations. It comes in 2″ wide rolls that are 30′ long. The tape is $\frac{1}{8}$″ thick. If stored or used in temperatures under 90° F. [32.2°C], the lifetime is indefinite.

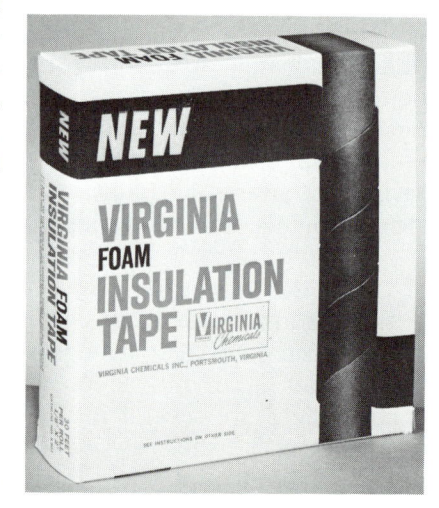

Virginia Chemicals
20-3. Foam insulation tape.

510

CHAPTER 20—LOAD ESTIMATION AND INSULATION

20-4. Slugs of insulation material and cords are workable into locations where sheet material cannot fit.

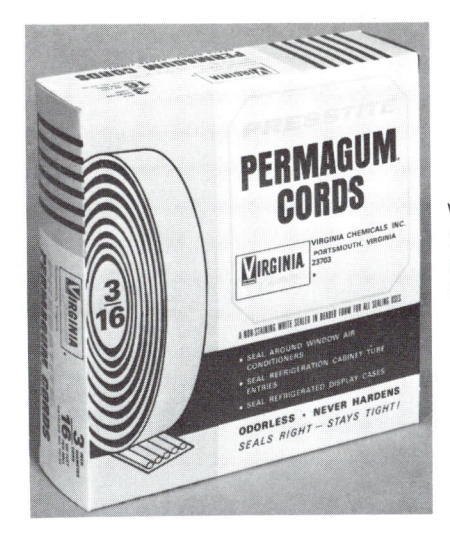

Virginia Chemicals

Foam insulation tape is made specifically for wrapping cold pipes to prevent pipe sweat. Fig. 20-3. It can be used to hold down heat loss on hot pipes below 180° F. [82.2°C]. It can be cut in pieces and easily molded around fittings and valves. It adheres to itself and clean metal surfaces. It is wrapped over pipes with about 1/4" overlap on each successive lap. Remember one precaution: Never wrap two or more parallel runs of tubing or pipe together, leaving air voids under the tape. Fill the voids between the pipes with Permagum® before wrapping. This will prevent moisture from collecting in the air spaces. This foam insulation tape has a unicellular composition. The *k* factor is 0.26 at 75° F. [23.9°C].

Permagum® is a nonhardening, water-resistant sealing compound. It is formulated to be nonstaining, nonbleeding, and to have excellent adhesion to most clean surfaces. It comes in containers in either slugs or cords. Fig. 20-4.

This sealer is used to seal metal to metal joints in air conditioners, freezers, and coolers. It can seal metal to wood joints and set plastic and glass windows in wood or metal frames. It can be used to seal electrical or wire entries in air conditioning installations or in freezers. It can be worked into various spaces. It comes with a paper backing so that it will not stick to itself.

Extrusions are simple to apply. Unroll the desired length and smooth it into place. It is soft and pliable. The bulk slug material can be formed and applied by hand or with tools such as a putty knife.

Sealing compounds are sometimes needed to seal a joint or an entry location. These compounds can be purchased in small units in white, nonstaining compositions.

Pipe Insulation

Pipe fittings are insulated for a number of reasons. Methods of insulating three different fittings are shown in Fig. 20-5. In most cases it is advisable to clean all joints and waterproof them with cement. A mixture of hot crude paraffin and granulated cork can be used to fill the cracks around the fittings.

Figure 20-6 shows a piece of rock cork insulation. It is molded

20-5. Pipe fittings covered with cork-type insulation. On one of the valves the top section can be removed if the packing needs replacing.

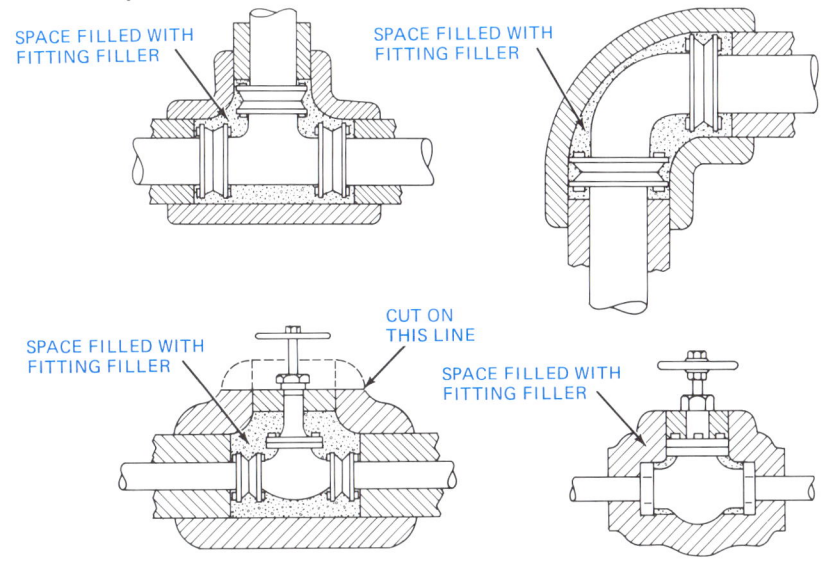

REFRIGERATION AND AIR CONDITIONING TECHNOLOGY

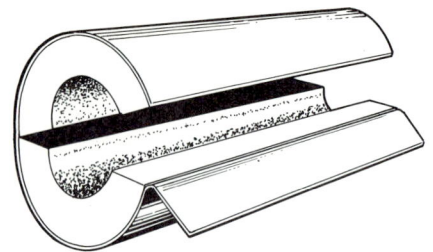

20-6. Pipe insulation.

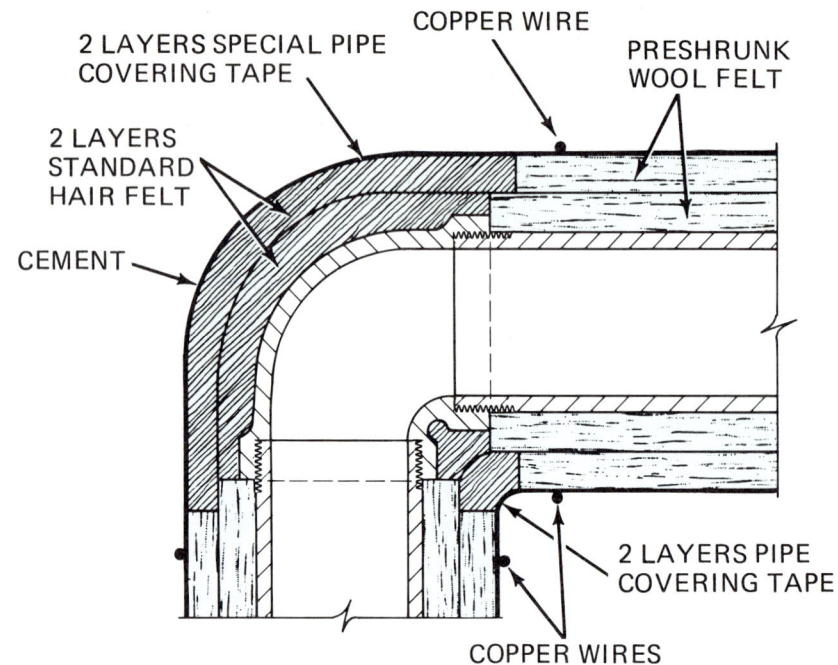

20-7. Insulated pipe fittings.

from a mixture of rock wool and waterproof binder. Rock wool is made from limestone that has been melted at about 3000° F. [1649°C]. It is then blown into fibers by high pressure steam. Asphaltum is the binder used to hold it into a moldable form. This insulation has approximately the same insulation qualities as cork. It can be made waterproof when coated with asphalt. Some more modern materials have been developed to give the same or better insulation qualities. The Vascocel® tubing can be used in the insulation of pipes. Pipe wraps are available to give good insulation and prevent dripping, heat loss, or heat gain.

Figure 20-7 shows a fitting insulated with preshrunk wool felt. This is a built-up thickness of pipe covering made of two layers of hair felt. The inside portion is covered with plastic cement before the insulation material is applied. After the application, waterproof tape and plastic cement should be added for protection against moisture infiltration. This type of insulation is used primarily on pipes located inside a building. If the pipe is located outside, another type of insulation should be used.

REVIEW QUESTIONS

1. Why is off-cycle defrost not used in commercial units?
2. Why does frost not form on an air conditioning unit?
3. Explain wall gain load, air change load, and product load.
4. How do you calculate heat leakage?
5. What are miscellaneous loads?
6. What does the capacity of a refrigerating compressor depend on?
7. What is the purpose of an air door?
8. What is the main component of an air door?
9. What is the basic purpose of Vascocel®?
10. Why is tubing insulation needed in a refrigerating system?

21

Wire Sizes and Voltage Requirements for Air Conditioning and Refrigeration Equipment

CHOOSING WIRE SIZE

There are two criteria for choosing wire size for installation of air conditioning or refrigeration equipment. The size of the electrical conductor wire recommended for a given appliance circuit depends upon two things: limitation on voltage loss and minimum wire size.

Limiting Voltage Loss

Proper operation of an electrical device must be under the conditions for which it was designed. The wire size selected must be low in resistance per foot of length. This will assure that the full load "line loss" of the total length of the circuit does not cause low voltage of the appliance terminals. Since the length of electrical feeders varies with each installation, wire sizing to avoid excessive voltage loss becomes the responsibility of the installing contractor. The National Electrical Code or local code should be followed.

Minimum Wire Size

To avoid field wiring being damaged by tensile stress or overheating, national and local codes establish minimum wire sizes. The maximum amperage permitted for a given conductor limits internal heat generation so that temperature will not damage its insulation. This assumes proper fusing that will limit the maximum current flow so that the conductor will always be protected.

Wire size and voltage loss go hand in hand, so to speak. The larger the wire, the more current it can handle without voltage loss along the lines. Each conductor or wire has resistance. This resistance, measured in ohms per unit of wire length, increases as the cross-sectional area of the wire decreases. The size of the wire is indicated by gage number. The higher the gage number, the smaller the wire. American Wire Gage (AWG) is the standard used for wire size. Each gage number has a resistance value in ohms per foot of wire length. The resistance of aluminum wire is 64% greater than that for copper of the same gage number.

Wire Selection

The wire size recommended for actual use should be the heavier of the two indicated by the procedures that will follow.

Local approval is usually necessary for any installation that has large current draws. The data presented here are based on the National Electrical Code. Much of the detail has been omitted in the interest of simplification. Thus, there may be areas of incompleteness not covered by a footnote or reference. In all cases it is recognized that final approval must come from the authority having local jurisdiction. The National Electrical Code sets forth minimum standards. It is an effort to establish some standard for safe operation of equipment.

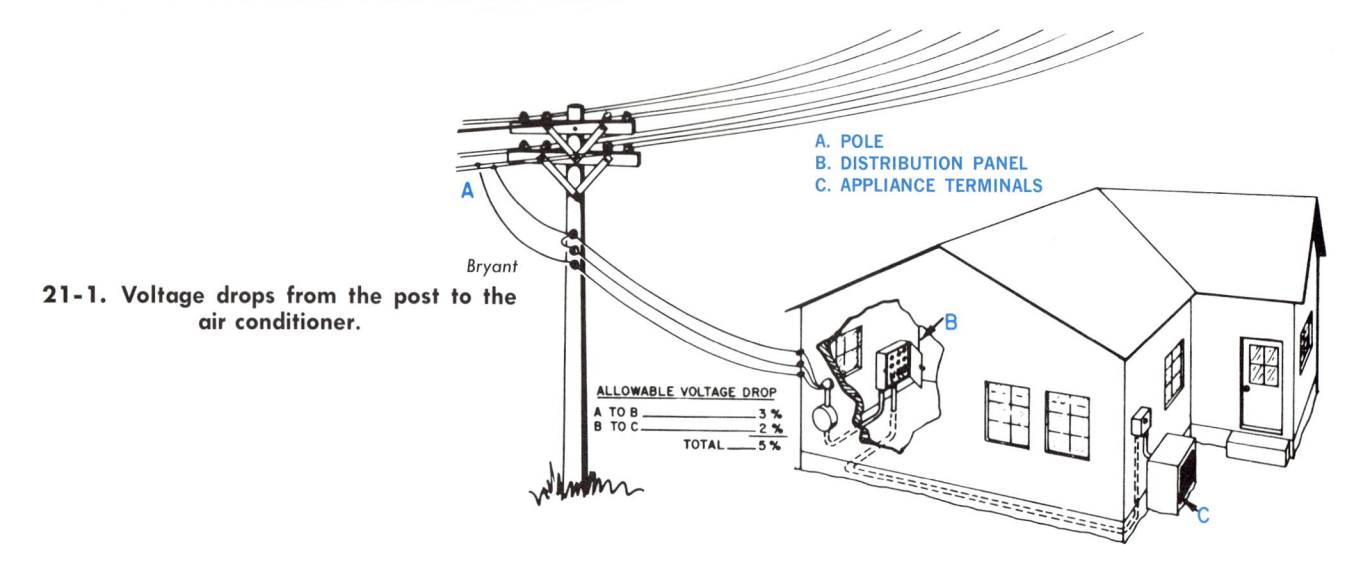

REFRIGERATION AND AIR CONDITIONING TECHNOLOGY

21-1. Voltage drops from the post to the air conditioner.

A. POLE
B. DISTRIBUTION PANEL
C. APPLIANCE TERMINALS

ALLOWABLE VOLTAGE DROP
A TO B ___ 3%
B TO C ___ 2%
TOTAL ___ 5%

WIRE SIZE AND LOW VOLTAGE

The voltage at which a motor or device should operate is stamped on the nameplate. This voltage indicates that the full capacity of the device is being utilized when that particular voltage is available. Motors operated at lower than rated voltage are unable to provide full horsepower without jeopardizing their service life. Electric heating units lose capacity even more rapidly at reduced voltages.

Low voltage can result in insufficient spark for oil burner ignition, reluctant starting of motors, and overheating of motors handling normal loads. Thus, it is not uncommon to protect electrical devices by selecting relays that will not close load circuits if the voltage is more than 15% below rating.

ARI (Air Conditioning and Refrigeration Institute) certified cooling units are tested to assure they will start and run at 10% above and 10% below their rating plate voltage. However, this does not imply that continuous operation at these voltages will not affect their capacity, performance, and anticipated service life. A large proportion of air conditioning compressor burnouts can be traced to low voltage. Because the motor of a hermetic compressor is entirely enclosed within the refrigerant cycle, it is important that it not be abused either by overloading or undervoltage. Both of these can occur during peak load conditions. A national survey has shown that the most common cause of compressor low voltage is the use of *undersized* conductors between the utility lines and the condensing unit.

The size of the wire selected must be one that, under full load conditions, will deliver acceptable voltages to the appliance terminals. The National Electrical Code requires that conductors be sized to limit voltage drop between the outdoor-pole service tap and the appliance terminals to not in excess of 5% of rated voltage under full load conditions. This loss may be subdivided, with 3% permissible in service drops, feeders, meters, and overcurrent protectors at the

Table 21-A.
Permissible Maximum Voltage Drops

For a line voltage of:	120	208	240	480
Feeders to distribution panel (3%)	3.6	6.24	7.2	14.4
Branch circuit to appliance (2%)	2.4	4.16	4.8	9.6
Total voltage drop fully loaded	6	10.4	12	24
Resultant* voltage at appliance	114	197.6	228	456

* Assumes full rated voltage where feeders connect to utility lines. If utility voltage runs low, the overall voltage drop should be further reduced so as to make available at the appliance terminals a voltage as close as practical to that specified on the appliance rating plate.

514

CHAPTER 21—WIRE SIZES AND VOLTAGE REQUIREMENTS FOR AIR CONDITIONING AND REFRIGERATION EQUIPMENT

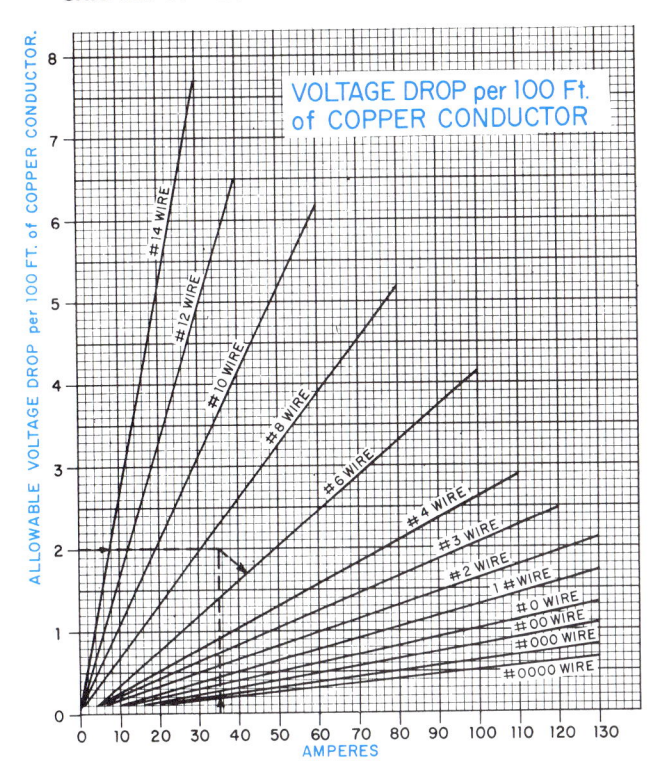

21-2. Conductor voltage drop ampere flow rate graph.

distribution panel and the appliance. See Fig. 21-1.

In a 240-V service, the wire size selected for an individual appliance circuit should cause no more than 4.8-V drop under full load conditions. Even with this 5% limitation on voltage drop, the voltage at the equipment terminals is still very apt to be below the rating plate values. Table 21-A.

Voltage Drop Calculations

Just as friction creates pressure loss in water flow through pipes, so does electrical resistance create voltage drop as current flows through a conductor. The drop increases with the length of the conductor (in feet), the current flow (in amperes), and the ohms of resistance per foot of wire. This relationship may be expressed as follows:

$$\text{Voltage drop} = \text{amperes} \times \text{ohms/foot} \times \text{length of conductor}$$

Figure 21-2 illustrates how voltage drop per 100′ of copper conductor will increase with the amount of current drawn through the conductor. The wire size is indicated on the straight line. Match the amperes with the wire size. Then follow over to the left column to determine the voltage drop. For instance, Fig. 21-2 shows that there will be a 2.04-V drop per 100′ of copper conductor for 20 A of current through a #10 wire.

THE EFFECTS OF VOLTAGE VARIATIONS ON AC MOTORS

Motors will run at the voltage varations already mentioned. This does not imply such operation will comply with industry standards of capacity, temperature rise, or normally anticipated service life. Figure 21-3 shows general effects. Such effects are not guaranteed for specific motors.

The temperature rise and performance characteristics of motors sealed within hermetic compressor shells constitute a special case. These motors are cooled by return suction gas—of varying quantity and temperature. Thus, Fig. 21-3 is not necessarily applicable to this specialized type of equipment.

The chart shows the approximate effect of voltage variations on motor characteristics. The reference base of voltage and frequency is understood to be that shown on the nameplate of the motor.

515

REFRIGERATION AND AIR CONDITIONING TECHNOLOGY

21-3. General effect of voltage variations on induction motor characteristics. For 1φ and 3φ squirrel cage induction motors. Not applicable to capacitor run or shaded pole motors. Assumes 60-hertz frequency is maintained.

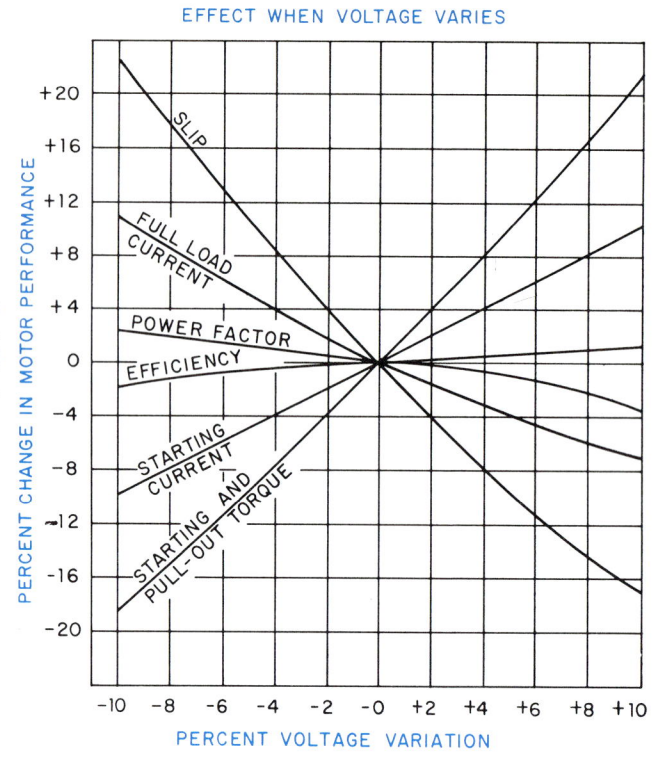

Some of the terms used in the chart are explained here.

Normal slip = synchronous speed—the rating plate speed. "Slip" in the graph indicates the change in normal slip.

Synchronous speeds for 60-Hz motors are:

2-pole—3600 r/min
4-pole—1800 r/min
6-pole—1200 r/min
8-pole—900 r/min

Table 21-B indicates the voltage drop that may be anticipated for various ampere flow rates through copper conductors of different gage size. Figure 21-2 provides the same data in graphic form.

These data are applicable to both single-phase and three-phase circuits. In each case, the wire length equals twice the distance from the power distribution panel to the appliance terminals, measured along the path of the conductors. This is twice the distance between B and C in Fig. 21-1, measured along the path of the conductors. For motorized appliances, particularly those that start under loaded conditions, the voltage at the appliance terminals should not drop more than 10% below rating plate values unless approved by the manufacturer. Thus, the voltage drop permissible in the load leads must anticipate any reduction below rated voltage that may be suffered under full load conditions at the point of power source connection (Point A in Fig. 21-1).

Troublesome voltage losses may also occur elsewhere if electrical joints or splices are mechanically imperfect and create unanticipated resistance. Such connections may exist in the distribution panel, the meter socket, or even where outdoor power drops are clamped to the feeder lines on poles. Where there is a wide variation between no-load voltage and operating voltage, sources of voltage drop can be determined by taking voltmeter readings at various points in the circuit. These points might be ahead of the meter, after the circuit disconnect switch, at the appliance terminals, and at other locations.

SELECTING PROPER WIRE SIZE

To provide adequate voltage at the appliance terminals, anticipate the minimum voltage that may exist at the distribution panel. Then determine the allowable voltage drop acceptable in the appliance circuit. This should not exceed 2% of rated voltage. It should, for example,

516

CHAPTER 21—WIRE SIZES AND VOLTAGE REQUIREMENTS FOR AIR CONDITIONING AND REFRIGERATION EQUIPMENT

Table 21-B.
Voltage Drop per 100 Feet of Copper Conductor of Wire Gage

Amperes*	#14	#12	#10	#8	#6	#4	#3	#2	#1	#0	#00	#000	#0000
5	1.29	0.81	0.51	0.32	0.21	0.13	0.11						
10	2.57	1.62	1.02	0.64	0.41	0.26	0.21	0.16	0.13	0.10			
15	3.86	2.43	1.53	0.96	0.62	0.39	0.31	0.24	0.19	0.15	0.12	0.10	
20	5.14	3.24	2.04	1.28	0.82	0.52	0.41	0.32	0.26	0.20	0.16	0.13	0.10
25	6.43	4.05	2.55	1.60	1.03	0.65	0.51	0.41	0.32	0.26	0.20	0.16	0.13
30	7.71	4.86	3.06	1.92	1.23	0.78	0.62	0.49	0.39	0.31	0.24	0.19	0.15
35		5.67	3.57	2.24	1.44	0.91	0.72	0.57	0.45	0.36	0.28	0.22	0.18
40		6.48	4.08	2.56	1.64	1.04	0.82	0.65	0.52	0.41	0.32	0.26	0.20
45			4.59	2.88	1.85	1.17	0.92	0.73	0.58	0.46	0.36	0.29	0.23
50			5.10	3.20	2.05	1.30	1.03	0.81	0.65	0.51	0.41	0.32	0.26
60			6.12	3.84	2.46	1.56	1.23	0.97	0.77	0.61	0.49	0.38	0.31
70				4.48	2.87	1.82	1.44	1.13	0.90	0.71	0.57	0.45	0.36
80				5.12	3.28	2.08	1.64	1.30	1.03	0.82	0.65	0.51	0.41
90					3.69	2.34	1.85	1.46	1.16	0.92	0.73	0.58	0.46
100					4.10	2.59	2.05	1.62	1.29	1.02	0.81	0.64	0.51
110						2.85	2.26	1.78	1.42	1.12	0.89	0.70	0.56
120							2.46	1.94	1.55	1.22	0.97	0.77	0.61
130								2.10	1.68	1.33	1.05	0.83	0.66
140									1.81	1.43	1.13	0.90	0.71
150										1.53	1.22	0.96	0.77
ohms/100ft copper	0.257	0.162	0.1018	0.064	0.041	0.0259	0.0205	0.0162	0.0128	0.0102	0.0081	0.0064	0.0051
aluminum	0.422	0.266	0.167	0.105	0.0674	0.0424	0.0336	0.0266	0.0129	0.0168	0.0133	0.0105	0.0084

* To determine voltage drop for aluminum conductors, enter the chart using 1.64 × actual amperes.
The conductor's length is twice the length of the branch leads, whether single or three phase.
Since resistance varies with temperature, it may be necessary to correct for wire temperature under load conditions if the ambient materially exceeds 80° F. If so, increase ampere values using the multiplier 1.0 + 0.002 × (ambient temperature −80° F.).
Example: If current flow and environment result in conductors reaching 140° F. under load conditions, the appliance ampere ratings should be increased by the multiplying factor 1.0 + 0.002 (140° −80°) = 1.0 + 0.12 = 1.12.

not exceed 4.1 volts for 208-V service. Table 21-A shows voltage drops for 120-, 208-, 240-, and 408-V service.

Determine the length of feed conductor. This is twice the length of the wire path from the source to the appliance. In Fig. 21-1, this is two times the distance from B to C measured along the path of the wire. If it is single-phase or three-phase, consider two conductors in establishing the total length of the circuit.

Determine the allowable drop per 100′ of conductor.

EXAMPLE: If for a 230-V installation, a 4.6-V drop is permissible, and the wire path is 115′ from the distribution panel to the appliance (this makes 230′ of conductor), then the allowable drop per 100′ will be:

$$4.6 \text{ V} \times \frac{100'}{230'} = 2.0 \text{ V per } 100'$$

Using either Table 21-B or Fig. 21-2, determine the gage wire required. When using the graph, select the gage number closest below and to the right.

EXAMPLE: The full load value is 35 A. The allowable voltage drop is 2.0 V per 100′. See Fig. 21-2.

Table Solution (Table 21-B): Select No. 6 wire. This results in a drop of 1.44 V per 100′.

Graph Solution (Fig. 21-2): Intersection lies between No. 6 wire and No. 8 wire. Select the larger of the two, in this case it would be No. 6.

UNACCEPTABLE MOTOR VOLTAGES

Occasionally it becomes neces-

REFRIGERATION AND AIR CONDITIONING TECHNOLOGY

Table 21-C.
Range of Locked Rotor Amperes per Motor Horsepower*

NEMA Code Letter	115 1ϕ	208 1ϕ	208 3ϕ	230 1ϕ	230 3ϕ	460 1ϕ	460 3ϕ
A	0–27.4	0–15.1	0–9.1	0–13.7	0–7.9	0–6.9	0–4.0
B	27.5–30.9	15.2–17.0	9.2–9.8	13.8–15.5	8.0–9.0	7.0–7.7	4.1–4.5
C	31.0–34.8	17.1–19.4	9.9–11.2	15.6–17.4	9.1–10.1	7.8–8.7	4.6–5.0
D	34.9–39.2	19.5–21.6	11.3–12.5	17.5–19.6	10.2–11.3	8.8–9.8	5.1–5.7
E	39.3–43.5	21.7–24.0	12.6–13.9	19.7–21.7	11.4–12.5	9.9–10.9	5.8–6.3
F	43.6–48.7	24.1–26.9	14.0–15.5	21.8–24.4	12.6–14.1	11.0–12.2	6.4–7.0
G	48.8–54.8	27.0–30.3	15.6–17.5	24.5–27.4	14.2–15.8	12.3–13.7	7.1–7.9
H	54.9–61.7	30.4–33.7	17.6–19.5	27.5–30.6	15.9–17.7	13.8–15.3	8.0–8.8
J	61.8–69.6	33.8–38.4	19.6–22.2	30.7–34.8	17.8–20.1	15.4–17.4	8.9–10.1
K	69.7–78.4	38.5–43.3	22.3–25.0	34.9–39.2	20.2–22.6	17.5–19.6	10.2–11.3
L	78.5–87.1	43.4–48.0	25.1–27.7	39.3–43.2	22.7–25.2	19.7–21.8	11.4–12.6
M	87.2–97.4	48.1–53.8	27.8–31.1	43.3–48.7	25.3–28.7	21.9–24.4	12.7–14.1
N	97.5–109	53.9–60.0	31.2–34.6	48.7–54.5	28.3–31.5	24.5–27.3	14.2–15.8
P	110–122	60.1–67.2	34.7–38.8	54.6–61.0	31.6–35.2	27.4–30.5	15.9–17.6
R	123–139	67.3–76.8	38.9–44.4	61.1–69.6	35.3–40.2	30.6–34.8	17.7–20.1
S	140–157	76.9–86.5	44.5–50.0	69.7–78.4	40.3–45.3	34.9–39.2	20.2–22.6
T	158–174	86.6–96.0	50.1–55.5	78.5–87.0	45.4–50.2	39.3–43.5	22.7–25.1
U	175–195	96.1–108	55.6–56.4	87.1–97.5	50.3–56.3	44.5–48.8	25.2–28.2
V	196 & up	109 & up	56.5 & up	97.6 & up	56.4 & up	48.9 & up	28.3 & up

*Locked rotor amperes appear on rating plates of hermetic compressors.
The NEMA Code Letter appears on the motor rating plate.
Multiply above values by motor horsepower.

sary to determine causes of unacceptable voltage conditions at motor terminals. Often this is necessary where excessive voltage drops are encountered as motors start. During this brief interval, the starting inrush current may approximate a motor's locked rotor amperage rating.

Table 21-C shows the range of locked rotor amperes per motor horsepower. LRA (locked rotor amperes) appears on the rating plates of hermetic compressors. Depending on the type of motor, its locked-rotor amperage may be two to six times its rated full load current. Motor starting torque varies as the square of the voltage. Thus, only 81% of the anticipated torque is available if the voltage drops to 90% of the rating during the starting period.

The full load amperage value must be considered in choosing the proper wire size and making sure the motor has acceptable

518

CHAPTER 21—WIRE SIZES AND VOLTAGE REQUIREMENTS FOR AIR CONDITIONING AND REFRIGERATION EQUIPMENT

Table 21-D.
Approximate Full Load Amperage Values for AC Motors

Motor	Single Phase*		3-Phase, Squirrel-Cage Induction		
HP	115 V	230 V	230 V	460 V	575 V
1/6	4.4	2.2			
1/4	5.8	2.9			
1/3	7.2	3.6			
1/2	9.8	4.9	2	1.0	0.8
3/4	13.8	6.9	2.8	1.4	1.1
1	16	8	3.6	1.8	1.4
1 1/2	20	10	5.2	2.6	2.1
2	24	12	6.8	3.4	2.7
3	34	17	9.6	4.8	3.9
5	56	28	15.2	7.6	6.1
7 1/2			22	11.0	9.0
10			28	14.0	11.0
15			42	21.0	17.0
20			54	27.0	22.0
25			68	34.0	27.0

*Does not include shaded pole.

voltages. These are shown in Table 21-D.

CALCULATING STARTING CURRENT VALUES AND INRUSH VOLTAGE DROPS

Single-Phase Current

Wire size and inrush voltage drop can be calculated. The following formula can be used for single-phase current.

EXAMPLE: If a single-phase 230-V condensing unit, rated at 22 A full load and having a starting current of 91 A, is located 125′ from the distribution panel and so utilizes 250′ of the No. 10 copper wire, the voltage drop expected during full load operation is calculated as follows:

Refer to the lower lines of Table 21-B. Note that the resistance of No. 10 copper wire is 0.1018 ohms/100′.

$$\text{Voltage drop} = 22 \text{ A} \times \frac{0.1018}{100'} \times 250' = 5.6 \text{ V}$$

(Note that 5.6 V exceeds the 2% loss factor, which is 4.6 V.)

If the full 3% loss (6.9 V) allowed ahead of the meter is present, then the voltage at the load terminal of the meter will be 223.1 V (230 − 6.9 = 223.1). Subtract the voltage drop calculated above and there will be only 217.5 V at the unit terminals during full load operation. Thus, 223.1 − 5.6 = 217.5. With a total loss of 5.4%, (230 − 217.5 = 12.5, or 5.4%), it is common practice to move to the next largest wire size. Therefore, for this circuit, AWG No. 8 wire should be used instead of No. 10.

Insofar as motor starting and relay operation are concerned, the critical period is during the initial instant of start-up when the inrush current closely approximates the locked rotor value. For the equipment described in the above example, the voltage drop

519

REFRIGERATION AND AIR CONDITIONING TECHNOLOGY

experienced at 91-A flow for No. 10 wire is again excessive, indicating the wisdom of using No. 8 wire.

For No. 10 wire:
$$\text{Inrush voltage drop} = 91 \text{ A} \times \frac{0.1018 \text{ ohms}}{100'} \times 250' = 23.1595 \text{ V},$$
or 23.2 V

For No. 8 wire:
$$\text{Inrush voltage drop} = 91 \text{ A} \times \frac{0.064 \text{ ohms}}{100'} \times 250' = 14.56 \text{ V}$$

For a 230-V circuit, the 23.2 V slightly exceeds a 10% drop between the meter and the appliance. To this must be added the voltage drop incurred in the lead-in wires from the outdoor power line. This total must then be deducted from the power line voltage on the poles, which may be less than 230 V during utility peak load periods. Although the inrush current may exist for only an instant, this may be long enough to cause a starting relay to open, thus cutting off current to the motor. Without current flow, the voltage at the unit immediately rises enough to reclose the relay, so there is another attempt to start the motor. While the unit may get underway after the second or third attempt, such "chattering relay" operation is not good for the relay, the capacitors, or the motor.

For electrical loads such as lighting, resistance heating, and cooking, inrush current may be considered the equivalent of normal current flow. In the case of rotating machinery, it is only during that initial period or rotation that the start-up current exceeds that of final operation. The same is true of relays during the instant of "pull-in."

Three-Phase Circuits

Calculating the inrush voltage drop for three-phase circuits is the same as calculating the drop for single-phase circuits. Again, the value for circuit length equals twice the length of an individual conductor. Since more conductors are involved, the normal current and the starting current per conductor are smaller for a motor of a given size. Thus, lighter wire may be used.

EXAMPLE: Using the same wire length as in the single-phase example and the lower values of 13.7 A full load and 61 A starting inrush per conductor for the three-phase rating of the same size compressor, the use of No. 10 conductor results in:

$$\text{Normal voltage drop} = 13.7 \text{ A} \times \frac{0.1018 \text{ ohms}}{100'} \times 250' = 3.5 \text{ V}$$
$$\text{Inrush voltage drop} = 61 \text{ A} \times \frac{0.1018 \text{ ohms}}{100'} \times 250' = 15.5 \text{ V}$$

Inrush Voltage Drop

The actual inrush current through an appliance usually is somewhat less than the total of locked-rotor current values. Locked-rotor current is measured with rated voltage at the appliance terminals. Since voltage drop in the feed lines reduces the voltage available at the terminals, less than rated voltage can be anticipated across the electrical components. Consequently, inrush currents and voltage drops are somewhat less. This fact is illustrated in the following, which is based on the same installation as that in the previous single-phase examples. However, here the actual locked-rotor current of 101 A is used. The formula can be found at the top of the next page.

CODE LIMITATIONS ON AMPERES PER CONDUCTOR

Varied mechanical conditions are encountered in field wiring. Thus, the National Electrical Code places certain limitations on the smallness of conductors installed in the field. Such limitations apply regardless of conductor length. They assure the following:

- That the wire itself has ample strength to withstand the stress of pulling it through long conduits and chases. With specific exceptions, no wire lighter than No. 14 copper is permitted for field wiring of line voltage power circuits.

- By stipulating the maximum amperage permissible for each wire gage, self-generated heat can be limited to avoid temperature damage to wire insulation. If wiring is installed in areas of high-ambient temperature, the amperage rating may need to be reduced.

- By stipulating the maximum amperage of overload pro-

520

CHAPTER 21—WIRE SIZES AND VOLTAGE REQUIREMENTS FOR AIR CONDITIONING AND REFRIGERATION EQUIPMENT

$$\text{Indicated Locked-Rotor Impedance} = \frac{\text{Rated Voltage}}{\text{Locked Rotor Amperes}} = \frac{230 \text{ V}}{101 \text{ A}} = 2.28 \text{ ohms}$$

$$\text{Resistance of No. 10 Leads} = \text{ohms}/100' \times \text{Length of Wire (in feet)} = \frac{0.1018 \text{ ohms}}{100'} \times 250' = 0.25 \text{ ohms}$$

$$\text{Total Indicated Load and Conductor Impedance} = (2.28 + 0.25) = 2.53 \text{ ohms}$$

$$\text{Inrush Current} = \frac{\text{Distribution Panel Voltage}}{\text{Total Indicated Impedance}}$$

$$\text{Inrush Current} = \frac{230 \text{ V}}{2.53 \text{ ohms}} = 91 \text{ A}$$

$$\text{Calculated Inrush Voltage Drop} = 91.0 \times \frac{0.1018 \text{ ohms}}{100'} \times 250' = 23.2 \text{ V}$$

$$\text{Locked Rotor Voltage Drop} = 101 \times \frac{0.1018 \text{ ohms}}{100'} \times 250' = 25.7 \text{ V}$$

tectors for circuits, current flow is limited to safe values for the conductor used. Some equipment has momentary starting currents that trip-out overload protectors sized on the basis of full load current. Here heavier fusing is permissible—but only under specific circumstances. Current flow limitations for each gage protect wire insulation from damage due to overheating.

HEAT GENERATED WITHIN CONDUCTORS

Heat generation due to current flow through the wire is important for the following two reasons.

1. Temperature rise increases the resistance of the wire and, therefore, the voltage drop in the circuit. Under most conditions of circuit usage, this added resistance generates additional heat in the wires. Finally, a temperature is reached where heat dissipation from the conductors equals the heat that they generate. It is desirable to keep this equilibrium temperature low.

2. Temperature also damages wire insulation. The degree of damage is dependent upon the insulation's ability to withstand temperature under varying degrees of exposure, age, moisture, corrosive environment, mechanical abuse, and thickness.

Estimating the probable operating temperature of a conductor and its insulation is difficult. The rate of heat dissipation from the wiring surfaces varies with the ambient temperature, the proximity of other heat-generating conductors, the heat conductivity of the insulation and jacket material, the availability of cooling air, and other factors. Free-standing individual conductors dissipate heat more effectively. However, the typical situation of two or three conductors, each carrying equal current and enclosed in a common jacket, cable, or conduit, anticipates

The number of Btu generated can be found by both of the following formulas.

$$\text{Btu generated} = \text{amperes}^2 \times \text{resistance in ohms} \times 3.4313^*$$

$$\text{Btu generated} = \text{amperes} \times \text{voltage drop} \times 3.4313^*$$

limitations as set forth by the National Electrical Code.

CIRCUIT PROTECTION

Circuits supplying power to appliances must incorporate some means for automatically disconnecting the circuit from the power source should there be abnormal current flow due to accidental grounding, equipment overload, or short circuits. Such overload devices should operate promptly enough to limit the buildup of damaging temperatures in conductors or in the electrical components of an appliance. However, devices selected to protect circuits feeding motors must be slow enough to permit the momentary inrush of heavy

---

*Conversion factor.

521

21-4. How a fuse works.

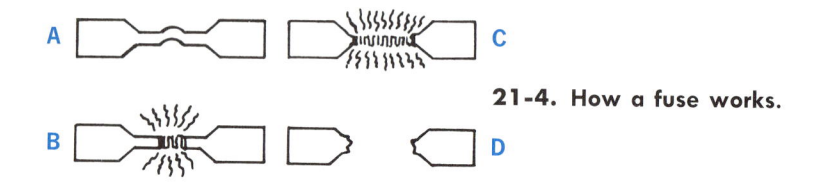

starting current. They must then disconnect the circuit if the motor does not start promptly, as can happen under low voltage conditions.

Devices heavy enough to carry continuously the motor starting current do not provide the overload protection desired. Likewise, heavily fused branch circuits do not adequately protect the low amperage components that cumulatively require the heavy fusing. For this reason some literature lists *maximum allowable* fuse sizes for equipment. While electrical components of factory-built appliances are individually safeguarded, the field combining of two or more units on one circuit may create a problem more complex than that normally encountered. Remember that the final authority is the local electrical inspector.

Standard Rule

With a few exceptions, the ampere capacity of an overload protector cannot exceed the ampacity values listed by wire size by the NEC. (Check the National Electrical Code for these exceptions.) If the allowable ampacity of a conductor does not match the rating of a standard size fuse or nonadjustable trip-circuit breaker, the device with the next largest capacity should be used. Some of the standard sizes of fuses and nonadjustable trip-circuit breakers are: 15, 20, 25, 30, 35, 40, 45, 50, 60, 70, 80, 90, 100, 110, 125, 150, 175, 200, 225, 250, and 300 amperes.

FUSES

One-Time Single-Element Fuses

If a current of more than rated load is continued sufficiently long, the fuse link becomes overheated. This causes the center portion to melt. The melted portion drops away. However, due to the short gap, the circuit is not immediately broken. An arc continues and burns the metal at each end until the arc is stopped because of the very high increase in resistance. The material surrounding the link tends to break the arc mechanically. The center portion melts first, because it is farthest from the terminals that have the highest heat conductivity. Fig. 21-4.

Fuses will carry a 10% overload indefinitely under laboratory-controlled conditions. However, they will blow promptly if materially overloaded. They will stand 150% of the rated amperes for the following time periods:

1 minute (fuse is 30 A or less).
2 minutes (fuse of 31 to 60 A).
4 minutes (fuse of 61 to 100 A).

Time-Delay Two-Element Fuses

Two-element fuses use the burnout link described previously. They also use a low-temperature soldered connection that will open under overload. This soldered joint has mass, so it does not heat quickly enough to melt if a heavy load is imposed for only a short time. However, a small but continuous overload will soften the solder so that the electrical contact can be broken.

With this type of protection against light overloads, the fusible link can be made heavier, yet blow quickly to protect against heavy overloads. This results in fewer nuisance burnouts and equipment shutdowns. Two types of dual-element fuses are shown in Fig. 21-5.

Types of Fuses

In addition to those fuses just described, there are three general categories based on *shape* and *size*.

The AG (automotive glass) fuse consists of a glass cylinder with metallic end caps between which is connected a slender metal element that melts on current overload. This fuse has a length of $1\frac{5}{16}''$ and a diameter of $\frac{1}{4}''$. It is available only for low amperages. While used in specific appliances, it is not used to protect permanently installed wiring.

Cartridge fuses are like AG fuses. However, they are larger. The cylindrical tube is fiber, rather than glass. The metallic end pieces may be formed as lugs, blades, or cylinders to meet a variety of fuse box socket re-

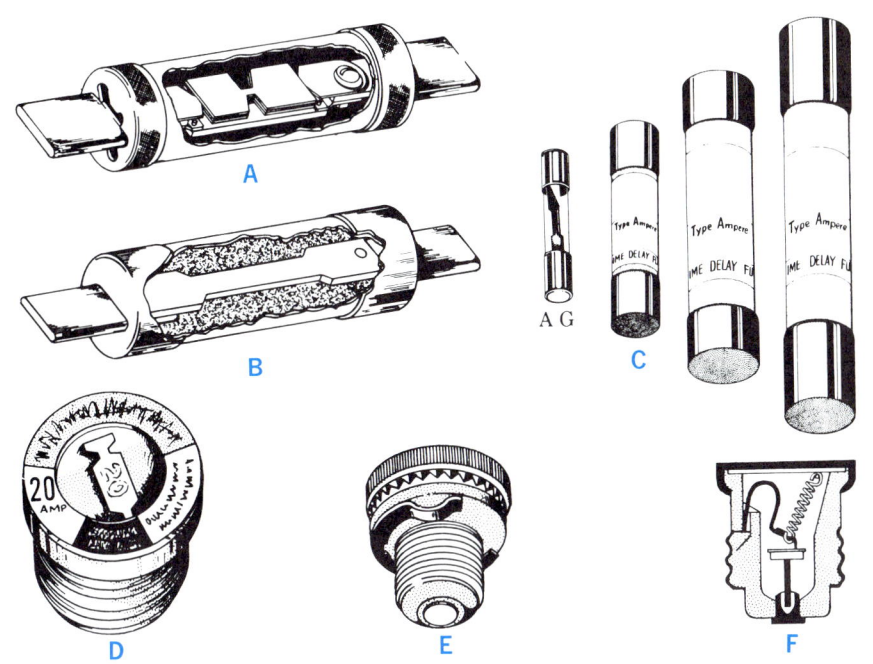

CHAPTER 21—WIRE SIZES AND VOLTAGE REQUIREMENTS FOR AIR CONDITIONING AND REFRIGERATION EQUIPMENT

quirements. The internal metal fusible link may be enclosed in sand or powder to quench the burnout arc.

Cartridge fuses are made in a variety of dimensions, based on amperage and voltage. Blade-type terminals are common above 60 A. Fuses used to break 600-V arcs are longer than those for lower voltages. Fuses are available in many capacities other than the listed standard capacities, particularly in the two-element, time-delay types. Often, they are so dimensioned as to not be interchangeable with fuses of other capacities.

Plug fuses are limited in maximum capacity to 30 A. They are designed for use in circuits of not more than 150 V above ground. Two-element time-delay types are available to fit standard screw lamp sockets. They are also available with nonstandard threads made especially for various amperage ratings.

21-5. Fuse types. (A) Two-element renewable fuse (cutaway view). The end caps screw off for replacement of the fusible element. **(B)** Ordinary one-time fuse (cutaway view). This type cannot be opened for replacement of the fusible element. **(C)** Cartridge-type fuses. AG stands for automotive glass. This type is used in electronic equipment and instruments. **(D)** Plug-type fuse used for a screw-in type receptacle. Old type used in home wiring circuits. **(E)** Fustat®. Special threads make this fit only a particular type of fuse holder. **(F)** Fusetron® is a dual-element plug-type fuse. The spring increases the possibility that it will trip at the correct rating.

REVIEW QUESTIONS

1. What are two criteria for choosing wire size for installation in air conditioning and refrigeration equipment?
2. What are some of the results of low voltage to a refrigerating system?
3. What is the most common cause of compressor low voltage?
4. What is synchronous speed?
5. How do you choose the proper size of wire for a job?
6. What does the abbreviation LRA mean in reference to motors?
7. How much torque is available from a motor if the voltage drops to 90% of its rated value?
8. What is the minimum number of wires needed to wire a three-phase compressor?
9. What causes heat generation in wire conductors?
10. What is the purpose of a fuse in a circuit?
11. What is the difference between a cartridge fuse and a plug fuse?

22

Careers

The field of air conditioning and refrigeration offers a variety of career opportunities. Air-conditioning and refrigeration mechanics install and service air-conditioning and refrigeration equipment. There are a number of different types of equipment that require service. Some systems are complex, but are easily broken down into smaller units for repair purposes. Some mechanics specialize in a particular part of a system.

INDUSTRIES THAT EMPLOY AIR-CONDITIONING AND REFRIGERATION MECHANICS

Approximately 200,000 persons work as air-conditioning and refrigeration mechanics. Most of these mechanics are employed by cooling contractors. Food chains, school systems, manufacturers, and other organizations use the services of air-conditioning and refrigeration mechanics. Large air-conditioning systems use many mechanics to keep the equipment operational. However, not all mechanics work for other people. About one in every seven mechanics is self-employed.

Manufacturers use refrigeration equipment for a variety of processes. Meat packers and chemical makers use refrigeration in some form. Temperature control is very important for many manufacturing processes.

Mechanics work in homes, office buildings, and factories. They work anywhere there is air conditioning. They bring to the job sites the tools and parts they need. During the repair season, mechanics may do considerable driving. Radios may be used to dispatch them to the jobs. If major repairs are needed, mechanics will transport parts or inoperative machinery to a repair shop.

Mechanics work in buildings that often are uncomfortable. This is because the air-conditioning or refrigeration system has failed. The mechanic may have to work in a cramped position. Rooftop units are a common practice in keeping smaller installations cool. Many of the systems have at least one unit on the roof of the building. Cooling towers are usually mounted on top of a building. Thus, the mechanic may be called upon to work in high places. In summer, the rooftop may be very hot. This trade does require some hazardous work. For instance, there are the dangers of electrical shock, torch burns, muscle strains, and other injuries from handling heavy equipment.

System installation calls for work with motors, compressors, condensing units, evaporators, and other components. These devices must be installed properly. This calls for the mechanic to be able to follow the designer's specification. In most instances, blueprints are used to indicate where and how the equipment is to be installed. The ability to read blueprints is essential for the air-conditioning and refrigeration mechanic. Such ability will help ensure that the ductwork, refrigerant lines, and electrical service is properly connected. Fig. 22-1.

After making the connections, it is then necessary to charge the system with the proper refrigerant. Proper operation must be assured before the mechanic is through with the job.

Equipment installation is but one of the jobs the mechanic must perform. If the equipment fails, then the mechanic must diagnose the cause of the trouble and make the proper repairs. The mechanic must find defects, inspect parts, and be able to know if thermostats and relays are working correctly. During the winter, air-conditioning mechanics inspect parts such as relays and thermostats. They also perform required maintenance. Overhauling may be included if compressors need attention or recharging. They may also adjust

CHAPTER 22—CAREERS

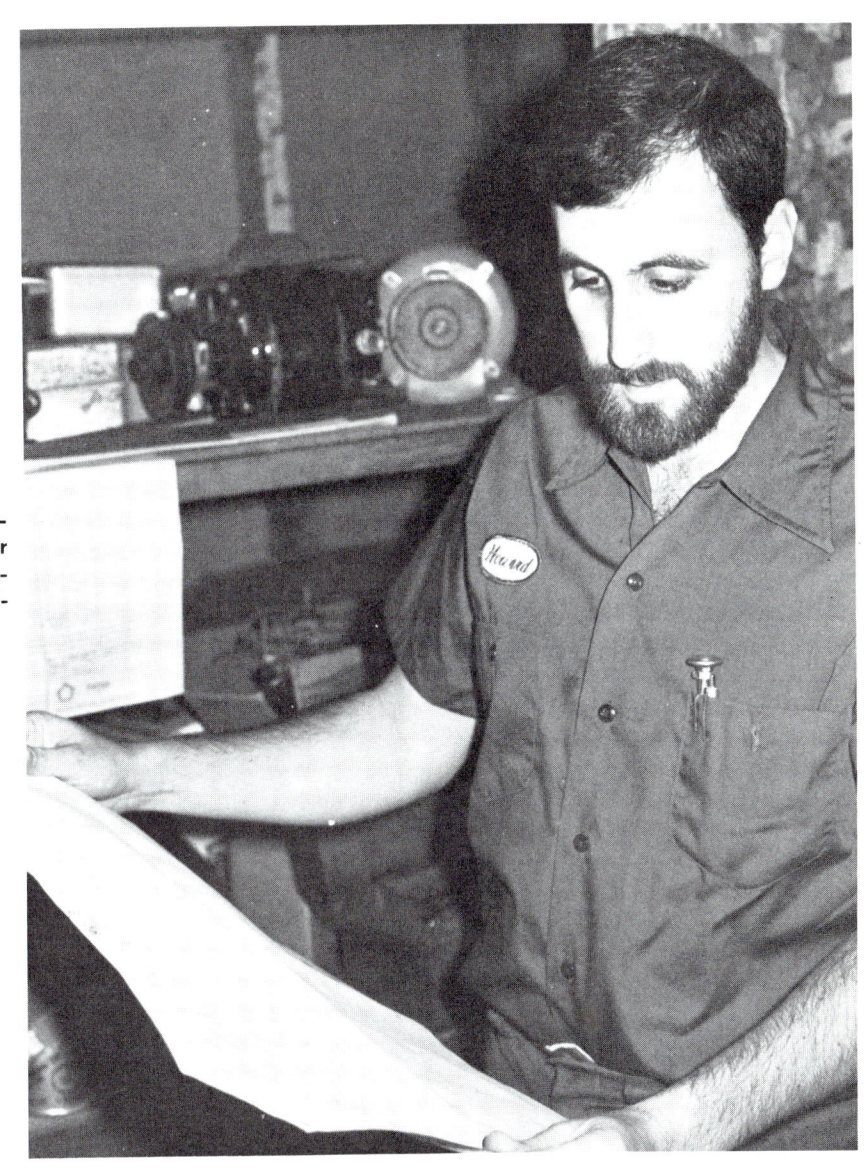

22-1. The air-conditioning and refrigeration mechanic must be familiar with blueprint symbols. Reading blueprints is essential to the proper installation of systems.

the airflow ducts for the change of seasons.

Air-conditioning and refrigeration mechanics use a number of special tools. They also use more common tools such as hammers, wrenches, metal snips, electric drills, pipe cutters and benders, and acetylene torches. Air ducts and refrigeration lines require more specialized tools. Voltmeters, ammeters, ohmmeters, and manometers are also part of the mechanic's toolbox. Electrical circuits and refrigeration lines must be tested and checked. Testing of electrical components is also required. A good background in electricity is necessary for any mechanic. Fig. 22-2.

Cooling systems sometimes are installed or repaired by other craft workers. For example, on a large air-conditioning installation job, especially where people are covered by union contracts, duct work might be done by sheet-metal specialists. Electrical work will be done by electricians. Installation of piping will be done by pipe fitters.

JOB QUALIFICATIONS

Most air-conditioning and refrigeration mechanics start as helpers. They acquire their skills

525

REFRIGERATION AND AIR CONDITIONING TECHNOLOGY

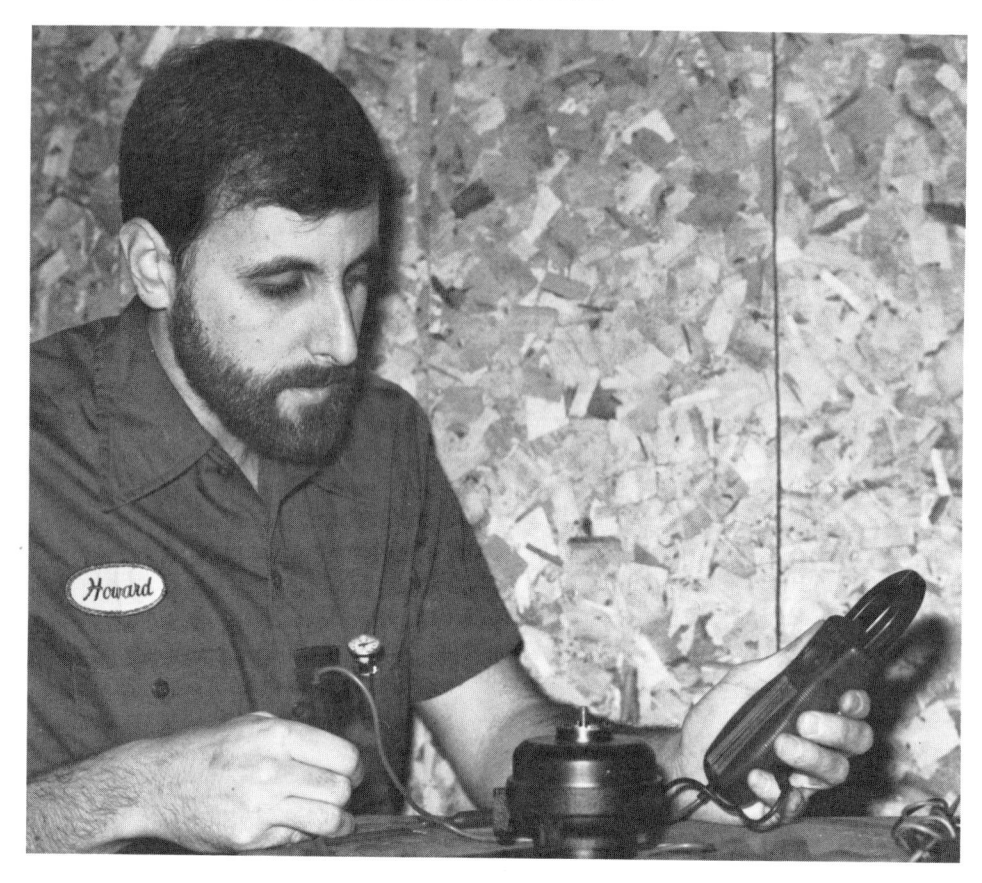

22-2. The air-conditioning and refrigeration mechanic must be able to test the effectiveness of a component. This demands skill in using a variety of instruments.

by working for several years with experienced mechanics. New people usually begin by assisting experienced mechanics. They do the simple jobs at first. They may carry materials or insulate refrigerant lines. In time, they do more difficult jobs, such as cutting and soldering pipes and sheet metal and checking electrical circuits. In four or five years the new mechanics are capable of making all types of repairs and installations.

Apprenticeship programs are run by unions and air-conditioning contractors. In addition to on-the-job training, apprentices receive 144 hours of classroom instruction each year. This instruction is in related subjects. Such subjects include the use and care of tools, safety practices, blueprint reading, and air-conditioning theory. Applicants for apprenticeships must have a high school diploma. They are given a mechanical aptitude test. Apprenticeship programs last four years.

Many high schools, private vocational schools, and junior colleges offer programs in air conditioning and refrigeration. Students study air-conditioning and refrigeration theory and equipment design and construction. They also learn the basics of installation, maintenance, and repair. Employers may prefer to hire graduates of these programs because they require less on-the-job training.

High school graduates are preferred for helpers. If they have mechanical aptitude and have had courses in mathematics, mechanical drawing, electricity, physics, and blueprint reading, they have a better opportunity to be hired. Good physical condition is also necessary. Mechanics sometimes have to lift and move heavy equipment.

To keep up with technological change and to expand their skills, experienced mechanics may take courses offered by a number of sources. The Refrigeration Service Engineers Society and the Air-Conditioning Contractors of America offer updating courses for mechanics. There are a number of trade magazines that help keep the mechanic up-to-date in the latest equipment and troubleshooting procedures.

Mechanics can advance. They can become supervisors. Some open their own contracting businesses. However, it is becoming difficult for one-person operations to operate successfully.

THE FUTURE

Employment of air-conditioning and refrigeration mechanics is expected to increase. The increase is expected to be about as fast as the average of all occupations during the 1980s. Many openings will occur as experienced mechanics transfer to other fields of work. As experienced mechanics retire or die they have to be replaced. The numbers will vary from year to year. However, in the United States, as people move toward the Sunbelt there will be more jobs for air-conditioning mechanics. Fig. 22-3.

Opportunities for air-conditioning and refrigeration mechanics are expected to follow trends in residential and commercial construction. Even during periods of slow growth, many mechanics will be needed to service existing systems. Installations of new energy-saving air-conditioning systems also will create new jobs. In addition, more refrigeration equipment will be needed in the production, storage, and marketing of food and other perishables. Because these trades have attracted many people, beginning mechanics may face competition for jobs as helpers or apprentices.

PAY AND BENEFITS

Most mechanics work for hourly wages. These skilled air-conditioning and refrigeration specialists receive pay frequently higher than those who work in similar specialties.

Apprentices receive a percentage of the wage paid experienced mechanics. Their percentage is about 40% at the beginning of their training. They receive 80% during their fourth year.

U.S. Department of Labor

22-3. The temperature in data processing centers must be carefully controlled if the data processing equipment is to operate properly. Such equipment often generates a great deal of heat. Thus, air conditioning is essential for the equipment to function properly. This operator works in an air-conditioned room.

Mechanics usually work a 40-hour week. However, during seasonal peaks they often work overtime or irregular hours. Most employers try to provide a full workweek for the entire year. By doing this, they have mechanics when they need them most—during the summer months. However, they may temporarily

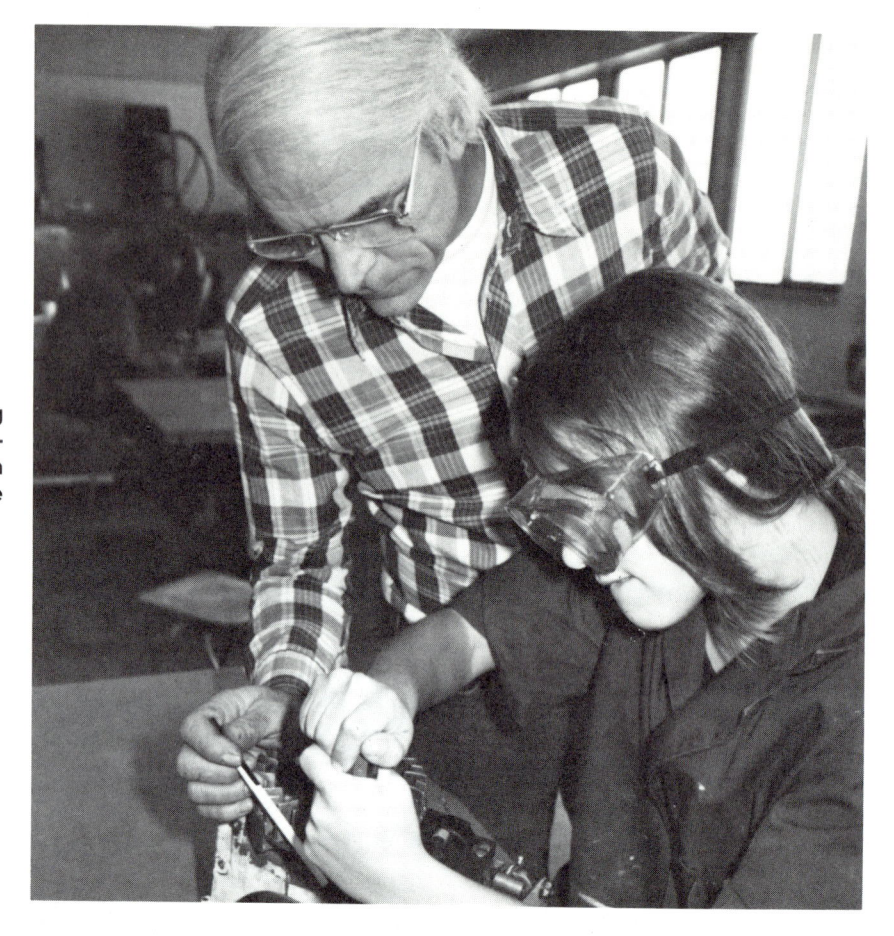

REFRIGERATION AND AIR CONDITIONING TECHNOLOGY

22-4. The field of air conditioning and refrigeration offers many opportunities to qualified teachers. Such teachers can work in private trade schools, as well as public schools.

reduce hours or lay off some mechanics when seasonal peaks end. In most shops that service air-conditioning and refrigeration equipment, employment is fairly stable throughout the year.

TEACHING AS A CAREER

A person interested in passing on to others his or her knowledge of air conditioning might be interested in teaching the subject in vocational schools. The public schools also offer classes in air-conditioning and refrigeration. Teachers of such courses often come from the trade itself. Once they have secured a position in the school, they may have to take certain college-level courses. These courses will deal with teaching methods and other subjects related to education.

Private trade schools are usually in need of good people with experience in the trade. They are needed to organize and teach apprentices. These schools may be sponsored by air-conditioning and refrigeration contractors or by unions. Fig. 22-4.

Pay and benefits are the same as for any other teacher in the public or the private schools. Working conditions are similar throughout the country. The de-

528

mand varies with the temperature. Therefore, the climate has much to do with the demand for air-conditioning and refrigeration specialists.

MORE INFORMATION

For more information about employment and training opportunities, contact the local office of the State Employment Service or firms that employ air-conditioning and refrigeration mechanics.

For pamphlets on career opportunities and training write to:

Air-Conditioning and Refrigeration Institute
1815 North Fort Myer Drive
Arlington, VA 22209

(The Institute prefers not to receive individual requests for large quantities of pamphlets.)

Air-Conditioning Contractors of America
1228 17th Street, NW
Washington, DC 20036

REVIEW QUESTIONS

1. How many hours of classroom instruction are received in an air conditioning and refrigeration apprenticeship program?
2. What business employs most air conditioning and refrigeration mechanics?
3. What percentage of the wage paid experienced workers is paid to apprentices?

Index

A

Accumulators, 291–293
Acids
  corrosive, 218
  in refrigeration systems, 259–260
Acoustical treatment, 457–458
Acrolein, 146
Actuators, 104–106
Addition, 11, 15
Air
  circulation of, in refrigerators, 338
  properties of, 427
Air conditioner. *See also* Cooling unit
  console, 483–487
  gas-fired, 488–494
  rooftop, 455–458
  solar, 495–500
  window unit, 448–451
Air conditioning
  and air flow distribution, 437–439
  careers in, 524–529
  chilled water, 477–481
  comfort, general discussion, 448–467
  commercial systems, 468–487
  compressors for, 162–201
  condensers for, 162–163, 202–211, 388
  evaporators, 226–253
  fittings for, 263–264
  gas, 488–494
  and humidity, 422, 427–429
  and psychometrics, 424–447
  reclaiming, 417–422
  troubleshooting. *See* Troubleshooting
Air Conditioning and Refrigeration Institute (ARI), 450–451, 508, 514, 529
Air-Conditioning Contractors of America, 529
Air conditioning systems
  absorption type, 476–477
  commercial, 468–487
  console-type, 483–487
  voltage requirements for, 513–523
Air distribution, in frozen food aisles, 390
Air flow distribution, 437–439
Air measurement, 51, 433–434
Air movement, 424–447
Air shutters, 338–339
Algae, 219–220
Algaecides, 219

American Society of Heating, Refrigeration, and Air-Conditioning Engineers (ASHRAE), 436, 508
Ammeter, 45–47
Ammonia
  to detect leaks, 160
  general discussion, 78, 146
Ammonia refrigerant, 489
Amplifiers, differential, 102–103
Antifoam reagent, 221
Antifreeze, needed in a chilled-water cooler (table), 490
Anti-slug feature, 173
Antismudge ring, 445
Apprenticeship programs, 526–527
Approach, in cooling towers, 212
Area
  effective, 435
  finding, 16–18
  free, 435
  outlet, 435
ARI. *See* Air Conditioning and Refrigeration Institute
Armature reaction, 112
Armature speed, 111
ASHRAE. *See* American Society of Heating, Refrigeration, and Air-Conditioning Engineers
Aspiration, 434–435
Atom, 71–72
Auxiliary devices, for control devices, 106
Azeotropic mixtures, 155

B

Bandwidths, in controllers, 103–104
Barometers, 73, 425
Bellows, 73
Biological organisms, 219–220
Bits, drilling, 26–27
Bleed-off, in cooling towers, 212
Blower
  ceiling, 239–240
  flooded ceiling, 236–238
  flooded floor-type, 237–239
  floor, 241
  low-temperature ceiling, 242
Bourdon tube, 33, 73
Brazing
  of crankcase pressure regulating valves, 270

INDEX

general discussion, 65, 67
of head pressure control valves, 276
Breakdown voltage, 87
Bridge
   balanced, 99
   Wheatstone, 98–99
Btu meters, 51
Burnouts
   compressor motor, 349–350
   hermetic motor, 263

C

Cabinet, refrigerator
   general discussion, 295
   sealing, 319
Cable cutter, armored, 28
Cables, measuring insulation resistance of, 130
Calcium chloride, 147
Cam, icemaker timing, 307–308
Capacitance, 85–87
Capacitive reactance, 91
Capacitors
   general discussion, 85–92
   ratings of, 362
   run, 320–321, 362–363
   start, 362
   testing, 126–127
   uses of, 92–93
Capacitor symbols, 86
Capacity, of refrigeration machines, 78
Cap-Check®, 53
Cap-Gage®, 53
Capillary tube, refrigerator
   general discussion, 294, 344
   inside and outside diameters, 59–60
Carbon dioxide
   to detect leaks, 160
   general discussion, 146–147
Carbon tetrachloride, 67–68
Cardiac sensitization, and fluorocarbons, 152–153
Careers, 524–529
Cases. *See also* Refrigeration units, commercial
   dehumidification of, 418–419
   ice-cream, 412–417
   meat-display, 405–412
   open produce, 403–405
   troubleshooting, 404
Cellulose acetate, and Freon®, 156
Central nervous system (CNS), effects on of fluorocarbon vapors, 152
Centrifugal switch, split-phase motor, testing, 126

Charging cylinder, 58
Charging stations, mobile, 59
Chaser kit, 52
Chiller-heater, 490–494
Chillers
   gas-fired, 489–490
   general discussion, 207–211, 481–483
   shell-and-tube, 226–229
Chisels, wood, 27
Circle, finding area of, 17
Circuit breakers, testing insulation resistance of, 131
Circuits
   AC, and capacitor, 90–95
   bridge, 98–99
   DC, power in, 84
   defined, 80
   parallel, 82–83
   protection of, 521–522
   series, 81–82
   three-phase, 520
Clippers, wire, 25
Clock, flushing, 422
Coils
   finned tube, 226
   refrigerator evaporator, 318
   testing insulation resistance of, 131
Comfort conditions, 434–436
Comfort control panel, 421–422
Compounds, 71
Compression ratio, 74–75
Compressor
   connections for, 195–198
   circuit breaker for, 364
   efficiency of, 361
   in expansion valve system, 468
   for freezer, replacing, 376–377
   fuses for, 364
   general discussion, 162–201
   hermetic, field testing, 363–368
   hermetic, general discussion, 163–200
   hermetic, troubleshooting (table), 464–466
   locked rotor volts, 364
   motors for, 184–195
   refrigerating, capacity of, 508–509
   refrigerator, 343–344
   replacement of, 344–349
   rotary, 198–200
   safety devices for, 472
   servicing (table), 346–348
   test stand for, 366

531

INDEX

troubleshooting (table), 346–348, 365
types of, 321
wire size for, 364
Compressor electrical components, testing, 367–368
Compressor systems, hermetic, moisture in (table), 131
Compressor tubes, 195–198
Condensers
  air-cooled, 202–203
  capacities (table), 509
  coaxial water-cooled, 206
  evaporative, 213
  evaporative, cleaning, 221–223
  in expansion valve system, 468
  freezer, repairing, 377
  general discussion, 162–163, 202–211, 213–214, 388
  refrigerator, 294, 344
  refrigerator, replacing, 352
  shell, cleaning, 223–224
  shell-and-coil, 206–207
  water-cooled, 203–207
  wrapped freezer, 370
Condensing units, outdoor, 392
Conduction, 433–434
Conductors
  and amperage limitations, 520–521
  heat generated in, 521
Contactors, 133
Controllers, 101–104
Controls. *See also* Control system
  in chillers, 210
  electric, general discussion, 107–144
  for hot-gas defrost of ammonia evaporators, 229–253
  lever-master, 285–290
Control system. *See also* Controls
  of chilled water air conditioning, 480–481
  constant liquid pressure, 242–245
Convection, 433–434
Conversions, metric-customary (table), 22
Coolers
  sectional walk-in storage, 396–400
  self-contained, walk-in, 392–393
Cooling cycle
  absorption, 488–489
  in direct-expansion system, 230–231
Cooling range, of cooling towers, 212
Cooling system, evaporative, 475–476
Cooling towers
  cleaning, 221–223
  general discussion, 163, 165, 211–214
Cooling unit
  packaged, 469–473
  rooftop, 470
Cooling water problems, 215–225
Copper losses, 95
Corrosion
  electrolytic, 219
  galvanic, 218–219
  general discussion, 215–216, 218–220
  oxygen cell, 219
Coulomb, 80
Criteria, noise, 435
CRO. *See* Valves, crankcase pressure regulating cross-controlling
Crossrail, freezer, divider, 333
Current
  full load (FLC), 131–132
  general discussion, 80
  locked rotor (LRC), 131
  in a parallel circuit, 82
  single-phase, 519–520
Current values, starting, calculating, 519–520
Cybertronic device, 99–100
Cycle
  pulldown, 362
  warm-up, 362
Cylinders, gas
  handling of, 341
  volume of, 18
Cylinder (geometrical form), area of, 18

D

Dampers
  fire and smoke, 443
  volume, 458
D'Arsonval meter movement, 45–46
DBV. *See* Valves, discharge bypass, 278–283
Decibels, 434
Decimal point, 15
Decimals, 15
Decomposition, thermal, of fluorocarbons, 153–154
Defrost controls
  general discussion, 138–140
  ice-cream case, 415
  refrigerator, 315–319
Defrost cycle
  in direct-expansion systems, 231–242
  heat pump, 503–504

INDEX

Defrost heater, refrigerator, 315–319
Defrost, hot-gas
   of ammonia-type evaporators, 246–249
   general discussion, 399–400
Defrosting
   automatic, 138–140
   of heat pumps, 502–504
   hot-gas, 139–140
   of meat cases, 408–412
   refrigerator, 313–319
Defrost system, refrigerator, parts of, 315–319
Defrost thermostats, refrigerator, 315–319
Defrost timer
   general discussion, 138–139
   termination, 362
Dehydrator cycle, in chillers, 209–210
Denominator, 14
Density
   general discussion, 72
   refrigerant, 158
Detergents, 225
Diaphragm, 72–73
Dielectric, 86
Dielectric constant, 87
Dielectric failure, 87
Diffusers
   air channel, 445
   air, selecting, 439–442
   ceiling, 444–447
   defined, 435
   luminaire, 445
Diodes, 96–97
Dirt, in refrigeration systems, 259
Dividend, defined, 12
Division, 12–14
Divisor, defined, 12
Doors
   air, 509
   evaporator, 295–296
   refrigerator, adjusting, 296–297
   refrigerator, aligning, 296–297
Drain system, freezer, 375
Drier coil
   freezer, installing, 377–378
   refrigerator, 332–333
Driers, 259–260
Drift, in cooling towers, 212
Drilling equipment, 26–27
Drills, masonry, 27
Drip pipes, for ice cream cases, 416–417
Drip tray, refrigerator, 296

Drop, 434–435
Drum, surge, 287
Ducturns, 439
Duty rating, 132

E

Eddy current losses, 95
Elastomers, and Freon®, 156
Electrical connections, for display cases, 401–402
Electrical controls
   for ice cream cases, 416
   for packaged cooling units, 470–471
Electrical systems, compressor motor, 184–195
Electricity
   current, 80
   and refrigeration controls, 80–106
   static, 80
Electrolytic, 87
Electrons, 71–72
Elements
   CAB resistive, 100–101
   chemical, 71
   humidity-sensing, 100–101
   temperature-sensing, 99–100
Enclosures, control equipment, standards for, 132
Enthalpy, 158, 433
Equations, 19–20
Ethyl chloride, 147
Evacuation, 57–58
Evaporator coil
   freezer, 376
   general discussion, 388
Evaporator-heat exchanger, replacing, 356–357
Evaporators
   ammonia-type 246–249
   coiled, 226–229
   flooded-ceiling, 236
   general discussion, 226–253
   ice-blocked, 360
   refrigerator, 294
   for residential use, 451–455
Expansion-valve air conditioning system, 468–469
Expansion valve kit, 468–469
External equalizer connection, 268

F

Fan assemblies, 337–338
Farad, 87
Faraday, Michael, 87
Feeder, bypass, 223
File, mill, 27

INDEX

Fill trough, icemaker, 308
Filter drier, refrigerator
  general discussion, 344
  replacing, 350–351
Filter, line, 260–261
Firestats, 457
Fittings
  refrigerant, 255–259
  refrigeration and air conditioning, 263–264
Flammability
  and Freon®, 151
  and refrigerants, 158
Flaring, of copper tubing, 62–63
Flooded liquid systems, 234–235
Floors, for walk-in coolers, 397–398
Fluorocarbons, potential hazards of, 151–154
Force, 72
Forming, of refrigerant tubing, 64–65
Formulas, mathematical, 20
Fouling, 215–216
Fractions, 13–14
Freezers
  general discussion, 369–387
    portable, 378–386
    troubleshooting, 378–386
Freon® refrigerants
  applications of, 154–155
  flammability of, 151
  general discussion, 147–161
  physical properties (table), 148–150
  pressures, absolute and gage, 151
  reaction with other materials, 155–156
Friction drop correction factors (table), 463
Frost removal, in refrigerator, 314–315
Fungi, 219–220
Fuse puller, 25
Fuses, 522–523

G

Gages
  accuracy of, 32–33
  air filter, efficiency of, 50–51
  compound, 73–74
  draft, 51
  general discussion, 32–52
  pressure, 32
  recalibration of, 35–36
Galvanic series (table), 218
Gasket, refrigerator door, 297
Gear, icemaker timing, 308
Graphics, reading, 21–22

Grilles
  air, selecting, 439–442
  and air capacity (table), 438
  ceiling supply, 443–444
  linear, 447
  return air, 437, 442–443
Grounding, of refrigerator/freezers, 322–323
Grounds, testing for, 124

H

Hammers, 28
Handles, refrigerator door, 297
Hardware
  refrigerant, 255–259
  refrigeration and air conditioning, 263–264
Heat
  latent, 75–76
  methods of recovering, 421
  radiant, measuring, 407
  reclaiming of, 417–422
  sensible, 75–76
  specific, 75
Heat content, 75
Heaters
  antisweat, 401
  crankcase, 183–184
  defrost, 335–337
  drain-trough, 339
  refrigerator defrost, length of operation, 319
  serpentine, 143
  tube, 331
Heat exchanger
  refrigerator, general discussion, 344
  refrigerator, replacing, 352
Heating units, rooftop, 470
Heat load, in cooling towers, 212
Heat pumps, 500–504
Heat reclaimer, 419–422
Hermetic compressor systems, moisture in, 131
High side, 388–389
Hinges, freezer, 372
Humidistat, 422
Humidity
  control of, 422
  relative, 427–429
Hydrolysis, 158–159
Hysteresis losses, 95

I

Ice ejector, icemaker, 306

INDEX

Icemaker
   general discussion, 301–313
   troubleshooting, 311–313
Ice stripper, icemaker, 306
Identification labels, icemaker, 305
Indicators, liquid, 261–263
Inductance, 92–93
Inductive reactance, 93
Insulation
   general discussion, 509–512
   for pipes, 510–512
   refrigerator, 295
Insulation resistance, 129–131

J

Jogging, 132

K

Knives, 28

L

Law, Perfect Gas, 424
Leakage, heat, calculating, 507
Leak detectors
   general discussion, 263–358
   halide, 43–45
Leaking, of refrigerants, 159
Leaks
   high-side, 358
   low-side, 357–358
   refrigerant, detecting, 159–161
   refrigerant, testing for, 378
   testing for, 67
Lid, freezer, 373
Lifting, proper procedure for, 342
Light, extension, 28
Line
   liquid, sizing, 460–463
   refrigerant, 256–257
   suction, sizing, 460–463
Linebreakers, 190–191
Liners, refrigerator inner, 295
Lithium bromide, 488
Load
   air change, 506
   cooling, calculating, 506–507
   miscellaneous, 506–507
   product, calculating, 506–508
   pull-down, 506
   refrigeration and air conditioning, 505–512
   wall gain, 506

Load estimation, 505–512
Lock assembly, freezer, 372
Low-side, 389
Lubrication cycle, in chillers, 210
Lucite®, and Freon®, 156

M

Magnetic field, rotating, 113–116
Magnetic flux, 93–94
Maintenance
   of packaged units, 472–473
   of refrigerators, 323
   of vacuum pumps, 56
   of window air conditioners, 450
Make-up, in cooling towers, 212
Manometer, 425–426
Markers, wire code, 28–29
Mass, 72
Mathematics, 11–24
Matter, structure of, 70–72
Mechanical centers, 391–392
Mechanics, air-conditioning and refrigeration, 524–529
Megger. *See* Megohmmeter
Megohmmeter, 128–129
Metals, reaction with Freon®, 155–156
Metering devices, refrigerant, 254–255
Meters
   sound, 51–52
   vibration, 51
Methyl chloride
   general discussion, 146
   to detect leaks, 160–161
Metric system, 20–23
Mobile homes, air conditioning units for, 463–466
Moisture
   in hermetic compressor systems, 131
   and refrigerants, 158–159
   in refrigeration systems, 259
Moisture analyzers, 51
Mold heater, icemaker, 306
Molecular weight
   general discussion, 147–151
   of selected refrigerants (table), 159
Motor assemblies, 337–338
Motor burnout cleanup, 140–141
Motor control, AC, 131–135
Motor controller, 131
Motor cooling cycle, in chillers, 209
Motor mounts, 182–183
Motors

INDEX

AC, full load amperage values—approximate (table), 519
AC, general discussion, 113–124
AC, squirrel-cage, 131–132
  capacitor-start, 118–120
  compound, 112–113
  compressor, burnout of, 349–350
  compressor, general discussion, 184–195
  compressor, troubleshooting, 184–195
  condenser fan, 320
  cooling, 120–122
  DC, general discussion, 107–113
  electric, general discussion, 107–144
  electric, measuring insulation resistance of, 130
  electric, starting, 110–111
  for hermetic compressors, 176–180
  icemaker, 308
  induction, construction of, 116–117
  mounting, 120–122
  multiloop, 110
  permanent split-capacitor (PSC), 363; troubleshooting (table), 365
  protection of, 133
  PSC. *See* Motors, permanent split-capacitor
  refrigerator, 343–344
  refrigerator compressor, electrical check of, 326–327
  series, 112
  shaded-pole, 117
  shunt, 110
  single-phase, 117
  sizes of, 120
  split-phase, 117–118
  synchronous, 122–123
Motor service factor, 132
Motor speed, 132
Motor starters, 111
Multimeter, 48–49
Multiplication, 12, 14
Multizone system, direct, 473–475

N

Nameplate, refrigerator, 297
National Electrical Code (NEC), 25, 132–133, 513, 514
National Electrical Manufacturer's Association (NEMA), 132
NC. *See* Noise criteria
NEC. *See* National Electrical Code
NEMA. *See* National Electrical Manufacturer's Association

Neutrons, 71
Nitrate, and Freon®, 156
Noise
  air, 441–442
  casing radiated, 441–442
Noise criteria, 441
Nucleus, 71
Numbers, mixed, 14
Numerator, 13
Nylon, and Freon®, 156

O

Odor, refrigerant, 159
Ohm, 80
Ohmmeter, 48
Ohm's Law, 80–81
Oil
  changing, 59
  charging, 58–59
  excess, and liquid indicators, 263
  mixing with refrigerant, 158
Oil problems, with vacuum pumps, 56–57
Opens, testing for, 125
Oral toxicity, and Freon®, 152
ORD. *See* Valve, head pressure control
ORI. *See* Valve, evaporator pressure regulating
ORI. *See* Valve, head pressure control
ORIT. *See* Valve, evaporator pressure regulating
ORO. *See* Valve, head pressure control.
Orlon, and Freon®, 156
Outdoor installation, of display case components, 417
Outlets, air, 436–447
Overcurrents, 133
Overload protection, refrigerator, 325–326
Overloads, on compressor motors, 184–195
Oxygen, corrosion by, 218

P

Parallelogram, area of, 17
Pascal's Law, 73
Perimeter system, 435–439
Perimeter tube, fiberglass insulated, 352–356
Permagum®, 511
Phase, 84
Pi, 17
Pilot light assembly, 230–231
Pipes
  insulation for, 510–512
  outside diameter of, 61
Piping, refrigerant, 458–463

INDEX

Plastics, reaction with Freon®, 156
Pliers, 25
Plugging, 132
Plugs, electrical, 450
Polychlorotrifluoroethylene, and Freon®, 156
Polyethylene, and Freon®, 156
Polystyrene, and Freon®, 156
Polyvinyl alcohol, and Freon®, 156
Positive temperature coefficient, 99–100
Power
  AC, 83–85
  computing, 85
  DC, 83–85
Power rating, 85
Pressure
  back, measuring, 403
  of crankcase pressure regulating valves, 270–271
  for evaporator pressure regulating valves, 273
  gage, 73–74
  general discussion, 72–75
  head, measuring, 403
  of head pressure control valves, 276
  line, 33–35
  measuring, 425–427
  refrigeration system, 157
  static, 435
  total, 436
  velocity, 436
Pressure indicating devices, 72–73
Pressure loss, for certain refrigerants (table), 267
Pressure requirement, for grilles, 440–441
Pressurizing, of cylinders, 341–342
Process tubes, 195–198
Product, defined, 12
Protector, motor overload, 133
Protons, 71
Psychrometer, sling, 51
Psychrometer, stationary, 51
Psychrometric chart, 429–430
Psychrometrics, 424–447
P-traps, suction-line, 256–257
Pump head, cooling tower, 212
Punches, 27

Q

Quotient, defined, 12

R

Radiation, 433–434
Radical sign, 15

Rapid Electrical Diagnosis unit (RED), 319–320
Receiver, 388
Recirculating systems, liquid refrigerant, 239–241
Recirculator, flooded, 241
Recorders
  general discussion, 52
  voltage and current, 32
Rectangle, area of, 17
Rectangular solid, volume of, 18
Rectifiers
  general discussion, 96–97
  silicon controlled (SCR), 97–98
RED. See Rapid Electrical Diagnosis unit
Refinishing, of refrigerator, 323–325
Refrigerant
  adding, 357
  boiling temperatures of (table), 157
  characteristics of (table), 145
  classifications of, 145–146
  defined, 145
  density of, 158
  enthalpy of, 158
  flammability of, 158
  flow control of, 254–293
  freezer, complete recharge of, 378
  freezer, overcharge of, 378
  general discussion, 78, 145–161
  high-side leak of, 358
  hydrolysis rate in water (table), 159
  for ice-cream cases, 413–414
  line sizes for (table), 257
  liquid density of, at 86°F. (table), 158
  low-side leak of, 357–358
  mixing with oil, 158
  and moisture, 158–159
  moisture content (table), 262
  molecular weights of (table), 159
  and odor, 159
  operating pressures of (table), 157
  overcharge of, 358
  pressure drop of (table), 266
  pressure of, 157
  properties of, 156–159
  shortage of, 361
  specific volumes at 5°F. (table), 157
  subcooling for (table), 267
  temperature of, 157
  tendency to leak, 159
  testing for leaks of, 358–359
  and toxicity, 159
  undercharge of, 357–358

INDEX

vertical lift of (table), 266
volume of, 157
Refrigerant 12 (R-12)
  general discussion, 78
  working with, 342
Refrigerant 22 (R-22)
  general discussion, 78
  working with, 342
Refrigerant 40, general discussion, 78
Refrigerant charge, 319
Refrigerant cycle
  for chilled-water air conditioning, 478–480
  in chillers, 207–209
  general discussion, 77–78, 203, 388–389
Refrigerant holding capacity, of accumulator, 292
Refrigeration
  careers in, 524–529
  commercial systems, 388–423
  and condensers, 202–207
  controls for, 80–106
  and evaporators, 226–253
  general discussion, 70–79
  and insulation, 509–512
  and load estimation, 505–509
  and psychometrics, 424–447
  and refrigerants, 145–161
  supermarket, 391
  tools for, 25–69
Refrigeration charge, improper, detecting, 402–403
Refrigeration equipment, voltage requirements for, 513–523
Refrigeration machine, absorption, 494–495
Refrigeration Service Engineers Society and the Air Conditioning Contractors of America, 526
Refrigeration systems, general discussion, 76–78, 388–423
Refrigeration units. *See also* Cases
  commercial, 391–422
  corner, 393–396
  usage factors and interior capacity (table), 509
  water-cooled, valve inspection, 395
Refrigerator/freezer
  cycle defrost, 297–299
  grounding, 322–323
  installation of, 322–323
  no-frost, top-mount, 298–300
  no-frost, side-by-side, 300–301
Refrigerators
  air circulation in, 338
  general discussion, 294–340
  servicing, 341–368
  single-door, 295–297
Registers
  and air capacity (table), 438
  ceiling supply, 443–444
  general discussion, 435
Regulator
  back-pressure, 249–251
  differential pressure-relief, 251–253
  dual-pressure, 245
Relative humidity, 428–432
Relays
  general discussion, 133–134, 180–181
  testing insulation resistance, 131
Remote systems, 455
Research centers, solar cooling, 497–500
Resins, and Freon®, 156
Resistance
  checking on hermetic compressor, 367
  general discussion, 80
  in a parallel circuit, 82–83
  relationship to current, 23
Resistor
  bleeder, 362
  variable, 99
Restrictions, refrigerant, 361–362
Restrictor, 34
Return air systems, 457
Returns, air, 437
Rock wool, 511–512
Room air motion, 445–446
Rotation, direction of motor, 122
Rotors
  general discussion, 116
  squirrel-cage, testing, 125–126
Rule
  folding, 28
  left-hand, 107
  right-hand, 109
Running time, 505–506

S

Safety
  electrical, 342–343
  in using fluorocarbons, 151–154
  in using scale remover, 222, 224
  in servicing refrigeration units, 341–343
  in soldering, 67
  in using solvents, 67–68
Safety devices, compressor, 472

INDEX

Saws, 27
Scale
   field testing of, 217–218
   formation of, 221
   general discussion, 215–221
   identification of, 217
Scale remover, 221–222
Scaling, 215–221
Schematic, reading, 141–143
SCR. *See* Rectifier, silicon controlled
Screwdrivers, 25–26
Self-inductance, 93
Self-leveling feature, 493–494
Semiconductors, 95–96
Sensible heat factor, 431–432
Sensors, 99–101
Series rheostat, 111–112
Servicing
   of console air conditioners, 486–487
   of crankcase pressure regulating valves, 271
   of refrigerators, 341–368
   of refrigerator section, 343
Service log, 486
Short circuit (in split-phase motor), testing for, 126
Shorts, checking for, 125
Signal arm, icemaker, 307
Sine waves, 83–84
Skin effects, and Freon®, 151–152
Slime, 219
Slimicides, 219
Slip, 116
Sludge, in refrigerant systems, 260
Slug (unit of mass), 72
Smoke detectors, 456–457
Sockets, light-bulb, 339
Solar energy, 495–500
Solder
   general discussion, 26
   for refrigerant lines, 256
Soldering, 65–67
Soldering equipment, 26
Soldering paste, 26
Solvents, 67–68, 225
Sound power level, 435
Specific gravity, 72
Splicing, of power cords, 353–354
Spread, 435
Square root, 15–16
Starters, 133
Stator, 116

Strainer, for crankcase pressure regulating valve, 270
Strainers, line, 260–261
Strippers, wire, 28
Subcooling, for certain refrigerants (table), 267
Subtraction, 12, 14
Suction line, 294, 388, 461
Sulfur dioxide
   in detecting leaks, 160
   as a refrigerant, 78, 146
Sump, amount of water in, 221
Superheat, measuring, 39–43
Swaging, of copper tubing, 64
Sweating (in coolers), 396–397
Switches
   centrifugal in split-phase motor, testing, 126
   energy saver, 320
   icemaker water valve, 309–310
   power saver, on refrigerators, 323
   refrigerator light, 330–331
   testing insulation resistance, 131
   timing, icemaker, 307
Synchronous speed, 132
Systems, direct-expansion, 229–234

T

Tables
   activity-heat relationships, 429
   air-cooled condensing unit capacities, 509
   amperage values for AC motors, 519
   antifreeze in a chilled-water cooler, 490
   capacitive value conversion table, 88
   capillary tubing inside and outside diameters, 60
   compressor troubleshooting and service, 346–348
   copper conductor voltage drop, 517
   copper tubing outside diameters, wall thicknesses, and weights, 60–61
   copper tubing suction line, 461
   copper wire current-carrying ability, 81
   enthalpy (selected refrigerants), 158
   fitting losses in equivalent feet of pipe, 462
   fluorocarbons, potential hazards of, 153
   fluoronated products molecular weight and boiling point, 152
   Freon® physical properties, 148–150
   friction drop correction factors—liquid and suction lines, 463
   galvanic series, 218
   gas-fuel air conditioning unit specifications, 491

INDEX

locked rotor ampere range per motor horsepower, 518
meat case checklist, 410–411
metal combinations in water-cooled condensers, 206
metric-customary conversions, 22
metric prefixes, 21
moisture in hermetic compressor systems, 131
NC criteria, 441
outlet velocity ratings, 439
pipe size and outside diameter, 61
pressure loss and required subcooling for refrigerant condensing, 267
PSC compressor motor troubles and corrections, 365
refrigerant characteristics, 145
refrigerant boiling temperatures, 157
refrigerant hydrolysis rate in water, 159
refrigerant line sizes, 257
refrigerant liquid density at 86°F., 158
refrigerant moisture content, measuring, 262
refrigerant molecular weights, 159
refrigerant operating pressures, 157
refrigerant specific volumes at 5°F., 157
refrigeration unit, usage factors and interior capacity, 509
register or grille size related to air capacities, 438
relative humidity permissible (in winter), 429
resistance and current relationship, 23
saturated vapor per pound of dry air, 429
SI metric base units, 21
solenoid valve service suggestions, 137
thermostat altitude adjustments, 328
troubleshooting console model air conditioners, 482–483
troubleshooting differential pressure relief regulator, 251
troubleshooting discharge bypass valves, 282
troubleshooting freezers: chest models, 382–384
troubleshooting freezers: upright models, 379–381
troubleshooting head pressure control valves, 277
troubleshooting hermetic compressor type air conditioning equipment, 464–466
troubleshooting refrigerated cases, 404
valve losses in equivalent feet of pipe, 461
valve sizing for high-temperature system, 231
valve sizing for low-temperature system, 233
vertical lift and pressure drop of refrigerants, 266
voltage drops, permissible, 514
wire size and current-carrying capacity, 81
Tape, measuring, 28
Teaching, as a career, 528–529
Teflon®, and Freon®, 156
Temperature
  ceiling, effects of, 407
  converting, 22–23, 424–425
  dewpoint, 427
  dry bulb, 427, 431
  freezer compartment, 360
  and gage performance, 35
  general discussion, 424–425
  and heat, 75–76
  measuring, 21–22
  minimum evaporator, at accumulator, 292
  provision compartment, 360
  refrigerant, 157
  room ambient, 319
  of suction gas at accumulator, 292
  wet bulb, 427, 431–432
Temperature controls, 136–138
Temperature differential, 435–436
Temperature rise, 132
Terminal boxes, locating, 442
Terminals, compressor, 181–182
Test cycling, of icemaker, 309
Tester, manual start, 327
Testing
  of insulation resistance, 129–131
  for refrigerant leaks, 358–359
Test prods, 29–30
TEV. *See* Valve, thermostatic expansion
Thermistor, 38
Thermometers, 36–43
  resistance, 38–39
  superheat, 39
  thermocouple, 38
Thermostats
  bimetallic, 137–138
  of comfort control panel, 422
  freezer, 373–375
  ice-cream cases, 415–416
  icemaker, 307
  outdoor, 502
  refrigerator, 327–330
  temperatures, 359–360
Thermostat construction, 138
Thermostat wiring, 138

540

INDEX

Throw, 436
Throw requirement, 440
Timers, defrost, 333–335, 399
Tolerances, of capacitors, 90
Tool kits, 30–32
Tools
    general discussion, 25–69
    insulation-stripping, 28
    masonry working, 27
    metalworking, 27
    woodworking, 27
Torque, 132
Touch-up, of refrigerator, 323–325
Toxicity
    and Freon®, 151
    of refrigerants, 159
Transformers, 93–95
Transistors, 97
Transposing, 19
Triangle, area of, 17
Trim, cold ban, 370–371
Troubleshooting
    of compressor motors, 184–195
    of compressors, 345–349, 365
    of console air conditioners, 482–483 (table), 487
    of direct multizone system, 475
    of discharge bypass valves (table), 282
    of electric motors with a volt-ammeter, 123–124
    of evaporators for residential use, 455, 464–466
    of expansion valve air conditioning system, 469
    of freezers, 378–386
    of head pressure control valves (table), 277
    of hermetic compressors (table), 464–466
    of icemakers, 311–313
    with the megohmmeter, 128–131
    of mobile home air conditioners, 464–466
    of refrigerated cases (table), 404
    of refrigerator components, 345–349
    of refrigerator electrical components, 325–327
    of relief regulators, differential pressure (table), 251
    of single-package rooftop units, 458, 464–466
    of window air conditioners, 451, 464–466
Tube bender, 65
Tube, perimeter, 332–333
Tubing
    capillary, 255, 283–284
    constricting, 63–64
    copper, cutting, 60–62
    copper, fitting by compression, 65
    copper, general discussion, 255–256
    copper, hard-drawn, 60
    copper, swaging, 64
    cutter, 63
    flaring, 62–63
    forming, 64–65
    general discussion, 59–65
    insulation for, 510–511
    outside diameters, wall thicknesses, and weights (table), 60–61
    refrigerant, forming, 64–65

U

Usage conditions, refrigerator, 319
Usage factors, 508–509

V

Vacuum, 73–74
Vacuum pumps, 54–57
Vacuum release, on freezers, 371–372
Valve gradient, 271
Valves
    automatic expansion, 254
    check, 290–291
    check, for ammonia-type evaporators, 246–249
    compressor, 257–258
    crankcase pressure regulating (CRO), 269–271
    for direct-expansion systems, 229–230
    discharge bypass (DBV), 278–283
    discharge bypass, troubleshooting (table), 282
    evaporator pressure regulating, 271–273
    expansion, in ice-cream cases, 414–415
    float, 255, 284
    gate, 290–291
    globe, 290–291
    hand expansion, 254
    head pressure control, 273–278
    head pressure control, troubleshooting (table), 277
    level control, 283–285
    line, 258–259
    losses in equivalent feet of pipe (table), 461
    pilot, 249
    receiver, 291
    refrigeration, 134–135
    regulator, for ammonia-type evaporators, 246–249
    relief, for ammonia-type evaporators, 247
    service, in sealed units, 290
    sizing for high-temperature systems (table), 231
    sizing for low-temperature systems (table), 233
    solenoid, 134–135, 137

INDEX

solenoid, in ammonia-type evaporators, 246–249
stop, 290–291
suction pressure regulating. *See* Valves, crankcase pressure regulating
thermostatic expansion (TEV), 254–255, 263–269
troubleshooting, 251–253
vacuum-release, for walk-in coolers, 397
water, 290–291
Vaporization, refrigeration from, 76
Vapor, saturated, per pound of dry air (table), 429
Varnish, in refrigerant systems, 260
Vectrol®, 437–439
Velocity
  face, 435
  outlet, 435
  residual, 435
  terminal, 436
Velometer, 434
Vinyl, and Freon®, 156
Vizi-Vapr®, 54
Voltage
  effect of variations on AC motors, 515–516
  general discussion, 80
  line, 360–361
  motor, unacceptable, 517–519
Voltage drop
  calculations, 515
  inrush, 519–520
  permissible maximum (table), 514
Voltage loss, limiting, 513
Volt-ammeter
  hook-on, 29–30
  split-core AC, to troubleshoot motors, 124–128
  in troubleshooting electric motors, 123–124
Voltmeter
  general discussion, 47–48
  indicating, 29
Voltmeter probe, 31
Volume
  finding, 18–19
  refrigerant, 157–158
Vortex shedders, 442
Vortex shedding, 442

W

Water
  determining volume of, 221–222
  for use in cooling equipment, 215–225
Water fill volume, icemaker, 309–310
Water supply line, icemaker, 308–309
Water valve assembly, icemaker, 306–307
Wattage, measuring, 361
Wattmeter, 49–50
Weight, 72
Wire
  copper, current-carrying ability (table), 81
  selecting proper size, 513–515, 516–517
  size, and current-carrying capacity (table), 81
  size of, and low voltage, 514–515
Wiring, icemaker, 308
Wrapper condenser, freezer, 375–376
Wrenches, 26

X, Y, Z

Zone, occupied, 435